Springer Proceedings in Complexity

Springer Proceedings in Complexity publishes proceedings from scholarly meetings on all topics relating to the interdisciplinary studies of complex systems science. Springer welcomes book ideas from authors. The series is indexed in Scopus.

Proposals must include the following:

- name, place and date of the scientific meeting
- a link to the committees (local organization, international advisors etc.)
- scientific description of the meeting
- list of invited/plenary speakers
- an estimate of the planned proceedings book parameters (number of pages/articles, requested number of bulk copies, submission deadline)

Submit your proposals to: Hisako.Niko@springer.com

More information about this series at https://link.springer.com/bookseries/11637

Samsul Ariffin Abdul Karim ·
Mohd Fadhlullah Abd Shukur · Chong Fai Kait ·
Hassan Soleimani · Hamzah Sakidin
Editors

Proceedings of the 6th International Conference on Fundamental and Applied Sciences

ICFAS 2020

Springer

Editors
Samsul Ariffin Abdul Karim
Fundamental and Applied Sciences
Universiti Teknologi PETRONAS
Seri Iskandar, Perak, Malaysia

Mohd Fadhlullah Abd Shukur
Fundamental and Applied Sciences
Universiti Teknologi PETRONAS
Seri Iskandar, Perak, Malaysia

Chong Fai Kait
Universiti Teknologi PETRONAS
Seri Iskandar, Perak, Malaysia

Hassan Soleimani
Universiti Teknologi PETRONAS
Seri Iskandar, Perak, Malaysia

Hamzah Sakidin
Universiti Teknologi PETRONAS
Seri Iskandar, Perak, Malaysia

ISSN 2213-8684 ISSN 2213-8692 (electronic)
Springer Proceedings in Complexity
ISBN 978-981-16-4512-9 ISBN 978-981-16-4513-6 (eBook)
https://doi.org/10.1007/978-981-16-4513-6

This Springer imprint is published by the registered company Springer Nature Singapore Pte Ltd.
The registered company address is: 152 Beach Road, #21-01/04 Gateway East, Singapore 189721, Singapore

Organization

The 6th International Conference on Fundamental and Applied Sciences (ICFAS 2020), 13–15 July 2021

Virtual Event

Organizing Committee

Chairman

Assoc. Prof. Dr. Hanita Daud

Co-chairman

Assoc. Prof. Dr. Maizatul Shima Shaharun
Assoc. Prof. Dr. Vijanth Sagayan

Secretary

Dr. Nooraini Zainuddin

Treasury

Dr. Mohana Sundaram Muthuvalu
Assoc. Prof. Dr. Ibrahima Faye

Technical

Dr. Khairulazhar Jumbri
Assoc. Prof. Dr. Balbir Singh Mahinder Singh
Assoc. Prof. Dr. Chong Fai Kait
Dr. A'fza Shafie
Assoc. Prof. Dr. Hassan Soleimani
Dr. Hayyiratul Fatimah Mohd Zaid

Dr. Hamzah Sakidin
Dr. Lee Kean Chuan
Dr. Lim Jun Wei
Dr. Zulkifli Merican Aljunid Merican

Publication

Dr. Samsul Ariffin Abdul Karim
Assoc. Prof. Dr. Anita Ramli
Assoc. Prof. Dr. Cecilia Devi Wilfred
Assoc. Prof. Dr. Mahmod Othman
Dr. Beh Hoe Guan
Dr. Khe Cheng Seong
Dr. Mohd Fadhlullah Abd Shukur
Dr. Normawati Mohamad Yunus

Sponsorship

Assoc. Prof. Dr. Noor Asmawati Mohd Zabidi
Prof. Dr. Norani Muti Mohamed

Publicity and Protocol

Dr. Dennis Ling Chuan Ching

Event

Dr. Noraini Abd Ghani
Noor Akmal Mohamad Saleh

Logistic

Mohd Faisal Taha

IT and Media

Muhammad Hanif Zahari
Dr. Rajalingam A/L Sokkalingam

Preface

The Fundamental and Applied Sciences Department, Universiti Teknologi PETRONAS, is honoured to host the Virtual Event of 6th International Conference on Fundamental and Applied Sciences (ICFAS 2020) from 13 to 15 July 2021. It is our pleasure to welcome delegates from Malaysia, Indonesia, Kazakhstan, Nigeria, Vietnam, Pakistan, Ghana, New Zealand and Brunei to share and exchange ideas on the latest advancement in mathematics, chemistry and physics with the theme of "Innovative Science towards Sustainability and Industrial Revolution 4.0". This conference has certainly provided a platform for sharing and discussion on recent advances in sustainable chemistry, nano and sustainable technology, and mathematical and analytical methods in science and technology. The conference topics include green materials, molecular modelling, catalysis, nanodevices and nanosystems, smart materials applications, solar cell technology, computational mathematics, data analysis and visualization, but not limited to numerical analysis. It is our hope that the contents of this book will benefit researchers, postgraduate students and industrial practitioners in the areas of mathematics, physics and chemistry as most of the topics are in line with IR 4.0. On behalf of the sponsors and the Conference Committees, we would like to gratefully acknowledge all the keynote and invited speakers, presenters and participants in contributing to the success of ICFAS 2020. Special appreciation and thank you to all the reviewers for providing their expertise in improving the quality of the papers, without which this book will not be possible. Finally, thank you to Springer for publishing our conference in Springer Proceedings in Complexity.

Seri Iskandar, Malaysia

Samsul Ariffin Abdul Karim
Mohd Fadhlullah Abd Shukur
Chong Fai Kait
Hassan Soleimani
Hamzah Sakidin

Contents

Editors and Contributors

About the Editors

Samsul Ariffin Abdul Karim is Senior Lecturer at Universiti Teknologi PETRONAS (UTP), Malaysia. He has published more than 140 papers in journals and conferences including three edited conference volumes and 40 chapters. He was the recipient of Effective Education Delivery Award and Publication Award (Journal and Conference Paper), UTP Quality Day 2010, 2011 and 2012, respectively. He was Certified WOLFRAM Technology Associate, Mathematica Student Level. He also has published nine books with Springer Publishing including five books with Studies in Systems, Decision and Control (SSDC) series.

Mohd Fadhlullah Abd Shukur is Senior Lecturer at the Fundamental and Applied Sciences Department, Universiti Teknologi PETRONAS (UTP). His research interest is in the field of ionic conductor including the development and characterization of polymer electrolytes for application in various energy storage devices. He has received several research grants, acting as the head researcher or co-researcher for such grants. He has been invited as reviewer for various journals and has published 41 articles in ISI/Scopus Indexed Journals. Currently, his H-index is 21 with 1347 citations in Google Scholar Citations.

Chong Fai Kait is Associate Professor at Universiti Teknologi PETRONAS. She graduated with B.Sc. (Hons) in Industrial Chemistry from Universiti Putra Malaysia in 1993. In 1998, she received her Ph.D. from University of Dundee, Scotland, in the area of heterogeneous catalysis for naphtha reforming process. She has more than 20 years of academic and research experience, and her main research interests are in photocatalysis and its integration with ionic liquids or other techniques for sustainable environmental applications.

Hassan Soleimani is Associate Professor at Universiti Teknologi PETRONAS, Malaysia. He has obtained a Ph.D. in wave propagation from Universiti Putra Malaysia. He has published more than 120 research articles in peer-reviewed international. He is Committee Member and Organizer of international conference of Diffusion in Solid and Liquids (DSL) since 2011 in Europe. He has been Visiting Scientist at many universities including University of Cambridge, UK; Wright State University, USA; University of Patras, Greece; and School of Engineering and Griffith University, Australia.

Hamzah Sakidin is Senior Lecturer at Universiti Teknologi PETRONAS (UTP), Seri Iskandar, Perak, since September 2013. His research interest is in the field of applied mathematics and mathematical modelling. He has received several research grants, acting as the head researcher or co-researcher for some research grants. He is also Author of several mathematics books for tertiary level of education. He has more than 20 years of experience teaching in secondary and tertiary level of education.

Contributors

Summaira Abbasi Centre of Innovative Nanostructures and Nanodevices (COINN), Universiti Teknologi PETRONAS, Seri Iskandar, Perak Darul Ridzuan, Malaysia

Nadia Syazana Abd Halim Universiti Teknologi PETRONAS, Seri Iskandar, Perak, Malaysia

Zhubanov Amin Abdirasululy Kazakh-British Technical University, Almaty, Kazakhstan

G. Abduakhitova Al-Farabi Kazakh National University, Almaty, Kazakhstan

Samsul Ariffin Abdul Karim Fundamental and Applied Sciences Department and Centre for Systems Engineering (CSE), Institute of Autonomous System, Universiti Teknologi PETRONAS (UTP), Seri Iskandar, Perak Darul Ridzuan, Malaysia

Ninie Suhana Abdul Manan Department of Chemistry, Faculty of Science, Universiti Malaya, Kuala Lumpur, Malaysia

Bashir Abubakar Abdulkadir Fundamental and Applied Sciences Department, Universiti Teknologi PETRONAS, Seri Iskandar, Perak Darul Ridzuan, Malaysia

Lazim Abdullah School of Informatics and Applied Mathematics, Universiti Malaysia Terengganu, Kuala Terengganu, Terengganu, Malaysia

Nor Ain Fathihah Abdullah Department of Fundamental and Applied Sciences, Universiti Teknologi PETRONAS, Seri Iskandar, Perak, Malaysia

Muslim Abdurrahman Department of Petroleum Engineering, Universitas Islam Riau, Pekanbaru, Indonesia

Akbar Abu Seman PETRONAS Research Sdn.Bhd, Kajang, Selangor, Malaysia

Dauda Abubakar Department of Physics, Bauchi State University Gadau, Bauchi, Nigeria

Muhammad Adil Department of Fundamental and Applied Sciences, Universiti Teknologi PETRONAS, Seri Iskandar, Perak Darul Ridzuan, Malaysia

Iman Eslami Afrooz Fundamental and Applied Sciences Department, Universiti Teknologi PETRONAS, Seri Iskandar, Perak, Malaysia

Farkhanda Afzal MCS, National University of Science and Technology (NUST), Islamabad, Pakistan

Mohd Arif Agam Faculty of Applied Science and Technology, Universiti Tun Hussein Onn Malaysia, UTHM Pagoh Campus, Muar, Johor, Malaysia

Farooq Ahmad Department of Electrical Engineering, University of Engineering and Technology, Lahore, Pakistan

Abdelazim Abbas Ahmed Institute of Hydrocarbon Recovery, Universiti Teknologi PETRONAS, Seri Iskandar, Perak Darul Ridzuan, Malaysia

Naser M. Ahmed School of Physics, Universiti Sains Malaysia, Pulau Pinang, Malaysia

Abdullah Al-Yaari Fundamental and Applied Sciences Department, Universiti Teknologi PETRONAS, Seri Iskandar, Perak, Malaysia

Hassan Ali Fundamental and Applied Science Department, Universiti Teknologi PETRONAS, Seri Iskandar, Perak Darul Ridzuan, Malaysia;
Department of Physics, University of Narowal, Narowal, Pakistan

Labiyana Hanif Ali Faculty of Science and Natural Resources, Universiti Malaysia Sabah, Sabah, Malaysia

Imtias Amir Centralized Analytical Laboratory, Universiti Teknologi PETRONAS Seri Iskandar, Perak, Malaysia

Sorfina Amran Department of Fundamental and Applied Sciences, Universiti Teknologi PETRONAS, Seri Iskandar, Perak, Malaysia

Nur Akila Syakida Idayu Khairul Anuar Fundamental and Applied Sciences Department, HICoE Centre for Biofuel and Biochemical Research, Institute of Self-Sustainable Building, Universiti Teknologi PETRONAS, Seri Iskandar, Perak, Malaysia

Muhamad Fatikul Arif Institut Teknologi Sumatera, Lampung, Indonesia

Muhammad Naeim Mohd Aris Department of Fundamental & Applied Sciences, Universiti Teknologi PETRONAS, Seri Iskandar, Perak, Malaysia

Saba Ayub Research Center for Dielectric and Advanced Matter Physics, Pusan National University, Busan, Republic of Korea

Ninna Sakina Binti Azman Department of Fundamental and Applied Sciences, Universiti Teknologi PETRONAS, Seri Iskandar, Perak, Malaysia;
Centre of Research in Ionic Liquids, Universiti Teknologi PETRONAS, Seri Iskandar, Perak, Malaysia

Areej Babiker Electronic Engineering Department, Future University, Khartoum, Sudan

Mawardi Bahri Department of Mathematics, Hasanuddin University, Makassar, Indonesia

Mohd Nazari Abu Bakar Faculty of Applied Sciences, Universiti Teknologi MARA Perlis, Arau, Perlis, Malaysia

Abayev Talgat Bakytuly Kazakh-British Technical University, Almaty, Kazakhstan

Robabeh Bashiri Centre of Innovative Nanostructure and Nanodevices (COINN), Universiti Teknologi PETRONAS, Bandar Seri Iskandar, Perak, Malaysia;
Department of Fundamental & Applied Sciences, Universiti Teknologi PETRONAS, Bandar Seri Iskandar, Perak, Malaysia

Lyazzat Bekbayeva Department of Engineering Disciplines, Asfendiyarov Kazakh National Medical University, Almaty, Kazakhstan;
Biology Program, School of Distance Education, Universiti Sains Malaysia, Minden, Penang, Malaysia

Nor Hafizah Berahim Universiti Teknologi PETRONAS, Seri Iskandar, Perak, Malaysia;
Petronas Research Sdn Bhd, Kajang, Selangor, Malaysia

Yih Hui Boon Department of Chemistry, Faculty of Science, Universiti Malaya, Kuala Lumpur, Malaysia;
Integrative Medicine Cluster, Advanced Medical and Dental Institute, Universiti Sains Malaysia, Pulau Pinang, Malaysia

Azry Borhan Department of Chemical Engineering, HICoE, Centre for Biofuel and Biochemical Research, Institute of Self-Sustainable Building, Universiti Teknologi PETRONAS, Seri Iskandar, Perak, Malaysia

Noorazlenawati Borhan Hydrocarbon Recovery Technology, PETRONAS Research Sdn Bhd, Bandar Baru Bangi, Selangor, Malaysia

Mageswaran Ravi Chandran College of Engineering, Universiti Tenaga Nasional (UNITEN), Selangor, Malaysia

Albert Ling Sheng Chang Malaysian Cocoa Board, Kota Kinabalu, Sabah, Malaysia

K. Y. Cheong School of Material and Mineral Resources, Universiti Sains Malaysia, Nibong Tebal, Pulau Pinang, Malaysia

Xiao Yan Chew Department of Physics Education, Pusan National University, Busan, Republic of Korea

Dennis Ling Chuan Ching Fundamental and Applied Sciences Department, Universiti Teknologi PETRONAS, Bandar Seri Iskandar, Perak, Malaysia

Chong Fai Kait Fundamental and Applied Sciences Department, Universiti Teknologi PETRONAS, Seri Iskandar, Perak, Malaysia;
Centre of Innovative Nanostructures and Nanodevices (COINN), Institute of Autonomous System (IAS), Universiti Teknologi PETRONAS, Seri Iskandar, Perak, Malaysia

Sarat Chandra Dass Heriot-Watt University Malaysia, Putrajaya, Malaysia

Hanita Daud Fundamental and Applied Sciences Department, Universiti Teknologi PETRONAS, Seri Iskander, Perak, Malaysia

John Ojur Dennis Fundamental and Applied Sciences Department, Universiti Teknologi PETRONAS, Seri Iskander, Perak, Malaysia;
Fundamental of Applied Science, Universiti Teknologi PETRONAS, Seri Iskandar, Perak, Malaysia

Israf Ud Din Department of Chemistry, College of Science and Humanity, Prince Sattam bin Abdulaziz University, Al-kharj, Saudi Arabia

Nur Nabilah Che Draman Fundamental and Applied Sciences Department, Universiti Teknologi PETRONAS (UTP), Seri Iskandar, Perak, Malaysia

Iskandar Dzulkarnain Universiti Teknologi PETRONAS, Seri Iskandar, Perak, Malaysia

Khaled Abdalla Elraies Petroleum Engineering Department, Universiti Teknologi PETRONAS, Seri Iskandar, Perak, Malaysia

Muhammad Faheem Department of Chemical Engineering, University of Engineering and Technology, Lahore, Pakistan

Syeda Saba Fatima Department of Chemical Engineering, HICoE, Centre for Biofuel and Biochemical Research, Institute of Self-Sustainable Building, Universiti Teknologi PETRONAS, Seri Iskandar, Perak, Malaysia

Ibrahima Faye Centre for Intelligent Signal and Imaging Research (CISIR), Universiti Teknologi PETRONAS, Seri Iskandar, Perak, Malaysia;
Fundamental and Applied Science Department, Universiti Teknologi PETRONAS (UTP), Seri Iskandar, Perak, Malaysia

Noraini Abd Ghani Department of Fundamental and Applied Sciences, Universiti Teknologi PETRONAS, Seri Iskandar, Perak, Malaysia;
Centre of Research in Ionic Liquids, Universiti Teknologi PETRONAS, Seri Iskandar, Perak, Malaysia

Syed Muhammad Arslan Gilani Department of Electrical Engineering, University of Engineering and Technology, Lahore, Pakistan

Alla Goncharova Department of Biotechnology, Faculty of Biochemistry, Al-Farabi, Kazakh National University, Almaty, Kazakhstan

Beh Hoe Guan Centre of Innovative Nanostructure and Nanodevices, Universiti Teknologi PETRONAS, Seri Iskandar, Perak, Malaysia;
Department of Fundamental & Applied Sciences, Universiti Teknologi PETRONAS, Seri Iskandar, Perak, Malaysia

Mohamad Amin Bin Hamid Centre of Innovative Nanostructure and Nanodevices, Universiti Teknologi PETRONAS, Seri Iskandar, Perak, Malaysia

Mohammed Falalu Hamza Department of Pure and Industrial Chemistry, Bayero University Kano, Kano, Nigeria

Noraini Hamzah School of Chemistry and Environment, Faculty of Applied Sciences, Universiti Teknologi MARA (UiTM), Shah Alam, Selangor, Malaysia

Amiratul Liyana Mohamad Hanapi Fundamental and Applied Sciences Department, Universiti Teknologi PETRONAS, Seri Iskandar, Perak, Malaysia

Noor Adilla Harim Fundamental and Applied Sciences Department, Universiti Teknologi PETRONAS (UTP), Seri Iskandar, Perak, Malaysia

Ishak Hashim Department of Mathematical Sciences, Faculty of Science and Technology, Universiti Kebangsaan Malaysia, Bangi, Selangor, Malaysia;
School of Mathematical Sciences, Universiti Kebangsaan Malaysia, Bangi, Selangor, Malaysia

Muhammad Hariz Hasmi Faculty of Science and Technology, Universiti Sains Islam Malaysia, Nilai, Negeri Sembilan, Malaysia

Abdullah Husin Program Studi Sistem Informasi, Universitas Islam Indragiri, Riau, Indonesia

Aarij Mahmood Hussaan Computer Science Department, IQRA University, Karachi, Pakistan

Zawawi Ibrahim Engineering and Processing Division, Malaysian Palm Oil Board (MPOB), Kajang, Selangor, Malaysia

Mudassar Iqbal Department of Fundamental and Applied Sciences, Universiti Teknologi PETRONAS, Seri Iskandar, Perak Darul Ridzuan, Malaysia;
Department of Mathematical Sciences, Balochistan University of Information Technology, Engineering and Management Sciences (BUITEMS), Quetta, Pakistan

Ariful Islam Department of Fundamental and Applied Sciences, Universiti Teknologi PETRONAS, Seri Iskandar, Perak Darul Ridzuan, Malaysia

Nur Ezzati Mohd Izudin Fundamental and Applied Sciences Department, Universiti Teknologi PETRONAS, Perak, Malaysia

Muhammad Izzatullah Seismic Modeling and Inversion Group (SMI), King Abdullah University of Science and Technology, Thuwal, Saudi Arabia

Ahmad Kamil Mohd Jaaffar Malaysian Cocoa Board, Kota Kinabalu, Sabah, Malaysia

Khairul Anuar Jamaluddin Fundamental of Applied Science, Universiti Teknologi PETRONAS, Seri Iskandar, Perak, Malaysia

A. Jati Faculty of Engineering, Universiti Teknologi Malaysia, Johor Bahru, Johor DT, Malaysia

Easter Joseph Department of Fundamental and Applied Sciences (FASD), Universiti Teknologi PETRONAS, Seri Iskandar, Perak, Malaysia

Mahdalena Julia Universitas Negeri Malang, East Java, Indonesia

Khairulazhar Jumbri Department of Fundamental and Applied Sciences, Universiti Teknologi PETRONAS, Seri Iskandar, Perak, Malaysia;
Centre of Research in Ionic Liquids (CORIL), Universiti Teknologi PETRONAS, Seri Iskandar, Perak, Malaysia

Hisyam Jusoh Civil and Environmental Engineering, Universiti Teknologi PETRONAS, Seri Iskandar, Perak, Malaysia

Donenov Beisen Kainarbaevich Kazakh-British Technical University, Almaty, Kazakhstan

Kozhanova Kaldanay Department of Engineering Disciplines, Asfendiyarov Kazakh National Medical University, Almaty, Kazakhstan

Samsul Ariffin Bin Abdul Karim Department of Fundamental and Applied Sciences and Centre for Systems Engineering (CSE), Institute of Autonomous System, Universiti Teknologi PETRONAS, Seri Iskandar, Perak, Malaysia

Tatyana Karpenyuk Department of Biotechnology, Faculty of Biochemistry, Al-Farabi, Kazakh National University, Almaty, Kazakhstan

Mohd Azlan Kassim School of Science and Technology, Sunway University, Bandar Sunway, Selangor, Malaysia

Yew Weng Kean School of Engineering and Physical Science, Heriot-Watt University Malaysia, Putrajaya, Malaysia

Lam Man Kee Department of Chemical Engineering, Universiti Teknologi PETRONAS, Seri Iskandar, Perak, Malaysia

Noran Nur Wahida Khalili Fundamental and Applied Sciences Department, Universiti Teknologi PETRONAS, Seri Iskandar, Perak, Malaysia

Ilyas Khan Department of Mathematics, Majmaah University, Majmaah, Saudi Arabia

Lai Weng Kin Fundamental and Applied Sciences Department, HICoE Centre for Biofuel and Biochemical Research, Institute of Self-Sustainable Building, Universiti Teknologi PETRONAS, Seri Iskandar, Perak, Malaysia

U. Kusherbayeva Al-Farabi Kazakh National University, Almaty, Kazakhstan

Lawal Adebayo Lanre Fundamental and Applied Science Department, Universiti Teknologi PETRONAS, Seri Iskandar, Perak, Malaysia

Kar Mun Lee Fundamental and Applied Sciences Department, Universiti Teknologi PETRONAS, Seri Iskandar, Perak, Malaysia

Wai Hong Leong Department of Fundamental and Applied Sciences, HICoE—Centre for Biofuel and Biochemical Research, Institute of Self-Sustainable Building, Universiti Teknologi PETRONAS, Seri Iskandar, Perak Darul Ridzuan, Malaysia

Jun Wei Lim Fundamental and Applied Sciences Department, Universiti Teknologi PETRONAS, Seri Iskandar, Perak, Malaysia;
Department of Fundamental and Applied Sciences, HICoE—Centre for Biofuel and Biochemical Research, Institute of Self-Sustainable Building, Universiti Teknologi PETRONAS, Seri Iskandar, Perak Darul Ridzuan, Malaysia

Lee Siew Ling Center for Sustainable Nanomaterials, Ibnu Sina Institute for Scientific and Industrial Research, Universiti Teknologi Malaysia, Johor Bahru, Malaysia;
Chemistry Department, Faculty of Science, Universiti Teknologi Malaysia, Johor Bahru, Malaysia

Afzan Mahmad Fundamental and Applied Sciences Department, Universiti Teknologi PETRONAS, Seri Iskandar, Perak, Malaysia;
Laboratory Department, Royal College of Medicine Perak, Universiti Kuala Lumpur, Perak, Malaysia

Hermiza Mardesci Fakultas Pertanian, Universitas Islam Indragiri, Riau, Indonesia

Siti Nur Athirah Mazlan Department of Fundamental & Applied Sciences, Universiti Teknologi PETRONAS, Seri Iskandar, Perak, Malaysia

Nur Najihah Md Zuki School of Chemical Sciences, Universiti Sains Malaysia, Pulau Pinang, Malaysia

Onn Chiew Mey Universiti Teknologi PETRONAS, Seri Iskandar, Perak, Malaysia

Xu Ming Ming Faculty of Science and Natural Resources, Universiti Malaysia Sabah, Sabah, Malaysia

Mardawani Mohamad Faculty of Bioengineering and Technology, Universiti Malaysia Kelantan, Jeli Campus, Jeli, Kelantan, Malaysia

Sharifah Mohamad Department of Chemistry, Faculty of Science, Universiti Malaya, Kuala Lumpur, Malaysia

Norani Muti Mohamed Centre of Innovative Nanostructure and Nanodevices, Universiti Teknologi PETRONAS, Seri Iskandar, Perak, Malaysia;
Fundamental and Applied Sciences Department, Universiti Teknologi PETRONAS, Seri Iskandar, Perak, Malaysia

S. M. Mohamed Esivan Faculty of Engineering, Universiti Teknologi Malaysia, Johor Bahru, Johor DT, Malaysia

Muhammad Naeim Mohd Aris Universiti Teknologi PETRONAS, Seri Iskandar, Perak, Malaysia

Khairul Arifin Mohd Noh Department of Geosciences, Universiti Teknologi PETRONAS, Seri Iskandar, Perak, Malaysia

Nur Diyan Mohd Ridzuan Fundamental and Applied Sciences, Universiti Teknologi PETRONAS, Seri Iskandar, Perak, Malaysia

Noor Asmawati Mohd Zabidi Universiti Teknologi PETRONAS, Seri Iskandar, Perak, Malaysia

Asad Mumtaz Centre of Innovative Nanostructures and Nanodevices (COINN), Universiti Teknologi PETRONAS, Seri Iskandar, Malaysia;
Center of Contaminant Control and Utilization (CenCoU), Institute of Contaminant Management for Oil and Gas, Universiti Teknologi PETRONAS, Perak, Malaysia

Mohamed Mahmoud Elsayed Nasef Malaysia-Japan International Institute of Technology, Universiti Teknologi Malaysia, Kuala Lumpur, Malaysia

Mohammad Nazhan Nasir College of Engineering, Universiti Tenaga Nasional (UNITEN), Selangor, Malaysia

El-Sayed Negim Laboratory of Advanced Materials and Technology, Kazakh-British Technical University, Almaty, Kazakhstan

Van Thien Nguyen Department of Mathematics, FPT University, Education Zone, Hanoi, Vietnam

Khairul Ariffin Mohd Noh Department of Geosciences, Universiti Teknologi PETRONAS, Seri Iskandar, Perak, Malaysia

Teh Ubaidah Noh Institute of Bioproduct Development (IBD), Universiti Teknologi Malaysia (UTM), Johor Bahru, Malaysia

Mohd Sofi Numin Department of Fundamental and Applied Sciences, Universiti Teknologi PETRONAS, Seri Iskandar, Perak, Malaysia

Ernee Sazlinayati Othman Department of Fundamental and Applied Sciences, Centre for Intelligent Signal and Imaging Research (CISIR), Universiti Teknologi PETRONAS, Seri Iskandar, Perak, Malaysia

Irwan Othman Centralized Analytical Laboratory, Universiti Teknologi PETRONAS, Seri Iskandar, Perak, Malaysia

Mahmod Othman Fundamental and Applied Sciences Department, Universiti Teknologi PETRONAS, Seri Iskandar, Perak, Malaysia

Derrick Asamoah Owusu Mathematics Department, Kwame Nkrumah University of Science and Technology, Kumasi, Ghana

Timotius Pasang Oregon Institute of Technology, Klamath Falls, USA

Avita Ayu Permanasari Universitas Negeri Malang, East Java, Indonesia

Yeo Wee Ping Faculty of Science, Universiti Brunei Darussalam, Bandar Seri Begawan, Brunei Darussalam

Poppy Puspitasari Mechanical Engineering Department, State University of Malang, Malang, East Java, Indonesia;
Universitas Negeri Malang, East Java, Indonesia

Ayesha Rafique Department of Electrical Engineering, University of Engineering and Technology, Lahore, Pakistan

Norhana Abd. Rahim Faculty of Science and Technology, Universiti Sains Islam Malaysia, Nilai, Malaysia

Nurul Yani Rahim School of Chemical Sciences, Universiti Sains Malaysia, Pulau Pinang, Malaysia

Noor A.'in A. Rahman Electrical and Electronics Engineering Department, Universiti Teknologi PETRONAS, Seri Iskandar, Perak, Malaysia

Nurul Jannah Abd Rahman Fundamental and Applied Sciences Department, HICoE Centre for Biofuel and Biochemical Research, Institute of Self-Sustainable Building, Universiti Teknologi PETRONAS, Seri Iskandar, Perak, Malaysia

Muggundha Raoov Ramachandran Department of Chemistry, Faculty of Science, Universiti Malaya, Kuala Lumpur, Malaysia

Haya Ramba Cocoa Research and Development Center, Kota Samarahan, Sarawak, Malaysia

Anita Ramli Fundamental and Applied Sciences Department, HICoE Centre for Biofuel and Biochemical Research, Institute of Self-Sustainable Building, Universiti Teknologi PETRONAS, Seri Iskandar, Perak, Malaysia;
School of Science and Technology, Sunway University, Bandar Sunway, Selangor, Malaysia

Kamilah Ramly Fundamental and Applied Sciences Department, Centre of Innovative Nanostructure and Nanodevices (COINN), Universiti Teknologi PETRONAS, Seri Iskandar, Perak, Malaysia

R. Rashid Faculty of Engineering, Universiti Teknologi Malaysia, Johor Bahru, Johor DT, Malaysia

Hemamalini Rawindran Department of Fundamental and Applied Sciences, HICoE—Centre for Biofuel and Biochemical Research, Institute of Self-Sustainable Building, Universiti Teknologi PETRONAS, Seri Iskandar, Perak Darul Ridzuan, Malaysia

Mohamad Amirul Ashraf Mohd Razip Department of Fundamental and Applied Sciences, Universiti Teknologi PETRONAS, Seri Iskandar, Perak, Malaysia

Makhmetova Aliya Ruslanovna Kazakh-British Technical University, Almaty, Kazakhstan

Bahruddin Saad Fundamental and Applied Sciences Department, Universiti Teknologi PETRONAS, Seri Iskandar, Perak, Malaysia

Nabilah Saafie Universiti Teknologi PETRONAS, Perak, Malaysia

Maziyar Sabet Department of Petroleum and Chemical Engineering, Jalan Tungku Link, Gadong, Universiti Teknologi Brunei (UTB), Darussalam, Brunei

Amir Reza Sadrolhosseini Department of Physics, Faculty of Science, University Putra Malaysia, Selangor, Malaysia

Mohamed Shuaib Mohamed Saheed Centre of Innovative Nanostructure and Nanodevices (COINN), Universiti Teknologi PETRONAS, Seri Iskandar, Perak, Malaysia;
Department of Fundamental & Applied Sciences, Universiti Teknologi PETRONAS, Seri Iskandar, Perak, Malaysia

Nurul Tasnim Sahrin Department of Fundamental and Applied Sciences, Universiti Teknologi PETRONAS, Seri Iskandar, Perak, Malaysia

Hamzah Sakidin Fundamental and Applied Sciences Department, Universiti Teknologi PETRONAS, Seri Iskandar, Perak, Malaysia

Fitri Norizatie Salehin Department of Fundamental and Applied Sciences, Universiti Teknologi PETRONAS, Seri Iskandar, Perak, Malaysia

Suhaida Salleh Cocoa Research and Development Center, Jengka, Pahang, Malaysia

Nur Afiqah Mohamad Saman Department of Fundamental and Applied Sciences, Universiti Teknologi PETRONAS, Seri Iskandar, Perak Darul Ridzuan, Malaysia

Kozhabekov Serik Samsalykovich Kazakh-British Technical University, Almaty, Kazakhstan

Wasan Saphanuchart PETRONAS Research Sdn. Bhd, Bangi, Malaysia

Richard M. N. Y. Sarpong-Streetor Fundamental and Applied Sciences Department, Universiti Teknologi PETRONAS, Seri Iskander, Perak, Malaysia

Azali Saudi Faculty of Computing and Informatics, Universiti Malaysia Sabah, Sabah, Malaysia

Gowri Selvaraj Fundamental and Applied Sciences Department, Universiti Teknologi PETRONAS, Seri Iskander, Perak, Malaysia

Afza Shafie Department of Fundamental and Applied Sciences, Universiti Teknologi PETRONAS, Seri Iskandar, Perak Darul Ridzuan, Malaysia

Maizatul Shima Shaharun Fundamental and Applied Sciences Department, Universiti Teknologi PETRONAS, Seri Iskandar, Perak, Malaysia

Muhammad Umair Shahid Department of Fundamental and Applied Sciences, Centre of Innovative Nanostructure and Nanodevices (COINN), Universiti Teknologi PETRONAS, Bandar Seri Iskandar, Perak, Malaysia;
Department of Materials Science and Engineering, Institute of Space Technology, Islamabad, Pakistan

Nurul Natasha Shahrom School of Chemistry and Environment, Faculty of Applied Sciences, Universiti Teknologi MARA (UiTM), Shah Alam, Selangor, Malaysia

Nadeem Ahmad Sheikh Fundamental and Applied Sciences Department, Universiti Teknologi PETRONAS, Seri Iskandar, Perak, Malaysia

Ewe Lay Sheng College of Engineering, Universiti Tenaga Nasional (UNITEN), Selangor, Malaysia

Mohd Fadhlullah Abd Shukur Fundamental and Applied Sciences Department, Universiti Teknologi PETRONAS, Seri Iskandar, Malaysia

Nadiah Sidek Department of Chemistry, Faculty of Science, Universiti Malaya, Kuala Lumpur, Malaysia

Surajudeen Sikiru Department of Fundamental and Applied Sciences, Universiti Teknologi PETRONAS, Seri Iskandar, Perak, Malaysia

Balbir Singh Mahinder Singh Department of Fundamental and Applied Sciences, Centre of Innovative Nanostructure and Nanodevices (COINN), Universiti Teknologi PETRONAS, Seri Iskandar, Perak, Malaysia;
Centre for Foundation Studies, Universiti Teknologi PETRONAS, Seri Iskandar, Perak, Malaysia

Rajalingam Sokkalingam Fundamental and Applied Sciences Department, Universiti Teknologi PETRONAS, Seri Iskandar, Perak, Malaysia

Hassan Soleimani Fundamental and Applied Science Department, Universiti Teknologi PETRONAS, Seri Iskandar, Perak, Malaysia;
Institute of Hydrocarbon Recovery, Universiti Teknologi PETRONAS, Seri Iskandar, Perak Darul Ridzuan, Malaysia

Nur Amirah Suhaimi Universiti Teknologi PETRONAS, Seri Iskandar, Perak, Malaysia

Jumat Sulaiman Faculty of Science and Natural Resources, Universiti Malaysia Sabah, Sabah, Malaysia

Mohd Faisal Taha Department of Fundamental and Applied Sciences, Universiti Teknologi PETRONAS, Seri Iskandar, Perak, Malaysia;
Centre of Research in Ionic Liquids (CORIL), Universiti Teknologi PETRONAS, Seri Iskandar, Perak, Malaysia

Nazrizawati Ahmad Tajuddin School of Chemistry and Environment, Faculty of Applied Sciences, Universiti Teknologi MARA (UiTM), Shah Alam, Selangor, Malaysia

Geok Bee Teh Faculty of Engineering & Information Technology, Southern University College, Skudai, Skudai, Johor DT, Malaysia

Farman Ullah Department of Fundamental and Applied Sciences, Centre of Innovative Nanostructure and Nanodevices (COINN), Universiti Teknologi PETRONAS, Seri Iskandar, Malaysia

Fahad Usman Fundamental and Applied Sciences Department, Universiti Teknologi PETRONAS, Seri Iskandar, Perak, Malaysia

Norhidayah Ahmad Wazir PETRONAS Research Sdn. Bhd, Bangi, Malaysia

Cecilia Devi Wilfred Fundamental and Applied Sciences Department, Universiti Teknologi PETRONAS, Seri Iskandar, Perak, Malaysia

Mohd Dzul Hakim Wirzal Department of Chemical Engineering, Universiti Teknologi PETRONAS, Seri Iskandar, Perak, Malaysia

Noorhana Yahya Fundamental and Applied Science Department, Universiti Teknologi PETRONAS, Bandar Seri Iskandar, Perak, Malaysia

Nor Hafizah Yasin PETRONAS Research Sdn.Bhd, Kajang, Selangor, Malaysia

N. A. S. Yuharmon Department of Fundamental and Applied Science, Universiti Teknologi PETRONAS, Seri Iskandar, Perak, Malaysia

Normawati M. Yunus Fundamental and Applied Sciences Department, Centre for Research in Ionic Liquids, Institute of Contaminant Management, Universiti Teknologi PETRONAS, Seri Iskandar, Perak, Malaysia

Yumn Suhaylah Yusoff Faculty of Science and Technology, Universiti Sains Islam Malaysia, Nilai, Negeri Sembilan, Malaysia

Jemilat Yetunde Yusuf Fundamental and Applied Science Department, Universiti Teknologi PETRONAS, Seri Iskandar, Perak, Malaysia

Mudasar Zafar Fundamental and Applied Science Department, Universiti Teknologi PETRONAS, Seri Iskandar, Perak, Malaysia

Nur Fatin Syazwanie Zahari Fundamental and Applied Sciences Department and Centre for Biofuel and Biochemical Research, Institute of Self- Sustainable Building, Universiti Teknologi PETRONAS, Seri Iskandar, Perak, Malaysia

N. A. Zaharudin Faculty of Engineering, Universiti Teknologi Malaysia, Johor Bahru, Johor DT, Malaysia

Hasnah Mohd Zaid Department of Fundamental and Applied Sciences, Universiti Teknologi PETRONAS, Seri Iskandar, Perak, Malaysia

Siti Noor Azella Zaine Department of Chemical Engineering, Universiti Teknologi PETRONAS, Seri Iskandar, Perak, Malaysia

Nooraini Zainuddin Department of Fundamental and Applied Sciences, Universiti Teknologi PETRONAS, Seri Iskandar, Perak Darul Ridzuan, Malaysia

Zuraini Zakaria Biology Program, School of Distance Education, Universiti Sains Malaysia, Minden, Penang, Malaysia

Zakariyya Uba Zango Fundamental and Applied Sciences Department, Universiti Teknologi PETRONAS, Perak, Malaysia;
Department of Chemistry, College of Natural and Applied Sciences, Al-Qalam University Katsina, Katsina, Nigeria

Zh. Zhanabekov Al-Farabi Kazakh National University, Almaty, Kazakhstan

Nur Insyirah Zulkifli Center of Contaminant Control and Utilization (CenCoU), Institute of Contaminant Management for Oil and Gas, Universiti Teknologi PETRONAS, Universiti Teknologi PETRONAS, Seri Iskandar, Perak, Malaysia

Sustainable Chemistry

Acoustic Performance of Mixing EFB and OPF Low-Density Fibreboards in Different Thickness

Mageswaran Ravi Chandran, Ewe Lay Sheng, Mohammad Nazhan Nasir, Yew Weng Kean, and Zawawi Ibrahim

Abstract Fast growth of worldwide oil palm industry and nation's economics have affected the environment for the mishandling of oil palm waste, which has endangered the flora and fauna that surrounds it. The wastes of oil palm can be turned into good source of material for sound absorption or other usages. Furthermore, the awareness about the harmful effects of fibreglass as sound absorbing material has certainly increased and the demand of replacing it with natural fibres are growing day by day. This research highlights the acoustic properties of mixing empty fruit bunch (EFB) and oil palm frond (OPF) at mixing ratios of 50% EFB-50% OPF, 40% EFB-60% OPF, 30% EFB-70% OPF and 20% EFB-80% OPF in thickness of 12, 14, 16 and 18 mm with density of 120 kg/m^3. The sound absorption coefficient, SAC (α) test was conducted using the impedance tube method (ITM) and the morphology of the samples were examined using scanning electron microscope, SEM. The results showed no trend for the sac values with increasing of thickness and OPF contents. Nevertheless, it is noteworthy that the frequency range with 0.8 sac values and above are increased with thickness, it may be attributed to tortuosity effects. The morphology of EFB and OPF mixture may play a crucial role in determining the sound absorption.

Keywords Sound absorption coefficient · Empty fruit bunch · Oil palm frond

---

M. R. Chandran · E. L. Sheng (✉) · M. N. Nasir
College of Engineering, Universiti Tenaga Nasional (UNITEN), Putrajaya Campus, Jalan IKRAM-UNITEN, 43000 Selangor, Malaysia
e-mail: laysheng@uniten.edu.my

Y. W. Kean
School of Engineering and Physical Science, Heriot-Watt University Malaysia, No. 1, Jalan Venna P5/2, Precinct 5, 62200 Putrajaya, Malaysia
e-mail: w.yew@hw.ac.uk

Z. Ibrahim
Engineering and Processing Division, Malaysian Palm Oil Board (MPOB), No. 6, Persiaran Institusi, Bandar Baru Bangi, 43000 Kajang, Selangor, Malaysia

S. A. Abdul Karim et al. (eds.), *Proceedings of the 6th International Conference on Fundamental and Applied Sciences*, Springer Proceedings in Complexity,
https://doi.org/10.1007/978-981-16-4513-6_1

1 Introduction

Malaysia being one of the largest producers of oil palm industry in the world. Based on research from 2016, there are 5.76 million hectares of oil palm plantation across Malaysia [1]. But rough estimation shows about 84 million tons of oil palm biomass wastes are available throughout Malaysia [2]. The Malaysian government is facing a hard time to dispose this oil palm biomass wastes because the amount of waste is huge. The oil palm trunk (OPT), empty fruit bunch (EFB), oil palm frond (OPF), palm kernel shell (PKS), etc. all belong to the palm oil industry. Due to the moisture content of EFB is 67%, that natural fibre can be considered as a very good source for fuel. Given that EFB undergoes a proper pre-pressing process [3]. EFB is usually subjected to open burning once the oil is extracted from them. Instead of open burning and polluting the environment, the heat energy from the burning could be directed to boilers in oil palm mills [4]. Even in power generation sector, EFB natural fibre is considered as valuable natural resource [5]. Besides that, natural fibres from palm oil tree can be used to produce daily products such as soap, cosmetic, sound absorbing material for acoustic purposes and etc. [6].

Technology around us are evolving at a very fast rate day by day. So, does technology on enhancing sound absorption characteristic, which have been studied over the years by researchers around the world. The motivation that drives these researchers is because the fibreglass and asbestos are infamous to have health effects of humans and could also harm animals as well [7]. Synthetic fibre, mineral wool, polyester are the base for sound absorbing panels that currently in production [8]. Therefore, researchers are opting to test out natural fibres as a decent replacement for fibreglass as it is environmentally friendly and have good acoustic properties. Based on investigation by past researchers, it is said that both fibrous and non-fibrous natural fibres are proven to be a sustainable acoustical material and it can be commercialized [9]. Besides that, waste fibre from paddy has good sound absorption, as good as synthetic glass wool. Waste fibre from paddy achieved sound absorption coefficient up to 0.80 at frequency of 2500 Hz [10]. Another research done on coir natural fibre, where the absorption coefficient average result of 0.80 was obtained [11].

For few years now, researchers have been testing various natural fibres on their acoustic properties. Natural fibres such as oil palm trunk [12], empty fruit bunch [13], oil palm frond [14], kapok [15], bamboo [16], arenga pinnata [17], paddy straw [18], coconut coir [19] and etc. Currently, OPT natural fibre has proved to have the best sound absorption coefficient thus far. The material was able to achieved SAC of 0.99 at frequency of 3000–6000 Hz for sample thickness 12 mm [12]. Sugarcane bagasse showed an average sound absorption of 0.80 at frequency range of 2000–4500 Hz [20].

Tortuosity is the expansion of pathway trough the pores in the material. Other than that, tortuosity also influences the internal structure of porous material on its acoustic performance [21]. A tight space structure will be small and there will also be less volume of air, resulting in a narrow passage for the sound wave to travel [22]. Hence, sound waves will travel in longer distance and the tortuosity will increase.

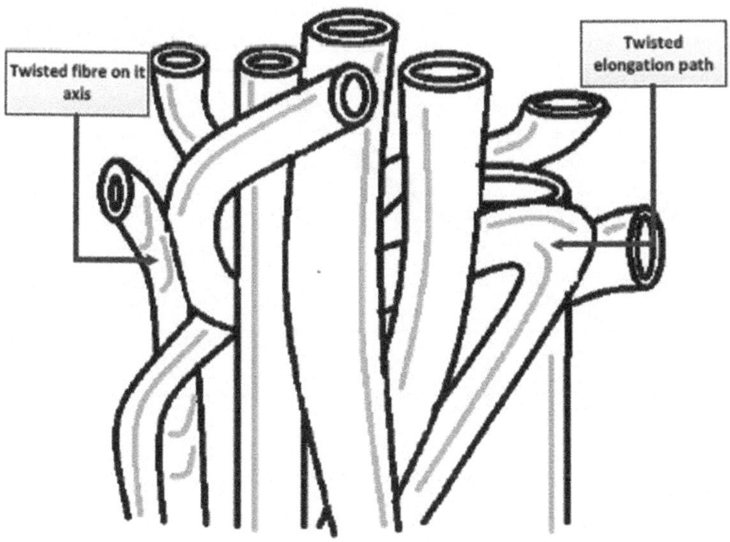

Fig. 1 Pathway of fibres (tortuosity)

High tortuosity will produce good sound absorption at high range frequency. Figure 1 shows how the pathway of fibres look like [23].

This research mainly focusses on the mixture of empty fruit bunch and oil palm frond natural fibres and their acoustic properties in thickness of 12, 14, 16 and 18 mm. By completing this research, the environmental problems and the noise pollution problem will be able to be solved at one go.

2 Methodology

The mixture of EFB and OPF low density fibreboard (LDF) at different mixing ratio were prepared. The fabrication process involved chipping, refining, glue bending, mat forming, pre-press, hot press and cool down. For further interpretation, can refer to my previous publication and other publication [24, 25].

3 Results and Discussions

Sound Absorption Coefficient, SAC (α) is the term that is used to measure the sound absorption rate of a material. SAC (α) is the ratio of sound transmitted through a material and the incident sound that emitted from the material. For instance, when a material achieved SAC (α) value of 0.95 at certain frequency. It means at that

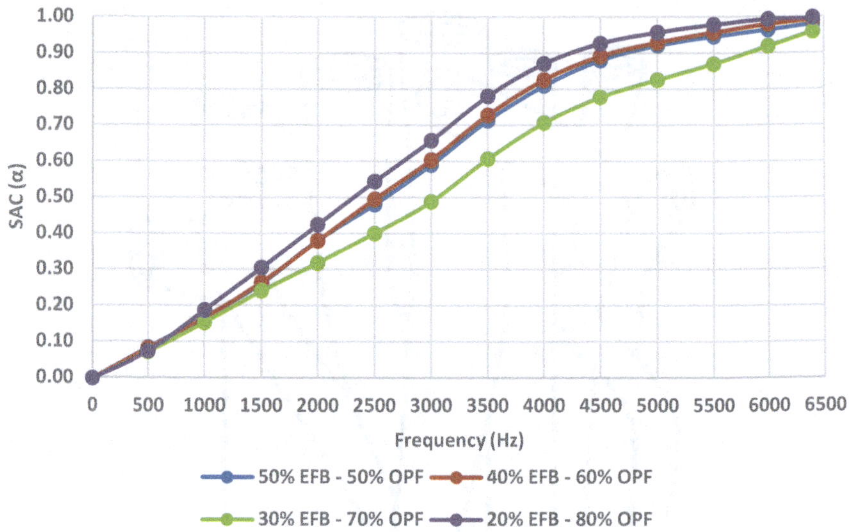

Fig. 2 SAC, α values versus frequency (Hz) in four different mixing ratios with thickness of 12 mm

frequency; the material is able to absorb 95% of the sound and only 5% of the remaining sound is reflected back to the surrounding as heat energy.

Figure 2 shows the SAC, α values of four different mixing ratios with the same thickness of 12 mm. SAC, α values are found to increases as the frequency increase from 0 to 6400 Hz for all samples. It can be clearly seen that the sac values for all samples increased with increasing in OPF contents except for sample with mixing ratio of 30% EFB-70% OPF. All samples performed outstandingly (sac values, $\alpha >$ 0.8) at high range of frequency 5000 Hz and above. The SAC results are comparable with OPT because of SAC, α values are more than 0.80 at frequency of 4000 Hz [8]. Compared to mixture of EFB and OPF sample thickness 12 mm, the OPT performed better at low frequency meanwhile mixture of EFB and OPF performs better at higher frequency. The average SAC, α for thickness 12 mm is higher when compared to other thickness.

Based on Fig. 3 the best SAC, α values obtained from sample 50% EFB-50% OPF, 40% EFB-60% OPF and 20% EFB-80% OPF were 0.98, 0.99 and 1.00 respectively at frequency 6400 Hz. Sample 30% EFB-70% OPF only achieved SAC, α more than 0.90 at frequency of 6000 Hz, whereas the other sample did it at frequency of 5000 Hz. The SAC, α values for all samples are found to increase linearly from 5000 to 6400 Hz. The average acoustic performance of sample 30% EFB-70% OPF may be affected by sample fabrication process [26].

Figure 4 shows the SAC, α values of four different mixing ratios with the same thickness of 14 mm. All the samples show very consistency in the increase of sound absorption as the frequency increases. Sample 40% EFB-60% OPF has a superior acoustic property at the frequency range of 1500–5000 Hz compared to other

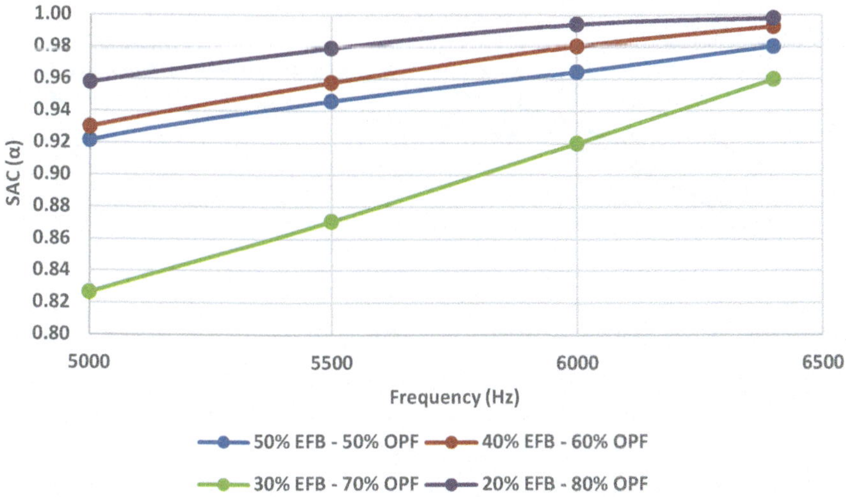

Fig. 3 SAC, α values versus frequency (Hz) in four different mixing ratios with thickness of 12 mm from frequency range of 5000 to 6400 Hz

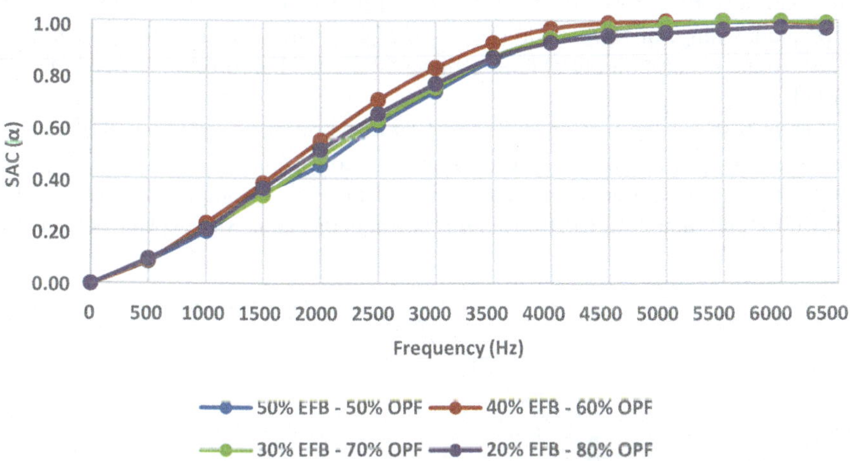

Fig. 4 SAC, α values versus frequency (Hz) in four different mixing ratios with thickness of 14 mm

samples. Whereas, sample 50% EFB-50% OPF has a marginally low SAC, α values at frequency 1000–3500 Hz contrasted to other samples.

Figure 5 shows all the LDF samples of thickness 14 mm with different mixing ratios reached the SAC values of 0.8 and above from the frequency range of 3500 to 6400 Hz. The SAC, α values are found to increase with frequency. Sample mixture of 40% EFB-60% OPF has the SAC, α of 0.90 at the frequency of 3500 Hz while the other samples only in the range of 0.80. Sample 50% EFB-50% OPF, 40% EFB-60%

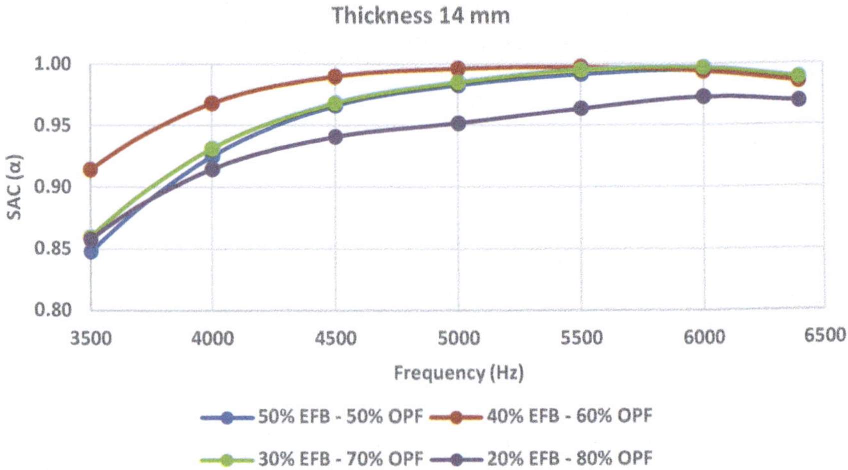

Fig. 5 SAC, α values versus frequency (Hz) in four different mixing ratios with thickness of 14 mm from frequency range of 3500 to 6400 Hz

OPF and 30% EFB-70% OPF all achieved unity, which is 100% sound absorption at frequency range of 5000–6000 Hz. Sample 20% EFB-80% OPF did not achieved 100% sound absorption but able to achieve 97% sound absorption at frequency of 6400 Hz.

Figure 6 shows the SAC, α values of four different mixing ratios with the same thickness of 16 mm. All the samples show very consistency in increasing of sound absorption as the frequency increased. Based on Fig. 5, sample mixture of 40% EFB-60% OPF and 20% EFB-80% OPF portraits the same SAC, α values from the

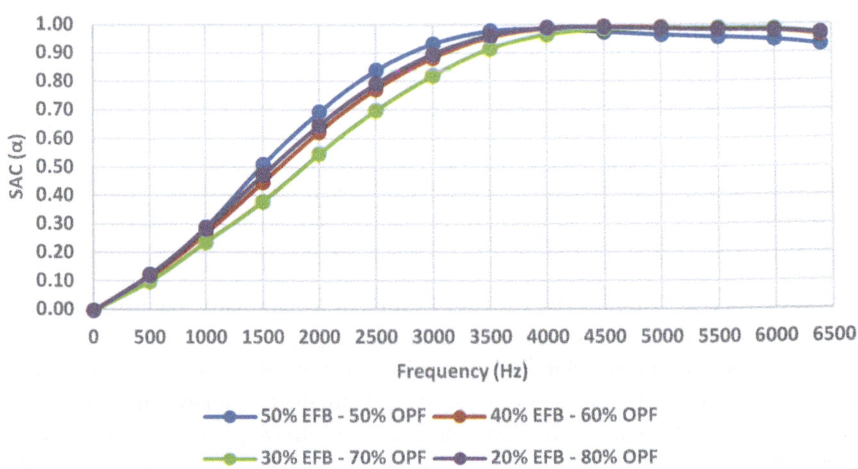

Fig. 6 SAC, α values versus frequency (Hz) in four different mixing ratios with thickness of 16 mm

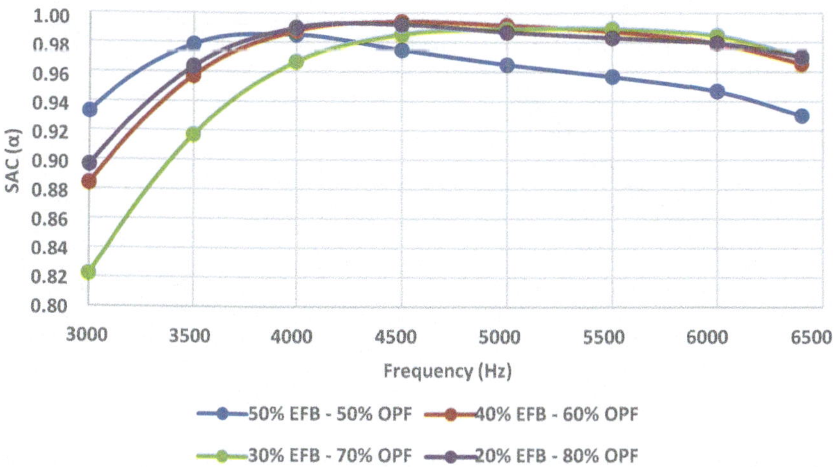

Fig. 7 SAC, α values versus frequency (Hz) in four different mixing ratios with thickness of 16 mm from frequency range of 3000 to 6400 Hz

frequency range of 3500 to 6400 Hz. Comparing this result with rice straw sound absorption, rice straw performs better at mid-range frequency. SAC, α of rice straw at frequency 2000 Hz is 0.70, but low at high frequency of 8000 Hz with SAC, α of 0.50 [27]. Although mixture of EFB and OPF did not performs as well as rice straw at mid-range frequency, mixture of EFB and OPF did very well at high frequency.

Figure 7 shows all the LDF of thickness 16 mm in different mixing ratios obtained the SAC values of 0.8 and above from the frequency range of 3000 to 6400 Hz. Sample with a mixture of 50% EFB-50% OPF and 20% EFB-80% OPF have the SAC, α values exceeded 0.90 at the frequency of 3000 Hz while samples 40% EFB-60% OPF and 30% EFB-70% OPF only exceeded SAC, α value more than 0.80 at frequency of 3000 Hz. Sample 50% EFB-50% OPF performed better than other three samples at frequency range of 1500–4000 Hz. All the samples except 50% EFB-50% OPF, almost achieved unity with SAC, α value of 0.99 with only 1% of remaining sound reflected.

Figure 8 shows the SAC, α values of four different mixing ratios with the same thickness of 18 mm. All the sample shows high consistency in increasing of sound absorption as the frequency increased. Based on Fig. 7, its observed that all the samples have very similar trend of SAC, α values from frequency range of 0 to 6400 Hz except for sample mixture 50% EFB-50% OPF, where it has a marginally lower SAC, α values from frequency range of 1500 to 3000 Hz. When compare to hollow bamboo, which performs excellently at mid-range frequency of 3600 Hz SAC, α of 0.95 [28]. However, mixture of EFB and OPF samples have an average SAC, α of 0.97 at frequency of 3500 Hz.

Figure 8 shows the LDF of thickness 18 mm with different mixing ratios that possessed the SAC values of 0.8 and above from the frequency range of 2500 to 6400 Hz and 0.9 and above from the frequency range of 3000 to 6400 Hz (Fig. 9).

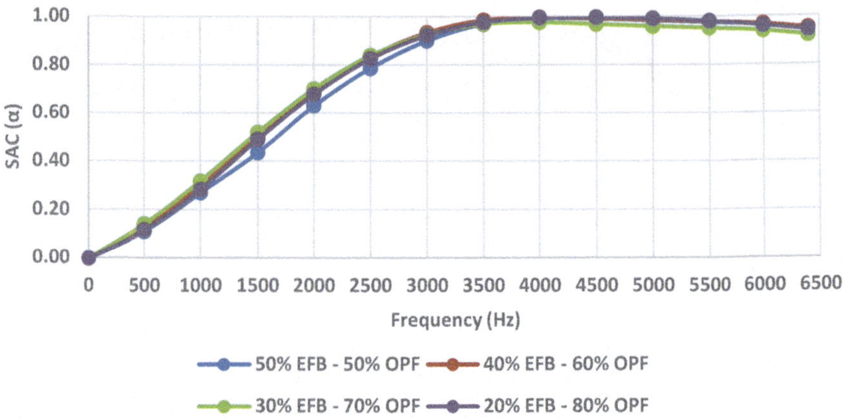

Fig. 8 SAC, α values versus frequency (Hz) in four different mixing ratios with thickness of 18 mm

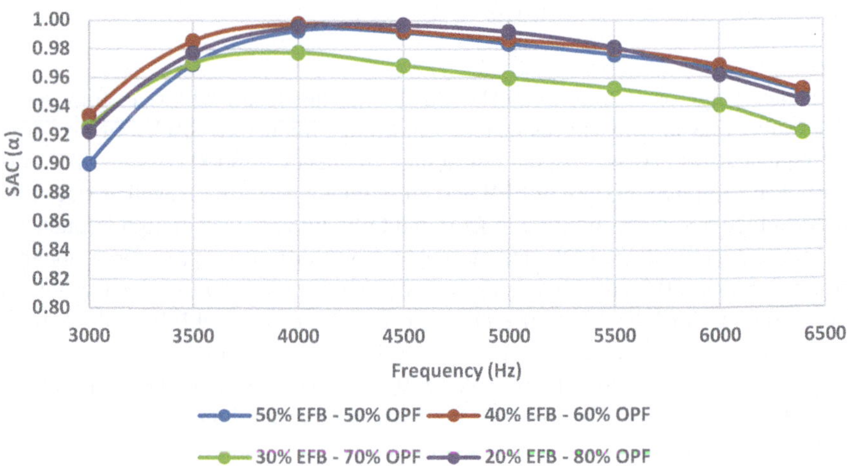

Fig. 9 SAC, α values versus frequency (Hz) in four different mixing ratios with thickness of 18 mm from frequency range of 3000 to 6400 Hz

Sample 40% EFB-60% and 20% EFB-80% achieved 100% sound absorption at frequency of 4000 Hz. After 4000 Hz, all the sample experience a decrease in SAC, α values. However, that is not a big issue because the SAC, α values are still more than 0.90 which means more than 90% sound absorption.

Tortuosity plays a large role in determining the acoustic performance of the natural fibre. The more fibre used to fabricate the fibreboard, the more tortuous the path will become [29, 30]. This will lead to a higher flow of resistivity. Thus, the time taken for the sound wave to strike is longer and more sound can be absorbed. The path of the natural fibre is more tortuous for all the thickness and it may be due to the combination of two different natural fibres. All the samples contain a certain porous percentage, which allows the reflection of sound wave travel and hence increases the absorption rate.

4 Conclusion

In a nutshell, by conducting this research using natural fibres for sound absorption material able to solve the two trending main issues. First, the waste from the natural fibres were able to reduce and put the waste into good use. Second, the sound absorption material will be able to replace synthetic fibres in the market. Natural fibres does not only show good acoustic properties but also plays a vital role in design ergonomics. Since natural fibre is renewable and biodegradable, they will produce lower emission during production compare to synthetic fibres [31]. The sound absorption coefficient test SAC, α was carried out for EFB and OPF mixed fibreboard with four different ratios in four different thickness of 12, 14, 16 and 18 mm. All the LDF samples are able to achieve unity ($\alpha = 1.00$) at high frequency. It's also notable that by increasing the thickness from 12 to 18 mm for all mixing ratios, it is able to widen the frequency range for SAC values of 0.8 and above from 5000–6400 Hz to 2500–6400 Hz. The mixture of EFB and OPF have never been tested for acoustic properties, hence therefore this research study is a new beginning into determining natural fibres as a substitute of synthetic fibre.

Acknowledgements The authors would like to thank Universiti Tenaga Nasional's BOLD grant project code: RJO10517844/060 and Malaysian Oil Palm Board (MPOB) for supporting the research work.

References

1. Nambiappan, B., et al.: Malaysia: 100 years of resilient palm oil economic performance. J. Oil Palm Res. **30**, 13–25 (2018)
2. Sulaiman, O., Salim, N., Nordin, N.A., Hashim, R., Ibrahim, M., Sato, M.: The potential of oil palm trunk as biomass as an alternative source for compressed wood **7**, 2688–2706 (2012)
3. Wicke, B., Sikkema, R., Dornburg, V., Faaij, A.: Exploring land use changes and the role of palm oil production in Indonesia and Malaysia. Land Use Policy **28**(1), 193–206 (2011)
4. Loh, S.K.: The potential of the Malaysian oil palm biomass as a renewable energy source. Energy Convers. Manag. **141**, 285–298 (2017)
5. Hansen, U.E., Nygaard, I.: Sustainable energy transitions in emerging economies: the formation of a palm oil biomass waste-to-energy niche in Malaysia 1990–2011. Energy Policy **66**, 666–676 (2014)
6. Choong, C.G., McKay, A.: Sustainability in the Malaysian palm oil industry. J. Clean. Prod. **85**, 258–264 (2014)
7. Kilburn, K.H., Powers, D., Warshaw, R.H.: Pulmonary effects of exposure to fine fibreglass : irregular opacities and small airways obstruction, pp. 714–720 (1992)

8. Kalaivani, R., Ewe, L.S., Zaroog, O.S., Woon, H.S., Ibrahim, Z.: Acoustic properties of natural fiber of oil palm trunk. Int. J. Adv. Appl. Sci. **5**(6), 88–92 (2018)
9. Oldham, D.J., Egan, C.A., Cookson, R.D.: Sustainable acoustic absorbers from the biomass. J. Appl. Acoust. **72**(6), 851–856 (2011)
10. Putra, A., Abdullah, Y., Efendy, H., Mohamad, W.M.F.W., Salleh, N.L.: Biomass from paddy waste fibers as sustainable acoustic material. Adv. Acoust. Vib. **2013** (2013)
11. Hosseini Fouladi, M., Ayub, M., Jailani Mohd Nor, M.: Analysis of coir fiber acoustical characteristics. Appl. Acoust. **72**(1), 35–42 (2011)
12. Kalaivani, R., Ewe, L.S., Chua, Y.L., Ibrahim, Z.: The effects of different thickness of oil palm trunk (Opt) fiberboard on acoustic properties. Am. J. Environ. Eng. Sci. **29**(5), 1105–1108 (2017)
13. Hee, O.K., Putra, A., Mohd Nor, M.J., Selamat, M.Z., Ying, L.Z.: Sound absorption performance of oil palm empty fruit bunch fibers. In: 23rd International Congress Sound Vibration, pp. 1–8 (2016)
14. Ibrahim Z., Aziz, A.A., Mokhtar, A., Ramli, R., Mamat, R.: Production of medium density fibreboard (mdf) from oil palm fronds and its admixture, pp. 1511–7871 (2012)
15. Veerakumar, A., Selvakumar, N.: A preliminary investigation on kapok/polypropylene nonwoven composite for sound absorption A. Indian J. Fibre Text. Res. **37**(2), 385–388 (2012)
16. Koizumi, T., Tsuijuchi, N., Adachi, A.: The development of sound absorbing materials using natural bamboo fibers and their acoustic properties. In: Inter-Noise Noise Conference Congress Proceedings vol. **2002** (5), pp. 713–718 (2002)
17. Ismail, L.: Acoustic and durability performance of arenga pinnata panel. Faculty Mechanical Manufacturing Engineering University, Tun Hussein Onn Malaysia (2012)
18. Abdullah, Y., Putra, A., Effendy, H., Farid, W.M., Ayob, R.: Dried paddy straw fibers as an acoustic absorber: a preliminary study. http://eprints2.utem.edu.my pp. 52–56 (2011)
19. Mahzan, S., Zaidi, A.M., Arsat, N., Hatta, M.N.M., Ghazali, M.I., Mohideen, S.R.: Study on sound absorption properties of coconut coir fibre reinforced composite with added recycled rubber. Int. J. Integr. Eng. Mech. Mater. Manuf. pp. 1–6 (2009)
20. Putra, A., Abdullah, Y., Efendy, H., Farid, W.M., Ayob, M.R., Py, M.S.: Utilizing sugarcane wasted fibers as a sustainable acoustic absorber. Procedia Eng. **53**, 632–638 (2013)
21. Jiang, S., Xu, Y., Zhang, H., White, C.B., Yan, X.: Seven-hole hollow polyester fibers as reinforcement in sound absorption chlorinated polyethylene composites. Appl. Acoust. **73**(3), 243–247 (2012)
22. Jayamani, E., Hamdan, S., Ezhumalai, P., Bakri, M.K.: Investigation on dielectric and sound absorption properties of banana fibers reinforced epoxy composites. J. Teknol. **78**(6–10), 97–103 (2016)
23. Markiewicz, E., Paukszta D., Borysiak, S.: Acoustic and dielectric properties of polypropylene-lignocellulosic materials composites. In: Polypropylene InTech, pp. 193–217 (2012)
24. Mageswaran, R., Sheng, E.L., Ibrahim, Z.: The thickness effects on acoustic properties of oil palm trunk natural fiber in density of 170 Kg/m3. Sci. Int. **30**(6), 845–849 (2018)
25. Ibrahim, Z.: Production of medium density fibreboard (MDF) from oil palm trunk. J. Appl. Sci. **11** (2014)
26. Ibrahim, Z., Aziz, A.A., Ramli, R., Mokhtar, A., Lee, S.: Effect of refining parameters on medium density fibreboard (MDF) properties from oil palm trunk (Elaeis guineensis), vol. 2013, pp. 127–131 (2013)
27. Yang, H.-S., Kim, D.-J., Kim, H.-J.: Rice straw wood particle for sound absorbing wooden construction materials. Bioresour. Technol. **86**, 117–121 (2003)
28. Putra, A., Khair, F.A., Nor, M.J.M.: Utilizing hollow-structured bamboo as natural sound absorber. Arch. Acoust. **40**(4), 601–608 (2015)
29. Vallabh, R., Banks-lee, P., Seyam, A.: New approach for determining tortuosity in fibrous porous media. J. Eng. Fiber. Fabr. **5**(3), 7–15 (2010)

30. Umnova, O., Attenborough, K., Shin, H., Cummings, A.: Deduction of tortuosity and porosity from acoustic reflection and transmission measurements on thick samples of rigid-porous materials, vol 66, pp. 607–624 (2005)
31. Begum, K., Islam, M.A.: Natural fiber as a substitute to synthetic fiber in polymer composites: a review. Res. J. Eng. Sci. **2**(3), 46–53 (2013)

Modified Alginate Adsorbent Using Ionic Liquid for Manganese Removal from Aqueous Solution

Noor A.'in A. Rahman, Gowri Selvaraj, and Cecilia Devi Wilfred

Abstract Removal of heavy metals from industrial effluent have been a subject of interest for researchers all around the globe, due to the industrial pollution which become a major cause of environmental degradation. The use of absorbent to extract those heavy metals can be enhanced with the addition of ionic liquids. The objective of this study is to improve the extraction efficiency of manganese (Mn) from industrial effluent using alginate beads loaded with imidazolium based ionic liquid. The strength of the beads was enhanced with the addition of polyvinyl alcohol. An extraction efficiency of 90% was achieved. The 'hard' beads have exceptional mechanical strength making them highly suitable to used in harsh conditions during the extraction process. Out of the two adsorption isotherms applied, the Langmuir model provides the best fit compared to the Freundlich model. The modified beads can meet variety of requests/applications and exhibits high metal adsorption ability.

Keywords Alginate · Ionic liquids · Manganese ions

1 Introduction

Industrial pollution is a major cause of environmental degradation which in turn is a key factor in turning extreme weather events into natural disasters. Numerous studies from all over the world have reported the trace contaminants pollute their land and surface water from time to time [1–6]. Frequent incidents of Malaysian raw water being contaminated with metals are of concern. Toxic metal such as manganese, Mn normally contributes to the rusty taste and reddish colour of water. Unlike organic contaminates, heavy metals are not biodegradable and known to be toxic and/or

N. A. A. Rahman (✉)
Electrical and Electronics Engineering Department, Universiti Teknologi PETRONAS, 32610 Seri Iskandar, Perak, Malaysia

G. Selvaraj · C. D. Wilfred
Fundamental and Applied Sciences Department, Universiti Teknologi PETRONAS, 32610 Seri Iskandar, Perak, Malaysia
e-mail: cecili@utp.edu.my

S. A. Abdul Karim et al. (eds.), *Proceedings of the 6th International Conference on Fundamental and Applied Sciences*, Springer Proceedings in Complexity,
https://doi.org/10.1007/978-981-16-4513-6_2

carcinogenic. Because of their high solubility in the aquatic environments, heavy metals can be absorbed by living organisms, thus effecting our food chain.

Innovative processes for treating industrial effluent containing heavy metal involves several technologies. Numerous approaches namely ion-exchange, water softening, adsorption by natural low-cost adsorbent, liquid extraction and so forth. Adsorption, especially by low-cost materials has become one of the alternative treatment method. Ionic liquids (ILs) act as a versatile extracting agent of metal in aqueous media or so-called "green extractants" due to their well-known versatile properties [7–9]. Recent advances and developments in ILs have helped to broaden its potential in various applications. ILs could be designed for specific applications in the extraction of metal ions [10–15]. Most of the studies showed that hydrophobic ILs is generally favourable for metal extraction due to their ability to extract metal ions in aqueous phase [16, 17].

Alginate is one of the most abundant, cheapest and environmentally friendly natural materials as adsorbent. Unmodified alginate is too soft, fragile and brittle for harsh experimental conditions. Alginate itself exhibits poor metal adsorption performance. By a combination of ionic liquids, alginate and polyvinyl alcohol (PVA) network, they are able to enhance the sorbent mechanical strength during harsh services. This new sorbent enhances the extraction performance and provides a simpler extraction process compared to current technique. The sorbent material is easy to handle and can be recycled for few times. A better understanding can be achieved regarding the adsorption process based on adsorption equilibrium studies. The most frequently used models to describe the equilibrium data of two-parameter isotherm are Langmuir and Freundlich. These models describe the adsorption process and how adsorbates interact with the absorbent [18]. This paper presents the capability of our adsorbent in removing manganese from contaminated water. The sorbent comprises of 10% (w/v) ionic liquid, alginate and PVA. The sorbent beads offer good mechanical strength toward harsh environment and can be recycled, hence reducing material cost since minimal ionic liquid would be needed.

2 Materials and Methods

2.1 Ionic Liquids

1-Hexylimidazolium-3-propionitrile chloride was synthesized and the structure of the ionic liquid was confirmed from 1H NMR spectra. 0.15 mol of imidazole salt and 0.19 mol of chlorohexane were mixed and stirred in the presence of dimethyl sulfoxide and potassium hydroxide at room temperature for 24 h. Solvent and unreacted reagents from this precursor were subsequently removed after the reaction. Next, 0.12 mol of 3-chloropropionitrile chloride was introduced dropwise into the precursor. The reactants were allowed to stir for 48 h at 55°C. The sample was purified to remove unreacted reagents. The resultant ionic liquid was concentrated under reduced pressure using rotary evaporator.

2.2 *Preparation of Alginate Modified ILs*

4 g of PVA and 0.675 g of sodium alginate were dissolved in 50 mL of Millipore water and stirred at 80°C for 8 h. IL (10% w/v) was added to this mixture. The mixture was stirred at 30°C for 6 h at a stirring rate of 500 rpm to obtain a homogeneous gel blend which was then extruded into a gently stirred saturated 3% w/v $CaCl_2$-boric acid solution using a syringe, which gave spherical hydrogel beads. After 24 h, the beads were washed with Millipore water to remove any impurities. The beads were dried in an oven at 35°C for 48 h. An estimated 50% shrinkage in bead size was observed after drying. The surface characteristics of the beads were analyzed using Scanning Electron Microscopy (SEM) after curing.

2.3 *Respond Surface Method (RSM)*

Factors influencing the extraction efficiency of Mn (VII) using the proposed IL/aqueous system was predicted using RSM. The factors investigated are pH of waste water, contact time, and weight of beads. Central composite design (CCD is a more popular rotatable and orthogonal technique applicable for modeling the main factors and interaction factors with a response [19]. The polynomial regression equation is most prominent relationship for analysis of correlation between factors and response. These factors were investigated by using preliminary experiments, and then the significant factors were modelled by central composite design (CCD).

Experimental design On the basis of the preliminary experiments, the factors which had the greatest influence on extraction efficiency (EE%) of Mn (VII) were selected. The CCD was used as an approach for optimizing the main variables which are pH of wastewater, contact time, and weight of beads. The initial concentration of the wastewater is 44 ppm. Independent variables in the range of -1 and $+ 1$ were coded according to classical methods Table 1. The independent variables and level are designed to evaluate the effects on the response, removal efficiency (Y) of manganese (Mn) from waste water. Design–expert (Stat-Ease 8.0.7.1) is the software used for designing the experiment statistical studies and response surface studies. The parameters are shown in Table 1.

Batch sorption experiment The optimization experiments were carried out based on the batch sorption studies. A 1000 dilution factor (DF) of wastewater was prepared by dissolving in 100 mL of Millipore water. Based on the experimental design, an

Table 1 Parameters and their level in CCD model

Variables	Factor	Unit	Range and levels (coded)		
			(low) −1	(central point) 0	(high) + 1
Contact time	X_1	min	60	180	300
Weight	X_2	g	0.2	0.35	0.5
pH	X_3	–	2	5	8

amount of weight PVA/alginate/ILs beads were shaken in 10 mL of wastewater at 400 rpm for a range of contact time by orbital shaker. All experiments were performed in duplicates and the result were given as average. Concentration of metal solution after absorption test was measured using Atomic Absorption Spectroscopy (AAS). The percentages of extraction efficiency (EE%) are calculated using the following equations respectively.

$$EE\% = \frac{C_i - C_f}{C_i} \times 100 \quad (1)$$

where C_i and C_f are the initial and final concentration of Mn (VII) in mg L^{-1}.

2.4 *Adsorption Isotherm*

The wastewater was diluted by using distilled water for this study. The pH of the diluted wastewater was between 6 and 7. The test matrix for the experiment is shown in Table 2.

Based on Table 3, the adsorption isotherm was evaluated. Langmuir and Freundlich isotherm models are used to describe the adsorption process that occurs on heterogonous surface of the absorbent.

Table 2 Test matrix for adsorption studies

Parameter	Minimum	Maximum
Contact time/(min)	30	300
Initial concentration/(ppm)	14	44

Table 3 Test matrix for isotherm study

Isotherm model	Equation	Indication
Langmuir isotherm	Linear form: $\frac{C_e}{q_e} = \frac{1}{q_m K_L} + \frac{C_e}{q_m}$	C_e: concentration of adsorbate at equilibrium (mg/g) q_e: adsorbed quantity (mmol/g) qm: monolayer maximal adsorption capacity (mmol/g) K_L: Langmuir constant related to adsorption capacity (mg/g) K_F: Freundlich constant related to adsorption capacity (mg/g)
Freundlich isotherm	Linear form: $\log q_e = \log K_F + \frac{1}{n} \log C_e$	

3 Result and Discussion

3.1 *Ionic Liquid*

1-Hexylimidazolium-3-propionitrile chloride (22.50 g, 75%); 1H NMR δ (DMSO-d6) 1H NMR (500 MHz, DMSO): δ 0.85 (t, 3H, CH3), 1.23–1.27 (m, 6H, CH2), 1.77–1.83 (m, 2H, CH2), 3.07 (t, 2H, CH2), 3.84 (t, 2H, CH2), 4.20 (t, 2H, CH2), 7.76 (d, 2H, imidazole), 9.25 (s, 1H, imidazole). This compound contains 20 hydrogen atoms, integration of the assigned peaks confirmed the amount of hydrogen atoms as shown in Fig. 1. The three protons in the imidazole ring gives rise to peaks above 7.70 ppm due to the anisotropy of induced magnetic field produced by the π ring current which has a deshielding effect on the aromatic protons. The hexyl chain protons are found between 0.84 and 3.85 ppm. Hydrogen atoms on the carbon-1 gives rise to a triplet peak whereas protons on carbons 2–4 give rise to a single peak due to their equivalent nature. Protons on carbon-6 appear further downfield due to the neighboring highly electronegative nitrogen atom. The protons on the propionitrile group appear at 3.07 and 4.20 ppm. The proton closer to aromatic ring experiencing greater deshielding.

3.2 *Appearance of the Modified Beads*

Basically, the produced beads in an almost sphere-like shape after pass through the drying process. The beads also harder towards the end of curing process. Figure 2 shows the surface of the modified beads taken using scanning electron microscope (SEM) at different magnification (80 X, 1000 X, 5000 X, 10,000 X), which showed the presence of pores on the surface of beads. Pore sizes between 155 and 250 nm are tabulated almost homogenously on the surface of beads. The diameter of dry beads is around 1.22 mm.

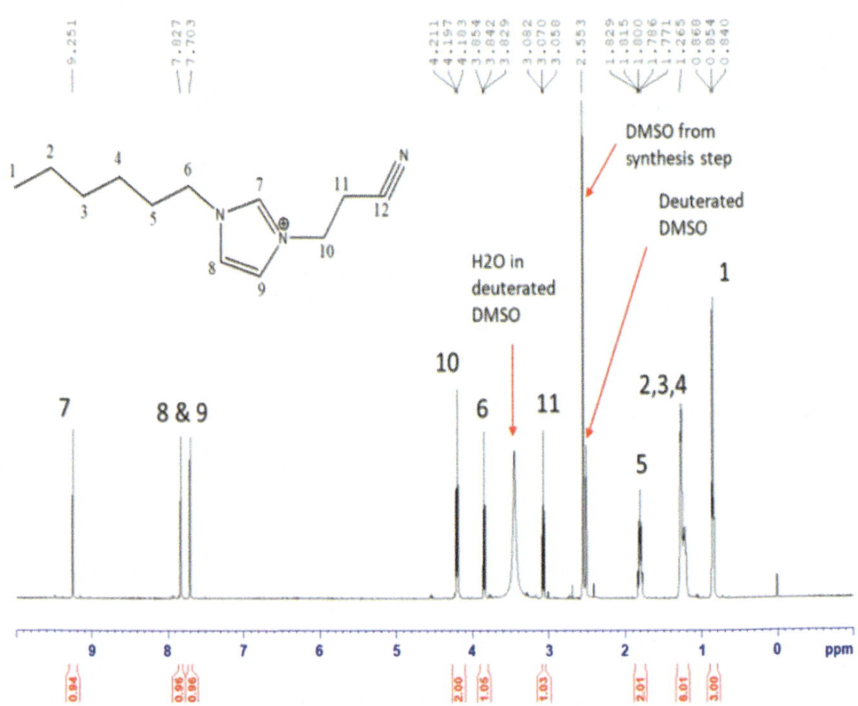

Fig. 1 NMR spectrum of 1-hexylimidazolium-3-propionitrile chloride

3.3 Optimization of Beads Towards Extraction of Mn Ions

A total 15 experiments were conducted as per Central Composite Design (CCD) surface statistical design. The ANOVA results indicates a coefficient of determination (R^2) for the extraction of Mn ions from waste water solution to be 0.9952, shown from the predicted against actual values in Fig. 3. Three-dimensional response surface is plotted plots as graphical representations, based on regression equation given by the model. The extraction efficiency of Mn ions increased by contact time and weight of absorbent up to the highest limit at pH 5.5 of wastewater. Longer contact time allowed the longer interaction between beads and Mn ions which would result in increased of the extraction efficiency. From the statistical analysis, it can be concluded that, the optimum values of the variables according to the quadratic model are wastewater at pH 3.54, contact time at 225 min and weight of absorbent is 0.49 g with desirability up to 1.0000 at the highest extraction efficiency, 95.95%.

Figure 4 depicts the dimensional response surface plot for the selected parameters in this work. Each parameter affects the EE% values. This interaction established a dependence of the degree of EE% on the mutual interaction between those three factors.

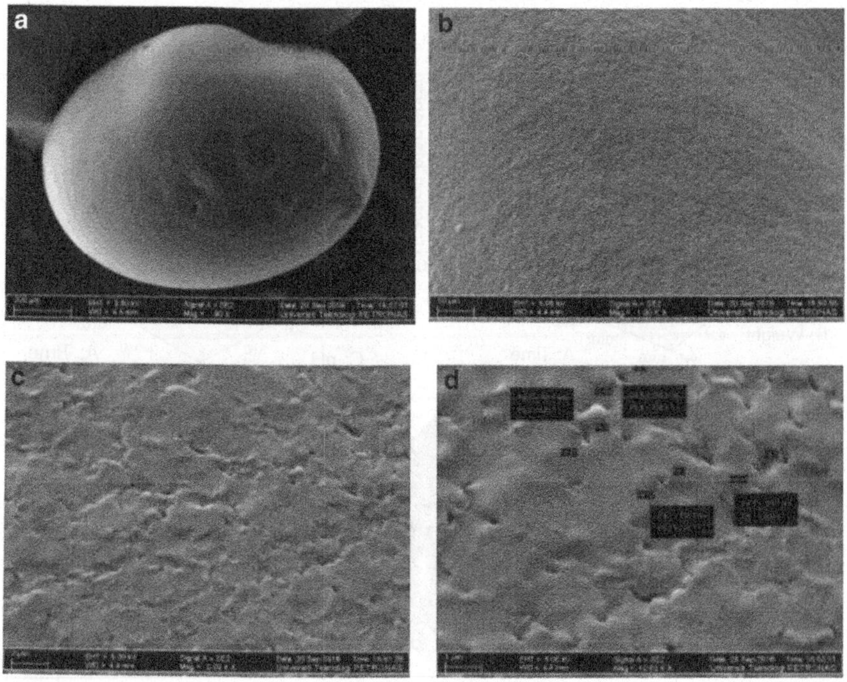

Fig. 2 Image of modified ILs beads taken by SEM at different magnification **a** 80 X **b** 1000 X **c** 5000 X **d** 10, 000 X

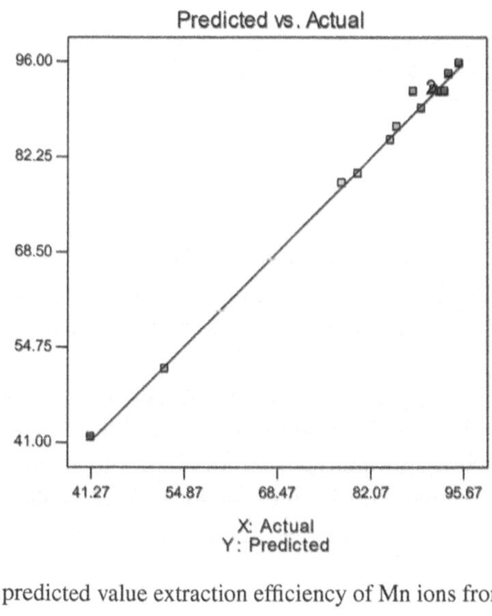

Fig. 3 Actual against predicted value extraction efficiency of Mn ions from wastewater solution

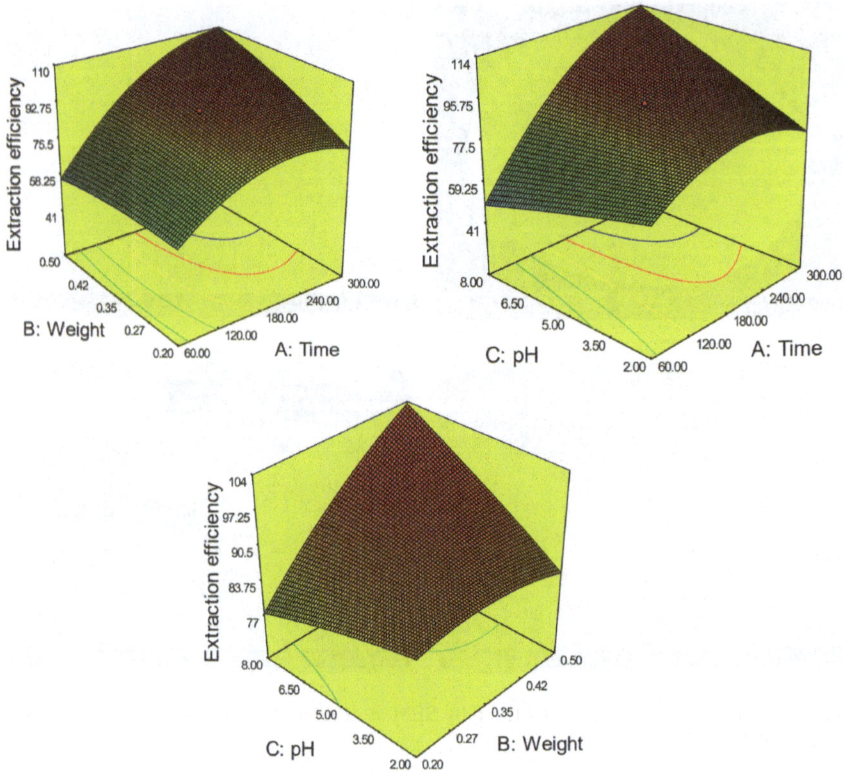

Fig. 4 Three-dimensional response surface plot for the effect of **a** weight and contact time at pH 5, **b** contact time and pH of wastewater with absorbent weight 0.35 g, and **c** pH and weight of modified beads at 180 min contact time

3.4 Adsorption Capacity

Figure 5 depicts the variation of Mn ions adsorbed on the absorbent with shaking time (0–300 min) at varied initial concentration of wastewater (14–44 ppm). Figure 5 indicates that the adsorption of Mn ions was quite rapid initially and reached an equilibrium after 60 min for all set. The initial faster rate may due to the availability of the uncovered surface area of the adsorbent. This type of absorbent (hydrogel) which are crosslinked hydrophilic polymers, are capable of expanding their volumes due to their high swelling in water [20]. In this case, the swelling process complete after 60 min of contact time, thus limit the adsorption capacity if the metal ion.

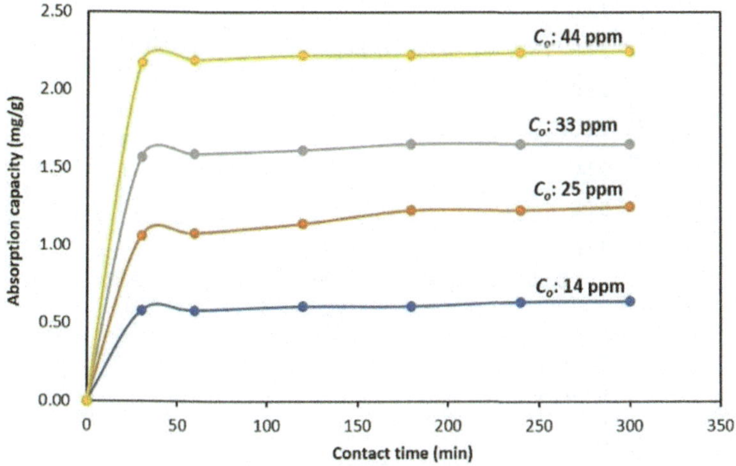

Fig. 5 Adsorption capacity of the absorbent at different initial concentration

3.5 *Absorption Isotherm Models*

The Langmuir and Freundlich isotherms are shown in Fig. 6. Isotherm parameters were obtained by linear regression analysis for each model. Based on the regression method, the Langmuir model describes the adsorption of metal ion onto Alginate/PVA/IL better than that of Freundlich isotherm model. Meanwhile, Table 4 listed the coefficient values for both isotherm models. According to this model, the maximal adsorption capacity increase between 14 and 44 ppm of initial concentration. It was generally attributed to the change with initial concentration. As initial concentration increase, the number of metal ion increase, which favor the adsorbent. The Langmuir model assumes that the uptake of metal ions is monolayer sorption on a homogeneous surface and without any interaction between adsorbed ions [21].

4 Conclusion

The present study aims at the optimizing the usage modified adsorbent based on ionic liquid and alginate to extract Mn ions from wastewater solution. The ionic liquids (1-hexyl-3-propionitrile imidazolium chloride) owns the functional group that have affinity to extract metal ions. The ILs having functional moiety will offer capability of adsorbent to extract metal ions. In this work, the incorporation of this ionic liquid with alginate have offer synergistic effect in order to serve better as metal extractant. The addition of PVA in this sorbent's preparation have enhance the physical strength of beads which make it applicable in harsh process (shaking and mechanical stirring) during extraction process. The preparation of this sorbent had

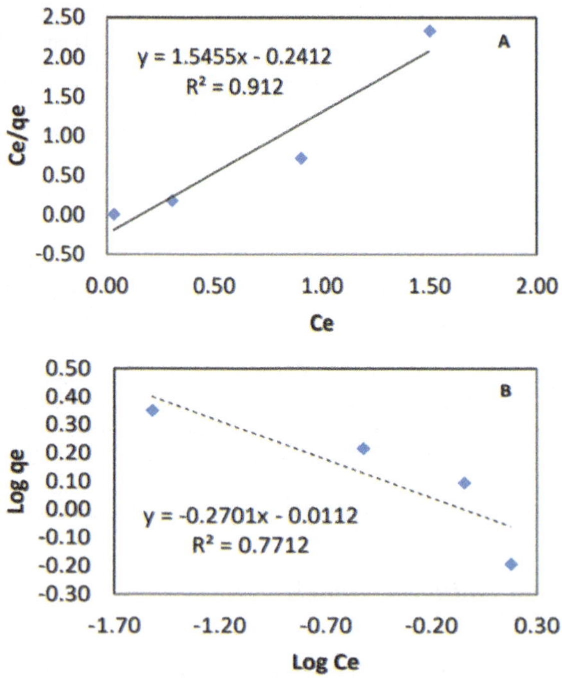

Fig. 6 Adsorption Isotherm models of Mn ions onto Alginate/PVA/ILs presented by **a** Langmuir model and **b** Freundlich model

Table 4 Coefficient values for both isotherms model

Langmuir model			Freundlich model		
q_m (mg g^{-1})	K_L (L mg^{-1})	R^2	1/n	K_F	R^2
0.64	0.745	0.912	2.207	0.975	0.771

been considered as a convenient and inexpensive production process. The adsorbent material is easy to handle during or after adsorptive removal of manganese and capable to serve an efficient adsorption of manganese ions, depending on the weight, concentration and pH of the solution during extraction. The predicted optimize values are proposed in order to gain the maximum EE% during the process. This sorbent has high metal adsorption ability. Furthermore, the adsorption process was found to be initial concentration dependent. The Langmuir isotherm can be correlated to experimental data better than Freundlich model. This combination sorbent (modified beads) can meet variety of request/application and offer high metal adsorption ability. Impregnation ionic liquid with alginate and PVA provide advantages in term of economic value.

Acknowledgements The authors would like to thank PRF (0153AB-A30) and PRGS (015MA0-049) for funding and facilities.

References

1. Khodami, S., Surif, M., Wan, W.M., Daryanabard, R.: Assessment of heavy metal pollution in surface sediments of the Bayan Lepas area, Penang Malaysia. Mar. Pollut. Bull. **114**(1), 615–622 (2017)
2. Orisakwe, O.E., Blum, J.L., Sujak, S., Zelikoff, J.T.: Metal pollution in Nigeria: a biomonitoring update. J. Health Pollut. **4**(6), 40–52 (2014)
3. Akpor, O.K.: Heavy metal pollutants in wastewater effluents: sources, effects and remediation. Adv. Biosci. Bioeng. **2**(4), 37–43 (2014)
4. Nyingi, B., Gitahi, K.J., Kiptoo, K., Jackson, K.: Heavy metas concentrations in water and selected fish species (tilapia, catfish and lungfish) from Lake Baringo, Kenya. Int. J. Sci. Enviro. Technol. **5**, 4288–4295 (2016)
5. Hu, H., Jin, Q., Kavan, P.: A study of heavy metal pollution in China: current status, pollution-control policies and countermeasures. Sustainability **6**(9), 5820–5838 (2014)
6. Armid, F.A., Shinjo, R., Ruslan, R.: Distributions and pollution assessment of heavy metals Pb, Cd and Cr in the water system of Kendari Bay, Indonesia. In: IOP Conference Series: Materials Science Engineering 172(1), 1–8 (2017)
7. Sankara Narayanan, T.S.N., Park, I.S., Lee, M.H.: Surface modification of magnesium and its alloys for biomedical applications: opportunities and challenges. Elsevier Ltd (2015)
8. Ríos, D.L., Irabien, A., Hollmann, F., José, F., Fernández, H.: Ionic liquids: green solvents for chemical processing. J. Chem. pp. 2–4 (2013)
9. Nelson, W.M: Are ionic liquids green solvents?. In: ACS Symposium Series pp. 30–41 (2002)
10. Zhou, Y., Boudesocque, S., Mohamadou, A., Dupont, L.: Extraction of metal ions with task specific ionic liquids: influence of a coordinating anion. Sep. Sci. Technol. **50**(1), 38–44 (2015)
11. Budak, T.B.: Removal of heavy metals from wastewater using synthetic ion exchange resin. Asian J. Chem. **25**(8), 4207–4210 (2017)
12. Leyma, R., Platzer, S., Jirsa, F., Kandioller, W., Krachler, R., Keppler, B.K.: Novel thiosalicylate-based ionic liquids for heavy metal extractions. J. Hazard. Mater. **314**, 164–171 (2016)
13. Platzer, S., Kar, M., Leyma, R., Chib, S., Roller, S., Jirsa, F., Krachler, R., MacFarlane, D.R., Kandioller, W., Keppler, B.K.: Task-specific thioglycolate ionic liquids for heavy metal extraction: synthesis, extraction efficacies and recycling properties. J. Hazard. Mater. **324**, 241–249 (2017)
14. Ríos, D.L., HernándezFernández, A.P., Lozano, F.P., Sánchez, L.J., Moreno, S., Moreno, J.I., Godínez, C.: Removal of metal ions from aqueous solutions by extraction with ionic liquids. J. Chem. Eng. Data **55**(2), 605–608 (2010)
15. Case, M.E., Fox, F.V., Baek, D.L., Mincher, B.J., Wai, C.M.: Extraction behavior of selected rare earth metals from acidic chloride media using tetrabutyl diglycolamide. Solvent Extr. Ion Exch. **35**(7), 496–506 (2017)
16. Wei, G.T., Yang, Z., Chen, C.J.: Room temperature ionic liquid as a novel medium for liquid/liquid extraction of metal ions. Anal. Chim. Acta **488**(2), 183–192 (2003)
17. Wang, J., Luo, J., Feng, S., Li, H., Wan, Y., Zhang, X.: Recent development of ionic liquid membranes. Green Energy Environ. **1**(1), 43–61 (2016)
18. Ayawei, N., Ebelegi, A.N., Wankasi, D.: Modelling and interpretation of adsorption isotherms. J. Chem. **2017**, 3039817 (2017). https://doi.org/10.1155/2017/3039817
19. Sadeghi, S., Moghaddam, A.Z.: Chromium speciation using task specific ionic liquid/aqueous phase biphasic system combined with flame atomic absorption spectrometry. J. Mol. Liquid **221**, 798–804 (2016)
20. Barakat, M.A.: New trends in removing heavy metals from industrial wastewater. Arab. J. Chem. **4**(4), 361–377 (2011)
21. Zhang, Y., Zhao, J., Jiang, Z., Shan, D., Lu, Y.: Biosorption of Fe (II) and Mn (II) ions from aqueous solution by rice husk ash. Biomed. Res. Int. 973095 (2014). https://doi.org/10.1155/2014/973095

Synthesis and Modification of Pour Point Depressant (PPD) Based on Copolymers of α-Olefins and Maleic Anhydride for Waxy Crude Oil

Kozhabekov Serik Samsalykovich, Zhubanov Amin Abdirasululy, Donenov Beisen Kainarbaevich, Makhmetova Aliya Ruslanovna, and Abayev Talgat Bakytuly

Abstract In this work, a synthesis and modifications of copolymers based on α-olefin and maleic anhydride was carried out. The influence of copolymers on the rheological behavior of waxy crude oil of the Akshabulak field (Kazakhstan) was investigated. The synthesis of copolymer is based on α-olefin: octadecene-1 with maleic anhydride by the method of initiated radical polymerization. Subsequently, this copolymer was modified with primary amines: butylamine (BA), octadecylamine (ODA). The copolymer was modified in a xylene solution with a Dean-Stark trap. The ratio of the ODC-MA copolymer to amines was 1:1.2 mol. The degree of completion of the reaction was estimated by the amount of water condensed in the Dean-Stark trap. The modified copolymers of ODC-MA amines (ODC-MA-BA, ODC-MA-ODA) were characterized by Fourier transform IR spectroscopy, which indicates the conversion of the starting anhydride cycles (1780–1850 cm^{-1}) to imide (1535, 1690–1700 cm^{-1}). Modified copolymers exhibit the properties of depressant additives-they reduce the viscosity and pour point temperature of crude oil, relative to crude oil without additives. The microscopy method was used to determine the morphology of wax crystals formed in oil in the process of lowering the temperature. It was shown that the resulting modified wax crystals have a smaller size compared to wax crystals in crude oil without additives, which shows the dispersing effect of the modified copolymers based on ODC-MA.

Keywords Copolymers · α-olefins · Maleic anhydride · Pour point depressant · Rheology · Crude oil

1 Introduction

The transportation of crude waxy oil through pipelines and storage in tanks is a huge problem due to wax deposits on the walls of the pipelines, which leads to an

K. S. Samsalykovich (✉) · Z. A. Abdirasululy · D. B. Kainarbaevich · M. A. Ruslanovna · A. T. Bakytuly
Kazakh-British Technical University, 59 Tole-Bi, 050005 Almaty, Kazakhstan

S. A. Abdul Karim et al. (eds.), *Proceedings of the 6th International Conference on Fundamental and Applied Sciences*, Springer Proceedings in Complexity,
https://doi.org/10.1007/978-981-16-4513-6_3

increased risk of pipeline stoppages. To prevent these negative consequences, pour point depressants (PPD) are used in world practice-polymer surfactants of various compositions [1, 2]. Global PPD production is projected to reach 1.9 billion by 2027. In the context of the COVID-19 crisis, the global PPD market is estimated at US 1.6 billion in 2020 [3]. There is a wide range of PPDs that are widely used in oil transportation technology, such as copolymers of ethylene and vinyl acetate [4–6] copolymers of α-olefin [7, 8] and their polar nitrogen-containing compounds [9–11]. One of the promising areas is the synthesis of copolymers with amphiphilic properties. Analysis of the literature has demonstrated that amphiphilic copolymers exhibit depression activity in relation to waxy crude oil [12]. The depression properties of amphiphilic copolymers are achieved due to the content of the polar part and linear alkyl groups in the polymer structure. Linear alkyl groups are incorporated into the wax structure and co-crystallize with decreasing temperature, while the polar part of the copolymer repels crystals, thereby preventing agglomeration [13–15]. In [16, 17], α-olefins and maleic anhydride act as reagents for the synthesis of amphiphilic copolymers. The use of copolymers based on maleic anhydride and α-olefins as PPDs and fluidity improvers makes it possible to control the oil fluidity depending on the length of the grafted pendant group [18]. Moreover, copolymers of maleic anhydride with hydrophilic functional groups attract the attention of researchers due to the possibility of regulating the hydrophobichydrophilic balance of polymers. Chen et al. [19] synthesized a copolymer of maleic anhydride and methylbenzyl acrylate and modified them with amines and alcohols. Esterified and amidated copolymers exhibited PPD properties to improve the fluidity of the oil. Xu et al. Synthesized a series of comb-type copolymers of methacrylate and maleic anhydride with various side groups amidated with long carbon chains from C14 to C16 [10].

However, PPD concentrations in excess of 4000 ppm are not commercially acceptable. Moreover, it has been found that the degree of imidization has a significant effect on performance as a flow improver. Flow improvers with a higher degree of imidization were found to be more effective [20]. Modification of the anhydride fragments of the copolymer with a primary amine makes it possible to increase the depressant activity of the copolymers [21].

2 Material and Methods

2.1 Materials

Maleic anhydride (98%), octadecene-1 (95%), butylamine (99%) and octadecylamine (99%) from Sigma-Aldrich, used without further purification. Toluenesulfonic acid (99%), benzoyl peroxide (75%, remainder water), o-xylene (95%), toluene (95%), and THF (99.9%) from Sigma-Aldrich, used without further purification. Methanol (98%) was used to purify the copolymer without further purification.

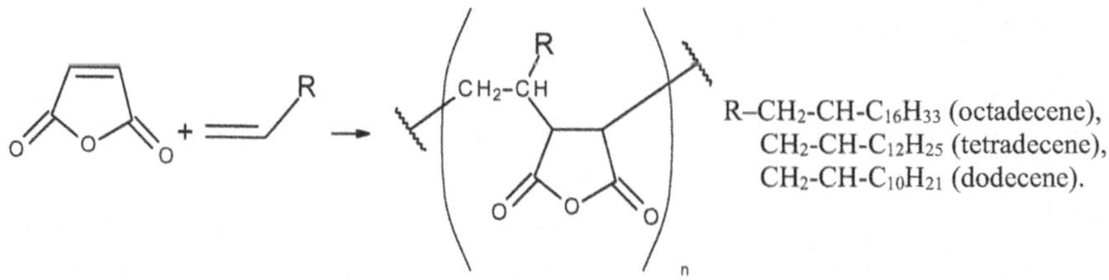

Fig. 1 Structure diagram of copolymers based on α-olefins (ODC-MA) and maleic anhydride (MA)

2.2 *Copolymerization of α-Olefins with Maleic Anhydride*

The synthesis of α-olefin octadecene-1 with maleic anhydride was conducted by free radical polymerization (see Fig. 1) The molar ratio of α-olefin to maleic anhydride is 1:1. The reagents were mixed in a three-necked flask under nitrogen atmosphere with constant stirring. Copolymerization was conducted in a toluene solution at 100 °C for 4 h. The initiator was dibenzoyl peroxide 4% (wt.) by weight of the starting monomers. The copolymer was precipitated in an excess volume of methanol, filtered, and dried in vacuum at 60 °C. The copolymer yield is ODC-MA 83%.

2.3 *Amidation of a Copolymer of Octadecene-1 and Maleic Anhydride*

The modification of the ODC-MA copolymer with primary amines was conducted according to the general procedure (see Fig. 2). ODC-MA copolymer and primary amine in a molar ratio of 1:1.2 in o-xylene solution were loaded into a two-necked flask equipped with a Dean-Stark attachment with a reflux condenser, stirrer, and thermometer. Toluenesulfonic acid, 0.5% by weight of the reagents, was used as a catalyst. The mixture was heated stepwise to a temperature of 150 °C. The reaction mixture was kept at this temperature with stirring for 12 h until the evolution of reaction water ceased. The resulting mixture was reprecipitated in methanol, filtered, and dried in vacuum to constant weight at 60 °C.

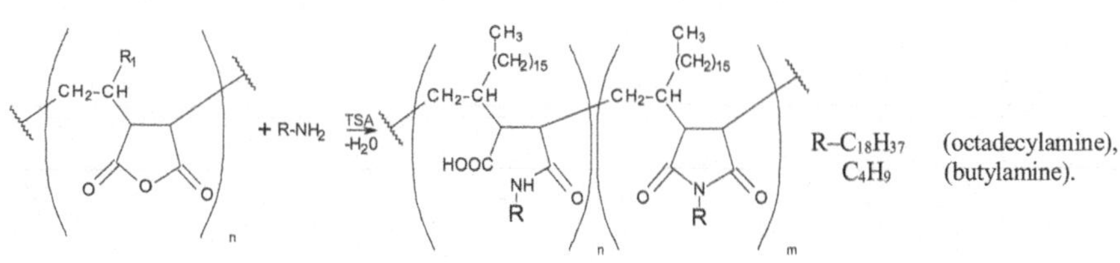

Fig. 2 A Structure of modified copolymers R = $C_{18}H_{37}$ (octadecylamine) and C_4H_9 (butylamine)

3 Results and Discussion

3.1 Characterization of the Copolymers

The structure of the copolymers was analyzed using ^{1}H and ^{13}C NMR spectroscopy (see Fig. 3) and FTIR spectrometry in the range of 400–4000 cm^{-1} (see Fig. 4).

The ^{1}H NMR spectra of the ODC-MA copolymer exhibits multiplet signals at 0.84–0.87 ppm, with an integrated proton intensity 3.0 N of the terminal methyl groups of the octadecene-1 fragment. The methylene protons of the octadecene-1 fragment appear as a broadened signal at 1.10–1.42 ppm with an integrated intensity of 22.1 N. The methine protons of maleic anhydride fragments resonate at an integrated intensity of 0.4 N 2.5–3.5 ppm. Traces of proton signals at 4.09–4.15 and

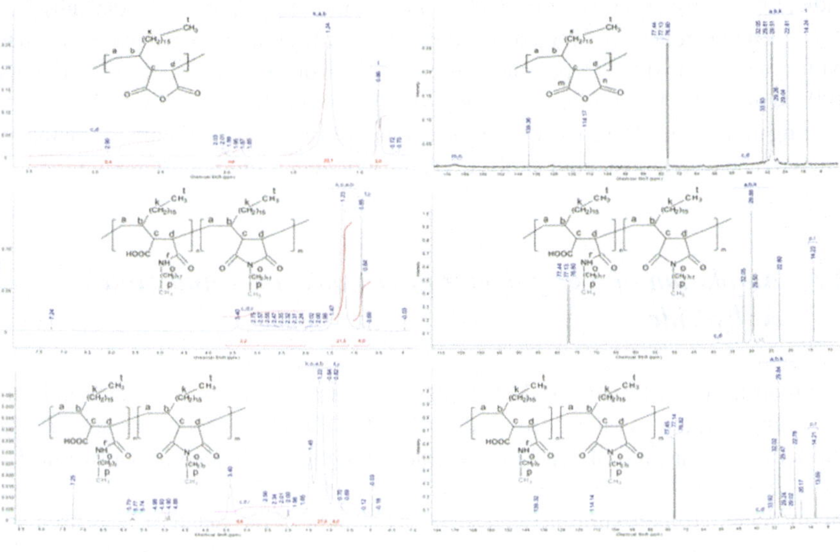

Fig. 3 ^{1}H and ^{13}C NMR spectra of the ODC-MA copolymer and the modified copolymers ODC-MA-ODA, and ODC-MA-BA

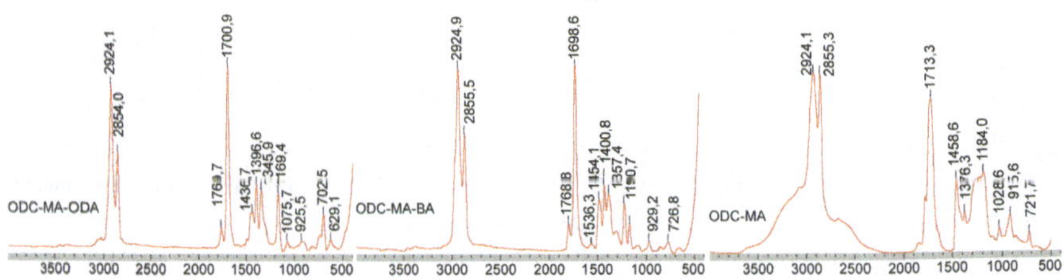

Fig. 4 IR spectrum of the synthesized copolymers ODC-MA and modified copolymers ODC-MA-BA and ODC-MA-ODA

4.89–4.99 ppm indicate the absence or insignificant amount of unsaturation in the copolymer.

In the ^{13}C NMR spectra of ODC-MA, the carbon atoms of the terminal methyl groups of the octadecene-1 fragments are observed at 14.27 ppm. The carbon atoms of the methylene and methine copolymer resonate at 22.81, 29.26, 29.86, 31.88 and 33.94 ppm. The methine carbon atoms of the maleic anhydride ring show up as broadened signals in the 41–44 ppm range. The carbonyl protons of this cycle are also manifested by broadened signals in the region of 170–175 ppm. In the ^{13}C spectra of the initial ODC-MA copolymer, the carbon of the anhydride group resonates at 114.17–139.36 ppm, but practically does not appear or practically disappears in the spectra of copolymers modified with primary amines. The IR spectra of the ODC-MA copolymer exhibit bands of stretching vibrations of the carbonyl group $C = O$ in the ester region of 1710–1713 cm^{-1}. The characteristic vibrations in the region (1770–1775 and 1850–1855 cm^{-1}) refer to the stretching vibrations of $C = O$ in the units of cyclic anhydride [22].

In the modified copolymers ODC-MA with amines (ODC-MA-BA and ODC-MA-ODA), the 1H NMR spectra of the copolymers exhibits multiplet signals in the 0.82–0.89 ppm with an integrated proton intensity from 3.0 to 6.0 of the terminal methyl groups of the octadecene-1 and alkylamide fragments. The methine protons of the modified copolymers were located in two singlet signals at 1.22–1.25 and 1.45 ppm with a total integral intensity of 21.0 H to 27.8 N. Methine protons of maleic anhydride fragments, as well as protons of the amide group and an adjacent methylene fragment, resonating at 2.0–3.5 ppm, with an integral intensity from 1.0 H to 2.2 N. The opening of the cycles of maleic anhydride and the formation of cycles with the participation of nitrogen atoms affect the picture of the spectra in this region: it increases by 3.40 ppm. Traces of proton signals of unsaturated bonds are practically not observed.

In the ^{13}C NMR spectrum of ODC-MA-BA, the carbon atoms of the terminal methyl groups of the octadecene-1 and alkylamine fragments are observed at 14.21–14.28 and 13.69 ppm. The carbon atoms of the methylene and methine copolymer resonate at 20.17, 22.78, 27.62, 29.85, 32.02, 33.92 and 38.50 ppm. Two broadened signals at 177.85 and 179.8 ppm correspond to the carbonyl carbon atoms of the copolymer, characterized by FTIR spectroscopy, indicating the transformation of the initial anhydride rings (1780–1850 cm^{-1}) into the imide ring (1535, 1690–1700 cm^{-1}, 1359 cm^{-1}) [22]. The presented data indicate the presence of the main functional groups in the structure of the copolymers.

3.2 *Molecular Weight Measurements*

The absolute molecular weight of the obtained PPDs was determined by dynamic light scattering on a Malvern Zetasizer Nano S. The results are presented in Table 1. The kilodalton (kDa) of the synthesized and modified polymers was found to be in the range from 9.6 (kDa) to 20.1 (kDa), and the correlation coefficient (R) ranges

Table 1 Absolute molecular weight (MW) of copolymers

Copolymer	ODC-MA	ODC-MA-BA	ODC-MA-ODA
MW (kDa)	12, 8	15, 1	20, 1

from 0.868 to 0.981. The molecular weight results for the copolymers produced with the Malvern Zetasizer Nano S are within the predicted molecular weight range. Thus, it can be assumed that the synthesized modified copolymers can be used as potential PPDs.

3.3 Pour Point and Rheological Measurements

The pour point determination of the waxy crude oil was conducted in accordance with ASTM D 5853. The results are demonstrated in Table 2. At 300, 500, and 1000 ppm copolymer dosage, the pour point of the Akshabulak crude oil is reduced by 9–15 °C in comparison with heat-treated oil, indicating the depression activity of the copolymers. The greatest depression effect was observed for the ODC-MA-BA copolymer. The presented data show that the ODC-MA copolymer modified with butylamine (ODC-MA-BA) shows the best result among the modified copolymers with a pour point of –6 °C at a concentration of 300 ppm. Further increasing the concentration of the PPD obtained does not improve the low temperature properties of the crude oil. Thus, the optimum concentration for the obtained PPDs was determined to be 0.03%.

Despite the fact that in most works an improvement in oil fluidity with an increase in the length of the pendant group [13] has been demonstrated, in this case the ODC-MA copolymer modified with short-chain amines shows a better result compared to long-chain pendant groups. The reason for the high efficiency of copolymers with a short pendant group is possibly associated with the availability of a hydrophilic group during the formation of wax crystals on the copolymer surface. Short pendant groups do not shield the access of the hydrophilic group of the copolymer to waxes, while long pendant groups in the ODC-MA-ODA copolymer shield the wax crystals from interaction with the hydrophilic group.

Table 2 Dependence of the pour point of Akshabulak crude oil with and without the addition of PPD

Sample	Pour point temperature		
	300 ppm (°C)	500 ppm (°C)	1000 ppm (°C)
Crude oil without PPD		+ 9	
Crude oil with ODC-MA	0	+ 3	+ 6
Crude oil with ODC-MA-BA	−6	−3	0
Crude oil with ODC-MA-ODA	−3	0	+ 3

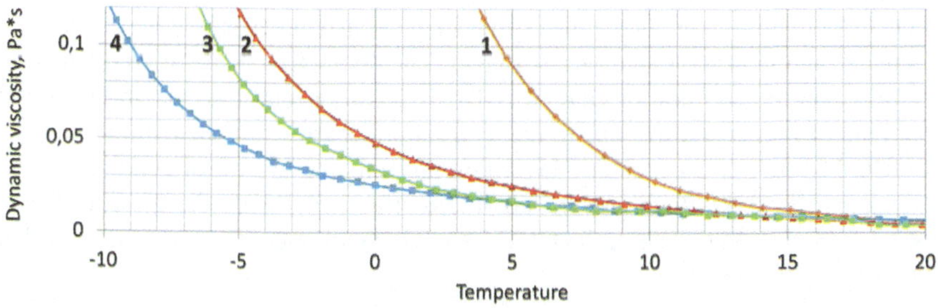

Fig. 5 The curve of dependence of the dynamic viscosity of Akshabulak crude oil on temperature with heat treatment at 60 °C, at shear rate (ɣ) 10 s^{-1} at a dosage of 300 ppm depressants with the addition of PPDs: ODC-MA[2], ODC-MA-ODA[3], ODC-MA-BA[4], and blank crude oil[1]

The rheological curve results are the same as the pour point results (see Fig. 5). From the presented data, we can observe that the most effective copolymer is modified with butylamine.

3.4 Microscopic Measurements

Presented are micrographs of blank crude oil (see Fig. 6), with the addition of ODC-MA copolymer (see Fig. 6) and modified ODC-MA-BA copolymer (see Fig. 6). From the obtained micrographs of crystals in crude oil Akshabulak it can be seen that at the same temperature (0 °C) wax crystals with the addition of ODC-MA modified with butylamine have a shape and size different from those of ODC-MA wax crystals and are even more different from crystals wax in blank crude oil. When ODC MA copolymer modified with butylamine is added to the oil, wax crystals resemble spherical in shape with smoother and more even surfaces with sizes ranging from 8 to 20 μm, while blank crude oil has a plate-like crystal shape with sizes ranging from 35 to 50 μm, a similar situation with crude oil wax crystals with the addition

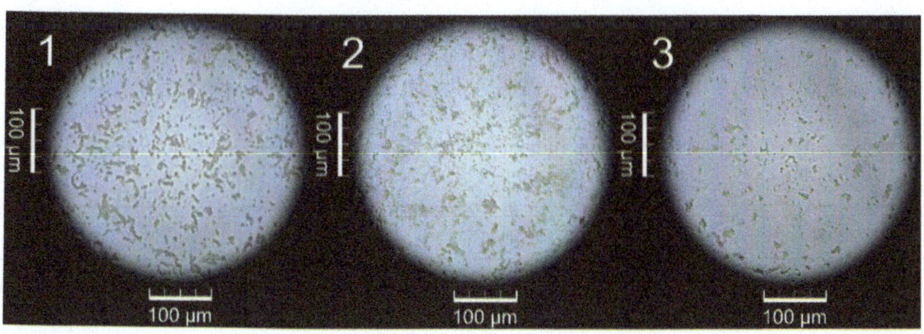

Fig. 6 Micrograph of Akshabulak crude oil without PPD[1], with PPD (ODC-MA)[2], and with modified PPD (ODC-MA-BA)[3]

of ODC MA copolymer, where the crystal shape is also lamellar, and the crystal size is smaller and is in the range of 25–40 μm. The distinctive shape of crude oil wax crystals with the addition of ODC-MA-BA is possibly related to the ability of the copolymer to co-crystallize due to the hydrophobic part, while the hydrophilic part of ODC-MA-BA creates a steric layer that prevents the growth of copolymer wax crystals, while short-chain side groups more effectively block the growth of wax crystals [23, 24]. In this case, the ODC-MA-BA copolymer acts not only as a PPD, but also as a dispersant.

4 Conclusion

The octadecene-1 maleic anhydride copolymer was synthesized and modified with different primary amines (octadecylamine and butylamine) of various lengths. The structure of copolymers based on α-olefins and MA was confirmed by ^{1}H ^{13}C NMR and FTIR spectroscopy. According to the data of rheological tests and optical microscopy, it was demonstrated that the effectiveness of the PPD action is largely influenced by the length of the side groups introduced by the anhydride group. In addition, microscopic studies have shown that the ODC-MA-BA copolymer, which has a dispersing effect during the crystallization of waxes. Studies on the determination of the pour point of crude oil and rheological tests demonstrated that the ODC-MA copolymer modified with short-chain butylamine shows high efficiency compared to ODC-MA copolymer modified with longer amine chain octadecylamine.

Acknowledgements This work was supported by the Science Committee of the Ministry of Education and Science of the Republic of Kazakhstan [IRN AP08855445, No. 335 dated November 24, 2020]

References

1. Van Engelen, G.P., Kaul, C.L., Vos, B.: Study of flow improvers for transportation of Bombay high crude oil through submarine pipeline. J. Petrol. Technol. **33**(12), 539–545 (1971)
2. Rønningsen, H.P.: Rheological behaviour of gelled, waxy North Sea crude oils. J. Petrol. Sci. Eng. **7**(3–4), 177–213 (1992)
3. Global Pour Point Depressants Industry. https://www.reportlinker.com/p05959752/Global-Pour-Point-Depressants-Industry.html?utm_source=GNW. Cited April 2021
4. Machado, A.L.C., Lucas, E.F.: Influence of ethylene-co-vinyl acetate copolymers on the flow properties of wax synthetic systems. J. Appl. Polym. Sci. **85**, 1337–1348 (2002)
5. Ashbaugh, H.S., Guo, X., Schwahn, D., Prud'homme, R.K., Richter, D., Fetters, L.J.: Interaction of paraffin wax gels with ethylene vinyl acetate co-polymers. Energy Fuels **19**, 138–144 (2005)
6. Machado, A.L.C., Lucas, E.F., González, G.: Poly (ethylene-co-vinyl acetate)(EVA) as wax inhibitor of a Brazilian crude oil: oil viscosity, pour point and phase behavior of organic solutions. J. Petrol. Sci. Eng. **32**, 159–165 (2001)

7. Ghosh, P., Das, M., Das, T.: Polyacrylates and acrylate-α-olefin copolymers: synthesis, characterization, viscosity studies, and performance evaluation in lube oil. Pet. Sci. Technol. **32**, 804–812 (2014)
8. Jun, X., Huiqin, Q., Shili, X., Li, L., Xuhong, G.: Synthesis of poly (maleic acid alkylamide-co-α-olefin-costyrene) co-polymers and their effect on the yield stress and morphology of waxy gels with asphaltenes. Energy Fuels **25**, 573–579 (2010)
9. Song, Y., Ren, T., Fu, X., Xu, X.: Synthesis and evaluation of some C14MC-MCNR2 copolymers and C14MC-MA-MCNR2 terpolymers as cold flow improvers in the diesel fuels. Pet. Sci. Technol. **23**, 669–679 (2005)
10. Xu, G.W., Xue, Y., Zhao, Z.C., Lian, X., Lin, H.L., Han, S.: Influence of poly (methacrylate-co-maleic anhydride) pour point depressant with various pendants on low-temperature flowability of diesel fuel. Fuel **216**, 898–907 (2018)
11. Zhou, M., He, Y., Chen, Y., Yang, Y., Lin, H., Han, S.: Synthesis and evaluation of terpolymers consist of methacrylates with maleic anhydride and methacrylic morpholine and their amine compound as pour point depressants in diesel fuels. Energy Fuels **29**, 5618–5624 (2015)
12. Xiu, Z., Dufils, P., Zhou, J., Cadix, A., Hatchman, K., Decoster, T., Ferlin, P.: Amphiphilic wax inhibitor for tackling crude oil wax deposit challenges. In: Paper presented at the Proceedings-SPE International Symposium on Oilfield Chemistry 10.2118 (2019)
13. Xu, J., Zhang, X., Sun, J., Li, L., Guo, X.: How comb-type poly(maleic acid alkylamide-co-α-olefin) assemble in waxy oils and improve flowing ability. Asia-Pac. J. Chem. Eng. **4**(5), 551–556 (2009)
14. Soni, H.P., Kiranbala, Bharambe, D.P.: Performance-based designing of wax crystal growth inhibitors. Energy Fuels **22**(6), 3930–3938 (2008)
15. Wei, B.: Recent advances on mitigating wax problem using polymeric wax crystal modifier. J. Pet. Explor. Prod. Technol. **5**, 391–401 (2015)
16. Ivchenko, P.V., Nifant'ev, I.E.: Polymer depressor additives: synthesis, microstructure, efficiency. Polym. Sci.-Ser. A **60**(5), 577–593 (2018)
17. Nifant'ev, I.E., Vinogradov, A.A., Bondarenko, G.N., Korchagina, S.A., Shlyakhtin, A.V., Roznyatovskii, V.A., Ivchenko, P.V.: Copolymers of maleic anhydride and methylene alkanes: synthesis, modification, and pour point depressant properties. Polym. Sci.-Ser. B **60**(4), 469–480 (2018)
18. Elganidi, I., Elarbe, B., Abdullah, N., Ridzuan, N.: Synthesis of a novel terpolymer of (BA-co-SMA-co-MA) as pour point depressants to improve the flowability of the Malaysian crude oil. Mater. Today **42**, 28–32 (2021)
19. Chen, F., Liu, J., Yang, T., Yin, S., Su, B., Xie, M., Dai, B., Han, S., Xue, Y.: Influence of maleic anhydride-co-methyl benzyl acrylate copolymers modified with long-chain fatty amine and long-chain fatty alcohol on the cold flow properties of diesel fuel. Fuel **268**, 117392 (2020)
20. Cao, K., Zhu, Q., Wei, X., Yao, Z.: Study on the influence of the imidization degree of poly(styrene-co-octadecyl maleimide) as a flow improver in waxy crude oils with asphaltenes. Energy Fuels **29**(2), 993–1000 (2015)
21. Wu, Y., Ni, G., Yang, F., Li, C., Dong, G.: Modified maleic anhydride co-polymers as pour-point depressants and their effects on waxy crude oil rheology. Energy Fuels **26**(2), 995–1001 (2012)
22. Kuptsov, A.H., Zhizhin, G.N.: Handbook of Fourier Transform Raman and Infrared Spectra of Polymers. Elsevier Academic, pp. 1–571 (1998)
23. Kudaibergenov, S., Shakhvorostov, A., Gussenov, I., Seilkhanov, T., Nuraje, N.: Application of novel hydrophobically modified polybetaines based on alkylaminocrotonates and methacrylic acid as pour point depressants and ASP flooding agent. Polym. Bull. (2019)
24. Kozhabekov, S.S., Zhubanov, A.A., Toktarbay, Z.: Study the rheological properties of waxy oil with modified pour point depressants for the South Turgai oil field in Kazakhstan. Oil Gas Sci. Technol. **74**, 28 (2019)

Effects of NPK Fertilizers on the Growth, Yield and Chemical Content of Tomato (*Lycopersicon esculentum* L. Mill)

Lyazzat Bekbayeva, Zuraini Zakaria, Tatyana Karpenyuk, Alla Goncharova, El-Sayed Negim, and Kozhanova Kaldanay

Abstract This paper investigates the effect of NPK fertilizers with different content of nitrogen on growth, yield and chemical content of tomato (*Lycopersicon esculentum* L. Mill). NPK fertilizers were used based on polyvinyl alcohol (P) blended with three different contents of urea (U) (PU35, N: 8.45%; PU50, N: 34.1%, and PU65, 44.04%) and K_3PO_4. NPK35, NPK50 and NPK65 fertilizers were prepared by mixing 90% PU with 10% K_3PO_4. All treatments were applied at the rate of 500 ml (1% w/v) of NPK directly to the soil. The first treatment was given at 15-day-old seedlings. Thereafter, the treatments with 5 replications were given at intervals of 15 days (6 s) each until 90 days. Final harvest was at 112 days. Data was analyzed using variance analysis (ANOVA) and compared using control without fertilizer. The results indicate that the NPK fertilizers cause improvement in the vegetative growth, yield as well as chemical content of tomato plants. Meanwhile, the application of NPK fertilizer (NPK50) with nitrogen content 31.7% increased the above parameters more than NPK35, NPK65 and control.

Keywords NPK · Fertilizers · Tomato · Growth · Yield · Nitrogen

---

L. Bekbayeva (✉) · K. Kaldanay
Department of Engineering Disciplines, Asfendiyarov Kazakh National Medical University, 88 Tole bi Street, Almaty, Kazakhstan

K. Kaldanay
e-mail: kozhanova.k@kaznmu.kz

L. Bekbayeva · Z. Zakaria
Biology Program, School of Distance Education, Universiti Sains Malaysia, 11800 Minden, Penang, Malaysia
e-mail: zuraini@usm.my

T. Karpenyuk · A. Goncharova
Department of Biotechnology, Faculty of Biochemistry, Al-Farabi, Kazakh National University, 050040 Almaty, Kazakhstan

E.-S. Negim
Laboratory of Advanced Materials and Technology, Kazakh-British Technical University, 59 Tole bi St., 050000 Almaty, Kazakhstan

S. A. Abdul Karim et al. (eds.), *Proceedings of the 6th International Conference on Fundamental and Applied Sciences*, Springer Proceedings in Complexity,
https://doi.org/10.1007/978-981-16-4513-6_4

1 Introduction

Tomato (*Lycopersicon esculentum* Mill.) is a very popular vegetable grown in the world [1]. A number of researchers focus on increment tomato yield by introducing new NPK fertilizers [2]. The application of NPK fertilizers to tomato plants at proper times with suitable rates could increase yield of tomatoes. Authors found that increasing NPK levels resulted in higher growth performance in tomato varieties [3]. A high concentration of NPK in the nutrient solution gave higher total yield and tomato fruit weight than the control nutrient solution [4]. It was reported that fertilizing tomato with NPK at deferent levels increased the vegetative growth characters, for example moderate dose of NPK (N: 26.7 g, P: 10.0 g and K: 53.3 g/m^2) significantly increased tomato growth and yield [5, 6]. The highest plant height, the maximum number of primary and secondary branches, number of flowers and fruits per plant as well as the greatest fruit size, fruit yield per plant and fruit tomato yield per ha were obtained when NPK fertilizer was applied at the rates of 120 kg N, 80 kg P and 75 kg K/ha [7]. The application of NPK at the rate of 20:10:10 has a highly significant effect on the growth and yield of tomato varieties and enhanced the soil chemical parameters [8]. Pansare et al. [9] studied the effect of NPK on yield and quality of tomato and evaluated the maximum yield of high-quality tomato when N, P and K fertilizers were added at a ratio of 3:1:2 (150 kg N/ha, 50 kg P/ha, and 100 kg K/ha). However, the application of NPK at rates of 60 kg N/ha, 50 kg P_2O_5/ha and 33 kg K_2O/ha gave the highest growth, and fruit yield of tomato as compared to control without fertilizers [10]. Felipe and Casanova [11] investigated the effects of N (0, 90, 180 and 270 kg/ha), P (P_2O_5, 0, 135, 270 and 405 kg/ha), and K (K_2O, 0, 90, 180 and 270 kg/ha) on the yield and number of fruits of tomato in Venezuela [11]. The best treatment, with the highest yield and number of fruits per plant, was 180 kg N, 270 kg P_2O_5, and 180 kg K_2O/ha. The present study was undertaken to improve growth, yield and chemical Content of Tomato (*Lycopersicon esculentum* L. Mill) fertilized with NPK at different rates NPK35 (N: 7.8, P: 5.7, K: 4.8%), NPK50 (N: 31.7, P: 4.05, K: 4.6%) and NPK65 (N: 40.3, P: 4.5, K: 5.1%).

2 Material and Method

2.1 Synthesis of Nitrogen Fertilizer (PU)

The nitrogen fertilizers based on polyvinyl alcohol (PVA) with urea (U) were synthesized with composition ratios i.e. 65:35 (PU35), 50:50 (PU50) and 35:65 (PU65) respectively by blending polymerization in presence of acetic acid as catalyst. The preparation of nitrogen fertilizer and the method of analysis (^{1}H NMR, FTIR, SEM, DSC and TGA) have been previously described in a previous investigation by Negim et al. [12]. The elemental analysis was carried out for the determination of carbon,

Table 1 Elemental composition of nitrogen fertilizer (PU)

Elements	C (%)	N (%)	O (%)
PU35	49.41	8.45	42.15
PU50	28.79	34.09	37.58
PU65	21.54	44.04	34.42

Table 2 Elemental composition of NPK fertilizers

Elements	C (%)	N (%)	O (%)	P (%)	K (%)
NPK35	39.05	7.8	42.7	5.7	4.8
NPK50	26.8	31.7	33.1	4.05	4.6
NPK65	19.9	40.3	31.09	4.5	5.1

nitrogen and oxygen content in the fertilizers as shown in Table 1. The analysis was performed on a Vario Micro Elemental Analyzer (Elementar, Germany).

2.2 *Preparation of NPK Fertilizer*

NPK fertilizers were prepared by mixing 90% PU35, PU50 and PU65 with 10% K_3PO_4. The elemental analysis of NPK fertilizers was performed on a Vario Micro Elemental Analyzer (Elementar, Germany) as shown in Table 2.

NPK fertilizers were used with a constant concentration of 1% w/v using distilled water.

2.3 *Experimental Design and Treatments*

The crop plant selected for the present study was *Lycopersicon esculentum* (tomato). The hybrid tomato seeds (Pearl-F1) were purchased from the local market and kept for one hour in a glass beaker with fresh water. Only the seeds that settled at the bottom of the beaker were used for conducting the experiment. The seeds were carefully sowed in plastic trays and regular sprinkling of water was done to keep the compost soil moist. After two weeks, the germinated seedlings were transferred and planted in plastic pots. The seedlings were pushed 5 cm deep into the soil and the depression was then loosely covered back by the soil. The soil was air-dried, sieved and packed (13.5 kg/pot), and was properly filled in 15 pots. Each pot was labelled with the pot number and the date of sowing of the seeds were recorded to determine the offset date for analysis. The day on which the seedlings were planted

Fig. 1 Tomato plants after two weeks of germination

in the pot was treated as day zero (Fig. 1). The plants were watered every day or on alternate days depending on the requirement. All 3 sets were prepared in five replicates. Nitrogen fertilizer treatment was given to the plants namely PU and a set of control pants. In each of the treatment, 500 mL (1% w/v) of PU was applied directly to the soil. The first treatment was given at 15-day-old seedlings. Thereafter, the treatments were given at intervals of 15 days each until 90 days. The control set was watered only with tap water without any fertilizers.

2.4 *Physical and Chemical Properties of the Soil*

In this study, the soil physical and chemical properties were analyzed before the addition of the N-fertilizer (PU) in different concentrations to the experimental soil to know the type and properties of the soil. The result is as presented in Table 3.

Table 3 Physical and chemical properties of the soil

Physical properties	
Sand	56.63 (%)
Silt	24.15 (%)
Clay	14.22 (%)
Soil texture	Sandy loam
Chemical properties	
pH	7.8
Ec	1.4 (mhos/cm^3)
Available N	81.0 (ppm)
Available P	3.04 (ppm)
Available K	40.8 (ppm)
Organic matter	0.6 (%)

The ingredients of the experimental soil were a mixture of clay (56.63%), fine sand (14.22%), and silt (24.15%). The chemical properties of the soil were 1.4 mhos/cm^3, 81.0 ppm N, 3.04 ppm P, 40.8 ppm K, 0.6 ppm of organic matter and pH was 7.8.

2.5 Data Recorded

Vegetative growth Plant height, number of main lateral branches, number of leaves, leaf area as well as fresh and dry weights of shoots were recorded at 4 and 8 weeks.

Chemical composition Leaves disks were taken at 4 and 8 weeks after transplanting to determine chlorophyll a, b according to the method described by Sartory and Grobbelaar [13]. Total carbohydrate content in dry matter of leaves was determined spectrophotometrically method described by Dubois et al. [14]. Nitrogen, phosphorus and potassium elements were determined in the leaves of tomato plants via digestion procedure according to Piper [15]. Nitrogen content was determined by modified micro-Kjeldahl method as described by Pregl [16]. Phosphorus and potassium contents in the sample were estimated using ammonium molybdate and flame photometer methods respectively, according to Chapman and Pratt [17].

Flowering and fruit yield Node number bearing the first flower, number of flower clusters per plant, number of flowers per cluster, number of flowers per plant, weight and number of fruits per plant were recorded.

Physical characteristics of fruits Fruit shape index was calculated using the ratio of vertical to horizontal diameters. Fruit volume was determined by using immersion method.

Chemical characteristics of fruits Soluble solids content (SSC) was determined by hand refractometer according to the method described by AOAC [18]. Titratable acidity was determined using the method described by AOAC [19] and AOAC [20]. Ascorbic acid content (vitamin C) was determined as described by AOAC [20]. Lycopene in the tomato samples was extracted with hexane: ethanol: acetone (2:1:1) (v/v) mixture following the method of Sharma and Le Maguer [21].

2.6 Statistical Analysis

The data calculated on different variables were subjected to analysis of variance (ANOVA) to observe the differences among the treatments and their interactions. Means were separated using Least Significant Difference (LSD at 5%) test. Statistical computer software "Statistix 8.1" was used for computing the ANOVA and LSD Mead et al. [22].

Table 4 The effect of NPK fertilizers on the growth vegetative characteristics of tomato plant

Treatments	Plant height (cm)		Number of lateral branches/plant		Number of leaves/plant	
	4 weeks	8 weeks	4 weeks	8 weeks	4 weeks	8 weeks
Control	84.83[d]	118.32[d]	18.88[c]	32.70[d]	43.50[c]	97.70[d]
NPK35	96.35[b]	139.76[b]	19.70[c]	47.42[c]	50.63[b]	111.42[b]
NPK50	101.05[a]	142.50[a]	25.65[a]	57.50[a]	60.96[a]	117.53[a]
NPK65	90.53[c]	131.65[c]	22.58[b]	51.65[b]	48.50[b]	101.15[c]
LSD 0.05	1.7886	1.61125	1.0898	0.9882	2.3742	1.7421

a, b, c and d = Statistical Analysis

3 Results and Discussion

3.1 *Plant Height (cm)*

The effect of NPK fertilizer treatment with varied concentrations of NPK on tomato plant height is shown in Table 4. The results indicated that the NPK fertilizers had a positive effect on the plant height compared to control. The plants fertilized with NPK50 gave the tallest plants with 101.05 and 142.5 cm at 4 and 8 weeks respectively, while plants fertilized with NPK65 showed height of 90.53 and 131.65 cm at 4 and 8 weeks respectively. The increase of plant height of tomatoes by NPK might be due to tomato plants being feeders for macronutrient elements including potassium (K), nitrogen (N) and phosphorus (P) [23]. Authors demonstrated that tallest plants of tomato were obtained from the plots treated with NPK fertilizer compared to the control [24, 25].

3.2 *Number of Lateral Branches*

The effect of different concentrations of NPK fertilizers on the number of lateral branches per tomato plant at 4 and 8 weeks is shown in Table 4. Number of lateral branches per plant fertilized with NPK was higher than control at 4 and 8 weeks of treatment. However, Plants fertilized with NPK50 (N: 31.7, P: 4.04, K: 4.6%) increased number of lateral branches to 57.5% at 8 weeks, while plants fertilized with NPK35 (N: 7.8, P: 5.7, K: 4.8%) increased lateral branches per plant to 45.0.02% at 8 weeks. The increase in number of lateral branches could be attributed to increased nitrogen content in NPK fertilizer. This result is in agreement with Manoj et. al. [7], whereby increasing levels of nitrogen in NPK resulted in 30% increase of lateral branches per tomato plant.

3.3 Number of Leaves Per Plant

The greatest number of leaves per plant was observed when level of N increased from 7.8% to 31.7% in NPK fertilizers. At 4 and 8 weeks of treatment, the NPK50 (N: 31.7, P: 4.04, K: 4.6%) resulted in the greatest number of leaves per plant (60.9 and 117.5 respectively) followed by NPK35 (N: 7.8, P: 5.7, K: 4.8%) (50.6 and 111.4 respectively) and NPK65 (N: 40.3, P: 4.5, K: 5.1%) (48.5 and 101.2 respectively) (Table 4). These values indicated significant response to the NPK fertilizers treatment and difference of nitrogen levels in the NPK had considerable influence on the number of leaves per plant [26]. These results were in agreement with those of Adekiya and Agbede [27]. The lowest value of leaf number was calculated for the control plants at 4 and 8 weeks (43.5 and 97.7 respectively).

3.4 Leaf Area (cm^2)

The effect of NPK fertilizer on leaf area of tomato at 4 and 8 weeks is shown in Table 5 whereby the fertilizer highly induced an increase in the leaf area over the control plants. These results indicated that NPK had a favourable effect on leaf area. The highest record of leaf area was obtained from plants fertilized with NPK50 (N: 31.7, P: 4.04, K: 4.6%), while control without fertilizer gave the lowest leaf area at 4 and 8 weeks. This attributed to role of nitrogen in NPK fertilizer which increasing cytokinin in the shoots and increasing leaf area. Similar results were recorded by Singh et al. [28] who found that NPK treatments significantly increased leaf tomato area.

Table 5 The effect of NPK fertilizers on the growth vegetative characteristics of tomato plant

Treatments	Leaf area (cm^2)		Shoots fresh weight (g/plant)		Shoots dry weight (g/plant)	
	4 weeks	8 weeks	4 weeks	8 weeks	4 weeks	8 weeks
Control	138.53^d	151.90^d	227.50^d	350.40^d	60.38^c	85.75^d
NPK35	220.35^c	250.00^c	272.18^b	625.20^b	69.10^b	131.38^b
NPK50	259.20^a	277.78^a	298.37^a	877.78^a	75.90^a	201.43^a
NPK65	245.40^b	263.00^b	252.23^c	543.10^c	61.35^c	111.20^c
LSD 0.05	1.4744	1.5031	1.0508	0.6722	1.0873	1.2683

a, b, c and d = Statistical Analysis

3.5 Fresh and Dry Weight of Shoots (g/Plant)

All applications of NPK fertilizers had a significant effect on the fresh and dry weights of shoots than control at 4 and 8 weeks. The fresh and dry weights of shoots tomato plants treated with NPK50 (N: 31.7%) were higher than NPK35, NPK65 and control at both 4 and 8 weeks (Table 5). NPK50 increased fresh shoot weight and dry shoot weight up to 150.15% and 134.8% respectively at 8 weeks, compared to control. While, NPK65 increased fresh shoot weight and dry shoot weight up to 54.99% and 29.67% respectively at 8 weeks. The increase in fresh and dry shoot weight might be due to that NPK improved root growth, which consequently promoted shoot growth. A similar result has been results reported by Etissa et al. [29] who demonstrated that fresh and dry weights of tomato shoots were affected by the application of NPK fertilizer.

3.6 Chlorophyll (a, b) Content (mg/dm^2)

The effect of NPK fertilizers containing different ratios of nitrogen on the chlorophyll a and b contents in the tomato leaves at 4 and 8 weeks is presented in Table 6. The results showed that chlorophylls content increased in all the tomato plants treated with NPK fertilizers. The maximum content of chlorophylls a and b (2.915 and 1.873 mg/dm^2 respectively) was recorded in plants that were fertilized with NPK50 (N: 31.7, P: 4.04, K: 4.6%) at 8 weeks while the minimum content of chlorophylls a and b content (2.2 and 1.4 mg/dm^2 respectively) was recorded in plants that were fertilized with NPK65 (N: 40.3, P: 4.5, K: 5.1%) at 8 weeks. Leaf chlorophyll content was found to be affected by different factors including types and concentration of fertilizers, nutrient concentration, distribution of chlorophyll in leaves and plant genotype [30].

Table 6 The effect of NPK fertilizers on chlorophyll a, b (mg/dm^2) and carbohydrate content (%) in leaves of tomato plant

Treatments	Chlorophyll a (mg/dm^2)		Chlorophyll b (mg/dm^2)		Carbohydrate (%)	
	4 weeks	8 weeks	4 weeks	8 weeks	4 weeks	8 weeks
Control	1.20^d	1.44^d	0.64^d	0.84^d	9.55^d	12.56^c
NPK35	1.94^b	2.55^b	0.78^c	1.19^c	11.32^b	13.75^b
NPK50	2.52^a	2.92^a	1.47^a	1.87^a	12.65^a	16.75^a
NPK65	1.72^c	2.16^c	0.89^b	1.39^b	10.63^c	12.63^c
LSD 0.05	0.0843	0.0631	0.0283	0.0401	0.2999	0.2222

a, b, c and d = Statistical Analysis

3.7 *Carbohydrate Content (%)*

Table 6 shows the response of total carbohydrate of tomato leaves to NPK fertilizers. It can be seen that NPK fertilizers significantly increased the total carbohydrate percentage in the treated plants over the control. The highest carbohydrate contents in tomato leaves were obtained with NPK50 fertilizer (16.75%), while the lowest carbohydrates content was observed in plants fertilized with NPK65 fertilizer (12.63%) at 8 weeks. This is attributed to the increase in the rate of photosynthesis process and construction of organic compounds mainly carbohydrates as reported by Xiukang and Yingying [31].

3.8 *Nitrogen, Phosphorus and Potassium Content (%)*

The nutrient percentage in leaves of tomato plants as influenced by NPK fertilizers in both 4 and 8 weeks is shown in Table 7. All NPK fertilizer treatments enhanced the absorption of N, P and K, hence increased the concentrations in the leaves as compared to the control plants. The highest leaf nitrogen content was the plant fertilized with NPK65 (N: 40.3, P: 4.5, K: 5.1%) whereas the highest concentrations of P and K were detected in the leaves treated with NPK 50 (N: 31.7, P: 4.04, K: 4.6%) at 4 and 8 weeks. The highest P content (0.65%) and K content (2.85%) in leaves of tomato plants were observed in NPK50 fertilizer and the lowest P content (0.44%) in leaves was found in NPK 65 fertilizer at 8 weeks, while the lowest K content (2.4%) was in NPK35 fertilizer at 8 weeks. The increase in nutrient percentage in leaves of tomato plants maybe due to improved absorption of N, P and K at higher rates of application and is consistent with the use of NPK as fertilizers for tomato production [27].

Table 7 The effect of NPK fertilizers on the chemical composition in leaves tomato at 4 and 8 weeks

Treatments	Nitrogen, (%)		Phosphorus, (%)		Potassium, (%)	
	4 weeks	8 weeks	4 weeks	8 weeks	4 weeks	8 weeks
Control	2.27^{d}	3.29^{d}	0.23^{c}	0.41^{d}	1.41^{c}	1.94^{d}
NPK35	3.90^{c}	4.42^{c}	0.44^{b}	0.53^{b}	1.82^{b}	2.43^{c}
NPK50	4.21^{b}	5.75^{b}	0.49^{a}	0.56^{a}	2.00^{a}	2.85^{a}
NPK65	5.13^{a}	6.67^{a}	0.43^{b}	0.44^{c}	1.87^{b}	2.66^{b}
LSD 0.05	0.165	0.2593	0.0296	0.0312	0.0602	0.0446

a, b, c and d = Statistical Analysis

Table 8 The effect of NPK fertilizers on flowering characteristics of tomato plant

Treatments	Node number bearing	Number of flower clusters/plant	Number of flowers/cluster	Number of flowers/plant
Control	5.50^{d}	14.11^{d}	4.35^{d}	75.40^{c}
NPK35	7.24^{c}	21.65^{b}	6.19^{c}	87.55^{b}
NPK50	8.63^{a}	28.69^{a}	8.21^{a}	99.64^{a}
NPK65	8.08^{b}	19.96^{c}	7.40^{b}	84.40^{b}
LSD 0.05	0.2754	0.1994	0.2846	3.3503

a, b, c and d = Statistical Analysis

3.9 *Characteristics of Flowers*

The influences of NPK fertilizer on flowering of tomato plants are shown in Table 8. Data revealed that the NPK improved the flowering characters. Increasing N content in NPK from 7.8% (NPK35) to 31.7% (NPK50) increased the node number bearing the first flower from 7.24 to 8.63, number of flower clusters per plant from 21.65 to 28.69, number of flowers per cluster from 6.19 to 8.21, and number of flowers per plant from 87.55 to 99.64. It might be due to the highest level availability nutrients to plant that forced towards the growth of vegetative parts then bloom as compared to control without fertilizers. Similar results were obtained by authors [3, 28] who detected similar flowering characteristics of tomato plants treatment with NPK fertilizers.

3.10 *Characteristics of Fruits*

Number of fruits per plant Tomato plants treatment with NPK produced a higher number of fruits than the control as shown in Table 9. The maximum fruit numbers were produced by plants fertilized with NPK50 (N: 31.7%) (35.7) and NPK35 (N: %) (31.4) compared to plants fertilized with NPK 65 (N: 40.3%) (28.8) and control

Table 9 Effect of NPK fertilizers on fruit numbers and physical characteristics of tomato fruits

Treatments	Fruit number/plant	Fresh weight (g/fruit)	Shape index	Fruit volume (cm^3)
Control	24.25^{d}	37.24^{d}	1.10^{d}	32.08^{d}
NPK35	31.38^{b}	75.45^{b}	1.25^{b}	40.53^{b}
NPK50	35.65^{a}	82.93^{a}	1.41^{a}	44.94^{a}
NPK65	28.80^{c}	71.60^{c}	1.81^{c}	37.74^{c}
LSD 0.05	0.3408	0.4701	0.0582	0.2017

a, b, c and d = Statistical Analysis

(24.3). These results agreed with Zekri and Obreza [32] who stated that lower concentrations of NPK limited plant growth, flower and fruit production of citrus. Increasing the nitrogen content in NPK increased the number of fruits due to the increased vegetative growth under abundance of nitrogen content in NPK for photosynthesis activity.

Weight and volume of fruits The data presented in Table 9 demonstrates that NPK treatment of tomato plants had significant effects on fresh weight, volume and shape index of the tomato fruit. Plants treated with NPK50 (N: 31.7, P: 4.05, K: 4.6%) showed the highest fruit weight (82.9 gm), volume (44.9 cm^3) and shape index (1.4), while plants fertilized with NPK65 (N: 40.3, P: 4.5, K: 5.1%) showed the lowest values of the above parameters (71.6 gm, 37.7 cm^3 and 1.2 respectively) but higher than the control. The findings of Fandi et al. [4] and Schon et al. [33] supported these results whereby increased weight of the tomato fruit parameters was attributed to the nutrient potential of the NPK fertilizer.

3.11 Chemical Composition of Fruits

The SSC, acidity, ascorbic acid and lycopene of fruits tomato plants treated with NPK is shown in Table 10. The SSC, titratable acidity of fruits increased with increasing nitrogen content in NPK up to 31.7% (NPK50). Beyond 31.7% (NPK65) the SSC, titratable acidity began to decrease but still higher than control fruits as shown in Table 10. Table 10 indicates that NPK treated plants produced fruits with higher ascorbic acid and lycopene than the control plants. Treatment with NPK 50 (N: 31.7, P: 4.05, K: 4.6%) gave the highest values for all the chemical characteristics of the tomato fruits. The increase in SSC, titratable acidity, ascorbic aci, lycopene of tomato fruits might be attributed to the effect of NPK in supplying the plants with various nutrients, and producing auxins (Mansour et al. [35]). These results were consistent with those of Xiukang and Yingying [31], Salama et al. [34], and Mansour et al. [35]. who illustrated that plant grown at high nitrogen content will result in fruits

Table 10 Effect of NPK fertilizers on some chemical characteristics of tomato fruits

Treatments	Soluble solids content (%)	Titratable acidity (%)	Ascorbic acid content (mg/100 g)	Lycopene (mg/100 g)
Control	3.60^{d}	0.65^{d}	17.18^{d}	25.08^{d}
NPK35	5.71^{b}	0.76^{b}	20.30^{b}	35.81^{b}
NPK50	6.10^{a}	0.87^{a}	22.91^{a}	37.48^{a}
NPK65	5.19^{c}	0.71^{c}	18.77^{c}	32.90^{c}
LSD 0.05	0.3292	0.0425	0.265	0.0593

a, b, c and d = Statistical Analysis

with high SSC, titratable acidity, ascorbic aci, lycopene and tomato fruits responded mainly to NPK fertilization (15: 15: 15).

4 Conclusions

NPK fertilizer treatments were carried with the aim of improving tomato growth and yield, and leaf chemical composition. NPK fertilizers were based on mixing 90% PU fertilizer with 10% K_3PO_4, with different nitrogen content NPK35 (N: 7.8, P: 5.7, K: 4.8%), NPK50 (N: 31.7, P: 4.05, K: 4.6%) and NPK65 (N: 40.3, P: 4.5, K: 5.1%). Data on plant height, number of lateral branches, number of leaves per plant, leaf area, fresh and dry weights of shoots, chlorophylls a and b; carbohydrate content, nutrient content, node number bearing the first flower, number of flower clusters per plant, number of flowers per cluster, number of fruits, weight, volume and shape index of fruits, soluble solid content (SSC), acidity, vitamin C, and lycopene pigment were recorded and statistically analyzed to evaluate the treatment effects. Results revealed that growth in tomato was more highly influenced by NPK treatments. However, the highest vegetative growth, fruit yield, nutrient content in the fruit content were found in tomato plants fertilized with NPK50 (N: 31.7%) followed by NPK35 (N: 7.8%) and NPK65 (N: 40.3).

References

1. Saravanan, S., Thamburaj, S., Veeraragavathatham, D., Subbiah, A.: Effects of seaweed extract and chlormequat on growth and fruit yield of tomato (*Lycopersicon esculentum* Mill.). Indian J. Agric. Res. **37**, 79–87 (2003)
2. Makela, P., Jokinen, K., Kontturi, M., Peltonen-Sainio, P., Pehu, E., Somersalo, S.: Foliar application of glycinebetaine—a novel product from sugar beet—as an approach to increase tomato yield. Ind. Crop Prod. **7**, 139–148 (1998)
3. Jayasinghe, H.A.S., Alwis, L.M.H., Gunadasa, K.S.G.: Effect of different NPK levels on the growth and yield of three tomato (Solanum lycopersicum) varieties in Sri Lanka. Asian Res. J. Agric. **2**(1), 1–6 (2016)
4. Fandi, M., Muhtaseb, J., Hussein, M.: Effect of N, P, K concentrations on yield and fruit quality of tomato (Solanum lycopersicum l.) in tuff culture. J. Cent. Eur. Agric. **11**(2), 179–184 (2010)
5. Noha, E.E., Ayoub, Z.A., Mohamed, O.W., Ibrahim, H.B.: Impact of N, P and K and organic fertilizer on tomato (*Lycopersicon Esculentum* Mill) growth and yield under cool plastic tunnel conditions. Int. J. Appl. Pure Sci. Agric. **3**(9), 38–45 (2017)
6. Chaurasia, S.N.S., Singh, K.P., Mathura, R.: Effect of foliar application of water-soluble fertilizers on growth. Yield and quality of tomato (*Lycopersicon esculentum* L.). Sri Lankan J. Agric. Sci. **42**, 66–70 (2005)
7. Manoj Kumar, M.L.M., Sanjay, K., Sutanu, M., Devendra, K.: Effect of nitrogen, phosphorus and potassium fertilizers on the growth, yield and quality of tomato var. Azad T-6. Asian J. Hort. **8**(2), 616–619 (2013)
8. Nnabude, P.C., Nweke, I.A., Nsoanya, L.N.: Response of three varieties of tomatoes (Lycopersicon esculentum) to liquid organic fertilizers (alfa life) and inorganic fertilizer (NPK 20: 10: 10) and for soil improvements. Eur. J. Pure Appl. Chem. **1**(1), 32–41 (2014)

9. Pansare, P.D., Desai, B.B., Chavan, U.D.: Effects of different nitrogen, phosphorus and potassium ratio on yield and quality of tomato. J. Maharashthra Agril. Univ. **19**(3), 462-463 (1994)
10. Tswanya, M.N., Olaniyi, J.O.: Effects of mineral fertilizers on growth and fruit yield of tomato variety (Lycopersicon lycopersicum Mill) in the Southern Guinea Savanna Zone of Nigeria. J. Agric. Sci. Food Technol. **2**(9), 147–153 (2016)
11. Felipe, E.F., Casanova, O.E.: Nitrogen, phosphorus and potassium fertilization in tomato (*Lycopersicon esculentum* Mill.) in alluvial bank soils of the Guarico river. Revista-Unellez-de-Ciencia-y-Tecnologia, Produccion Agricola **17**(1), 21–44 (2000)
12. Negim, E.-S., Bekbayeva, L., Hanan, A., Yeligbayeva, G.: The effect of blend ratios on physico-mechanical properties and miscibility of cross-linked poly (vinyl alcohol)/urea blends. J Physiscs Conf. Seri. **1123**(1), 1–9 (2018)
13. Sartory, D.P., Grobbelaar, J.U.: Extraction of chlorophyll a, from freshwater phytoplankton for spectrophotometric analysis. Hydrobiologia **114**, 177–187 (1984)
14. Dubois, M., Gilles, K.A., Hamilton, J., Roberts, R., Smith, F.: Colorimetric method for determination of sugar and related substances. Ann. Chem. **28**, 350–356 (1956)
15. Piper, C.S.: Soil and Plant Analysis. University of Adelaide, pp.258–275 (1947)
16. Pregl, F.: Quantitative organic microanalysis, 4th edn. J. and A. Churchill Ltd., London (1945)
17. Chapman, H.D., Pratt, P.F.: Methods of Analysis for Soil, Plant and Water. University of Californa Division Agriculture Science, U.S.A (1961)
18. AOAC.: Official Method of Analysis of Association of Agricultural Chemists 10th (ed.) (1965)
19. AOAC.: Official Method of Analysis of Association of Agricultural Chemists 15th (ed.) (2005)
20. AOAC.: Official Method of Analysis of Association of Agricultural Chemists 10th (ed.) (1970)
21. Sharma, Maguer L.: Lycopene in tomatoes and tomato pulp fractions. Ital. J. Food Sci. **2**, 107–113 (1996)
22. Mead, R., Currnow, R.N., Harted, A.M.: Statistical Methods in Agricultural and Experimental Biology 2nd (ed.), pp. 54–60 (1993)
23. Eliakira K., Peter H.: Effect of poultry manure and NPK (23:10:5) fertilizers on tomato variety grown on selected soil of Morogoro Region, Tanzania. Asian J. Crop Sci. 1–11 (2014)
24. Nafu, A.K., Togun, A.O., Abiodun, M.O., Chude, V.O.: Effects of NPK fertilizer on growth, dry matter production and yield of eggplant in southwestern Nigeria. Agric. Biol. J. N. Am. **2**(7), 1117–1125 (2011)
25. Isah, A.S., Amans, E.B., Odion, E.C., Yusuf, A.A.: Growth rate and yield of two tomato varieties (*Lycopersicon esculentum* Mill) under green manure and NPK fertilizer rate Samaru Northern Guinea Savanna. Int. J. Agronomy 1–8, (2014)
26. Samia, O.Y., Wigdan, M.A.A., Mariod, A.A.: Effect of urea, NPK and compost on growth and yield of soybean (Glycin max L.), in Semi-Rid region of Sudan. Int. Scholarly Res. Netw. pp. 1–6 (2012)
27. Adekiya, A.O., Agbede, T.M.: Growth and yield of tomato (*Lycopersicon esculentum* Mill) as influenced by poultry and NPK fertilizer. Emir. J. Food Agric. **21**(1), 10–20 (2009)
28. Singh, R., Kohli, U.K., Sharma, S.K.: Effect of nitrogen, phosphorous and potassium combinations on yield of tomato hybrids. Ann. Agric. Res. **21**, 27–31 (2000)
29. Etissa, E., Dechassa, N., Alamirew, T., Alemayehu, Y., Desalgen, L.: Growth and yield components of tomato as influenced by nitrogen and phosphrous fertilizer application in different growing seasons. Ethiop. J. Agric. **23**, 57–77 (2013)
30. Uddling, J., Gelang-Alfredsson, J., Piikki, K., Pleijel, H.: Evaluating the relationship between leaf chlorophyll concentration and SPAD-502 chlorophyll meter readings. Photosynth. Res. **91**, 37–46 (2007)
31. Xiukang, W., Yingying, X.: Evaluation of the effect of irrigation and fertilization by drip fertigation on tomato yield and water use efficiency in greenhouse. Int. J. Agronomy 1–10 (2016)
32. Zekri, M., Obreza, A.: Plant nutrients for citrus trees, soil and water science department, Institute Food Agric. Sci. **200** (2003)

33. Schon, M., Compton, M., Bell, E., Burns, I.: Nitrogen concentrations affect pepper yield and leachate nitrate-nitrogen from rockwool culture. HortSci. **29**(10), 1139–1142 (1994)
34. Salama, A.A., Nashwa, A.I.A., El-Sayed, H.E.K.: Effect of deficit irrigation levels and NPK fertilization rates on tomato growth, yield and fruits quality. Middle East J. Agric. Res. **6**(3), 587–604 (2017)
35. Mansour, A.F.M., Ahmed, F.F., Abdel Aal, A.M.K., Cimpoies, G.P.: Use mineral, organic, slow release and biofertilizers for anna apple trees in a sandy soil. In: African Crop Science Conference Proceedings **8**, 265–271 (2007)

Production of Greener Biodiesel from a Low-Cost CaO Catalysts of Waste Freshwater Shells

Nazrizawati Ahmad Tajuddin, Nurul Natasha Shahrom, and Noraini Hamzah

Abstract In this study, a low-cost and environment-friendly catalysts from waste freshwater shells of *Polymesoda erosa (kepah), Paratapes textilis (lala),* and *Pholas orientalis (mentarang)* are used in the production of biodiesel. All the samples were collected at Sungai Besar, Selangor, Malaysia is a source of calcium carbonate ($CaCO_3$) which changes to calcium oxide (CaO) in calcination temperatures 900 °C for 3 h. The thermogravimetric, morphology, structure, surface area and functional group of catalyst were studied by using Thermogravimetric Analysis (TGA), X-ray Diffraction (XRD), Scanning Electron Microscopy (SEM), *Brunauer–Emmett–Teller* (BET), and Fourier Transform Infrared Spectroscopy (FTIR) respectively. Transesterification reaction was done in the presence of waste cooking oil, methanol and freshwater shells catalyst at a temperature of 65 °C. Under the reaction condition (temperature 65 °C, methanol/oil molar ratio 30:1, reaction time 12 h, and catalyst 5 wt.% of oil), the biodiesel yield of *Polymesoda erosa (kepah), Paratapes textilis (lala)* and *Pholas orientalis (mentarang)* were 30.98%, 28.64%, and 11.11% respectively.

Keywords Biodiesel · Freshwater shells · CaO catalyst · Kepah · Lala · Mentarang

1 Introduction

Nowadays, the finding of an alternative energy source is the main green environmental focus due to the increment of energy, depletion of fossil fuel resources, expensive price of diesel fuel, global warming, and environmental pollution [1–3]. In concomitant with the rapidly increasing world population, the energy demand is predicted to increase by 50% of its present by 2040 [4]. Biodiesel is frequently generated in the presence of the catalyst through the method of transesterification of

N. A. Tajuddin (✉) · N. N. Shahrom · N. Hamzah
School of Chemistry and Environment, Faculty of Applied Sciences, Universiti Teknologi MARA (UiTM), 40450 Shah Alam, Selangor, Malaysia
e-mail: nazriza@uitm.edu.my

N. Hamzah
e-mail: pnoraini@uitm.edu.my

S. A. Abdul Karim et al. (eds.), *Proceedings of the 6th International Conference on Fundamental and Applied Sciences*, Springer Proceedings in Complexity,
https://doi.org/10.1007/978-981-16-4513-6_5

oil with alcohol [5] where the glycerol is a by-product [6–8]. Normally, methanol is used as a co-reactant for the transformation of triglyceride to generate methyl ester fatty acid because it is low-priced, easily available, and easy to separate from the product mixture [9, 10]. Biodiesel could produce from renewable sources where it is biodegradable, non-toxic, low sulfur content, inexpensive, environmental-friendly, and reduces the greenhouse gas emission effect [10–12]. Waste cooking oils are a suitable option for biodiesel production according to the situation where the quantity of waste cooking oil produced, waste management issues and environmental hazards, crude oil, and vegetable oil import expenses, also high-speed diesel imports [13]. Vegetable oils are consumed primarily for the manufacturing of biodiesel for many centuries since they can directly use in the engine without changing their properties. Edible oil including long-chain alkyl ester fatty acids that are transformed into short-chain viscous alkyl esters depicted as biodiesel [14, 15].

In general, $Ca(NO_3)_2$, $CaCO_3$, or $Ca(OH)_2$ is the raw material to produce CaO catalysts which react greatly in the transesterification of biodiesel reaction. Unfortunately, preparing highly efficient solid catalysts is complex, costly, and requires certain skills to operate them [2]. The catalyst's manufacturing costs could be a major factor in its industrial applications. Biodiesel production, therefore, needs an efficient and cheap catalyst to make the process cost-effective and environmentally friendly, by reducing its price and be competitive with petroleum diesel. To discourse these issues, the attentiveness of using waste material such as $CaCO_3$ from numerous natural calcium sources as raw resources for catalyst synthesis such as mollusk shell, eggshell, meretrix, snail shell, and clamshell could exterminate the wastes and at the same time formed a competence heterogeneous catalyst for transesterification [2].

It has been reported that catalyst synthesized from the freshwater waste shells fit as an alternative for renewable catalyst and at the same time very beneficial in term of recycles the waste generated [1, 2, 8, 16]. Many studies have been conducted on the growths of heterogeneous catalysts from waste in the usage of shells, bones, or skin of marine animals due to their rich $CaCO_3$ to be decomposed into CaO catalyst by calcination process [17]. Waste shells were mainly composed of $CaCO_3$ but when it was heated at 750–800 °C, it was converted to CaO [18]. It has been reported, CaO derived from a calcined *Polymesoda erosa (kepah), Paratapes textilis (lala) and Pholas orientalis (mentarang)* have the potential to be used as a novel heterogeneous catalyst for the esterification [2]. To date, no extension study has been done on transesterification reaction using these types of the catalyst with waste cooking oil. Hence, the main aim of this study is to prepare the mixed oxide catalyst using waste freshwater seashells of *Paratapes textilis (lala), Polymesoda erosa (kepah),* and *Pholas orientalis (mentarang)* as the catalyst before being applied in waste cooking oil to convert into biodiesel via transesterification reaction.

2 Materials and Methods

2.1 *Preparation of Low-Cost Catalyst from Freshwater Seashell*

The waste freshwater shells as a source of CaO were collected from the household waste at Sungai Besar, Selangor, Malaysia. The collected samples of waste freshwater shells were cleaned and wash several times with distilled water to remove impurities. Then, the shells were dried in an oven at 100 °C for 4 h to remove water and were crushed using a pestle and mortar followed by subsequent grinding until it becomes powdered [16]. Samples were further calcined in a furnace at 900 °C for 3 h at atmospheric pressure and was stored in a vacuum desiccator.

2.2 *Catalysts Characterization*

Catalysts were characterized by Powder X-ray Diffraction (PXRD) using X'pert-Pro PANanalytical diffractometer with a Cu Kα X-ray radiation ($\lambda = 1.54056$) with 2θ range of 5°–80°. The specific surface area, pore size and pore volume distribution of all the catalyst were determined by using the *Brunauer-Emmet-Teller* (BET) using a Micromeritics ASAP 2060 instrument. The surface morphological structures of the catalyst samples before and after calcined were scanned by using Scanning Electron Microscope (SEM) by Hitachi Tabletop Microscope TM3030Plus. The infrared spectra were obtained from Perkin Elmer (Model: Spectrum One) FTIR Spectrometer at wavenumbers from 4000 to 450 cm^{-1} using the KBR pellet technique. The samples were analyzed using a High-performance modular simultaneous TGA & DTA/DSC thermal analyzer (ambient/2400 °C) analyzer under a nitrogen flow rate of dry air at 30 ml/min condition with a temperature range from 25 to 1000 °C with 10 °C/min of heating rate to determine thermal transition of the sample.

2.3 *Pre-treatment of Waste Cooking Oil (WCO)*

Waste cooking oil (WCO) was collected at a stall from Dataran Cendekia at Universiti Teknologi MARA (UiTM), Shah Alam, Selangor, Malaysia. The collected WCO has been filtered to separate suspended solid material and debris. The water content was removed from WCO through drying at 120 °C for 1 h. The pre-treatment process was carried out to remove excess free fatty acid content in waste cooking oil by adding 0.5 g activated carbon to 100 ml of waste cooking oil. Then, the mixture was kept in a shaker flask at 150 rpm for 30 min under room temperature. After that, the oil was filtered before biodiesel production. To determine the free fatty acid (FFA) content, 1 g of oil was dissolved in 25 ml of ethanol and 25 ml of diethyl ether with a few drops of phenolphthalein was added as an indicator. The titration process was stopped when the solution turned into light pink color. If the FFA obtain is higher than 1%, the FFA value is not suitable for transesterification using basic catalysts [24].

2.4 *Transesterification Process*

25 g waste cooking oil, methanol, and catalyst were introduced in a round bottom flask with a water condenser attached to the water cooler. The mixture was heated for 12 h at 65 °C. The reaction mixture was allowed to cool down and then the solid catalyst is separated using a centrifuge. The biodiesel and glycerol combination were allowed to settle in a separating funnel overnight and the top layer of biodiesel and the bottom layer of glycerol will form [19]. The transesterification product is crude ester, fatty acid methyl ester (FAME), and glycerol. Crude biodiesel was purified using distilled water. Otherwise, the contaminants such as residual methanol, glycerol, catalyst, glycerides, and FFA, could decrease biodiesel quality and adversely affect engine performance on application [20]. The separatory funnel was washed with distilled water at 50 °C for three times until the solution is neutral [21]. The composition of FAME was determined by using gas chromatography equipped with a flame ionization detector (GC-FID). The percentage of FAME content was calculated as shown in Eqs. (1) and (2) by GC analysis based on the standard method EN 14103.

$$\text{Purity }(\%) = \frac{Area\ of\ all\ FAME}{Area\ of\ Reference} \times \frac{Wt.\ of\ reference}{Wt.\ of\ biodiesel\ sample} \times \frac{Wt.\ of\ reference}{Wt.\ of\ biodiesel\ sample} \times 100 \quad (1)$$

The biodiesel yield was defined using the following equation:

$$\text{Yield }(\%) = \frac{Area\ of\ all\ FAME}{Area\ of\ Reference} \times \frac{Wt.\ of\ reference}{Wt.\ of\ biodiesel\ sample} \times \frac{Wt.\ of\ biodiesel\ produced}{Wt.\ of\ oil\ used} \times 100 \quad (2)$$

The programmed column oven temperature increased by 40 °C min^{-1} from 190 to 210 °C and 40 °C min^{-1} to 230 °C with a 3 min holding time. The carrier gas was nitrogen gas at a flow rate of 2.2 ml/min. Both the inlet and detector temperatures were maintained at 250 °C.

3 Results and Discussion

3.1 Catalysts Characterization

The results of TGA analyses of *Polymesoda erosa (kepah), Paratapes textilis (lala)* and *Pholas orientalis (mentarang)* shells are shown in Table 1. There are two decomposition peaks, one is around 80–250 °C belong to dehydration of water and around 800–900 °C which is belong to decomposition of carbonate compound. These results confirmed that the experimental conditions used in the calcinations 900 °C are suitable for the decomposition of the $CaCO_3$ present in the shells [22].

The FTIR analyses of the fresh *Polymesoda erosa (Kepah), Paratapes textilis (Lala),* and *Pholas orientalis (Mentarang)* shells were carried out. Based on the result in Table 1, the major absorption bands observed were 2521.06 cm^{-1}, 1788.15 cm^{-1}, 1475.95 cm^{-1}, 1082.80 cm^{-1}, 860.87 cm^{-1}, 712.74 cm^{-1}, and 699.52 cm^{-1}, which are attributed to asymmetric stretch, out of the plane bend and in-plane bend vibration modes for CO_3^{2-} molecules on the catalyst surface. A broad absorption band of 3393.69 cm^{-1} is also observed due to the stretching vibration of structural water molecules.

Figure 1 depicts the XRD of the fresh and calcined shells of *Polymesoda erosa (Kepah)* (1a), *Paratapes textilis (Lala)*(1b), and *Pholas orientalis (Mentarang)* (1c). The peaks obtained were compared with the standard peaks in the Joint Committee of Powder Diffraction Standards (JCPDS) files. The peaks at 2θ of 32.40°, 37.65°,

Table 1 TGA decomposition temperature and FTIR characteristic bands and wavelengths

Catalyst	TGA	FTIR	
	Decomposition (°C)	Characteristic band	Wavenumber (cm^{-1})
Polymesoda erosa (kepah)	893.18	O–H stretching	3393.69
		C–H scissoring	1475.95
		=C–H out of plane bending	860.87
		C-H	712.24
Paratapes textilis (lala)	892.70	O–H stretching	3396.32
		C–H scissoring	1474.49
		=C–H out of plane bending	862.01
		C–H	712.66
Pholas orientalis (mentarang)	890.24	O–H stretching	3413.35
		C–H scissoring	2533.20
		=C–H out of plane bending	1475.90
		C–H	862.63

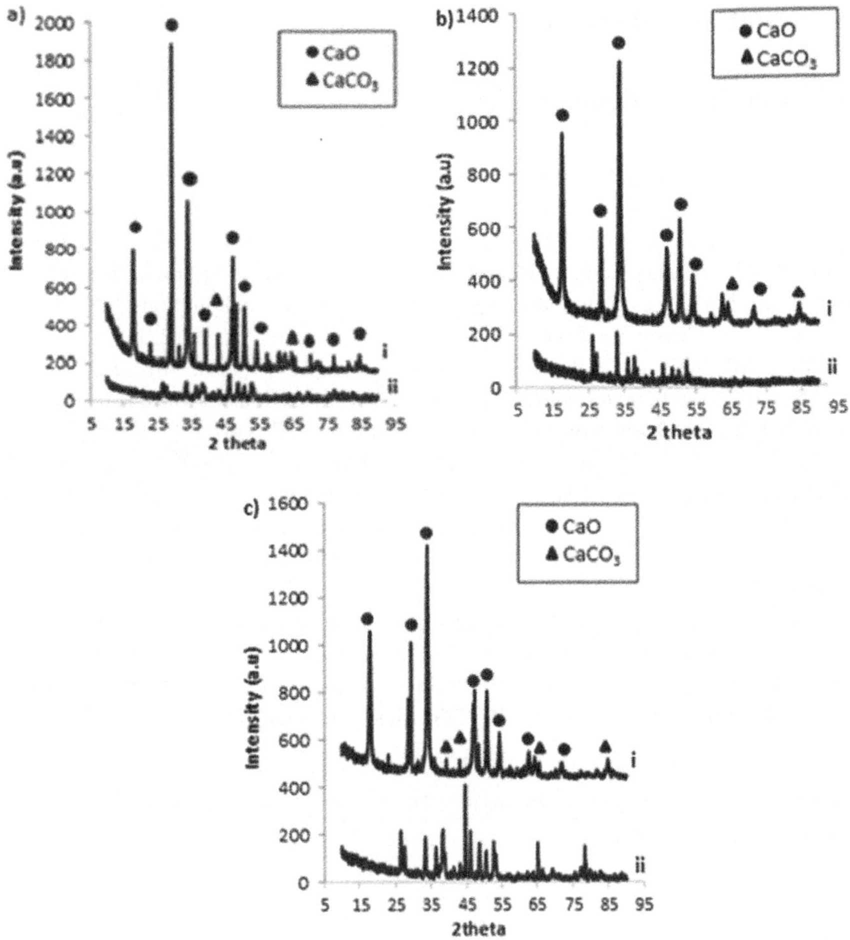

Fig. 1 XRD diffraction patterns of *Polymesoda erosa (Kepah)* (**a**), *Paratapes textilis (Lala)* (**b**) and *Pholas orientalis (Mentarang)* (**c**) of fresh (i) and calcined (ii)

54.13° correspond to CaO and those at 18.22°, 47.20°, 50.92°, 54.24°, 64.56° correspond to $CaCO_3$. The peaks of CaO (JCPDS No. 00-037-1497) are observed at the catalyst after being calcined. Narrow, sharp, and high intense peaks of the calcined catalyst define the well-crystallized structure of the CaO catalyst. However, $CaCO_3$ still remaining after calcination at 900 °C for 3 h. This finding further confirmed that at higher calcination temperature, most of the $CaCO_3$ compound has been completely transformed to CaO by evolving the CO_2, while some remain unchanged [22]. This result is similar to a study done by Ngamcharussrivichai and co-workers [23] where $CaCO_3$ peaks disappeared simultaneously with the increment of CaO peaks. Hence, to ensure all the carbonate compound fully decompose the calcination time need to be prolonged.

The surface area of the calcined freshwater shells was further investigated by using the N_2 adsorption/desorption technique. These catalyst exhibit type IV isotherms and type H1 hysteresis loops indicate microporous-mesoporous material with narrow-slit pore [17]. These are also associated with solids consisting of nearly cylindrical channels or agglomeratesor compacts of near-uniform spheres. The BET surface area of *Polymesoda erosa (kepah)*after calcined was 351 m^2/g. Associated with this, the higher calcination temperature will enhance the surface area and basicity sites of the catalyst due to segregation happened as CaO is dispersed on the surface [24]. By this, it is firmed that a higher temperature (900 °C) is the optimum calcination temperature of these freshwater shells. Segregation formed from calcination generates very fine powders, providing a higher surface area, which later will be expected to accelerate the mass transport and significantly increase the reaction rates and yield.

The morphology of the calcined catalyst was examined by Scanning Electron Microscope (SEM). Figure 2 shows the surface morphology observed under 1000× magnification. The shells calcined at 900 °C showed a decrease in the agglomeration size and segregation of particles, particularly due to the large amount of CO_2 that has been escaped from the decomposition of $CaCO_3$ molecules from the decomposition of the carbonate groups, liberating CO_2 concomitantly with the formation of small pores.

3.2 *Production of Biodiesel Using Waste Cooking Oil*

A pre-treatment of WCO was done to reduce the FFA contents (the initial FFA value was 5.33 mg KOH/g). The acid value after the esterification process was 0.98 mg KOH/g and the FFA percentage obtained was 0.0098%. Characterization of methyl ester compounds in biodiesel samples was analyzed using GC-FID. Based on the results, there were three FAME presents in all samples of biodiesel which are methyl myristate, methyl palmitate, and methyl linoleate.

Data shows 30.98% FAME was obtained from *Polymesoda erosa* (kepah), 28.64% FAME of *Paratapes textilis (lala), and 11.11% from Pholas orientalis (mentarang)* at 900 °C calcined temperature. The reaction of triglycerides with methanol proceeds in a stepwise manner to produce the intermediates of diglyceride and monoglyceride, with subsequent reactions with methanol producing glycerol and biodiesel. In summary, solid base catalysts specifically that CaO, which is the natural sources from $CaCO_3$ obtained from freshwater shells of *Polymesoda erosa (kepah), Paratapes textilis (lala), and Pholas orientalis (mentarang)*, is a good heterogeneous base catalyst for the transesterification of waste cooking oil with methanol. It was demonstrated that the high temperature of calcination has segregated the morphology, is expected to induce towards the higher surface area, and later enhanced the catalytic performance.*)*.

Fig. 2 SEM images for *Polymesoda erosa (Kepah)* (**a**): uncalcined (left) calcined (right), *Paratapes textilis (Lala)* (**b**): uncalcined (left) calcined (right) and *Pholas orientalis (Mentarang)* (**c**): uncalcined (left) and calcined (right)

Acknowledgements Authors would like to express a sincere heartfelt to the Research Management Centre (RMC), Universiti Teknologi MARA (UiTM) for financing the project under GPK grant (600-RMC/GPK 5/3 (117/2020).

References

1. Safieddin Ardebili, S.M., Solmaz, H., İpci, D., Calam, A., Mostafaei, M.: A review on higher alcohol of fusel oil as a renewable fuel for internal combustion engines: applications, challenges, and global potential. Fuel, 118516 (2020)
2. Syazwani, O.N., Teo, S.H., Islam, A., Taufiq-Yap, Y.H.: Transesterification activity and characterization of natural CaO derived from waste venus clam (*Tapes belcheri* S.) material for enhancement of biodiesel production. Process Saf Environ Prot. Institution of Chemical Engineers **05**, 303–315 (2017)
3. Toledo Arana, J., Torres, J.J., Acevedo, D.F., Illanes, C.O., Ochoa, N.A., Pagliero, C.L.: One-step synthesis of CaO-ZnO efficient catalyst for biodiesel production. Int. J. Chem. Eng. (2019)
4. Dhawane, S.H., Kumar, T., Halder, G.: Recent advancement and prospective of heterogeneous carbonaceous catalysts in chemical and enzymatic transformation of biodiesel. Energy Convers. Manag., 176–202 (2018)
5. Ambat, I., Srivastava, V., Sillanpää, M.: Recent advancement in biodiesel production methodologies using various feedstock: a review. Renew. Sustain. Energy Rev., 356–69 (2018)
6. Davoodbasha, M., Pugazhendhi, A., Kim, J.-W., Lee, S.-Y., Nooruddin, T.: Biodiesel production through transesterification of *Chlorella vulgaris*: synthesis and characterization of CaO nanocatalyst. Fuel **300**, 121018 (2021)
7. Jung, S., Kim, M., Lin, K.Y.A., Park, Y.K., Kwon, E.E.: Biodiesel synthesis from bio-heavy oil through thermally induced transesterification. J. Clean. Prod. **94**, 126347 (2021)
8. Almazrouei, M., Elagroudy, S., Janajreh, I.: Transesterification of waste cooking oil: quality assessment via thermogravimetric analysis. Energy Procedia, 2070–2076 (2019)
9. Yusuff, A.S., Popoola, L.T.: Optimization of biodiesel production from waste frying oil over alumina supported chicken eggshell catalyst using experimental design tool. Acta Polytech. Czech Technical University **59**, 88–97 (2019)
10. Michael, A.T., Ajibola, V.O., Agbaji, E.B., Yusuf, J.: Methanolic synthesis of fatty acid methyl esters (FAME) from Waste Materials. Chem. Sci. Int. J. **26**, 1–14 (2019)
11. Kaewdaeng, S., Sintuya, P., Nirunsin, R.: Biodiesel production using calcium oxide from river snail shell ash as catalyst. Energy Procedia, 937–942 (2017)
12. Piker, A., Tabah, B., Perkas, N., Gedanken, A.: A green and low-cost room temperature biodiesel production method from waste oil using egg shells as catalyst. Fuel **182**, 34–41 (2016)
13. Khan, H.M., Ali, C.H., Iqbal, T., Yasin, S., Sulaiman, M., Mahmood, H.: Current scenario and potential of biodiesel production from waste cooking oil in Pakistan: an overview. Chemical Industry Press, Chinese J. Chem. Eng (2019)
14. Kirubakaran, M., Arul Mozhi Selvan, V.: A comprehensive review of low cost biodiesel production from waste chicken fat. Renew. Sustain. Energy Rev., 390–401 (2018)
15. Ahmad Tajuddin, N., Lee, A.F., Wilson, K.: Chapter 6—Production of biodiesel via catalytic upgrading and refining of sustainable oleagineous feedstocks. Handb Biofuels Prod [Internet], pp. 121–64. Woodhead Publishing (2016)
16. Hadiyanto, H., Afianti, A.H., Navi'A, U.I., Adetya, N.P., Widayat, W., Sutanto, H.: The development of heterogeneous catalyst C/CaO/NaOH from waste of green mussel shell (Perna varidis) for biodiesel synthesis. J. Environ. Chem. Eng. **5**, 4559–4563 (2017)
17. Boonyuen, S., Smith, S.M., Malaithong, M., Prokaew, A., Cherdhirunkorn, B., Luengnaruemitchai, A.: Biodiesel production by a renewable catalyst from calcined Turbo jourdani (Gastropoda: Turbinidae) shells. J. Clean. Prod. **177**, 925–929 (2018)

18. Boro, J., Deka, D., Thakur, A.J.: A review on solid oxide derived from waste shells as catalyst for biodiesel production. Renew. Sustain. Energy Rev, 904–910 (2012)
19. Mansir, N., Teo, S.H., Rashid, U., Saiman, M.I., Tan, Y.P., Alsultan, G.A.: Modified waste egg shell derived bifunctional catalyst for biodiesel production from high FFA waste cooking oil. A review. Renew. Sustain. Energy Rev, 3645–3655 (2018)
20. Tshizanga, N., Aransiola, E.F., Oyekola, O.: Optimisation of biodiesel production from waste vegetable oil and eggshell ash. S. Afr. J. Chem. Eng. **23**, 145–156 (2017)
21. Yaakob, Z., Mohammad, M., Alherbawi, M., Alam, Z., Sopian, K.: Overview of the production of biodiesel from Waste cooking oil. Renew. Sustain. Energy Rev. **18**, 184–193 (2013)
22. Marinković, D.M., Stanković, M.V., Veličković, A.V., Avramović, J.M., Miladinović, M.R., Stamenković, O.O.: Calcium oxide as a promising heterogeneous catalyst for biodiesel production: Current state and perspectives. Renew. Sustain. Energy Rev., 1387–1408 (2016)
23. Lertpanyapornchai, B., Ngamcharussrivichai, C.: Mesostructured Sr and Ti mixed oxides as heterogeneous base catalysts for transesterification of palm kernel oil with methanol. Chem. Eng. J. **264**, 789–796 (2015)
24. Sun, H., Wu, C., Shen, B., Zhang, X., Zhang, Y., Huang, J.: Progress in the development and application of CaO-based adsorbents for CO_2 capture—a review. Mater. Today Sustain., 1–27 (2018)

CO_2 Hydrogenation to Methanol: Effects of Reaction Temperatures and Pellet Crushing on the Catalytic Performance

Nor Hafizah Berahim, Noor Asmawati Mohd Zabidi, Nadia Syazana Abd Halim, Akbar Abu Seman, and Nor Hafizah Yasin

Abstract Methanol synthesis from CO_2 hydrogenation is a potent clean energy solution that can address both climate change and depletion of natural resources. Nevertheless, the development of effective catalyst capable of providing adequate activity and stability remains one of the most significant barriers to the real implementation of such a reaction. In view of this, Cu/ZnO tri-promoted with transition metals from Group IV (GIV), V (GV) and VII (GVII) supported on SBA-15 pellet has been synthesized using the impregnation technique. The synthesized catalyst was characterized using temperature-programmed reduction (TPR), field emission scanning electron microscope (FESEM), N_2 adsorption/desorption and X-ray fluorescence (XRF) techniques. The catalytic performance of the catalyst in a CO_2 hydrogenation reaction was evaluated using a fixed-bed reactor and the products were determined using online GC. Effects of reaction temperatures and crushing of the pellet were also investigated in this study. The catalyst resulted in highest MeOH selectivity of 54.59% at 250 °C. Crushing the pellet into smaller size did not affect the catalytic performance significantly.

Keywords CO_2 · Methanol · Cu catalyst · SBA-15 pellet

N. H. Berahim (✉) · N. A. Mohd Zabidi · N. S. Abd Halim
Universiti Teknologi PETRONAS, 32610 Seri Iskandar, Perak, Malaysia
e-mail: nor_19001638@utp.edu.my

N. A. Mohd Zabidi
e-mail: noorasmawati_mzabidi@utp.edu.my

A. Abu Seman · N. H. Yasin
PETRONAS Research Sdn.Bhd, 43000 Kajang, Selangor, Malaysia
e-mail: akbar_seman@petronas.com

N. H. Yasin
e-mail: norhafizah.yasin@petronas.com

S. A. Abdul Karim et al. (eds.), *Proceedings of the 6th International Conference on Fundamental and Applied Sciences*, Springer Proceedings in Complexity,
https://doi.org/10.1007/978-981-16-4513-6_6

1 Introduction

Excess CO_2 concentrations in the atmosphere, which contributes to the global warming phenomena, has become a serious attention worldwide. Due to the need for mitigating CO_2 emissions and the increasing energy demands, current research activities are focusing to develop new and efficient technologies to cater the issues [1, 2]. In this respect, conversion of CO_2 into useful chemicals such as methanol (MeOH) has been the subject of many CO_2 utilization studies over the past few decades. Currently there is an existing demonstration plant in Iceland for MeOH synthesis using CO_2 as feed while the source of hydrogen is mainly coming from geothermal energy [3]. MeOH was mainly used as a starting feedstock in the chemical industries and also an important fuel which is easy to store and transport [4]. CO_2 is thermodynamically stable, thus highly active and selective catalyst for methanol synthesis are yet to be explored.

Industrially, $Cu/ZnO/Al_2O_3$ has been employed in methanol synthesis using syngas ($H_2/CO/CO_2$) as a feedstock. The commercial catalyst for methanol synthesis consists of 50–70% CuO, 20–50% ZnO, and 5–20% of metal promoters [5]. Unfortunately, such catalyst is neither active nor selective for pure CO_2 system [6]. Seeking a better catalyst for methanol synthesis from CO_2 hydrogenation remains an intensive research subject. This is due to poor thermal stability of the conventional catalyst which it suffered from rapid deactivation [7, 8]. However, most researchers believe that the copper phase activity has been linked to the Cu/ZnO system [9, 10]. That is why the majority of modified catalysts still contain Cu–Zn oxides and other components can be added to alter the catalytic properties [11]. Various activators or other metals (Zr, Si, La, Ti, Cr, Ga, Ce, Fe, Nb, Pd, etc.) has been investigated to improve the methanol synthesis catalyst from H_2/CO_2 feeding gas [12]. The effect of the support was also extensively studied. The types of support used plays an important role to improve catalytic and physical properties such as increase the surface area, stabilize the active phase and stabilize against sintering [13]. In general, basic oxides such as La_2O_3, Sm_2O_3, Nb_2O_5, In_2O_3, and ThO_2 were used as supports to favor methanol formation [14]. Mesoporous silica such as SBA-15 is well-known for its well-ordered structures and high surface area that will allow high accessibility of the active metals. The versatility in synthesizing different shapes of meso-structures also make it interesting to be implied as one of the promising catalyst support [15].

2 Methodology

2.1 Preparation of Catalyst Pellet Support (SBA-15)

SBA-15 pellet was pretreated in the tube furnace under Ar flow at temperature 400 °C for 5 h to remove moisture and impurities.

2.2 Preparation of Cu/ZnO Based Catalyst with the Addition of Tri-Promoters

Cu/ZnO with fixed metal loading of 15 wt% at a ratio of 7:3 and 3 wt% of promoters (GVII/GV/GIV) at a ratio 1:1:1 was synthesized using wetness impregnation method. The amount of each precursor and promoter added were calculated based on the amount of total catalyst mass being prepared. Firstly, desired amount of copper (II) nitrate trihydrate ($Cu(NO_3)_2.3H_2O$), zinc nitrate hexahydrate ($Zn(NO_3)_2.6H_2O$), Group VII salt, Group V salt, and Group IV salt were dissolved in a desired amount of deionized water to produce an aqueous solution. The solution was stirred to ensure homogeneity. Then, the prepared aqueous precursor solution was added dropwise into a beaker containing the pellet support (SBA-15). The catalyst sample Cu/ZnO/GVII/GV/GIV/SBA-15 was denoted as CZ(M)S. M represents the tri-promoters (GVII/GV/GIV).

The impregnated sample was dried at room temperature and calcined at 350 °C for 4 h.

2.3 Catalyst Characterization

TPR analysis was conducted using a Thermo Finnigan TPD/R/O 110 CE equipped with a thermal conductivity detector (TCD). 40 mg catalyst was placed in the quartz tube and was pre-treated under the flow of pure N_2 at 250 °C with a hold time of 1 h to remove moisture and impurities. The analysis was then continued by switching the gas flow to 5% H_2/Ar (20 mL/min) to a maximum temperature of 950 °C with ramping 10 °C/min and hold for an hour. The hydrogen consumption is monitored by a thermal conductivity detector (TCD).

The morphology of the synthesized catalyst was observed using Hitachi SU8000 field emission scanning electron microscope (FESEM).

Textural analysis has been carried out on a Micromeritics ASAP 2020 analyzer by determining the nitrogen adsorption/desorption isotherms at −196 °C. Prior to analysis, the samples was degassed at 350 °C (heating rate 10 °C/min). The Brunauer-Emmett-Teller (BET) specific surface area and pore volume are assessed from the adsorption data. The mean pore diameter is determined by applying the Barrett-Joyner-Halenda (BJH) model to the isotherm desorption branch [16].

The bulk compositions of active metals impregnated onto the pellet support was determined using XRF method. 3.0 g of calcined catalyst sample was put into the sample holder and the analysis were carried out based on calibration curves obtained using certified method.

2.4 Catalyst Evaluation

The CO_2 hydrogenation reactions to methanol were carried out in a tubular, stainless steel micro-activity fixed bed reactor (PID Eng and Tech). Prior to a reaction, the calcined catalyst was reduced in 5 vol% H_2 in He at 400 °C for 2 h under atmospheric pressure. The reactor tube was pressurized with He to 22.5 bar, then switched to feed gas containing CO_2 and H_2 with ratio of 1:3. The reaction temperature was set at 200–300 °C for 5 h and the reactor effluents were analyzed using an on-line gas chromatograph equipped with TCD and FID detectors. The catalytic performance was evaluated based on the MeOH selectivity.

3 Results and Discussion

3.1 Reduction Behavior of Tri-Promoted Cu/ZnO Based Catalyst

The characterization of catalyst by H_2-TPR is used to explore the ability of the catalyst to be activated via H_2 consumption. As shown in Fig. 1, the CZ(M)S exhibited broad reduction peak at temperature 200–500 °C region which corresponded to the reduction of Cu species and other species from the precursor salts since this sample was not washed and filtered. The result from Table 1 indicates a main reduction peak at 383 °C and a shoulder at 631 °C, thus the catalyst was activated at 400 °C prior to the reaction.

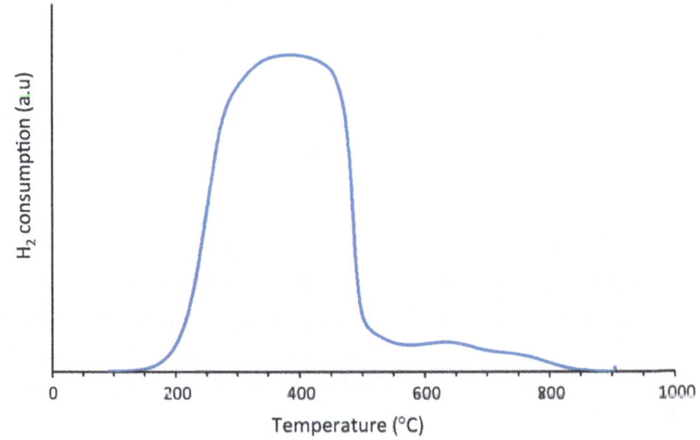

Fig. 1 TPR profile of CZ(M)S

Table 1 TPR quantification data for CZ(M)S

Catalyst	H_2 consumption (μmol/g)	Reduction temperature peak (°C)	
		T1	T2
CZ(M)S	1258.75	383	631

Fig. 2 FESEM images **a** SBA-15 **b** CZ(M)S

Table 2 Textural properties of the samples

Sample	S_{BET} (m^2/g)	V_P ($cm^{3/}g$)	D_{BJH} (nm)
SBA-15	305	0.40	5.6
CZ(M)S	260	0.35	4.15

3.2 *Morphology of Catalyst*

The morphology of the bare pellet support (SBA-15) and CZ(M)S catalyst were investigated using FESEM, and the acquired images are presented in Fig. 2.

SBA-15 pellet support, Fig. 2a had an irregular shape. The impregnated catalyst, CZ(M)S shows that the particles were distributed over the support which were responsible to the decreasing surface area of the support (see Table 2) [17].

3.3 *Textural and Physical Properties*

The textural properties such as the catalyst surface area (S_{BET}), pore volume (V_P) and pore diameter (D_{BJH}) are shown in Table 2. Surface area and pore volume of SBA-15 support are reduced to some degree after the metal oxide loading, suggesting the active phase has been successfully confined into the mesoporous channels [17].

Table 3 Bulk compositions of the sample

Catalyst	Composition of active metals (wt%)		Composition of promoters (wt%)		
			Tri-promoters (M)		
	Cu	Zn	GVII	GV	GIV
CZ(M)S	11.65	3.35	0.52	1.15	1.33

Table 3 shows the bulk compositions of active metals impregnated based on the XRF analysis. The prepared catalyst was loaded with 15 wt% at Cu to Zn ratio 7:3 respectively, while all the tri promoters (M) were added at 1 wt% each. The result of XRF analysis was different from the expected value whereby theoretically, the amount of Cu should be around 10.5 wt%, Zn 4.5 wt%, tri-promoters (M) each at 1 wt% which suggested that the impregnation method did not result in homogeneous deposition of metals onto the surface of the pellet.

3.4 Catalytic Performance

The catalytic performance of CZ(M)S catalyst was tested for CO_2 hydrogenation reaction using a micro-activity fixed bed reactor at pressure 22.5 bar, GHSV 4320 ml/g.h and H_2/CO_2 molar ratio of 3. The effect of different reaction temperatures (200–300 °C) has been evaluated. Time-on-stream plot for CZ(M)S in terms of MeOH selectivity is depicted in Fig. 3. It was observed that the reactions were stable throughout the 5 h time-on-stream. The highest MeOH selectivity, 54.59% was achieved at the reaction temperature 250 °C. However, MeOH selectivity slightly dropped to 47.73% when the reaction temperature increased from 250 to 300 °C. Figure 4 shows the products distribution for the reaction in terms of selectivity. It can

Fig. 3 Time-on stream profile for MeOH selectivity at different reaction temperatures

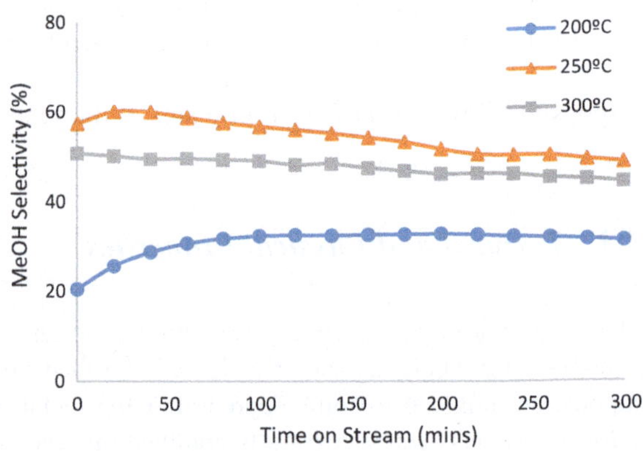

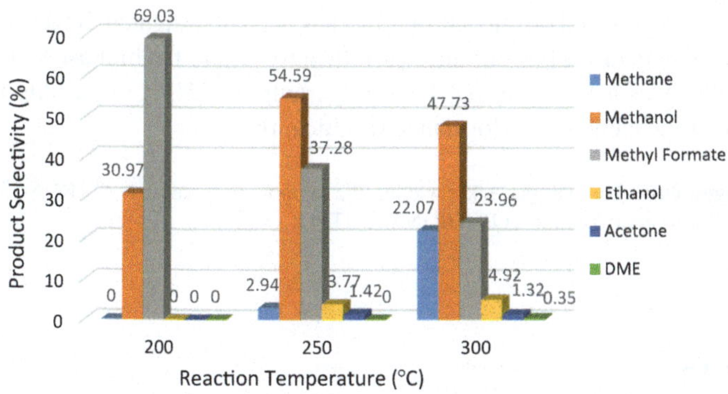

Fig. 4 Products distribution at different reaction temperatures

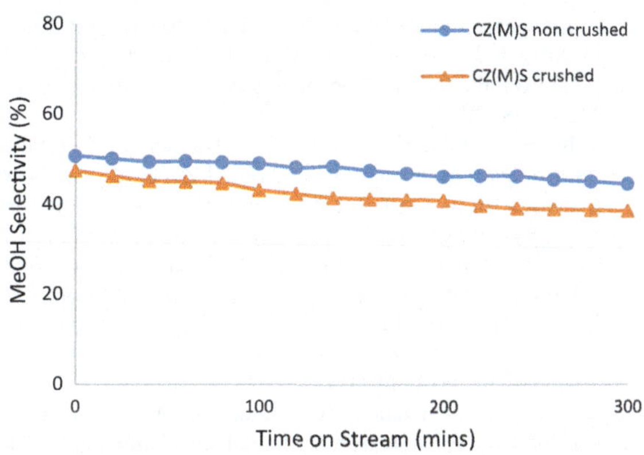

Fig. 5 MeOH selectivity time-on-stream profile for crushed versus non-crushed CZ(M)S catalyst at 300 °C

be seen that, at lower reaction temperature of 200 °C, the reaction favors the methyl formate formation.

The effect of catalyst crushing on the MeOH selectivity has been evaluated at reaction temperature 300 °C. The results are summarized in Fig. 5, which shows that crushing the pellet into smaller size did not affect the catalyst performance significantly.

4 Conclusion

Cu/ZnO based catalyst supported on SBA-15 pellet with the addition of tri-promoters (M) was synthesized via impregnation method. The catalyst reducibility, morphology, textural and physical properties has been investigated. Reaction temperature affects the catalyst performance significantly where the highest MeOH selectivity, 54.59% was achieved at reaction temperature of 250 °C. Crushing the pellet did not affect the catalytic performance significantly.

Acknowledgements The authors acknowledge the support provided by PETRONAS Research Sdn.Bhd. and Universiti Teknologi PETRONAS (UTP).

References

1. Kameyama, H., Yoshizaki, K., Yasuda, I.: Carbon capture and recycle by integration of CCS and green hydrogen. Energy Procedia **4**, 2669-2676 (2011)
2. Ibram, G.: Conversion of carbon dioxide into methanol-a potential liquid fuel: fundamental challenges and opportunities (a review). Renew. Sust. Energ. Rev. **31**, 221–257 (2014)
3. Da Silva, R.J., Pimentel, A.F., Monteiro, R.S., Mota, C.J.A.: Synthesis of methanol and dimethyl ether from the CO_2 hydrogenation over Cu·ZnO supported on Al_2O_3 and Nb_2O_5. J. CO_2 Util. **15**, 83–88 (2016)
4. Velu, S., Suzuki, K., Okazaki, M., Kapoor, M.P., Osaki, T., Ohashi, F.: Oxidative steam reforming of methanol over CuZnAl(Zr)-oxide catalysts for the selective production of hydrogen for fuel cells: catalyst characterization and performance evaluation. J. Catal. **194**(2), 373–384 (2000)
5. Choi, Y., Futagami, K., Fujitani, T., Nakamura, J.: Role of ZnO in Cu/ZnO methanol synthesis catalysts-morphology effect or active site model? Appl. Catal. A **208**(1–2), 163–167 (2001)
6. Inui, T., Hara, H., Takeguchi, T., Kim, J.B.: Structure and function of Cu-based composite catalysts for highly effective synthesis of methanol by hydrogenation of CO_2 and CO. Catal. Today **36**(1), 25–32 (1997)
7. Samei, E., Taghizadeh, M., Bahmani, M.: Enhancement of stability and activity of Cu/ZnO/Al_2O_3 catalysts by colloidal silica and metal oxides additives for methanol synthesis from a CO_2-rich feed. Fuel Process. Technol. **96**, 128–133 (2012)
8. Szizybalski, A., Girgsdies, F., Rabis, A., Wang, Y., Niederberger, M., Ressler, T.: In situ investigations of structure-activity relationships of a Cu/ZrO_2 catalyst for the steam reforming of methanol. J. Catal. **233**(2), 297–307 (2005)
9. Waugh, K.C.: Methanol synthesis. Catal. Lett. **142**(10), 1153–1166 (2012)
10. Behrens, M., Studt, F., Kasatkin, I., Kühl, S., Hävecker, M., Abild-Pederson, F., Zander, S., Girgsdies, F., Kurr, P., Kniep, B., Tovar, M., Fischer, R.W., Nørskov, J.K., Schlögl, R.: The active site of methanol synthesis over Cu/ZnO/Al_2O_3 industrial catalysts. Sci. **336**(6083), 893–897 (2012)
11. Allam, D., Cheknoun, S., Hocine, S.: Operating conditions and composition effect on the hydrogenation of carbon dioxide performed over CuO/ZnO/Al_2O_3 catalysts. Bull. Chem. React. Eng. Catal. **14**(3), 604–613 (2019)
12. Huang, C., Chen, S., Fei, X., Liu, D., Zhang, Y.: Catalytic hydrogenation of CO_2 to methanol: study of synergistic effect on adsorption properties of CO_2 and H_2 in CuO/ZnO/ZrO_2 system. Catal. **5**(4), 1846–1861 (2015)
13. Wang, L., Yang, L., Zhang, Y., Ding, W.: Promoting effect of an aluminum emulsion on catalytic performance of Cu-based catalysts for methanol synthesis from syngas. Fuel Process. Technol. **91**(7), 723–728 (2010)
14. Ramaroson, E., Kieffer, R., Kiennemenn, A.: Reaction of CO-H_2 and CO_2-H_2 on copper-zinc catalysts promoted by metal oxides of groups III and IV. Appl. Catal. **4**(3), 281–286 (1982)
15. Naik, S.P., Elangovan, S.P., Okubo, T., Sokolov, I.: Morphology control of mesoporous silica particles. J. Phys. Chem. C **111**(30), 11168–11173 (2007)

16. Barrett, E.P., Joyner, L.G., Halenda, P.P.: The determination of pore volume and area distributions in porous substances i. computations from nitrogen isotherms. J. Am. Chem. Soc. **73**(1), 373–380 (1951)
17. Tasfy, S.F.H., Zabidi, N.A.M., Shaharun, M.S., Subbarao, D.: The role of support morphology on the performance of Cu/ZnO-catalyst for hydrogenation of CO_2 to methanol. In: AIP Conference Proceedings **1669**(1) (2015)

Catalytic Conversion of CO_2 into Alcohols: Comparison of Supports

Nur Insyirah Zulkifli and Noor Asmawati Mohd Zabidi

Abstract Promoted Bimetallic Cu/ZnO catalysts were synthesized on Al_2O_3, Al_2O_3-ZrO_2 and zeolite supports. The catalytic performances in CO_2 hydrogenation reaction were evaluated in a fixed-bed reactor system at H_2:CO_2 (3:1), 22.5 bar and 250 °C for 5 h. The catalyst supported on Al_2O_3–ZrO_2 resulted in the highest CO_2 conversion (15.81%) and methanol selectivity (68.77%). In general, all the tested catalysts show low selectivity towards ethanol.

Keywords CO_2 Hydrogenation · Cu–ZnO Bimetallic Catalyst · Alumina · Zeolite · Alumina Zirconia

1 Introduction

There has been an intense interest from researchers towards CO_2 valorization and conversion to valuable products and chemicals [1, 2]. In response to the need of energy supply, the researchers explore the challenges to utilize CO_2 gas to form methanol which is the main precursor to synthesize valuable chemicals for liquid fuels production such as di-methyl-ether (DME) as well as an alternative oil derivative and several bulk chemicals for example methyl- tert-butyl-ether (MTBE), formaldehyde and acetic acid.

Unfortunately, the activation of CO_2 requires high energy input as it exists in its most stable state. Commercially, methanol is produced from syngas feedstock using Cu/ZnO/Al_2O_3 catalyst [3]. Besides the commercialized alumina support, zirconia

N. I. Zulkifli · N. A. Mohd Zabidi (✉)
Center of Contaminant Control and Utilization (CenCoU), Institute of Contaminant Management for Oil and Gas, Universiti Teknologi PETRONAS, Universiti Teknologi PETRONAS, 32610 Seri Iskandar, Perak, Malaysia
e-mail: noorasmawati_mzabidi@utp.edu.my

N. I. Zulkifli
e-mail: nur_19000323@utp.edu.my

S. A. Abdul Karim et al. (eds.), *Proceedings of the 6th International Conference on Fundamental and Applied Sciences*, Springer Proceedings in Complexity,
https://doi.org/10.1007/978-981-16-4513-6_7

(ZrO_2) is chosen as the support/promoter because zirconia is a strong and thermal-resistant material that can withstand harsh reducing or oxidizing atmosphere. Findings by Witoon et al., proved that CO_2 hydrogenation is strongly influenced by the zirconia phase, Cu-ZrO_2 interaction of amorphous-ZrO_2 and tetragonal-ZrO_2 is stronger than that of monoclinic-ZrO_2 which may enhance the spillover of atomic hydrogen from copper surface to the zirconia surface [4]. A study conducted by Zhang et al. suggests that the presence of ZrO_2 in the commercial CO_2 hydrogenation catalyst has increased the CO_2 conversion up to 25.9–24.7% at 250 °C and 5 MPa using 10% of alumina sol in a slurry bed reactor [5].

Zeolite has well-defined pores in molecular size, good porosity, and high thermal and chemical stability [6]. These characteristics attribute to excellent ability of zeolite to perform in a harsh reaction condition and prevent the deactivation of copper which has low thermal stability. Heating of Cu can cause sintering, agglomeration of copper-based catalysts that limits its usage to a relatively low and small temperature range [7]. Therefore, zeolite might serve as a suitable support to increase copper thermal stability. Previous findings show that added silica suppress the agglomeration of copper and zinc sites, and improved the stability of catalyst [8]. Silica support had been widely studied in CO_2 hydrogenation and it was found to increase the production of methanol compared to commercial industrial catalyst support, alumina. Hence, it is a great idea to test the activity of zeolite, which contain alumina and silica, as a catalyst support for methanol synthesis via CO_2 hydrogenation reaction. A theoretical study using ab initio calculation conducted by Chan et al., found that zeolite can significantly reduce energy barriers of catalytic CO_2 hydrogenation with suitable chemical modifications [9].

CO_2 conversion increased to 17.1% on the Nb-promoted catalyst compared to 14.2% over the unpromoted catalyst from a CO_2 hydrogenation reaction [8]. According to Mohd Zabidi et al., the methanol yield of the unpromoted Cu-based catalyst was reported at 51.4 g/h gcat and it increased to 143 g/h gcat over the Nb-promoted catalyst. In addition, Ud-Din et al. also studied the effect of niobium on carbon nanofibers for Cu/ZrO_2 based catalysts for liquid phase hydrogenation of CO_2 to methanol and reported that Nb_2O_5 was able to enhance Cu-based catalyst activity. When niobium oxide was incorporated into the catalyst, small XRD peak around 31° appear indicates that zirconia crystal was shifted to amorphous phase at 0.4 to 1.2 wt.% Nb_2O_5 loading. This finding agrees with Zhang et al., which claim that amorphous zirconia phase results in high interfacial area improved methanol formation [10, 11]. Furthermore, the TPR peak also shifted to lower temperature suggesting that the addition of Nb_2O_5 facilitated the reduction of dispersed Cu [10].

Besides Nb, the activity of Mn as a catalyst promoter for CO_2 hydrogenation had also been reported. It was found that the synergistic effect of Mn and Zr were able to inhibit the agglomeration of metallic oxide particles and resulted in an increase in BET surface area and the dispersion of smaller sized nanoparticles [11]. Besides that, triple promoter effect of Mn, Nb, Zr showed increasing activity of Cu/ZnO/Al_2O_3 where CO_2 conversion and methanol selectivity increased to 12.47% and 60.28% from 7.67% and 42.17%, respectively[12].

This study aims to find the catalyst support which resulted in the highest CO_2 conversion and alcohol selectivity. Effect of catalyst reduction temperature on the catalytic performance in CO_2 hydrogenation reaction were investigated.

2 Experimental Section

2.1 Catalyst Synthesis

The catalyst was prepared using incipient wetness impregnation method with 15 wt% of metal loading using Cu to ZnO ratio of 7:3. Combination of Mn/Nb/Zr metals were incorporated onto the catalyst as the promoter. For production of 5 g of the catalyst, 0.002 g of $Mn(NO_3)_2 \cdot 4H_20$ [Merck], zirconia (IV) oxynitrate hydrate [Sigma Aldrich], $C_4H_4NNbO_9$ [Merck], 1.997 g of copper nitrate, $Cu(NO_3)_2$ [Merck], and 0.822 g of zinc nitrate, $Zn(NO_3)_2$ [Sigma Aldrich], were dissolved in deionized water and stirred then added to alumina oxide [Sasol] or zeolite [Sigma Aldrich] support in dropwise manner. The solution was continuously stirred for 24 h and the pH was maintained at 7 using 10% ammonia solution, NH_4OH [Merck]. Then, the solution was filtered and the sample was washed using deionized water before being dried at 120 °C for 12 h and calcined in air at 350 °C for 4 h. The same procedure was applied to synthesis of Cu–ZnO on Al_2O_3–ZrO_2 support except that only Mn and Nb promoters were added to the solution. The catalyst on Al_2O_3, zeolite and Al_2O_3–ZrO_2 support was denoted as mnzcza, mnzczz and mnczaz, respectively.

2.2 Catalyst Evaluation

Catalyst evaluation was performed in a micro-activity fixed bed reactor (Microactivity Reference, PID Eng Tech). Prior to the hydrogenation reaction, a 0.2 g of fresh catalyst was reduced in H_2 at 20 mL min^{-1} for 2 h. The CO_2 hydrogenation reaction was performed at 22.5 bar, 250 °C, and the gas feed ratio of CO_2 to H_2 is 1:3 with total flow ratef 36 mL/min for 5 h. The reactor effluents were analyzed on-line using a gas chromatograph (Agilent 6890) equipped with a TCD detector for analysis of H_2 and CO_2, and with FID detector for alcohol and other hydrocarbons analysis [8]. CO_2 conversion and alcohol selectivity were calculated using Eqs. 1 and 2, respectively.

$$CO_2\ conversion(\%) = \frac{\text{Mole of } CO_2 \text{ in} - \text{Mole } of\ CO_2 \text{ out}}{Mole\ of\ CO_2\ in} \times 100 \quad (1)$$

$$Alcohol\ selectivity(\%) = \frac{\text{Mole of alcohol produced}}{Totalmoleofproduct} \times 100 \quad (2)$$

2.3 Catalyst Characterization

The catalyst surface morphology was observed on Hitachi-8020 field emission scanning electron microscope (FESEM). The reduction profile of the catalyst was studied via hydrogen temperature-programmed reduction using a TPR equipment (1100 CE Instrument). Meanwhile, the catalyst elemental composition is studied using X-Ray Fluorescence, XRF. For catalyst components and phase studies, X-Ray diffraction (XRD) D2 Phase from Bruker was employed and the phase identification were determined using PANanalytical High Score Plus Software.

3 Results and Discussion

3.1 Morphological Properties of Samples

FESEM images with 5000 magnification of mnzcza, mnczaz and mnzczz are shown in Fig. 1. Al_2O_3, Al_2O_3–ZrO_2 and zeolite exhibit irregular shape. However, it has lesser agglomeration on Al_2O_3–ZrO_2 support compared to those on Al_2O_3 and zeolite support. This is in agreement with literatures that found that Cu/ZnO catalyst on alumina support tend to agglomerate on its surface [4, 13, 14].

Figure 2 shows the H_2-TPR profiles of mnzcza, mnzczz and mnczaz. Almost similar reduction pattern was observed for both mnzcza and mnczaz samples. Mnzcza reduction starts higher at 200 °C with a peak at 250 °C while mnczaz started lower at 186 °C with highest peak at 300 °C and both reactions finished at 310 °C and 324 °C, respectively. This peak can be assigned to reduction of Cu^{2+} species to Cu^0 [15]. Different reduction profile was obtained for Cu–ZnO/Zeolite/Mn/Nb/Zr, mnzczz, where a broad peak was observed ranging from 230 °C until 513 °C with a maximum peak at 430 °C. A higher reduction temperature of mnzczz suggests that a stronger interaction between CuO and zeolite support compared to those observed for Al_2O_3 and Al_2O_3–ZrO_2 supports. Ud-Din et al. and Tasfy et al. observed the same reduction behavior in the Cu-based catalyst [10, 16] Nevertheless, all the samples' reduction temperature started between 250 and 300 °C thus the catalyst reduction temperature was carried out at a reasonable temperature range (250 °C).

Elemental composition and textural properties were studied using x-ray fluorescence (XRF) and Surface Area Analyzer and Porosimetry Analyzer (SAP), respectively, as shown in Table 1. As observed, mnzcza has the highest BET area followed by mnczaz and mnzczz with 165.39, 139.38 and 12.81 m^2/g, respectively. The increasing trend of pore volume may be explained during calcination, some of the smaller pores merge with each other causing the change in pore diameter. On the other hand, the elemental composition of the synthesized catalysts is in correlation with theoretical value calculated during the catalyst preparation (70:30 of 15 wt% with promoter load at 0.1 wt%), with small deviation on Cu percentage range from 0.40 to 0.60%.

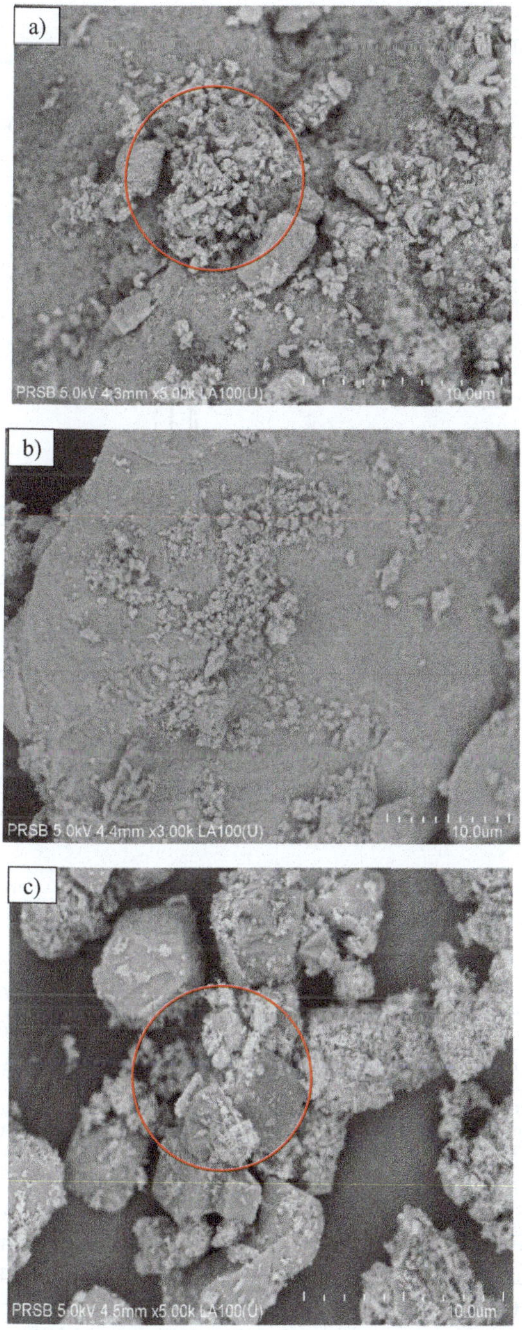

Fig. 1 FESEM images of **a** mnzcza, **b** mnczaz and **c** mnzczz, red circle indicate agglomeration spot

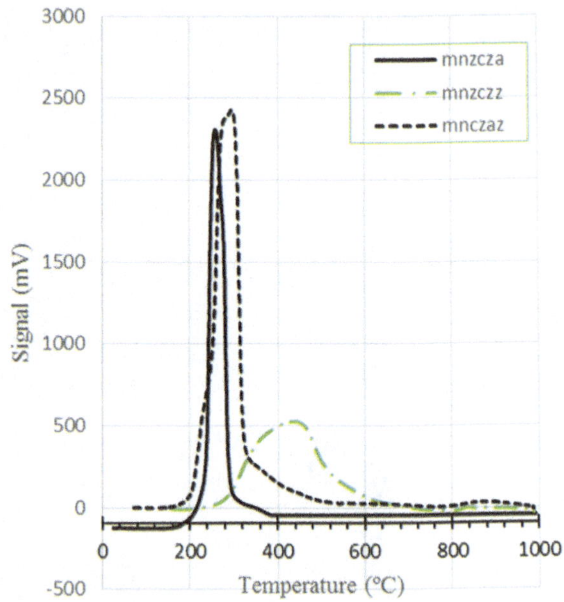

Fig. 2 H_2-TPR profiles of mnzcza, mnczaz and mnzczz

Table 1 Elemental composition and textural properties of the catalysts

Catalyst label	mnzcza	mnczaz	mnzczz
S_{BET} (m^2/g)	165.39	139.38	12.81
Vp (cm^3/g)	0.42	0.31	0.06
D_{BJH} (nm)	6.97	6.18	22.92
Weight (%)			
Cu	11.02	11.11	10.90
Zn	3.77	3.87	3.90
Mn	0.015	0.022	0.015
Nb	0.03	0.082	0.026
Zr	0.25	0	0.27
Total: ~ 15.09%			

S_{BET} is BET surface area; Vp is pore volume; and D_{BJH} is the BJH pore diameter; textural and elemental data were quantified via Surface Area Analyzer and Porosimetry Analyzer (SAP) and X-ray Fluorescence XRF, respectively

XRD analysis was conducted to determine the catalyst phase composition and crystallographic of the catalyst using peak matching with International Centre for Diffraction Data (ICCD) and Inorganic Crystal Structure Database (ICSD). All the loaded Cu/ZnO promoted catalysts showed the same pattern of diffraction peaks. The only noticeable difference amongst the catalyst is the peak intensity of each catalyst. The low peak intensity suggest the presence of amorphous phase or microcrystalline structure such as the one that present in mnczaz while the rest remains invariance. The XRD profile of the catalyst is illustrated below with their respective support for comparison purpose in Fig. 3.

Referring to Fig. 3, four significant peaks can be detected at 2θ values 37.2°, 45.5°, 60.4° and 66.7° which indicate the presence of gamma aluminium oxide, γ-Al2O3 (ICDD 00-001-1307) and two CuO peaks at 75.2° and 43.3° (ICSD 01-078-5773). Meanwhile, ZnO presence is detected to be overlapped with CuO and alumina peak at 37.2° and 66.7° [14]. The result obtained is in correspondence with the control study conducted by Halim (2019) with the same preparation method [12]

As shown in Fig. 4, four significant peaks of alumina zirconia support was detected at 19.8°, 32.2°, 60.4° and 60.9° (ICSD 03-065-6868 and ICSD 01–076-8886) and the peaks are visible in mnczaz diffraction peak which confirming the existence of

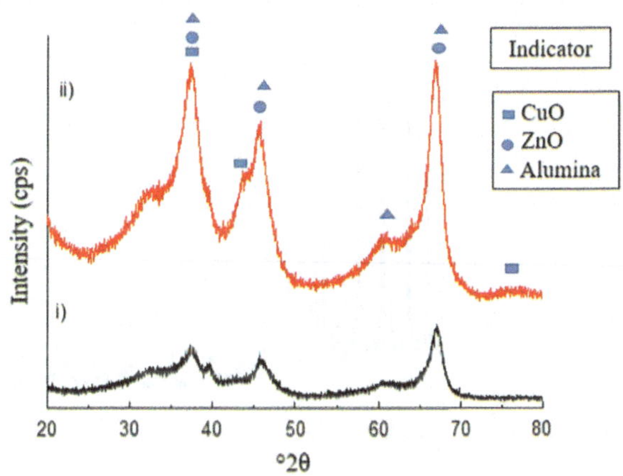

Fig. 3 XRD profiles of mnzcza (top) and alumina support (below)

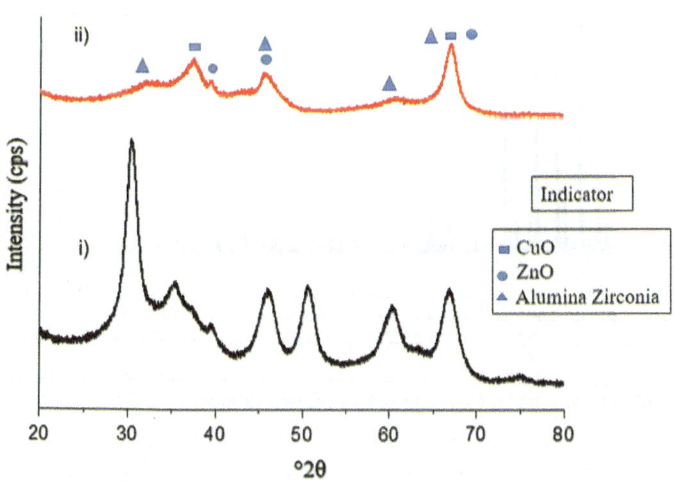

Fig. 4 XRD profiles of mnczaz (top) and alumina zirconia support (below)

alumina zirconia as a catalyst support. CuO peak is detected at, 39.7° and 45.6° meanwhile ZnO at 37.6° and 67.4° (ICSD 01-070-6827, ICDD 00-013-0311).

The crystallinity of alumina zirconia support decrease when CuO and ZnO is incorporated to the support. According to Witoon et al. and Abbas et al., amorphous zirconia is a better catalyst for CO_2 hydrogenation compared to tetragonal and monoclinic zirconia. This may be the reason why the mnczaz activity is the best among other two catalysts with alumina and zeolite support [4].

Referring to Fig. 5, six significant peaks can be detected at 2θ values 22°, 23°, 27°, 30°, 42° and 44° which indicate the presence of zeolite (ICSD 98-020-1188) and five CuO peaks at 34.8°, 38°, 59°,61.5° and 65.9° (ICSD 03-065-2309). Meanwhile, ZnO presence is detected to be overlapped with CuO peak at 38° and 59° [17]. Other than that, ZnO peaks are also visible at 42°, 44°, 47°, 54.7°, and 69.7° (ICDD 00-021-1486). All the promoters added, MnO, ZrO_2 and Nb_2O_5 diffraction peaks were undetectable due to the low amount of species added to the sample.

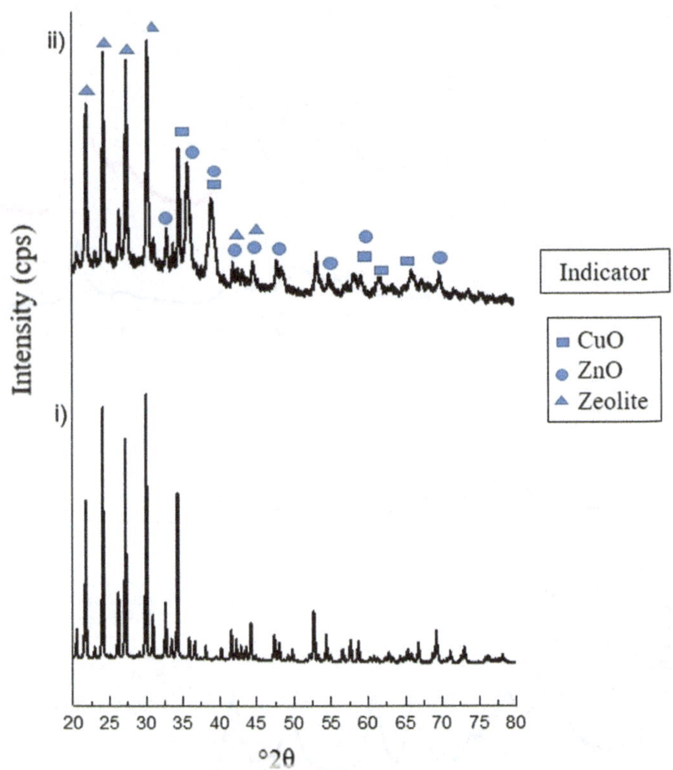

Fig. 5 XRD profiles of mnzczz (top) and zeolite support (bottom)

3.2 Catalyst Performance

The catalytic performance was evaluated at reaction temperature of 250 °C, pressure of 22.5 bar, and CO_2 to H_2 in ratio 1:3. The activity of the catalysts are represented by the values of conversion of the CO_2 as shown in Fig. 6a while the methanol and ethanol selectivity are as shown in Fig. 6b. Table 2 summarized the findings of the catalyst performance with same reduction temperature, 250 °C. In addition to methanol and ethanol, other products detected include methyl formate and methane.

As shown in Fig. 6a, mnzcza resulted in the highest CO_2 conversion (13.76%), compared to those obtained using mnczaz and mnzczz catalysts with CO_2 conversion values of 10.48% and 10.00%, respectively. However, different pattern was observed in methanol selectivity where the Al_2O_3–ZrO_2 supported catalyst had resulted in the highest methanol selectivity (66.99%), compared to 45.91% and 51.95% methanol selectivity obtained over the Al_2O_3 and zeolite-supported catalysts, respectively. However, all the catalysts resulted in low selectivity towards ethanol with selectivity ranging only from 5.23 to 7.06%. This may be due to unsuitable reaction condition for ethanol production. According to literatures, the optimum temperature for Cu-based catalysts to produce ethanol using syngas is at 280 to 310 °C and pressure range of 40–100 bar. Besides that, the H_2/CO is more favourable to produce ethanol between 0.45 and 2.33 [18–22]. In contrast, this study utilized lower pressure at 25 bar and higher feedstock ratio.

To improve the catalysts' performance, the effect of reduction temperature of the catalyst, prior to the catalytic reaction, was investigated. Samples were reduced according to their TPR profiles, namely, the mnczaz sample was reduced at 300 °C and the mnzczz sample at 430 °C. Table 3 summarized their catalytic performance and Fig. 7a, b show the catalytic activity of the catalysts.

Table 3 shows that the Al_2O_3–ZrO_2 supported catalyst (mnczaz) resulted in higher CO_2 conversion (15.81%) compared to 10.22%, obtained from the zeolite-supported catalyst (mnzczz). Similar trend was observed in methanol selectivity, where mnczaz catalyst resulted in the higher methanol selectivity of 68.77%. The improved catalytic activity could be due to greater fraction of Cu species being reduced at 300 °C compared to that at 250 °C. The lower activity exhibited by catalyst supported on zeolite (mnzczz) could be due to sintering of Cu particles caused by the high reduction temperature of 430 °C. However, the mnzczz catalyst reduced at 430 °C still exhibited some activity towards ethanol formation with selectivity of 3.16% whereas ethanol was not detected over the Al_2O_3–ZrO_2 supported catalyst that was reduced at 300 °C. The reduction temperature of Al_2O_3–ZrO_2 supported catalyst had a significant effect on the ethanol formation over the catalyst possibly due to sintering of copper particles at 300 °C since ethanol selectivity of 7% was obtained when the reduction temperature was kept at 250 °C.

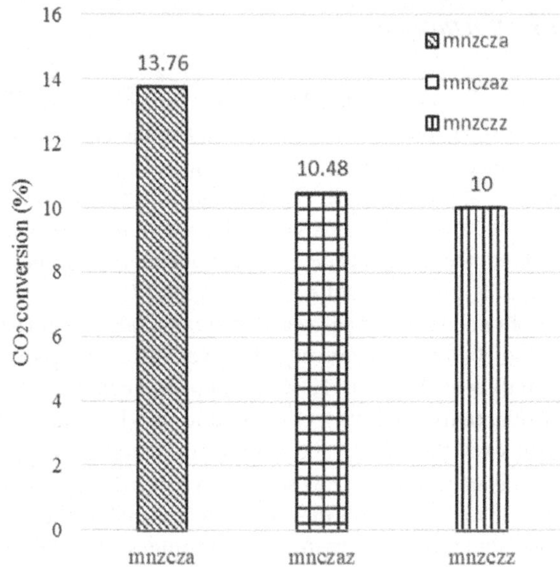

(a)

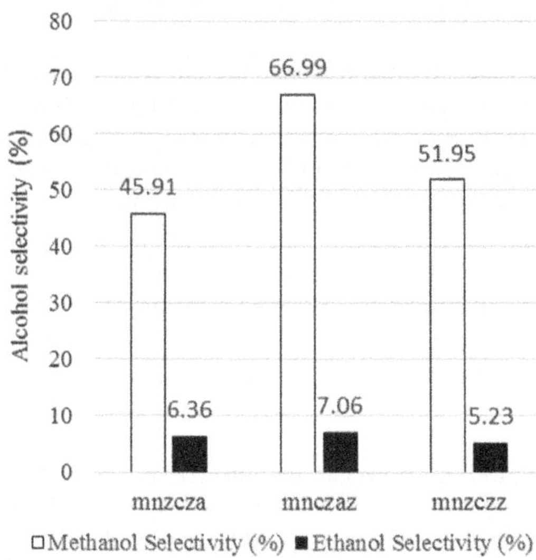

(b)

Fig. 6 Performance of catalysts at 250°C reduction temperature, reaction temperature of 250 °C, pressure of 22.5 bar, and CO_2:H_2 (1:3)

Table 2 CO_2 hydrogenation activity of the catalysts using reduction temperature of 250 °C, reaction temperature of 250 °C, pressure of 22.5 bar and CO_2:H_2 (1:3)

Catalyst label	mnzcza	mnczaz	mnzczz
Catalyst support	Alumina oxide (Al_2O_3)	Alumina zirconia (Al_2O_3-ZrO_2)	Zeolite
Promoter	Mn, Nb, Zr	Mn, Nb	Mn, Nb, Zr
CO_2 conversion (%)	13.76	10.48	10.00
Methanol selectivity (%)	45.91	66.99	51.95
Ethanol selectivity (%)	6.36	7.06	5.23

Table 3 CO_2 hydrogenation activity of mnczaz and mnzczz at their respective reduction temperature

Catalysts	mnczaz	mnzczz
Reduction temperature (°C)	300	430
CO_2 conversion (%)	15.81	10.22
Methanol selectivity (%)	68.77	64.53
Ethanol selectivity (%)	0	3.16

Reaction condition: reaction temperature of 250 °C, pressure of 22.5 bar, and CO_2:H_2 (1:3)

4 Conclusion

Cu-based catalysts were prepared on Al_2O_3, Al_2O_3–ZrO_2 and zeolite supports and their catalytic activities in CO_2 hydrogenation reaction were evaluated. Based on the results, alumina–zirconia supported catalyst resulted in the CO_2 conversion of 15.87%, followed by alumina and zeolite-supported catalysts with 13.76% and 10.22%, respectively. Alumina–zirconia supported catalyst also exhibited the highest methanol selectivity of 68.77%, which is 20.91% higher than that obtained using alumina-supported catalyst. However, all the synthesized catalysts resulted in low selectivity towards ethanol due to unsuitable reaction condition for ethanol production.

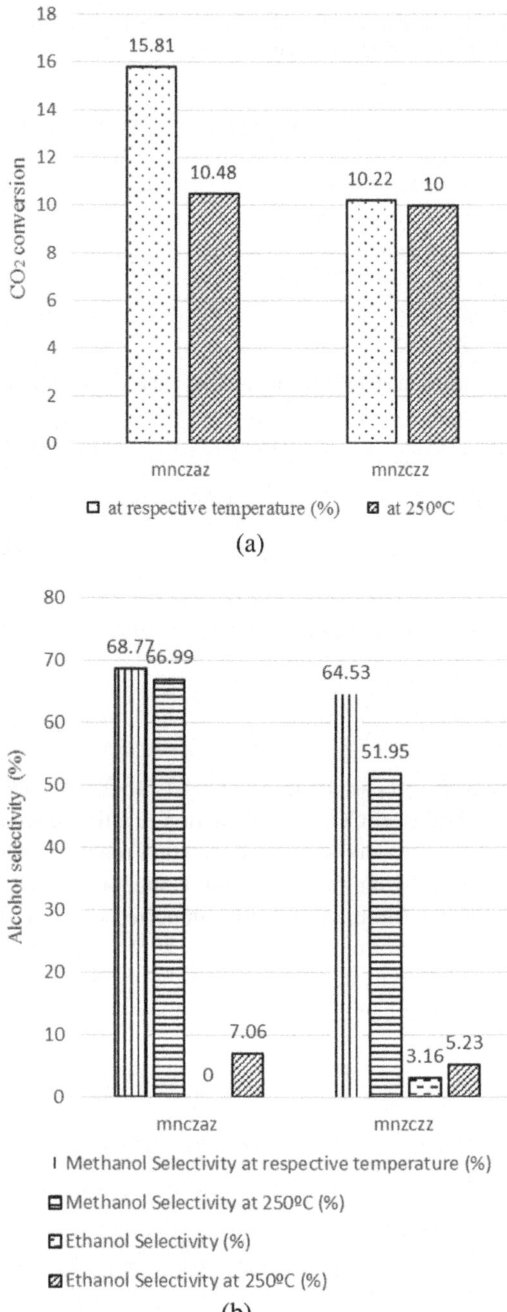

Fig. 7 Performance of mnczaz, and mnzczz catalyst at 250 °C reduction temperature compared with their respective reduction temperature via micro-fixed bed reactor: **a** CO_2 conversion (%), **b** alcohol selectivity

Acknowledgements The authors gratefully acknowledge the financial support from YUTP-Fundamental Research Grant (Cost center 015LCO-111).

References

1. Li, J., Wang L., Cao, Y., Zhang, C., He, P., Li, H.: Recent advances on the reduction of CO_2 to important C^{2+} oxygenated chemicals and fuels. Chin. J. Chem. Eng. **26**(11), 2266–2279 (2018)
2. Vidal, A.B., et al.: CO_2 activation and methanol synthesis on novel Au/TiC and Cu/TiC catalysts. J. Phys. Chem. Lett. **3**(16), 2275–2280 (2019)
3. Prasad, P.S., Bae, J.W., Kang, S., Lee, Y., Jun, K.: Single-step synthesis of DME from syngas on Cu–ZnO–Al_2O_3/Zeolite bifunctional catalysts: the superiority of ferrierite over the other zeolites. Fuel Process. Technol. **89**(12), 1281–1286 (2008)
4. Witoon, T., Jiraporn, C., Porntipar, D.: CO_2 Hydrogenation to methanol over Cu/ZrO_2 catalysts: effects of zirconia phases. Chem. Eng. J. **293**, 327–336 (2016)
5. Zhang, Y., et al.: Catalytic performance of spray-dried Cu/ZnO/Al_2O_3/ZrO_2 catalysts for slurry methanol synthesis from CO_2 hydrogenation. J. CO_2 Utilization **15**, 72–82 (2016)
6. Makertiharta, I.G.B.N., Dharmawijaya, P.T., Wenten, I.G.: Current progress on zeolite membrane reactor for CO_2 hydrogenation. In: International Conference on Engineering, Science and Nanotechnology (ICESNANO 2016), pp 04001. American Institue of Physics, 040001, Bandung (2017)
7. Liu, X.M., Lu, G.Q., Yan, Z., Beltramini, J.: Recent Advances in Catalysts for Methanol Synthesis via Hydrogenation of CO and CO_2, pp. 6518–6530. American Chemical Society (2003)
8. Mohd Zabidi, N.A., Tasfy, S., Shaharun, S.M.: Effects of Nb promoter on the properties of Cu/ZnO/SBA-15 catalyst and performance in methanol production. Key Eng. Mater. **708**, 94–97 (2019)
9. Chan, B., Radom, L.: Design of effective zeolite catalysts for the complete hydrogenation of CO_2. J. Am. Chem. Soc. **128**(16), 5322–5323 (2006)
10. Din, I.U., Shaharun, M.S., Subbarao, D., Naeem, A., Hussain, F.: Influence of niobium on carbon nanofibres based Cu/ZrO_2 catalysts for liquid phase hydrogenation of CO_2 to methanol. Catal. Today **259**, 303–311 (2015)
11. Tasfy, S., Mohd Zabidi, N.A., Shaharun, M.S., Subbarao, D.: The influence of Mn, Zr and Pb promoters on the performance of Cu/ZnO/SBA-15 catalyst for hydrogenation of CO_2 to methanol. Defect and Diffusion Forum **365**, 178–182 (2015)
12. Halim, N.S.A.: Synthesis, characterization and performance of copper-zinc oxide with mixed promoter catalyst for CO_2 hydrogenation to methanol. Universiti Teknologi PETRONAS, Perak (2019)
13. Mohd Zabidi, N.A., Tuan Sulong, T.S., Sardar, S.: Synthesis and characterization of Cu/ZnO catalyst on carbon nanotubes and Al_2O_3 support. Mater. Sci. Forum **916**, 139–43 (2018)
14. Li, Z., Yan, S., Fan, H.: Enhancement of stability and activity of Cu/ZnO/Al_2O_3 catalysts by microwave irradiation for liquid phase methanol synthesis. Fuel **106**, 178–186 (2013)
15. Luo, M.F., Ping, F., Mai, H., Yun, L.X.: In situ XRD, Raman, and TPR studies of CuO/Al_2O_3 catalysts for CO oxidation. J. Mol. Catal. A: Chem. **239**(1–2), 243–248 (2005)
16. Tasfy, S.F.H., Mohd Zabidi, N.A., Shaharun, M.S., Subbarao, D.: Effect of Mn and Pb promoters on the performance of Cu/ZnO-catalyst in CO_2 hydrogenation to methanol. Appl. Mech. Mater. **625**, 289–292 (2014)
17. Huang, C., Chen, S., Fei, X., Liu, D., Zhang, Y.: Catalytic hydrogenation of CO_2 to methanol: study of synergistic effect on adsorption properties of CO_2 and H_2 in CuO/ZnO/ZrO_2 system. Catalysts **5**(4), 1846–1861 (2015)

18. Wang, N., Fang, K., Lin, M., Jiang, D., Li, D., Sun, Y.: Synthesis of higher alcohols from syngas over Fe/K/β-MO_2C catalyst. Catal. Lett. **136**(1–2), 9–13 (2010)
19. Pan, W.X., Cao, R., Griffin, L.G.: Direct alcohol synthesis using copper/cobalt catalysts. J. Catal. **114**(2), 447–456 (1998)
20. Slaa, J.C., Weierink, G.J., Ommen, J.G., Ross, J.R.H.: TPR and infrared measurements with Cu/Zno/Al_2O_3 based catalysts for the synthesis of methanol and higher alcohols from CO + H_2. Catal. Today **12**(4), 481–490 (1992)
21. Calverley, E.M., Kevin, T.S.: The effects of carbon dioxide, methanol, and alkali promoter concentration on the higher alcohol synthesis over a Cu/ZnO/Cr_2O_3 catalyst. J. Catal. **130**(2), 616–26 (1991)
22. Gupta, M., Miranda, L.S., James, J.S.: Heterogeneous catalytic conversion of dry syngas to ethanol and higher alcohols on Cu-based catalysts. ACS Catal. **1**(6), 641–656 (2011)

Raman Spectroscopy of TiO_2 Nanoparticles Synthesized by Hydrolysis of $TiCl_4$: Effect of Sulfate Ions Concentration

Kar Mun Lee, Chong Fai Kait, Jun Wei Lim, and Geok Bee Teh

Abstract Titanium dioxide (TiO_2) nanoparticles were synthesized via precipitation method. The objective of the work is to investigate the influence of ammonium sulfate concentration on the properties of as-synthesized and calcined TiO_2 nanoparticles at room temperature. The concentrations were varied based on molar fraction of $TiCl_4$:ammonium sulfate ranging from 20:1 to 0.5:1. The morphology and types of phases present in the samples were characterized using TEM, XRD and Raman spectroscopy. Titanium dioxide particles with pure anatase phase were successfully synthesized at 350 °C for 2 h with crystallite size ranging from 22 to 67 nm. Typical spherical shaped TiO_2 particles were observed from TEM images. As the concentration of sulfate ions increased, the average crystallite size was found to decrease. Based on the Raman spectra, shifting of peak position (E_g mode) to lower wavenumber with decreased crystallite size was found to be different from other reported works. This may be caused by defects formed due to traces of sulfate ions coordinated to Ti(IV) surface atoms. The major anatase peaks such as 635 cm^{-1} (E_g mode) and 397 cm^{-1} (B_{1g} mode) were found to be predominant in all uncalcined samples and the peaks intensity increased after calcination.

Keywords Titanium dioxide · Sulfate · Raman spectra · Oxygen defects

K. M. Lee (✉) · C. F. Kait · J. W. Lim
Fundamental and Applied Sciences Department, Universiti Teknologi PETRONAS, Seri Iskandar, 32610 Seri Iskandar, Perak, Malaysia

C. F. Kait
e-mail: chongfaikait@utp.edu.my

J. W. Lim
e-mail: junwei.lim@utp.edu.my

C. F. Kait
Centre of Innovative Nanostructures and Nanodevices (COINN), Institute of Autonomous System (IAS), Universiti Teknologi PETRONAS, 32610 Seri Iskandar, Perak, Malaysia

G. B. Teh
Faculty of Engineering & Information Technology, Southern University College, Skudai, 81300 Skudai, Johor DT, Malaysia
e-mail: gbteh@sc.edu.my

S. A. Abdul Karim et al. (eds.), *Proceedings of the 6th International Conference on Fundamental and Applied Sciences*, Springer Proceedings in Complexity,
https://doi.org/10.1007/978-981-16-4513-6_8

1 Introduction

TiO_2 is the most versatile and prominent metal oxide semiconductor photocatalyst which has been widely used in many environmental and energy applications such as hydrogen production as renewable energy [1], dielectric material for ultrathin-film capacitors, chemical sensors, self-cleansing surfaces, pigment of dyes and paints [2], solar cells [3], CO_2 conversion to hydrocarbon fuels, desulfurization [4], photoacoustic signaling [5] and environmental remediation, specifically for wastewater treatment and air purification under ultraviolet and visible light radiation [6]. Its application is highly motivated by the exceptional electronic and optical properties, chemical and thermal stability, low production cost, non-toxicity, rapid reaction rates and strong oxidizing ability towards organic pollutants [7, 8].

In order to develop highly effective TiO_2 photocatalysts, the properties such as particle size, shape and morphology phase are critical in determining the final properties of TiO_2. Several state-of-arts review the impact of preparation method of TiO_2 towards its physical and optical properties related to its applications [9, 10]. Over the past three decades, a number of methods such as hydrothermal, sol–gel, solvothermal, chemical vapor deposition (CVD), direct oxidation, electro-deposition (ED), micro emulsion, microwave, sono-chemical, and precipitation method have been reported for the preparation TiO_2 [11, 12]. Synthesis of TiO_2 by precipitation method is easy and able to obtain pure nanoparticles phase at ambient operating conditions, in addition with benefits of low cost, and minimal chemicals usage. It has high potential to produce the photoactive material on industrial scale. The photocatalytic efficiency depends on the preparation method, post-treatment conditions, and ions concentration.

Synthesis of TiO_2 by precipitation from $TiCl_4$ sol in the presence of ammonium sulfate was first investigated by Zhang and co-workers [13–15]. Ultrafine titania powders with primary particle size of 3.5 nm and pure anatase phase was obtained at high hydrolysis temperatures (70–98 °C). TiO_2 with mixed phases (anatase–rutile) was produced in the absence of sulfate ions. The addition of sulfate ions plays an important role to promote the formation of nano-size crystallites and stabilizing the anatase phase even at high temperature. Li et al. [16] showed that the preparation of TiO_2 using sulfuric acid as additive favors formation of fine crystallite size anatase phase even at high treatment temperature. Similar method was used by Zou et al. [17] to synthesize sulfated titania by hydrolysis of $TiCl_4$ in the presence of H_2SO_4 under reflux condition. Another study in 2010 investigated the influence of different sulfate precursors. Their study shows that the employment of ammonium sulfate results in reduced surface area but increased acidic properties of titania compared to sulfuric acid [18].

In this paper, we reported the synthesis of TiO_2 NPs by hydrolysis of $TiCl_4$ in an aqueous medium with the addition of sulfate ions at room temperature. Anatase TiO_2 nanocrystals were obtained and the effect of the sulfate ion concentration was investigated. The objective of the paper is to investigate the effect of post treatment calcination temperature on the photocatalyst properties using TEM, XRD and Raman spectroscopy. Raman spectroscopy is a powerful tool for the investigation of the structural properties of nanoparticles. It was reported that Raman spectroscopy is very sensitive to the surface phase of a solid sample when the sample absorbs UV light [19, 20]. In addition, Raman spectroscopy could be used to detect and identify sulfate ions on the titania surface [21].

2 Material and Methods

2.1 Chemicals

Titanium(IV) chloride ($TiCl_4$) (purity: 99.9%), ammonium sulfate ($(NH_4)_2SO_4$, 99.9% purity) and ammonium hydroxide (NH_4OH, 30% v/v) were purchased from Merck. All the chemicals were used without further purification.

2.2 TiO_2 Synthesis

A modified chemical precipitation method described by Zhang et al. [13] was referred for the synthesis of TiO_2 nanoparticles by hydrolyzing $TiCl_4$ in 0.5 M ammonium sulfate $(NH_4)_2SO_4$ solution. Approximately 18 mL (0.1 mol) of $TiCl_4$ was added drop-wise into aqueous solution containing different concentration of $(NH_4)_2SO_4$. The concentration of sulfate was varied for different molar ratio of Ti^{4+}:SO_4^{2-} as shown in Fig. 1. Precipitation was carried out by drop-wise addition of 2.5 M NH_4OH until pH 10. White precipitates of TiO_2 were washed several times with distilled water until chloride-free. The precipitates were dried overnight in an oven at 80 °C. Then the dried powder was manually ground into fine powder using pestle and mortar. The as-synthesized TiO_2 samples were calcined at 350 °C for 2 h at a heating rate of 10 °C/min separately, producing yellow powder. The synthesis scheme is shown as in Table 1. For the ease of referencing, all the samples were named TiO_2_X, while calcined samples were labeled as TiO_2_XC where 'X' refers to mol of $TiCl_4$ added based on 1 mol of SO_4^{2-} while 'C' refers to calcined samples. For example, TiO_2_20C denoted calcined TiO_2 with sulfate ions added according to Ti^{4+}:SO_4^{2-} molar ratio of 20:1.

2.3 TiO_2 Characterization

The morphology of TiO_2 nanoparticles was characterized using a HT-7800 (Hitachi, Japan) transmission electron microscope (TEM). The phase composition of the samples was determined using X-ray diffaction (PANalytical X'Pert[3] Powder, AA

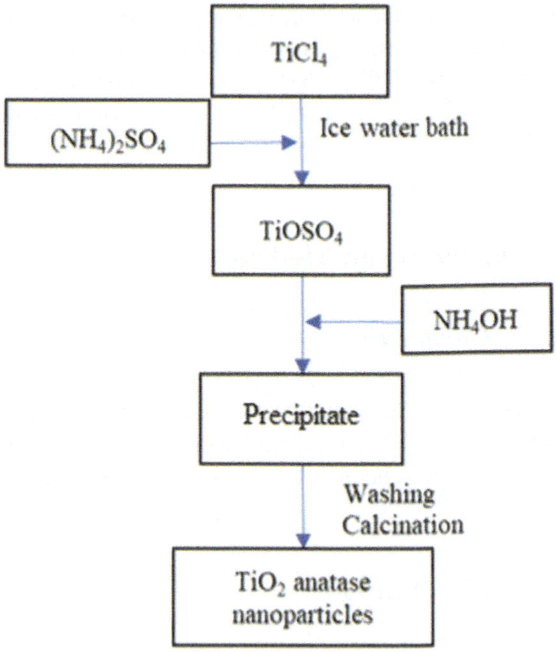

Fig. 1 Synthesis scheme of titanium dioxide nanoparticles

Table 1 Sample label of synthesized titanium dioxide

As-synthesized samples	Calcined samples	Ti^{4+}:SO_4^{2-} molar ratio
TiO_2_20	TiO_2_20C	20:1
TiO_2_13	TiO_2_13C	13:1
TiO_2_2	TiO_2_2C	2:1
TiO_2_0.5	TiO_2_0.5C	0.5:1

Almelo, Almelo, The Netherlands) with Cu K_α radiation (40 kV, 40 mA) at a 2θ angle from 10° to 70° with the step size of 0.01°. The average crystallite sizes of the TiO_2 were estimated from the most intense characteristic diffraction peak at 2θ = 25° which represent the (101) plane of anatase TiO_2 using Scherrer's formula: $D = K\lambda/\beta\cos\theta$ where K is the Scherrer's constant (0.9), λ is the X-ray wavelength (0.15418 nm), β is the full-width at half maximum of the selected diffraction peak (in radian), and θ is the Bragg's angle. Raman spectra were recorded at room temperature using HORIBA LabRAM HR Evolution system equipped with a CCD (Andor) detector. The excitation source was a 514 nm laser. The power of the laser beam was set to below 4 mW to avoid sample damage and the spectral resolution was 3–4 cm^{-1}.

3 Results and Discussion

The morphology of calcined titanium dioxide samples was determined from TEM micrographs as shown in Fig. 2, all the samples showed typical spherical-shaped TiO_2 nanoparticles of size ranging from 7 to 22 nm. The agglomeration of particles were observed which may be caused by vigorous stirring during hydrolysis of $TiCl_4$ aqueous solution [17].

Figure 3 displays the XRD patterns of samples calcined at 350 °C for 2 h. All the samples show the dominant anatase diffraction peak at $2\theta = 25.3°$, which was assigned to the crystallographic plane (101). This is in agreement with previous work of producing anatase TiO_2 nanoparticles [17, 18]. A series of weaker peaks around $2\theta = 37.8°$, 48.0°, 53.9°, 55.1° and 62.7° were assigned to the (004), (200), (105), (211) and (204) planes, respectively [22]. There is no characteristic peak assigned to sulfur compounds, suggesting that the sulfate ions were highly dispersed on the catalysts [23]. From Fig. 3, it can be observed for the peak at $2\theta = 25.3°$ that its intensity became stronger with increasing sulfate ions concentration. The average crystallite sizes of anatase TiO_2 samples are listed in Table 2, showing that the sizes gradually increased from 22 to 67 nm with decreasing amount of sulfate ions added during synthesis.

Raman spectroscopy is used to observe the crystalline phase of synthesized TiO_2 samples. The technique is sensitive to metal-oxide vibrations thus variations of Raman spectra can be easily detected with decrease in particle size [24]. Figure 4 shows the Raman spectra of TiO_2 prepared at different molar ratio of $Ti^{4+}:SO_4^{2-}$ before calcination. Due to the limitation of the Raman system that has been used

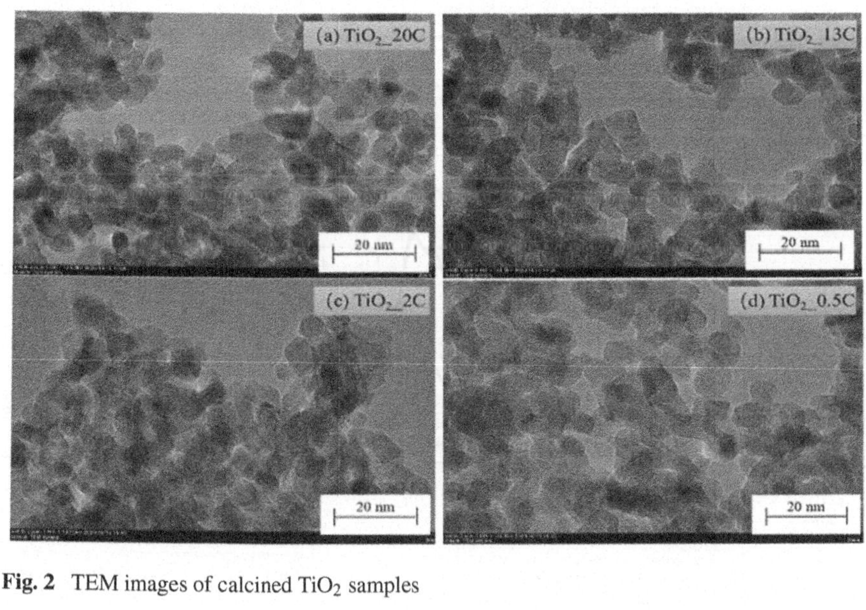

Fig. 2 TEM images of calcined TiO_2 samples

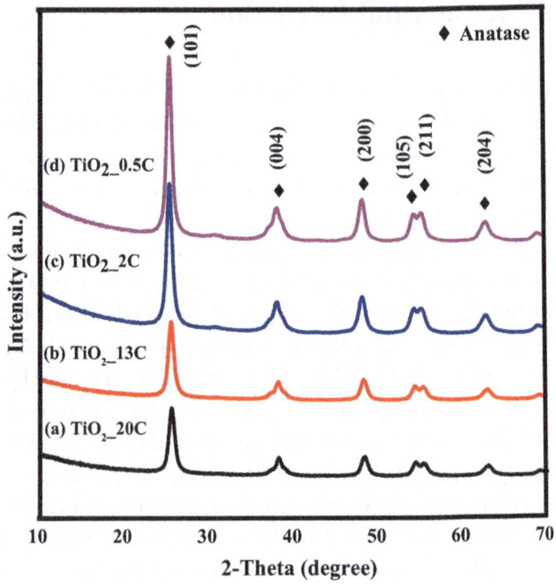

Fig. 3 XRD pattern of calcined titanium dioxide prepared by with different molar ratio of Ti^{4+}:SO_4^{2-}

Table 2 Crystallite sizes of calcined samples

Calcined samples	Average crystallite size (nm)
TiO_2_0.5C	22
TiO_2_2C	28
TiO_2_13C	37
TiO_2_20C	67

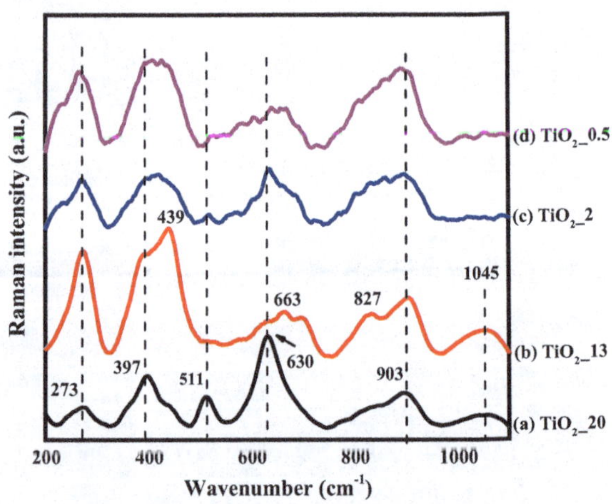

Fig. 4 Raman spectra of the uncalcined titania prepared by using different molar ratio of Ti^{4+}:SO_4^{2-}

in the present study, the range below 200 cm^{-1} is not detectable. Typical intense peak at 144 and 197 cm^{-1} are not able to be investigated. Figure 4 shows Raman spectra of uncalcined samples. The characteristic peaks match fairly consistently with the structure of amorphous TiO_6 octahedron. It was reported that the bands near 273 and 903 cm^{-1} are attributed to Ti–OH bonds [25]. Peaks at about 439 cm^{-1} are assigned to Ti–O bending vibration involving the three-fold oxygen. Peaks at 663 and 827 cm^{-1} are ascribed to Ti–O bending and stretching vibration involving two-fold oxygen [26]. Zooming into sample TiO_2_20, three major peaks of anatase phase at 397 (B_{1g}), 511 (A_{1g}) and 630 (E_g) cm^{-1} were present but with poor crystallinity. This shows that the arrangement of nanocrystals were in zig-zag arrangement promoting growth of anatase phase [27]. Pure anatase phase titanium dioxide was obtained after calcination at 350 °C for 2 h as shown in Fig. 5. In addition to determining TiO_2 phases, Raman spectroscopy can detect the presence of sulfate ions. The weak band at 1045 cm^{-1} is related to the stretching and splitting mode of SO_4^{2-} ions [28]. It was observed in low sulfate molar fractions samples which may indicate that SO_4^{2-} ions enter the nearest coordination environment of the Ti(IV) atom. The weak band was observed to become weaker in samples with higher sulfate concentration. This might be due to sulfate ions incorporated into TiO_2 bulk region [23].

The Raman spectra in Fig. 6 for all the calcined samples labelled as TiO_2_20C, TiO_2_13C, TiO_2_2C and TiO_2_0.5C display three major absorption peaks: 395, 515, and 637 cm^{-1} which are attributed to the Raman-active modes of anatase phase with the symmetries of B_{1g}, A_{1g}, and E_g, respectively [23]. All the samples were successfully transformed into pure anatase phase after heat treatment was introduced at 350 °C for 2 h. This is in agreement with reported data by our research group previously in which anatase TiO_2 can be easily transformed at calcination temperature as low as 350 °C for 2 h in the presence of small amount of sulfate ions [29]. Compared to work reported by Li and co-worker, anatase TiO_2 was obtained at calcination temperature of 400 and 600 °C for 2 h [16]. With increasing sulfate ions concentration added to the precursor, the Raman peak of E_g mode observed drastic decrease in

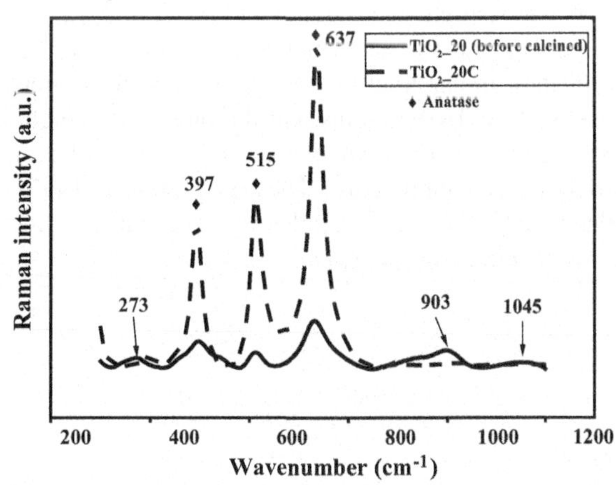

Fig. 5 Raman spectra of TiO_2_20 before and after calcination at 350 °C for 2 h

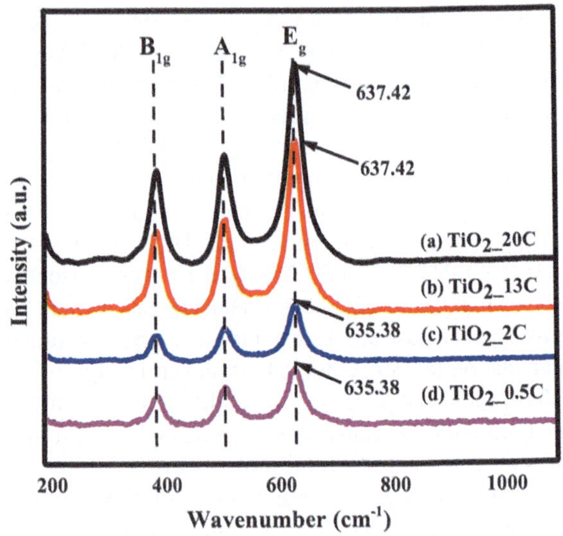

Fig. 6 Raman spectra of calcined samples

intensity. The reduction in scattering intensity might be due to the breakdown of long-range translational crystal symmetry caused by the incorporated defects [30]. It has been reported that the Raman peak position is highly correlated with the crystallite size. The peak position of E_g tends to shift towards lower wavenumbers with increasing crystallite size [24]. However, the Raman spectra in Fig. 6 showed the opposite trend: TiO_2_20C and TiO_2_13C with larger crystallite sizes displayed higher peak position at 637 cm^{-1} compared to the smaller crystallite sizes of TiO_2_2C and TiO_2_0.5C. This can be explained based on the occurrence of oxygen defects on the surface of titania [31]. It was shown that TiO_2_20C and TiO_2_13C displayed the presence of a weak band assigned to sulfate group at 1045 cm^{-1} before calcination. Upon heat treatment, detachment of sulfate ions from TiO_2 surface led to formation of oxygen vacancies or Ti(III) defect sites [32]. Thus, oxygen defects result in shifting of E_g band to a higher wavenumber [31]. TiO_2_2C and TiO_2_0.5C contain less oxygen defects on the surface which resulted in a lower E_g peak position. For samples with high molar fraction of sulfate ions, these ions were not found coordinated near surface Ti(IV) atoms. On the other hand, high molar fraction of sulfate ions promote formation of smaller crystallite sizes of anatase TiO_2 during hydrolysis of $TiCl_4$ at room temperature. These precipitates adsorbed sulfate ions resulting in both electrostatic and steric repulsion to prevent agglomeration, thus resulting in finer sized titania particles [15].

4 Conclusion

The increasing concentration of sulfate ions present in the precursor solution during synthesis led to the formation of smaller crystallite size titania. Anatase TiO_2 with high crystallinity was obtained after calcination at 350 °C for 2 h. The XRD analysis and Raman spectroscopy showed the formation of anatase crystalline phase. Raman spectra has shown more details about the metal-oxide vibration and structural description based on the Raman peak position and intensity. It was found that Raman peak intensity decrease as increase of sulfate ion concentration added, however TiO_2_20C with the largest crystallite size (67 nm) showed a red-shift which may be due to more sulfate ions being incorporated to Ti(IV) surface before calcination. Optimization of the sulfate ions incorporated in the TiO_2 lattice could be a new approach in modifying the properties of anatase nanoparticles.

Acknowledgements The authors would like to express their gratitude to Yayasan Universiti Teknologi PETRONAS (YUTP 015LC0-276) and The MURATA Science Foundation (015ME0-188) for financial support. Research facilities from the Centre of Innovative Nanostructures and Nanodevices (COINN) and Photocatalyst Laboratory at Catalyst Research (CARE) Laboratory in Universiti Teknologi PETRONAS are greatly acknowledged.

References

1. Bazri, B., Lin, Y., Lu, T., Chen, C., Kowsari, E., Hu, S., Liu, R.: A heteroelectrode structure for solar water splitting: Integrated cobalt ditelluride across a TiO_2-passivated silicon microwire array. Catal. Sci. Technol. **7**(7), 1488–1496 (2017)
2. Janus, M., Zatorska, J., Czyzewski, A., Bubacz, K., Kusiak-Nejman, E., Morawski, A.W.: Self-cleaning properties of cement plates loaded with N, C-modified Tio2 photocatalysts. Appl. Surf. Sci. **330**, 200–206 (2015)
3. Lee, Y.S., Gopi, C.V.V.M., Reddy, A.E., Nagaraju, C., Kim, H.J.: High performance of TiO_2/CdS quantum dot sensitized solar cells with a Cu–ZnS passivation layer. New J. Chem. **41**(5), 1914–1917 (2017)
4. Zaid, H.F.M., Chong, F.K., Mutalib, M.I.A.: Extractive deep desulfurization of diesel using choline chloride-glycerol eutectic-based ionic liquid as a green solvent. Fuel **192**, 10–17 (2017)
5. Fan, Z., Huang, X., Tan, C., Zhang, H.: Thin metal nanostructures: Synthesis, properties and applications. Chem. Sci. **6**(1), 95–111 (2015)
6. Agbe, H., Nyankson, E., Raza, N., Dodoo-Arhin, D., Chauhan, A., Osei, G., Kumar, V., Kim, K.H.: Recent advances in photoinduced catalysis for water splitting and environmental applications. J. Ind. Eng. Chem. **72**, 31–49 (2019)
7. Dong, H., Zeng, G., Tang, L., Fan, C., Zhang, C., He, X., He, Y.: An overview on limitations of TiO_2-based particles for photocatalytic degradation of organic pollutants and the corresponding countermeasures. Water Res. **79**, 128–146 (2015)
8. Kumar, S.G., Rao, K.S.R.K.: Comparison of modification strategies towards enhanced charge carrier separation and photocatalytic degradation activity of metal oxide semiconductors (TiO_2, WO3 and ZnO). Appl. Surf. Sci. **391**, 124–148 (2017)
9. Gupta, P., Paul, S.: Solid acids: Green alternatives for acid catalysis. Catal. Today **236**, 153–170 (2014)

10. Cargnello, M., Gordon, T.R., Murray, C.B., Cargnello, M., Gordon, T.R., Murray, C.B.: Solution-phase synthesis of titanium dioxide nanoparticles and nanocrystals. Chem. Rev. **114**(114), 9319–9345 (2014)
11. Chen, X., Mao, S.S.: Titanium dioxide nanomaterials: Synthesis, properties, modifications and applications. Chem. Rev. **107**(7), 2891–2959 (2007)
12. Li, L., Chen, Y., Jiao, S., Fang, Z., Liu, X., Xu, Y., Pang, G., Feng, S.: Synthesis, microstructure, and properties of black anatase and B phase TiO_2 nanoparticles. Mater. Des. **100**, 235–240 (2016)
13. Zhang, Q., Gao, L., Guo, J.K.: Preparation of nanosized TiO_2 powders from hydrolysis of $TiCl_4$. J. Inorg. Mater. **15**(1), 21–25 (2000)
14. Zhang, Q., Gao, L., Guo, J.K.: Effects of calcination on the photocatalytic properties of nanosized TiO_2 powders prepared by $TiCl_4$ hydrolysis. Appl. Catal. B Environ. **26**(3), 207–215 (2000)
15. Gao, L., Zhang, Q.: The promoting effect of sulfate ions on the nucleation of TiO_2 (Anatase) nanocrystals. Mater. Trans. **42**(8), 1676–1680 (2001)
16. Li, W., Zeng, T.: Preparation of TiO_2 anatase nanocrystals by $TiCl_4$ hydrolysis with additive H_2SO_4. PLoS ONE **6**(6), 21082 (2011)
17. Zou, J., Gao, J., Wang, Y.: Synthesis of highly active H_2O_2-sensitized sulfated titania nanoparticles with a response to visible light. J. Photochem. Photobiol. A Chem. **202**(2–3), 128–135 (2009)
18. Ropero-Vega, J.L., Aldana-Pérez, A., Gómez, R., Niño-Gómez, M.E.: Sulfated titania [TiO_2/SO_4^{2-}]: A very active solid acid catalyst for the esterification of free fatty acids with ethanol. Appl. Catal. A Gen. **379**(1–2), 24–29 (2010)
19. Zhang, J., Li, M., Feng, Z., Chen, J., Li, C.: UV Raman spectroscopic study on TiO_2. I. Phase transformation at the surface and in the bulk. J. Phys. Chem. B **110**(2), 927–935 (2006)
20. Ilie, A.G., Scarisoareanu, M., Morjan, I., Dutu, E., Badiceanu, M., Mihailescu, I.: Principal component analysis of Raman spectra for TiO_2 nanoparticle characterization. Appl. Surf. Sci. **417**, 93–103 (2017)
21. Hardcastle, F.: Raman spectroscopy of titania (TiO_2) nanotubular water-splitting catalysts. J. Ark. Acad. Sci. **65**, 43–48 (2011)
22. Dabbawala, A.A., Alhassan, S.M., Mishra, D.K., Jegal, J., Hwang, J.S.: Solvent free cyclodehydration of sorbitol to isosorbide over mesoporous sulfated titania with enhanced catalytic performance. Mol. Catal. **454**, 77–86 (2018)
23. Zhao, H., Jiang, P., Dong, Y., Huang, M., Liu, B.: A high-surface-area mesoporous sulfated nano-titania solid super acid catalyst with exposed (101) facets for esterification: facile preparation and catalytic performance. New J. Chem. **38**(9), 4541 (2014)
24. Ekoi, E.J., Gowen, A., Dorrepaal, R., Dowling, D.P.: Characterisation of titanium oxide layers using Raman spectroscopy and optical profilometry: Influence of oxide properties. Results Phys. **12**, 1574–1585 (2019)
25. Qian, L., Du, Z.L., Yang, S.Y., Jin, Z.S.: Raman study of titania nanotube by soft chemical process. J. Mol. Struct. **749**(1–3), 103–107 (2005)
26. Kim, H.M., Miyaji, F., Kokubo, T., Nakamura, T.: Effect of heat treatment on apatite-forming ability of Ti metal induced by alkali treatment. J. Mater. Sci. Mater. Med. **8**(6), 341–347 (1997)
27. Zhang, J., Wang, X., Wang, J., Wang, J., Ji, Z.: Effect of sulfate ions on the crystallization and photocatalytic activity of TiO_2/diatomite composite photocatalyst. Chem. Phys. Lett. **643**, 53–60 (2016)
28. Allodi, V., Bruti, S., Giarola, M., Sgambetterra, M., Navarra, M., Panelo, S., Mariotto, G.: Structural and spectroscopic characterization of a nanosized sulfated TiO_2 filler and of nanocomposite nafion membranes. Polymers (Basel) **8**(3), 68–81 (2016)
29. Chandran, K., Chong, F.K., Mohd Zaid, H.F.: Morphology of titanium dioxide synthesized via precipitation technique: Effect of calcination and reflux on the surface morphology. J. Phys. Conf. Ser. **1123**(1), 012059 (2018)
30. Pal, M., Pal, U., Jiménez, J.M.G.Y., Pérez-Rodríguez, F.: Effects of crystallization and dopant concentration on the emission behavior of TiO_2: Eu nanophosphors. Nanoscale Res. Lett. **7**, 1 (2012)

31. Choudhury, B., Choudhury, A.: Local structure modification and phase transformation of TiO_2 nanoparticles initiated by oxygen defects, grain size, and annealing temperature. Int. Nano Lett. **3**, 55 (2013)
32. Lebedeva, I.I., Sizeneva, I.P., Kisel'kov, D.M., Val'tsifer, V. A.: Study of the effect of ammonium sulfate additives on the structure and photocatalytic activity of titanium dioxide. Russ. J. Appl. Chem. **87**(5), 547–554 (2014)

Growth of *Lactobacillus casei* and *Propionibacterium jensenii* in Different Glucose Concentration and Incubation Temperature

S. M. Mohamed Esivan, R. Rashid, A. Jati, and N. A. Zaharudin

Abstract The effects of two different parameters on the growth rate of *L. casei* and *P. jensenii* were discussed in this study. The impacts of glucose concentration 5, 10, 15, 18, 20 g/L) and incubation temperature (30, 35, 37, 40, 42 °C) on the growth of both microorganisms in co-culture were investigated. The fermentation process was carried out for seven days in a 500 mL screw-capped bottle with 300 mL sterile MRS broth. The viable cell counts of bacteria, pH drops, and glucose concentration were analysed during the incubation period. The findings show that glucose concentration and incubation temperature had a significant impact on the production of the cell. The highest specific growth rate of *L. casei* and *P. jensenii* are 0.030/h and 0.041/h, respectively. Based on this study, the most suitable initial glucose concentration and incubation temperature for the fermentation of *L. casei* and *P. jensenii* in co-culture fermentation are 18 g/L and 37 °C, respectively.

Keywords Co-culture fermentation · *Lactobacillus casei* · *Propionibacterium jensenii* · The effect of glucose · Incubation temperature

1 Introduction

Probiotics are feed supplement of live microorganism that will confer beneficial effects to the host [1, 2]. The U.S. Food and Drug Administration (FDA) refer to these live microorganisms as direct-fed microbial in animal feeds. This term is suitable for animal feed as the definition given a narrower application as microbial-based feed additives [3]. From the descriptions given by Fuller [1] and FAO/WHO [2], the live microorganisms must be viable during storage and consumptions. In addition, the concentration of the direct-fed microbial must be adequate in the range of 7 and 8 log CFU/mL upon consumption [4]. Direct-fed microbial must be non-pathogenic, technologically suitable for industrial processes, resistant to acid and bile, capable of producing antimicrobial substances, modulating immune responses, and influential

S. M. Mohamed Esivan · R. Rashid (✉) · A. Jati · N. A. Zaharudin
Faculty of Engineering, Universiti Teknologi Malaysia, 81310 Johor Bahru, Johor DT, Malaysia
e-mail: r-roslina@utm.my

S. A. Abdul Karim et al. (eds.), *Proceedings of the 6th International Conference on Fundamental and Applied Sciences*, Springer Proceedings in Complexity,
https://doi.org/10.1007/978-981-16-4513-6_9

to the gut's activities. However, Applegate et al. [5] stated that the stringent considerations of direct-fed microbial attributes could not be achieved using only a single strain.

Co-culture is defined by Bader et al. [6] as incubation of anaerobic or aerobic of different specified microbial strains under aseptic conditions. Co-culture fermentation offered efficient carbon utilisation in the presence of lignocellulose (substrates) [7], improved the growth and survival populations, and subsequently increased the biomass yield [8]. Hence, a co-culture system proposed a satisfactory compromise in developing direct-fed microbial due to its capability to achieve the desired aspect of microbiological, technological performance promoting, and registration capabilities of the strains tested.

Lactobacillus and *Propionibacterium* are commonly utilised in co-culture fermentation. Both species have found their usage in various co-culture processes with different microorganisms to produce various substances or products. *Lactobacillus* sp., one of the lactic acid bacteria (LAB), is often co-incubated with the microorganisms such as *Saccharomyces* sp., *Bifidobacterium* sp., *Aspergillus* sp., *Bacillus* sp., and *Propionibacterium* sp. *Propionibacterium* sp., also known as propionic acid bacteria (PAB), is co-incubated with other species such as *Streptococcus* sp., *Lactococcus* sp., *Bifidobacterium* sp., and *Veillonella* sp. Both species can be found used in probiotics production, food and beverages production, microbial chemical and organic acid production, antimicrobial substances and others such as natural folate and vitamin B12.

Hugenschmidt et al. [9] reported that the co-cultivation of LAB and PAB species have a stimulatory effect on the growth rate of the cultures and metabolic dependency between two species, as LAB species produce lactate and PAB species use this. This effect is because lactate is much more preferred than sugars by PAB. Wu et al. [10] reported that the co-culture fermentation of *Bifidobacterium* sp. (LAB bacterium) with *Propionibacterium freudenreichii* increased the growth and survival of *Bifidobacterium* sp. To six times compared to its mono-culture growth and survival in the same medium. The same result was reported by Ranadheera et al. [11], significant growth and survival of *L. acidophilus* co-culture with *P. jensenii* during storage at 4 °C was observed.

In this study, a co-culture fermentation of *L. casei* and *P. jensenii* was carried out at various glucose concentration and incubation temperature to evaluate each microbes' cell growth in co-culture, pH reduction and glucose consumption.

2 Material and Methods

2.1 *Microorganism*

The microorganisms used in this study were bought from American Type Cell Culture (ATCC Virginia, USA). The species used in this study were *Lactobacillus casei* ATCC 393 and *Propionibacterium jensenii* ATCC 4871.

2.2 *Fermentation Conditions*

Co-culture fermentations were carried out in MRS (de Man Rogosa and Sharpe) broth. The fermentation medium was prepared as follows 10.0 g peptone from casein, 8.0 g meat extract, 4.0 g yeast extract, 2.0 g di-potassium hydrogen phosphate, 1.0 g TWEEN® 80, 2.0 g di-ammonium hydrogen citrate, 5.0 g sodium acetate, 0.2 g magnesium sulphate and 0.04 g manganese sulphate) were added into one-litre distilled water with the concentration for glucose was varied accordingly.

The effect of glucose concentration on co-culture fermentation was investigated by fermentation at various glucose concentrations ranging from 5 to 20 g/L. The medium was inoculated with *L. casei* and *P. jensenii* at 1:4 (% v/v) inoculum ratio, respectively. Fermentation was carried out statically at 30 °C for 120 h. The samples were collected at 0, 3, 6, 9, 12, 24, 48, 72, 96 and 120 h. These samples were then analysed for viable cells of *L. casei* and *P. jensenii*, pH and glucose concentration.

Selected glucose concentration then was used to study the effect of incubation temperature on co-culture fermentation. The same inoculation ratio was used. The fermentation was carried out at five different incubation temperatures, 30, 35, 37, 40 and 42 °C The samples were collected at 0, 3, 6, 9, 12, 24, 48, 72, 96 and 120 h. These samples were then analysed for viable cells of *L. casei* and *P. jensenii*, pH and glucose concentration.

2.3 *Enumeration of Viable Cell*

The enumerations of the viable cell for both microbes were carried out using the spread plate method. 100 μL from serial decimal dilutions were plated on each suitable agar, MRS agar for *L. casei,* while sodium lactate agar (SLA) for *P. jensenii.* The plates with visible colonies in the range of 30–300 colonies were calculated and recorded as colony-forming units per millilitre (CFU/mL). MRS agar was prepared by mixing 68.2 g of MRS agar (Merck) with one litre of distilled water. SLA agar was prepared by adding 10 g of enzymatic digest of casein, 10 g of yeast extract, 2.5 g potassium phosphate monobasic, 16.67 g of sodium lactate, 0.005 g of magnesium sulphate and 15 g agar into one litre of distilled water. Both agars were heated until boiled before autoclaved. Afterwards, the agars were poured into the petri dishes inside a safety cabinet to avoid contamination on the agar.

2.4 Measurement of pH and Glucose Concentration

The pH of the culture was directly measured using a pH metre at room temperature. The determination of total sugar consumption by *L. casei* and *P. jensenii* in co-culture fermentation was carried out using the 3,5-dinotrosalicyclic acid (DNS) method. First, one millilitre culture and one millilitre distilled water were pipetted into two different test tube. Then, 2 mL DNS was added into the test tubes, and the mixture was thoroughly mixed before heated at 90 °C for 5 min. Afterwards, the mixture was cooled down, and 7 mL of distilled water was added. Then the absorbance was measured at 540 nm.

3 Result and Discussion

3.1 The Effect of Glucose Concentration on the Growth of L. casei and P. jensenii and pH Changes During Co-culture Fermentation

As glucose was the primary carbon source used in this study, different initial glucose concentrations were varied in the range of 5 and 20 g/L. The initial population of *L. casei* and *P. jensenii* was 7.77 and 6.62 log CFU/mL. The growth of *L. casei* and *P. jensenii* during co-culture fermentation in five different glucose concentrations are shown in Fig. 1. At the same time, the pH changes of the co-culture in different glucose concentrations are shown in Fig. 2.

The increase in numbers of cell density was measured from time to time, and a growth curve was plotted. Several distinct growth phases can be observed in Fig. 1: exponential, stationary, and death phase. Figure 1 shows the unnoticeable lag phase due to the cell's ability to adjust rapidly with the environment resulting in a shorter lag phase. At the time the second sampling was taken, the cell already reached its exponential phase. Each microbe's specific growth rate in different glucose concentrations was calculated from the growth curve and tabulated in Table 1.

The optimum glucose concentration for the growth of co-culture fermentation was found at a glucose concentration of 18 g/L, with a specific growth rate for *L. casei* and *P. jensenii* was 0.021 h^{-1}. Although the specific growth rate of *P. jensenii* was higher at a glucose concentration of 20 g/L, the specific growth rate of *L. casei* was lower than at 18 g/L. Table 1 shows that the growth rate of *L. casei* reached its optimum at 18 g/L but decreasing at a glucose concentration higher than that. However, in *P. jensenii* case, the growth rate increased as the glucose concentration increased.

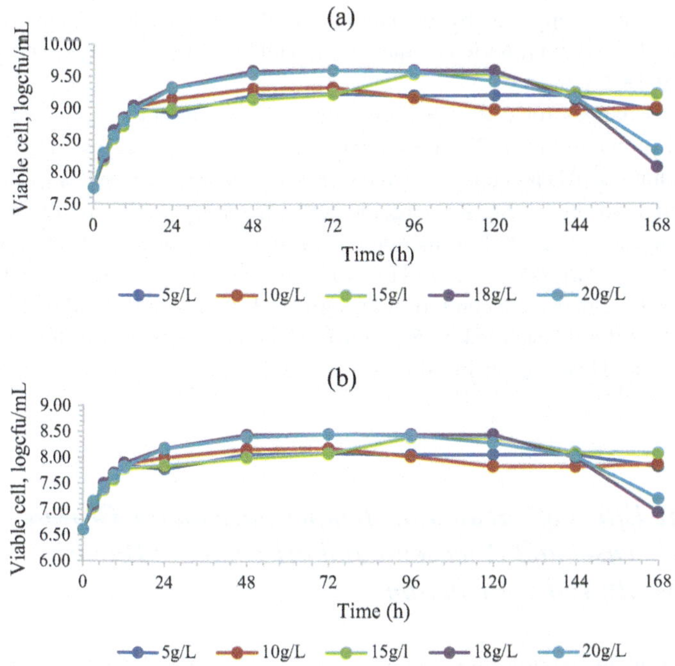

Fig. 1 The growth of *L. casei* **a** and *P. jensenii* **b** during co-culture fermentation in different glucose concentration at the temperature of 30 °C, inoculation ratio of 1:4% v/v (*L. casei*: *P. jensenii*)

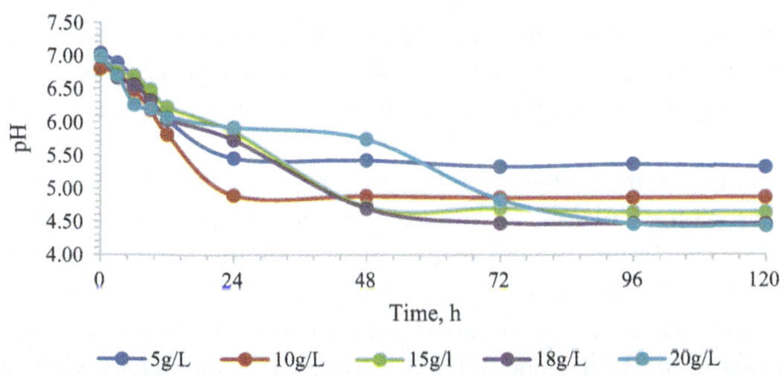

Fig. 2 The variances of pH in co-culture of *L. casei* and *P. jensenii* in various glucose concentration

Table 1 The specific growth rate of *L. casei* and *P. jensenii* at a different initial glucose concentration

Glucose concentration (g/L)		5	10	15	18	20
Specific growth rate (μ h^{-1})	*L. casei*	0.011	0.011	0.016	0.021	0.010
	P. jensenii	0.013	0.013	0.018	0.021	0.023

This result can be supported by the outcomes of the study made by Khosravi-Darani and Zoghi [12], which reported an increase in yield and productivity of fermentation products at 18% w/w.

The pH changes for different initial glucose concentrations were recorded for 120 h and shown in Fig. 2. The initial pH of the medium was adjusted to 7.0 before the fermentation. pH changes during the fermentation of *Lactobacillus* sp. and *Propionibacterium* sp. indicating that organic acid was being produced. The lowest pH recorded at the end of 120 h fermentation was 4.44 for glucose concentration of 18 and 20 g/L. A sharp decrease in pH can be observed for glucose concentration of 10 g/L. At the glucose concentration of 5 g/L, pH culture at the end of 120 h incubation time did not reach below 5 compared to other batches with higher glucose concentration. This is probably due to the limited glucose supply available in the medium.

3.2 The Effect of Incubation Temperature on the Growth of L. casei and P. jensenii and pH Changes During Co-culture Fermentation

Five different incubation temperatures, 30, 35, 37, 40 and 42 °C, were studied. The initial cell concentrations were kept the same as during the study of glucose concentration, while the initial glucose concentration was at 18 g/L. The growth of *L. casei* and *P. jensenii* during co-culture in five different incubations temperature is shown in Fig. 3. At the same time, the pH changes of the co-culture in different incubation temperature are depicted in Fig. 4. The specific growth rate of each species incubated at different temperature was calculated from the growth curve and tabulated in Table 2.

As seen in Fig. 3, the highest viability of cell obtained was 9.62 log CFU/mL for *L. casei* and 8.48 log CFU/mL for *P. jensenii* at an incubation temperature of 37 °C. The specific growth rates for both species were the highest at 37 °C with 0.030 h^{-1} and 0.041 h^{-1} for *L. casei* and *P. jensenii*, respectively, as depicted in Table 2. The specific growth rate for *L. casei* and *P. jensenii* were the lowest at a temperature higher than 40 °C, 0.017 h^{-1} and 0.019 h^{-1} for *L. casei* and *P. jensenii*, respectively. The decrease in growth rate might be probably due to the decreasing activity of cellular enzymes. These enzymes are most active at the optimum temperature, thus directly affect the growth rate of cells [13].

This study shows that 37 °C was suitable for obtaining high viability cells for both *L. casei* and *P. jensenii*. This finding is per the previous study by Patrick and McDowell [14], for the optimum growth of Propionibacterium sp. was recorded between 30 and 37 °C, and the optimum growth of *L. casei* in pumpkin waste was reported at 37 °C [15]. The findings might be varied because the value of the controlled parameters was different, and the microbes used were not the same. This

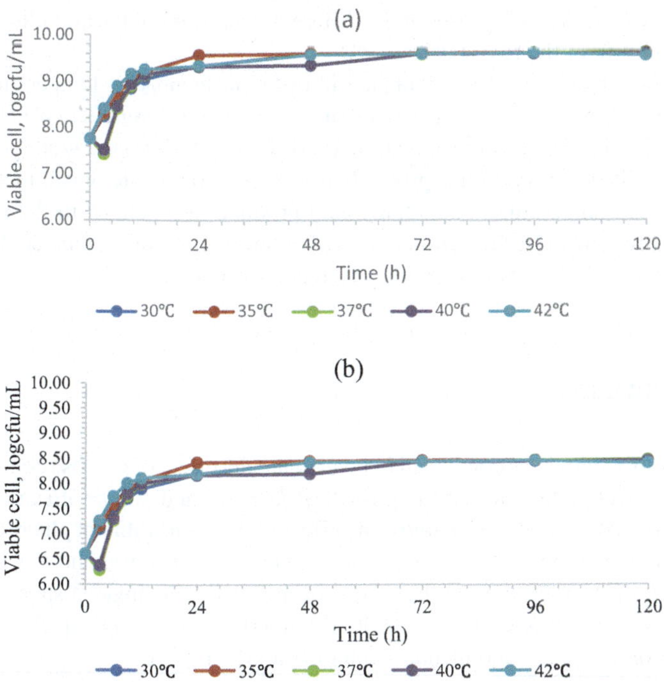

Fig. 3 The growth of *L. casei* (**a**) and *P. jensenii* (**b**) during co-culture fermentation in different incubation temperature at a glucose concentration of 18 g/L, inoculation ratio of 1:4% v/v (*L. casei*: *P. jensenii*)

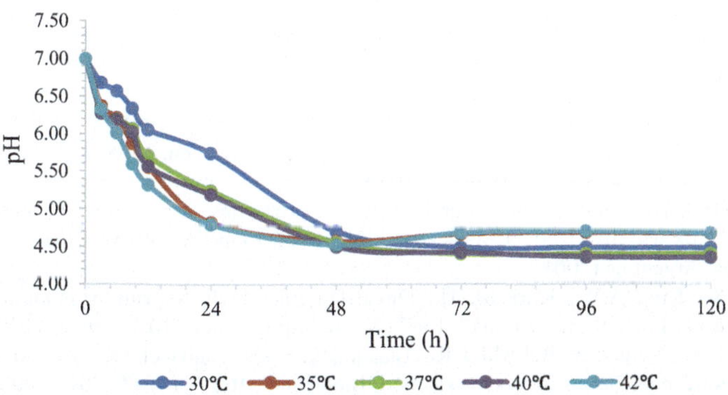

Fig. 4 The variances of pH in co-culture of *L. casei* and *P. jensenii* in various incubation temperature

Table 2 The specific net growth rate of *L. casei* and *P. jensenii* in different incubation temperature

Incubation temperature (°C)		30	35	37	40	42
Specific growth rate (μ h^{-1})	*L. casei*	0.021	0.026	0.030	0.029	0.017
	P. jensenii	0.021	0.020	0.041	0.034	0.019

different result might be because of the different conditions of the co-culture and the inoculation ratio of the two species.

The pH changes for 120 h incubated in different incubation temperatures were recorded and shown in Fig. 4. There were no significant differences in pH changes in all the batches except during incubation at 30 °C which showed a gradual decrease in pH value. This observation is probably due to the growth rate of cell that affected by incubation temperature. As organic acid production is related to the growth of the cell, slow growth of the cell indicates slow production of organic acid. The pH culture at the end of the fermentation was recorded at 4.35.

4 Conclusion

In general, this study was conducted to investigate the effect of glucose concentration and incubation temperature on the growth of *L. casei* and *P. jensenii* in co-culture fermentation. At the end of this study, the effect of different value of glucose concentration and incubation temperature on the growth rate of *L. casei* and *P. jensenii* has been determined. According to the results of this study, the highest specific growth rate of *L. casei* and *P. jensenii* was 0.030 h^{-1} and 0.041 h^{-1}, respectively, in glucose concentration of 18 g/L and incubation temperature of 37 °C.

Acknowledgements The authors would like to express their appreciation to Universiti Teknologi Malaysia for the support under the UTM Fundamental Research grant with Project No. 21H31.

References

1. Fuller, R.: Probiotics in man and animals. J. Appl. Bacteriol. **66**, 1365–1378 (1989). https://doi.org/10.1111/j.1365-2672.1989.tb05105.x
2. FAO/WHO: Evaluation of health and nutritional properties of probiotics in food including powder milk with live lactic acid bacteria. Report of a Joint FAO/WHO Expert Consultation. Córdoba, Argentina (2001)
3. Seo, J.K., Kim, A.W., Kim, M, H.: Direct-fed microbials for ruminant animals. Asian-Australasian J. Anim. Sci. 23:1657–1667 (2010). https://doi.org/10.5713/ajas.2010.r.08
4. Makinen, K., Nerger, B., Bel-Rhlid, R., Anaanta, E.: Science and technology for the mastership of probiotic applications in food products. J. Biotechnol. **162**, 356–365 (2012). https://doi.org/10.1016/j.jbiotec.2012.07.006
5. Applegate, T.J., Klose, V., Steiner, T., Ganner, A., Schatzmayr, G.: Probiotics and phytogenics for poultry: myth or reality? In: Informal Nutrition Symposium at the Poultry Science Association's 98th annual meeting in Raleigh, North Carolina, 20–23 July (2009)
6. Bader, J., Mast-Gerlach, E., Popović, K., Bajpai, R., Stahl, U.: Relevance of microbial co-culture fermentations in biotechnology. J. Appl. Microbiol. **109**, 371–387 (2010)
7. Tesfaw, A., Assefa, F.: Co-culture: A great promising method in single cell protein production. Biotechnol. Mol. Biol. Rev. **9**(2), 12–20 (2014)

8. Aghababaie, M., Khanahmadi, M., Beheshti, M.: Developing a kinetic model for co-culture of yogurt starter bacteria growth in pH controlled batch fermentation. J. Food Eng. **166**, 72–79 (2015)
9. Hugenschmidt, S., Schwenninger, S.M., Lacroix, C.: Concurrent high production of natural folate and vitamin B12 using a co-culture process with Lactobacillus plantarum SM39 and Propionibacterium freudenreichii DF13. Process Biochem. **46**, 1063–1070 (2011). https://doi.org/10.1016/j.procbio.2011.01.021
10. Wu, Q.Q., You, H.J., Ahn, H.J., Kwon, B., Ji, G.E.: Changes in growth and survival of *Bifidobacterium* by co-culture with Propionibacterium in soy milk, cow's milk, and modified MRS medium. Int. J. Food Microbiol. **157**, 65–72 (2012)
11. Ranadheera, C.S., Evans, C.A., Adams, M., Baines, S.K.: Co-culturing of probiotics influences the microbial and physico-chemical properties but not sensory quality of fermented dairy drink made from goats' milk. Small Rumin. Res. **136**, 104–108 (2016)
12. Khosravi-Darani, K., Zoghi, A.: Comparison of pretreatment strategies of sugarcane baggase: Experimental design for citric acid production. Biores. Technol. **999150**, 6986–6993 (2008)
13. Panesar, P.S., Kennedy, J.F., Knill, C,J., Kosseva, M.: Production of L(+) lactic acid using *Lactobacillus casei* from Whey. Brazilian Arch. Biol. Technol. 53, 219–226 (2010). https://doi.org/10.1590/S1516-89132010000100027
14. Patrick, S., McDowell, A.: Genus I. Propionibacterium Orla-Jensen 1909. In: Goodfellow, M., Kämpfer, P., Busse, H.-J., Trujillo, M.E., Suzuki, K., Ludwig, W., William B, W. (eds.) Bergey's Manual of Systematic Bacteriology vol. 5, 2nd edn., pp. 1138–1188. Springer New York Dordrecht Heidelberg London, Athens, USA (2012)
15. Genevois, C., Flores, S., de Escalada Pla, M.: Byproduct from pumpkin (*Cucurbita moschata* Duchesne ex poiret) as a substrate and vegetable matrix to contain *Lactobacillus casei*. J. Funct. Foods. **23**, 210–219 (2016). https://doi.org/10.1016/j.jff.2016.02.030

A Comparison of Two Methods for the Development of Low-Cost Carbonaceous Adsorbent from Rubber-Seed Shell (RSS)

Syeda Saba Fatima, Azry Borhan, and Muhammad Faheem

Abstract Biomass-derived porous carbons have recently gained much attention for carbon dioxide (CO_2) adsorption. In this work, RSS were used as a precursor to prepare highly porous activated carbon (AC). RSSAC was synthesized through the conventional two-stage chemical activation as well as through a new three-stage chemical activation process. RSS were activated with K_2CO_3 under different ratios and carbonized at 500 °C, 600 °C, and 700 °C for different activation times. AC was carefully characterized using thermogravimetric analysis (TGA), elemental composition analysis, scanning electron microscopy (SEM), Fourier transform infrared spectroscopy (FTIR), and CO_2 adsorption capacity using HPVA II to obtain AC with higher adsorption capacity towards CO_2 capture. The SEM images of RSSAC revealed a highly porous structure. From the elemental composition analysis, carbon (77%) and oxygen (18%) were the major elements found in RSSAC. The yield of the activation process decreased with a higher ratio of K_2CO_3. A maximum yield of 81.84% was obtained at 500 °C and 1:1 ratio. The highest CO_2 adsorption capacity of 60.06 mg/g was obtained for AC3(700) prepared by three-stage activation. From these findings, it can be concluded that the three-stage activation is an effective method for producing AC with higher adsorption capacity towards CO_2 and that RSSAC can serve as a suitable adsorbent for CO_2 capture.

Keywords Activated carbon · CO_2 capture · Adsorption

S. S. Fatima · A. Borhan (✉)
Department of Chemical Engineering, HICoE, Centre for Biofuel and Biochemical Research, Institute of Self-Sustainable Building, Universiti Teknologi PETRONAS, 32610 Seri Iskandar, Perak, Malaysia
e-mail: azrybo@utp.edu.my

S. S. Fatima
e-mail: syeda_18003267@utp.edu.my

M. Faheem
Department of Chemical Engineering, University of Engineering and Technology, Lahore 54890, Pakistan
e-mail: faheem@uet.edu.pk

S. A. Abdul Karim et al. (eds.), *Proceedings of the 6th International Conference on Fundamental and Applied Sciences*, Springer Proceedings in Complexity,
https://doi.org/10.1007/978-981-16-4513-6_10

1 Introduction

In recent years, adsorption has emerged as a promising technique for the removal of CO_2. However, the success of adsorption lies in the development of durable sorbents with high CO_2 adsorption capacity, low cost, and ease of regeneration [1]. AC derived from biomass has gained much attention because of abundantly available precursors, high thermal stability, better CO_2 adsorption capacity, low cost, ease of preparation and regeneration, control of pore structure and, low selectivity to water because of its hydrophobic nature [2, 3]. One of the precursors which has gained much attention recently is the RSS because of its high carbon content, abundant availability, low cost, and effective utilization in the adsorption process [4, 5]. Malaysia is the fifth largest producer of natural rubber. It contributes about 46% of the world's total rubber production besides Thailand and Indonesia [4].

AC can be synthesized from both physical and chemical activation processes. Chemical activation has the advantage of shorter activation times and a lower temperature range for pyrolysis (typically 500–800 °C). It also gives better control of textural properties and higher yield [6] and is generally preferred over physical activation. Chemical activation is usually conducted in two steps: (1) decomposition of the precursor by pyrolysis, and (2) activation by suitable chemical reagent. Recently, a new three-stage activation process is introduced for the preparation of AC. For three-stage activation, the first stage is the hydrothermal decomposition of precursor to form hydrochar, which is then activated with suitable chemical reagent and pyrolyzed to obtain hydrochar derived biochar (Fig. 1).

Although KOH is widely used for chemical activation of AC it is not suitable for large-scale applications because of its corrosive nature which can easily damage the equipment, particularly at high temperatures. Furthermore, it is difficult to remove it from waste streams and can lead to serious environmental issues [7]. To overcome these issues, a different activating agent of environmental benign nature is desired. K_2CO_3 has a mild activating nature compared to KOH and is frequently used as

Fig. 1 Rubber-seed shell washed and dried

a food additive [6]. In addition, K_2CO_3 also shows similar activating behavior as KOH. In this work, RSS is used as a precursor and K_2CO_3 as an activating agent. Although K_2CO_3 has been used to produce AC using various biomass precursors [6, 7], this is the first example of RSS-derived AC produced via K_2CO_3 activation using three-stage activation method for CO_2 capture.

2 Experimental

2.1 *Materials and Pre-treatment*

Approximately 1 kg of RSS was collected from Tersusun Tanah Hitam, 32100 Chemor Perak. The shells were knocked with a hammer to remove the inner white part of the seeds. Then, they were soaked in water and washed thoroughly to remove the dirt and impurities. Aftermath, the washed shells were dried, grounded and sieved to collect the particles with size <500 μm and stored in an airtight container.

Chemicals used were purchased from Sigma Aldrich located at Petaling Jaya, Selangor D.E, Malaysia.

2.2 *Preparation of AC*

The RSS were activated via two-stage activation and the improved three-stage activation process. For two-stage activation, approximately 30 g of raw RSS were soaked in 100 ml aqueous K_2CO_3 solution in various ratios and left overnight to ensure that K_2CO_3 is completely adsorbed into RSS. The impregnated RSS were dried at 110 °C and were carbonized in a muffle furnace under limited oxygen conditions. The temperatures studied for carbonization were 500 °C, 600 °C, and 700 °C for one and two hours. The carbonization temperature range was selected by the TGA analysis.

For three-stage activation, 20 g of RSS were mixed with 100 ml distilled water in a Teflon-lined autoclave, subsequently used hydrothermal carbonization to produce hydrochar. After that samples were activated with K_2CO_3 and carbonized at the same temperature range After carbonization, samples were washed repetitively with deionized water to remove ash and inorganic salts and dried at 110 °C to remove moisture. Dried AC samples were placed in labeled bottles and stored in a desiccator for characterization and further use.

2.3 Characterization of AC

The structure and morphology of raw and treated RSS were studied using SEM (with an SEM Zeiss SUPRA 55-VP). All samples were analyzed under different resolutions to compare their structure. To find the elemental composition of the untreated RSS and prepared AC samples, ultimate analysis was carried out using an elemental analyzer (Perkin-Elmer 2400 Series II CHNS/O).

FTIR spectroscopy was used to investigate the surface functional groups of both fresh and treated RSS. The CO_2 adsorption capacity of prepared AC samples was observed using High Pressure Volumetric Analyzer (HPVA II).

3 Results and Discussion

3.1 Yield Analysis

The nomenclature, operating conditions, impregnation ratio, and yield of AC samples are summarized in Table 1.

Table 1 Nomenclature, impregnation ratio, operating conditions, and yield of RSSAC

Sample name	IR	Carbonization temperature (°C)	Activation time (min)	Weight after impregnation (g)	Weight after washing/drying (g)	Yield (%)
AC2 (500)	1:1	500	120	5	4.0980	81.84
AC2 (500)	1:2	500	120	5	2.6501	53.00
AC2 (500)	1:3	500	120	5	1.6350	32.70
AC2 (600)	1:1	600	120	5	3.6244	72.48
AC2 (600)	1:2	600	120	5	1.5982	31.96
AC2 (600)	1:3	600	120	5	0.7965	15.93
AC2 (700)	1:1	700	120	5	2.5220	50.44
AC2 (700)	1:2	700	120	5	0.5280	10.56
AC2 (700)	1:3	700	120	5	0.4080	8.16

Table 2 Elemental composition of raw and treated RSS

Elements	Raw RSS	After carbonization	
		Two-stage activation	Three-stage activation
	Weight (%)	Weight (%)	Weight (%)
Carbon	48.4038	77.716	77.1438
Nitrogen	0.7437	0.3818	0.6582
Hydrogen	6.8622	2.9452	2.7496
Oxygen[a]	43.9903	18.957	19.4484

[a] Calculated by the difference

The results clearly show that increasing the impregnation ratio of K_2CO_3 decreases the yield of the activation process. Like most of the lignocellulosic materials, RSS has hemicellulose, cellulose, and lignin as the main components. During activation, these components decompose and most of the non-carbon elements like nitrogen, oxygen, and hydrogen are released in the form of volatile matter leaving behind solid residue rich in carbon. As the impregnation ratio increases, the evolution of the gaseous product in the form of volatile matter increases which ultimately reduces the yield of AC formed. As more volatile matter is released, AC samples with a higher impregnation ratio are expected to be more porous with a richer carbon matrix.

3.2 *Elemental Composition Analysis*

To find the composition of the samples, ultimate analysis was carried out using an elemental analyzer. Table 2 shows the corresponding elemental composition of raw and treated RSS samples. Two main elements detected in the raw and treated samples were carbon and oxygen. From studies, the acceptable range of carbon content for a good biomass precursor should be more than 40% [8].

3.3 *Surface Morphology and Structure*

The surface morphology of raw RSS and some selected samples of both activation methods were studied, and results are shown in Fig. 2. The samples were analyzed under different magnification scales for better comparison. The SEM image of raw RSS in Fig. 2a indicated a rough and uneven surface with no noticeable pores. However, the presence of well-developed pores is seen in activated samples in Fig. 2b, c. During the activation process, K_2CO_3 decomposes to K_2O and CO_2, and potassium metal is formed after series of reactions:

$$K_2CO_3 + C \rightarrow K_2O + 2CO \quad (1)$$

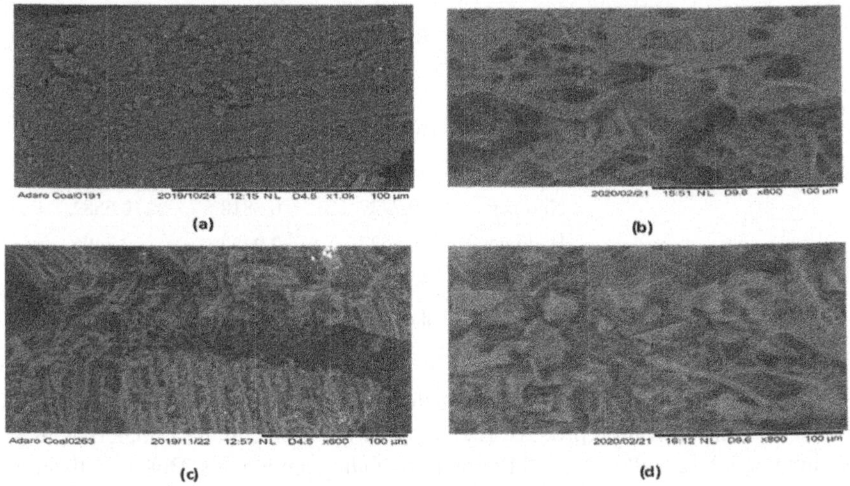

Fig. 2 SEM images of selected samples. **a** Fresh RSS, **b** RSS treated by two step activation at 500 °C, **c** RSS treated by 3 step activation at 500 °C, **d** RSS treated by two-step activation at 700 °C

$$K_2O + C \rightarrow 2K + CO \quad (2)$$

The highly porous structure of the carbon matrix is the result of a chemical reaction between K_2CO_3 and carbon. The carbon matrix expands by metallic potassium intercalation, resulting in the formation of fine pores [7]. Distribution of fine pores appears on carbon structure which acts as active sites for adsorption. A comparison of carbon structure before and after activation is made between Fig. 2a and b, c. It shows that K_2CO_3 works very well as activating agent. The structure of AC shown in Fig. 2c has an array of very fine pores and is expected to have higher S_{BET} and higher adsorption capacity.

3.4 FT-IR Spectroscopy

Figure 3 shows the FT-IR spectra of raw and treated RSS samples. Although the IR spectra are somewhat similar for both samples, the intensity of some peaks is reduced in the treated samples because some temperature sensitive groups are diminished after carbonization. The wide peak at 3336 cm^{-1} for the treated sample and at 3309 cm^{-1} for raw RSS is assigned to –OH stretching vibration due to hydroxyl groups. The intensity of this peak is significantly reduced in the treated sample due to the removal of surface-bound water at a high temperature. The peak observed at 1729 cm^{-1} is due to –C=O stretching of aldehydic and acetyl functionalities in hemicellulose which is very common in biomass. For RSS, a sharp peak at 1574 cm^{-1} corresponds to C=C stretch due to cyclic alkene which acts as the backbone of carbon structure.

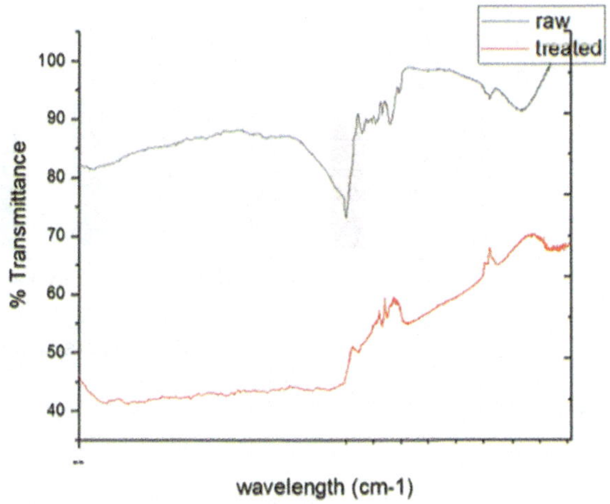

Fig. 3 FT-IR spectra of raw and treated RSS

The significant decrease in peak intensity of carboxyl and hydroxyl groups suggests dehydration and decarboxylation were the main reactions that occurred during heat treatment of the sample.

3.5 *Adsorption Test*

The promising samples prepared from both activation methods were selected for the CO_2 adsorption study. Test of adsorption capacity of AC was carried out by batch adsorption on High Pressure Volumetric Analyzer (HPVA II) at 25 °C. For CO_2 adsorption studies, low temperature is desired due to the exothermic nature of the process.

The results show that AC produced from three-stage activation has a higher CO_2 adsorption capacity. A maximum adsorption capacity of 60.06 mg/g was obtained for sample AC3(700).

Figure 4 presents a comparison of adsorption capacities for AC derived from various biomass precursors and presented in this work. AC3(700) presents good adsorption capacity and has the potential to be used as a green adsorbent for large-scale applications.

A comparison of CO_2 adsorption capacity of prepared AC with other carbonaceous and non-carbonaceous adsorbents is presented in Table 3. Some adsorbent material such as CNT and KIT-6 present very high value of CO_2 adsorption capacity compared to AC, but these adsorbents are quite expensive and have some limitations which hinder their large-scale applications. The obtained results of CO_2 adsorption capacity are comparable with those reported earlier.

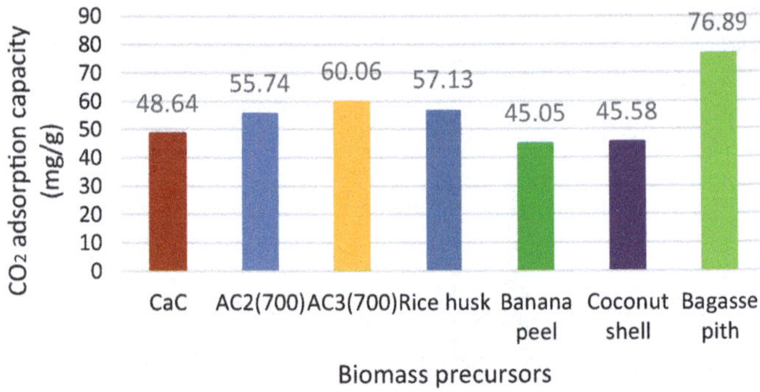

Fig. 4 Comparison of CO_2 adsorption capacity of AC from different biomass precursors

Table 3 Comparison of different adsorbents for CO_2 capture

Adsorbent	P_{CO_2} (atm)	T (K)	CO_2 adsorption capacity (mmol/g)	References
RSS activated carbon	1	298	1.365	This work
Amorphous silica	0.17	303	1.364	[9]
HMS	0.9	293	1.205	[10]
MgO	–	673	0.200	[11]
Basic alumina	1	293	1.000	[12]
Zr pillared clay	1	293	0.682	[13]
4A	1	298	2.500	[14]
CNT	1	343	2.530	[15]
Nanoporous carbon	1	348	1.091	[16]
MSU-J	1	378	0.645	[17]
KIT-6	1	343	1.682	[18]

4 Conclusion

To summarize, RSSAC synthesized using K_2CO_3 exhibits good structural properties with well-developed pores. The structure of the samples carbonized at 500 °C was better. From the SEM images, the well-developed porous structure of RRSAC can be seen. The carbon content of the prepared AC sample is 77%, making it a suitable adsorbent material for CO_2 capture. The yield of the activation process decreases with higher impregnation ratios of K_2CO_3 but porosity increases. The maximum yield of 81.84% was obtained at 500 °C with an impregnation ratio (1:1). A maximum CO_2 adsorption capacity of 60.06 mg/g was obtained for AC prepared by three-stage activation. The low-cost widely available RSS precursor and mild K_2CO_3 used in the synthesis of AC exhibited promising results and indicates that this type of adsorbent can be used in industrial applications.

Acknowledgements This work was supported by Yayasan Universiti Teknologi PETRONAS (YUTP-FRG 015LCO-068). The authors also gratefully acknowledge Universiti Teknologi PETRONAS for the technical and facilities support. Support from the Ministry of Education Malaysia through the HICOE award to Centre for Biofuel and Biochemical Research (CBBR) is also acknowledged.

References

1. Sethia, G., Sayari, A.: Comprehensive study of ultra-microporous nitrogen-doped activated carbon for CO_2 capture. Carbon **93**, 68–80 (2015)
2. Plaza, M.G., Garcia, S., Rubiera, F., Pis, J.J., Pevida, C.: Post-combustion CO_2 capture with a commercial activated carbon: comparison of different regeneration strategies. Chem. Eng. J. **163**, 41–47 (2010)
3. Wickramaratne, N.P., Jaroniec, M.: Importance of small micropores in CO_2 capture by phenolic resin-based activated carbon spheres. J. Mater. Chem. A **1**, 112–116 (2013)
4. Eka, H.D., Tajul Aris, Y., Wan Nadiah, W.A.: Potential use of Malaysian rubber (Hevea brasiliensis) seed as food, feed and biofuel. Int. Food Res. J. **17**, 527–534 (2010)
5. Borhan, A., Yusup, S., Lim, J.W., Show, P.L.: Characterization and modelling studies of activated carbon produced from rubber-seed shell using KOH for CO_2 adsorption. Processes **7**, 855 (2019)
6. Tran, H.N., You, S.J., Chao, H.P.: Fast and efficient adsorption of methylene green 5 on activated carbon prepared from new chemical activation method. J. Environ. Manage. **188**, 322–336 (2017)
7. Yue, L., Xia, Q., Wang, L., Wang, L., DaCosta, H., Yang, J., Hu, X.: CO_2 adsorption at nitrogen-doped carbons prepared by K_2CO_3 activation of urea-modified coconut shells. J. Colloid Interface Sci. **511**, 259–267 (2018)
8. Borhan, A., Taha, M.F., Hamzah, A.A.: Characterization of activated carbon from wood sawdust prepared via chemical activation using potassium hydroxide. Adv. Mater. Res. **832**, 132–137 (2014)
9. Park, J.H., Celedonio, J.M., Seo, H., Park, Y.K., Ko, Y.S.: A study on the effect of the amine structure in CO_2 dry adsorbents on CO_2 capture. Catal. Today **265**, 68–76 (2016)
10. Knowles, G.P., Chaffee, A.L.: Aminopropyl-functionalized silica CO_2 adsorbents via sonochemical methods. J. Chem. **2016**(Art. ID 1070838) (2016)
11. Xu, X., Song, C., Andresen, J.M., Miller, B.G., Scaroni, A.W.: Novel polyethylenimine-modified mesoporous molecular sieve of MCM-41 type as high-capacity adsorbent for CO_2 capture. Energy Fuels **16**, 1463–1469 (2002)
12. Yong, Z., Mata, V., Rodrigues, A.E.: Adsorption of carbon dioxide on basic alumina at high temperatures. J. Chem. Eng. Data **45**, 1093–1095 (2000)
13. Pereira, P.R., Pires, J., Bortaas de Carvalho, M.: Zirconium pillared clays for carbon dioxide/methane separation. 1. Preparation of adsorbent materials and pure gas adsorption. Langmuir **14**, 4584–4588 (1998)
14. Siriwardane, R.V., Shen, M.S., Fisher, E.P., Poston, J.A.: Adsorption of CO_2 on molecular sieves and activated carbon. Energy Fuels **15**, 279–284 (2001)
15. Liu, Q., Shi, J., Wang, Q., Tao, M., He, Y., Shi, Y.: Carbon dioxide capture with polyethylenimine-functionalized industrial-grade multiwalled carbon nanotubes. Ind. Eng. Chem. Res. **53**, 17468–17475 (2014)
16. Tang, Z., Han, Z., Yang, G., Yang, J.: Polyethylenimine loaded nanoporous carbon with ultra-large pore volume for CO_2 capture. Appl. Surf. Sci. **277**, 47–52 (2013)

17. Jiao, J., Cao, J., Xia, Y., Zhao, L.: Improvement of adsorbent materials for CO_2 capture by amine functionalized mesoporous silica with worm-hole framework structure. Chem. Eng. J. **306**, 9–16 (2016)
18. Kishore, R., Ghoshal, A.K.: Polyethylenimine functionalized As-synthesized KIT-6 adsorbent for highly CO_2/N_2 selective separation. Energy Fuels **30**, 9635–9644 (2016)

Adsorptive Removal of Bisphenol A Using Zeolitic Imidazolate Framework (ZIF-8)

Afzan Mahmad, Maizatul Shima Shaharun, Zakariyya Uba Zango, Teh Ubaidah Noh, and Bahruddin Saad

Abstract Bisphenol A is one of the compounds capable of causing endocrine disruption and is also a pollutant in the environment. Even in low concentrations, these pollutants have significant effects on humans, animals and the environment. This work's main goal was to study the performance of adsorption of BPA using a Metal Organic Framework (MOF), ZIF-8. Textural analysis has shown the porous nature of the MOF as possessing surface areas of 1299 m^2/g. Other characterizations were conducted using Fourier Transformed Infrared (FTIR) spectroscopy, X-ray Photoelectron Spectroscopy (XPS), Field Emission Scanning Electron Microscope (FESEM), and powdered X-ray diffraction (XRD). The MOF was evaluated for application as an adsorbent for (BPA) removals in an aqueous solution. Adsorption isotherms using Langmuir and Freundlich models were investigated, while kinetics was studied using pseudo-first-order and pseudo-second-order. In conclusion, this MOF exhibited favorable features of ideal adsorbents for BPA removal in wastewater.

Keywords Adsorption · Bisphenol-A · Metal-organic frameworks · Removal

A. Mahmad · M. S. Shaharun (✉) · Z. U. Zango · B. Saad
Fundamental and Applied Sciences Department, Universiti Teknologi PETRONAS, Seri Iskandar, Perak, Malaysia
e-mail: maizats@utp.edu.my

A. Mahmad
Laboratory Department, Royal College of Medicine Perak, Universiti Kuala Lumpur, Perak, Malaysia

Z. U. Zango
Department of Chemistry, Al-Qalam University Katsina, Katsina, Nigeria

T. U. Noh
Institute of Bioproduct Development (IBD), Universiti Teknologi Malaysia (UTM), Johor Bahru, Malaysia

S. A. Abdul Karim et al. (eds.), *Proceedings of the 6th International Conference on Fundamental and Applied Sciences*, Springer Proceedings in Complexity,
https://doi.org/10.1007/978-981-16-4513-6_11

1 Introduction

Metal organic frameworks (MOFs) are a class of crystalline organic–inorganic hybrid compounds formed by coordination of metal clusters or ions with organic linkers, in which bivalent or trivalent aromatic carboxylic acids or N-containing aromatics are commonly used to form frameworks with zinc, copper, chromium, aluminium, zirconium, and other elements [1]. Since MOFs have large surface areas and high pore volumes in uniformly sized pores as well as high metal content, they have emerged as interesting materials for various applications in energy storage [2], CO_2 adsorption [3], hydrocarbon adsorption/separation [4], catalysis [5], sensor [6], magnetism [7], drug delivery [8], luminescence [9], and others.

Zeolitic imidazolate frameworks (ZIFs) are a family of microporous materials in which Zn or Co atoms are linked through N atoms of ditopic imidazolates to form a range of neutral framework structures as shown in Fig. 1 [10]. The frameworks of ZIF compounds can be represented by T(Im)2 (Im = imidazolate and its derivative, T = tetrahedrally coordinated metal ion) similar to the (Al)SiO_2 frameworks of (alumino) silicate zeolites; in particular, the T–Im–T angle of 145° is close to the Si–O–Si angle typically found in zeolites [10]. ZIFs materials can have structures analogous to the standard zeolite topology, such as rho, sod, gme, lta, and ana. Compared to all the ZIFs, ZIF-8 has been a widely investigated structure due to its high thermal and chemical stability [10].

In 2002, the family of MOFs was extended to imidazolate-based compounds that are nowadays known as zeolitic imidazole frameworks (ZIFs) [11]. Nine imidazolate type linkers and mixtures thereof were reacted with zinc or cobalt nitrate in mixtures of DMF and DEF in various concentrations and molar ratios of metal to linker in the

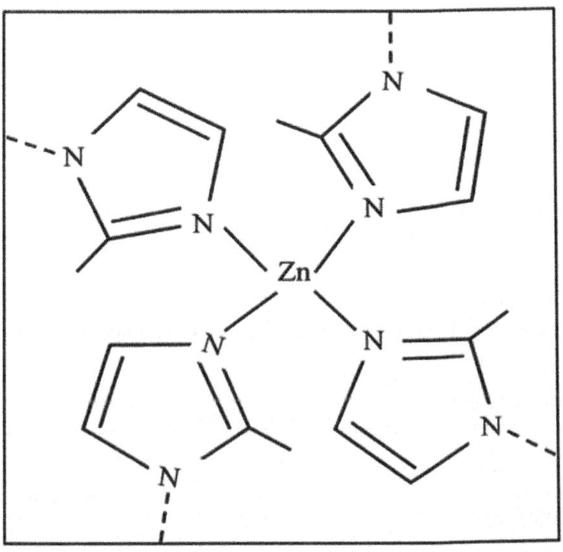

Fig. 1 Chemical structure of ZIF-8

temperatures ranging from 65–150 °C [10]. Among the products obtained, ZIF-8 is the most widely-investigated structure; Zn atoms coordinated tetrahedrally with 2-methylimidazolate (HMeIm), leading to the formula, $Zn(MeIM)_2$ [10]. ZIF-8 exhibits a sod topology formed by four- and six-membered ring ZnN_4 clusters with internal cavities, 1.16 nm in diameter, connected by 0.34 nm windows. ZIF-8 has been tested for gas adsorption and storage of hydrogen [12, 13], and as a heterogeneous catalyst [14–16].

2 Materials and Method

All materials used in this work were of analytical grade and used as received without further purification. BPA analytical standards (>99% purity) and Basolite Z1200@ZIF-8 (99% purity) were purchased from Sigma Aldrich, USA.

2.1 *Characterizations*

The X-ray diffraction analysis was conducted using a Bruker D8 Advance X-ray diffractometer, operated at 40 kV with a current of 40 mA, using Ni-filtered Cu Kα radiation with a graphite monochromator ($\lambda = 1.5406$ Å) in a 2θ angle of 5–50°. Fourier transformed infrared spectroscopic measurements were recorded using an FTIR spectrophotometer (Perkin Elmer Spectrum 65) within the wavenumber range 400–4500 cm^{-1}. The surface morphology was examined using field emission scanning electron microscopy (FESEM, Zeiss Supra 55 VP), operated at 200 kV, with the samples sputter-coated with gold metal.

2.2 *Preparation of the BPA Solution*

The BPA stock solution was freshly prepared in distilled water by dissolving 100 mg in a 100 mL volumetric flask to make a solution of 1000 mg/L. The working solution is prepared daily by diluting a certain amount of the stock solution.

2.3 *BPA Removal Experiment*

Effect of Contact Time

The batch experiment for BPA adsorption onto the ZIF-8 was conducted using 30 mg/L of the 100 mL of BPA solution in the Erlenmeyer flask. The amount of adsorbent was maintained at 0.2 g, and the mixing solution was agitated using an orbital shaker (Daihan), at 200 rpm [17], and room temperature (25 °C). At a predetermined time, about 2 mL of the supernatant was collected, filtered with a nylon syringe membrane (0.45 μm), and the absorbance read at 276 nm using a UV spectrophotometer (Shimadzu, UV-1800) fitted with a quartz cell [18].

The percentage of removal was determined using the following formula:

$$\%R = \frac{C_o - C_e}{C_o} \times 100 \quad (1)$$

Similarly, the quantity of the analytes adsorbed onto the MOF at a certain time (q_t) was determined using the formula:

$$q_t = \frac{(C_o - C_t)V}{w} \quad (2)$$

And the equilibrium quantity adsorbed (q_e) is calculated as;

$$q_e = \frac{(C_o - C_e)V}{w} \quad (3)$$

where C_o and C_t are the BPA's initial and equilibrium concentrations (mg/L), w is the mass of BPA (g), and V is the solution volume (L).

Effect of Agitation Speed

The effect of different agitation speeds (50, 100, 150, 200, and 250 rpm) was investigated. The analyte concentration (30 mg/L) and MOF dosage (0.2 g) were kept constant. The percentage of removals was calculated against each temperature from the absorbance results.

2.4 *Adsorption Isotherms*

According to the Langmuir and Freundlich models, the effect of equilibrium interactions between the BPA and MOF was studied.

Langmuir Isotherms

The Langmuir model assumes that the adsorption process does not proceed beyond monolayer coverage. All the sites available on the adsorbent surface are equivalent, and the surface is uniform. Essentially, once the adsorbate is attached on the site, and no further adsorption can take place at that site, the Langmuir isotherm equation is given below:

$$\frac{C_e}{q_e} = \frac{1}{Q_{maxK_L}} + \frac{C_e}{Q_{max}} \quad (4)$$

where C_e is the equilibrium concentration of the adsorbate (mg/L), q_e is the amount of adsorbate adsorbed per unit mass of adsorbent (mg/g), Q_o is the maximum monolayer adsorption capacity of the adsorbent (mg/g), and K_L is the Langmuir adsorption constant related to the free energy adsorption (L/mg). The Langmuir equation's essential characteristics can be expressed in terms of dimensionless separation factor, R_L, defined as:

$$R_L = \frac{1}{1 + C_0 K_L} \tag{5}$$

where K_L is the Langmuir constant, and C_0 is the highest initial dye concentration (mg/L). The value of R_L indicates the type of the isotherm to be either unfavorable ($R_L > 1$), linear ($R_L = 1$), favorable ($0 < R_L < 1$), or irreversible ($R_L = 0$).

Freundlich Isotherms

The Freundlich model assumes that the heat of adsorption is decreased in logarithmic magnitude with an increase in the extent of adsorption. The adsorption sites are distributed exponentially concerning adsorption energy. This isotherm does not indicate the adsorption limit when coverage of adsorbate is sufficient to fill a monolayer. The equation that describes the isotherm is given as below:

$$\ln q_e = \frac{1}{n_F} \ln C_e + \ln K_F \tag{6}$$

where q_e is the amount of adsorbate adsorbed at equilibrium, (mg/g), C_e is the equilibrium concentration of adsorbate (mg/L), K_F is the Freundlich constant, (mg/g (L/mg)1/n), and n_F is the Freundlich heterogeneity factor. The value of n indicates favourable adsorption when $1 < n < 0$. A value of n below one indicates a normal Freundlich isotherm. The value of n closer to zero indicates a more heterogeneous surface, while the value of n above one indicates cooperative adsorption.

2.5 *Kinetics Study*

Adsorption kinetic studies are essential to describe the adsorbate uptake rate, which controls the residencies in the adsorbent-solution interface. The kinetics of adsorption uptake is required to select optimum operating conditions for a full-scale batch process. Various kinetic models have been developed, such as the pseudo-first-order, pseudo-second-order, and intra-particle kinetic models.

Pseudo-First-Order-Kinetic Model

A linear form of the pseudo-first-order model is described as:

$$\ln(q_e - q_t) = \ln q_e - k_1 t \tag{7}$$

The q_e and q_t (mg/g) represent the amount of BPA adsorbed at equilibrium and time t, respectively. The k_1 (1/min) is the pseudo-first-order rate constant. It is obtained from the plot $\ln(q_e - q_t)$ against t.

Pseudo-Second-Order Kinetic Model

The pseudo-second-order equation is described as:

$$\frac{t}{q_t} = \frac{1}{k_2 q_e^2} + \frac{t}{q_e} \tag{8}$$

where q_e and q_t are the adsorption capacities at equilibrium and at time t, respectively (mg/g) and k_2 (g/mg h) is the rate constant of pseudo-second-order sorption. By plotting t/q_t versus, the values q_e and k_2 can be determined from the slope and intercept. This model suggests that the rate-limiting step is chemical adsorption, concerning valency forces through the sharing or exchange of electrons in the adsorption process.

3 Results and Discussions

3.1 *Characterizations of ZIF-8*

The chemical structure of the adsorbent is vital to understanding the adsorption process. The FTIR technique is an important tool to identify the characteristic of functional groups, which are instrumental in the BPA compound's adsorption. The spectra showed the vibrations of imidazole and Zn^{+2} ions due to their bond origin. Thus, it informs on the basic or acidic nature of the sample. It presented remarkable bands at 3433.23, 3115.68, 1677.31, 1592.08, 1171.75, 1143.37, and 426 cm^{-1} for the ZIF-8 sample. These FTIR bands were consistent with those previously reported by Ordonez [19]. The band 3115.68–3433.23 cm^{-1} are attributed to the N–H stretching vibration of the ZIF-8.

Another signal around 1592.08 cm^{-1} arose from the C=C stretch mode, while the band at 1677.31 cm^{-1} corresponded to the C=N stretch vibration. The signals 1143.37–1171.75 cm^{-1} were band derived from aromatic C–N stretching mode. The unique stretching Zn–N stretching vibration band was observed at position 426 cm^{-1}, suggesting that zinc ions combined chemically with nitrogen atoms of the methylimidazole groups to form the imidazolate. The N–H stretching vibration, C=C stretch mode, C=N stretch vibration, C–N stretching mode, C=C bending mode, and Zn–N stretching vibration signals that appeared in ZIF-8 still could be found in ZIF-8, suggesting the existence of a functional group on ZIF-8.

In this study, the ZIF-8 obtained was characterized for surface area, morphology, chemistry, pore size and volume. The optimized ZIF-8 was then used for BPA removal (Table 1).

Table 1 Surface area characterization for ZIF-8

Properties	ZIF-8
BET surface area ($m^2\ g^{-1}$)	1299
Micropores surface area ($m^2\ g^{-1}$)	761
Micropores volume ($cm^3\ g^{-1}$)	0.60
Total pore volume ($cm^3\ g^{-1}$)	0.06
Average pore diameter (nm)	8.30
Microporosity (%)	0.99

The morphological characteristic of adsorbent was observed using the Field Emission Scanning Electron Microscope (FESEM) analysis. From the FESEM images (Fig. 2), it is observed that the external surfaces of the ZIF-8 are relatively smooth with fine pores. The morphology of the ZIF-8 is a hexagonal cage and shows a cubic structure (Fig. 2).

The ZIF-8 evolution was followed by XPS from the Zn 2p, C 1s, O 1s, and N 1s photoelectron peaks (Fig. 4). One of the key features in the C 1s photoelectron spectra of the ZIF-8 is the component at 285.98 eV assigned to graphitic-like compounds with a marked asymmetry toward high binding energies due to the presence of C=C bonds from hydrocarbon. In the 1021.98–1045.48 eV ranges, the contributions were from the imidazole and Zn^{+2} ions due to their bond origin. In this study, the binding

Fig. 2 FESEM images of ZIF-8

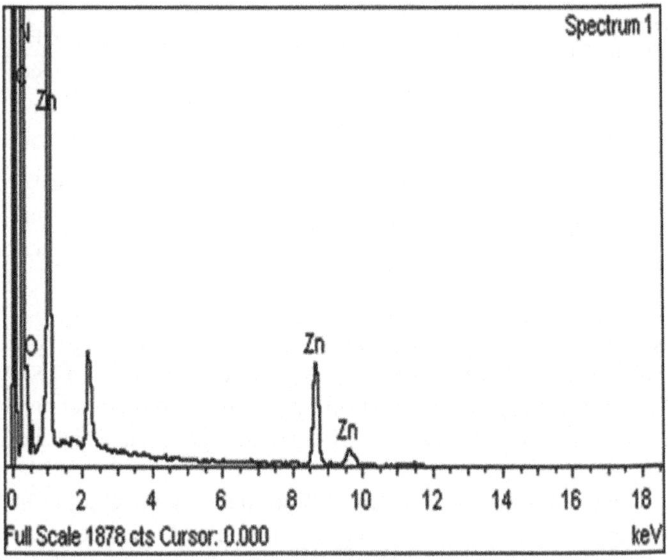

Fig. 3 EDX image of ZIF-8

energy uses an internal standard in the following sections where this binding energy is estimated to be 531.88 eV for O (1s). The results show that the N (1s) binding energy of ZIF-8 is in good agreement with 399.38 eV. These observations are consistent with the almost complete existence of a functional group of ZIF-8 data.

3.2 *Removal Studies*

The contact time was studied in this experiment. Contact time could explain the influence of systems parameters on the rate extent of adsorption of ZIF-8 onto BPA. It was clear from Fig. 5 that for 30 ppm concentration of BPA, it takes 10–30 min to achieve the full equilibrium amount (q_e). However, for ZIF-8, it takes the lowest contact time with 35 min to achieve full equilibrium. This is due to the availability of active sites for the small amount of BPA molecules to adsorb on the surface of the ZIF-8. The process is also rapid, indicating the adsorption probably occurs primarily in the presence of functional groups of the ZIF-8.

At this point, the amount of BPA being adsorbed onto ZIF-8 is in a state of dynamic equilibrium with the amount of ZIF-8 desorbing. The time required to attain this state of equilibrium is termed the equilibrium time, and the amount of ZIF-8 adsorbed at the equilibrium time reflects the maximum adsorption capacity of the BPA under those operating conditions. Therefore, it can be concluded that the contact time is much less for BPA when using ZIF-8.

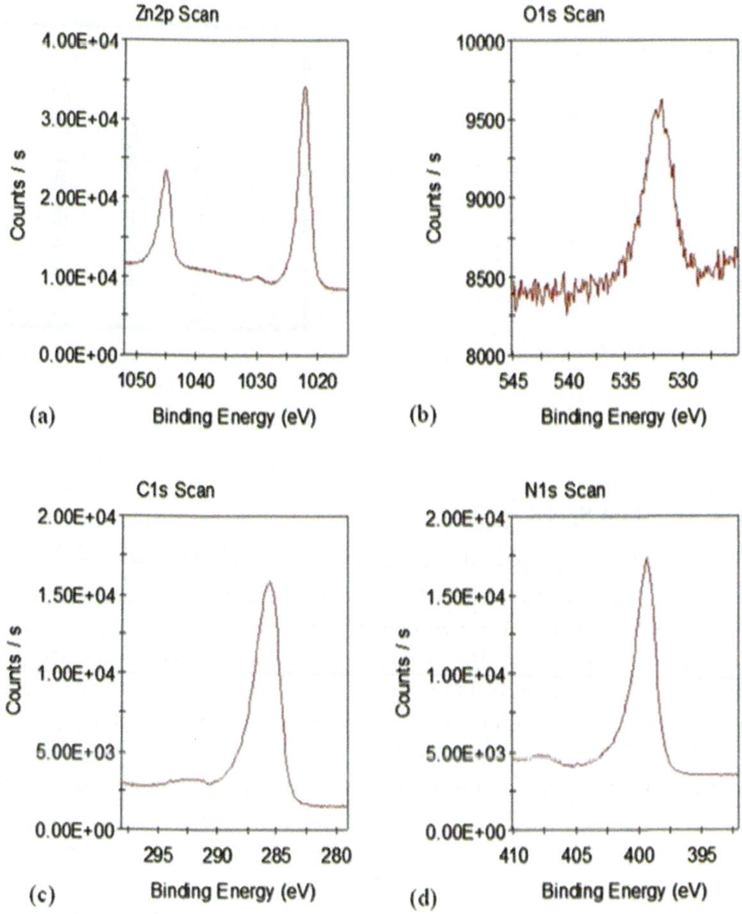

Fig. 4 XPS spectra obtained from ZIF-8 where **a** Zn (2p), **b** O (1s), **c** C (1s) and **d** N (1s) regions, respectively

Fig. 5 Plot of quantity absorbed versus time

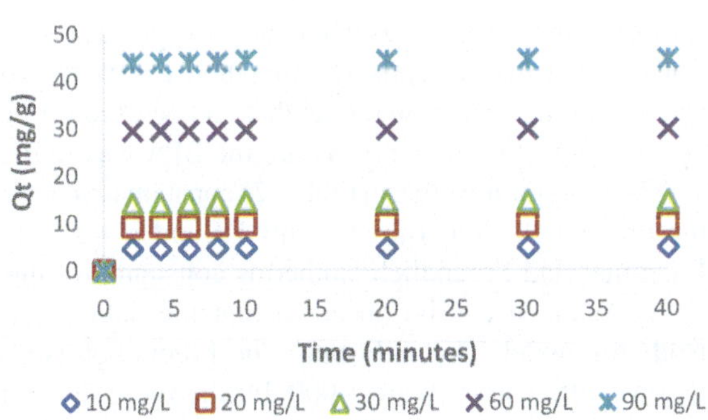

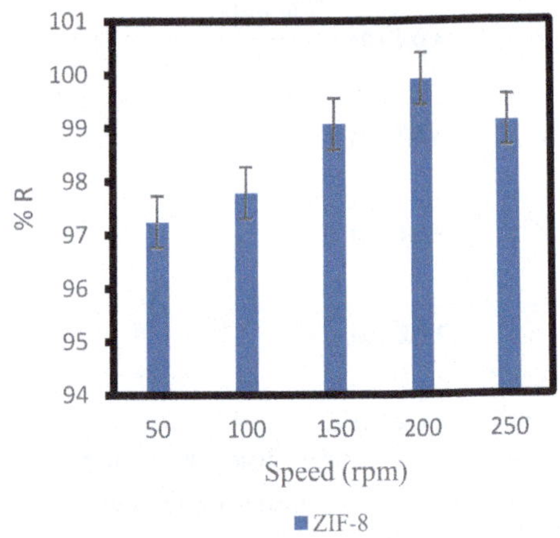

Fig. 6 Percentage removal of BPA using different speed

Effect of Agitation Speed

The effect of agitation speed for the adsorption of BPA onto the ZIF-8 was studied by conducting the batch adsorption experiment at different shaking speeds (50–250 rpm) of the orbital shaker. The removal efficiency was found to increase by increasing the speed of the shaker (Fig. 6), which is attributed to the increase in the rate at which BPA diffuses from the bulk of the solution to the boundary layer surrounding the adsorbent, hence why the contact between the BPA molecules and the active pores of the MOF is high. The optimum removal was found at 200 rpm and achieved a percentage removal of 99.90. Hence, 200 rpm has been maintained throughout the work.

3.3 *Adsorption Isotherms*

The adsorption isotherm indicates how the adsorbate molecules distribute between the liquid and solid phases when the adsorption process reaches an equilibrium state. The isotherm data analysis by fitting them to different isotherm models is an important step in finding a suitable model that can be used for design purposes. Therefore, the adsorption isotherm of ZIF-8 onto the BPA was studied and observed.

The Langmuir isotherm (Table 2) constants qmax and K_L were determined. While the Freundlich isotherm constants Kf and n were also determined, Table 2 shows Langmuir and Freundlich isotherms constants for the adsorption of the ZIF-8 onto BPA. Again, it can be observed that the data fits very well with the Freundlich isotherm model. This is because the Freundlich isotherm model's values of correlation coefficient (R2) are 1.000. Furthermore, it can be observed that the values of qmax are increasing for ZIF-8. Thus it can be concluded that the adsorption process using ZIF-8 is exothermic.

Table 2 Isotherms study on ZIF-8

Langmuir isotherm model

MOF	Dosage (g)	Q_m (mg/g)	K_L (L/mg)	Correlation coefficient (R^2)	Separation factor (R_L)
ZIF-8	0.05	345	0.0064	0.9065	0.9402
	0.1	1667	0.0006	0.8173	0.9940
	0.2	714	0.0007	0.9162	0.9930
	0.3	1667	0.0002	0.9199	0.9982
	0.4	357	0.0007	0.8936	0.9930

Freundlich isotherm model

MOF	Dosage (g)	n	K_f (mg/g(L/mg)$^{1/n}$)	Correlation coefficient (R^2)
ZIF-8	0.05	1.19775	3.1074	0.9875
	0.1	1.01947	1.0524	0.9998
	0.2	1.02375	1.8889	0.9998
	0.3	1.00604	3.3060	1
	0.4	1.02333	3.7818	0.99998

The increase of q_{max} suggests that the adsorption capacity is enhanced for ZIF-8. This effect may be because, at higher surface areas, an increase in free volume occurs due to the increased mobility of the ZIF-8. It may also be due to the increasing interaction between the ZIF-8 and the BPA in the endothermic process. Separation factors, R_L, was calculated and tabulated in Table 2. It can be concluded that the adsorption process is a favourable process since the values of R_L are less than unity ($0 < R_L < 1$).

3.4 Adsorption Kinetics

In order to optimize the design of an adsorption system to remove BPA, it is important to establish the most appropriate correlation for the equilibrium data for each system. For this reason, several kinetic models, including the pseudo-first-order and pseudo-second-order models, were tested to find out the adsorption mechanisms.

Figure 7 shows the plot of the pseudo-second-order for ZIF-8 with BPA. Table 3 lists the calculated results of the pseudo-first-order and pseudo-second-order models for ZIF-8. The calculated q_e values obtained from the first-order kinetic model did not provide reasonable values and are too low compared to the experimental q_e values. This finding suggests that the adsorption of ZIF-8 onto BPA is not a pseudo-first-order model.

The values of k_2 parameters, and the experimental $q_{e2,}$ calculated, and the correlation coefficients are also presented in Table 3 ZIF-8 within 60 min. The correlation

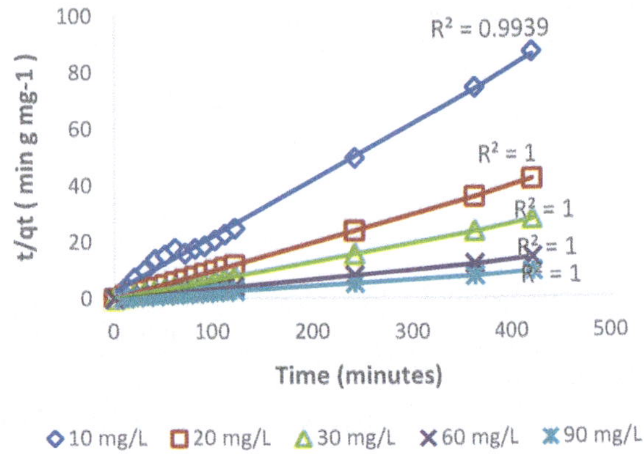

Fig. 7 Second order kinetic plot of BPA adsorption (variety dosage of ZIF-8)

Table 3 Comparison between pseudo-first order and pseudo-second order kinetic models for ZIF-8 adsorption on BPA

Model	Parameters	Value
Pseudo-first order	q_e, exp (mg/g)	9.9390
	q_e, cal (g/mg)	0.2819
	K1 (min^{-1})	10.9094
	R^2	0.5015
Pseudo-second order	q_e, cal (g/mg)	0.1861
	K1 (min^{-1})	10.0502
	R^2	1.000

coefficients for the second-order kinetic model were greater than 0.9 for ZIF-8. The calculated q_{e2} values also agreed very well with the experimental data in the pseudo-second-order kinetics. The calculated q_e values obtained from the second-order kinetic model did give reasonable values and is almost the same compared to the experimental q_e values. This finding suggests that the adsorption of ZIF-8 onto BPA is a pseudo-second-order model.

4 Conclusion

In conclusion, the ZIF-8 has shown a good potential of the adsorbent to remove BPA from aqueous solution. The equilibrium data fits well with Langmuir's and Freundlich's isotherm models. The kinetic study showed that the second-order kinetic model could best describe the experimental data.

Acknowledgements This work was sponsored by a grant provided through Financial Aid from Universiti Teknologi PETRONAS (UTP) and the Ministry of Higher Education Malaysia, FRGS No: FRGS/1/2011/SG/UTP/02/13 is gratefully acknowledged. The authors acknowledge the use of the facilities within the UTP Centralized Analytical Laboratory (CAL).

References

1. Zhao, D., Timmons, D.J., Yuan, D., Zhou, H.C.: Tuning the topology and functionality of metal–organic frameworks by ligand design. Acc. Chem. Res. **44**, 123–133 (2011)
2. Rowsell, J.L., Spencer, E.C., Eckert, J., Howard, J.A., Yaghi, O.M.: Gas adsorption sites in a large-pore metal–organic framework. Science **309**, 1350–1354 (2005)
3. Kitaura, R., Seki, K., Akiyama, G., Kitagawa, S.: Porous coordination-polymer crystals with gated channels specific for supercritical gases. Angew. Chem. **115**, 444–447 (2003)
4. Haque, E., Lee, J.E., Jang, I.T., Hwang, Y.K., Chang, J.S., Jegal, J.: Adsorptive removal of methyl orange from aqueous solution with metal–organic frameworks, porous chromium–benzenedicarboxylates. J. Hazard. Mater. **181**, 535–542 (2010)
5. Kim, J., Kim, S.N., Jang, H.G., Seo, G., Ahn, W.S.: CO_2 cycloaddition of styrene oxide over MOF catalysts. Appl. Catal. A **453**, 175–180 (2013)
6. Achmann, S., Hagen, G., Kita, J., Malkowsky, I.M., Kiener, C., Moos, R.: Metal–organic frameworks for sensing applications in the gas phase. Sensors **9**, 1574–1589 (2009)
7. Yoon, J.W., Jhung, S.H., Hwang, Y.K., Humphrey, S.M., Wood, P.T., Chang, J.S.: Gas-sorption selectivity of CUK-1: a porous coordination solid made of cobalt (ii) and pyridine-2, 4-dicarboxylic acid. Adv. Mater. **19**, 1830–1834 (2007)
8. Horcajada, P., Chalati, T., Serre, C., Gillet, B., Sebrie, C., Baati, T.: Porous metal–organic-framework nanoscale carriers as a potential platform for drug delivery and imaging. Nat. Mater. **9**, 172–178 (2010)
9. Rocha, J., Carlos, L.D., Paz, F.A.A., Ananias, D.: Luminescent multifunctional lanthanides-based metal–organic frameworks. Chem. Soc. Rev. **40**, 926–940 (2011)
10. Banerjee, R., Phan, A., Wang, B., Knobler, C., Furukawa, H., O'Keeff, M.: High-throughput synthesis of zeolitic imidazolate frameworks and application to CO_2 capture. Science **319**, 939–943 (2008)
11. Tian, Y.Q., Cai, C.X., Ji, Y., You, X.Z., Peng, S.M., Lee, G.H.: [Co5 (im) 10· 2 MB]∞: a metal-organic open-framework with zeolite like topology. Angew. Chem. **114**, 1442–1444 (2002)
12. Zhou, W., Wu, H., Yildirim, T.: Enhanced H_2 adsorption in isostructural metal−organic frameworks with open metal sites: strong dependence of the binding strength on metal ions. J. Am. Chem. Soc. **130**, 15268–15269 (2008)
13. Assfour, B., Leoni, S., Seifert, G.: Hydrogen adsorption sites in zeolite imidazolate frameworks ZIF-8 and ZIF-11. J. Phys. Chem. C **114**, 13381–13384 (2010)
14. Cho, H.Y., Kim, J., Kim, S.N., Ahn, W.S.: High yield 1–L scale synthesis of ZIF-8 via a sonochemical route. Microporous Mesoporous Mater. **169**, 180–184 (2013)
15. Tran, U.P., Le, K.K., Phan, N.T.: Expanding applications of metal−organic frameworks: zeolite imidazolate framework ZIF-8 as an efficient heterogeneous catalyst for the knoevenagel reaction. ACS Catal. **1**, 120–127 (2011)
16. Miralda, C.M., Macias, E.E., Zhu, M., Ratnasamy, P., Carreon, M.A.: Zeolitic imidazole framework-8 catalysts in the conversion of CO_2 to chloropropene carbonate. ACS Catal. **2**, 180–183 (2012)
17. Wang, L.C., Ni, X.J., Cao, Y.H., Cao, G.: Adsorption behavior of bisphenol A on CTAB–modified graphite. Appl. Surf. Sci. **428**, 165–170 (2018)
18. Park, E.Y., Hasan, Z., Khan, N.A., Jhung, S.H.: Adsorptive removal of bisphenol-A from water with a metal–organic framework, a porous chromium–benzenedicarboxylate. J. Nanosci. Nanotechnol. **13**, 2789–2794 (2013)
19. Ordonez, M.J.C., Balkus, K.J., Jr., Ferraris, J.P., Musselman, I.H.: Molecular sieving realized with ZIF-8/Matrimid® mixed–matrix membranes. J. Membr. Sci. **361**, 28–37 (2010)

Inclusion Complexes of β-Cyclodextrin/Pioglitazone and β-Cyclodextrin-Ionic Liquid/Pioglitazone: Spectroscopic Methods Combined with Molecular Docking Studies

Nurul Yani Rahim, Nur Najihah Md Zuki, Khairulazhar Jumbri, Sharifah Mohamad, Ninie Suhana Abdul Manan, Nadiah Sidek, Yih Hui Boon, and Muggundha Raoov Ramachandran

Abstract The complexes of β-cyclodextrin/pioglitazone and β-cyclodextrin-ionic liquid/pioglitazone were prepared by kneading method. Characterization using FT-IR, thermogravimetric analysis and NMR had proven the formation of the complexes. From ^{1}H NMR determination, pioglitazone is proven to form hydrogen bonding and inclusion complex with native β-cyclodextrin and β-cyclodextrin-ionic liquid. The molecular docking also reveals the same interaction involved for the formation of complexes β-cyclodextrin/pioglitazone and β-cyclodextrin-ionic liquid/pioglitazone. The binding energy obtained from molecular studies for the complex of β-cyclodextrin-ionic liquid/pioglitazone is higher than the binding energy of β-cyclodextrin/pioglitazone.

N. Y. Rahim (✉) · N. N. Md Zuki
School of Chemical Sciences, Universiti Sains Malaysia, 11800 Pulau Pinang, Malaysia
e-mail: nurulyanirahim@usm.my

K. Jumbri
Department of Fundamental and Applied Sciences, Universiti Teknologi PETRONAS, 32610 Seri Iskandar, Perak, Malaysia
e-mail: khairulazhar.jumbri@utp.edu.my

S. Mohamad · N. S. Abdul Manan · N. Sidek · Y. H. Boon · M. R. Ramachandran
Department of Chemistry, Faculty of Science, Universiti Malaya, 50603 Kuala Lumpur, Malaysia
e-mail: sharifahm@um.edu.my

N. S. Abdul Manan
e-mail: niniemanan@um.edu.my

M. R. Ramachandran
e-mail: muggundha@um.edu.my

Y. H. Boon
Integrative Medicine Cluster, Advanced Medical and Dental Institute, Universiti Sains Malaysia, 13200 Pulau Pinang, Malaysia

S. A. Abdul Karim et al. (eds.), *Proceedings of the 6th International Conference on Fundamental and Applied Sciences*, Springer Proceedings in Complexity,
https://doi.org/10.1007/978-981-16-4513-6_12

Keywords β-cyclodextrin · Pioglitazone · Ionic liquid · Inclusion complex · Hydrogen bonding · Binding energy

1 Introduction

Pioglitazone (PZN) is a widely prescribed oral anti-diabetic drug with low solubility. As such the absorption through oral route of pioglitazone is limited rate of dissolution and it needs to expansion in solubility as well as rate of dissolution for increasing in bioavailability via oral route. Several techniques such as micronization, cyclodextrins complexation, use of surfactants, solid dispersion in water soluble and dispersible carriers, use of salts, micro emulsions and self-emulsifying micro and nano disperse systems can be used to achieve better solubility [1]. Among the different techniques of solubility improvement, complexation techniques are one of the best suitable and economical method, which eventually helps to increase the dissolution rate of the drug and oral bioavailability.

Cyclodextrins (CDs) are cyclic oligosaccharides derived from starch, and contain six (α-CD), seven (β-CD), eight (γ-CD) (α-1, 4)-linked α-D-glucopyranose units [2]. The major property of CDs to adjustment the physicochemical as well as biological properties of drugs which are low aqueous soluble in nature by forming of drug and CDs inclusion complexes. Due to the chair conformation of the glucopyranose units, the CDs take the shape of a truncated cone or torus, rather than a perfect cylinder. The interior cavity of the cone is highly hydrophobic, and the exterior is hydrophilic owing to hydroxyl group [3]. The hydroxyl group combined with the hydrogen atoms outside surface of CDs build up a polar exterior to compatible with polar environments. The central cavity of the CD molecule is lined with skeletal carbons and ethereal oxygens of the glucose residue, which gives it a relatively lipophilic character [2]. As a consequence of these features, CDs can encapsulate a variety of hydrophobic molecule, or part of it, inside their cavity through non-covalent interactions to form inclusion complexes of host–guest type [3]. Compared with the α- and γ-CDs, β-CD is more widely investigated due to their high chemical stability and low cost. In addition, β-CD also has the special size of its hydrophobic cavity (cavity size: α-CD < β-CD < γ-CD) which affords to form inclusion complexes with numbers of organic and inorganic compounds [4]. However, the application of native β-CD was not always satisfactory [5]. Therefore, recently, researches have been focused on the preparation of modified β-CD to be used as complexation with drugs [6].

The addition of different substituent groups onto the rim of β-CD provides multiple interactions such as π-π, dipole–dipole, electrostatic interaction and hydrogen bonding which contributes significantly to the complexation formation. Ionic liquids (ILs) are examples of new substituent groups that are been used to modify β-CD [6–9]. ILs are salt-liked materials which are liquid below 100 °C and even below room temperature [10]. As salts they are by essence made of cation and anion. The term ILs covers inorganic as well as organic molten salt. The advantage of modified β-CD with IL is an additional electrostatic and π-π interactions can be derived and affect the formation of inclusion complex [11].

In this present research, the complexes of β-CD/PZN and βCD-IL/PZN are prepared by kneading method. The modified βCD with IL which is 3-benzylimidazoliym tosylate is prepared producing (β-CD-IL) as previous research [11–13]. To the best of our knowledge, most of the researches only study the inclusion complexes of native β-CD with PZN [2, 3, 14–16]. By using spectroscopic method such as 1H NMR and UV-Vis will give the insight inclusion behaviour of PZN complexes with native β-CD and modified β-CD experimentally. Theoretical method was studied to calculate the possible binding energy as well as its potential to predict the interactions among the macromolecules and guest via molecular docking.

2 Experimental

2.1 *Materials*

β-CD was purchased from Acros (Geel, Belgium) (99%). Pioglitazone hydrochloride was purchased from Aldrich (St. Louis, MO, USA). The analytical solvents (methanol and ethanol) were purchased from Merck (Darmstadt, Germany).

2.2 *Instrumentations*

A Perkin-Elmer 2000 FT-IR (Waltham, USA) spectrophotometer was used to record all infrared (IR) spectra. IR data were recorded from 400 to 4000 cm^{-1}. Absorption spectra measurements were carried out with a Shimadzu UV 2600 (Kyoto, Japan) spectrophotometer in the range of 190–800 nm. All NMR spectra were recorded using Bruker Avance III 500 MHz.

2.3 *Preparation of β-CD/PZN and β-CD-IL/PZN Complexes*

The complexes of β-CD/PZN and β-CD-IL/PZN were prepared using the conventional kneading method [17, 18]. PZN and β-CD with the molar ratio 1:1 were kneaded thoroughly with a minimal amount of ethanol to form a homogenous paste for 30 min. The precipitate was dried to constant mass and kept in a desiccator. The product was characterized using FT-IR, TGA and 1H NMR. The same procedure was applied for preparation complexes of β-CDIL/PZN using β-CD-IL replacing β-CD.

2.4 Determination of Absorption Spectra of β-CD/PZN and β-CDIL/PZN Complexes

The solution of β-CD/PZN complex was prepared by adding 2.0 mL of 0.01 mM PZN aliquot and 3.2 mL of 0.003 M β-CD solution into a 10.0 mL standard volumetric flask and diluted to the mark with ultra-pure water. The absorption spectra of βCD/PZN complex were recorded against blank reagent. This blank reagent was prepared with the same reagent concentration but without the addition of PZN. The absorption spectra of PZN and β-CD alone were also recorded. The same procedure was applied to determine the absorption spectra of complex of β-CD-IL/PZN by replacing β-CD with β-CD-IL.

2.5 Molecular Docking Procedure

The preliminary β-CD and PZN configurations were retrieved from the ATB server which were optimized under semi-empirical and DFT (B3LYP/6-31G*) methods, respectively. On the other hand, the construction of the modified β-CD (β-CD-IL) molecular structure was built by using Pymol from the native β-CD in which the hydroxyl group at the primary rim in one of the glucopyranose unit was substituted with imidazole group.

This study aims to rationalize the experimental finding with the docking analysis. The docking procedure was executed by using AutoDockTools software version 1.5.6 where PZN was introduced into the β-CD cavity. Lamarckian Genetic Algorithm (LGA) was applied to identify the possible binding modes of PZN with β-CD through the entire process. A three-dimensional grid box of 50 Å size (x, y, z) with 0.375 Å spacing centred on β-CD was developed using Autogrid program. Its size was appropriately designed to sufficiently enclose not only the β-CD hollow region but also its exterior part. β-CD and PZN were set as rigid macromolecule and flexible guest molecule, respectively. The interaction energy (ΔG_{bind}) was then calculated by the following expression:

$$\Delta G_{bind} = \Delta G_{VDW} + \Delta G_{electrostatic} + \Delta G_{HBond} + \Delta G_{desolv} + \Delta G_{tor} \quad (1)$$

The inhibition constant (K_i) of the host–guest interaction was determined from the following term:

$$K_i = e^{\left(\Delta G_{bind}/RT\right)} \quad (2)$$

The same procedure was repeated for β-CD-IL. Finally, the lowest binding free energies were chosen for further analysis.

3 Results and Discussion

3.1 FT-IR Characterization

The formation of the complexes β-CD/PZN and β-CDIL/PZN were confirmed using FT-IR. Figure 1a–e were illustrated the FT-IR spectra of β-CD, β-CD-IL, PZN, complexes of β-CD/PZN and β-CD-IL/PZN, respectively. The absorption bands at 3300 cm^{-1} observed in Fig. 1a, b, d, e were dedicated to the OH stretching of β-CD. The IR spectrum of PZN in Fig. 1c showed an absorption peak at 3085 cm^{-1} corresponding to NH stretching. The IR spectra of the complexes in Fig. 1d, e had some similarities in general. The IR spectra of complexes β-CD/PZN and β-CD-IL/PZN in Fig. 1d, e, respectively show that the amide-NH stretching band of PZN was not clearly detected might because of the co-occurrence with the OH intensified band at 3300 cm^{-1} [2]. The absorption band at 1684 cm^{-1} showed in Fig. 1c was attributed to carbonyl group in PZN had been broadened and shifted to a higher wave number at 1692 cm^{-1} as illustrated in Fig. 1d, e after forming complexes with β-CD [14] and β-CD-IL.

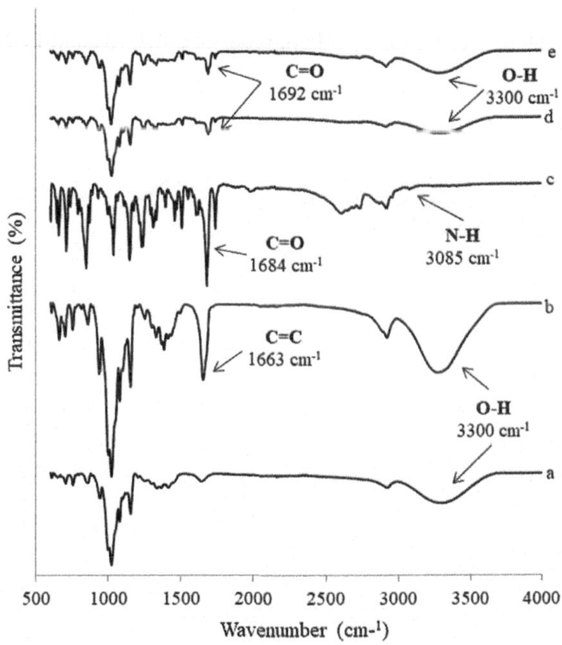

Fig. 1 The Fourier transform infrared (FT-IR) spectra of **a** β-CD; **b** β-CDIL; **c** PZN; **d** complex of β-CD/PZN; and **e** complex of β-CD-IL/PZN

3.2 *TGA Characterization*

Thermogravimetric analysis (TGA) was done on samples to identify the changes in weight percent with respect to temperature change [19]. TGA was performed on β-CD, β-CDIL, PZN, complexes of β-CD/PZN and β-CD-IL/PZN in the temperature range of 30–900 °C. Figure 2a showed β-CD obtained 12% of weight loss at 67 °C due to loss of water molecules in the cavity of β-CD that stabilizes the ring structure [20]. Then, the weight loss at second stage which found at 329 °C was related to decomposition of the β-CD structure due to the transition from solid to liquid phase [19]. The thermogravimetric profile for PZN in Fig. 2b showed a maximum rate of weight loss (92%) at 248 °C due to melting of organic compound.

The thermogravimetric profile for complex of β-CD/PZN (Fig. 2d) obtained 8% weight loss of water molecules at 61 °C. The weight of complex β-CD/PZN continued to lose 71% of original weight at 224 °C. The decomposition temperature at second stage for complex of β-CD/PZN was lesser than decomposition temperature for β-CD. This phenomenon suggests that formation of complex decrease the thermal stability of β-CD [21]. The thermal behaviours of β-CD-IL in Fig. 2c obtained 6% of weight loss at the first stage due to loss of water. The second decomposition stage of β-CD-IL at 231 °C with 45% of weight loss. The complex of β-CD-IL/PZN (Fig. 2e) lost 67% of original weight at 218 °C. The results showed complex formation induced changes in the thermal behaviour of β-CD-IL and PZN [14].

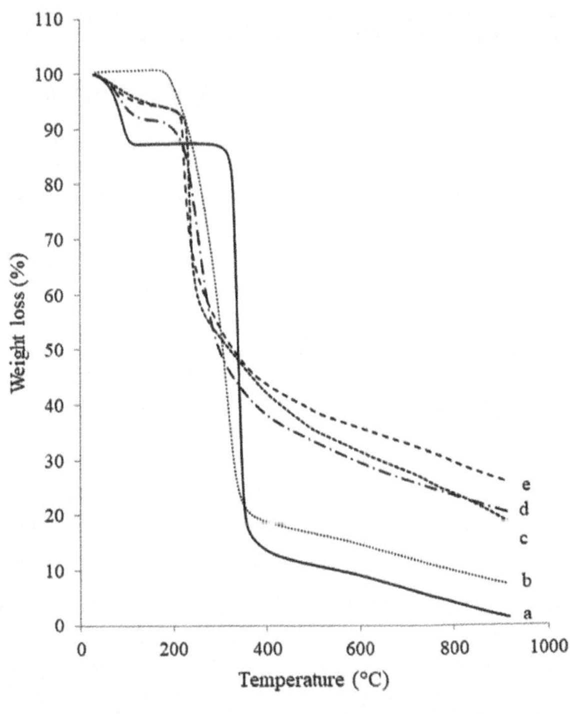

Fig. 2 Thermogravimetric profiles of **a** β-CD; **b** PZN; **c** β-CD-IL; **d** complex of β-CD/PZN and **e** complex of β-CD-IL/PZN

3.3 Evaluation on the Inclusion Behaviour for Complexes of βCD/PZN and β-CD-IL/PZN

The spectroscopic technique which is 1H NMR was used to study the inclusion behavior between native β-CD and modified β-CD with targeted compound, PZN. The values of the chemical shifts (δ) for different protons in β-CD, β-CD-IL, complexes of β-CD/PZN and β-CD-IL/PZN are listed in Tables 1 and 2. The deduced structures of the β-CD, β-CD-IL, PZN and complexes of β-CD/PZN and β-CD-IL/PZN are shown in Figs. 3 and 4. Normally, the inclusion of a nonpolar region of a compound into the hydrophobic cavity would affect the inner protons of the glucose units of β-CD, namely, H3 and H5, whereas the protons on the exterior torus of β-CD (H1, H2 and H4) would remain unaffected [22]. The induced shift (Δδ) is defined as the difference in chemical shift in the presence or absence of compound in cavity of β-CD. The induced shift is calculated by using Eq. (3):

$$\Delta\delta = \delta(\text{complex}) - \delta(\text{free}) \quad (3)$$

Table 1 Chemical shifts (Δ) of β-CD, pioglitazone, and complex β-CD/pioglitazone

Protons	β-CD δ	Pioglitazone δ	β-CD/pioglitazone	
			δ	Δδ
H1	4.8347	–	4.8380	0.0033
H2	3.3348		3.3321	−0.0027
H3	3.6362	–	3.6441	**0.0079***
H4	3.3767	–	3.3572	**−0.0190***
H5	3.5742	–	3.5798	0.0056
H6*	–	–	–	–
Ha′	–	1.2388	1.2246	−0.0142
Hb′	–	2.7857	2.7403	- 0.0454
Hc′	–	8.7010	8.6212	−0.0798
Hd′	–	8.3537	8.1669	−0.1868
He′	–	7.9289	7.7733	- 0.1556
Hf′	–	3.0684	–	–
Hg′	–	4.8821	4.8785	- 0.0036
Hh′	–	6.8837	6.8798	−0.0039
Hi′	–	7.1446	7.1403	−0.0043
Hj′	–	3.2869	–	–
Hk′	–	4.3799	4.3580	−0.0219
Hl′	–	12.0054	12.0009	−0.0045

* Values in bold refer to the highest induced shift of that particular proton

Table 2 Chemical shifts (Δ) of β-CD-IL, pioglitazone, and complex β-CD-IL/pioglitazone

Protons	β-CD-IL δ	Pioglitazone δ	β-CD-IL/pioglitazone	
			δ	Δδ
H1	4.8374	–	4.8397	0.0023
H2	3.3127	–	3.3158	0.0031
H3	3.6395	–	3.6417	0.0022
H4	3.3681	–	3.3571	**−0.0110***
H5	3.5742	–	3.5776	**0.0034***
H6*	3.9201	–	3.9191	−0.0010
H7	–	–	–	–
H8	7.2565	–	7.2762	0.0197
H9	7.1281	–	7.1410	0.0129
H10	–	–	–	–
H11	2.2940	–	2.2936	−0.0004
Ha	–	–	–	–
Hb	7.4784	–	7.4726	−0.0058
Hc	7.4293	–	7.4220	−0.0073
Hd	–	–	–	–
He	7.7497	–	7.7518	0.0021
Hf	7.9545	–	8.1414	0.1869
Hg	–	–	–	–
Hh	–	–	–	–
Ha′	–	1.2388	1.2248	−0.0140
Hb′	–	2.7857	–	–
Hc′	–	8.7010	8.6119	−0.0891
Hd′	–	8.3537	–	–
He′	–	7.9289	7.9587	0.0298
Hf′	–	3.0684	–	–
Hg′	–	4.8821	4.8985	0.0164
Hh′	–	6.8837	6.8799	−0.0038
Hi′	–	7.1446	–	–
Hj′	–	3.2869	–	–
Hk′	–	4.3799	4.3544	−0.0255
Hl′	–	12.0054	11.9942	−0.0112

* Values in bold refer to the highest induced shift of that particular proton

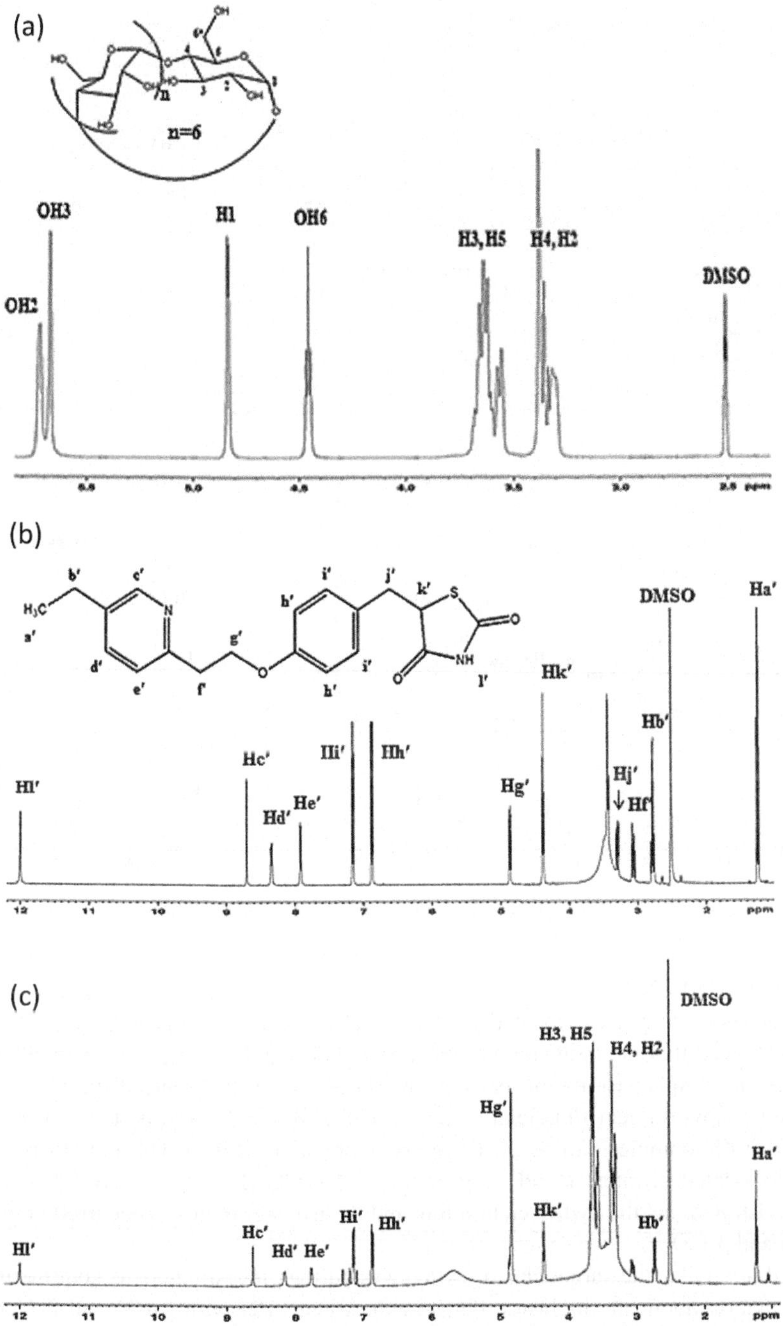

Fig. 3 1H NMR spectrum of **a** β-CD; **b** pioglitazone; and **c** complex of βCD/pioglitazone

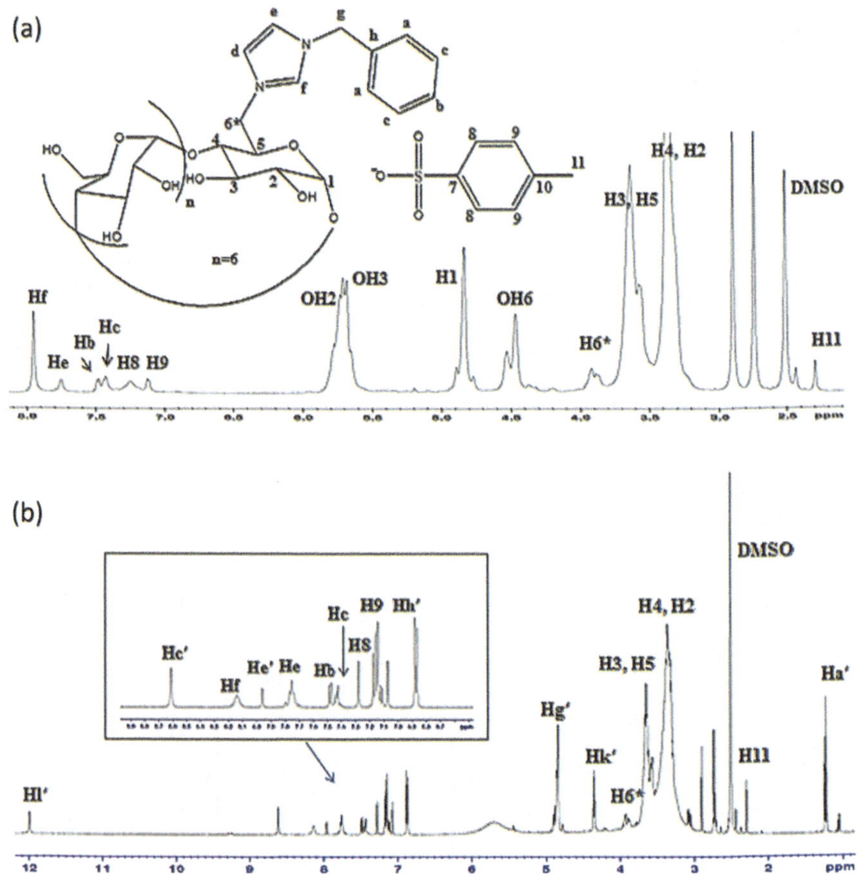

Fig. 4 1H NMR spectrum of **a** β-CD-IL and **b** complex of β-CDIL/pioglitazone

For β-CD/PZN complex (Table 1), the significant change was observed on Δδ at H3 and H4 protons which located at interior and exterior cavity of β-CD, respectively. This showed that the inclusion complex and hydrogen bonding were formed. This inclusion complex formation between β-CD and PZN had caused the change of the microenvironment thus lead to the upfield shifts in PZN protons [23]. For β-CD-IL/PZN complex (Table 2), there are appreciable shift at H4 and H5 protons which located at interior and exterior torus of β-CD-IL, respectively. This result demonstrated that the hydrogen bonding and inclusion complex is occurred between PZN and β-CD-IL.

Additionally, the molecular docking was further investigated to acquire more information about the inclusion behavior between PZN with β-CD and β-CD-IL. Figure 5a–d illustrate the most stable host–guest complexes for β-CD/PZN and β-CD-IL/PZN. β-CD and β-CD-IL are rendered in line model while PZN is displayed in stick model. As can be seen in the figure, the docking of PZN into each macromolecule

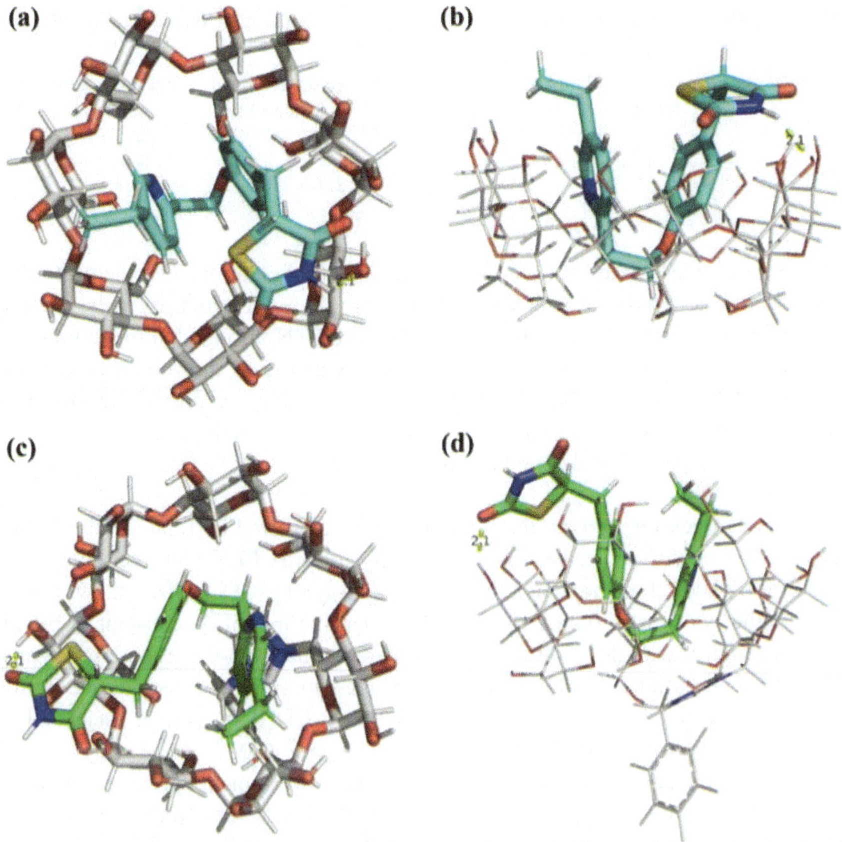

Fig. 5 The molecular structure of **a** β-CD/PZN complex (top view), **b** βCD/PZN complex (side view), **c** β-CD-IL/PZN complex (top view), **d** βCD-IL/PZN complex (side view) obtained from molecular docking calculation

generates different conformations of the complex. Nevertheless, the bent shape of the PZN observed inside both cavities are almost comparable to a certain degree. For β-CD, PZN fitted into its cavity with the thiazolidinedione and ethyl groups protruded outside the wider secondary rim. Apart from this outer projection of the thiazolidinedione ring, it can be seen in the figure that a hydrogen bonding was successfully established between the amide's hydrogen attached to the ring with the O_3 atom of βCD. This phenomenon is in accordant with the other study where this thiazolidinedione ring showed its ability to form hydrogen bonding with the hydrophilic site of peroxisome proliferator-activated receptor [24].

On the other hand, its aliphatic linker acted as the bridge by directing the position of both pyridine and phenyl rings to face the interior region of the β-CD glucopyranose unit. Moreover, the phenyl ring specifically offers supplementary hydrophobic interactions [25]. Similarly, the bending structure of docked PZN into the β-CD-IL

Table 3 Hydrogen bonds formed between β-CD and β-CD-IL with PZN as obtained from molecular docking

Complexes	Hydrogen bond donor	Hydrogen bond acceptor	Hydrogen bond distance (Å)	Energy (kcal/mol)
β-CD/PZN	H (amine)	O_3	2.065	−0.255
β-CD-IL/PZN	H_2	O (amide)	2.072	−1.074

cavity reveals that the thiazolidinedione ring and ethyl chain favored the protruded configuration across the broader frame of the cavity. However, contrary to the binding mode of native β-CD, the affinity of β-CD-IL with PZN through hydrogen bonding was generated via the oxygen of the amide moiety with the H_2 atom on the glucopyranose unit. Additionally, PZN preferably aligned its pyridine and phenyl rings towards the glucopyranose ring, a similar behavior was discovered in the formation of native β-CD/PZN complex. The detailed distances and energies of the hydrogen bonds are tabulated in Table 3.

From the docking studied, the binding energy tabulated in Table 4 reveals that native β-CD possessed weaker binding affinity towards PZN with the value of −6.07 kcal/mol. On the other hand, the binding energy of β-CD-IL/PZN shows a relatively higher value of −6.28 kcal/mol. Moreover, the inhibition constant of β-CD-IL/PZN acquired a slightly lower value compared to native β-CD complex in which the smaller inhibition constant value signifies the superiority of the binding ability. Thus, the docking procedure demonstrated that β-CD-IL possesses an excellent binding potency towards PZN compared to the native β-CD.

Overall, the results of molecular docking studies were meet agreement with the experimental data where it was found that the PZN was form inclusion complex and hydrogen bonding with β-CD and β-CD-IL. Additionally, the UV-Vis absorption spectra of β-CD/pioglitazone and β-CD-IL/pioglitazone complexes were further investigated to acquire more information on the inclusion behavior between pioglitazone with β-CD and β-CD-IL. The plots of UV-Vis absorption for βCD, pioglitazone and complexes of β-CD/pioglitazone and βCD-IL/pioglitazone are presented in Figs. 6 and 7.

In Fig. 6, it was recorded that β-CD has almost no absorption throughout the wavelength; hence, its absorbance can be neglected [19]. Absorption spectra for PZN and the complex β-CD/PZN were similar at most of the points along the wavelength

Table 4 Binding energy, final intermolecular energy, torsional free energy and inhibition constant for all the complexes

Complexes	Binding energy, ΔG_{bind} (kcal/mol)	VDW + HBond + desolv energy	Electrostatic energy	Torsional free energy	Inhibition constant, K_i (μM)
β-CD/PZN	−6.07	−8.49	+0.03	+2.39	35.31
β-CD-IL/PZN	−6.28	−8.58	−0.09	+2.39	24.93

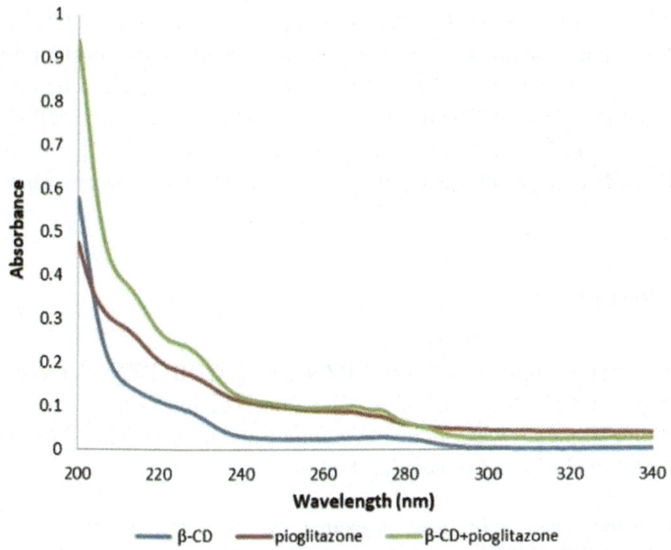

Fig. 6 Absorption spectra of β-CD, PZN and complex of β-CD/PZN with [βCD]: 0.003 M [PZN]: 0.01 mM; T = 25 °C

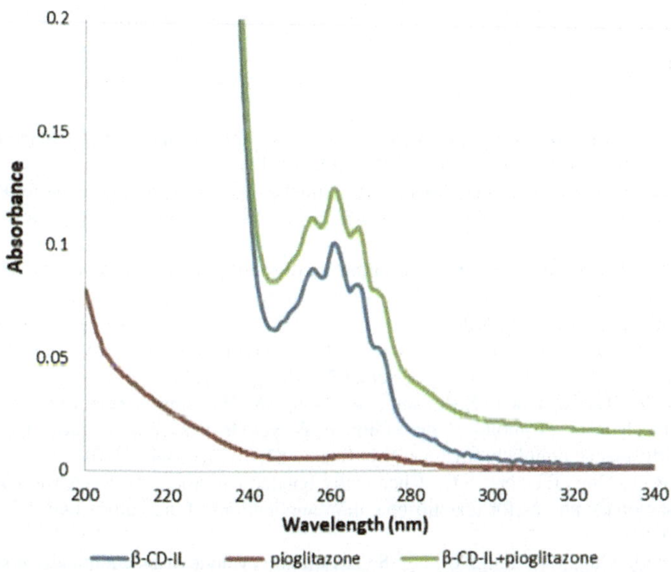

Fig. 7 Absorption spectra of β-CD-IL, PZN and β-CD-IL/PZN [β-CD-IL]: 0.003 M [PZN]: 0.01 mM; T = 25 °C

recorded. The absorbance of β-CD/PZN is higher than absorbance of PZN alone. The increased of intensity at all points of wavelength was due to the formation inclusion phenomena between β-CD and PZN. The same phenomena have been observed by most of researches [19, 23]. The results obtained in Fig. 7 revealed that β-CD-IL had λ_{max} in the range of 230–260 nm as stated in previous studies [11–13]. The absorbance of β-CD-IL/PZN underwent the hyperchromic effect (increase in absorbance).

4 Conclusion

In this study, the complexes of β-CD/PZN and β-CDIL/PZN were successfully prepared and compared their inclusion behaviour. The interactions that contributed for the both of complexes were inclusion complex and hydrogen bonding determined from 1H-NMR analysis and molecular docking.

Acknowledgements This work was supported by Ministry of Education Malaysia (FRGS/1/2018/STG01/USM/03/1) and Universiti Sains Malaysia for Short-Term Research Grant (304/PKIMIA/6315106).

References

1. Kour, P., Kataria, M.K., Bilandi, A.: Dissolution rate enhancement of pioglitazone by solid dispersion. Indo Am. J. Pharm. **5**(7), 2664–2681 (2015)
2. Beloshe, S.P., Chougule, D.D., Shah, R.R., Ghodke, D.S., Pawar, N.D., Ghaste, R.P.: Effect of method of preparation on pioglitazone HCl-β-cyclodextrin inclusion complexes. Asian J. Pharm. **4**(2), 168–172 (2014)
3. Ali, S.M., Upadhyay, S.K.: Complexation studies of pioglitazone hydrochloride and β-cyclodextrin: NMR (1H, ROESY) spectroscopic study in solution. J. Incl. Phenom. Macrocycl. Chem. **62**(1–2), 161–165 (2008)
4. Valente, A.J., Söderman, O.: The formation of host–guest complexes between surfactants and cyclodextrins. Adv. Colloid Interface Sci. **205**, 156–176 (2014)
5. Zhou, Z.M., Li, X., Chen, X.P., Fang, M., Dong, X.: Separation performance and recognition mechanism of mono (6-deoxy-imino)-β-cyclodextrins chiral stationary phases in high-performance liquid chromatography. Talanta **82**(2), 775–784 (2010)
6. Wang, R.Q., Ong, T.T., Ng, S.C.: Chemically bonded cationic β-cyclodextrin derivatives as chiral stationary phases for enantioseparation applications. Tetrahedron Lett. **53**(18), 2312–2315 (2012)
7. Wang, R.Q., Ong, T.T., Tang, W., Ng, S.C.: Cationic cyclodextrins chemically bonded chiral stationary phases for high-performance liquid chromatography. Anal. Chim. Acta. **718**, 121–129 (2012)
8. Wang, R.Q., Ong, T.T., Ng, S.C.: Chemically bonded cationic β-cyclodextrin derivatives and their applications in supercritical fluid chromatography. J. Chromatogr. A **1224**, 97–103 (2012)
9. Wang, R.Q., Ong, T.T., Ng, S.C.: Synthesis of cationic β-cyclodextrin derivatives and their applications as chiral stationary phases for high-performance liquid chromatography and supercritical fluid chromatography. J. Chromatogr. A **1203**(2), 185–192 (2008)

10. Yao, X., Tan, T.T.Y., Wang, Y.: Thiolene click chemistry derived cationic cyclodextrin chiral stationary phase and its enhanced separation performance in liquid chromatography. J. Chromatogr. A **1326**, 80–88 (2014)
11. Rahim, N.Y., Tay, K.S., Mohamad, S.: β-cyclodextrin functionalized ionic liquid as chiral stationary phase of high-performance liquid chromatography for enantioseparation of β-blockers. J. Incl. Phenom. Macrocycl. Chem. **85**(3–4), 303–315 (2016)
12. Rahim, N.Y., Tay, K.S., Mohamad, S.: Chromatographic and spectroscopic studies on β-cyclodextrin functionalized ionic liquid as chiral stationary phase: enantioseparation of flavonoids. Chromatographia **79**(21–22), 1445–1455 (2016)
13. Rahim, N.Y., Tay, K.S., Mohamad, S.: Chromatographic and spectroscopic studies on β-cyclodextrin functionalized ionic liquid as chiral stationary phase: enantioseparation of NSAIDs. Adsorpt. Sci. Technol. **36**(1–2), 130–148 (2018)
14. Elbary, A.A., Kassem, M.A., Abou Samra, M.M., Khalil, R.M.: Formulation and hypoglycemic activity of pioglitazone-cyclodextrin inclusion complexes. Drug Discov. Ther. **2**(2), 94–107 (2008)
15. Rao, V.K., Singh, G., Porwal, P.K., Kaur, S.: Enhancement of pioglitazone hydrochloride solubility via inclusion complexation with β-cyclodextrins-optimization and evaluation studies. Asian J. Biomater. Res. **4**(2), 39–51 (2017)
16. Abdulraheem, A., El-Tohamy, M.F., Alnajjar, H.A., Alsaffar, N.: Electrochemical studies of modified 2-hydroxypropyl beta-cyclodextrin and modified carbon nanotubes sensors for determination of anti-diabetic pioglitazone hydrochloride. World J. Pharm. Pharm. Sci. **7**(2), 197–217 (2017)
17. Cwiertnia, B., Hladon, T., Stobiecki, M.: Stability of diclofenac sodium in the inclusion complex with β-cyclodextrin in the solid state. J. Pharm. Pharmacol. **51**(11), 1213–1218 (1999)
18. Daruházi, Á.E., Szente, L., Balogh, B.: Utility of cyclodextrins in the formulation of genistein: part 1 preparation and physicochemical properties of genistein complexes with native cyclodextrins. J. Pharm. Biomed. Anal. **48**(3), 636–640 (2008)
19. Sambasevam, K.P., Mohamad, S., Sarih, N.M., Ismail, N.A.: Synthesis and characterization of the inclusion complex of β-cyclodextrin and azomethine. Int. J. Mol. Sci. **14**(2), 3671–3682 (2013)
20. Furó, I., Pócsik, I., Tompa, K., Teeäär, R., Lippmaa, E.: CP-DD-MAS 13C-NMR investigations of anhydrous and hydrated cyclomalto-oligosaccharides: the role of water of hydration. Carbohydr. Res. **166**(1), 27–33 (1987)
21. Chen, M., Diao, G., Zhang, E.: Study of inclusion complex of β-cyclodextrin and nitrobenzene. Chemosphere **63**(3), 522–529 (2006)
22. Zhang, D.D., Zhao, P.Y., Huang, N.J., Wu, Y.L., Zhai, Y.M.: Study of α-cyclodextrin or dimethylcyclodextrin/toluene in CF COOD/DO. In: The 5th International Symposium on Cyclodextrins, Editions de Santé, pp. 146–149 (1990)
23. Wang, H.Y., Han, J., Feng, X.G.: Spectroscopic study of orange G-β-cyclodextrin complex and its analytical application. Spectrochim. Acta A Mol. Biomol. Spectrosc. **66**(3), 578–585 (2007)
24. Thangavel, N., Al Bratty, M., Akhtar Javed, S., Ahsan, W., Alhazmi, H.A.: Targeting peroxisome proliferator-activated receptors using thiazolidinediones: strategy for design of novel antidiabetic drugs. Int. J. Med. Chem. 1069718 (2017)
25. Yekta, R., Dehghan, G., Rashtbari, S., Sheibani, N., Movahedi, A.A.M.: Activation of catalase by pioglitazone: multiple spectroscopic methods combined with molecular docking studies. J. Mol. Recogn. **30**(12), 1099–1352 (2017)

Mechanism of Alkaline Surfactant Polymer in Oil-Water Interface: Physicochemical of Fluid Rheology, Interfacial Tension and Molecular Dynamics Simulation

Mohd Sofi Numin, Khairulazhar Jumbri, Anita Ramli, and Noorazlenawati Borhan

Abstract Chemical's injection such as alkaline, polymer, and surfactant in Chemical Enhanced Oil Recovery (CEOR) can improve oil recovery by modifying the injected fluid characteristics to enhance interaction with oil in the reservoir. The alkaline surfactant polymer (ASP), which consists of 0.3 M S672 cocamidopropyl betaine derivative surfactant, 1.0 M sodium carbonate (Na_2CO_3), and 1000 ppm hydrolyzed polyacrylamide (HPAM) polymer has good rheological and most shear-thinning properties, better viscoelastic, and most stable emulsion with the highest viscosity. Interactions were investigated using the molecular dynamics (MD) simulation approach by simulating surfactant molecules at the oil-water interface to determine interfacial tension (IFT). Experimentally measured IFT values demonstrate that 0.3 M S672 concentration has the lowest surfactant value of 0.04 mN/m and predicted IFT was calculated using the GROMACS software. The behaviors of surfactant molecules at the interface were captured after the simulation to investigate the interface behaviors. The density and IFT values of pure octane with SPC water were 712.95 kg/m^3 and 47.072 mN/m, respectively, showing excellent agreement with the experimental findings. Increasing surfactant concentration at the oil-water interface has lowered IFT values from 37.20 to 19.07 mN/m, indicating that surfactant molecules adhered correctly at the interface.

M. S. Numin · K. Jumbri (✉) · A. Ramli
Department of Fundamental and Applied Sciences, Universiti Teknologi PETRONAS, Seri Iskandar, Perak, Malaysia
e-mail: khairulazhar.jumbri@utp.edu.my

M. S. Numin
e-mail: mohd_18003507@utp.edu.my

A. Ramli
e-mail: anita_ramli@utp.edu.my

N. Borhan
Hydrocarbon Recovery Technology, PETRONAS Research Sdn Bhd, Bandar Baru Bangi, Selangor, Malaysia
e-mail: noorazlenawati_borhan@petronas.com.my

S. A. Abdul Karim et al. (eds.), *Proceedings of the 6th International Conference on Fundamental and Applied Sciences*, Springer Proceedings in Complexity,
https://doi.org/10.1007/978-981-16-4513-6_13

Keywords Enhanced oil recovery · Rheology · Interfacial tension · Molecular dynamics simulation

1 Introduction

Choosing the right chemicals in ASP/SP flooding plays a significant role in ensuring the reservoir fluids are conductive for oil extraction. In ASP and SP flooding, the surfactant acts as a surface acting agent by lowering the interfacial tension (IFT) value between the oil and the injected fluid and mobilizing the trapped oil between the reservoir's rock surface. The hydrophobic group of surfactants will interact with the oil and changes its properties, thus making it easier to flow. Betaine is the type of surfactant that usually used in CEOR. Its structural properties contain both hydrophilic cation and hydrophilic anion groups that allow low irritation, excellent strong water resistance, and broad applicability [1]. The performance of ASP and SP in flooding applications can be determined through various tests, including phase behavior analysis, solubility analysis, and recovery factor determination via direct injection of ASP/SP into the simulation reservoir system. However, the mechanism and interaction of the injected CEOR fluid with the reservoir oil require further investigation using more precise techniques such as rheology tests, interfacial tension (IFT) measurement, and molecular dynamics simulation.

An alkaline substance is added to the CEOR flooding to hinder surfactant adsorption to the reservoir surface, thus reducing the IFT value through ionic strength. According to Leitenmueller et al. [2], carbonate-based alkali is frequently used in CEOR flooding due to its buffer capacity, which prevents a pH value from changing dramatically. The wettability of rock is partially altered by an in-situ emulsion formed by alkali reacting with naturally occurring acid species of oil. By increasing the viscosity and decreasing the effective permeability of the injected fluid, the presence of polymer improves the mobility ratio and sweeps efficiency in the reservoir [3]. HPAM (partially hydrolyzed polyacrylamide) has been used as a mobility control polymer since the 1960s. The molecule is made up of a flexible chain structure called a random coil, and as a polyelectrolyte, it interacts with the ions present in the solution. Additionally, Diaconu et al. 1998 demonstrated that HPAM could increase the oil recovery factor by thickening the viscosity of injected water-flooded oil and increasing sweep efficiency [4]. According to a study published in 2017 by Sudhir Kumar et al. and Seung Jun Lee in 2009, measuring rheological properties is an excellent tool for describing the parameter for mobility control in chemical EOR. It will determine the apparent viscosity of EOR polymers as a function of shear rate, concentration, and temperature, as viscosity is the primary factor to consider when controlling mobility in EOR processes [5, 6]. Jang et al. [7] demonstrated the rheological properties of two polymer solutions, xanthan gum and HPAM, by comparing their viscosities at various salinities, concentrations, and temperatures [7]. It was discovered that injecting xanthan gum is more effective in reservoirs with higher salinity. However, one of xanthan gum's disadvantages is that it cannot penetrate low-permeability rocks [8]. In this work, the IFT analysis and MD simulation were employed to determine the surfactant's interfacial reaction directly at the interface and the mechanism by which the IFT value is reduced and its relationship to viscosity properties. In 2016, Jian C. et al. used MD simulation to investigate the effect of asphaltene concentrations on the IFT of the oil/water interface [9].

2 Materials and Methods

2.1 Emulsion Rheology Test

The rheological measurement of the emulsion is a valuable tool for understanding the flow properties of the emulsion during the storage and injection process [10]. The emulsion rheology is also related to emulsion stability. The emulsion produced by an emulsion tendency test was analyzed by using a rheometer equipped with an interfacial rheology system (IRS) to study the behavior of emulsion that is related to rheology properties such as bulk viscosity, viscoelastic, and shear properties. The IRS consists of a bicone (radius of 34.12 mm and cone angle of 5°) and a glass of cup (inner radius of 40 mm and height of 20 mm) surrounded by an insulation jacket and covered by a metallic cap. A 3 mL of emulsion sample was used to fill the measuring cup for a test. The frequency sweep was performed from 0.00628 to 62.8 angular frequency at 0.1–100% shear rates. For viscosity, the sample was analyzed at temperatures varied from 30 to 80 °C, and 1 to 70 s^{-1} shear rates.

2.2 IFT Analysis

The IFT measurement was done to determine the best composition of ASP in reducing the IFT value. The analysis started by measuring IFT values of a single betaine surfactant solution in five different concentrations (0.1, 0.2, 0.3, 0.4, and 0.5 M) at temperatures of 30 and 60 °C. The IFT values between the crude oil and surfactant solution were measured for 60 min using spinning drop video tensiometer instrument (Dataphysics, Germany).

2.3 Molecular Dynamics Simulation

The MD simulation method predicts surfactant mechanisms by lowering the IFT value in the oil-water interface. The crude oil model used in this method was octane. A total of 6 simulation systems were designed to probe the IFT trend with varying surfactant concentrations. The first system was the octane-water interface (200–2100 number of molecules) without any chemicals in the interface. Then, the other simulation systems contained different number of surfactant molecules at the interface (10, 20, 30, 40 and 50) to investigate the effect of concentration on the behavior of surfactant injected in the oil-water interface. The initial structures of all molecules were first built using *Chemspace*, and the initial coordinates were submitted to the Automated Topology Builder (ATB) and Repository Version 3.0 server [11] to obtain the chemical topology files. The initial configuration was constructed for all systems using *Packmol* with a box dimension of $4 \times 4 \times 8$ nm. To begin the simulation, the energy minimization was performed using steepest descent followed by conjugate gradient methods for 5000 steps. The simulation proceeded with 10 ns NVT MD to establish proper orientation in the system's temperature, followed by NPnAT ensemble simulation for 10 ns. Pn and A are the isonormal pressure perpendicular to the z-direction interface and the isointerfacial area. NPnAT ensemble is capable of predicting the IFT value in MD simulation [9]. In the NPnAT ensemble, the pressure in only z-direction is controlled. Throughout this study, all the simulations were computed using GROMACS software. The temperature and pressure were controlled by the velocity rescaling thermostat [12] and Parrinello-Rahman barostat [13], respectively at 303 K, and 1.0 bar. In all simulations, periodic boundary condition, full electrostatics with particle-mesh Ewald method [14], a cut-off distance of 1.4 nm for van der Waals, SETTLE algorithm [15] to constrain all bonds for water molecules, LINCS algorithm [16] to constrain all bonds for solute as well as organic solvent molecules with a time step of 2.0 fs were used [9].

The IFT was calculated by using the following equation [17]:

$$\gamma = \frac{1}{2}\left(Pz - \frac{Px + Py}{2}\right)Lz \tag{1}$$

where Px, Py, and Pz are the diagonal components of the pressure tensor, and Lz is the box length in the z-direction. The effectiveness of this equation and calculation methodology were proved by calculating the IFT value of the octane-water interface and compared with the experimental data. The results showed that the density and IFT values calculated by MD simulation and experimental data from literature showed good agreement. By MD simulation, the IFT value calculated for pure octane with SPC water was 47.07 mN/m, whereas 50.70 mN/m was obtained experimentally [18]. While the equilibrium density of the system from NPT simulation at 303 K and 0.1 MPa was 712.95 kg/m^3 compared to 698.39 kg/m^3 from experimental finding [19]. This proves that the density and IFT calculation method by MD simulation was in excellent agreement with the experimental with percentage error of 2.04% for density and 7.16% for IFT.

3 Results and Discussions

3.1 Rheological Properties of ASP and SP Emulsion

The previous work on the rheological properties of ASP and SP emulsions are summarized in Table 1 [20]. The ASP and SP compositions gave efficient rheological properties in terms of shear and viscoelastic properties, and stability at different temperatures making it favorable towards CEOR application. A 20 WC ASP has optimum emulsion properties with the highest consistency index value (K) interpreted from the power-law equation representing more shear thinning behavior. A 20 WC also gave a better viscoelastic property with a lower relaxation time and more stable emulsion with higher viscosity at high temperatures. The results showed that the polymer can function efficiently in surfactant, and alkaline also helps to produce better rheological properties that increases the sweep efficiency. The study by Narendra Kumar and Ajay Mandal proved the importance of injected fluid that has a stable viscosity and viscoelastic properties with high modulus and SF value that is able to increase the oil recovery [21]. During the fluid injection process in CEOR flooding, the surfactant reduced the IFT value and mobilize the trapped crude oil between the rock in the reservoir. The oil mobilizes and forms an oil bank, which cause the decreasing of the pressure drop due to the greater mobility of water than trapped oil. To overcome this problem, the good rheological properties of injected fluid, such as viscosity stability, shear, and viscoelastic properties, are needed to increase the pressure drop that in turn increases sweep and oil displacement efficiency [21]. Thus, the combination of S672, Na_2CO_3, and HPAM in ASP flooding was further analyzed by IFT measurement to determine the best composition in lowering the IFT value. On the other hand, MD simulation was used to explain the mechanism in reducing the IFT value.

Table 1 Rheological properties of ASP and SP emulsion

System	K value from power law equation	Relaxation time, s	Viscosity at 80 °C
ASP (60 crude: 40 brine)			
20 WC	1.40	0.0054	8.2
40 WC	0.85	0.067	5.2
50 WC	0.64	0.105	5.7
60 WC	0.79	0.095	8.5
80 WC	0.50	0.090	1.7
SP (60 crude: 40 brine)			
20 WC	0.55	0.095	5.4
40 WC	0.44	0.095	7.8
50 WC	0.61	0.099	2.7
60 WC	0.43	0.117	3.4

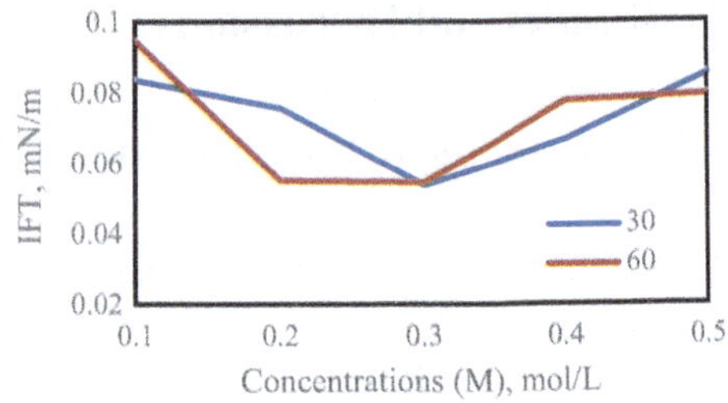

Fig. 1 The IFT values measured with different concentrations of surfactant solution at temperature of 30 and 60 °C for 60 min

3.2 *Interfacial Tension (IFT) Analysis*

The IFT values of S672 betaine surfactant at different concentrations of 0.1, 0.2, 0.3, 0.4, and 0.5 mol/L were measured. Figure 1 illustrates the IFT values measured for all surfactant concentration at temperatures of 30 and 60 °C for 60 min. The IFT values indicated that all concentrations of S672 surfactant from 0.1 to 0.5 M was able to mobilize the oil trapped in the pore structure of the reservoir with an IFT value less than 10^{-2} mN/m [22]. The figure for both temperatures showed that the IFT value decreases as the surfactant concentration increases. When the surfactant concentration was above 0.3 M, the IFT value increased. The phenomenon was caused when the surfactant solution reached a critical micelle concentration, CMC. At CMC, all the surfactant molecules were saturated at the interface or surface of the solution. The addition of more surfactants enhanced the formation of micelles in the solution and increased the IFT value [23]. From the findings, 0.3 M of S672 surfactant was the most efficient concentration in reducing the IFT value called optimum concentration. The temperature did not affect the efficiency of the surfactant in reducing the IFT value. The IFT value remained less than 10^{-2} mN/m at temperature of 60 °C indicating that the S672 surfactant has a good heat-resistant ability. As stated by Negin et al. [24], a surfactant with both positive and negative hydrophilic heads has a good temperature resistance and thermostability. Zhang et al. [25] also studied the heat-resistant ability of betaine surfactant and found that the surfactant has a good resistant-ability [25]. The range of CMC also increased at 60 °C, where 0.2–0.3 M S672 surfactant gave the lowest IFT values. Therefore, the temperature tolerance properties of the S672 surfactant are suitable for EOR application and can be used in a wide temperature range.

3.3 *The Mechanism in Reducing IFT Value by MD Simulation*

IFT Calculation and Direct Visualization. All the calculated IFT values are shown in Fig. 2. The different composition of ASP has different trends as the concentration increases. In surfactant system, the IFT values decreased from 37.20 to 19.07 mN/m

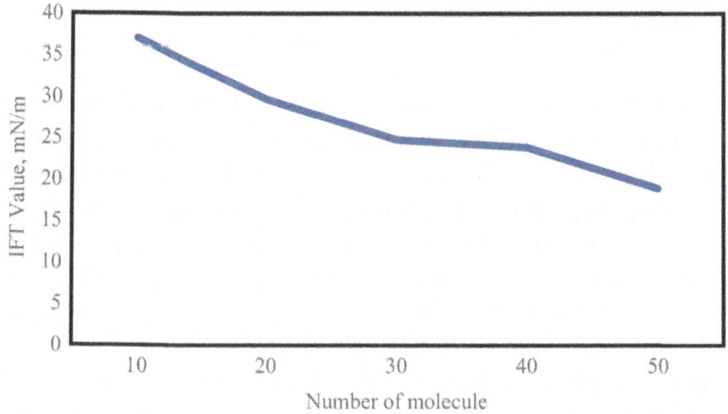

Fig. 2 Interfacial tension value of surfactant at the different number of molecules

as the surfactant concentration (molecules) increased indicating that more surfactants adsorbed in the oil/water interface [26]. The snapshot of surfactant systems (10 and 50 molecules) at temperature of 300 K and pressure of 0.1 MPa after 10 ns simulation are shown in Fig. 3. It can be observed that all the surfactants adsorbed perfectly at the interface, where more surfactant molecules were adsorbed for 50 molecules surfactant system. The adsorption of surfactant in the interface played the most prominent role in lowering the IFT value of the interface system. The higher the surfactant adsorption, the lower the IFT value until the surfactant reached the highest concentration, where the IFT value decreased [27].

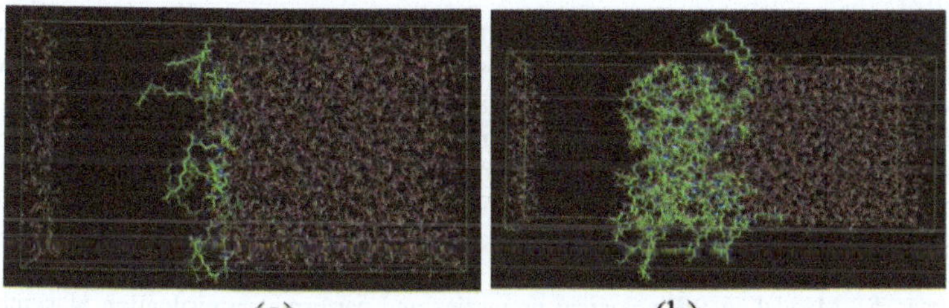

Fig. 3 Snapshot of surfactant systems for **a** 10 and **b** 50 surfactant molecules at 303 K and 0.1 MPa after 10 ns NPnAT simulation. Green represents the surfactant, and red represents water molecules. The octane molecules were removed for clarity

4 Conclusion

In conclusion, the rheological properties of 20 WC ASP composition were the most efficient in mobilizing the oil. The IFT values for the surfactant systems were successfully determined at five different concentrations, with 0.3 M serving as the CMC. The MD simulation technique was successfully performed to calculate the IFT values for the surfactant systems, and the results were in agreement with experimental findings. The IFT value decreased as the surfactant concentration increased. The adsorption of surfactant molecules at the interface was observed that showed perfect correlation with the IFT values. In addition, the IFT value decreased as the adsorption of surfactant molecules at the interface increased. The future works will focus on conducting additional analysis on the IFT surfactant solution, both experimentally and through MD simulation, with the addition of alkaline and polymer, in order to determine the optimal ASP composition in lowering the IFT value and investigating its mechanism, respectively.

Acknowledgements The author thanks PETRONAS Group Research and Technology (GR&T) for the research grant to conduct this research.

References

1. Li, N., Zhang, G., Ge, J., Zhang, L., Liu, X., Wang, J.: Ultra-low interfacial tension between heavy oil and betaine-type amphoteric surfactants. J. Dispersion Sci. Technol. **33**(2), 258–264 (2012)
2. Leitenmueller, V., Toumi, O., Hofstaetter, H., Leoben, M.: Microemulsion Formation & Its Effect on Rheology Using Carbonate-Based Alkalis for AP or ASP Floods in the Matzen Field. Society of Petroleum Engineers (2018)
3. Amirbayov, T.: Simulation study of the polymer flooding applied to the Norne field E-segment. Dissertation, Norwegian University of Science and Technology (2014)
4. Chelaru, C., Diaconu, I., Simionescu, I.: Polyacrylamide obtained by plasma-induced polymerization for a possible application in enhanced oil recovery. Polym. Bull. **40**, 757–764 (1998)
5. Kumar, S., Mandal, A.: Rheological properties and performance evaluation of synthesized anionic polymeric surfactant for its application in enhanced oil recovery. Polymer **120**, 30–42 (2017)
6. Lee, S., Kim, D.H., Huh, C., Pope, A.G.: Development of a comprehensive rheological property database for EOR polymers. In: Proceedings—SPE Annual Technical Conference and Exhibition, vol. 5, pp. 3161–3174 (2009)
7. Jang, H.Y., Zhang, K., Chon, B.H., Choi, H.J.: Enhanced oil recovery performance and viscosity characteristics of polysaccharide xanthan gum solution. J. Ind. Eng. Chem. **21**, 741–745 (2015)
8. Xu, L., Xu, G., Liu, T., Chen, Y., Gong, H.: The comparison of rheological properties of aqueous Welan gum and Xanthan gum solutions. Carbohydr. Polym. **92**(1), 516–522 (2013)
9. Jian, C., Poopari, M.R., Liu, Q., Zerpa, N., Zeng, H., Tang, T.: Reduction of water/oil interfacial tension by model asphaltenes: the governing role of surface concentration. J. Phys. Chem. **120**(25), 5464–5654 (2016)

10. Qiu, F.: The potential applications in heavy oil EOR with the nanoparticle and surfactant stabilized solvent-based emulsion. In: Society of Petroleum Engineers—Canadian Unconventional Resources and International Petroleum Conference 2010, vol. 1, pp. 157–168 (2010)
11. Stroet, M., Caron, B., Visscher, K.M., Geerke, D.P., Malde, A.K., Mark, A.E.: The automated topology builder version 3.0 (ATB3.0): prediction of solvation free enthalpies in water and hexane. J. Chem. Theor. Comput. **14**(11), 5834–5845 (2018)
12. Bussi, G., Donadio, D., Parrinello, M.: Canonical sampling through velocity rescaling. J. Chem. Phys. **126**(014101) (2007)
13. Parrinello, M., Rahman, A.: Polymorphic transitions in single crystals: a new molecular dynamics method. J. Appl. Phys. **52**, 7182–7190 (1981)
14. Essmann, U., Perera, L., Berkowitz, M.L., Darden, T., Lee, H., Pedersen, L.G.: A smooth particle mesh Ewald method. J. Chem. Phys. **103**, 8577 (1995)
15. Miyamoto, S., Kollman, P.A.: Settle: an analytical version of the SHAKE And RATTLE algorithm for rigid water models. J. Comput. Chem. **13**(8), 952–962 (1992)
16. Hess, B.: P-LINCS: a parallel linear constraint solver for molecular simulation. J. Chem. Theor. Comput. **4**, 116–122 (2008)
17. Kunieda, M.: Molecular dynamics study of oil–water interfacial equilibrium in petroleum engineering. Ph.D Dissertation, Kyoto University Research Information Repository (2012)
18. Donahue, D.J., Bartell, F.E.: The boundary tension at water-organic liquid interfaces. J. Phys. Chem. **56**(4), 480–484 (1952)
19. Lemmon, E., Mclinden, M., Friend, D.: In: Linstrom, P.J., Mallard, W.G. (eds.) Thermophysical Properties of Fluid Systems in NIST Chemistry Webbook. NIST Standard Reference Database Number 69 (2005)
20. Numin, M.S., Jumbri, K., Ramli, A., Borhan, N.: Microemulsion rheological analysis of alkaline, surfactant, and polymer in oil-water interface. Processes **8**(7), 762 (2020)
21. Kumar, N., Mandal, A.: Surfactant stabilized oil-in-water nanoemulsion: stability, interfacial tension, and rheology study for enhanced oil recovery application. Energy Fuels **32**, 6452–6466 (2018)
22. Rosen, M.J., Wang, H., Shen, P., Zhu, Y.: Ultralow interfacial tension for enhanced oil recovery at very low surfactant concentrations. Langmuir **21**, 3749–3756 (2005)
23. Lee, K.S., Lee, J.H.: Hybrid chemical EOR using low-salinity and smart waterflood. In: Hybrid Enhanced Oil Recovery using Smart Waterflooding. Hanyang University, Seoul, South Korea, Gulf Professional Publishing, pp. 65–110 (2019)
24. Negin, C., Ali, S., Xie, Q.: Most common surfactants employed in chemical enhanced oil recovery. Petroleum **3**(2), 197–211 (2017)
25. Zhang, F., Ma, D., Wang, Q., Zhu, Y., Luo, W.: A novel hydroxypropyl sulfobetaine surfactant for high temperature and high salinity reservoir. In: International Petroleum Technology Conference, Beijing, China (2013)
26. Wu, Z., Yue, X., Cheng, T., Yu, J., Yang, H.: Effect of viscosity and interfacial tension of surfactant-polymer flooding on oil recovery in high-temperature and high-salinity reservoirs. J. Pet. Explor. Prod. Technol. **4**, 9–16 (2014)
27. Kong, L., Saar, K.L., Jacquat, R., Hong, L., Levin, A., Gang, H., Ye, R., Mu, B., Knowles, T.P.J.: Mechanism of biosurfactant adsorption to oil/water interfaces from millisecond scale tensiometry measurements. Interface Focus **7**(6) (2017)

Free Energy Calculation of CO_2 in Cholinium-Based Amino Acid Ionic Liquids

Fitri Norizatie Salehin, Khairulazhar Jumbri, Anita Ramli, Mohd Azlan Kassim, and Mohd Faisal Taha

Abstract Solvation free energy of CO_2 in six cholinium based ionic liquids (ILs) was predicted by using Bennet Acceptance Ratio (BAR) method employed in molecular dynamics (MD) simulation. From solvation free energy, the degree removal of acid gases can be measured by gas solubility in liquids and quantified by excess chemical potential (μ^{ex}) and Henry's law constant (k_{H}) . In this work, the predicted solvation free energies of CO_2 in six [Chl][AA] ILs showed the highest value of k_H was 78.89 atm obtained by [Chl][SRI] and the lowest value gained by [Chl][MEH] was 45.35 atm. Radial distribution function (RDF) for specific atom of CO_2 and pure ILs was evaluated to assess our understanding on the interaction between the gas and the ILs. There were several characteristics that affected the solubility of CO_2 in these six cholinium-based ILs such as the type of amines, functional group, free volume of ILs, and length of alkyl chain of anions.

Keywords Carbon dioxide · Ionic liquids · Molecular dynamics (MD) simulation · Solubility · Henry's law constant

F. N. Salehin · K. Jumbri · A. Ramli · M. F. Taha (✉)
Department of Fundamental and Applied Sciences, Universiti Teknologi PETRONAS, 32610 Seri Iskandar, Perak, Malaysia
e-mail: faisalt@utp.edu.my

F. N. Salehin
e-mail: fitri_17008492@utp.edu.my

K. Jumbri
e-mail: khairulazhar.jumbri@utp.edu.my

A. Ramli
e-mail: Anita_ramli@utp.edu.my

K. Jumbri
Centre of Research in Ionic Liquids (CORIL), Universiti Teknologi PETRONAS, 32610 Seri Iskandar, Perak, Malaysia

A. Ramli · M. A. Kassim
School of Science and Technology, Sunway University, 47500 Bandar Sunway, Selangor, Malaysia
e-mail: azlanka@sunway.edu.my

S. A. Abdul Karim et al. (eds.), *Proceedings of the 6th International Conference on Fundamental and Applied Sciences*, Springer Proceedings in Complexity,
https://doi.org/10.1007/978-981-16-4513-6_14

1 Introduction

CO_2 is a colorless and odorless gas that exist in the natural sources such as volcanos, petroleum and natural gas. CO_2 also can be found in industrial like petrochemicals, chemical manufacturing plants, coal fired power plants, and cement manufacturing [1]. The increasing of CO_2 emission in the atmosphere had caused some serious environmental problems such as greenhouse effect, global warming, ocean acidification and others. Consequently, many technologies were developed to reduce and control the CO_2 emissions. The current technologies that are being used to reduce CO_2 emissions are carbon capture and storage, adsorption, absorption, scrubbing, and cryogenic separation. The option of methods generally based on the process which is pre-combustion, post combustion or oxy fuel combustion [2]. The most technology that is being used for acid gases removal such as CO_2 and H_2S is amine gas treating. In this process, several of amine-based solvents are used, for example diethanolamine (DEA), monoethanolamine (MEA), and methyldiethanolamine (MDEA), to remove the acid gases [3]. Even though these aqueous amine solutions are useful, but this process still suffers serious drawbacks like high energy consumption, costly, solvent loss, corrosion, and degradation by oxidation [4]. Therefore, to solve these problems, the suitable solvents with an outstanding performance were developed such as ionic liquids (ILs). ILs are molten organic salts that consists of cation and anion with the melting point below 100 °C. The use of ILs as solvents for acid gas removal receives great attention from researchers due to its unique properties such as low vapor pressure, ability to dissolve a variety of chemical compounds, and high thermal and chemical stability [5]. Furthermore, the physical and chemical properties of ILs also can be tuned by altered the structure of cations or anions [6]. These unique characteristics of ILs make them suitable for wide range of applications in energy management, gas separation, catalysis and others [7, 8].

For the past few decades, it was reported that the solubility of acid gases in ILs was rapidly increased. In 2001, the first solubility of CO_2 in [bmim][PF_6] at temperature of 313 K was investigated by Blanchard et al. [9]. Their findings showed that the solubility of CO_2 was about 0.6 mol fraction at ambient pressure up to 8 MPa. However, when the pressure was below ambient pressure, the highest solubility of CO_2 only 0.035 mol fraction. The usage of conventional ILs in industrial to absorb CO_2 can only be applied at high pressure (89 atm or higher) and it would take a longer time to achieve equilibrium [10]. Hence, functionalized ILs was designed to enhance CO_2 capture by incorporating imidazolium cation with amine [11]. Primary amine was used in amine functionalized ILs for absorption of CO_2 and the highest molar uptake of CO_2 per mole of [pNH_2bim][PF_6] at normal pressure and temperature was found to be 0.5 mol fraction. Due to these findings, many of amino acid based ILs have been developed for acid gases removal [12, 13]. In addition, some other researchers have also studied CO_2 solubility or absorption through computer simulation. In 2010, Gurkan et al., studied the experimental and theoretical absorption of CO_2 under 298 K and the pressure from 1 to 3 atm [14]. The results showed that by using amine tethered at the anions component compared to the cations, the equimolar uptake can be simply achieved which in agreement with other previous study [15]. Besides, they also proposed to eliminate the viscosity issues by synthesizing ILs with aprotic heterocyclic anions (AHA). These AHA ILs have provided improvement in amino ILs structural properties that led to lower viscosity of ILs. Other than that, the predicted solubility of CO_2 in terms of Henry's constant in [bmim][PF_6] were calculated using test particle insertion method [16]. The results showed underestimation of Henry's constant due to overestimation of the interaction strength between solute and solvent. These findings indicated that interaction of anions with solute was one of the factors that can affect the solubility of gas compared to cations [16].

Even though ILs can act as a favourable choice for acid gas capture but some of the ILs are not environmentally friendly since some of them may contain long alkyl chains that consequently increase their toxicities such as imidazolium, pyridinium, pyrrolidinium and morpholinium based ILs [17]. Hence, to solve this problem, the cation should be based on the natural derivatives, low toxicity, and biocompatibility such as cholinium based ionic liquids. Cholinium is a *N,N,N*-trimethylethanolammonium cation which is quaternary ammonium salt and this cation exhibits low toxicity due to the presence hydroxyl group. Figure 1 shows the structure of cholinium cation in 3D.

Cholinium based ILs are able to dissolve almost all insoluble materials due to their high ability to form hydrogen bond. Therefore, cholinium based ILs are widely being used in the industrial applications [18]. Besides, Yuan et al. successfully demonstrated the ability of aqueous cholinium-based ILs as CO_2 absorbents with the absorption capacity of 1.454 mol CO_2/kg absorbent at 308.2 K [19]. Furthermore, other functionalized cholinium based amino acid ILs exhibited an excellent CO_2 capture capacity ranging from 1.62 to 4.31 mol of CO_2/kg and the mechanism of reaction was supported by DFT calculation [20]. Hence, to contemplate the low toxic and environmentally friendly of the cholinium cation, it is important to incorporate with the competent anion pair in order to develop more favorable absorbent such as amino acids anions. According to Aparicio et al., choline that tethered with amino acid anions can be used in acidic gas separation from biogas. This is due to the strong polar-polar interaction in the protein that can contribute to gas capture [18].

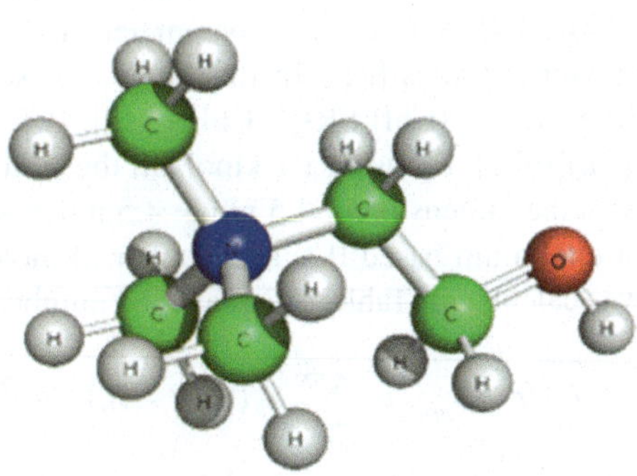

Fig. 1 Structure of cholinium cation [Chl]$^+$

It is important to understand the microscopic and structure characteristics of ILs in designing ILs for specific purposes. Although, the study of dynamic properties of ILs through computer simulation is increasing, but there have been little reported about the theoretical study on functionalized ILs especially amino acid ILs in molecular dynamics simulation viaOPLS force field. This project was focused on the analysis of the solubility of CO_2 in cholinium based amino acid, [Chl][AA], ILs. In this piece of work, cholinium based ILs were selected to be paired with six amino acids anions and the density of these ILs were predicted at 298 K and 1 bar. The solubility and thermodynamic properties such as excess chemical potential (μ^{ex}), Henry's law constant (k_H), enthalpy (h^{ex}), entropy (s^{ex}) and RDF were also calculated. The predicted results were compared with the available experimental data from other literature studies.

2 Computational Details

2.1 *Structure Optimization*

The geometry structures of solute and solvents were optimized using density functional theory (DFT) performed with the basic set B3-LYP in Turbomole (TmoleX) [20].

2.2 *Force Fields Validation*

In computer simulation, it is important to validate the force fields to obtain the accurate results. The ILs topology was prompted using antechamber and then was converted to Groningen Machine for Chemical Simulation (GROMACS) topology using Antechamber Python Parser interface (ACPYPE) Phython script. In this work, force fields used was Optimized Potentials for Liquids Simulations (OPLS) and Eq. 1 shows the summaries of the OPLS force field's equation.

The OPLS force field parameters for six [Chl][AA] ILs were obtained from previous studies [21]. In this study, the selected six [Chl][AA] ILs, as shown in Table 1, are [Chl][SRI], [Chl][GLZ], [Chl][MEH], [Chl][PHY], [Chl][ALN] and [Chl][PRL]. By using Packmol, all the simulations were simulated in the cubic box with the dimension of 4.5 nm × 4.5 nm × 4.5 nm [22]. Based on the actual density of cholinium based ILs, the number of molecules that can fit in the simulation box was calculated. Table 1 displays the number of molecules in the simulation box.

$$U\left(\left|r_{ij}\right|\right)_{total} = \sum_{bonds} k_b\left(\left|r_{ij}\right| - r_0\right)^2 + \sum_{angles} k_\theta\left(\theta_{ijk} - \theta_0\right)^2$$

Table 1 Number of molecules used in the simulation box for MD simulation

Entry	Cholinium based amino acids ILs	Number of molecules	
		Cation	Anion
1	[Chl][SRI]	316	316
2	[Chl][GLZ]	352	352
3	[Chl][MEH]	249	249
4	[Chl][PHY]	234	234
5	[Chl][ALN]	318	318
6	[Chl][PRL]	282	282

$$+\sum_{dihedrals}\sum_{n=1}^{N}k_x[1+\cos(nx_{ijkl}-\delta_n)$$
$$+\sum_{impropers}k_\varphi(\varphi-\varphi_0)^2$$
$$+\sum_{i=j}^{N-1}\sum_{j>1}^{N}\left\{4\varepsilon_{ij}\left(\left(\frac{\sigma_{ij}}{|r_{ij}|}\right)^{12}-\left(\frac{\sigma_{ij}}{|r_{ij}|}\right)^{6}\right)+\frac{1}{4\pi\varepsilon_0^2}\frac{q_iq_j}{|r_{ij}|}\right\} \quad (1)$$

The number of steps used was 2.0 fs with PBC for all directions (x, y, z). Lennard Jones and Coulomb (non-bonded) interactions was calculated up to 1.2 nm and the electrostatic interactions was calculated using PME [23, 24] with a grid spacing of 0.12 nm and fourth order interpolation. In every 5 steps, the neighbor searching of 1.2 nm was updated. Pressure and temperature were controlled by using Berendsen barostat and Berendsen thermostat. The bond lengths were constrained with LINCS [25]. Two heat baths were supplied with heat and the temperature coupling constant was 0.1 ps. 1.0 bar of pressure and 2.0 ps of relaxation time were applied. The isothermal compressibility for pressure control was set as 4.5×10^{-5} bar^{-1}. Every 500 steps the energies, forces, velocities, and trajectories were recorded.

Each system underwent energy minimization using steepest descent then followed by conjugate gradient with 5000 steps each. After minimization, the simulations were then subjected to pre-equilibrated canonical ensemble (NVT) with 5 ns implemented to control the temperature. The equilibration process was continued with the isothermal-isobaric ensemble (NPT) with 50 ns time run. Both simulations of NVT and NPT were carried out with different temperatures starting from 293 to 363 K at 1 bar.

2.3 Solvation Free Energy

Solubility of CO_2 can be predicted by using solvation free energy calculation. Solvation free energy of CO_2 in six [Chl][AA] ILs was calculated theoretically using Bennett Acceptance Ratio (BAR) method. Bennett equation is shown in Eq. 2.

$$\Delta G_{ji}^{BAR} = k_B T \left(\text{In} \frac{f(H_i - H_j + C)j}{f(H_j - H_i - C)i} \right) + C \tag{2}$$

According to Bennett [26], Eq. 2 is the initial steps to evolve free energy calculation i.e. by knowing how many lambda, λ, points needed to explain the free energy difference between states i and j. k_B is the Boltzmann constant, T is temperature while for H_i and H_j represent the Hamiltonians at states i and j, respectively. Value of C is determined repetitively to get $f(H_i - H_j + C)j = f(H_j - H_i + C)i$ where f is the Fermi function (Eq. 3) [27].

$$f(x) = \frac{1}{1 + \exp\left(\frac{x}{k_B T}\right)} \tag{3}$$

The free energy deviation between the states is acquired from Eqs. 4 and 5.

$$\Delta G_{ji}^{BAR} = -k_B T In \frac{N_j}{N_i} + C \tag{4}$$

$$\Delta G_{ji}^{BAR} = \sum_{i=1}^{n-1} \Delta G_{i+1,i}^{BAR} \tag{5}$$

where N_i and N_j represent the number of coordinates frames at λ_i and λ_j. Generally, if there is sufficient overlap between the energy differences in forward and backward state, the convergence of this operation can be fulfilled. Besides, this part is required for BAR method. These can be noticed by the overlap integral that can be obtained in normalized histograms of energy differences [28]. According to our previous work, in order to obtain valid results, the integral should be at least 0.01. However, if there is an insufficient overlap, the iterative procedure does not converge nor produce invalid results [26].

$$\Delta U_l^{FW} = U(\lambda_{l+1}) - U(\lambda_l)|_{\lambda_i} \tag{6}$$

$$\Delta U_{l+1}^{BW} = U(\lambda_l) - U(\lambda_{l+1})|_{\lambda_i} \tag{7}$$

The λ symbol in Eqs. 6 and 7 shows the λ-value where the trajectory was prompted. When the histograms of U_l^{FW} and U_{l+1}^{BW} are overlaying, the overlap is not easily spotted because the exact amount of the overlapping is difficult to estimate by visual

Table 2 Number of molecules used in simulation box for solvation free energy

Entry	[Chl][AA] ILs	Number of molecules	
		Cation	Anion
1	TIP4P	907	
2	[Chl][SRI]	92	92
3	[Chl][GLZ]	104	104
4	[Chl][MEH]	73	73
5	[Chl][PHY]	69	69
6	[Chl][ALN]	94	94
7	[Chl][PRL]	83	83

inspection. Thus, by using the overlap integral $OI_{l,l+1}$ as numerical guide, it can enhance the existing free energy. The interval λ_l, λ_{l+1} is defined as in Eq. 8.

$$OI_{l,l+1}^{\mathrm{BAR}} = 2.\sum_{b=1}^{B} \frac{h_b\left(\Delta U_l^{\mathrm{FW}}\right).h_b\left(\Delta U_{l+1}^{\mathrm{BW}}\right)}{h_b\left(\Delta U_l^{\mathrm{FW}}\right) + h_b\left(\Delta U_{l+1}^{\mathrm{BW}}\right)} \tag{8}$$

where $h_b(X)$ is a normalized histogram of quantity X with a total number of B bins. According to the definition, the configurations in the overlap zone will displays non-zero contributions to $OI_{l,l+1}^{\mathrm{BAR}}$. This means that 0 indicates no overlapping while 1 indicates perfect overlapping. In every system, CO_2 molecules was placed at the centre of a 3.0 nm × 3.0 nm × 3.0 nm in the simulation box and solvated in ILs and aqueous system using the TIP4P water model. Table 2 shows the listed number of molecules and [Chl][AA] ILs used in the simulation box for solvation free energy.

The solvation free energy of CO_2 was conducted by using the same conditions and parameters as in pure ILs except for the additional free energy parameters in this simulation. This additional is 21 equidistant sets of λ, i.e. from $\lambda = 0$ to $\lambda = 1$, with 5 ns simulation time for each λ. Meanwhile, the soft-core energy parameters were set to αLJ = 0.5, λ power reliance was set to 2 and the soft-core sigma was = 0.3. The post-processing of BAR which are coordinates, $\delta H/\delta\lambda$, and velocities were saved to disk at every 5 steps. Every node needs to undergo energy minimization which were 5000 steps of steepest descent and accompanied with 5000 steps of conjugate gradient. Then, the position restrains was applied in the system, and the system required to pre-equilibrate in in canonical ensemble (NVT) for 100 ps [29].

Afterwards, the system was continued for 2 ns in isobaric-isothermal (NPT) at 298 K and the data was recorded at the last 1 ns. For the calculation of free energy with BAR, it is essential to gain the post-processing data. Based on Eq. 2, it is important to calculate the energy difference between the system at H_i and H_j over the trajectories simulated at H_i and H_j separately. Hence, bonded, non-bonded and kinetic energies required a re-calculation for the post-processing steps. For this study, OPLS force field was used to implement the MD simulations and BAR calculations. For successful treatment for all atoms in the system, the GROMACS packages version 5.1

was applied. This is because this system provided highly optimized and parallelized algorithms [29].

After simulations were accomplished, the free energy deviation was gained from the output data using GROMACS BAR tool to obtain the solvation free energy. The solubility of CO_2 in liquids solvents can be gained by MD simulation by determining the excess chemical potential (μ^{ex}) and Henry's law constant (k_H). Basically, Henry's law constant is implemented to estimate gas solubility in a liquid at definite dilution. The solubility of gas is greater when the value of Henry's law constant is the lower. In addition, Henry's constant can be attained from excess chemical potential as shown in Eq. 9.

$$k_H = \frac{\mathrm{RT}\rho}{M}\exp\left(\frac{\mu^{ex}}{\mathrm{RT}}\right) \tag{9}$$

where μ^{ex} is excess chemical potential, R is gas constant, T represents temperature (K), ρ and M are the density and molecular weight of pure ILs, respectively. In order to calculate the excess properties such as enthalpy and entropy (h^{ex} and s^{ex}), numerical temperature derivatives were carried out for the excess chemical potential based on Eqs. 10 and 11 [30].

$$s^{ex} = -\left[\frac{\partial\mu^{ex}}{\partial \mathrm{T}}\right] \tag{10}$$

$$h^{ex} = \mu^{ex} + s^{ex} \tag{11}$$

3 Results and Discussion

3.1 Solvation Free Energy of CO_2 in Water

Solvation free energy also can be used for the validation of the OPLS force fields, particularly to determine the possibility of empirical parameters for instance partial atomic charges [31]. Water is one of the conventional solvents that has been used to evaluate the dependability of the physical model in estimating the solvation free energy due to its significant biological relevance and accessibility of experimental results [32]. The solubility of CO_2 in water was predicted by using the same force field used in our previous paper [21]. CO_2 was contemplated as most soluble when the value of μ^{ex} is in negative value compared when the value of μ^{ex} is in positive value. To attain the reliable results, the predicted solvation free energy obtained in this work was compared with other literature studies using different force fields and methods. According to Table 3, Entry 1 to 3 used the same water model, i.e. TIP4P, and the values of k_H for Entry 1, 2, and 3 were 2446.91 atm, 2388.35 atm and

Table 3 Excess chemical potential (μ^{ex}) and Henry's law constant (k_H) for CO_2 in water at 298 K and 1.0 bar

Entry	Method	μ^{ex} simulation (kJ/mol)	μ^{ex} experiment (kJ/mol)	(k_H) simulation (atm)	(k_H) experiment (atm)	Refs.
1	BAR	1.49	–	2446.91	–	This work
2	Widom	1.48	–	2388.35	–	[33]
3	FEP	1.28	–	–	–	[33]
4	Experiment	–	1.24	–	2358.75	[34]
5	Experiment	–	1.00	–	2011.90	[35]

2187.02 atm, respectively [33]. There was slightly difference in the k_H value which was possibly due to the different conditions and methods performed in simulation. The solubilities of CO_2 in Entry 2 and 3 were estimated at 298 K and the pressure of 1.0 bar. Meanwhile, for Entry 4 and 5, the solubilities of CO_2 were obtained from experimental studies with the k_H values of 2358.75 atm and 2011.90 atm, respectively [34]. Overall, based on the data in Table 3, it can be concluded that the k_H value of CO_2 in this work was the highest compared to other studies indicating that CO_2 was least soluble in water.

3.2 *Solvation of Free Energy of CO_2 in [Chl][AA] ILs*

Shivakumar et al. stated that solvation free energy is a significant indicator in validating the solvent models and force fields [36]. Besides, it is also vital to use other than water as the solvents to ascertain the accuracy of the solvation free energy of CO_2 [37]. The predicted solvation free energies of CO_2 in [Chl][AA] ILs are presented in Table 4.

Based on data in Table 4, there was a large difference in terms of k_H for CO_2 in water and in ILs systems. The k_H value of CO_2 in water was higher compared to CO_2 in [Chl][AA] ILs indicating that the CO_2 was more soluble in [Chl][AA] ILs compared to in aqueous system. Solubility of CO_2 can be expressed in terms of excess chemical potential (μ^{ex}) and Henry's law constant values (k_H) [38]. In this study, the pressure range was not high enough to affect Henry's law constant to be a strong function of pressure. Besides, under specific conditions, Henry's law constant is less dependent on the pressure [39]. CO_2 was least soluble in [Chl][SRI] showed by the smallest value of μ^{ex} and the highest value of k_H. Meanwhile, CO_2 was most soluble in [Chl][MEH] showed by the highest value of μ^{ex} and the smallest value of k_H. In general, the estimated free energies of CO_2 in [Chl][AA] ILs were found to be in the range of 45.35–78.89 atm which was close to the values of the previous studies [40, 41].

Table 4 Excess chemical potential (μ^{ex}), and Henry's law constant (k_H) of CO_2 in six [CHL][AA] ILs at 298 K and 1.0 bar

System	Excess chemical potential, μ^{ex} (kJ/mol)	Henry's law constant, k_H (atm)	Enthalpy, h^{ex} (kJ/mol)	Entropy, s^{ex}($\times 10^{-3}$ kJ/K mol)
CO_2-water system	1.49 ± 0.00	2446.91	−2.28	−1.54
[Chl][MEH]	−2.19 ± 0.05	45.35	−5.90	−12.46
[Chl][PHY]	−1.36 ± 0.03	58.83	−0.21	3.85
[Chl][PRL]	−1.82 ± 001	59.67	−3.70	−6.31
[Chl][ALN]	−2.02 ± 0.03	62.66	−3.81	−6.0
[Chl][GLZ]	−1.88 ± 0.09	73.56	−2.84	−3.23
[Chl][SRI]	−1.39 ± 0.02	78.89	−0.43	3.23

The solubility of CO_2 can be affected by several factors such as type of functional group, type of amine, free volume effects and length of alkyl chain of anions. In this study, based on Henry's law constant, the solubility of CO_2 in [Chl][AA] ILs with different anions increased in the following order: [Chl][SRI] < [Chl][GLZ] < [Chl][ALN] < [Chl][PRL] < [Chl][PHY] < [Chl][MEH]. The existence of (OH–) group in $[SRI]^-$ anion created stronger hydrogen bonding interaction of cation and anion which in turn decreased the free volume of ILs resulting in low solubility of CO_2 [42]. CO_2 solubility in [Chl][MEH], [Chl][PHY], [Chl][GLZ], and [Chl][ALN] was affected by the length of alkyl chain of anions. Ma et al. stated that by decreasing the length of alkyl chain of anions, the molar density of ILs increased, hence decreasing the free volume of ILs [10]. Therefore, it is difficult for CO_2 to accommodate the void space in the ILs which in turn reduces the solubility of CO_2. This could explain the lower solubility of CO_2 in [Chl][GLZ] and [Chl][ALN] compared to [Chl][PHY] and [Chl][MEH] as $[GLZ]^-$ and $[ALN]^-$ have shorter alkyl chain length than $[PHY]^-$ and $[MEH]^-$.

According to Xing et al., aminate-anion ILs were more dominant in dissolving CO_2 compared to aminate-cation ILs [43, 44]. Aminate-cation ILs have stronger interaction between anion and cation which makes amine group less accessible for CO_2 interactions resulting in the decrease of CO_2 solubility [44]. Although [Chl][PRL] has a long alkyl chain length due to its pentagon ring, the solubility of CO_2 in [Chl][PRL] was found to be lower than in [Chl][PHY] and [Chl][MEH]. This could be due to prolinate anion ($[PRL]^-$), that caused the steric hindrance effect resulting in the difficulty of CO_2 to access and interact with nitrogen atom in $[PRL]^-$ [43].

The solvation of enthalpy is the measurement for the strength of interaction between the ILs and solute. Meanwhile, the solvation of entropy is used to determine the degree of randomness or ordering in the solution which correlate with gas dissolution [45]. Table 4 displays the estimated solvation entropy and enthalpy for CO_2-ILs systems at 1.0 bar and 298 K. In general, low value of h^{ex} (less negative value) indicates weak interaction between gas and ILs, while high value of s^{ex} (less negative value) indicates high degree of ordering. All the predicted h^{ex} values were low which was in good agreement with the literature [46]. Solvation enthalpy of CO_2 increased in following order: [Chl][PHY] < [Chl][SRI] < [Chl][GLZ] < [Chl][ALN] < [Chl][PRL] < [Chl][MEH]. It can be concluded that CO_2 has stronger interaction with $[Chl]^+$ and $[MEH]^-$ compared to other ILs.

The solvation entropy of CO_2 in cholinium based ILs decreased in the following order: [Chl][PHY] > [Chl][SRI] > [Chl][GLZ] > [Chl][ALN] > [Chl][PRL] > [Chl][MEH]. This can be elucidated that [Chl][MEH] has the lowest of randomness due to strong interaction between solute and solvent resulting in higher solubility of CO_2 in [Chl][MEH]. Besides, the negative value of s^{ex} was due to finite space arrangement for solute in solvent which creates strong hydrogen bonding interaction between the carbonyl or hydroxyl group in IL with CO_2 [38]. From our previous work, it was found that solvation of enthalpy displayed more dominant effect in term of gas solubility compared to the solvation of entropy [21].

3.3 Radial Distribution Function (RDF)

Radial distribution function (RDF) is a qualitative analysis used to measure the probability of a pair of atoms or molecules at a certain distance, *r*, with the same density and same probability in random distribution function as shown in Eq. 12.

$$RDFg(r) = \frac{\rho(r)}{\langle\rho\rangle} \tag{12}$$

Figure 2a, b illustrate the RDF graphs for CO_2-cation, CO_2-anion and cation–anion. The solvation shell of CO_2-cation for [Chl][SRI] and [Chl][MEH] was spotted at the same distance around 0.50 nm. For cation–anion peaks, [Chl][SRI] and [Chl][MEH] were located at different distance which were about 0.51 nm and 0.59 nm, respectively. Although [Chl][MEH] has longer distance than [Chl][SRI], but [Chl][MEH] had broader peak than [Chl][SRI]. This can be explained in term of length of alkyl chain of anions and interaction energy between the ILs and gas. Methionate anion ($[MEH]^-$) has longer alkyl chain than serinate anion ($[SRI]^-$) which led to decrease in molar density of ILs. Safavi et al. mentioned that molar density of ILs was directly proportional to the interaction strength between cation and anion. As the molar density of ILs decreases, the interaction between anion and cation become weaker thus increases the free volume of ILs resulting in higher solubility of CO_2 [47]. The CO_2-anion solvation peaks of CO_2-MEH and CO_2-SRI were discovered at ~0.45 nm and ~0.43 nm, respectively. Even though, CO_2-MEH has slightly longer distance than CO_2-SRI, but CO_2-SRI had broader peak compared to CO_2-MEH. This indicates that CO_2 has weak interaction with $[SRI]^-$ causing low CO_2 solubility which was in good agreement with the predicted h^{ex}.

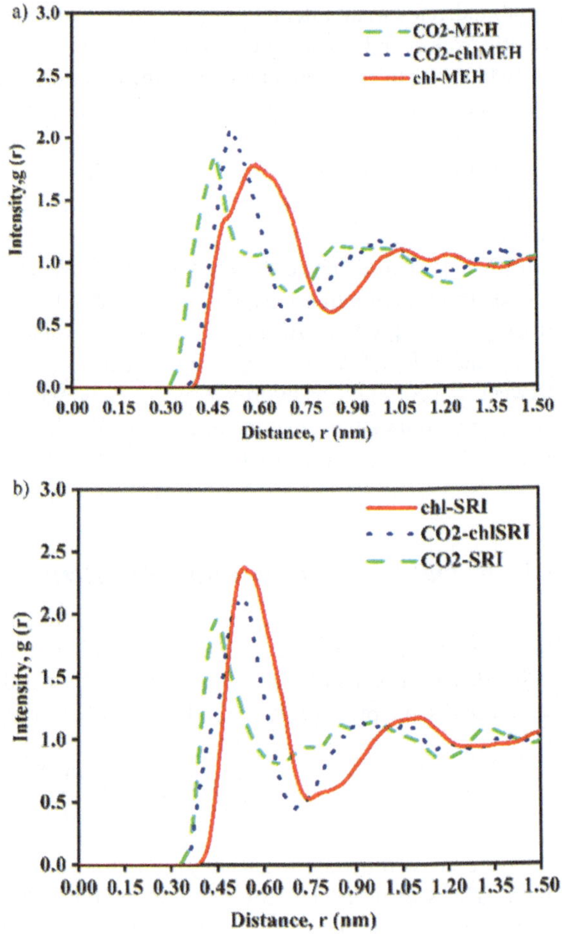

Fig. 2 Centre of mass RDF for the cation–anion, cation-CO_2 and anion-CO_2 interactions in **a** [Chl][MEH] and **b** [Chl][SRI]

Figure 3 exhibits the RDF graphs of CO_2 solvated in [Chl][AA] ILs. Figure 3a represents the CO_2-anions RDF using hydrogen or oxygen atom for anions and carbon atom for CO_2. The highest peaks of C-MEH and C-PHY were found to be located at near ~0.36 nm and these peaks decreased to the distance at about ~0.35 nm for C-GLZ, C-ALN and C-PRL, whereas, for C-GLZ the peak was discovered at around ~0.34 nm. These findings showed great correspondence to the work carried out by Harun et al. [48]. Meanwhile, Fig. 3b depicts the RDF graphs for anions and oxygen atoms in CO_2. The O-MEH, O-ALN, O-PRL, and O-PHY showed the peaks located at ~0.33 nm while the O-GLZ and O-SRI spotted the peak around ~0.32 nm. Based on Fig. 3a, b, it can be said that oxygen from CO_2 is more oriented towards anions compared to carbon from CO_2.

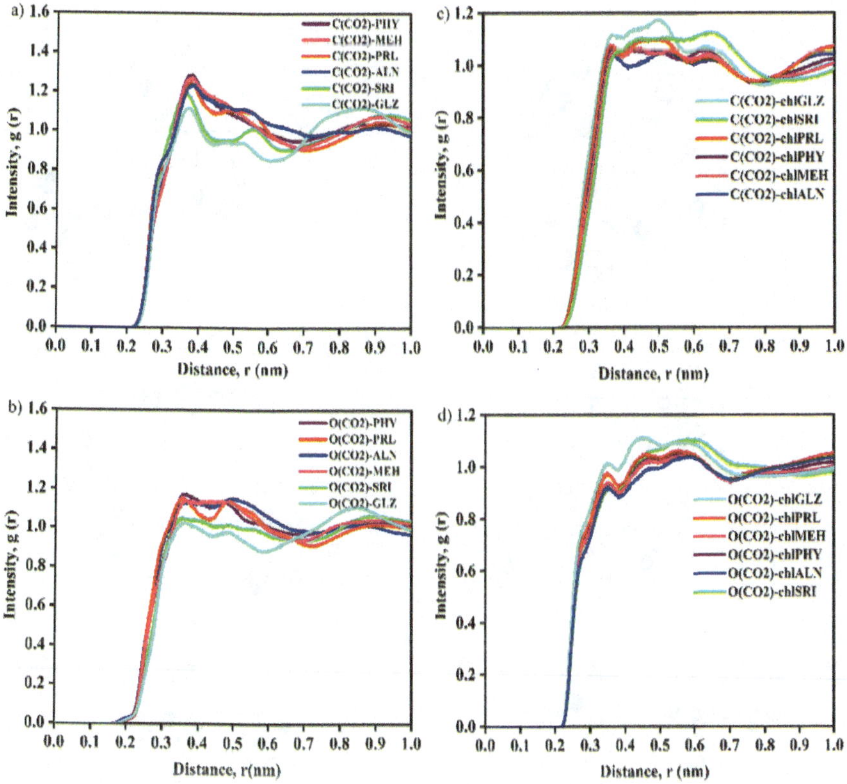

Fig. 3 Atom–atom pair radial distribution function (RDF) for CO_2 in ILs. **a** Carbon atom of CO_2 with anions, **b** Oxygen atom of CO_2 with anions, **c** Carbon atom of CO_2 with hydrogen atom as the centre of cations and **d** Oxygen atom of CO_2 with hydrogen atom as a centre of cations

Figure 3c, d display the RDF of CO_2-cations by using oxygen and carbon from CO_2 and hydrogen atom as the centre of cation. It can be observed that C(CO_2)-cations have longer distance at ~0.37 nm than O(CO_2)-cations at distance ~0.35 nm. From Fig. 3, generally, it can be concluded that oxygen atoms from CO_2 were more oriented towards cations and anion compared to carbon atom from CO_2 and these results are in good agreement with previous work [35]. Furthermore, it also indicates that CO_2 well interacted with the ILs due to intermolecular interaction [48]. This would contribute to high CO_2 solubility in [Chl][AA] ILs. Figure 4 exemplifies the interaction of CO_2 molecules with cations and anions in MD simulation.

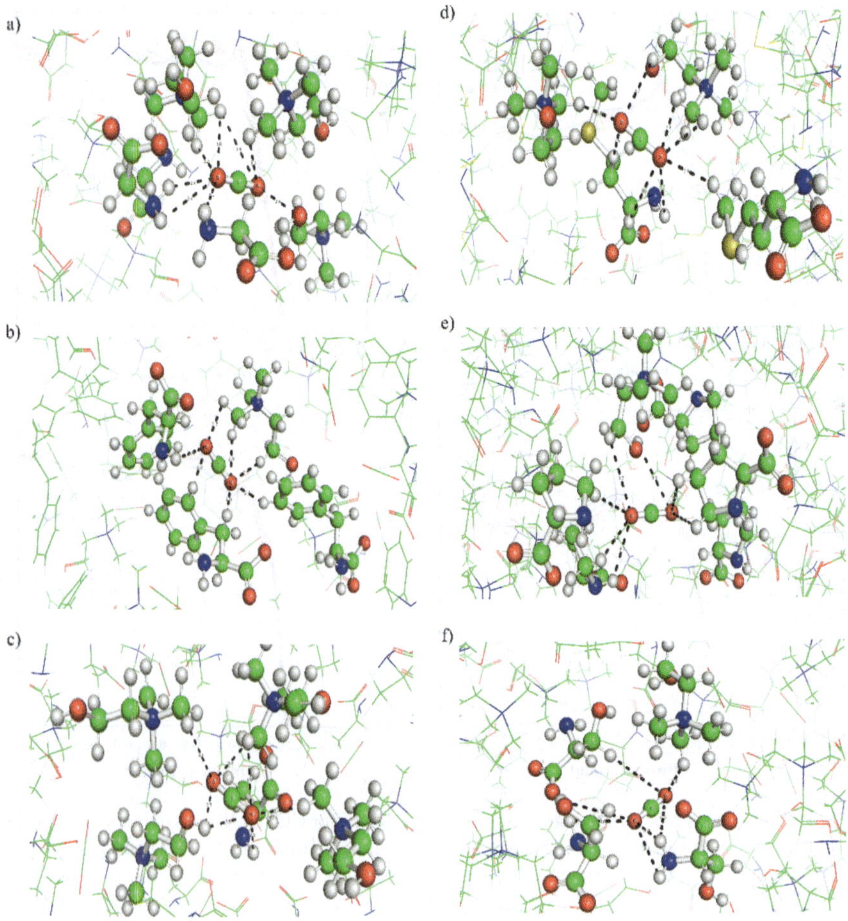

Fig. 4 MD snapshots of CO_2 and [Chl][AA] ILs. **a** CO_2-[Chl][GLZ], **b** CO_2-[Chl][PHY], **c** CO_2-[Chl][ALN], **d** CO_2-[Chl][MEH], **e** CO_2-[Chl][PRL] and **f** CO_2-[Chl][SRI]

4 Conclusion

In conclusion, the solvation free energies of CO_2 in six [Chl][AA] ILs were successfully reproduced using the BAR method with the values of Henry's law constant (k_H) ranging from 45.35 to 78.89 atm, at 298 K and 1.0 bar. The lower the value of k_H, the greater the solubility of CO_2 in the ILs. CO_2 showed the highest solubility in [Chl][MEH] and was least soluble in [Chl][SRI]. From the radial distribution function (RDF) results, it showed that oxygen atom was more oriented towards anions and cations compared to the carbon atom from CO_2. The longer alkyl chain of anions reduced the interactions between cation and anion, thus provided more space (free volume) for CO_2 to occupy.

Acknowledgements All authors gratefully acknowledge the financial support by Yayasan Universiti Teknologi PETRONAS Grant (015LC0-074).

References

1. Yang, Q., Wang, Z., Bao, Z., Zhang, Xing, H., Dai, S.: New insights into CO_2 absorption mechanisms with amino-acid ionic liquids. ChemSusChem **9**(8), 806–812 (2016)
2. Li, X., Hou, M., Zhang, Z., Wang, X., Zou, L.: Absorption of CO_2 by ionic liquid/polyethylene glycol mixture and the thermodynamic parameters. Green Chem. **10**(8), 879–888 (2008)
3. Lv, B., Jing, G., Qian, Y., Zhou, Z.: An efficient absorbent of amine-based amino acid-functionalized ionic liquids for CO_2 capture: high capacity and regeneration ability. Chem. Eng. J. **289**, 212–218 (2016)
4. Silveira, A.J., Pereda, S., Tavares, F.W., Abreu, C.R.A.: A molecular dynamics study of the solvation of carbon dioxide and other compounds in the ionic liquids [emim][B(CN)$_4$] and [emim][NTf$_2$]. Fluid Phase Equilib. **491**, 1–11 (2019)
5. Mattedi, S., Carvalho, P.J., Coutinho, J.A.P., Alvarez, V.H., Iglesias, M.: High pressure CO_2 solubility in N-methyl-2-hydroxyethylammonium protic ionic liquids. J. Supercrit. Fluids **56**(3), 224–230 (2011)
6. Zhang, J., Sun, J., Zhang, X., Zhao, Y., Zhang, S.: The recent development of CO_2 fixation and conversion by ionic liquid. Greenhouse Gases Sci. Technol. **1**(2), 142–159 (2011)
7. Wishart, J.F.: Energy applications of ionic liquids. Energy Environ. Sci. **2**(9), 956–961 (2009)
8. Han, D., Row, K.H.: Recent applications of ionic liquids in separation technology. Molecules **15**(4), 2405–2426 (2010)
9. Blanchard, L.A., Gu, Z., Brennecke, J.F.: High-pressure phase behavior of ionic liquid/CO_2 systems. J. Phys. Chem. B **105**(12), 2437–2444 (2001)
10. Ma, J.-W., Zhou, Z., Zhang, Zhang, Z.-B., Li, A.-M.: Ditetraalkylammonium amino acid ionic liquids as CO_2 absorbents of high capacity. Environ. Sci. Technol. **45**(24), 10627–10633 (2011)
11. Bates, E.D., Mayton, R.D., Ntai, I., Davis, J.H., Jr.: CO_2 capture by a task-specific ionic liquid. J. Am. Chem. Soc. **124**(6), 926–927 (2002)
12. Feng, Z., Cheng-Gang, F., You-Ting, Ai-Min, L., Zhi-Bing, Z.: Absorption of CO_2 in the aqueous solutions of functionalized ionic liquids and MDEA. Chem. Eng. J. **160**(2), 691–697 (2010)
13. Xue, Z., Zhang, Z., Han, J., Chen, Y., Mu, T.: Carbon dioxide capture by a dual amino ionic liquid with amino-functionalized imidazolium cation and taurine anion. Int. J. Greenhouse Gas Control **5**(4), 628–633 (2011)
14. Gurkan, B.E., De La Fuente, J.C., Mindrup, E.M., Schneider, W.F., Brennecke, J.F.: Equimolar CO_2 absorption by anion-functionalized ionic liquids. J. Am. Chem. Soc. **132**(7), 2116–2117 (2010)
15. Zhang, Y., Zhang, S., Lu, X., Fan, W., Zhang, X.: Dual amino-functionalised phosphonium ionic liquids for CO_2 capture. Chem. Eur. J. **15**(12), 3003–3011 (2009)
16. Shah, J.K., Maginn, E.J.: Monte Carlo simulations of gas solubility in the ionic liquid 1-n-butyl-3-methylimidazolium hexafluorophosphate. J. Phys. Chem. B **109**(20), 10395–10405 (2005)
17. Santos, J.I., Gonçalves, A.M.M., Pereira, J.L., Ventura, S.P.M., Gonçalves, F.: Environmental safety of cholinium-based ionic liquids: assessing structure-ecotoxicity relationships. Green Chem. **17**(9), 4657–4668 (2015)
18. Aparicio, S., Atilhan, M.: A computational study on choline benzoate and choline salicylate ionic liquids in the pure state and after CO_2 adsorption. J. Phys. Chem. B **116**(30), 9171–9185 (2012)

19. Yuan, S., Chen, Y., Ji, X., Yang, Z., Lu, X.: Experimental study of CO_2 absorption in aqueous cholinium-based ionic liquids. Fluid Phase Equilib. **445**, 14–24 (2017)
20. Bhattacharyya, S., Shah, F.U.: Ether functionalized choline tethered amino acid ionic liquids for enhanced CO_2 capture. ACS Sustain. Chem. Eng. **4**(10), 5441–5449 (2016)
21. Salehin, F.N.M., Jumbri, K., Ramli, A., Daud, S., Abdul Rahman, M.B.: In silico solvation free energy and thermodynamics properties of H_2S in cholinium-based amino acid ionic liquids. J. Mol. Liq. **294**, 111641 (2019)
22. Steffen, C., Thomas, K., Huniar, U., Rubner, O., Schroer, A.: Software news and updates TmoleX—a graphical user interface for TURBOMOLE. J. Comput. Chem. **31**(16), 2967–2970 (2010)
23. Sambasivarao, S.V., Acevedo, O.: Development of OPLS-AA force field parameters for 68 unique ionic liquids. J. Chem. Theory Comput. **5**(4), 1038–1050 (2009)
24. Martinez, L., Andrade, R., Birgin, E.G., Martínez, J.M.: PACKMOL: a package for building initial configurations for molecular dynamics simulations. J. Comput. Chem. **30**(13), 2157–2164 (2009)
25. Essmann, U., Perera, L., Berkowitz, M.L., Lee, H., Pedersen, L.G.: A smooth particle mesh Ewald method. J. Chem. Phys. **103**(19), 8577–8593 (1995)
26. Bennett, C.H.: Efficient estimation of free energy differences from Monte Carlo data. J. Comput. Phys. **22**(2), 245–268 (1976)
27. Darden, T., York, D., Pedersen, L.: Particle mesh Ewald: an N·log(N) method for Ewald sums in large systems. J. Chem. Phys. **98**, 10089–10092 (1993)
28. Hess, B., Bekker, H., Berendsen, H.J.C., Fraaije, J.G.E.M.: LINCS: a linear constraint solver for molecular simulations. J. Comput. Chem. **18**(12), 1463–1472 (1997)
29. Jumbri, K., Micaelo, N.M., Abdul Rahman, M.B.: Solvation free energies of nucleic acid bases in ionic liquids. Mol. Simul. **43**(1), 19–27 (2017)
30. Sánchez-Badillo, J., Gallo, M., Alvarado, S., Glossman-Mitnik, D.: Solvation thermodynamic properties of hydrogen sulfide in $[C_4mim][PF_6]$, $[C_4mim][BF_4]$, and $[C_4mim][Cl]$ ionic liquids, determined by molecular simulations. J. Phys. Chem. B **119**(33), 10727–10737 (2015)
31. Hansen, N., Van Gunsteren, W.F.: Practical aspects of free-energy calculations: a review. J. Chem. Theory Comput. **10**(7), 2632–2647 (2014)
32. Shivakumar, D., Williams, J., Wu, Y., Shelley, J., Sherman, W.: Prediction of absolute solvation free energies using molecular dynamics free energy perturbation and the OPLS force field. J. Chem. Theory Comput. **6**(5), 1509–1519 (2010)
33. Lísal, M., Smith, W.R., Aim, K.: Analysis of Henry's constant for carbon dioxide in water via Monte Carlo simulation. Fluid Phase Equilib. **226**(1–2), 161–172 (2004)
34. Dhima, A., De Hemptinne, J.-C., Jose, J.: Solubility of hydrocarbons and CO_2 mixtures in water under high pressure. Ind. Eng. Chem. Res. **38**(8), 3144–3161 (1999)
35. Jiao, D., Rempe, S.B.: CO_2 solvation free energy using quasi-chemical theory. J. Chem. Phys. **134**(22), 224506 (2013)
36. Shivakumar, D., Harder, E., Damm, W., Friesner, R.A., Sherman, W.: Improving the prediction of absolute solvation free energies using the next generation OPLS force field. J. Chem. Theory Comput. **8**(8), 2553–2558 (2012)
37. Zhang, J., Tuguldur, B., Van Der Spoel, D.: Force field benchmark of organic liquids. 2. Gibbs energy of solvation. J. Chem. Inf. Model. **55**(6), 1192–1201 (2015)
38. Shokouhi, M., Adibi, M., Jalili, A.H., Hosseini-Jenab, M., Mehdizadeh, A.: Solubility and diffusion of H_2S and CO_2 in the ionic liquid 1-(2-hydroxyethyl)-3-methylimidazolium tetrafluoroborate. J. Chem. Eng. Data **55**(4), 1663–1668 (2010)
39. Jalili, A.H., Mehdizadeh, A., Shokouhi, M., Sakhaeinia, H., Taghikhani, V.: Solubility of CO_2 in 1-(2-hydroxyethyl)-3-methylimidazolium ionic liquids with different anions. J. Chem. Thermodyn. **42**(6), 787–791 (2010)
40. Jalili, A.H., Mehrabi, M., Zoghi, A.T., Shokouhi, M., Taheri, S.A.: Solubility of carbon dioxide and hydrogen sulfide in the ionic liquid 1-butyl-3-methylimidazolium trifluoromethanesulfonate. Fluid Phase Equilib. **453**, 1–12 (2017)

41. Shaikh, A.R., Karkhanechi, H., Kamio, E., Yoshioka, T., Matsuyama, H.: Quantum mechanical and molecular dynamics simulations of dual-amino-acid ionic liquids for CO_2 capture. J. Phys. Chem. C **120**(49), 27734–27745 (2016)
42. De Santis, S., Masci, G., Casciotta, F., Campetella, M., Gontrani, L.: Cholinium-amino acid based ionic liquids: a new method of synthesis and physico-chemical characterization. Phys. Chem. Chem. Phys. **17**(32), 20687–20698 (2015)
43. Sistla, Y.S., Khanna, A.: CO_2 absorption studies in amino acid-anion based ionic liquids. Chem. Eng. J. **273**, 268–276 (2015)
44. Xing, H., Yan, Y., Yang, Q., Yang, Y., Ren, Q.: Effect of tethering strategies on the surface structure of amine-functionalized ionic liquids: inspiration on the CO_2 capture. J. Phys. Chem. C **117**(31), 16012–16021 (2013)
45. Jalili, A.H., Safavi, M., Ghotbi, C., Hosseini-Jenab, M., Taghikhani, V.: Solubility of CO_2, H_2S, and their mixture in the ionic liquid 1-octyl-3-methylimidazoliumbis(trifluoromethyl)sulfonylimide. J. Phys. Chem. B **116**(9), 2758–2774 (2012)
46. Chen, J.-J., Li, W.-W., Li, X.-L., Yu, H.-Q.: Carbon dioxide capture by aminoalkyl imidazolium-based ionic liquid: a computational investigation. Phys. Chem. Chem. Phys. **14**(13), 4589–4596 (2012)
47. Safavi, M., Ghotbi, C., Taghikhani, V., Jalili, A.H., Mehdizadeh, A.: Study of the solubility of CO_2, H_2S and their mixture in the ionic liquid 1-octyl-3-methylimidazoliumhexafluorophosphate: experimental and modelling. J. Chem. Thermodyn. **65**, 220–232 (2013)
48. Harun, N., Masiren, E.E.: Molecular dynamic simulation of CO_2 absorption into mixed aqueous solutions MDEA/PZ. Chem. Eng. Res. Bull. **19**, 1–11 (2017)

Microwave-Assisted Solvothermal Liquefaction of Kenaf Stalks for Bio-oil Production

Anita Ramli, Lai Weng Kin, Nur Akila Syakida Idayu Khairul Anuar, Normawati M. Yunus, and Nurul Jannah Abd Rahman

Abstract In this study, microwave-assisted solvothermal liquefaction was investigated to produce bio-oil from Kenaf stalks. Optimization of process parameters was performed using concentrated sulfuric acid as catalyst at biomass to solvent ratio of 1:10, catalyst loading of 10% and microwave power of 300 W with constant stirring at 200 rpm. The maximum bio-oil yield of 87% was achieved at temperature of 180 °C for duration of 20 min, in which ethanol was used as solvent. Different types of catalysts were also tested to produce bio-oil at the optimum process conditions. It was found that synthesized ionic liquid [BMIM][Cl] catalyst could give the highest quality of bio-oil in terms of heating value (26.55 MJ/kg) with considerable bio-oil yield.

Keywords Bio-oil · *Kenaf stalks* · Solvothermal liquefaction · Catalyst · Ionic liquid · Microwave

A. Ramli (✉) · L. W. Kin · N. A. S. I. K. Anuar · N. J. A. Rahman
Fundamental and Applied Sciences Department, HICoE Centre for Biofuel and Biochemical Research, Institute of Self-Sustainable Building, Universiti Teknologi PETRONAS, 32610 Seri Iskandar, Perak, Malaysia
e-mail: anita_ramli@utp.edu.my

L. W. Kin
e-mail: lai.weng_23349@utp.edu.my

N. A. S. I. K. Anuar
e-mail: nur_18003521@utp.edu.my

N. J. A. Rahman
e-mail: nurul_g03700@utp.edu.my

N. M. Yunus
Fundamental and Applied Sciences Department, Centre for Research in Ionic Liquids, Institute of Contaminant Management, Universiti Teknologi PETRONAS, 32610 Seri Iskandar, Perak, Malaysia
e-mail: normaw@utp.edu.my

S. A. Abdul Karim et al. (eds.), *Proceedings of the 6th International Conference on Fundamental and Applied Sciences*, Springer Proceedings in Complexity,
https://doi.org/10.1007/978-981-16-4513-6_15

1 Introduction

The rapid depletion of fossil fuels, as well as the aftermath of fossil fuel usage have led researchers around the world to search for a promising renewable energy. Bio-oil derived from plants and animals is one of the potential alternatives, which has gained widespread recognition around the globe. Bio-oil is a mixture of various hydrocarbons compounds, usually containing acids, alcohols, aldehydes, esters, ketones, phenols and lignin-derived oligomers. Bio-oil is commonly used as substitute for fuel in engines, boilers and turbines. According to Mohan et al. [1], bio-oils emits negligible amount of SO_x and relatively lower amount of NO_x, as compared to the fossil fuels. In addition, bio-oil can also be used for production of adhesives, surfactants, pharmaceuticals and bio-degradable polymers.

In the recent years, various technologies have been developed to produce liquid bio-oil products from lignocellulosic biomass. These biomass conversion technologies are divided into two main categories, biochemical conversion and thermochemical conversion [2]. As compared to biochemical conversion, thermochemical conversion is processed at relatively higher temperature. Furthermore, the process is usually performed under pressurized condition and in inert atmosphere. The thermochemical conversion can be further classified into different methods namely, pyrolysis, liquefaction, gasification and combustion. Among the technologies, the solvothermal liquefaction seems to be a more promising method due to its low operating temperature (523–647 K) and pressure (40–220 bar) as compared to the pyrolysis method [3].

The liquefaction of biomass to produce bio-oil often contained many shortcomings, as compared to the conventional fossil fuels, such as high oxygen content, high acid value and low heating value [3]. Therefore, the production of bio-oil requires the process parameters to be optimized to obtain maximum yield and high quality of final product. It is interesting to note that, the yield and the quality of bio-oil synthesized via microwave irradiation have not been investigated thoroughly. The microwave irradiation offers the advantage of extremely rapid heating throughout the volume of the reaction mixture because it penetrates and produces a volumetrically distributed heat source.

In this study, the optimization of process parameters for liquefaction of Kenaf stalks was investigated using solvothermal liquefaction process under the irradiation of microwave to maximize the energy efficiency, as well as reducing the process duration. The optimum reaction conditions where a maximum bio-oil yield was achieved were then utilised to further study the effect of catalyst types on liquefaction yield and bio-oil quality. The catalysts used include homogeneous sulfuric acid, heterogeneous ZSM-5 and ionic liquid [BMIM][Cl] catalysts. The synthesis and characterization of the ionic liquid have been covered in this study.

2 Methodology

2.1 Synthesis of Ionic Liquid [BMIM][Cl]

20 g of 1-methylimidazole was transferred into a 3-neck round-bottom flask. 22.5 g of 1-chlorobutane was added into the round-bottom flask. The mixture was then heated to 80 °C under reflux for 48 h in a silicon oil bath. After the reaction, the sample was cooled to room temperature and washed five times with ethyl acetate. The remaining ethyl acetate in the solution was further removed using rotary evaporator at 50 °C under vacuum for 5 h.

2.2 Characterization of Ionic Liquid [BMIM][Cl]

The molecular structure and conformation of the synthesized ionic liquid of 1-Butyl-3-methylimidazolium Chloride ([BMIM][Cl]) was analysed using 1H-NMR spectrometer (Bruker Aviance III NMR) at 500 MHz. 25 mg of the ionic liquid was mixed with 0.65 mL of dimethyl sulfoxide (DMSO) as solvent and transferred into NMR tube for the analysis.

The moisture content was measured by titration (Metler Toledo V30 Karl Fischer Titrator). 0.3 g of the ionic liquid was added into a vial and tightly sealed. The vial was then heated up at 110 °C for 10 min. A blank (empty vial) was also prepared. The result obtained for the [BMIM][Cl] sample was in relevance of the blank sample.

2.3 Optimization of Solvothermal Liquefaction Process

Kenaf stalks were purchased from Lembaga Kenaf dan Tembakau Negara (LKTN), Malaysia. The Kenaf stalks were grinded and sieved to <0.4 mm particle size. The solvothermal liquefaction process was carried out using Milestone Advanced Microwave Synthesis (MA143). The liqucfaction process was started off by loading Kenaf stalks and ethanol at ratio of 1:10. Then, 0.2 g of concentrated sulfuric acid (H_2SO_4) catalyst was added into vessel. A magnetic bar stirrer was then placed into the vessel. The vessel was tightly capped and placed into microwave.

The liquefaction process was subjected to different reaction temperature (140–200 °C), duration (10–25 min) and type of solvent [ethanol, water and ethanol–water (1:1)]. After the reaction, the products were allowed to cool down to room temperature. Then, the liquefied products were dissolved in 100 mL of solvent, under constant stirring for 4 h at 400 rpm. The products were then filtered through Whatman 1442–110 filter paper using a vacuum filtration pump, to separate the liquid and solid residues. The solid residues were further washed with solvent to ensure maximum liquid products has been extracted. The liquid products were then rotary evaporated at 50 °C under reduced pressure of 120 mbar to remove the solvent. The final product, which was a black viscous liquid was obtained.

2.4 Effect of Catalysts

The solvothermal liquefaction process was then carried out at optimum reaction condition using different type of catalysts of commercial heterogeneous Zeolite Socony Mobil-5 (ZSM-5; Zeolyst), and synthesized ionic liquid [BMIM][Cl].

2.5 Qualitative and Quantitative Analysis

The synthesized bio-oil was qualitatively analyzed using GC-MS (Shimadzu GCMS-QP2020) with a capillary polar wax column (30.00 m × 0.25 mm × 0.25 μm). The pressure was 83.50 kPa. The injection volume had the total flow of 34.50 mL/min. The oven temperature was fixed at 40 °C and was hold for 1 min, before ramping it up to 250 °C (6 °C/min) and hold for 24 min. The split mode with the ratio of 20:1 was employed, while the injection temperature was set at 250 °C. Helium gas was used as the carrier gas with the flow rate of 1.50 mL/min.

The carbon, hydrogen, nitrogen and sulfur contents were quantitatively measured using CHNS analyser (CHNS Elementar Variomicro Cube). The elemental composition obtained from the CHNS were used to compute the higher heating values (HHV) using Dulong formula:

$$\text{HHV (MJ/kg)} = 0.3383\text{ZC} + 1.422(\text{ZH} - \text{ZO}/8) \tag{1}$$

where ZC, ZH and ZO represent the weight percentage of carbon, hydrogen and oxygen content present in the bio-oil, respectively.

3 Result

3.1 Characterization of Catalyst

Figure 1 shows the NMR spectrum of synthesized ionic liquid [BMIM][Cl]. As can be seen, the result obtained was as follows: 1H-NMR (500 MHz, DMSO, ppm): δ 9.500 (s, 0H), 7.877–7.800 (d, 1H), 4.216–4.187 (t, 2H), 3.880 (s, 0H), 1.793–1.734 (p, 4H), 1.283–1.208 (m, 5H), 0.901 (t, 2H).

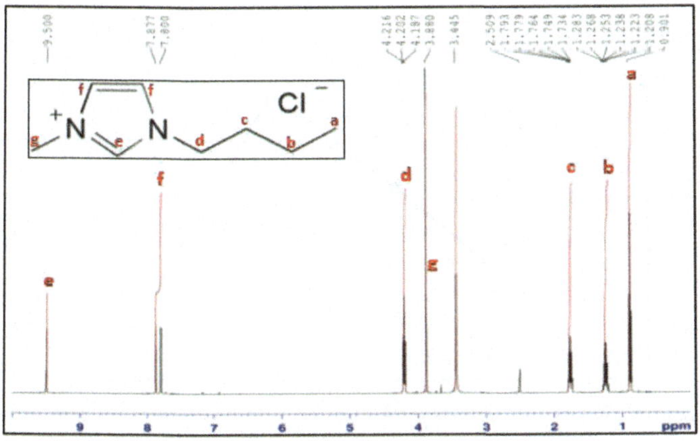

Fig. 1 ^{1}H-NMR spectrum of synthesized [BMIM][Cl]

From the result, there were few foreign peaks appear in the spectrum. At δ 3.445, the signal refers to the moisture present in the sample. This suggested that the synthesized [BMIM][Cl] may contain some moisture as impurity. In addition, another singlet (δ 2.509) was noticed which could be due to the impurity of DMSO solvent.

The moisture content of the ionic liquid [BMIM][Cl] was determined to be 2.505%. The low moisture content obtained is highly desirable as water acts as an impurity which may affect the results.

3.2 *Optimization Study*

Figure 2 shows the effect of temperature on bio-oil yield using ethanol as solvent for 20 min and catalysed by H_2SO_4. It was observed that the bio-oil yield increased as the temperature was increased from 140 to 180 °C. According to Reddy et al. [4], the higher the temperature, the easier the bonds were broken into fragments and subsequently, undergo hydrolysis and repolymerization reactions to form bio-oil. Also, at high temperature, the solid products depolymerized while the gas products aggregated to bio-oils [5] . However, when the reaction temperature was further increased to 200 °C, the bio-oil yield decreased significantly. This signified that the liquid bio-oil was converted into gaseous products at temperature above 180 °C, resulting in loss of bio-oil yield. It was also reported that higher temperature result in higher rate of reaction and favours free radical reaction which leads to formation of gaseous products [6].

Figure 3 shows the effect of liquefaction time on bio-oil yield using ethanol as solvent at optimum temperature (180 °C) and catalysed by H_2SO_4. From the result, it was observed that the bio-oil yield increases at prolonged reaction time, from 10 to 20 min. This phenomenon was occurred due to the competition between the

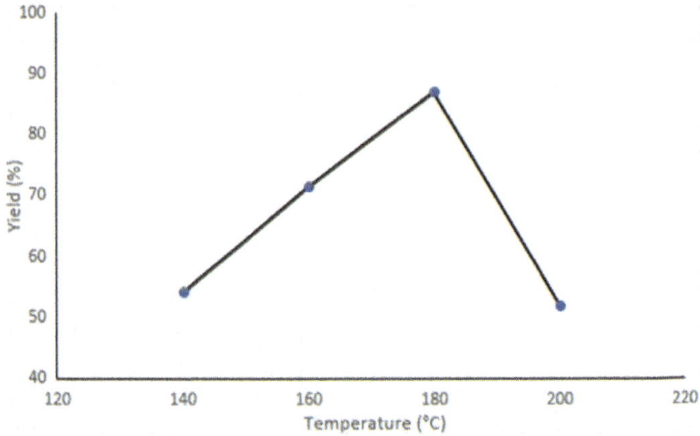

Fig. 2 The effect of temperature on bio-oil yield

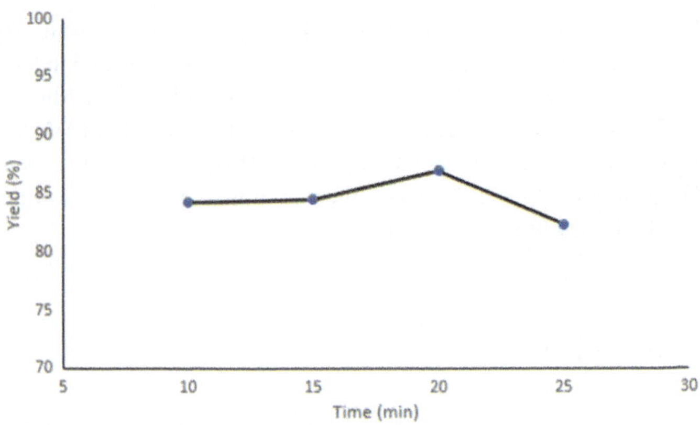

Fig. 3 The effect of liquefaction time on bio-oil yield

degradation of feedstock and repolymerization of bio-oil reactions during liquefaction process. At shorter liquefaction time, the degradation outweighs the repolymerization reaction. At higher liquefaction time, the repolymerization of bio-oil fragments dominates the liquefaction process, therefore producing liquified bio-oil [2]. However, when the liquefaction time was further increased to 25 min, significant decrease in bio-oil yield was observed.

Figure 4 shows the effect of different solvent system on bio-oil yield at optimum temperature (180 °C) for 20 min and catalysed by H_2SO_4. It was found that ethanol solvent resulted in the highest liquefaction yield, followed by ethanol-water and water. The high liquefaction yield from ethanol could be due to the present of hydrogen donor to hydrogenate the biomass fragments during liquefaction process [7]. This inhibits the condensation, cyclization or repolymerization of free radicals,

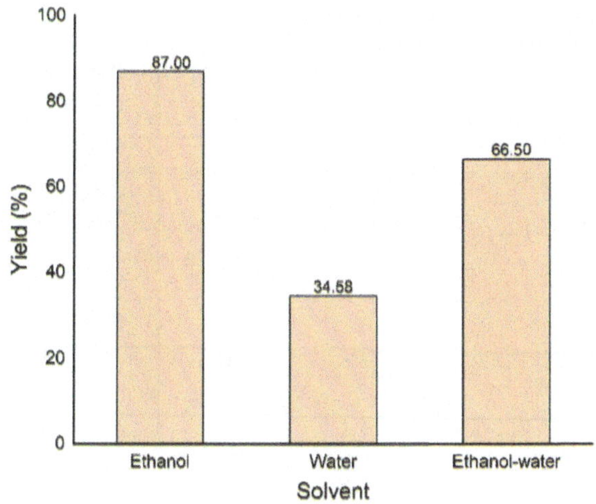

Fig. 4 The effect of solvent system on bio-oil yield

therefore reducing the formation of char. Ethanol also can react with acidic components in the bio-oil to produce esters [8]. Similarly, Liu and Zhang [9] also compared the solvent efficiency on liquefaction yield between ethanol, acetone and water and it was found that the ethanol solvent produced the highest liquefaction yield.

3.3 *Effect of Catalysts*

In this part, different types of catalysts were tested to produce bio-oil at optimum reaction condition of 180 °C for 20 min and ethanol as solvent system, as shown in Fig. 5. Maximum liquefaction yield of 87% was achieved using H_2SO_4 catalyst followed by the synthesised ionic liquid [BMIM][Cl] catalyst (43.17%). However, significant decrease in liquefaction yield was observed when heterogeneous catalyst of ZSM-5 was employed in the liquefaction process. According to Behrendt et al. [10], homogeneous catalysts in general are more active as compared to the heterogeneous catalysts, resulting in higher conversion of the biomass to bio-oil. In addition, H_2SO_4 is known for its high acidity and oxidization properties, allowing the catalyst to break down the molecules of the biomass, promoting the depolymerization reaction which results in higher yield of bio-oil [2]. On the other hand, Li et al. [11] reported that addition of ionic liquid will influence the solvent polarity and enhance the interaction between the solvent and biomass which resulted in increasing liquefaction yield. Also, ionic liquid could disrupt hydrogen bond structure and lead to the increase in number of active intermediates. These intermediates will be dissolved and stabilized in the solvent, which result in high bio-oil yield.

Table 1 shows the elemental composition of bio-oil produced at optimum reaction condition using different types of catalysts along with their heating value. It has been observed that the type of catalysts heavily influenced the elemental compositions and

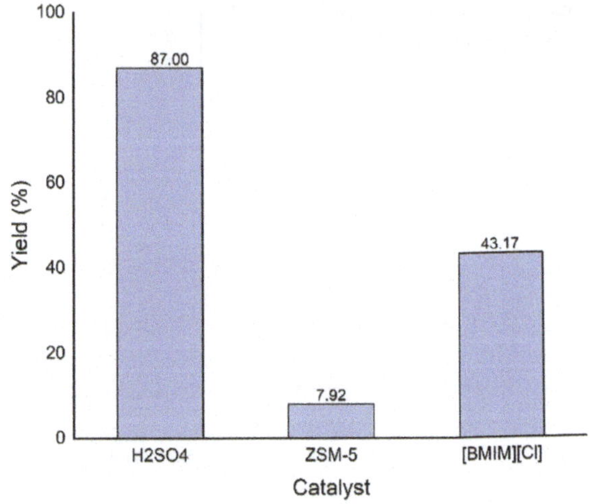

Fig. 5 The effect of catalyst types on bio-oil yield

Table 1 Elemental composition and heating value of bio-oil using different types of catalysts

Catalyst	Elemental compositions (wt%)					HHV (mJ/kg)
	C	H	N	O	S	
H_2SO_4	42.3	6.91	0.09	50.71	0.00	15.11
ZSM-5	52.05	8.28	0.53	38.59	0.45	22.50
[BMIM][Cl]	53.41	9.39	9.56	27.35	0.30	26.55

the heating value. The homogeneous H_2SO_4 catalyst resulted in the lowest heating value as compared to the other catalysts. The low heating value could be due to the reduced amount of carbon content contained in the bio-oil. On the other hand, when the liquefaction process was catalysed by ionic liquid [BMIM][Cl], the carbon content and the heating value of the bio-oil increased significantly. The heterogeneous ZSM-5 catalyst resulted in better quality of bio-oil in terms of heating value as compared to the H_2SO_4 catalyst. However, as previously mentioned, the yield of the ZSM-5 catalysed bio-oil was lower compared to the use of H_2SO_4 and [BMIM][Cl] as the catalyst.

From GC-MS analysis, it has been found that different types of catalysts resulted in the different compositions of bio-oil. The H_2SO_4-catalysed bio-oil produced mainly esters (41.43%), followed by ketones (16.18%), C5 sugars and its derivatives (14.52%), aldehydes (4.37%), nitrogen-containing compound (2.29%) and alkanes (1.76%). As for the ZSM-5-catalysed bio-oil, it produced a large fraction of ethers (33.42%), mostly due to the large yield of 1,2-dimethoxy-4-(1-methoxy-1-propenyl) benzene. It also produced a considerable number of phenolic compounds (13.82%), carboxylic acids (7.28%), aldehydes (5.83%) and nitrogen-containing compounds

(3.93%). As for the [BMIM][Cl]-catalysed bio-oil, it mainly consists of nitrogen-containing compounds (38.33%), mostly due to the formation of imidazole derivatives. This was further complemented with the high nitrogen content obtained from CHNS result, as shown in Table 1. The bio-oil also contained ethers (24.98%), ketones (6.6%), phenolic compounds (4.67%), aldehydes (4.18%), esters (3.29%) and carboxylic acids (2.3%).

4 Conclusion

Kenaf stalks have been successfully liquefied via solvothermal liquefaction method to produce bio-oil. The optimum process condition was determined under microwave irradiation using H_2SO_4 as the catalyst. At the optimum process condition, the synthesized ionic liquid [BMIM][Cl] was found to give considerable bio-oil yield with higher amount of carbon content and heating value as compared to H_2SO_4 and ZSM-5 catalysts.

Acknowledgements The research was funded by Yayasan Universiti Teknologi PETRONAS research grant (015LC0-145). The authors thank Universiti Teknologi PETRONAS for providing the necessary facilities to support this project.

References

1. Mohan, D., Pittman, C.U., Jr., Steele, P.H.: Pyrolysis of wood/biomass for bio-oil: a critical review. Energy Fuels **20**(3), 848–889 (2006)
2. Zhang, L., Xu, C., Champagne, P.: Overview of recent advances in thermo-chemical conversion of biomass. Energy Convers. Manage. **51**(5), 969–982 (2010)
3. Gallakota, A.R.K., Kishore, N., Gu, S.: A review on hydrothermal liquefaction of biomass. Renew. Sustain. Energy Rev. **81**, 1378–1392 (2018)
4. Reddy, H.K., Muppaneni, T., Rastegary, J., Shirazi, S.A., Ghassemi, A., Deng, S.: ASI: hydrothermal extraction and characterization of bio-crude oils from wet chlorella sorokiniana and dunaliclla tcrtiolccta. Environ. Prog. Sustain. Energy **32**(4), 910–915 (2013)
5. Yuan, X., Wang, J., Zeng, G., Huang, H., Pei, X., Li, H., Liu, Z., Cong, M.: Comparative studies of thermochemical liquefaction characteristics of microalgae using different organic solvents. Energy **36**(11), 6406–6412 (2011)
6. Kean, C.W., Sahu, J.N., Wan Daud, W.M.A.: Hydrothermal gasification of palm shell biomass for synthesis of hydrogen fuel. BioResources **8**(2), 1831–1840 (2013)
7. Akhtar, J., Amin, N.A.S.: A review on process conditions for optimum bio-oil yield in hydrothermal liquefaction of biomass. Renew. Sustain. Energy Rev. **15**(3), 1615–1624 (2011)
8. Isa, K.M., Abdullah, T.A.T., Ali, U.F.M.: Hydrogen donor solvents in liquefaction of biomass: a review. Renew. Sustain. Energy Rev. **81**, 1259–1268 (2018)
9. Liu, Z., Zhang, F.S.: Effects of various solvents on the liquefaction of biomass to produce fuels and chemical feedstocks. Energy Convers. Manage. **49**(12), 3498–3504 (2008)
10. Behrendt, F., Neubauer, Y., Oevermann, M., Wilmes, B., Zobel, N.: Direct liquefaction of biomass. Chem. Eng. Technol. **31**(5), 667–677 (2008)

11. Li, Q., Liu, D., Song, L., Hou, X., Wu, C., Yan, Z.: Efficient hydro-liquefaction of woody biomass over ionic liquid nickel based catalyst. Ind. Crops Prod. **113**, 157–166 (2018)

Interaction of TFSI-Imidazole and TFSI-Pyridinium ILs with MOFs from Molecular Docking Simulation

Nor Ain Fathihah Abdullah, Khairulazhar Jumbri, and Nurul Yani Rahim

Abstract Metal Organic Frameworks (MOFs) are porous material consisting of coordination bonds between metal cations and organic bridging ligands. Owing to its large surface area and tunable porosity, MOFs are an ideal material for various applications. The consolidation of ionic liquids (ILs) interior MOF might enhance the performance of pristine MOF, however, the depth study of fundamental interaction between IL and MOF are scarce. The interactions between the MOFs and ILs moiety, especially the types of interaction play an important role in determining the stability, compatibility, and selectivity of this hybrid material. This research will be centered on differing qualities of MOF-IL by utilizing computational approach to study the interaction of ILs and MOFs. Molecular docking calculation was carried out to identify the binding energy (ΔG_{bind}) between MOF5, ZIF8, UIO66 with TFSI-imidazole (EMIM, PMIM, BMIM) and TFSI-pyridinium ILs (BPYDM, EMPYDM, PMPYDM, BMPYDM). The initial 3D structures were retrieved from website database and were prepared using default parameters in docking software. The ILs was docked consecutively into the MOFs cavity inside the grid box dimension to identify binding sites of MOF. From the results, TFSI-imidazole and TFSI-pyridinium ILs exhibited good interaction with MOFs. In term of highest stability and strongest binding affinity, the PMPYDM-TFSI showed superior binding towards UIO66 with ΔG_{bind} of -8.60 kcal mol^{-1}.

Keywords Binding energy · Molecular docking · Metal organic framework · Ionic liquids

N. A. F. Abdullah · K. Jumbri (✉)
Department of Fundamental and Applied Sciences, Universiti Teknologi PETRONAS, 32610 Seri Iskandar, Perak, Malaysia
e-mail: khairulazhar.jumbri@utp.edu.my

N. Y. Rahim
School of Chemical Sciences, Universiti Sains Malaysia, 11800 Pulau Pinang, Malaysia
e-mail: nurulyanirahim@usm.my

S. A. Abdul Karim et al. (eds.), *Proceedings of the 6th International Conference on Fundamental and Applied Sciences*, Springer Proceedings in Complexity,
https://doi.org/10.1007/978-981-16-4513-6_16

Abbreviations

EMIM	1-ethyl-3-methylimidazolium
PMIM	1-propyl-3-methylimidazolium
BMIM	1-butyl-3-methylimidazolium
BPYDM	1-butylpyridinium
EMPYDM	1-ethyl-4-methylpyridinium
PMPYDM	1-propyl-4-methylpyridinium
BMPYDM	1-butyl-4-methylpyridinium
TFSI	Bis((trifluoro)sulfonyl)imide

1 Introduction

MOFs are a class of porous material first reported by Omar Yaghi's group in 1995 [1] and have gained a lot of attention in the recent decade because they combine the benefits of both organic and inorganic porous materials. MOFs consist of coordination bonds between metal cations and organic bridging ligands. MOFs have unique properties such as high porosity, high surface area, and tunable porosity which are able to create tunable host-guests interaction [2–4]. The pore size of MOFs can be tuned by appropriate selection of metal centers and organic linkers and usually their porosity remain permanent even after the guest species are removed [5]. A vast number of investigations on MOFs have been reported including gas storage [6], energy storage [7], catalysis [8], and water treatment [9].

Many researchers still work on the MOFs material as it offered exceptional porosity and wide range of potential uses [10–12]. Recent research has shown that the hybridization of MOFs with another component might give new materials that have similar characteristics with both the host MOFs and material added. This combination could enhance the host MOFs properties and may lead to a new type of 'designer hybrid material' in various applications. There are several methods that have been reported in preparing MOF hybrid material which is a covalent modification, non-covalent interaction and using MOF as a template or precursor. Among these methods, the non-covalent interaction by encapsulation is easier and convenient. This method allowed the other species to be entrapped inside MOF pores without the formation of a bond. The hybridization of MOF using the encapsulation method produces more stable and superior hybrid MOF with great functionality.

Ionic liquids (ILs) are another kind of organic–inorganic hybrid materials composed of organic cations and inorganic anions [13]. ILs are "green solvent" that have gained attention because of their unique physical and chemical properties. The synthesis of ILs is highly flexible, has the ability to tailor the components of ions and tune physicochemical properties and functionalities. It has been widely studied in various applications such as energy storage [14], catalysis [15], and separation and extraction [16]. Although ILs have several favorable qualities that make them attractive in certain research, their applicability in real operations is limited due to their liquid form. Toxicity, product purification, expensive operating system and high viscosity, for example, are the most pressing issues for a variety of applications. Therefore, the idea of many researchers by immobilizing ILs into solid porous materials is aspiring and may improve the properties of these hybrid materials and also could overcome the drawback depending on its application. These hybrid MOFs have shown promising results in various application such gas and separation [17, 18], energy storage [19], and electrochemical [20].

Understanding the basic fundamental of IL-IL and MOF-IL interactions using computational study is utmost important [21–24], as it allowed one to design a stable and remarkable performance of hybrid materials. Both MOFs and ILs bear a variety of interaction sites and functional groups. Consequently, more complicated interaction systems are involved such as Coulombic forces (electrostatic attraction or repulsion), polarization, van der Waals interactions, π-π interactions, and hydrogen bonding. Therefore, the understanding of these specific interactions between ILs and MOFs is of utmost importance to optimize the performance of the MOF-IL hybrid materials; however, it is still a big challenge.

From the findings, the hybridization of MOFs with ILs may provide new classes of adsorbent with the appropriate selection of ILs and MOFs. The different selection of ILs and MOFs might affect and improve the adsorption capacity of heavy metals and might improve the stability of MOFs in the aqueous solution. In pre-calculation screening, among the common cation tested, imidazole and pyridinium showed superior strength towards MOFs. The TFSI anion was fixed to maintain the hydrophobic character of ILs. Herein, molecular docking calculation was performed focusing on TFSI-imidazole and TFSI-pyridinium based ILs to determine the interaction energy between combinations of cation/anion with MOFs for further investigation.

2 Computational Details

2.1 System Preparation

The initial structures of MOF5, ZIF8 and UIO66 were retrieved from ChemTube3D [25] and Cambridge Crystallographic Data Centre (CCDC) [26]. The initial structure of cations and anions of ILs were retrieved from web database [27].

2.2 Docking Procedure

The molecular docking was performed to identify the binding energy and the binding affinity of the ILs that binds to a specific receptor binding sites of MOFs. The docking simulation was performed using AutoDock 4.2 software [28] by running the autogrid

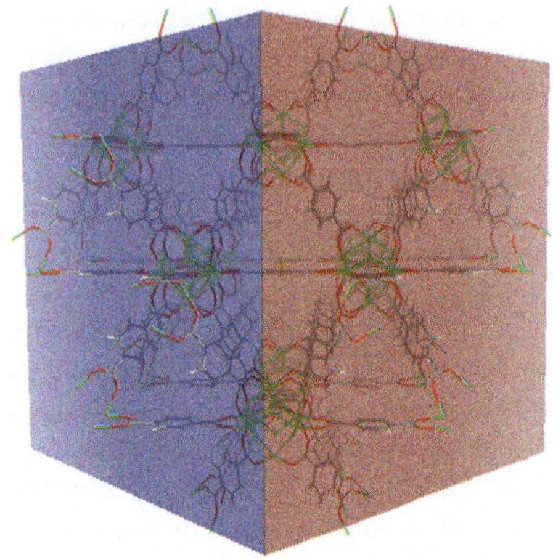

Fig. 1 Snapshot of grid box (90 × 90 × 90 Å) for MOF (e.g. for UIO66)

and autodock calculation to estimate the binding affinity using the scoring function to find the best binding mode. All the molecules were prepared and saved as PDBQT file format to be further used for docking calculation.

The MOF-IL system was prepared by introducing each cation and anion into the macromolecular receptors cavity of MOFs consecutively. A grid box of 90 × 90 × 90 Å with a grid spacing of 0.375 Å was generated adequate enough to occupy the whole MOFs surface to identify the best possible binding site as shown in Fig. 1.

In order to identify the appropriate binding modes and conformation for ILs, the Lamarckian Genetic Algorithm (LGA) [29] was employed. The binding energy (ΔG_{bind}) of the inclusion complex was then calculated by the following Eq. 1:

$$\Delta G_{bind} = \Delta G_{VDW} + \Delta G_{elect} + \Delta G_{Hbond} + \Delta G_{desolv} + \Delta G_{tor} \quad (1)$$

where ΔG_{VDW} is van der Waal energy, ΔG_{elect} refers to electrostatic energy, ΔG_{HBond} is hydrogen bonding energy, ΔG_{desolv} represents desolvation energy and ΔG_{tor} is torsional free energy. The inhibition constants (K_i) of all complexes were determined from the following Eq. 2:

$$K_i = e^{\left(\frac{\Delta G_{bind}}{RT}\right)} \quad (2)$$

3 Results and Discussion

3.1 *Screening of TFSI-Imidazole and TFSI-Pyridinium ILs onto Each MOFs*

Molecular docking calculation was carried out to find the free binding energy of the TFSI-imidazole (EMIM, PMIM, BMIM) and TFSI-pyridinium based ILs (BPYDM, EMPYDM, PMPYDM, BMPYDM) with three MOFs (MOF5, ZIF8, UIO66). The calculated binding energy of MOF5, ZIF8 and UIO66 were presented in Table 1. The negative value of binding energy (ΔG_{bind}) of the system describes the stability of the system. The high negative value of binding energy implies the highest stability of the complex. In general, ΔG_{bind} obtained for TFSI-imidazole ILs (as shown in Table 1) in the range of -1.91 to -1.99 kcal mol^{-1} which was slightly lower than ΔG_{bind} obtained for TFSI-pyridinium ILs (-1.91 to -2.10 kcal mol^{-1}). The higher negative ΔG_{bind} obtained for TFSI-pyridinium ILs compared to TFSI-imidazole ILS signifies that the superior of pyridinium cations to form strong binding affinity with MOF5. From the TFSI-imidazole results, it can be seen that the binding interaction

Table 1 The comparison of binding energies of TFSI-imidazole and TFSI-pyridinium for MOF5, ZIF8 and UIO66

MOFs	ILs	ΔG_{bind} (kcal/mol)
MOF5	EMIM/TFSI	−1.91
	PMIM/TFSI	−1.95
	BMIM/TFSI	−1.99
	BPYDM/TFSI	−2.02
	EMPYDM/TFSI	−1.91
	PMPYDM/TFSI	−2.12
	BMPYDM/TFSI	−2.10
ZIF8	EMIM/TFSI	−1.42
	PMIM/TFSI	−2.00
	BMIM/TFSI	−1.40
	BPYDM/TFSI	−2.08
	EMPYDM/TFSI	−2.17
	PMPYDM/TFSI	−2.21
	BMPYDM/TFSI	−2.37
UIO66	EMIM/TFSI	−4.25
	PMIM/TFSI	−4.18
	BMIM/TFSI	−4.18
	BPYDM/TFSI	−4.23
	EMPYDM/TFSI	−4.12
	PMPYDM/TFSI	−8.60
	BMPYDM/TFSI	−4.19

Table 2 The optimum binding energies (ΔG_{bind}) and inhibition constant (Ki) of TFSI-imidazole and TFSI-pyridinium ILs of MOF5, ZIF8 and UIO66

MOFs	ILs	ΔG_{bind} (kcal mol^{-1})	Ki (mM)
MOF5	BMIM-TFSI	−1.99	35.07
	PMPYDM-TFSI	−2.12	27.94
ZIF8	PMIM-TFSI	−2.00	34.46
	BMPYDM-TFSI	−2.37	18.17
UIO66	EMIM-TFSI	−4.25	7.72
	PMPYDM-TFSI	−8.60	4.9×10^{-4}

gets stronger with the addition of alkyl chain in imidazole cation that stabilizes the whole system. While in TFSI-pyridinium based ILs case, there is inconsistent ΔG_{bind} values with the addition of alkyl chain in pyridinium cation.

The similar case of MOF5 was observed in ZIF8 as well where the TFSI-pyridinium ILs exhibit higher negative value of ΔG_{bind} in the range of −2.08 to −2.37 kcal mol^{-1} compared to TFSI-imidazole (−1.40 to −2.00 kcal mol^{-1}). The TFSI-imidazole ILs showed inconsistent values of ΔG_{bind} with addition of alkyl chain in which was contrast to the TFSI-pyridinium. The addition of alkyl chain in pyridinium cation increases the strength and stability of the ZIF8 complex. For UIO66 system, in general, the ΔG_{bind} obtained for UIO66 was in the range of −4.12 to −8.60 kcal mol^{-1}. The ΔG_{bind} obtained for all systems have quite similar range and did not differ much from each other except for UIO66 bind with PMPYDM-TFSI which increased gradually to −8.60 kcal mol^{-1}. In the case of TFSI-imidazole, the addition of alkyl chain decreased the binding energy, meanwhile inconsistent ΔG_{bind} values were observed in TFSI-pyridinium systems when the alkyl chain was added to pyridinium cation.

The binding energies and the inhibition constant of the MOF-IL systems with optimum results were tabulated in Table 2. The inhibition constant (Ki) describes the binding affinity of the MOFs towards ILs. The lower inhibition constant obtained, the stronger the binding affinity of MOFs towards ILs. Generally, the low Ki obtained was found for pyridinium based ILs compared to imidazole ILs which showed the stronger interaction of pyridinium group with these three MOFs. Particularly, the lowest Ki is observed in UIO66 (4.9×10^{-4} mM) for pyridinium group (PMPYDM-TFSI) while the highest Ki was observed in MOF5 (35.07 mM) for imidazole group, BMIM-TFSI IL. The results indicated that, UIO66 exhibited the strongest binding affinity towards PMPYDM-TFSI compared to MOF5 and ZIF8.

From the binding energy obtained, it is pronounced that the TFSI-pyridinium based IL exhibits higher value than TFSI-imidazole based IL for all MOFs system. The high stability in TFSI-pyridinium is probably due to aromaticity of the pyridinium structure. It was reported that the harmonic oscillator measure of aromaticity (HOMA) values for pyridinium was closer to the unity value (~1) than imidazole [30]. The HOMA value closer to a unity indicates the stronger aromaticity of the ring, which affirming that pyridinium has strong aromaticity than imidazole. The stronger aromaticity of the ring could contribute to the higher stability of the pyridinium structure interacting with TFSI anion inside MOFs.

3.2 *The Most Stable Conformer of Molecular Docking Structure*

The corresponding lowest energy conformer of docking structures for MOF5 (PMPYDM/TFSI), ZIF8 (BMPYDM/TFSI) and UIO66 (PMPYDM-TFSI) were obtained and portrayed in Fig. 2, respectively. It was found that the ILs preferred to

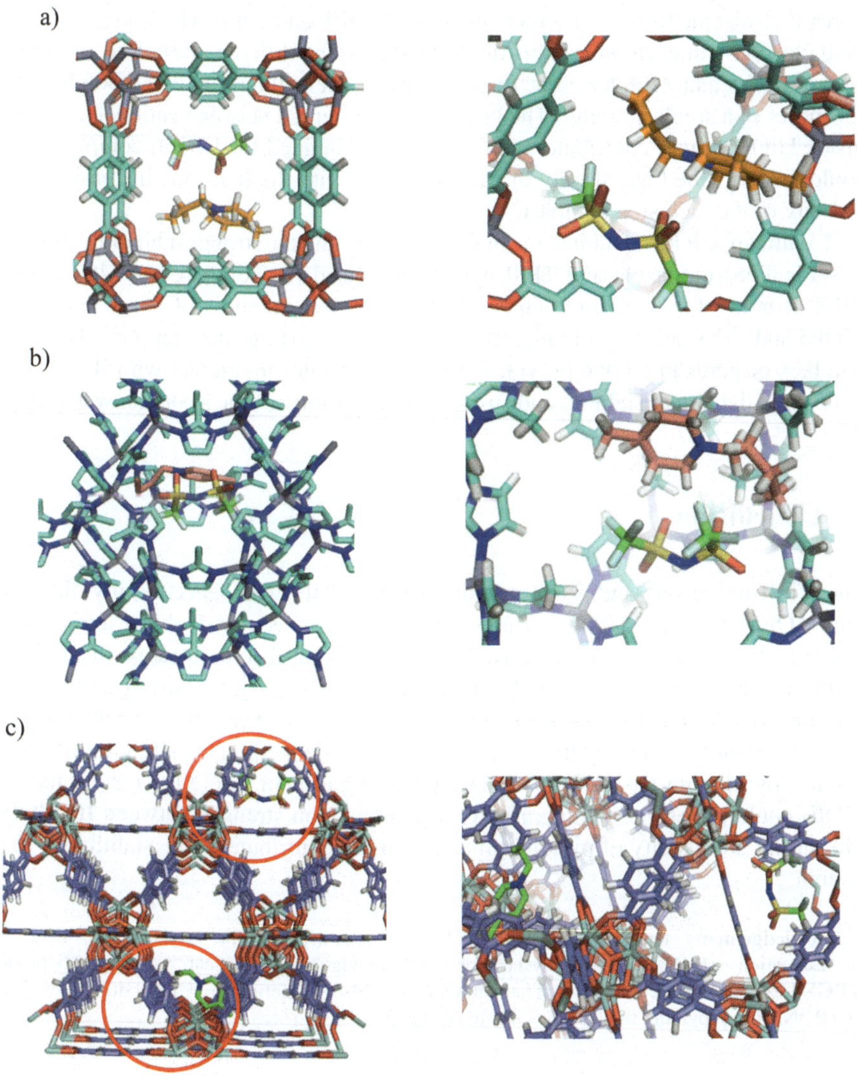

Fig. 2 The lowest conformer molecular docking structures of **a** PMPYDM/TFSI-MOF5, **b** BMPYDM /TFSI-ZIF8, **c** PMPYDM/TFSI-UIO66

bind at the corner of the MOFs pores. For MOF5 and ZIF8 complexes, both anion and cation preferred to bind around the corner of the pores near to the metal sites (as shown in Fig. 2a, b) which was in good agreement with previous studies [31, 32] where the authors used small size of ILs. Particularly, the small size of the ILs could promote the stability and strong interaction with MOFs [33]. Additionally, both cation and anion in MOF5 and ZIF8 complexes bind closely to TFSI anion around the metal sites of MOFs pores.

In contrast, the cation and anion in UIO66 both preferred to reside at different pores and distant from each other unlike in MOF5 and ZIF8 which preferred to bind closely with each other. This might be due to the different variations of pores where MOF5 and ZIF8 have one distinct pore in one cell unit while UIO66 have a few pores in a single cell unit. The large distant observed between cation and anion inside UIO66 might contribute to strong ΔG_{bind} obtained for UIO66 complex that could promote the high affinity of MOFs to bind with the ILs, thus increasing the stability of the inclusion complex.

To sum up, UIO66 exhibited the highest stability and the strongest binding affinity towards TFSI-imidazole and TFSI-pyridinium based ILs compared to MOF5 and ZIF8. It is notable that single unit of UIO66 has few numbers of pores unlike in MOF5 and ZIF8 (as shown in Fig. 2), which only consist one distinct pore. The few numbers of pores in UIO66 provides more reaction sites to interact with ILs which in resulted the higher stability and strong binding affinity of the system towards ILs.

4 Conclusion

The TFSI-imidazole and TFSI-pyridinium based IL were successfully docked into MOF5, ZIF8 and UIO66, consecutively using molecular docking simulation. These matching ILs and MOFs have shown promising results in terms of binding ability towards each other, with TFSI-pyridinium exhibiting the strongest interaction towards MOF5 (-2.12 kcal mol^{-1}), ZIF8 (-2.37 kcal mol^{-1}) and UIO66 (-8.60 kcal mol^{-1}). Among the three MOFs, UIO66 showed superior binding affinity towards pyridinium-based ILs. The size and the location of where ILs resides inside MOFs could affect the binding affinity and interaction strength between IL-MOF. The strong aromaticity of pyridinium cation might contribute to the stability of the system.

Acknowledgements The authors would like to acknowledge the support of Minister of Education (MOE) of Malaysia for the Fundamental Research Grant Scheme (FRGS/1/2020/STG04/UTP/02/3). The authors are also grateful to Universiti Teknologi PETRONAS in providing the facilities for this research.

References

1. Yaghi, O.M., Li, H.: Hydrothermal synthesis of a metal-organic framework containing large rectangular channels. J. Am. Chem. Soc. **117**(41), 10401–10402 (1995)
2. Lee, Y.-R., Kim, J., Ahn, W.-S.: Synthesis of metal-organic frameworks: a mini review. Korean J. Chem. Eng. **30**(9), 1667–1680 (2013)
3. Zhou, H.-C.J., Kitagawa, S.: Metal-organic frameworks (MOFs). Chem. Soc. Rev. **43**(16), 5415–5418 (2014)
4. Silva, P., Vilela, S.M.F., Tomé, J.P.C., Almeida Paz, F.A.: Multifunctional metal–organic frameworks: from academia to industrial applications. Chem. Soc. Rev. **44**(19), 6774–6803 (2015)
5. Li, J.R., Kuppler, R.J., Zhou, H.C.: Selective gas adsorption and separation in metal-organic frameworks. Chem. Soc. Rev. **38**(5), 1477–1504 (2009)
6. Ma, S., Zhou, H.-C.: Gas storage in porous metal–organic frameworks for clean energy applications. Chem. Commun. **46**(1), 44–53 (2010)
7. Wang, L., Han, Y., Feng, X., Zhou, J., Qi, P., Wang, B.: Metal–organic frameworks for energy storage: batteries and supercapacitors. Coord. Chem. Rev. **307**, 361–381 (2016)
8. Dhakshinamoorthy, A., Li, Z., Garcia, H.: Catalysis and photocatalysis by metal organic frameworks. Chem. Soc. Rev. **47**(22), 8134–8172 (2018)
9. Feng, M., Zhang, P., Zhou, H.-C., Sharma, V.K.: Water-stable metal-organic frameworks for aqueous removal of heavy metals and radionuclides: a review. Chemosphere **209**, 783–800 (2018)
10. Zango, Z.U., Jumbri, K., Sambudi, N.S., Hanif Abu Bakar, N.H., Fathihah Abdullah, N.A., Basheer, C., Saad, B.: Removal of anthracene in water by MIL-88(Fe), NH_2-MIL-88(Fe), and mixed-MIL-88(Fe) metal–organic frameworks. RSC Adv. (71), 41490–41501 (2019)
11. Isiyaka, H.A., Jumbri, K., Sambudi, N.S., Zango, Z.U., Saad, B., Mustapha, A.: Removal of 4-chloro-2-methylphenoxyacetic acid from water by MIL-101(Cr) metal-organic framework: kinetics, isotherms and statistical models. Roy. Soc. Open Sci. **8**(1), 201553 (2021)
12. Zango, Z.U., Sambudi, N.S., Jumbri, K., Abu Bakar, N.H.H., Abdullah, N.A.F., Negım, E.-S.M., Saad, B.: Experimental and molecular docking model studies for the adsorption of polycyclic aromatic hydrocarbons onto UiO-66(Zr) and NH2-UiO-66(Zr) metal-organic frameworks. Chem. Eng. Sci. **220**, 115608 (2020)
13. Dupont, J., de Souza, R.F., Suarez, P.A.Z.: Ionic liquid (molten salt) phase organometallic catalysis. Chem. Rev. **102**(10), 3667–3692 (2002)
14. Eftekhari, A., Liu, Y., Chen, P.: Different roles of ionic liquids in lithium batteries. J. Power Sources **334**, 221–239 (2016)
15. Steinrück, H.-P., Wasserscheid, P.: Ionic liquids in catalysis. Catal. Lett. **145**(1), 380–397 (2015)
16. Isiyaka, H.A., Jumbri, K., Sambudi, N.S., Zango, Z.U., Fathihah Abdullah, N.A., Saad, B., Mustapha, A.: Adsorption of dicamba and MCPA onto MIL-53(Al) metal–organic framework: response surface methodology and artificial neural network model studies. RSC Adv. **10**(70), 43213–43224 (2020)
17. Kumar, K.V., Preuss, K., Titirici, M.-M., Rodríguez-Reinoso, F.: Nanoporous materials for the onboard storage of natural gas. Chem. Rev. **117**(3), 1796–1825 (2017)
18. Herm, Z.R., Bloch, E.D., Long, J.R.: Hydrocarbon separations in metal–organic frameworks. Chem. Mater. **26**(1), 323–338 (2013)
19. Huang, W., Li, S., Cao, X., Hou, C., Zhang, Z., Feng, J., Ci, L., Si, P., Chi, Q.: Engineering: metal–organic framework derived iron sulfide–carbon core–shell nanorods as a conversion-type battery material. ACS Sustain. Chem. Eng. **5**(6), 5039–5048 (2017)
20. Wiers, B.M., Foo, M.-L., Balsara, N.P., Long, J.R.: A solid lithium electrolyte via addition of lithium isopropoxide to a metal–organic framework with open metal sites. J. Am. Chem. Soc. **133**(37), 14522–14525 (2011)
21. Ishak, M.A.I., Jumbri, K., Bustam@Khalil, M.A., Daud, S.: Molecular dynamic simulation on the stability of corporated metal organic framework and choline based ionic liquids. J. Comput. Theor. Nanosci. **17**(2–3), 1547–1556 (2020)

22. Ishak, M.A.I., Jumbri, K., Daud, S., Abdul Rahman, M.B., Abdul Wahab, R., Yamagishi, H., Yamamoto, Y.: Molecular simulation on the stability and adsorption properties of choline-based ionic liquids/IRMOF-1 hybrid composite for selective H_2S/CO_2 capture. J. Hazard. Mater. **399**, 123008 (2020)
23. Mohd Salehin, F.N., Jumbri, K., Ramli, A., Daud, S.: An insight into solubility of H_2S in choline based ionic liquids from molecular dynamics simulation. J. Comput. Theor. Nanosci. **17**(2–3), 1422–1431 (2020)
24. Ishak, M.A.I., Taha, M.F., Wirzal, M.D.H., Nordin, M.N., Abdurrahman, M., Jumbri, K.: Choline-based ionic liquids-incorporated IRMOF-1 for H_2S/CH_4 capture: insight from molecular dynamics simulation. Processes **8**(4), 412 (2020)
25. Chem3D Tube: https://www.chemtube3d.com/. Last accessed 2020/11/5
26. Groom, C.R., Bruno, I.J., Lightfoot, M.P., Ward, S.C.: The Cambridge structural database. Acta Crystallogr. Sect. B Struct. Sci. Cryst. Eng. Mater. **72**, 171–179 (2016)
27. Malde, A.K., Zuo, L., Breeze, M., Stroet, M., Poger, D., Nair, P.C., Oostenbrink, C., Mark, A.E.: An automated force field topology builder (ATB) and repository: version 1.0. J. Chem. Theor. Comput. **7**(12), 4026–4037 (2011)
28. Morris, G.M., Huey, R., Lindstrom, W., Sanner, M.F., Belew, R.K., Goodsell, D.S., Olson, A.J.: AutoDock4 and AutoDockTools4: automated docking with selective receptor flexibility. J. Comput. Chem. **30**(16), 2785–2791 (2009)
29. Fuhrmann, J., Rurainski, A., Lenhof, H.-P., Neumann, D.: A new Lamarckian genetic algorithm for flexible ligand-receptor docking. J. Comput. Chem. **31**(9), 1911–1918 (2010)
30. Wu, C., De Visscher, A., Gates, I.D.: Comparison of electronic and physicochemical properties between imidazolium-based and pyridinium-based ionic liquids. J. Phys. Chem. B **122**(26), 6771–6780 (2018)
31. Gupta, K.M., Chen, Y., Hu, Z., Jiang, J.: Metal–organic framework supported ionic liquid membranes for CO_2 capture: anion effects. Phys. Chem. Chem. Phys. **14**(16), 5785–5794 (2012)
32. Chen, Y., Hu, Z., Gupta, K.M., Jiang, J.: Ionic liquid/metal–organic framework composite for CO_2 capture: a computational investigation. J. Phys. Chem. C **115**(44), 21736–21742 (2011)
33. Li, Z., Xiao, Y., Xue, W., Yang, Q., Zhong, C.: Ionic liquid/metal–organic framework composites for H_2S removal from natural gas: a computational exploration. J. Phys. Chem. C **119**(7), 3674–3683 (2015)

Impact of Photoperiod on the Carbon Metabolic Pathways of *Chlorella Vulgaris* for Biomass Production and Nutrient Removal in Treating Nutrient-Rich Wastewater

Nur Afiqah Mohamad Saman, Wai Hong Leong, Hemamalini Rawindran, Mardawani Mohamad, Muslim Abdurrahman, and Jun Wei Lim

Abstract Cultivation of *Chlorella vulgaris* has been done in different photoperiod to study the effect of photoperiod cycles on the on the carbon metabolic pathways of the microalgae species in order to enhance its biomass and lipid productions. Autotrophic, heterotrophic and three different photoperiod cycles of mixotrophic growth mode were evaluated for its biomass growth, carbon and nutrient uptakes and lipid yield. The studied photoperiods for mixotrophic growth are 16:8, 12:12 and 8:16 h (light:dark hours). Heterotrophic condition produced much lower microalgal biomass and lipid yield compared to microalgae grown under autotrophic and mixotrophic condition. There is no significant difference in microalgal biomass yields observed under continuous illumination between 16:8 h, 12:12 h and 8:16 h photoperiods. All studied parameters showed a near complete removal of COD on the second day of cultivation. Higher nitrogen removal was observed at longer photoperiod conditions.

Keywords Microalgae · Photoperiod · Biomass · Carbon · Nitrogen

---

N. A. M. Saman
Department of Fundamental and Applied Sciences, Universiti Teknologi PETRONAS, 32610 Seri Iskandar, Perak Darul Ridzuan, Malaysia
e-mail: nur.afiqah_25213@utp.edu.my

W. H. Leong · H. Rawindran · J. W. Lim (✉)
Department of Fundamental and Applied Sciences, HICoE—Centre for Biofuel and Biochemical Research, Institute of Self-Sustainable Building, Universiti Teknologi PETRONAS, 32610 Seri Iskandar, Perak Darul Ridzuan, Malaysia
e-mail: junwei.lim@utp.edu.my

M. Mohamad
Faculty of Bioengineering and Technology, Universiti Malaysia Kelantan, Jeli Campus, 17600 Jeli, Kelantan, Malaysia
e-mail: mardawani.m@umk.edu.my

M. Abdurrahman
Department of Petroleum Engineering, Universitas Islam Riau, Jl. Kaharuddin Nasution No. 113, Pekanbaru 28284, Indonesia
e-mail: muslim@eng.uir.ac.id

S. A. Abdul Karim et al. (eds.), *Proceedings of the 6th International Conference on Fundamental and Applied Sciences*, Springer Proceedings in Complexity,
https://doi.org/10.1007/978-981-16-4513-6_17

1 Introduction

Microalgae is steadily gaining attraction as the preferred feedstock for sustainable biofuel productions termed as the third-generation biofuels. Moreover, microalgae have the ability in fixing atmospheric carbon dioxide for its growth and produce a variety of end products [1]. Production rate of biodiesel from microalgae is about 5000–20,000 gal/acre/year which is comparatively high than oil production from other feedstocks [2]. In addition, some microalgae strains are very versatile that they can be grown in extreme environments and conditions e.g. high nutrient wastewater loadings and saline water bodies [3–5].

While the merits of using microalgae as a potential biofuel feedstock are undeniably promising, the commercialization of microalgae-based biofuels is rather uncertain [6]. Thus, leading to extensive studies to further enhance the biomass and lipid productivities for eventual exploitations as microalgal biomass feedstock. These include manipulation of cultivation parameters and cultivation modes. The main obstacle that impedes the commercialization of microalgal biodiesel production is the low lipid productivities due to the low cell biomass and lipid production. Owing to the low biomass yield, the harvesting process become difficult which can subsequently lead to high unit cost of downstream process. Various studies into optimizing or modifying the cultivation conditions have been investigated to induce both the microalgal biomass and lipid productivities [7, 8]. As microalgae are known to be able to utilize various carbon metabolic pathways, namely, autotrophic, heterotrophic and mixotrophic, for its growth, manipulating its growth metabolic pathways perhaps could provide an insight into inducing the cellular biomass and lipid productions [9].

Other than that, photoperiod is one of the important physicochemical factors affecting the growth of microalgae as it regulates the cell division in microalgae reproduction. The influence of this photoperiod on biomass and lipid productivity of microalgae has not been clearly established. To further improve the biomass yields of microalgae, it is crucial to optimize the photoperiod to achieve high microalgal biomass growth productivities. Hence, this study is aimed to investigate the carbon metabolic pathways through varying the light and dark cycles i.e., photoperiod for optimizing the microalgal biomass production along with the nutrient removal from nutrient-rich wastewater. Thus, able to provide an insight on the dynamics of microalgal biomass growth in relation to the carbon and nitrogen metabolic pathways under different photoperiod conditions.

2 Materials and Methods

2.1 Wastewater Medium Preparation

Prior to the experiment, a synthetic wastewater medium simulating a high strength municipal wastewater was prepared based on a modified wastewater composition from Leong et al. [10] yielding concentrations of chemical oxygen demand, COD and ammonium-nitrogen, NH_4^+–N at 145 mg/L and 48 mg/L, respectively. The following were compositions (mg/L) for the wastewater: sucrose (109), $FeCl_3.6H_2O$ (10), $CaCl_2$ (42), $(NH_4)_2SO_4$ (226), K_2HPO_4 (180), KH_2PO_4 (35) and $MgSO_4$ (49).

2.2 Chlorella Vulgaris *Cultivation Using Synthetic Wastewater Medium*

Samples of the freshwater microalgae, *Chlorella vulgaris* was obtained from the culture collections belonging to the Centre for Biofuel and Biochemical Research (CBBR), Universiti Teknologi PETRONAS. The microalgae, *Chlorella vulgaris* was chosen in the study due to its ability to populate ponds naturally combined with its adaptability in tolerating high nutrient loadings in sewage effluents [11]. Stock culture of the microalgae, *Chlorella vulgaris* (500 mL) was inoculated in a 5-L laboratory clear glass bottle containing approximately 4.5 L of the prepared synthetic wastewater medium. The photobioreactor was continuously aerated with compressed air and illuminated with a white light-emitting diode (LED) light at the light intensity of 1200 lx. The initial cultivation pH was adjusted to 7.1 ± 0.1.

2.3 *Photoperiod Experimental Design*

The experimental setup consisted of culturing the microalgae, *Chlorella vulgaris* acclimated to the synthetic wastewater medium under five different photoperiod design settings in 500 mL Erlenmeyer flasks as functional bioreactors. Approximately 50 mL of the *Chlorella vulgaris* inoculum was introduced in each of the bioreactor containing 450 mL of the synthetic wastewater medium. The different photoperiod designs in the study with varying light:dark cycle (h:h) included 24:0 (autotrophic), 16:8, 12:12, 8:16 and 0:24 (heterotrophic) designated as BR-24L, BR-16L, BR-12L and BR-8L and BR-0L, respectively (Table 1). All the bioreactors except for the BR24-L (autotrophic) were supplemented with the additional organic

Table 1 Designated bioreactors under varying photoperiod cultivation conditions

Bioreactor	Photoperiod	
	Light (h)	Dark (h)
BR-24L (Autotrophic)	24	0
BR-16L	16	8
BR-12L	12	12
BR-8L	8	16
BR-0L (Heterotrophic)	0	24

carbon source (sucrose) from the compositions of the synthetic wastewater medium. Each of the bioreactor was continuously aerated with compressed air and the cultivation pH was regulated and maintained at 7.0 ± 0.1 using either 1 N of H_2SO_4 or 1 N of NaOH throughout the study. The experimental study was carried out until the stationary growth phase was achieved and performed in duplicates. Samplings were executed every two days for analyzing the microalgal biomass, COD and NH_4^+–N concentrations.

2.4 *Analytical Procedures*

The microalgal biomass concentration was determined via spectrophotometry using a UV-VIS spectrophotometer (Shimadzu UV-2600). An analysis of optical density was measured at the wavelength of 688 nm and the microalgal biomass concentration, B (g/L) was calculated using Eq. 1.

$$B = 0.3557 \times OD_{688nm}, R^2 = 0.988 \quad (1)$$

The concentration of COD was determined using the closed reflux, titrimetric method: 5520 following the Standard Methods for the Examination of Water and Wastewater [12]. The concentrations of NH_4^+–N species were also analysed following the Standard Methods [12], namely, the titrimetric method: 4500-NH_3.

3 Results and Discussion

3.1 *Microalgal Biomass Growth*

The growth patterns of *Chlorella vulgaris* were visualized in Fig. 1. The initial microalgal biomass in all bioreactors are relatively the same, which was around

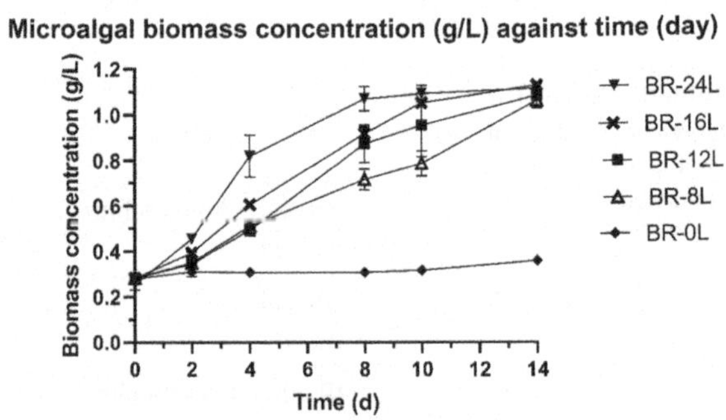

Fig. 1 Microalgal biomass growth cultured under different photoperiods

0.2 g/L. The first 2 days were considered as the lag phase period of microalgae cultured in all bioreactors as demonstrated by a relatively low biomass growth. BR-24L experienced a sudden increase in biomass concentrations (days 2–8) before entering the stationary phase at the 10th day of cultivation. BR-16L experienced a logarithmic growth phase in the 4th–8th day of cultivation whereas BR-12L and BR-8L experienced logarithmic growth phase later in the 8th–14th day and 10th–14th day of cultivation respectively. A trend was observed at which longer photoperiod induces faster microalgal biomass growth. Meanwhile, no logarithmic growth phase was observed in BR-0L indicating that the growth of *Chlorella vulgaris* was inhibited in heterotrophic culture.

At the end of cultivation study (14th day), the final biomass yields of 1.115 ± 0.003 g/L, 1.130 ± 0.001 g/L, 1.087 ± 0.040 g/L, 1.065 ± 0.020 g/L, and 0.359 ± 0.009 g/L were achieved in BR-24L, BR-16L, BR-12L, BR8-L, and BR-0L respectively. Although the final biomasses of BR-24L, BR-16L, BR-12L and BR-8L are comparable, BR-16L attained the highest biomass yield which suggested that under cyclic dark/hour conditions, cell production of biomass is higher.

3.2 Carbon (COD) Removal

COD measurement was carried out in the present study to estimate the carbon removal efficiencies in terms of COD removal from the synthetic wastewater media by *Chlorella vulgaris* at different photoperiods. The amount of organic carbon (sucrose) added to the bioreactors in this study generating an initial COD value of 103.63 mg/L. Figure 2 depicts the removal of COD under different photoperiods.

Rapid removal of COD was observed where about 80% COD removal was attained in all bioreactors on the second day of cultivation where similar COD removal patterns were noticed stipulating that there was no significant relationship between COD removal and cultivation of *Chlorella vulgaris* at different photoperiods. At the end of cultivation, the final concentrations of COD were found to be 6.72 ± 1.33, 8.64 ± 3.991, 13.44 ± 2.661, and 13.44 ± 2.661 mg/L in BR16-L, BR12-L, BR-8L

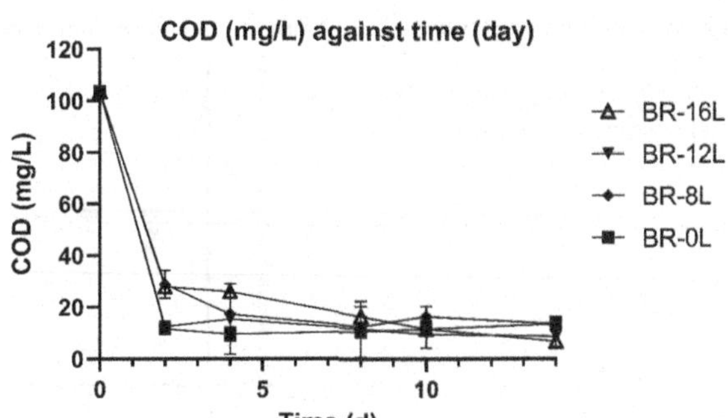

Fig. 2 COD removal by *Chlorella vulgaris* cultivated under various photoperiods

and BR-0L respectively. The almost complete removal of COD from microalgae culture suggested that complete biodegradable COD source was immediately being assimilated by *Chlorella vulgaris.*

3.3 Nitrogen Removal

Microalgae is known to be able to utilize nutrients in particular nitrogen and phosphorus for its growth [13]. Nitrogen could present in various form such as nitrite (NO_2^-), nitrate (NO_3^-), ammonium (NH_4^+–N), ammonia (NH_3), nitric acid (HNO_3) and nitrogen gas (N_2). In the present study, NH_4^+–N species was introduced in the synthetic wastewater medium. The removal of NH_4^+–N species efficiencies by *Chlorella vulgaris* cultured at different photoperiods is represented in Fig. 3.

From Fig. 3, high NH_4^+–N removal is observed in BR-24L followed by BR-12L, BR-16L and BR-0L with final concentration of 0, 0.38 ± 0.527, 3.15 ± 0.873, 15.75 ± 0.873 and 30.87 ± 0.873 respectively. BR-24L exhibited the fastest NH_4^+–N removal at which complete removal was observed at the 8th day of cultivation at which the microalgal growth has reached stationary growth phase. A similar trend was observed in BR-16L, BR-12L, and BR-8L where high removal of NH_4^+–N species is achieved as the microalgal growth entering stationary growth phase. Contrarily, the NH_4^+–N uptake by *Chlorella vulgaris* grown in BR-0L was impeded as no decline in NH_4^+–N concentrations is seen after second day of cultivation which assumed that NH_4^+–N was not assimilated by microalgae.

It is interesting to note that, although the intake of nitrogen was hindered in BR-0L, a prominent decrease in COD is seen in the second day of cultivation. An assumption was made that a complete consumption of the organic carbon source i.e. sucrose, occurred during the second day of cultivation. The notorious decrease in COD values caused by the complete degradation of organic carbon sources by the microalgae which explained the stagnant growth of microalgae in BR-0L (heterotrophic) even after 14 days of cultivation. The growth of microalgae grown under total dark condition is highly dependent on the presence of organic carbon source. Once the organic

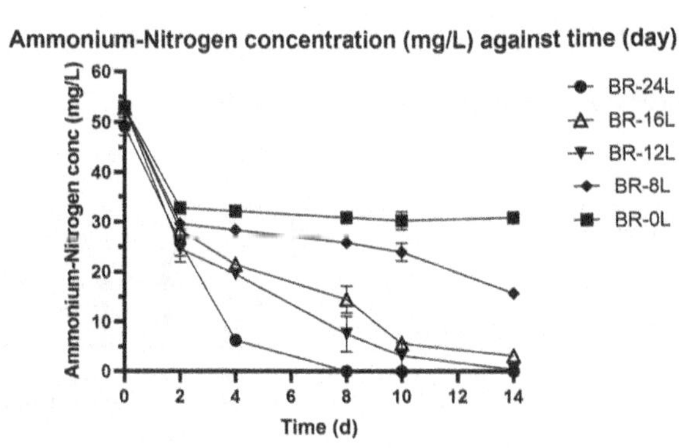

Fig. 3 Ammonium-nitrogen (NH_4^+–N) removal by *Chlorella vulgaris* cultivated under various photoperiods

carbon source was used up, the microalgal growth will be ceased as there is no source of energy left to support the cell metabolism for microalgal growth. The reasoning is further supported by the intake of NH_4^+–N species was prohibited as the assimilation of nitrogen into microalgal cell is largely associated with the availability of organic carbon source that serves as energy to convert NH_4^+–N into biodegradable nitrogen for microalgae assimilation [14]. Nonetheless, other bioreactors continue to demonstrate an increase in biomass growth and reduction in nitrogen concentration although organic compound was fully consumed. It can be inferred that after second day of cultivation, the growth of microalgae in other bioreactors were solely rely on the light energy via the photosynthetic pathway. It is also worth noting that higher NH_4^+–N removal efficiencies were achieved at longer photoperiods which is in tandem with the trend observed in the microalgal biomass growth.

4 Conclusion

This study aimed at finding the optimum photoperiods in enhancing biomass and lipid productions by investigating the effect of photoperiods on the biomass production, carbon and nutrient uptakes and lipid yields. Five different photoperiods were chosen to study in this study. Higher microalgal biomass yield accompanied by higher nitrogen removal by *Chlorella vulgaris* were observed under longer light photoperiod cycles. However, there were no remarkable differences observed in the light cycles of 24 h, 16 h and 12 h in the photoperiod study. Therefore, the 12 hphotoperiod duration is deemed to be the most ideal cultivation condition considering lower energy consumption for illumination viewing from an economic perspective. It can be deduced that the concentration of organic carbon source i.e. sucrose, introduced in the microalgal growth culture is relatively low leading to carbon deficit in the microalgal consortium. Dark/light cycle can counterbalance the shortage of organic carbon source by utilizing carbon dioxide as an inorganic carbon source in the presence of light for photosynthetic metabolic grow. As a result, high biomass growth was achieved under the 16 h, 12 h and 8 h light cycles of photoperiod whereas the growth of the heterotrophic condition (0 h) was totally inhibited.

Acknowledgements The financial supports from the Ministry of Higher Education Malaysia via Fundamental Research Grant Scheme (FRGS) with the cost center of 015MA0-110 (FRGS/1/2020/TK0/UTP/02/20), HICoE-Center for Biofuel and Biochemical Research with the cost center of 015MA0-052, Research Collaboration Grant UTP-UMP-UMT-UCTS with the cost center of 015MD0-019 and Yayasan Universiti Teknologi PETRONAS via YUTP-FRG with the cost center of 015LC0-126 are gratefully acknowledged.

References

1. Rashid, N., Rehman, M.S.U., Sadiq, M., Mahmood, T., Han, J.I.: Current status, issues and developments in microalgae derived biodiesel production. Renew. Sustain. Energy Rev. **40**, 760–778 (2014)
2. Khan, S.A., Rashmi, Hussain, M.Z., Prasad, S., Banerjee, U.C.: Prospects of biodiesel production from microalgae in India. Renew. Sustain. Energy Rev. **13**(9), 2361–2372 (2009)
3. Salama, E.S., Kurade, M.B., Abou-Shanab, R.A., El-Dalatony, M.M., Yang, I.S., Min, B., Jeon, B.H.: Recent progress in microalgal biomass production coupled with wastewater treatment for biofuel generation. Renew. Sustain. Energy Rev. **79**, 1189–1211 (2017)
4. Khoo, K.S., Chew, K.W., Yew, G.Y., Leong, W.H., Chai, Y.H., Show, P.L., Chen, W.H.: Recent advances in downstream processing of microalgae lipid recovery for biofuel production. Bioresour. Technol. **304**, 122996 (2020)
5. Vo, H.N.P., Ngo, H.H., Guo, W., Chang, S.W., Nguyen, D.D., Chen, Z., Wang, X.C., Chen, R., Zhang, X.: Microalgae for saline wastewater treatment: a critical review. Crit. Rev. Environ. Sci. Technol. **50**(12), 1224–1265 (2020)
6. Lam, M.K., Lee, K.T.: Microalgae biofuels: a critical review of issues, problems and the way forward. Biotechnol. Adv. **30**(3), 673–690 (2012)
7. da Silva Ferreira, V., Sant'Anna, C.: The effect of physicochemical conditions and nutrient sources on maximizing the growth and lipid productivity of green microalgae. Phycol. Res. **65**(1), 3–13 (2017)
8. Fields, M.W., Hise, A., Lohman, E.J., Bell, T., Gardner, R.D., Corredor, L., Moll, K., Peyton, B.M., Characklis, G.W., Gerlach, R.: Sources and resources: importance of nutrients, resource allocation, and ecology in microalgal cultivation for lipid accumulation. Appl. Microbiol. Biotechnol. **98**(11), 4805–4816 (2014)
9. Leong, W.H., Lim, J.W., Lam, M.K., Uemura, Y., Ho, Y.C.: Third generation biofuels: a nutritional perspective in enhancing microbial lipid production. Renew. Sustain. Energy Rev. **91**, 950–961 (2018)
10. Leong, W.H., Kiatkittipong, K., Kiatkittipong, W., Cheng, Y.W., Lam, M.K., Shamsuddin, R., Mohamad, M., Lim, J.W.: Comparative performances of microalgal-bacterial co-cultivation to bioremediate synthetic and municipal wastewaters whilst producing biodiesel sustainably. Processes **8**(11), 1427 (2020)
11. Leong, W.H., Lim, J.W., Lam, M.K., Uemura, Y., Ho, C.D., Ho, Y.C.: Co-cultivation of activated sludge and microalgae for the simultaneous enhancements of nitrogen-rich wastewater bioremediation and lipid production. J. Taiwan Inst. Chem. Eng. **87**, 216–224 (2018)
12. APHA, AWWA, WEP: Standard Methods for the Examination of Water and Wastewater, 22nd ed. American Public Health Association, Washington, DC (2012)
13. Perez-Garcia, O., Bashan, Y.: Microalgal heterotrophic and mixotrophic culturing for biorefining: from metabolic routes to techno-economics. In: Algal Biorefineries, pp. 61–131. Springer, Cham (2015)
14. Perez-Garcia, O., Escalante, F.M., de-Bashan, L.E., Bashan, Y.: Heterotrophic cultures of microalgae: metabolism and potential products. Water Res. **45**(1), 11–36 (2011)

Effect of Anodization Voltage on TiO_2 Nanotubes for Photodegradation of Formaldehyde

Nurul Tasnim Sahrin, Chong Fai Kait, Lee Siew Ling, and Mohd Dzul Hakim Wirzal

Abstract In this work, titanium dioxide nanotubes (TiO_2 NTs) were prepared via electrochemical anodization of titanium foil in choline chloride-glycerol ionic liquid as the electrolyte. The anodization voltage was varied from 5 to 50 V to investigate its effect on the properties of TiO_2 NTs. The NTs were characterized using FESEM and XRD analyses while the photocatalytic performance of the NTs was investigated for photodegradation of formaldehyde (HCHO) under visible light irradiation. Experimental results showed that the anodization voltage has an important impact on the surface morphology of TiO_2 NTs and the amorphous phase without heat treatment process. Higher anodization voltage led to pore rupture. TiO_2 NTs fabricated at anodization voltage of 30 V showed enhanced photocatalytic performance where 26.99 mg L^{-1} g^{-1} of removal after 120 min of reaction duration in the presence of visible light radiation.

N. T. Sahrin (✉) · C. F. Kait
Department of Fundamental and Applied Sciences, Universiti Teknologi PETRONAS, 32610 Seri Iskandar, Perak, Malaysia
e-mail: nurul_17007753@utp.edu.my

C. F. Kait
e-mail: chongfaikait@utp.edu.my

C. F. Kait
Centre of Innovative Nanostructures and Nanodevices (COINN), Institute of Autonomous System (IAS), Universiti Teknologi PETRONAS, 32610 Seri Iskandar, Perak, Malaysia

L. S. Ling
Center for Sustainable Nanomaterials, Ibnu Sina Institute for Scientific and Industrial Research, Universiti Teknologi Malaysia, Johor Bahru, Malaysia

Chemistry Department, Faculty of Science, Universiti Teknologi Malaysia, Johor Bahru, Malaysia

L. S. Ling
e-mail: lsling@utm.my

M. D. H. Wirzal
Department of Chemical Engineering, Universiti Teknologi PETRONAS, 32610 Seri Iskandar, Perak, Malaysia
e-mail: Mdzulhakim.wirzal@utp.edu.my

S. A. Abdul Karim et al. (eds.), *Proceedings of the 6th International Conference on Fundamental and Applied Sciences*, Springer Proceedings in Complexity,
https://doi.org/10.1007/978-981-16-4513-6_18

Keywords TiO_2 nanotubes · Ionic liquid · Anodization · Formaldehyde

1 Introduction

TiO_2 is the most important and versatile metal oxide semiconductor photocatalyst and finds an unequivocal prospect in environmental pollutant removal. However, its recovery from waste streams is the crucial drawback owing to its practical application for environmental pollutants removal. The fabrication of TiO_2 nanotubes (NTs) on a substrate is an effective approach to overcome this limitation. In addition, TiO_2 NTs offers several other benefits such as uniform morphologies, high surface-to-volume ratio, well-oriented growth as well as cost effective construction making them a promising functional material for environmental applications [1].

Recently, various techniques especially hydrothermal and electrochemical anodization have been employed to fabricate TiO_2 NTs [2]. Amongst the methods investigated so far, anodization has garnered great interest because of its ease and simplicity, greater control and reproducibility. Moreover, this technique produces strongly adherent nano porous TiO_2 layer on the substrate, which is more desirable for environmental pollutants removal. The TiO_2 NTs fabricated with anodization have been tested and proved useful in several applications including dye sensitized solar cells, photocatalysis, biomedical applications and gas sensing [3].

Several important synthesis parameters such as anodization voltage and time, electrolyte composition, post treatment temperature etc. are affecting the properties of TiO_2 NTs. It is reported that anodization voltage is one key parameter that effect the formation of TiO_2 nanotube arrays on the foil surface and straight channels against the foil [4]. The applied voltage can strongly influence the various properties of TiO_2 NTs namely, pore diameter, inter pore distance and film thickness [1]. Numerous research groups have investigated the effect of anodization voltage on the properties of TiO_2 NTs. However, there is limited literature related to the effect of anodization voltage on the photodegradation of environmental pollutants, especially formaldehyde (HCHO) removal from waste air streams.

The present study is aimed to fabricate TiO_2 NTs using glycerol mediated ionic liquid as the electrolyte. The IL electrolyte used in this study is greener, less toxic, has favourable electrochemical properties and more importantly it is cheaper and display high biocompatibility [5, 6]. The electrolyte was obtained by mixing choline chloride and glycerol in 1:2 molar ratio where glycerol acts as a complexing agent and hydrogen bond donor while choline chloride as hydrogen acceptor [7]. The formation of choline chloride-glycerol eutectic ionic liquid (EIL) was confirmed by the determination of its melting point which is lower than the melting point of the starting materials, mainly due to the formation of intermolecular hydrogen bonds during synthesis [8]. Different voltages from 5 to 50 V were applied for the fabrication of TiO_2 NTs via anodization and the properties including surface morphology and structural properties of the obtained TiO_2 NTs were determined using field emission scanning electron microscopy (FESEM) and X-ray diffraction

(XRD). The photocatalytic performance of the fabricated TiO_2 NTs was evaluated for the removal of gaseous HCHO under visible light irradiation.

2 Materials and Methodology

2.1 *Synthesis of Chloride-Based Electrolyte*

Choline chloride (ChCl, 98%, Sigma Aldrich) and glycerol ($C_3H_8O_3$, 98%, Fischer Scientific) were mixed at 1:2 molar ratio and heated for 30 min at 80 °C to form a colorless eutectic ionic liquid (EIL) [8, 9]. The EIL was used as the electrolyte for the anodization process of Ti foil to form TiO_2 NTs.

2.2 *Synthesis of TiO_2 NTs*

The substrate used for anodization process was technical grade titanium foil with thickness of 0.1 mm (Titanium, Ti Gr5/Tc4 Grade 5 ASTM B265 Thin Plate Sheet), was cut into 2 cm × 1 cm. The Ti substrate were ultrasonically washed in acetone for 10 min, followed by thorough rinsing with deionized water and dried in air. Anodization process was conducted using electrochemical set up and the details can be found in our previous work [10]. Briefly, the set up consisted of platinum rod as cathode and Ti substrate as anode 2.5 cm was the distance fixed between cathode and anode. The electrode was immersed in 35 mL of choline chloride-glycerol eutectic ionic liquid (ChCl-Gly EIL) as the electrolyte and the experiment was carried out at anodization voltages of 5, 10, 15, 20, 30, 40 and 50 V (DC power supply) for a fixed period of 1 h under ambient temperature and pressure. After anodization process, the obtained samples were removed instantly from the electrolyte solution and rinsed with deionized water. The samples were air-dried in ambient atmosphere. The as-synthesized TiO_2 NTs were further studied without undergoing any calcination process.

2.3 *Characterization of TiO_2 NTs*

In order to investigate the surface morphology of the synthesized TiO_2 NTs, the samples were characterized by Field emission scanning electron microscopy (FESEM) using Carl Zeiss instrument (SUPRA 55VP). The images were captured at 50 kX magnification at an acceleration voltage of 30 kV. The fabricated NTs were further analyzed using X-ray diffractometer (XRD, PANalytical X'Pert^3 Powder) with Cu Kα radiation (40 kV, 40 mA). The scanning was carried out at 2θ angle between 10° and 80° with the step size of 0.01°.

2.4 *Photocatalytic Degradation of Formaldehyde*

The photocatalytic activity for the removal of HCHO was evaluated in a batch mode. The titanium foil containing TiO_2 NTs was positioned in the 250 mL quartz photoreactor attached with a holder. HCHO (37% in aq. solution, Sigma Aldrich) of 3.22 ppm was stored in a tightly sealed stainless-steel container. HCHO vapor filled the quartz photoreactor by diffusion when the valves were opened. After 30 min of equilibration in the dark, the quartz photoreactor was illuminated with a halogen lamp (150 W) as the visible light source. The photocatalytic activity of HCHO was monitored by sampling at a regular interval of 30 min for a total of 120 min. A HCHO sensor meter (Hal Tech) connected at the outlet to measure the HCHO vapor concentration. The sensor can measure the HCHO concentration up to 10 ppm. During the experiments, the temperature of the photoreactor was maintained at 25 ± 1 °C using a cooling fan. At the end of the reaction, the remaining HCHO was purged with He. The photodegradation performance, X was investigated and calculated using Eq. (1):

$$X = \frac{C_o - C_t}{\mathrm{m}} \tag{1}$$

where, X denotes the concentration of HCHO (mg L^{-1}) removed per gram of TiO_2 NTs, C_o is the initial HCHO concentration (mg L^{-1}), C_t represents HCHO concentration at sampling time, t (mg L^{-1}) and m represents the mass of the TiO_2 NTs photocatalyst used (g).

3 Results and Discussion

3.1 *Characterization of TiO_2 NTs*

3.1.1 Morphology of TiO_2 NTs

The NTs morphology and structure are strongly influenced by the anodization condition, particularly the anodization voltage, as it is the important factor controlling the tube diameter [11]. Generally, NTs growth occurs proportional to the anodization voltage up to a voltage where dielectric breakdown of the oxide occurs. The effect of anodization voltage on the TiO_2 NTs formation was investigated with the objective to fabricate NTs with optimum pore diameter for improving photodegradation of HCHO under visible light illumination.

Figure 1 shows the FESEM images of anodized Ti foil in ChCl-Gly EIL at various anodization voltages. The evolution morphologies for TiO_2 NTs as a function of voltage are shown in Fig. 1a–h. No NTs formation was observed at a lower anodization voltages of 5 and 10 V as depicted in Fig. 1a, b. The results suggest that lower voltages are incapable of initiating NTs formation and growth in ChCl-Gly EIL electrolyte. Increasing the anodization voltage to 15 V produced a few poor-ordered TiO_2 NTs structures concentrated at one place on the surface of the Ti foil as can be seen in Fig. 1c. This is because the formation of oxide layer is too thin which retards the possibility of oxide layer dissolution to form the tubular structure [12].

Increasing the anodization voltage to 20 V, led to increased coverage of irregular shaped TiO_2 NTs on the surface of Ti foil as illustrated in Fig. 1d. Compared to the sample prepared at 20 V, the sample prepared at 30 V showed the formation of denser clusters of the TiO_2 NTs as presented in Fig. 1e. Nevertheless, it is important to note that further increase in anodization voltage beyond 30 V led to NTs rupture, disintegration, and less dense distribution of the tubular structure of the TiO_2 NTs which can be observed in Fig. 1f–h. The results suggest that anodization voltages of 30 V could produce TiO_2 NTs with well-defined pore structure and evenly distributed NTs on the surface of Ti foil. Apparently, the average inner tube diameter increased from 23.62 ± 3.52 nm to 35.31 ± 3.87 nm when the voltage was raised from 15 to 50 V as tabulated in Table 1. The results are consistent with those reported by Omidvar et al. [13].

In the present work, the NTs diameter was observed to increase with increasing anodization voltage which may be due to the severe electrical field dissolution which accelerates the formation of pore-like structures during the initial stage of anodization process [14]. Studies suggested that the difference in anodization voltage determines the strength of electrical field developed across the oxide layer and thus affects the tube diameter. Consequently, larger NTs were obtained at higher voltage [15]. Previous studies reported that NTs with diameter ranging between 30 and 90 nm can be produced by anodization in 0.25 wt% (NH_4F + ethylene glycol) solution + 2 vol% water at voltages between 30 and 60 V [4]. In addition, Gong et al. showed that an increase in TiO_2 NTs diameter could be obtained, ranging from 25 to 65 nm for anodization voltage ranging from 10 to 40 V where Ti foils were anodized in 0.5 wt% HF solution [16]. These studies showed that the NTs diameter was greatly influenced by the anodization voltage. In this work, the TiO_2 NTs with diameter from 23.62 ± 3.52 nm to 35.31 ± 3.87 nm were synthesized depending on the anodization voltage.

3.1.2 XRD Analysis

X-ray diffraction analysis was carried out to investigate the effect of anodization voltage on structural properties of the synthesized TiO_2 NTs. The XRD patterns of the TiO_2 NTs synthesized at various anodization voltages for 1 h duration are shown in Fig. 2. All the NTs samples were prepared without post-heat treatment or calcination process. In Fig. 2, all samples represent characteristic diffraction peaks

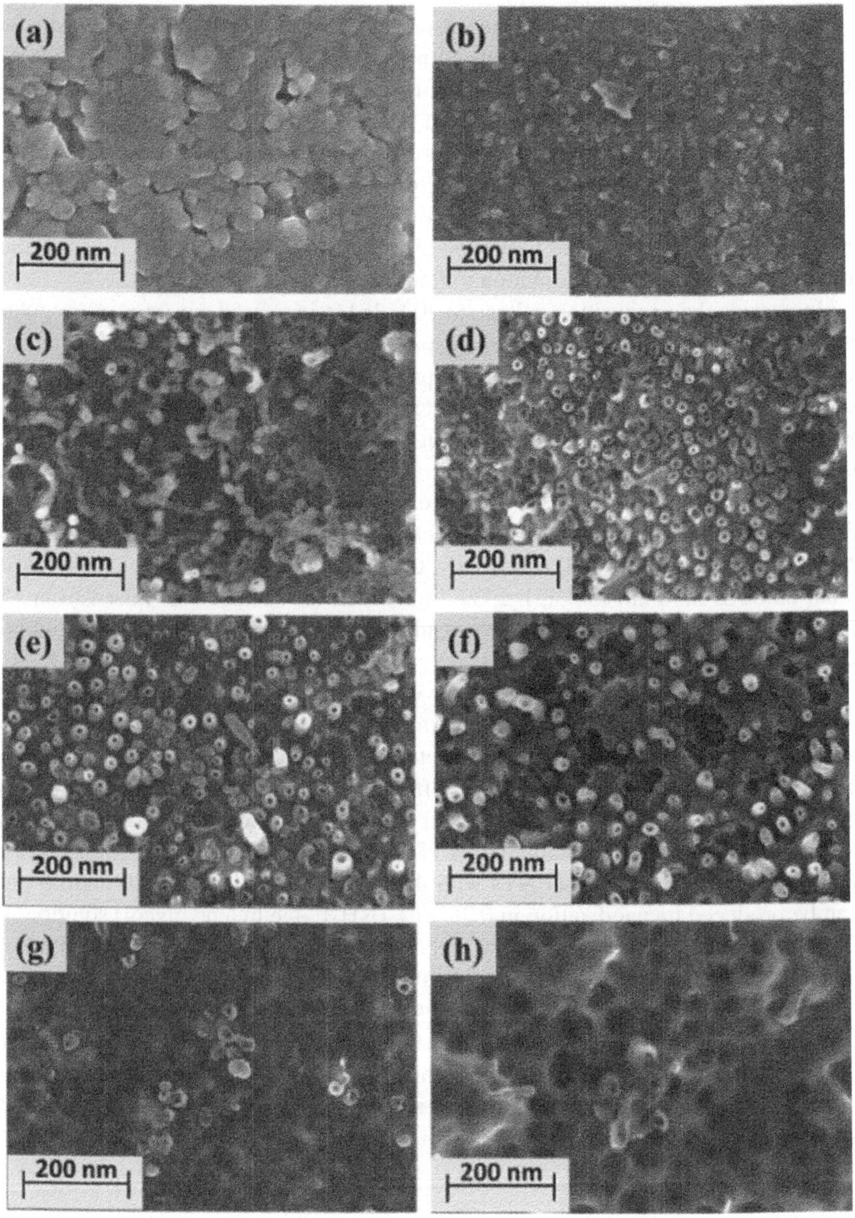

Fig. 1 FESEM images of TiO_2 NTs observed for 60 min at different anodization voltages: **a** 5 V, **b** 10 V, **c** 15 V, **d** 20 V, **e** 30 V, **f** 35 V, **g** 40 V and **h** 50 V at 50 kX magnification

Table 1 Average inner tube diameter of TiO_2 NTs anodized for 60 min at different anodization voltage

Sample	Anodization voltage (V)	Average inner tube diameter (nm)
(a)	5	–
(b)	10	–
(c)	15	23.62 ± 3.52
(d)	20	27.15 ± 4.69
(e)	30	33.87 ± 7.28
(f)	35	34.46 ± 6.11
(g)	40	35.72 ± 3.95
(h)	50	35.31 ± 3.87

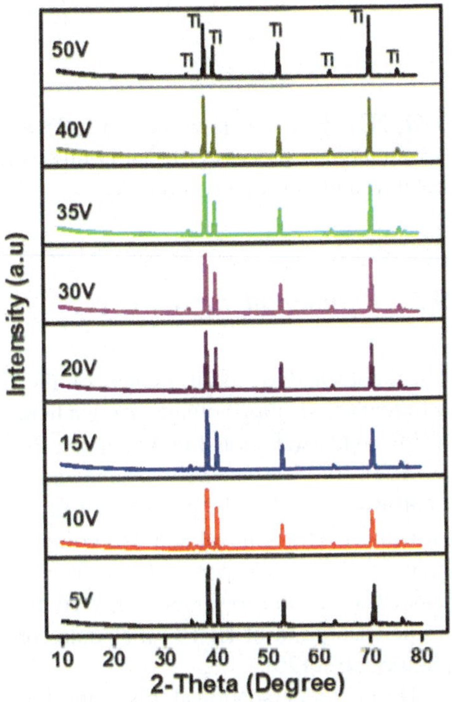

Fig. 2 XRD patterns of TiO_2 NTs anodized at different anodization voltages

for Ti metal. The diffraction peaks at 2θ of 34.9°, 35.8°, 40.3°, 52.9°, 62.9°, 70.6° and 76.3° correspond to (100), (002), (101), (102), (110), (103) and (112) planes, respectively [17, 18]. None of the diffraction peaks represent any of the crystalline structures of TiO_2. It can be observed that the anodization voltage has no significant effect on the structural properties of the synthesized TiO_2 NTs. The position and the intensity of the diffraction peaks showed no clearly observable change with the anodization voltage confirming the synthesis parameters have no significant effect on the structural properties especially the crystalline structure of the synthesized

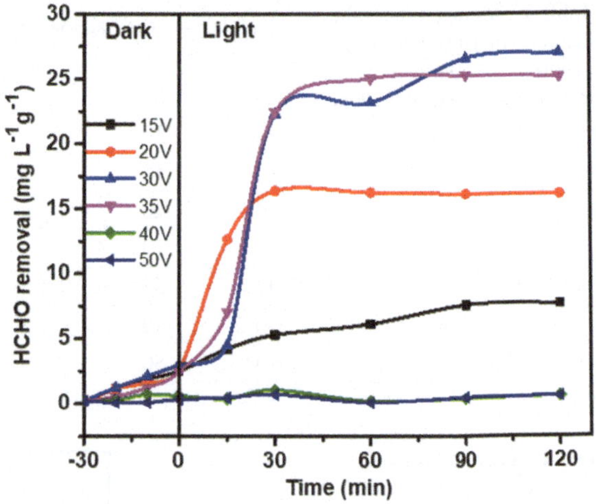

Fig. 3 Effect of anodization voltage on the photodegradation of formaldehyde

TiO_2 NTs. Based on previous work, TiO_2 NTs existed in amorphous phase without heat treatment process [19] but in this work the amorphous TiO_2 NTs were able to displayed photocatalytic property.

3.2 *Photocatalytic Activity*

The photocatalytic performance of the TiO_2 NTs photocatalyst synthesized at different anodization voltages was evaluated for the removal of gaseous HCHO under visible light irradiation. Figure 3 shows the removal of HCHO by TiO_2 NTs anodized at 15, 20, 30, 35, 40 and 50 V of anodization voltage for 1 h duration. Reaction without light irradiation but the presence of photocatalyst showed only ~3 mg L^{-1} g^{-1} of gaseous HCHO removal and was compared with the HCHO removal under visible light irradiation. The performance clearly shows that HCHO removal was enhanced under visible light irradiation. The HCHO removal without irradiation can be due to the adsorption of HCHO molecules on the in-wall of the reactor and/or on the photocatalyst [20].

The photodegradation ability of the TiO_2 NTs is strongly correlated to the average tube diameter and surface area. It can be seen from Fig. 3, sample anodized at 15 V showed only 7.70 mg L^{-1} g^{-1} of HCHO removal after 120 min of irradiation duration. However, further increase in the anodization voltage to 20 V led to an increment in the HCHO removal (16.17 mg L^{-1} g^{-1}). A noticeable improvement of HCHO removal can be seen for sample anodized at 30 V with 26.99 mg L^{-1} g^{-1} of removal after 120 min of reaction duration in the presence of light radiation, suggesting significant photocatalytic activity of the synthesized TiO_2 NTs. Nevertheless, further increase in the anodization voltage to 35, 40 and 50 V, the HCHO removal showed a decreasing trend which resulted in only 25.18, 0.65 and 0.64 mg L^{-1} g^{-1}, respectively. Overall,

HCHO removal shows increasing trends with increasing anodization voltage (15–35 V) from 29.37 to 75.94% of removal and decreases significantly to 11.25 and 10.94% for sample anodized at 40 and 50 V, respectively.

On the other hand, an increase in the anodization voltage led to production of larger average inner tube diameter. This further increased the surface area of the TiO_2 NTs, which facilitates in enhanced photodegradation of HCHO [17, 21]. Nonetheless, samples anodized at 35, 40 and 50 V result in decreased HCHO removal despite having large inner tube diameter. This could be due to the poor-ordered NTs structures which were unable to maintain the integrity of the tubular structure of the TiO_2 NTs synthesized at higher anodization voltages. The results suggest that the tubular structure integrity of the TiO_2 NTs is strongly correlated to the performance of HCHO removal. The TiO_2 NTs synthesized at 30 V displayed the highest performance towards HCHO removal due to their better ordered structure and large tube diameter.

4 Conclusion

TiO_2 NTs were successfully synthesized via electrochemical anodization of Ti foil in glycerol-mediated EIL based electrolyte. The effect of the anodization voltage on the surface morphology, structural properties and subsequent photocatalytic performance of the TiO_2 NTs for photodegradation of HCHO was investigated. The anodization voltage was found to greatly affect the surface morphology, pore size and distribution of the nanotubes on the surface of the Ti foil. At low anodization voltage, higher density and uniformly distributed TiO_2 NTs were formed while at higher anodization voltage the rupture and disintegration of the tubular structure was observed. All the synthesized photocatalyst maintained amorphous phase without heat treatment process. The optimum anodization voltage of 30 V was found to be the most effective for the removal of gaseous HCHO. The TiO_2 NTs fabricated at 30 V anodization voltage displayed 26.99 mg L^{-1} g^{-1} of removal of HCHO after 120 min of reaction duration in the presence of visible light radiation.

Acknowledgements The authors would like to thank the Centre of Research in Ionic Liquids (CORIL), Centre of Innovative Nanostructures and Nanodevices (COINN), and Photocatalyst Laboratory at Catalyst Research (CARE) Laboratory in Universiti Teknologi PETRONAS for providing the facilities. Research Grants awarded by the Ministry of Higher Education Malaysia (FRGS/1/2016/STG01/UTP/02/1), Yayasan Universiti Teknologi PETRONAS (YUTP 015LC0-276) and financial support from UTM-UTP CRG (015MC0-019) are greatly acknowledged.

References

1. Indira, K., Mudali, U.K., Nishimura, T., Rajendran, N.: A review on TiO_2 nanotubes: influence of anodization parameters, formation mechanism, properties, corrosion behavior, and biomedical applications. J. Bio Tribo Corros. **1**(4), 28 (2015)

2. Sreekantan, S., Saharudin, K., Wei, L.C.: Formation of TiO_2 nanotubes via anodization and potential applications for photocatalysts, biomedical materials, and photoelectrochemical cell. IOP Conf. Ser. Mater. Sci. Eng. **21**(1), 012002 (2011)
3. Regonini, D., Clemens, F.J.: Anodized TiO_2 nanotubes: effect of anodizing time on film length, morphology and photoelectrochemical properties. Mater. Lett. **142**, 97–101 (2015)
4. Qin, L., Chen, Q., Lan, R., Jiang, R., Quan, X., Xu, B., Zhang, F., Jia, Y.: Effect of anodization parameters on morphology and photocatalysis properties of TiO_2 nanotube arrays. J. Mater. Sci. Technol. **31**(10), 1059–1064 (2015)
5. Ju, Y.J., Lien, C.H., Chang, K.H., Hu, C.C., Wong, D.S.H.: Deep eutectic solvent-based ionic liquid electrolytes for electrical double-layer capacitors. J. Chin. Chem. Soc. **59**(10), 1280–1287 (2012)
6. Baokou, X., Anouti, M.: Physical properties of a new deep eutectic solvent based on a sulfonium ionic liquid as a suitable electrolyte for electric double-layer capacitors. J. Phys. Chem. C **119**(2), 970–979 (2015)
7. Zaidi, W., Boisset, A., Jacquemin, J., Timperman, L., Anouti, M.: Deep eutectic solvents based on N-methylacetamide and a lithium salt as electrolytes at elevated temperature for activated carbon-based supercapacitors. J. Phys. Chem. C **118**(8), 4033–4042 (2014)
8. Jaihindh, D.P., Fu, Y.P.: Facile synthesis of deep eutectic solvent assisted BiOCl/$BiVO_4$@AgNWs plasmonic photocatalysts under visible light enhanced catalytic performance. Catal. Today **297**, 246–254 (2017)
9. Shaabani, A., Afshari, R.: Magnetic Ugi-functionalized graphene oxide complexed with copper nanoparticles: efficient catalyst toward Ullman coupling reaction in deep eutectic solvents. J. Colloid Interface Sci. **510**, 384–394 (2018)
10. Sahrin, N.T., Nawaz, R., Chong, F.K., Lee, S.L., Wirzal, W.D.H.: Visible light photodegradation of formaldehyde over TiO_2 nanotubes synthesized via electrochemical anodization of titanium foil. Nanomaterials **10**(1) (2020)
11. Zhang, W., Liu, Y., Guo, F., Liu, J., Yang, F.: Kinetic analysis of the anodic growth of TiO_2 nanotubes: effects of voltage and temperature. J. Mater. Chem. C **7**(45), 14098–14108 (2019)
12. Ismail, S., Khairul, K.A., Hisham, N.A.A.N., Mamat, M.S., Azam, M.A.: Effect of voltage on TiO_2 nanotubes formation in ethylene glycol solution. Jurnal Teknologi **79**(5–2), 117–120 (2017)
13. Omidvar, H., Goodarzi, S., Seif, A., Azadmehr, A.R.: Influence of anodization parameters on the morphology of TiO_2 nanotube arrays. J. Superlattices Microstruct. **50**(1), 6–39 (2011)
14. Lai, C.W., Sreekantan, S.: Effect of applied potential on the formation of self-organized nanotube arrays and its photoelectrochemical response. J. Nanomater. 1–7 (2011)
15. Kapusta-Kołodziej, J., Syrek, K., Pawlik, A., Jarosz, M., Tynkevych, O., Sulka, G.D.: Effects of anodizing potential and temperature on the growth of anodic TiO_2 and its photoelectrochemical properties. J. Appl. Surf. Sci. **396**, 1119–1129 (2017)
16. Gong, D., Grimes, C.A., Varghese, O.K., Hu, W., Singh, R.S., Chen, Z., Dickey, E.C.: Titanium oxide nanotube arrays prepared by anodic oxidation. J. Mater. Res. **16**(12), 3331–3334 (2001)
17. Li, H., Wang, G., Niu, J., Wang, E., Niu, G., Xie, C.: Preparation of TiO_2 nanotube arrays with efficient photocatalytic performance and super-hydrophilic properties utilizing anodized voltage method. J. Results Phys. **14**, 102499 (2019)
18. Ng, S., Yam, F.K., Beh, K., Hassan, Z.: Titanium dioxide nanotubes in chloride based electrolyte: an alternative to fluoride based electrolyte. Sains Malaysiana **43**(6), 947–951 (2014)
19. Li, G., Liu, Z.Q., Lu, J., Wang, L., Zhang, Z.: Effect of calcination temperature on the morphology and surface properties of TiO_2 nanotube arrays. J. Appl. Surf. Sci. **255**(16), 7323–7328 (2009)
20. Yu, L., Wang, L., Sun, X., Ye, D.: Enhanced photocatalytic activity of rGO/TiO_2 for the decomposition of formaldehyde under visible light irradiation. J. Environ. Sci. China **73**, 138–146 (2018)
21. Yu, J., Wang, B.: Effect of calcination temperature on morphology and photoelectrochemical properties of anodized titanium dioxide nanotube arrays. J. Appl. Catal. B Environ. **94**(3), 295–302 (2010)

Molecular Dynamics Simulation of H_2S Solubility in Protic Ionic Liquids

Sorfina Amran, Mohamad Amirul Ashraf Mohd Razip, Khairulazhar Jumbri, and Mohd Faisal Taha

Abstract Amine solvent commonly used to remove H_2S especially in oil and gas industries. However, amine solvent has major drawbacks such as, high flammability, high volatility, and less efficiency in capturing H_2S. Hence, understanding the behaviour of molecules at molecular level and solubility of H_2S in protic ionic liquids (PILs) becomes the fundamental interest. In this work, molecular dynamics (MD) simulation with an aid of Bennet Acceptance Ratio (BAR) method was employed to unravel the key factors of the H_2S solubility in three PILs namely [2MAES], [2PAES], and [2BAES]. The predicted solvation free energy of H_2S in these PILs and water showed negative value with the lowest value obtained by 2BAES with ΔG of −19.42 kJ mol^{-1}. This indicated that H_2S was highly soluble in 2BAES compared to the other two PILs, having a value of −12.74 kJ mol^{-1} and −17.73 kJ mol^{-1} for 2MAES and 2PAES, respectively. To support the understanding, radial distribution function (RDF) and self-diffusion coefficient (D) were calculated. From these analyses, increasing the mobility of cation–anion resulted in an increase the H_2S solubility.

Keywords Solvation free energy · Hydrogen sulfide · Bennet acceptance ratio · Molecular dynamics · Protic ionic liquids

S. Amran · M. A. A. M. Razip · K. Jumbri · M. F. Taha (✉)
Department of Fundamental and Applied Sciences, Universiti Teknologi PETRONAS, 32610 Seri Iskandar, Perak, Malaysia
e-mail: faisalt@utp.edu.my

S. Amran
e-mail: sorfina_22386@utp.edu.my

K. Jumbri
e-mail: khairulzahar.jumbri@utp.edu.my

K. Jumbri · M. F. Taha
Centre of Research in Ionic Liquids (CORIL), Universiti Teknologi PETRONAS, 32610 Seri Iskandar, Perak, Malaysia

S. A. Abdul Karim et al. (eds.), *Proceedings of the 6th International Conference on Fundamental and Applied Sciences*, Springer Proceedings in Complexity,
https://doi.org/10.1007/978-981-16-4513-6_19

1 Introduction

Nowadays, the oil and gas industries are being more thoughtful to study effective removal of H_2S that has high toxicity and corrosive properties which is commonly produced from shale gas and desulfurization processes. Generally, amine solution is a well-known technology for removal of acid gases. Although the amine solution is very opportune, but it still has serious problems such as high flammability, high volatility and apart from that, the absorption process is very costly. Moving forward to solve these problems, ionic liquids (ILs) became a suitable candidate to replace amine solvent due to their good traits which are low volatility, less toxic and non-flammable. The solubility study of H_2S in ILs can be determined from wet-lab experiments and computational studies. However, the former is not preferable as it is expensive, time-consuming, and most importantly, needs proper equipment since H_2S is very toxic and corrosive resulting in scarce information on the H_2S solubility in ILs. Moreover, the effect of ILs' ions on the solubility of H_2S is still unanswered from the previous studies since wet experiment cannot provide an answer at the molecular level.

Literally, there is a lot of data available for a gas solubility in ILs such as CO_2, N_2 and etc. but not for H_2S. Pomelli et al. [1] investigated the influence of the interaction between H_2S and ILs on the H_2S solubility using NMR spectroscopy and quantum mechanics at 1400 kPa and 25 °C. They used [bmim]-based ILs with different anions. From the study they found that the solubility of H_2S decreased in the order of (expressed in mole fraction) [bmim][Cl] (0.86) > [bmim][BF_4] (0.79) > [bmim][TfO] (0.78) > [bmim][Tf_2N] (0.77) > [bmim][PF_6] (0.72). As for $[Tf_2N]^-$ anion with different cations, there was a slight cation effect on the solubility of H_2S ranging from 0.72 to 0.90 (express in mole fraction). Ab initio calculation showed that H_2S molecules made a good interaction with $[Cl]^-$, $[TfO]^-$, $[BF_4]^-$, and $[Tf_2N]^-$ since it showed a strong polar-polar activity with hydrogen bond. The interaction of H_2S with anion-ILs showed a comparable strength to the conventional hydrogen bond which is responsible to the high solubility of H_2S [1]. Sakhaeinia et al. [2] used different types of ILs which are 1-ethyl-3-methylimidazolium hexafluorophosphate ([emim][PF_6]) and 1-ethyl-3-methylimidazolium bis(trifluromethyl)sulfonylimide ([emim][Tf_2N]). In their study, the data were correlated using two models which were Krichevsky–Kasarnovsky (KK) equation, and the extended Henry's law combined with the Pitzer's virial expansion for the excess Gibbs energy. The result showed that the solubility of H_2S in [emim][Tf_2N] was higher than [emim][PF_6] due having a bigger anion size hence lower molar density. In other words, it has more volume to facilitate H_2S compared to $[PF_6]^-$ anion [2].

According to Jalili et al. [3], they found that increasing the temperature will decrease the solubility and by increasing the *n* of [$C_n$mim][Tf_2N], the solubility of H_2S in ILs also increase. Therefore, from this literature review, it shows that the length of alkyl chain, and temperature play a main role in order to increase the solubility of H_2S in ILs. Huang et al. [4] studied the absorption of H_2S and CO_2 in PILs using ammonium-based cation with formate and acetate as counter cations. They found that PILs encompassed a higher absorption capacity of H_2S than normal

ILs. In addition, the solubility of CO_2 showed a lower magnitude than H_2S. In terms of ideal selectivity, CO_2/H_2S showed greater and higher ideal selectivity in PILs as compared to common ILs [4]. Throughout this literature, the gas solubility data for H_2S is still scarce and only a few studies reported the H_2S solubility in ILs [5–7].

Thus, to answer these questions, the proper study at molecular level through molecular dynamics (MD) simulation is necessary in order to understand the mechanism of interaction between H_2S and ILs. Therefore, in this work, three PILs namely 2-methylamino ethanol salicylate [2MAES], propylamino ethanol salicylate [2PAES], and 2-butylamino ethanol salicylate [2BAES] were used to predict the solubility of H_2S through MD simulation for better understanding in terms of solvation behaviour.

2 Computational Modelling

2.1 Molecular Dynamics Simulation of Neat PILs

As for molecular dynamic (MD) simulation of neat PILs, the cubic simulation box with a dimension of 4.5 nm × 4.5 nm × 4.5 nm was created using Packmol. The number of molecules required in the simulation box implemented with periodic boundary condition (PBC) were calculated based on the PILs' density. Table 1 tabulates the number of ions used in MD simulation. Figure 1 shows the 3D structures of cations and anion of PILs.

MD simulations for all neat PILs were conducted using the GROMACS simulation package. Each system was energy minimized first using steepest descent followed by conjugate gradient with 5000 steps for both algorithms. Afterwards, the systems were subjected to pre-production canonical ensemble system (NVT) with 5 ns implemented with Berendsen thermostat to control the temperature. The equilibration process was continued in isobaric isothermal ensemble (NPT) with 30 ns time run implemented with Berendsen thermostat and Berendsen barostat, respectively. Both simulations of NVT and NPT run with different temperatures starting from 303 to 353 K.

Table 1 Number of molecules used in the simulation box for MD simulation

Entry	PILs	Number of molecules	
		Cation	Anion
1	[2MAES]	298	298
2	[2PAES]	256	256
3	[2BAES]	252	252

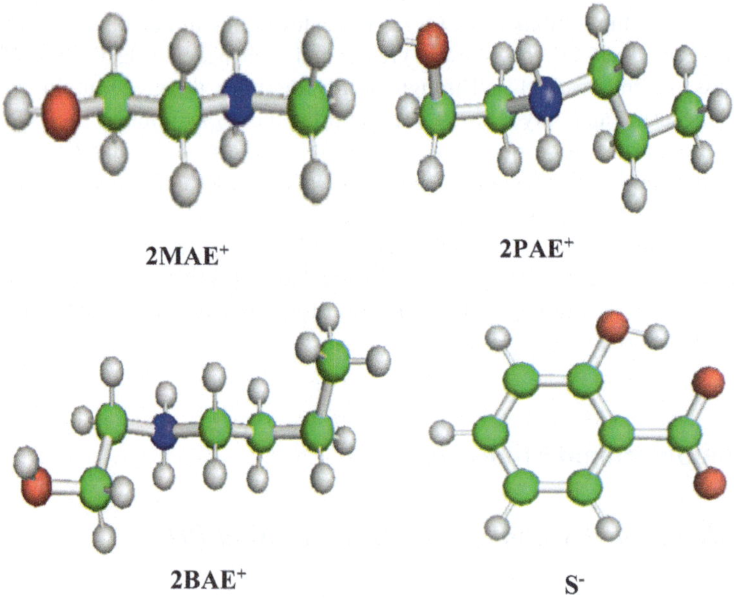

Fig. 1 The 3D structures of cations and anion [8]

2.2 *Free Energy Simulation*

The solvation free energy of H_2S in three PILs were theoretically calculated using Bennett Acceptance Ratio (BAR) method. The system was set up by using the same method as mentioned in previous section. However, a smaller 27 nm^3 cubic simulation box was used. Table 2 lists the number of molecules used in the simulation box for solvation free energy.

Technically, BAR method is more efficient than Thermodynamic Integration (TI) in terms of average and more robust due to less depending on the choice of intermediate states [9]. The details explanation of Bennett's equation is shown in Eq. 1 below:

Table 2 Number of molecules used in simulation box for solvation free energy

Entry	System	Number of molecules		
		Cation	Anion	Water
1	[2MAES]	88	88	–
2	[2PAES]	76	76	–
3	[2BAES]	70	70	–
4	SPC216	–	–	907

$$\Delta G_{ji}^{BAR} - K_B T\left(In\frac{f(H_i - H_j + C)j}{f(H_j - H_i - C)i}\right) + C \tag{1}$$

Above equation shows the first step by developing the free energy calculation to know how many lambda, λ points are required to explain the transformation from state i and j, where f is the fermi function, the K_B is the Boltzmann constant and T is temperature as shown in Eq. 2. The H_i andH_j represent the Hamiltonians at states i and j.

$$f(x) = \frac{1}{1 + \exp\left(\frac{x}{k_B T}\right)} \tag{2}$$

The value of C is determined repeatedly to achieve $f(H_i - H_j + C)j = f(H_j - H_i - C)i$ state. Later the difference free energy between the states is gained from Eqs. 3 and 4:

$$\Delta \mathrm{G}_{ji}^{\mathrm{BAR}} = -k_B T In\frac{N_j}{N_i} + C \tag{3}$$

$$\Delta \mathrm{G}_{ji}^{\mathrm{BAR}} = \sum_{i=1}^{n-1} \Delta \mathrm{G}_{i+1,i}^{\mathrm{BAR}} \tag{4}$$

where, N_i and N_j indicate the coordinates frames at λ_i and λ_j. Technically in this process the convergences can be reached if the energy difference between the forward (FW) and the backward (BW) give a sufficient overlap which is a pivotal part for BAR method, and all can be monitored by overlapping integral gained from normalized histograms of the energy differences [10]. To obtain a valid result, the overlap integral should be at least 0.01. However, if the overlap is insufficient, it will cause the iterative procedure to not converge nor produce an invalid result [10].

$$\Delta U_l^{\mathrm{FW}} - U(\lambda_{l+1}) - U(\lambda_l)|_{\lambda_l} \tag{5}$$

$$\Delta U_{l+1}^{\mathrm{BW}} = U(\lambda_l) - U(\lambda_{l+1})|_{\lambda_l} \tag{6}$$

The λ symbol in Eqs. 5 and 6 indicate the λ value where the trajectory is generated. By overlaying the histogram of ΔU_l^{Fw} and ΔU_{l+1}^{Bw} sometime the overlap cannot be detected due to the difficulty in assessing the correct amount of overlap by visual inspection. Thus, to improve this difficulty, the overlap integral $OI_{l,l+1}$ was used. Back to Eqs. 5 and 6, the λ_l and λ_{l+1} is defined as:

$$OI_{l,l+1}^{\mathrm{BAR}} = 2.\sum_{b=1}^{B} \frac{h_b(\Delta U_l^{\mathrm{FW}}).h_b(\Delta U_{l+1}^{\mathrm{BW}})}{h_b(\Delta U_l^{\mathrm{FW}}) + h_b(\Delta U_{l+1}^{\mathrm{BW}})} \tag{7}$$

where h_b (X) is a normalised histogram of quantity X with the sum up of b bins. In fact, non-zero contribution to $OI_{l,l+1}^{BAR}$ only happens for overlap zone which from the definition it stated that 0 means no overlap while 1 shows a perfect overlap [10]. As for this simulation, 21 equidistant set of λ values were used from $\lambda = 0$ to $\lambda = 1$ with 5 ns simulation time for each lambda. The soft-core energy parameters were set to $\alpha LJ = 0.5$ and soft-core sigma $= 0.3$, while the λ power dependency was set to 2 [10]. The post-processing of BAR which was coordinates, velocities and $\delta H/\delta\lambda$, were saved to a disk every 5 steps.

Steepest descent method was used for energy minimization with 5000 steps followed by 5000 steps of a conjugate gradient for each node. The system was subjected to pre-equilibrated in the canonical ensemble (NVT) for 100 ps with no position restraints applied [7]. Then the system was continued in isobaric-isothermal (NPT) with 2 ns at 298 K and 1.0 bar, respectively for all systems. At the last 1 ns, the data was recorded. Post-processing of the data is necessary and pivotal for the calculations of free energy with BAR. The energy difference between the system at H_i and H_j is essential to be calculated over the trajectories simulated at H_i and H_j separately. Thus, the non-bonded, bonded and kinetic energies required a re-calculation for the post-processing step. Figure 2 illustrates the cubic simulation box containing [2BAES] ions. In this simulation, the OPLS force field parameters were used to perform all simulations and BAR calculations. For the effective treatment of an all-atom system, GROMACS package version 4.5 was used since it offered highly optimized and parallelized algorithms.

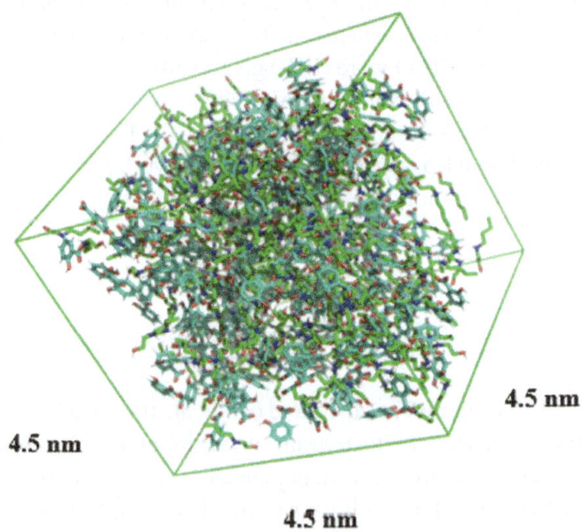

Fig. 2 A snapshot of [2BAES] molecules inside a cubic simulation box

3 Results and Discussion

3.1 Force Field Validation

In molecular dynamics (MD) simulation, force field parameters are very important in order to produce good results and accurately predict the solvation free energy. Most commonly, to validate the force field parameter is by validating the predicted properties such as density against experimental data. Thus, it will be a focus on the force field validation against experimental density. Technically, the properties that affect the PILs' density are size and shape of the ions, the interaction of cation–anion and molecular packing [11, 12]. In this research, the predicted densities of three PILs namely [2MAES], [2PAES], and [2BAES] at temperatures ranging from 303 to 353 K were obtained from NPT simulation and compared with the experimental results.

Generally, the force field parameters will be accepted when the error is less than 5%. The results tabulated in Table 3 shows the highest error value of 3.90% at 353 K and the lowest is 0.12% at 303 K, thus indicating that the OPLS force field used is acceptable and in agreement with experimental. The density decreased in the order of

Table 3 The list of predicted density obtained from MD simulation against experimental data

PILs	T/(K)	Exp, ρ (kg/m^3)[a]	Sim, ρ (kg/m^3)	% Error
[2MAES]	303	1151.7	1176.8	2.17
	313	1145.3	1174.5	2.55
	323	1138.9	1169.0	2.65
	333	1132.6	1168.8	3.19
	343	1126.2	1165.2	3.46
	353	1119.8	1163.5	3.90
[2PAES]	303	1126.6	1122.5	0.37
	313	1119.8	1117.3	0.22
	323	1113.3	1111.7	0.14
	333	1106.9	1113.3	0.58
	343	1100.3	1112.5	1.11
	353	1093.2	1105.9	1.17
[2BAES]	303	1103.7	1105.0	0.12
	313	1097.0	1098.8	0.17
	323	1090.4	1096.6	0.57
	333	1083.7	1093.7	0.92
	343	1077.1	1085.1	0.75
	353	1070.0	1084.6	1.36

The experimental density data were obtained from [a]Ahmad et al. [8]

Table 4 The solvation free energy obtained from free energy difference with excess chemical potential and Henry's constant

Entry	System	ΔG (kJ mol^{-1})	Std. dev	μ^{ex} (kJ mol^{-1})	k_H (atm)
1	Water	−4.52	±0.13	−2.37	519.9
				−2.31[a]	539.9[b]
2	[2MAES]	−12.74	±0.31	−2.86	42.49
3	[2PAES]	−17.73	±0.41	−3.48	27.92
4	[2BAES]	−19.42	±0.31	−3.81	22.76

The experimental data of H_2S in water was obtained from [a]Riahi and Rowley [16] and [b]Glew and Hames [17]

[2MAES] > [2PAES] > [2BAES]. As the chain length increased, the charge density of cation decreased and thus increase in molar volume [13]. According to Hawker et al. [14], this is because the ions are unable to pack perfectly due to the flexibility of the alkyl chain length resulting in the decrease of density.

3.2 *Free Energy of H_2S*

Solvation free energy is used to evaluate or estimate numerous physicochemical characteristics of a molecule [15]. The H_2S solubility in a liquid can be used to measure the degree of removal of the H_2S and is quantified by Gibbs free energy (ΔG) and Henry's constant (kH). In this research, the force field of H_2S was validated in water against experimental data since there is unavailable experimental data of the solubility of H_2S in these PILs. As tabulates in Table 4, the result shows that the solubility data of H_2S in water through MD simulation was in good agreement with the experimental and can be used to correctly calculate the solubility of H_2S in these PILs. Figure 3 shows the graph of free energy.

Generally, the solvation process is favorable and easy to occur if the total value of Gibbs free energy (ΔG) is negative [18]. From this term, it shows that the result obtained for PILs from MD simulation is in negative values which indicated that H_2S was soluble in PILs and slightly soluble in water. From these PILs, [2BAES] shows the most negative value of −19.42 kJ mol^{-1} indicating that H_2S was most soluble in [2BAES] as compared to the other PILs. Technically, when the alkyl length chain of cation increased, more negative values of ΔG was produced ranging from −12.74 to −19.42 kJ mol^{-1}. The possible reason behind this was due to the free volume effect which can be related to the interaction energy between the cation and anion. Fundamentally by having a weak interaction energy between cation and anion, it provides a space (free volume) to accommodate the H_2S in PILs and provide a greater freedom of movement in self-diffusion coefficient of PILs resulting in an increase of the H_2S solubility in PILs [19, 20].

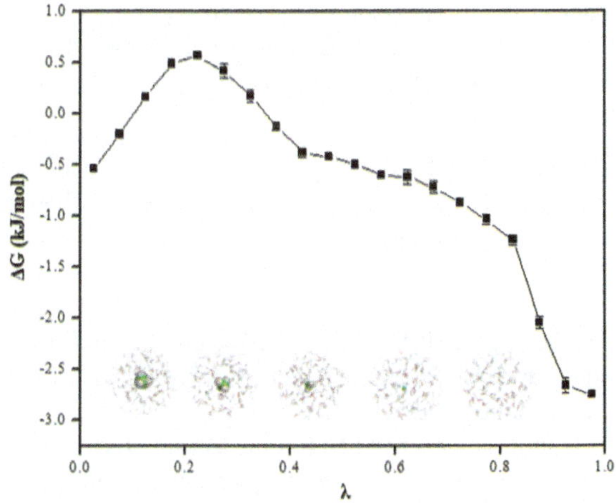

Fig. 3 The graph of free energy

The strength of interaction energy of cation–anion can be seen through the qualitative analysis of radial distribution function (RDF). Basically, RDF works to find the probability of a pair of atoms or molecules in a shell, $g(r)$ up to a distance of r with another atom chosen as references point. Referring to Fig. 4, the distance between cation and anion of [2MAES] is approximately about ~0.50 nm, while [2PAES] and [2BAES] were ~0.51 nm and ~0.52 nm, respectively. From these distances, it can be seen that [2MAES] formed a stronger interaction compared to the other PILs [21]. However, the distance between the cation and anion of [2MAES], [2PAES] and [2BAES] showed a less significant difference. Thus, to support this qualitative analysis, the quantitative analysis was provided by calculating the self-diffusion coefficient (D) of cation and anion obtained from the linear fitting slope of mean square

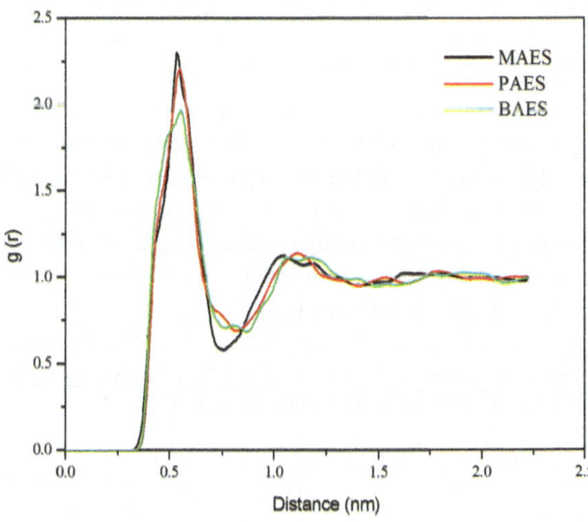

Fig. 4 The RDF graph of [2MAES], [2PAES] and [2BAES], respectively

Table 5 The self-diffusion coefficient (D) from a slope of MSD plot (with a standard deviation in parentheses) and cationic (t_+) and anionic (t_-) transferences numbers for cations and anion

System	Self-diffusion coefficient (D) ($\times 10^{-11}$ m^2 s^{-1})			t_+	t_-
	D_{cation}	D_{anion}	D_{system}		
[2MAES]	1.13 (0.008)	0.74 (0.004)	0.84 (0.006)	0.60	0.40
[2PAES]	1.24 (0.005)	0.88 (0.007)	1.03 (0.008)	0.58	0.41
[2BAES]	1.35 (0.004)	0.86 (0.006)	1.06 (0.005)	0.61	0.39

displacement (MSD). Self-diffusion coefficient (D) is used to measure and study the liquid dynamic in a dense fluid. Further details of the self-diffusion coefficient (D) are given in Table 5.

It shows that [2MAES] has the lower value of D_{system} in self-diffusion coefficient which was 0.84 × 10^{-11} m^2 s^{-1} in contrast with the other two PILs which were 1.03 × 10^{-11} m^2 s^{-1} for [2PAES] and 1.06 × 10^{-11} m^2 s^{-1} for [2BAES], respectively. This showed that [2MAES] formed a strong interaction between cation–anion that restricted its mobility to be slowly diffuse and reduce the space between the two resulting in less free volume. Ultimately, from these three PILs, the transport number of cation, t_+ are higher compared to the transport number of anion, t_-. The higher the value of the transport number, the faster the ions will diffuse. Thus, this indicates that the cations diffused faster than anions and contributed to a stronger ionic conductivity in the electrolyte.

4 Conclusion

In this work, the force field parameters for [2MAES], [2PAES], [2BAES] were successfully validated against experimental density with the percentage error less than 5%. The solvation free energy of H_2S in PILs were successfully reproduced using BAR method with the value of ΔG obtained for H_2S in these PILs were ranging from −12.74 to −19.42 kJ mol^{-1}, respectively at 298 K and 1.0 bar. The more the negative value of ΔG, the higher the solubility of H_2S in PILs. These results indicated that the solubility of the gas was highest in [2BAES]. To understand the solvation mechanism, qualitative and quantitative analyses were calculated and evaluated. The quantitative analysis showed that higher D_{system} value provide more free volume for H_2S to accommodate and form interaction between ions resulting in an increase H_2S solubility in PILs.

Acknowledgements All authors gratefully acknowledge the financial support by Yayasan Universiti Teknologi PETRONAS Grant (015LC0-074).

References

1. Pomelli, C.S., Chiappe, C., Vidis, A., Laurenczy, G., Dyson, P.J.: Influence of the interaction between hydrogen sulfide and ionic liquids on solubility: experimental and theoretical investigation. J. Phys. Chem. B **111**(45), 13014–13019 (2007)
2. Sakhaeinia, H., Jalili, A., Taghikhani, V., Safekordi, A.: Solubility of H_2S in ionic liquids 1-ethyl-3-methylimidazolium hexafluorophosphate ([emim][PF_6]) and 1-ethyl-3-methylimidazolium bis(trifluoromethyl)sulfonylimide ([emim][Tf_2N]). J. Chem. Eng. Data **55**(12), 5839–5845 (2010)
3. Jalili, A.H., Safavi, M., Ghotbi, C., Mehdizadeh, A., Hosseini-Jenab, M., Taghikhani, V.: Solubility of CO_2, H_2S, and their mixture in the ionic liquid 1-octyl-3-methylimidazolium bis(trifluoromethyl)sulfonylimide. J. Phys. Chem. B **116**(9), 2758–2574 (2012)
4. Huang, K., Zhang, X., Xu, Y., Wu, Y., Hu, X., Xu, Y.: Protic ionic liquids for the selective absorption of H_2S from CO_2: thermodynamic analysis. AIChE J. **60**(12), 4232–4240 (2014)
5. Salehin, F.N.M., Jumbri, K., Ramli, A., Daud, S., Abdul Rahman, M.B.: In silico solvation free energy and thermodynamics properties of H_2S in cholinium-based amino acid ionic liquids. J. Mol. Liq. **294**, 111641 (2019)
6. Iman Ishak, M.A., Taha, M.F., Hakim Wirzal, M.D., Nordin, M.N., Abdurrahman, M., Jumbri, K.: Choline-based ionic liquids-incorporated IRMOF-1 for H_2S/CH_4 capture: insight from molecular dynamics simulation. Processes **8**(4), 412 (2020)
7. Ishak, M.A.I., Jumbri, K., Daud, S., Abdul Rahman, M.B., Abdul Wahab, R., Yamagishi, H., Yamamoto, Y.: Molecular simulation on the stability and adsorption properties of choline-based ionic liquids/IRMOF-1 hybrid composite for selective H_2S/CO_2 capture. J. Hazard. Mater. **399**, 123008 (2020)
8. Ahmad, N.A., Jumbri, K., Ramli, A., Ahmad, H., Rahman, M.B.A., Wahab, R.A.: Design and molecular modelling of phenolic-based protic ionic liquids. J. Mol. Liq. **308**, 113062 (2020)
9. Gonçalves, P., Stassen, H.: Calculation of the free energy of solvation from molecular dynamics simulations. Pure Appl. Chem. **76**(1), 231–240 (2004)
10. Jumbri, K., Micaelo, N., Abdul Rahman, M.: Solvation free energies of nucleic acid bases in ionic liquids. Mol. Simul. **43**(1), 19–27 (2017)
11. Fareghi-Alamdari, R., Hatefipour, R., Rakhshi, M., Zekri, N.: Novel diol functionalized dicationic ionic liquids: synthesis, characterization and DFT calculations on H-bonding influence on thermophysical properties. RSC Adv. **6**(82), 78636–78647 (2016)
12. Ahmad, N.A., Jumbri, K., Ramli, A., Ghani, N.A., Ahmad, H., Kassim, M.A.: Synthesis, characterisation and antioxidant properties of ferulate-based protic ionic liquids: experimental and modelling approaches. J. Mol. Liq. **278**, 309–319 (2019)
13. Mandai, T., Masu, H., Imanari, M., Nishikawa, K.: Comparison between cycloalkyl- and n-alkyl-substituted imidazolium-based ionic liquids in physicochemical properties and reorientational dynamics. J. Phys. Chem. B **116**(7), 2059–2064 (2012)
14. Hawker, R., Haines, R., Harper, J.: Variation of the cation of ionic liquids: the effects on their physicochemical properties and reaction outcome. Cheminform **47**(19) (2016)
15. Choi, H., Kang, H., Park, H.: New solvation free energy function comprising intermolecular solvation and intramolecular self-solvation terms. J. Cheminformatics **5**(1), 8 (2013)
16. Riahi, S., Rowley, C.: Solvation of hydrogen sulfide in liquid water and at the water–vapor interface using a polarizable force field. J. Phys. Chem. B **118**(5), 1373–1380 (2014)
17. Glew, D., Hames, D.: Aqueous nonelectrolyte solutions. Part X. Mercury solubility in water. Can. J. Chem. **49**(19), 3114–3118 (1971)
18. Sánchez-Badillo, J., Gallo, M., Alvarado, S., Glossman-Mitnik, D.: Solvation thermodynamic properties of hydrogen sulfide in [C_4mim][PF_6], [C_4mim][BF_4], and [C_4mim][Cl] ionic liquids, determined by molecular simulations. J. Phys. Chem. B **119**(33), 10727–10737 (2015)
19. Damas, G., Dias, A., Costa, L.: A quantum chemistry study for ionic liquids applied to gas capture and separation. J. Phys. Chem. B **118**(30), 9046–9064 (2014)

20. Tsuzuki, S., Shinoda, W., Saito, H., Mikami, M., Tokuda, H., Watanabe, M.: Molecular dynamics simulations of ionic liquids: cation and anion dependence of self-diffusion coefficients of ions. J. Phys. Chem. B **113**(31), 10641–10649 (2009)
21. Haron, N., Lee, V., Sairi, N., Zain, S., Alias, Y., Aroua, M.: Microstructures, interactions, and dynamics properties studies of aqueous guanidinium triflate ionic liquid from molecular dynamics simulations. J. Mol. Liq. **227**, 184–193 (2017)

Synthesis and Characterization of Nickel–Magnesium Catalyst Supported on Reduced Graphene Oxide

Nur Diyan Mohd Ridzuan, Maizatul Shima Shaharun, Israf Ud Din, and Poppy Puspitasari

Abstract In this study, nickel–magnesium catalyst supported on reduced graphene oxide nanosheets was synthesized. rGO support was characterized by X-ray diffraction (XRD), surface area and porosity, and thermogravimetric analysis, and fourier-transform infrared (FTIR) and Raman spectroscopy to understand its physicochemical properties. The results revealed that the edge-oxidized rGO has high surface area (258.11 m^2g^{-1}) with approximately 36 multi graphitic layers where interlayer spacing is 0.336 nm each. High thermal stability of rGO makes it an excellent support to be used for high-temperature catalytic activity. Subsequently, rGO supported catalysts were synthesized via incipient wetness impregnation method. The amount of Ni was fixed at 20 wt% whilst Mg was varied at 0 wt % and 5 wt% relative to rGO content to assess its effect as second metal towards properties of the catalyst. X-ray diffractograms of Ni_{20}/rGO and $Ni_{20}Mg_5$/rGO demonstrated formation of new peaks due to presence of NiO and NiO MgO whereas (002) rGO peak at 26.5° does not show obvious changes, concluding the stability of rGO after the impregnation and calcination processes. Upon impregnation, the surface area and porosity of Ni_{20}/rGO is lower than that of rGO due to dispersion of Ni metal on the rGO surface. BET surface area of $Ni_{20}Mg_5$/rGO catalyst further decreases to 103.95 m^2g^{-1} due to the addition of metal content but its porosity is higher than Ni_{20}/rGO. Hence, it is postulated that Ni and Mg metal form solid solution (NiO–MgO) on rGO nanosheets

N. D. Mohd Ridzuan · M. S. Shaharun (✉)
Fundamental and Applied Sciences, Universiti Teknologi PETRONAS, 32610 Seri Iskandar, Perak, Malaysia
e-mail: maizats@utp.edu.my

N. D. Mohd Ridzuan
e-mail: diyanridzuan@gmail.com

I. U. Din
Department of Chemistry, College of Science and Humanity, Prince Sattam bin Abdulaziz University, Al-kharj 16278, Saudi Arabia

P. Puspitasari
Mechanical Engineering Department, State University of Malang, 5 Semarang Street, 65145 Malang, East Java, Indonesia
e-mail: poppy@um.ac.id

S. A. Abdul Karim et al. (eds.), *Proceedings of the 6th International Conference on Fundamental and Applied Sciences*, Springer Proceedings in Complexity,
https://doi.org/10.1007/978-981-16-4513-6_20

which has bigger particle size compared to NiO, lowering its penetration into rGO mesopores.

Keywords Reduced graphene oxide · Nickel catalyst · Magnesium

1 Introduction

Following the first isolation in 2004 by Novoselov and Geim [1] graphene has attracted interest both for fundamental aspects and application studies due to its exceptional mechanical, chemical and thermal properties [2, 3]. Graphene is a sp^2 hybridized carbon-based material with a hexagonal monolayer network forming two dimensional structure [4]. Solubility, dispersibility and chemical stability of this material can be enhanced through functionalization and modification of graphene depending on the properties necessary for the application [5, 6]. The high surface area of graphene as compared to other carbon allotropes makes it a suitable material for support in heterogeneous catalysis. In this work, reduced graphene oxide supported nanocatalyst (Ni and Mg) is proposed to be used for CO_2 methanation via Sabatier reaction from Eq. (1) [7]. This strategy can be explained by the fact that the support has significant influence on the catalyst properties in a few aspects. Firstly, presence of support can improve the dispersion of active phase. Apart from that, the addition of support onto catalyst can improve adsorption towards CO_2. It also decreases the formation of the inactive spinel phases and modify the reducibility of the oxides precursors through manipulating the interaction between the active phase and support [8]. Often, ceramics supports namely silica [9], aluminium oxide [10], and zeolite [11] are investigated. Reduced graphene oxide (rGO) can be seen as a potential support for CO_2 methanation reaction due to its high thermal and electronic conductivities. Then, rGO has special tunable physicochemical characteristic as well as controllable textures and microstructures. Thus, the property of rGO can be modified to adjust the deposition of and interaction with the metal nanoparticles [2]. Since rGO has high surface area [4], dispersion of metal nanoparticles can be enhanced, agglomeration can be avoided, sintering of carbon deposit can be resisted, and exposed active sites can be increased.

$$CO_2(g) + 4\,H_2(g) \leftrightarrow CH_4(g) + 2\,H_2O \tag{1}$$

Catalysts based on transition metals from group VIII (Pd, Ru, Rh, Ni and Co) were commonly employed for the methanation of CO_2. Among them, Ni-based catalyst has been extensively studied for industrial application due to their feasibility, low cost and ease of availability. However, under the severe reaction condition of CO_2 methanation, Ni catalyst may be deactivated due to the sintering of Ni particles, formation of mobile nickel sub-carbonyls and formation of carbon deposits [12, 13]. Apart from using support, these limitations could also be solved synergistically by adding promoters or second metal to the catalyst. Bimetallic catalyst system has attracted a great deal of attention to enhance the stability and catalytic activity of nickel [14]. Beforehand, various second metals namely Fe [15], Zr [8], Pd [16] and Mg [12, 17] have been added as promoters to Ni catalyst which will intervene in the reaction to capture carbon dioxide. To improve the efficiency of CO_2 methanation, magnesium oxide, MgO, has been used together with Ni as bimetallic catalyst. This is because MgO shows high activity for reverse water gas shift reaction at moderate temperatures, relatively cheap and abundant [18]. Most importantly, MgO was found to be an effective promoter to improve resistance towards carbon deposition and minimize Ni particles sintering [10]. Theoretically, the presence of MgO will produce solid solution with Ni on the support and can provide an alternative pathway by reacting with CO_2 to form magnesium carbonate, $MgCO_3$. $MgCO_3$ is the precursor to initiate reaction to produce methane whereas H_2 is activated by Ni nanoparticles [16]. Addition of second metal can also potentially minimize the CO byproduct by inhibiting CO desorption as intermediate. Earlier, MgO was utilized as promoter for nanoparticles supported on silica in several studies [9, 16, 17] and found success as enhanced CH_4 selectivity and CO_2 conversion at optimum temperature and H_2/CO_2 ratio was observed in the presence of MgO.

To the best of our knowledge, the use of reduced graphene oxide nanosheets (rGO) as support for Ni–Mg bimetallic catalyst system has not been reported in any open literature. Herein, we reported the synthesis of nickel and magnesium oxide nanoparticles supported on rGO via incipient wetness impregnation method with the expectation to improve metal dispersion on rGO and study the effect of Mg as second metal into the catalyst system. The physicochemical properties of rGO support and catalysts were thoroughly characterized by different spectroscopic, surface and thermal analysis.

2 Experimental

2.1 Materials

Reduced graphene oxide was purchased from Sigma Aldrich. Nickel nitrate hexahydrate ($Ni(NO_3)_2 \cdot 6H_2O$) and magnesium nitrate hexahydrate ($Mg(NO_3)_2 \cdot 6H_2O$) were both obtained from Merck. All the solutions were prepared using distilled water.

2.2 Catalyst Synthesis

Catalysts were synthesized from incipient wetness impregnation method. The amount of Ni was maintained at 20 wt% whilst Mg was varied at two different amounts (0 and 5 wt%) relative to rGO content to assess its effect as second metal towards properties of the catalyst. Firstly, respective amount of $Ni(NO_3)_2 \cdot 6H_2O$) and $Mg(NO_3)_2 \cdot 6H_2O$

were accurately weighed and dissolved into distilled water to prepare impregnating solution. Volume of solution used was equal to rGO volume. The salt precursors were added dropwise to rGO support in a beaker. The impregnated sample was left for 12 h before it was dried at 100 °C for another 12 h. After calcination at 400 °C for 2 h, Ni_{20}/rGO and $Ni_{20}Mg_5$/rGO catalysts were obtained.

2.3 *Characterization*

2.3.1 Thermogravimetric Analysis

Thermogravimetric analysis (TGA) of rGO support was conducted using METTLER TOLEDO TGA/DSC STARe thermogravimetric analyzer with Gas Controller GC10 over the temperature range of ambient temperature 25–900 °C (heating ramp: 20 °C/min) in N_2 atmosphere. The purpose of TGA was to characterize the thermal properties and analyse weight loss profile of rGO support which is used for determination of catalyst calcination temperature.

2.3.2 XRD Analysis

The analysis of interlayer spacing, crystallinity and phase identification of rGO support and catalysts were determined using X-ray diffraction recorded on powder diffractometer (X'Pert³ Powder and Empyrean, PANalytical) with Cu Kα radiation source between 2θ of 10° to 85° (scanning step of 0.01°/step) on continuous scanning. Interlayer spacing of rGO layers (d), crystal stack height (L_c), crystallite size (L_a) and estimated number of layers (N_{GP}) were obtained from Eq. (2)–(5) respectively.

$$\lambda = 2d \sin \theta \tag{2}$$

$$L_c = K\lambda / B \cos \theta \tag{3}$$

$$L_a = K_2\lambda / B \cos \theta \tag{4}$$

$$N_{GP} = L_c / d \tag{5}$$

where λ is wavelength of Cu Kα radiation (1.5418 nm), K is shape factor (0.89), K_2 is Warren Form Factor (1.84) and B is line broadening at half maximum of respective peak.

2.3.3 Raman Spectroscopy

Raman spectrum of rGO support catalyst were obtained using Horiba JobinYvonHR800 spectrometer with 514 nm green laser as excitation source in the Raman shift range of 400–4000 cm^{-1} to identify the structure and degree of disorder in carbon-based material of the rGO support. Crystallite size (L_a) of the rGO was calculated from Eq. (6).

$$L_a = (2.4 \times 10^{-10})\lambda^4(I_D/I_g)^{-1} \quad (6)$$

where λ is the wavelength of excitation source, I_D and I_G are the respective intensity of D and G band of rGO.

2.3.4 SAP Analysis

Surface area and porosity (SAP) analysis of support and catalysts were performed using Micromeritics ASAP 2020 analyzer with N_2 as adsorbate. The sample was degassed before being subjected to N_2 adsorption. Surface area was determined from BET model, pore size and volume were determined from BJH method and micropore analysis was carried out using t-plot analysis.

2.3.5 FTIR Spectroscopy

Fourier-transform infrared (FTIR) spectrophotometer (Pelkin Elmer, Spectrum One) was used to investigate chemical bonds and oxygen functional group of rGO support and catalysts in the range of 400–4000 cm^{-1} with a resolution of 4 cm^{-1}. Sample pellet was prepared by grinding with KBr.

3 Results and Discussion

3.1 Thermogravimetric Analysis

Thermogram of commercial rGO was obtained together with DTG curve as shown in Fig. 1. From the general view of the thermogram, rGO has excellent thermal stability at high temperature exposure because the heating of sample up to 900 °C resulted in only 15.14% weight loss. In the initial step, a steep lost caused by heating is observed and this is thought to be due to loss of physisorbed water on the sample where 97.72% residue is left after 127 °C. Removal of moisture adsorbed is commonly observed in porous materials as supported by other studies [4, 19, 20]. Then, the heating of the sample up to 500 °C resulted in another 4% degradation. After final heating at

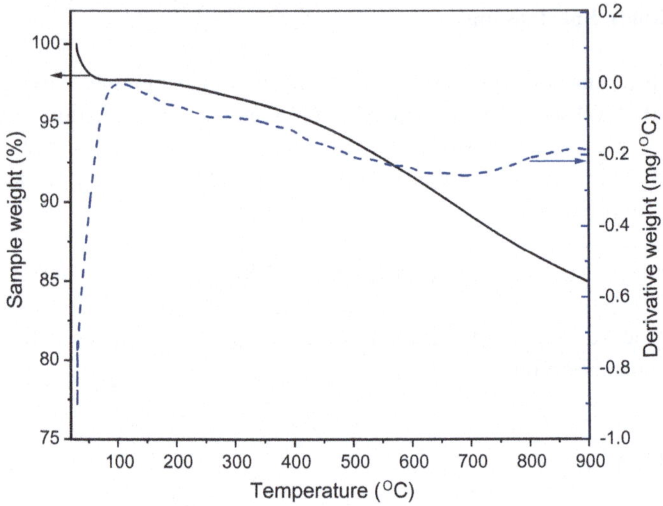

Fig. 1 Thermogram of rGO support

900 °C, 84.86% sample is left. Hence, it is suggested that calcination of catalyst to be carried out below 500 °C to avoid rGO decomposition. In the synthesis step, the catalyst was ideally calcined at 400 °C. This rGO is expected to be able to anchor the catalyst active phase without being decomposed at elevated temperature (max: 350 °C) during CO_2 methanation reaction. This feature is important because decomposition of support could result in metal sintering, reducing the conversion efficiency. The thermal property of this rGO contradicts to that of Deerattrakul et al. [19] in which it was found that upon heating up to 600 °C, the synthesized rGO was degraded by 60% beginning at 400 °C due to decomposition of oxygen functional groups and carbon backbone. Hence, the rGO used in this study has significantly higher thermal stability.

3.2 *XRD Analysis*

As depicted in Fig. 2, the diffraction pattern of rGO demonstrate a good agreement to hexagonal phase with major diffraction peak appear at 26.52° indexed to (002) facet of graphite crystal [21]. This peak is an indication for interlayer spacing (d) of graphitic planes (Fig. 3) that can be calculated from Bragg's law [22]. The interlayer spacing is 0.336 nm which is comparable to rGO synthesized from other studies [4, 20]. The crystal stack height (L_c) of the rGO nanosheets was found to be 12.27 nm; hence, the N_{GP} is 36 layers. The L_a is 25.37 nm as estimated from Eq. (5). The sharp and intense peak of (002) plane reflects high crystallinity of rGO which can be explained by the structure of commercial rGO that is only edge-oxidized; hence, making its stacking

Fig. 2 X-ray diffractogram

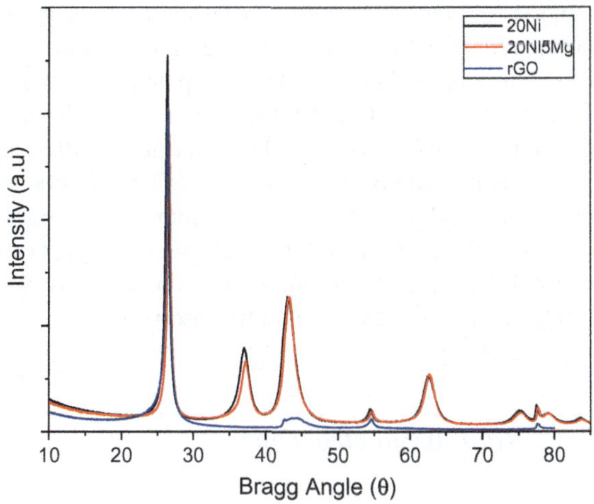

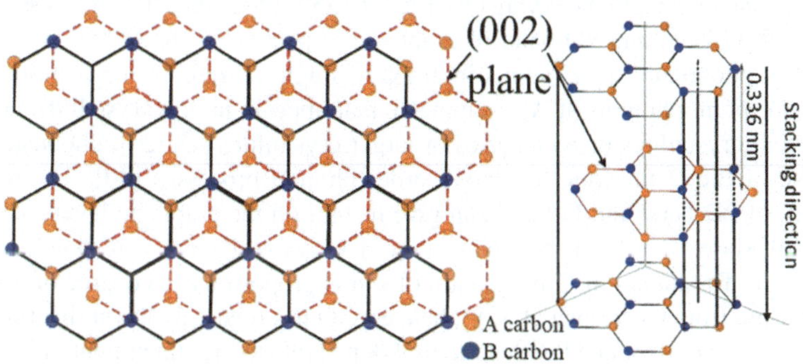

Fig. 3 Structure of rGO carbon backbone

to be more comparable to well-ordered graphite. This is different from other studies [19, 20, 23] in which XRD pattern of rGO commonly have broad nature due to poor ordering of the sheets along the stacking direction. The commercial rGO from Sigma Aldrich is a two-dimensional, layered carbon allotrope where layers are consists of carbon atoms arranged in a honeycomb fashion with oxygen functional groups at the edge whilst each layer is connected through van der wall's interaction [24] at interlayer spacing of 0.336 nm. The structure of rGO sheets that give rise to the XRD pattern is illustrated in Fig. 3. Furthermore, a small peak observed at 43.82° belong to (100) plane which gives intuition on the short-range order in graphene layer.

After metal impregnation, X-Ray diffractogram of catalysts (Ni_{20}/rGO and $Ni_{20}Mg_5$/rGO) were obtained. With the impregnation of 20% Ni onto rGO, reflections for the Ni phase were consistent with those for NiO at 37.07°, 43.27°, 62.72°, 75.16° and 79.13° which correspond to (111), (200), (220), (311) and (222) diffraction planes [25]. The presence of these peaks indicates success of wetness impregnation method to incorporate active metal onto rGO support. When 5% of Mg is added, no apparent changes on the diffractogram could be observed and this is likely to be due to loading being too low compared to Ni. This too could be attributed to the superposition of MgO and NiO peaks [17, 26]. The effect of Mg modifier towards the catalyst system could be concluded further as different amount of Mg is loaded onto the catalyst. XRD peaks of $Ni_{20}Mg_5$/rGO could be attributed to NiO–MgO mixed phases as solid solution and this is commonly suggested from other studies [9, 17, 27]. With the preliminary study of this material for synthesis at 40 °C, it confirms that rGO is able to withstand high temperature as there is no significant changes at (002) peaks of graphene could be observed.

3.3 *Raman Spectroscopy*

Figure 4 shows Raman spectrum of rGO support and $Ni_{20}Mg_5$/rGO catalysts. It is apparent that three major peaks are observed at 1352, 1572 and 2704 cm^{-1} associated to D-band, G-band and 2D-band of graphene respectively. The G-line is assigned to the E_{2g} phonon of sp^2 C atoms in graphite single crystal. D line is a breathing mode of κ-point phonons at A_{1g} symmetry that appears in polycrystalline samples from small crystallites or boundaries of larger crystallites [28, 29]. 2D band is the overtone of the D line due to second order phonon processes [30]. The ratio of intensity of 2D-band and G-band can give insight on the graphene layers [29, 31]. The result shows that $I_{2D}/I_g > 1$; hence, the rGO is multilayer, correspond with the XRD result. Broadness of G-band and D-band is proportional to degree of disorder as suggested from Gurrero et al. [28]; hence, rGO has relatively lower disorder. The disorder raised is associated with structural defect due to the attachment of hydroxyl

Fig. 4 Raman spectrum of rGO

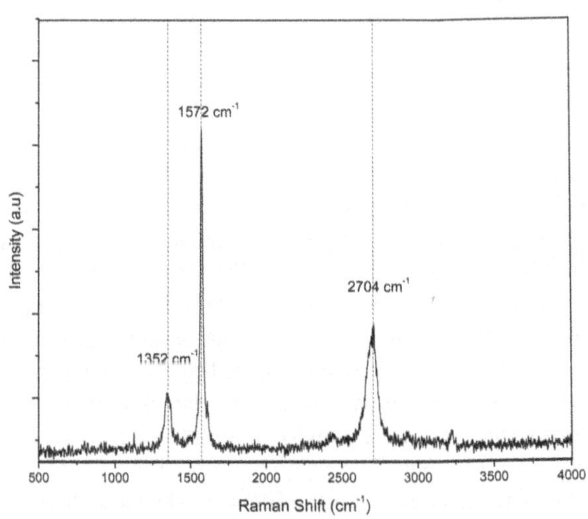

and epoxide groups on the carbon backbone [32]. The in-plane sp^2 crystallite size, L_a, was found to be 18.13 nm. The L_a is comparatively lower from L_a calculated from XRD analysis.

3.4 FTIR Spectroscopy

FTIR spectra of three samples are presented in Fig. 5. The presence of strong peak at 1079 cm^{-1} and weak absorbance at 670 cm^{-1} are respectively attributed to stretching mode of C–O and C–O–C functional group located over the edge of rGO sheets. The skeletal vibration from the aromatic C=C stretching band appear at 1578 cm^{-1} with medium intensity peak meanwhile an absorptive peak at 1720 cm^{-1} correspond to carbonyl function groups (C=O, COOH) stretching vibration band. In the range of wavenumbers 2000–3000 cm^{-1} a rather clean baseline is observed due to absence of C–H bond except for a peak at 2923 cm^{-1}. A broad medium absorption at 3420 cm^{-1} is assigned to antisymmetric and symmetric stretching of OH bond group very likely coming from oxygen functional group at the edge of the graphene sheets [24, 28]. Overall, the spectra of support and catalysts merge well depicting no degradation of rGO after undergoing synthesis and calcination processes. The notable reduction of overall absorption in catalysts is probably due to removal of physisorbed water after the calcination and impregnation as surface area and porosity decreases. In the view of these spectra as well, there is no formation of new peaks after metal loading suggesting that only physical anchoring occurred between metal and support, supposedly by filling of the pores as evidenced from Sect. 3.4.

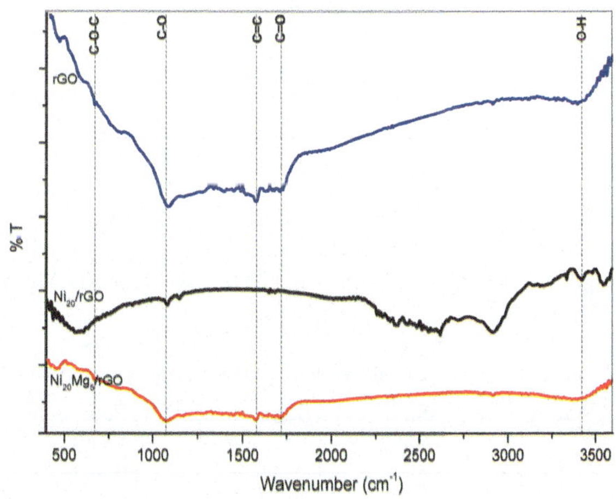

Fig. 5 FTIR spectra of rGO and catalysts

3.5 SAP Analysis

The isotherm of rGO support and catalysts are shown in Fig. 6, while comparative values of SAP result between rGO support, Ni_{20}/rGO and $Ni_{20}Mg_5$/rGO obtained are shown in Table 1. As presented in Fig. 6a, the isotherm of rGO exhibits Type IV and possess H3 hysteresis loop according to IUPAC classification [33]. This indicates that the sample has pores dominated in mesoporous range (2–50 nm) whilst the H3 type hysteresis loop represents aggregates of plate-like particles giving rise to slit-shaped pores [34]. The hysteresis loop occurs due to the capillary condensation

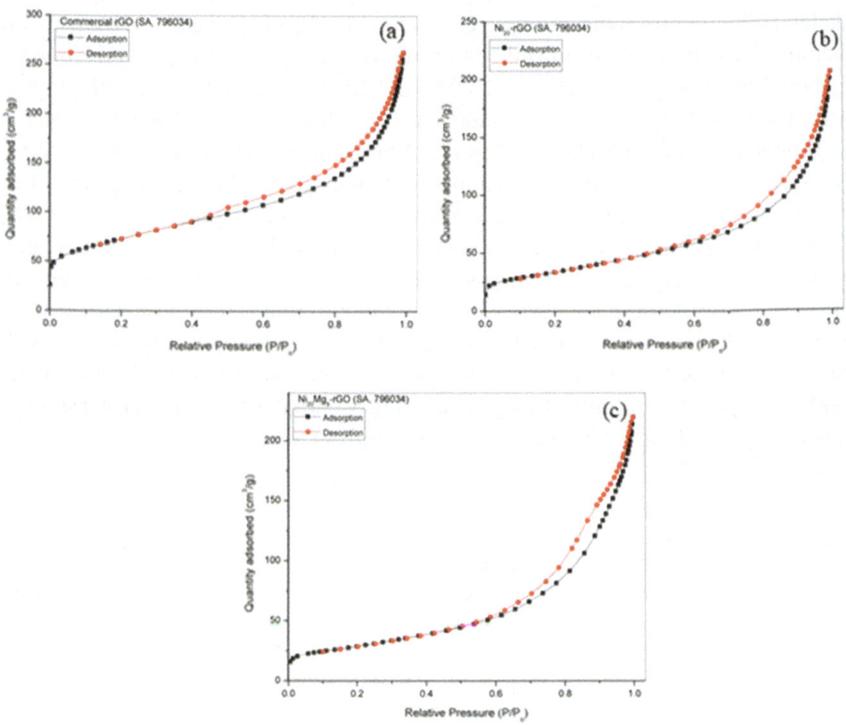

Fig. 6 N_2 sorption isotherm of **a** rGO **b** Ni_{20}/rGO **c** $Ni_{20}Mg_5$/rGO

Table 1 Surface area and porosity of rGO, Ni_{20}/rGO and $Ni_{20}Mg_5$/rGO

Catalyst	S_{BET} (m^2g^{-1})	S_{micro} (m^2g^{-1})	Pore volume (cm^3g^{-1})	Micropore volume (cm^3g^{-1})	Pore width (nm)
rGO	258.11	44.064	0.3858	0.0198	7.6483
Ni_{20}–rGO	122.01	3.512	0.2908	0.000766	10.1104
$Ni_{20}Mg_5$–rGO	103.953	4.436	0.3156	0.00137	10.6021

in the mesopores. The isotherm and hysteresis loop pattern after metal impregnation (Fig. 6b and c) are similar but the amount of N_2 adsorbed decreased due to loading of Ni and Mg which has much higher density than rGO [35]. This is correlated with the proportional decrease of S_{BET} when the amount of metal impregnated onto rGO support increase. Therefore, this provides complimentary evidence that the nanoparticles fill and block rGO pores. However, there is discrepancy between the trend for surface area and pore volume when amount of impregnated metal increases. The lower micropore surface area of Ni_{20}/rGO potentially suggest that NiO has smaller size that it penetrates up to micropores, correspond to micropore volume as well [36]. Hence, it is postulated that Ni and Mg metal form solid solution (NiO–MgO) on rGO nanosheets which has bigger size compared to NiO, lowering its penetration into rGO mesopores. It should be mentioned that this is highly probable because XRD analysis did not show peaks of pure MgO; hence, the result supported the postulation that addition of 5% Mg second metal produce NiO–MgO solid solution instead of forming two different oxides. This phenomenon is also in general agreement with other researchers studying bimetallic catalysts [10].

4 Conclusion

This work was devoted to study the capability of reduced graphene oxide to act as catalyst support via incipient wetness impregnation method. rGO physicochemical properties are well characterized and the results showed that rGO has high thermal stability, high surface area with ample porosity with multilayer edge-oxidized graphitic planes. rGO nanosheets has crystallite size along (002) stacking direction of 12.27 nm. Incipient wetness impregnation method was found to be a successful technique to physically anchor Ni and Mg on rGO where XRD analysis evidenced the presence of NiO and NiO–MgO in 2θ range of 30°–80°. It is supposed that the nanoparticles were penetrated into the rGO mesopores as BET surface area showed significant decrement proportional to amount of metal loading. The effect of Mg addition as second metal results in the formation of solid solution (NiO–MgO) based on the comparison between Ni_{20}/rGO and $Ni_{20}Mg_5$/rGO.

Acknowledgements The financial assistance from the Foundation of Universiti Teknologi PETRONAS (YUTP-FRG) with the cost center 015LC0-253 is gratefully acknowledged.

References

1. Novoselov, K.S., et al.: Electric field effect in atomically thin carbon films. Science **306**(5696), 666–669 (2004)
2. Julkapli, N.M., Bagheri, S.: Graphene supported heterogeneous catalysts: An overview. Int. J. Hydrogen Energ. **40**(2), 948–979 (2015)

3. Song, S., et al.: Biomedical application of graphene: from drug delivery, tumor therapy, to theranostics. Colloids Surf. B **185**, 110596 (2020)
4. Park, S., et al.: Hydrazine-reduction of graphite-and graphene oxide. Carbon **49**(9), 3019–3023 (2011)
5. Alkhouzaam, A., Qiblawey, H., Khraisheh, M., Atieh, M., Al-Ghouti, M.: Synthesis of graphene oxides particle of high oxidation degree using a modified Hummers method. Ceram. Int. **46**(15), 23997–24007 (2020)
6. Muniyalakshmi, M., Sethuraman, K., Silambarasan, D.: Synthesis and characterization of graphene oxide nanosheets. Mater. Today Proc. **21**, 408–410 (2020)
7. Feng, Y., Yang, W., Chen, S., Chu, W.: Cerium promoted nano nickel catalysts Ni–Ce/CNTs and Ni–Ce/Al_2O_3 for CO_2 methanation. Integr. Ferroelectr. **151**(1), 116–125 (2014)
8. Cai, M., Wen, J., Chu, W., Cheng, X., Li, Z.: Methanation of carbon dioxide on Ni/ZrO_2–Al_2O_3 catalysts: Effects of ZrO_2 promoter and preparation method of novel ZrO_2–Al_2O_3 carrier. J. Nat. Gas Chem. **20**(3), 318–324 (2011)
9. Guo, M., Lu, G.: The effect of impregnation strategy on structural characters and CO_2 methanation properties over MgO modified Ni/SiO_2 catalysts. Catal. Commun. **54**, 55–60 (2014)
10. Hu, D., et al.: Enhanced investigation of CO methanation over Ni/Al_2O_3 catalysts for synthetic natural gas production. Ind. Eng. Chem. Res. **51**(13), 4875–4886 (2012)
11. Borgschulte, A., et al.: Sorption enhanced CO_2 methanation. Phys. Chem. Chem. Phys. **15**(24), 9620–9625 (2013)
12. Meshkani, F., Rezaei, M.: Nanocrystalline MgO supported nickel-based bimetallic catalysts for carbon dioxide reforming of methane. Int. J. Hydrogen Energ. **35**(19), 10295–10301 (2010)
13. Su, X., et al.: Catalytic carbon dioxide hydrogenation to methane: A review of recent studies. J. Energ. Chem. **25**(4), 553–565 (2016)
14. Aziz, M., Jalil, A., Triwahyono, S., Ahmad, A.: CO_2 methanation over heterogeneous catalysts: Recent progress and future prospects. Green Chem. **17**(5), 2647–2663 (2015)
15. Hwang, S., et al.: Methanation of carbon dioxide over mesoporous nickel–M–alumina (M=Fe, Zr, Ni, Y, and Mg) xerogel catalysts: Effect of second metal. Catal. Lett. **142**(7), 860–868 (2012)
16. Park, J.-N., McFarland, E.W.: A highly dispersed Pd–Mg/SiO_2 catalyst active for methanation of CO_2. J. Catal. **266**(1), 92–97 (2009)
17. Li, Y., Lu, G., Ma, J.: Highly active and stable nano NiO–MgO catalyst encapsulated by silica with a core–shell structure for CO_2 methanation. RSC Adv. **4**(34), 17420–17428 (2014)
18. Sommerbauer, G.B., Lux, S., Siebenhofer, M.: AIChE Annual Meeting
19. Deerattrakul, V., Dittanet, P., Sawangphruk, M., Kongkachuichay, P.: CO_2 hydrogenation to methanol using Cu–Zn catalyst supported on reduced graphene oxide nanosheets. J. CO_2 Util. **16**, 104–113 (2016)
20. Shen, J., et al.: One step hydrothermal synthesis of TiO_2-reduced graphene oxide sheets. J. Mater. Chem. **21**(10), 3415–3421 (2011)
21. Ma, H., et al.: Graphene intercalated Ni–SiO_2/GO–Ni–foam catalyst with enhanced reactivity and heat-transfer for CO_2 methanation. Chemi. Eng. Sci. (2018)
22. Callister, W.D., Rethwisch, D.G.: Materials Science and Engineering 9th edn. Wiley (2014)
23. Li, Y., et al.: CO oxidation over graphene supported palladium catalyst. Appl. Catal. B **125**, 189–196 (2012)
24. Lavin-Lopez, M., Paton-Carrero, A., Sanchez-Silva, L., Valverde, J., Romero, A.: Influence of the reduction strategy in the synthesis of reduced graphene oxide. Adv. Powder Technol. **28**(12), 3195–3203 (2017)
25. Zhang, Z., et al.: Impacts of nickel loading on properties, catalytic behaviors of Ni/γ–Al_2O_3 catalysts and the reaction intermediates formed in methanation of CO_2. Int. J. Hydrogen Energ. **44**(18), 9291–9306 (2019)
26. Tan, J., et al.: Highly dispersed and stable Ni nanoparticles confined by MgO on ZrO_2 for CO_2 methanation. Appl. Surf. Sci. **481**, 1538–1548 (2019)

27. Guo, M., Lu, G.: The difference of roles of alkaline-earth metal oxides on silica-supported nickel catalysts for CO_2 methanation. RSC Adv. **4**(102), 58171–58177 (2014)
28. Guerrero-Contreras, J., Caballero-Briones, F.: Graphene oxide powders with different oxidation degree, prepared by synthesis variations of the Hummers method. Mater. Chem. Phys. **153**, 209–220 (2015)
29. Malard, L., Pimenta, M., Dresselhaus, G., Dresselhaus, M.: Raman spectroscopy in graphene. Phys. Rep. **473**(5–6), 51–87 (2009)
30. Yadav, N., Lochab, B.: A comparative study of graphene oxide: Hummers, intermediate and improved method. FlatChem **13**, 40–49 (2019)
31. Cançado, L., et al.: General equation for the determination of the crystallite size La of nanographite by Raman spectroscopy. Appl. Phys. Lett. **88**(16), 163106 (2006)
32. Bong, S., et al.: Graphene supported electrocatalysts for methanol oxidation. Electrochem. Commun. **12**(1), 129–131 (2010)
33. Lowell, S., Shields, J.E., Thomas, M.A., Thommes, M.: Characterization of porous solids and powders: Surface area, pore size and density, vol. 16. Springer Science and Business Media (2012)
34. Ramimoghadam, D., Hussein, M.Z.B., Taufiq-Yap, Y.H.: Hydrothermal synthesis of zinc oxide nanoparticles using rice as soft biotemplate. Chem. Cent. J. **7**(1), 136 (2013)
35. Aziz, M., et al.: Highly active Ni-promoted mesostructured silica nanoparticles for CO_2 methanation. Appl. Catal. B **147**, 359–368 (2014)
36. Mohd Ridzuan, N.D., Shaharun, M.S., Lee, K.M., Ud Din, I., Puspitasari, P.: Influence of nickel loading on reduced graphene oxide-based nickel catalysts for the hydrogenation of carbon dioxide to methane. Catalysts **10**(5), 471 (2020)

UiO-66 and ZIF-8 Metal-organic Frameworks for Acenaphthene Adsorption

Zakariyya Uba Zango, Anita Ramli, Khairulazhar Jumbri, and Muslim Abdurrahman

Abstract Comparative adsorption of acenaphthene onto UiO-66 and ZIF-8 metal-organic frameworks (MOFs) were investigated. Response surface methodology (RSM) was employed for the process optimizations according to the central composite design (CCD) with 5 inputs variables. The adsorption efficiency achieved were 99.7 and 60.7% for the UiO-66 and ZIF-8, respectively according to the RSM optimization conditions. The model was significantly described according to statistical analysis of variance (ANOVA). The adsorption process was well fitted by pseudo-second order kinetic model with R^2 values of 0.9947 and 0.97780 for the UiO-66 and ZIF-8, respectively.

Keywords Acenaphthene · Adsorption · Metal-organic frameworks · Response surface methodology

Z. U. Zango · A. Ramli (✉) · K. Jumbri
Department of Fundamental and Applied Sciences, Universiti Teknologi PETRONAS, 3210 Seri Iskandar, Perak, Malaysia
e-mail: anita_ramli@utp.edu.my

K. Jumbri
e-mail: khairulazharjumbri@utp.edu.my

Z. U. Zango
Department of Chemistry, Al-Qalam University Katsina, Katsina 2137, Nigeria

A. Ramli
HiCOE Centre for Biofuel and Biochemical Research (CBBR), Universiti Teknologi PETRONAS, 3210 Seri Iskandar, Perak, Malaysia

K. Jumbri
Centre of Research in Ionic Liquids (CORIL), Universiti Teknologi PETRONAS, 3210 Seri Iskandar, Perak, Malaysia

M. Abdurrahman
Department of Petroleum Engineering, Universitas Islam Riau, Jalan Kaharuddin Nasution No. 113, Pekanbaru 28284, Indonesia
e-mail: muslim@eng.uir.ac.id

S. A. Abdul Karim et al. (eds.), *Proceedings of the 6th International Conference on Fundamental and Applied Sciences*, Springer Proceedings in Complexity,
https://doi.org/10.1007/978-981-16-4513-6_21

1 Introduction

Pollution caused by the presence of organic contaminants in environmental waters has been a disturbing phenomenon affecting the peaceful existence of living organisms. Acenaphthene, a member of Polycyclic aromatic carbons (PAHs) heavily found in crude oil and coal [1, 2]. It consisted of naphthalene connected by the ethylene at position 1 and 8 (Fig. 1). It also occurs in small quantity as a result of biomass decay [3]. It has found variety of applications in industrial processing as a precursor and intermediate material for the production of products such as dyes production, plastics, insecticides, fungicides and other manufacturing processes [4, 5]. It is an important component in coal tar for road and highway constructions and as preservative in wood industries [6]. It is among the most prominent organic pollutant ubiquitously found in environmental waters [7]. It is resulted from oil spillage, effluents discharged from petroleum and petrochemical industries, coal conversion process, urban air due to automobile exhaust and cigarette smoke, as well as municipal wastewater due to the soil surface and street asphalt runoff [8, 9]. It is relatively insoluble in water (0.4 mg/100 mL), thus easily transported into the water bodies. Findings have shown the presence of acenaphthene in various ground and surface waters. It is recalcitrant to atmospheric degradation, thus undergoes bioaccumulation in the soil and sediments and subsequently consumed by the plants and aquatic organisms [10]. It was known to cause significant impact to both plants and animals due to its toxic and hazardous effects such as eye and skin irritations, lungs cancer and other respiratory complications [11, 12]. It was listed among the priority PAHs organic pollutant according to USEPA. Thus, essentially required to be eliminated from environmental waters.

Adsorption process have been recognized as an alternative wastewater remediation technology. It is a low-cost and environmental-friendly technique for variety of pollutants removal from the water. The availability of various natural and synthetic adsorbents materials for both organic and inorganic pollutants removal has been capitalized as the way forward compared to other water treatment technologies such as coagulation, reverse osmosis and enzymatic degradation processes [13, 14]. Thus, various adsorbents from natural and sources such as biomass [15], clays, soil and sand particles [16], etc., have been well reported for the acenaphthene adsorption from environmental waters. Despite showing lots of promise in water treatment, most of these adsorbents are not suitable for PAHs removal from water due to the low specific

Fig. 1 Chemical structure of acenaphthene

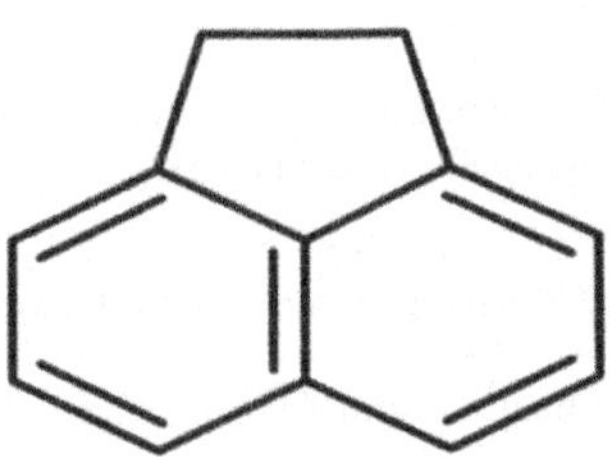

surface area and pore volumes of the adsorbents [17]. The use of synthetic materials such as activated carbons [18], biochar [19], carbon nanotubes, graphene [20], and silica have also been reported. However, AC is not suitable for organic pollutant removal at low concentrations.

Advancement in materials sciences have paved the way for the invention of new classes of materials with versatile applications. Metal-organic frameworks (MOFs) are among the most advanced materials that attracted the attention of researchers from scientific and engineering fields. They are comprised of coordination sphere made from interaction of metal-ion with organic moieties, forming extensive crystalline frameworks [21, 22]. They acquired unique properties such as exceptionally high porosities, mechanical and thermal strength [23, 24]. Thus, they have found tremendous applications in various fields such as carbon dioxide capture, energy storage, sensors, drug delivery, catalysis and wastewater remediations [25, 26]. They offered promising applications as super adsorbents for both organic and inorganic pollutants removal from water [27–29], due to their ultrahigh porosities and good stabilities.

This work is thus aimed at adsorption of acenaphthene from aqueous medium using UiO-66 and ZIF-8 MOFs. They were chosen due to their high BET surface area and pore volumes as well as their stabilities in both aqueous and organic phases. Response surface methodology (RSM) was used for the experimental design using central composite design (CCD). The kinetics of the adsorption process was evaluated, and the reusability of the MOFs was studied.

2 Materials and Methods

The materials purchased in this work were analytically graded and thus were used as received. Acenaphthene standard (99% purity), ZIF-8 (99% purity) zirconium tetrachloride (99.99% purity), terephthalic acid (97% purity) obtained from Sigma Aldrich (USA), and supplied by Avantis Laboratory, Malaysia.

2.1 *Synthesis and Characterizations of the MOFs*

UiO-66 was synthesized and characterized according to the procedure reported in our previous work [30]. While ZIF-8 was commercially purchased.

2.2 Preparation of the Stock Solution

The acenaphthene stock solution was prepared in acetone by using 10 mg of the standard in 100 ml volumetric flask to make a solution of 100 mg/L. Working solution was prepared from the stock solution in double deionized water using serial dilutions.

2.3 Synthesis of the MOFs

Batch adsorption experiment was designed by the RSM software (Design Expert 11) using full factorial CCD comprises of 5 input variables: contact time (min), dosage (mg), concentration (mg/L), pH and temperature (°C) with 5 center points. The adsorption was conducted using 30 mL of the acenaphthene solution in an incubator shaker (Incubator ES 20/60, bioSan) at 200 rpm. The sample was analyzed using UV-visible spectrophotometer (GENESYS 30) analysis at 220 nm. The responses were determined as the percentage adsorption efficiency achieved by the UiO-66 and ZIF-8 MOFs accordingly to the formula:

$$\%R = \frac{C_0 - C_e}{C_0} \times 100 \tag{1}$$

And the amount of the acenaphthene adsorbed onto the MOFs at certain time (qt) and equilibrium (qe) were determine from the formula:

$$q_t = \frac{(C_0 - C_t)V}{w} \tag{2}$$

$$q_e = \frac{(C_0 - C_e)V}{w} \tag{3}$$

where C_0, C_t and C_e are the initial, time and equilibrium concentrations (mg/L), respectively, w is the weight of adsorbent (g), and V is the volume of the solution (L).

3 Results and Discussion

3.1 RSM Optimizations for Acenaphthene Adsorption onto UiO-66 and ZIF-8

The synthesis of UiO-66 was achieved using solvothermal technique described in our previous studies [30], while ZIF-8 was commercially purchased. The BET surface area of the MOFs were 1421 m^2/g and 1299 m^2/g, while the pore volumes were 0.91 and 0.60 m^3/g for the UiO-66 and ZIF-8, respectively.

The adsorption studies for the removal acenaphthene were thus conducted using both UiO-66 and ZIF-8. The experimental conditions were determined by the CCD consisting of 5 parameters input variables with 47 number of experimental runs as described in Table 1. The adsorption efficiency of the two MOFs were determined according to each set of the experimental conditions given. The adsorption efficiency of UiO-66 was found to be much higher than that of ZIF-8, which was attributed the higher surface area and pore volume of the MOF [31, 32]. Thus, the highest removal efficiency achieved was 99.7% and 60.7% according to the described multi-variate conditions given in Fig. 2a,b for both UiO-66 and ZIF-8, respectively.

The corresponding 2-dimension (2D) and 3-dimension (3D) graphs for the RSM optimizations of the adsorption process. According to the Fig. 2, the adsorption efficiency was shown to increase with increase in the dosage of the MOFs due to the increase in the available number of adsorption sites on the surface of the MOFs. UiO-66 has shown to efficiently adsorbed larger organic molecules such as dyes and pharmaceuticals from aqueous solutions, attributed to its potential breathing capacity [33], high porosity [34], hydrophobicity and extensive stabilities in both organic and aqueous medium [35]. Similarly, the adsorption efficiency was increase with the increasing the contact time for the adsorption due to the increase in the interaction of the acenaphthene with the surface of the MOFs.

The model fittings were described by the analysis of variance (ANOVA) as presented in Table 2 according to the Fischer test (F-test) and p-values (probability > F) with 95% confidence level for the input variables. The overall model p-values were significant (< 0.0500), with F-value of 33.44 and 126.34 for the UiO-66 and ZIF-8, respectively. The significant terms for the adsorption and parameters combination were A, B, C, D, AD, A^2 and A, B, C, D, AB, BD, A^2 for the UiO-66 and ZIF-8 respectively. The significance of the model was also described by the reasonable agreements between the adjusted R^2 (0.9338 and 0.9820 for UiO-66 and ZIF-8, respectively) and predicted R^2 (0.8733 and 0.9638 for UiO-66 and ZIF-8, respectively) values with less than 20% differences. The lack of fit test for the replicated design points of the residual errors against the pure errors for the model were 0.1915 and 0.5840. The adequate precision of the model was 17.3082 and 32.6708 for the UiO-66 and ZIF-8, respectively, implying that the model can be used to navigate the design space for the adsorption studies. The scatter plots (Fig. 3) have shown significance of the model in terms of good agreement between the experimental and predicted adsorption efficiencies. Thus, RSM model was significant for the design optimizations of acenaphthene adsorption onto UiO-66 and ZIF-8.

3.2 *Contact Time and Kinetics Study*

The uptake of acenaphthene by the UiO-66 and ZIF-8 MOFs were studied. Adsorption experiment was studied using the optimized RSM data. Figure 4 highlighted the

Table 1 RSM optimizations for acenaphthene adsorption onto UiO-66 and ZIF-8 MOFs

Name	Units	Low	High	− Alpha	+ Alpha
A: Time	min	5	25	0	35
B: Dosage	mg	3	5	2	6
C: Concentration	mg/L	1	3	0	4
D: pH		2	6	0	8
E: Temperature	°C	25	35	20	40

Std	Run	Factor 1	Factor 2	Factor 3	Factor 4	Factor 5	Response 1	Response 2
		Time	Dosage	Concentration	pH	Temperature	UiO-66 (%R)	ZIF-8 (%R)
		min	mg	mg/L		°C		
1	1	5	3	1	2	25	67.3	30.2
3	2	5	5	1	2	25	77.1	33.5
22	3	25	3	3	2	35	84.2	50.2
24	4	25	5	3	2	35	97.5	53.6
14	5	25	3	3	6	25	96.1	51.2
16	6	25	5	3	6	25	98.4	55.2
36	7	15	6	2	4	30	87.4	45.7
45	8	15	4	2	4	30	85.2	44.3
35	9	15	2	2	4	30	72.2	42.1
5	10	5	3	3	2	25	58.5	29.6
17	11	5	3	1	2	35	63.2	31.5
20	12	25	5	1	2	35	98.5	57.4
42	13	15	4	2	4	40	79.3	44.4
7	14	5	5	3	2	25	58.6	33.8
46	15	15	4	2	4	30	79.1	43.6
31	16	5	5	3	6	35	56.3	32.4
33	17	5	4	2	4	30	54.2	32.1
26	18	25	3	1	6	35	95.5	56.3
28	19	25	5	1	6	35	97.1	57.2
32	20	25	5	3	6	35	92.4	55.4
43	21	15	4	2	4	30	77.3	43.7
39	22	15	4	2	10	30	76.2	44.2
18	23	25	3	1	2	35	94.3	56.3
25	24	5	3	1	6	35	55.7	33.5
34	25	35	4	2	4	30	99.7	60.7
2	26	25	3	1	2	25	94.1	56.4
41	27	15	4	2	4	20	76.5	44.5

(continued)

Table 1 (continued)

Std	Run	Factor 1	Factor 2	Factor 3	Factor 4	Factor 5	Response 1	Response 2
		Time	Dosage	Concentration	pH	Temperature	UiO-66 (%R)	ZIF-8 (%R)
		min	mg	mg/L		°C		
10	28	25	3	1	6	25	98.5	56.7
27	29	5	5	1	6	35	57.2	34.8
4	30	25	5	1	2	25	99.1	57.5
8	31	25	5	3	2	25	97.8	55.4
11	32	5	5	1	6	25	59.5	35.1
29	33	5	3	3	6	35	56.4	32.5
15	34	5	5	3	6	25	58.3	33.3
47	35	15	4	2	4	30	77.7	44.2
30	36	25	3	3	6	35	92.6	54.6
21	37	5	3	3	2	35	56.4	32.3
23	38	5	5	3	2	35	58.4	33.6
38	39	15	4	4	4	30	75.5	43.6
44	40	15	4	2	4	30	77.3	45.2
6	41	25	3	3	2	25	91.5	53.7
12	42	25	5	1	6	25	98.5	55.2
9	43	5	3	1	6	25	55.4	33.5
40	44	15	4	2	8	30	77.3	46.3
19	45	5	5	1	2	35	56.4	35.2
13	46	5	3	3	6	25	55.4	33.4
37	47	15	4	5	4	30	75.2	43.2

effect of contact time for the adsorption process. Sudden adsorption of the acenaphthene was observed at the onset of the experiment, attributed to the higher BET surface area of the MOFs [36]. The equilibrium of the adsorption was attained within 30 min with the MOFs achieving equilibrium adsorption capacities of 23.814 mg/g and 14.627 mg/g for UiO-66 and ZIF-8, respectively. The adsorption efficiency achieved at equilibrium was 99.23% and 60.95% for UiO-66 and ZIF-8, respectively, confirming the superior adsorption capacity of UiO-66 and its higher porosities as previously reported in our studies [37].

To understand the mechanism and the rate controlling step for the process, the adsorption data for the UiO-66 and ZIF-8 effect of contact time was kinetics models was treated by the kinetics models of pseudo-first order (Eq. 4), pseudo-second order (Eq. 5), intra-particle diffusion (Eq. 6) and Elovich (Eq. 7) models linearly expressed as.

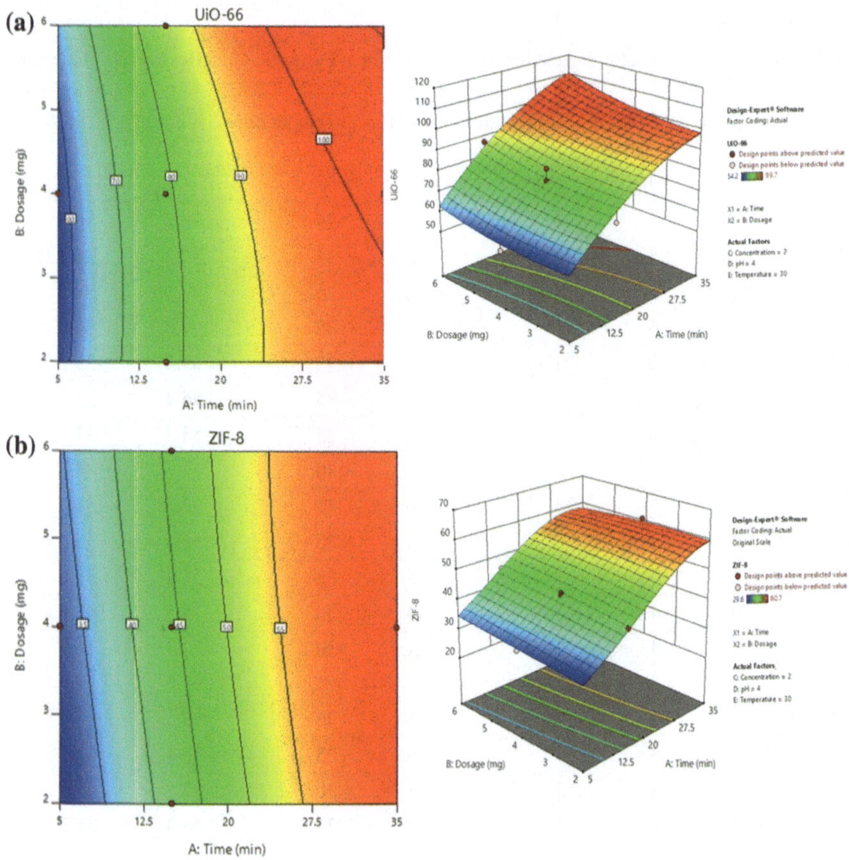

Fig. 2 2D and 3D RSM plots for the adsorption of acenaphthene onto **a** UiO-66 and **b** ZIF-8 MOFs under the condition: contact time of 35 min; dosage of 4 mg; concentrations of 2 mg/L, pH of 4 and temperature of 30 °C

$$\ln(q_e - q_t) = \ln q_e - k_1 t \tag{4}$$

$$\frac{t}{q_t} = \frac{1}{k_2 q_e^2} + \frac{t}{q_e} \tag{5}$$

$$q_t = K_P t^{\frac{1}{2}} + C \tag{6}$$

$$q_t = \frac{1}{\beta} In() + \frac{1}{\beta} Int \tag{7}$$

where q_e (mg/g) and q_t (mg/g) represent the amount of the acenaphthene adsorbed by the MOFs at equilibrium and t (mins). k_1 (1/min) k_2 (q/mg/min) represented the

Table 2 ANOVA for the adsorption of acenaphthene onto UiO-66 and ZIF-8 MOFs

	UiO-66						ZIF-8					
Source	SS	df	MS	F-value	p-value		SS	df	MS	F-value	p-value	
Model	0.0004	20	0.0000	33.44	< 0.0001	Significant	0.0014	20	0.0001	126.34	< 0.0001	Significant
A-Time	0.0004	1	0.0004	636.44	< 0.0001		0.0013	1	0.0013	2455.38	< 0.0001	
B-Dosage	3.434E−06	1	3.434E−06	5.77	0.0238		0.0000	1	0.0000	22.62	< 0.0001	
C-Concentration	3.811E−06	1	3.811E−06	6.40	0.0178		5.770E−06	1	5.770E−06	10.76	0.0029	
D-pH	2.554E−06	1	2.554E−06	4.29	0.0484		3.105E−06	1	3.105E−06	5.79	0.0235	
E-Temperature	1.980E−06	1	1.980E−06	3.33	0.0797		2.672E−07	1	2.672E−07	0.4985	0.4864	
AB	8.016E−09	1	8.016E−09	0.0135	0.9085		3.309E−06	1	3.309E−06	6.17	0.0197	
AC	8.144E−07	1	8.144E−07	1.37	0.2528		1.668E−07	1	1.668E−07	0.3112	0.5817	
AD	4.568E−06	1	4.568E−06	7.67	0.0102		2.068E−06	1	2.068E−06	3.86	0.0603	
AE	7.053E−07	1	7.053E−07	1.18	0.2865		4.333E−07	1	4.333E−07	0.8083	0.3769	
BC	4.427E−08	1	4.427E−08	0.0743	0.7873		2.146E−08	1	2.146E−08	0.0400	0.8430	
BD	2.858E−08	1	2.858E−08	0.0480	0.8283		3.603E−06	1	3.603E−06	6.72	0.0154	
BE	5.467E−07	1	5.467E−07	0.9181	0.3468		2.582E−07	1	2.582E−07	0.4816	0.4938	
CD	1.962E−06	1	1.962E−06	3.30	0.0810		9.847E−08	1	9.847E−08	0.1837	0.6718	
CE	4.490E−07	1	4.490E−07	0.7540	0.3932		2.190E−07	1	2.190E−07	0.4085	0.5283	
DE	7.972E−07	1	7.972E−07	1.34	0.2578		4.070E−07	1	4.070E−07	0.7592	0.3915	
A^2	0.0000	1	0.0000	33.65	< 0.0001		0.0000	1	0.0000	86.03	< 0.0001	
B^2	2.631E−08	1	2.631E−08	0.0442	0.8352		8.056E−08	1	8.056E−08	0.1503	0.7014	
C^2	1.937E−07	1	1.937E−07	0.3253	0.5734		4.969E−07	1	4.969E−07	0.9270	0.3445	
D^2	2.434E−07	1	2.434E−07	0.4088	0.5282		3.000E−07	1	3.000E−07	0.5595	0.4612	
E^2	9.380E−09	1	9.380E−09	0.0158	0.9011		1.998E−08	1	1.998E−08	0.0373	0.8484	

(continued)

Table 2 (continued)

	UiO-66						ZIF-8					
Source	SS	df	MS	F-value	p-value		SS	df	MS	F-value	p-value	
Residual	0.0000	26	5.955E−07				0.0000	26	5.361E−07			
Lack of fit	0.0000	22	6.564E−07	2.52	0.1915	Not significant	0.0000	22	6.146E−07	5.90	0.0484	Not significant
Pure error	1.043E−06	4	2.607E−07				4.170E−07	4	1.043E−07			
Cor total	0.0004	46					0.0014	46				

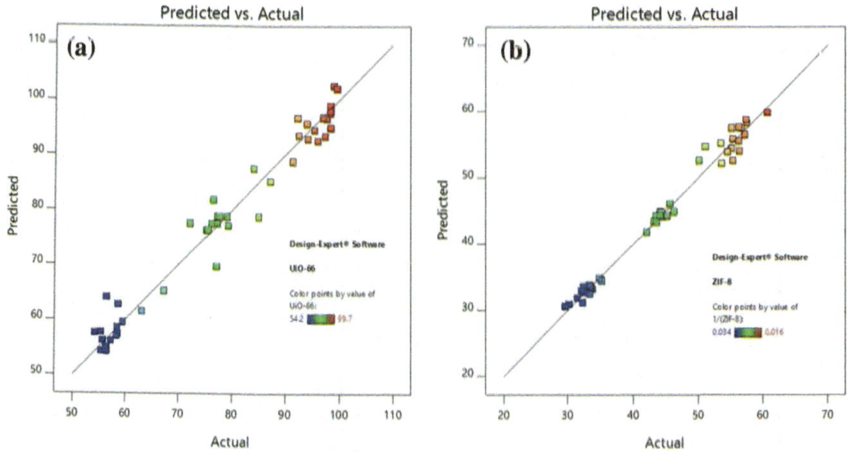

Fig. 3 Scatter plots for the RSM adsorption of acenaphthene onto UiO-66 and ZIF-8 MOFs

Fig. 4 Effect of contact time for the adsorption of acenaphthene onto UiO-66 and ZIF-8 MOFs

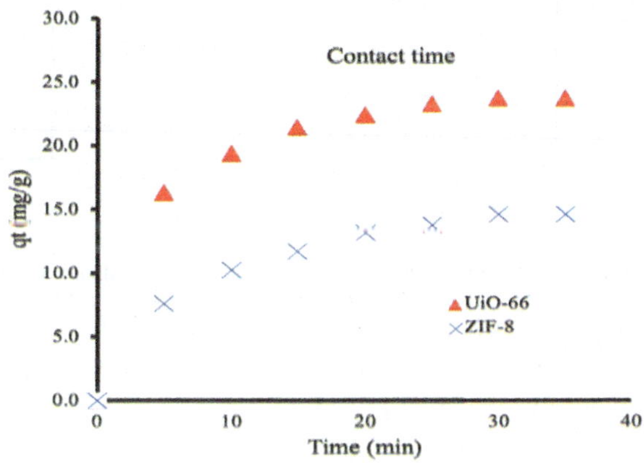

pseudo-first order and pseudo-second order rate constants, respectively. Kp represented the intraparticle rate constant and C is constant. The α (mg/gmin) and β (g/mg) were the initial desorption rate and desorption constant for the Elovich model, respectively.

Figure 5 depicted the plots of the kinetic models studied. The adsorption data was found to be precisely suited by the pseudo-second order model in comparison to other models evaluated. It is R^2 value was 0.9947 and 0.97780 for the UiO-66 and ZIF-8, respectively. Also, its calculated adsorption capacities were 24.80 and 15.87 mg/g for the UiO-66 and ZIF-8, respectively, very closer to the obtained experimental values of 23.814 and 14.627 mg/g for UiO-66 and ZIF-8, respectively, indicating the chemisorption is the rate limiting step of the process and the availability of the adsorption sites on the surface of the MOFs. Other reports have also described

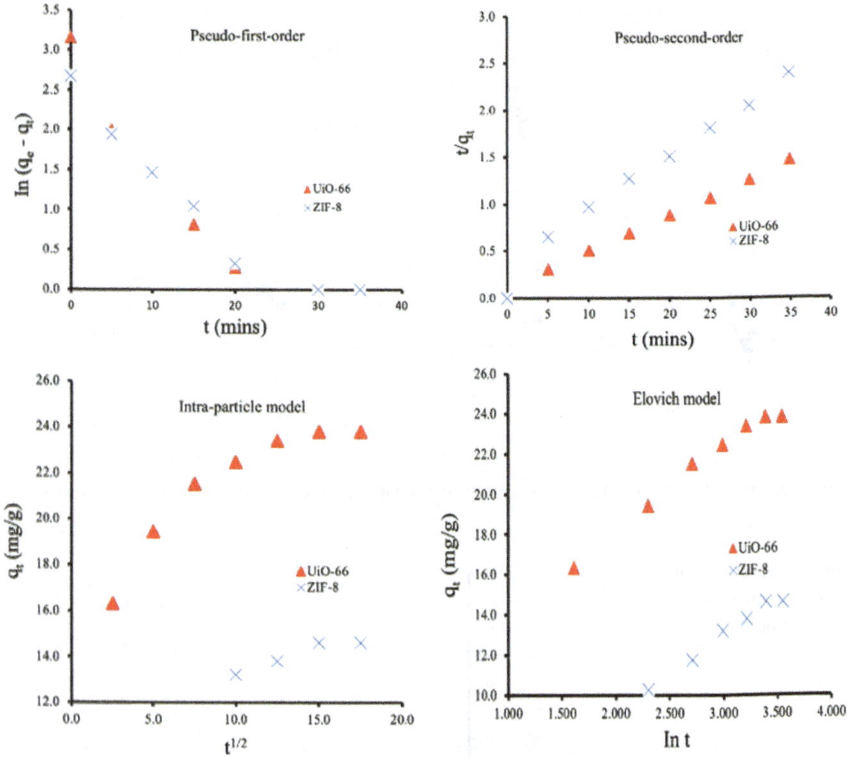

Fig. 5 Adsorption kinetics plots for acenaphthene adsorption onto UiO-66 and ZIF-8 MOFs

the adsorption of organic pollutants onto UiO-66 [38, 39] and ZIF-8 [40, 41] as pseudo-second order process.

4 Conclusion

RSM was used for the experimental design of acenaphthene adsorption onto UiO-66 and ZIF-8 MOFs. Five factor parameters were employed, and best performance of the MOFs was achieved for UiO-66 with adsorption efficiency of 99.7 due to its higher porosity in comparison to ZIF-8 which recorded adsorption efficiency of 60.7%, according to the experimental condition with contact time of contact time of 35 min, dosage of 4 mg, concentrations of 2 mg/L, pH of 4 and temperature of 30 °C. The overall model p-values were significant (< 0.0500) for both UiO-66 and ZIF-8 MOFs. The lack of fit test for the model was insignificant, which was desirable for both MOFs and the adequate precision was 17.3082 and 32.6708 for the UiO-66 and ZIF-8, respectively. Good agreement between the RSM and experimental findings.

Acknowledgements We would like to acknowledge UTP-UIR grant (015MEO-166) received from Universitas Islam Riau Indonesia and FRGS grant (FRGS/1/2020/STG04/UTP/02/3) received from Ministry of Higher Education Malaysia.

References

1. Guieysse, B., Bernhoft, I., Andersson, B.E., Henrysson, T., Olsson, S., Mattiasson, B.: Degradation of acenaphthene, phenanthrene and pyrene in a packed-bed biofilm reactor. Appl. Microbiol. Biotechnol. **54**(6), 826–831 (2000). https://doi.org/10.1007/s002530000442
2. Zango, Z.U., Jumbri, K., Sambudi, N.S., Abu Bakar, N.H.H., Abdullah, N.A.F., Basheer, C., Saad, B.: Removal of anthracene in water by MIL-88(Fe), NH_2-MIL-88(Fe), and mixed-MIL-88(Fe) metal-organic frameworks. RCS Adv. **9**, 41490–41501, (2019). https://doi.org/10.1039/c9ra08660arsc.li/rsc-advances
3. Jiang, L., Zhang, B., Wang, Y., Sun, J., Ma, X., Wang, G., Fu, S., Lin, C., Li, Y.: Three new acenaphthene derivatives from rhizomes of musa basjoo and their cytotoxic activity. Nat. Prod. Res. pp. 1–6 (2019). https://doi.org/10.1080/14786419.2019.1647422
4. Hirano, Y., Iba, Y., Kuroda, N., Kubota, Y., Inagaki, S.: Catalytic carbonization of acenaphthene for the preparation of ordered mesoporous carbon CMK-1 toward application as electrochemical double-layer capacitor electrode with ionic liquid electrolyte. Chem. Lett. **48**(6), 521–524 (2019). https://doi.org/10.1246/cl.190090
5. Bukharkina, T.V., Grechishkina, O.S., Digurov, N.G., Kon'kov, I.I.: Catalytic oxidation of acenaphthene and its derivatives in acetic acid. Org. Process Res. Dev. **6**(4), 394–400 (2002). https://doi.org/10.1021/op0100448
6. Ye, Y.F., Ma, F.Y., Wu, M., Wei, X.Y., Liu, J.M.: Increase of acenaphthene content in creosote oil by hydrodynamic cavitation. Chem. Eng. Process. Process Intensif. **104**, 66–74 (2016). https://doi.org/10.1016/j.cep.2016.03.001
7. Radwan, A.M.Y., Magram, S.F., Ahmed, Z.: Adsorption of acenaphthene using date seed activated carbon. J. Environ. Sci. Technol. **11**(1), 10–15 (2018). https://doi.org/10.3923/jest.2018.10.15
8. Alade, A.O., Amuda, O.S., Ibrahim, A.O.: Isothermal studies of adsorption of acenaphthene from aqueous solution onto activated carbon produced from rice (Oryza sativa) husk. Elixir Chem. Eng. **46**, 87–95 (2012). https://doi.org/10.1080/19443994.2012.677514
9. Rani, C.N., Karthikeyan, S.: Feasibility study of acenaphthene degradation in a novel slurry UV photocatalytic membrane reactor: Effect of operating parameters and optimization using response surface modeling. Chem. Eng. Process.-Process Intensif. pp. 108051 (2020). https://doi.org/10.1016/j.cep.2020.108051
10. Safitri, R., Handayani, S., Surono, W., Astika, H., Damayanti, R., Kusmaya, F.D., Rukiah, Balia, R.L.: Biodegradation of phenol, anthracene and acenaphthene singly and consortium culture of indigenous microorganism isolates from underground coal gasification area. In: IOP Conference Series Earth Environment Science **306**(1) (2019). https://doi.org/10.1088/1755-1315/306/1/012026
11. Gracht Van Der, H.: Potential of lost fishing gears for adsorption of PAHs.Centro Interdisplinar de Investigarcao Marinho e Ambiental (2020)
12. Patrick, U.A., Chiwuike, A.-O.: A column experiment showing adsorption dynamics and kinetics of selected PAH using plantain and cassava peels. Int. J. Adv. Sci. Technol. **111**, 129–146 (2018). https://doi.org/10.14257/ijast.2018.111.12
13. Smol, M., Włodarczyk-Makuła, M., Włóka, D.: Adsorption of polycyclic aromatic hydrocarbons (PAHs) from aqueous solutions on different sorbents. Civ. Environ. Eng. Reports. **13**(2), 87–96 (2014). https://doi.org/10.2478/ceer-2014-0017

14. Huang, Y., Fulton, A.N., Keller, A.A.: Simultaneous removal of PAHs and metal contaminants from water using magnetic nanoparticle adsorbents. Sci. Total Environ. **571**, 1029–1036 (2016). https://doi.org/10.1016/j.scitotenv.2016.07.093
15. Lu, L., Lin, Y., Chai, Q., He, S., Yang, C.: Removal of acenaphthene by biochar and raw biomass with coexisting heavy metal and phenanthrene. Colloids Surfaces A Physicochem. Eng. Asp. **558**, 103–109 (2018). https://doi.org/10.1016/j.colsurfa.2018.08.057
16. Mortazavi, M., Baghdadi, M., Seyed Javadi, N.H., Torabian, A.: The black beads produced by simultaneous thermal reducing and chemical bonding of graphene oxide on the surface of amino-functionalized sand particles: application for PAHs removal from contaminated waters. J. Water Process Eng. **31** (2019). https://doi.org/10.1016/j.jwpe.2019.100798
17. Zango, Z.U., Sambudi, N.S., Jumbri, K., Ramli, A., Hana, N., Abu, H., Saad, B., Nur, M., Rozaini, H., Isiyaka, H.A., Osman, A.M., Sulieman, A.: An overview and evaluation of highly porous adsorbent materials for polycyclic aromatic hydrocarbons and phenols removal from wastewater, pp. 1–40 (2020)
18. Alade, A.O., Amuda, O.S., Afolabi, A.O., Adelowo, F.E.: Adosrption of acenaphthene unto ativated carbon produced from agricultural wastes. J. Envionmental Sci. Technol. **5**(4), 192–209 (2012)
19. Lu, L., Li, A., Ji, X., Yang, C., He, S.: Removal of acenaphthene from water by Triton X-100-facilitated biochar-immobilized pseudomonas aeruginosa. RSC Adv. **8**, 23426–23432 (2018). https://doi.org/10.1039/C8RA03529F
20. Han, B., Li, Y., Qian, B., He, Y., Peng, L., Yu, H.: Adsorption and determination of polycyclic aromatic hydrocarbons in water through the aggregation of graphene oxide. Open Chem. **16**(1), 716–725 (2018). https://doi.org/10.1515/chem-2018-0078
21. Bueken, B., Vermoortele, F., Cliffe, M.J., Wharmby, M.T., Foucher, D., Wieme, J., Vanduyfhuys, L., Martineau, C., Stock, N., Taulelle, F., Van Speybroeck, V., Goodwin, A.L., De Vos, D.: A breathing zirconium metal-organic framework with reversible loss of crystallinity by correlated nanodomain formation. Chem.-A Eur. J. **22**(10), 3264–3267 (2016). https://doi.org/10.1002/chem.201600330
22. Connolly, B.M., Mehta, J.P., Moghadam, P.Z., Wheatley, A.E.H., Fairen-Jimenez, D.: From synthesis to applications: metal-organic frameworks for an environmentally sustainable future. Curr. Opin. Green Sustain. Chem. **12**, 47–56 (2018). https://doi.org/10.1016/j.cogsc.2018.06.012
23. Adesina Adegoke, K., Samuel Agboola, O., Ogunmodede, J., Oluyomi Araoye, A., Solomon Bello, O.: Metal-organic frameworks as adsorbents for sequestering organic pollutants from wastewater. Mater. Chem. Phys. **253**, p. 123246 (2020). https://doi.org/10.1016/j.matchemphys.2020.123246
24. Gomar, M., Yeganegi, S.: Adsorption of 5-fluorouracil, hydroxyurea and mercaptopurine drugs on zeolitic imidazolate frameworks (ZIF-7, ZIF-8 and ZIF-9). Microporous Mesoporous Mater. **252**, 167–172 (2017). https://doi.org/10.1016/j.micromeso.2017.06.010
25. Jin, Z., Yang, H.: Exploration of Zr-metal-organic framework as efficient photocatalyst for hydrogen production. Nanoscale Res. Lett. **12** (2017). https://doi.org/10.1186/s11671-017-2311-6
26. Zango, Z.U., Jumbri, K., Sambudi, N.S., Ramli, A., Bakar, N.H.H.A., Saad, B., Rozaini, M.N.H., Isiyaka, H.A., Jagaba, A.H., Aldaghri, O., Sulieman, A.: A critical review on metal-organic frameworks and their composites as advanced materials for adsorption and photocatalytic degradation of emerging organic pollutants from wastewater. Polymers (Basel). **12**, 1–42 (2020). https://doi.org/10.3390/polym12112648
27. Zango, Z.U., Ramli, A., Jumbri, K., Sambudi, N.S., Isiyaka, H.A., Abu Bakar, N.H.H., Saad, B.: Response surface methodology optimization and kinetics study for anthracene adsorption onto MIL-88(Fe) and NH_2-MIL-88(Fe) metal- organic frameworks. In: Series IOP Conference Science Materials **88**, pp. 0–12 (2021). https://doi.org/10.1088/1757-899X/1092/1/012035
28. Petit, C.: Present and future of MOF research in the field of adsorption and molecular separation. Curr. Opin. Chem. Eng. **20**, 132–142 (2018). https://doi.org/10.1016/j.coche.2018.04.004

29. Shahmirzaee, M., Hemmati-Sarapardeh, A., Husein, M.M., Schaffie, M., Ranjbar, M.: A review on zeolitic imidazolate frameworks use for crude oil spills cleanup. Adv. Geo Energy Res. **3**(3), 320–342 (2019). https://doi.org/10.26804/ager.2019.03.10
30. Zango, Z.U., Sambudi, N.S., Jumbri, K., Abu Bakar, N.H.H., Abdullah, N.A.F., Negim, E.S.M., Saad, B.: Experimental and molecular docking model studies for the adsorption of polycyclic aromatic hydrocarbons onto UiO-66(Zr) and NH_2-UiO-66(Zr) metal-organic frameworks. Chem. Eng. Sci. p. 115608 (2020). https://doi.org/10.1016/j.ces.2020.115608
31. Jamali, A., Shemirani, F., Morsali, A.: A comparative study of adsorption and removal of organophosphorus insecticides from aqueous solution by Zr-based MOFs. J. Ind. Eng. Chem. **80**, 83–92 (2019). https://doi.org/10.1016/j.jiec.2019.07.034
32. Yang, J.M., Ying, R.J., Han, C.X., Hu, Q.T., Xu, H.M., Li, J.H., Wang, Q., Zhang, W.: Adsorptive removal of organic dyes from aqueous solution by a Zr-based metal-organic framework: effects of Ce(III) doping. Dalt. Trans. **47**(11), 3913–3920 (2018). https://doi.org/10.1039/c8dt00217g
33. Bayazit, S.S., Şahin, S.: Acid-modulated zirconium based metal organic frameworks for removal of organic micropollutants. J. Environ. Chem. Eng. **8**(5) (2020). https://doi.org/10.1016/j.jece.2020.103901
34. Alamgir, Talha, K., Wang, B., Liu, J.H., Ullah, R., Feng, F., Yu, J., Chen, S., Li, J.R.: Effective adsorption of metronidazole antibiotic from water with a stable Zr(IV)-MOFs: Insights from DFT, kinetics and thermodynamics studies. J. Environ. Chem. Eng. **8**(1), p. 103642 (2020). https://doi.org/10.1016/j.jece.2019.103642
35. Shi, M., Huang, R., Qi, W., Su, R., He, Z.: Synthesis of superhydrophobic and high stable Zr-MOFs for oil-water separation. Colloids Surfaces A Physicochem. Eng. Asp. **602**, p. 125102 (2020). https://doi.org/10.1016/j.colsurfa.2020.125102
36. Yan, F., Liu, Z.Y., Chen, J.L., Sun, X.Y., Li, X.J., Su, M.X., Li, B., Di, B.: Nanoscale zeolitic imidazolate framework-8 as a selective adsorbent for theophylline over caffeine and diprophylline. RSC Adv. **4**, 33047–33054 (2014). https://doi.org/10.1039/c4ra05293e
37. Zango, Z.U., Ramli, A., Jumbri, K., Soraya, N., Ahmad, H.I., Hana, N., Abu, H., Saad, B.: Optimization studies and artificial neural network modeling for pyrene adsorption onto UiO-66(Zr) and NH_2-UiO-66(Zr) metal organic frameworks. Polyhedron **192**, p. 114857 (2020). https://doi.org/10.1016/j.poly.2020.114857
38. Zhang, W., Yang, J.M., Yang, R.N., Yang, B.C., Quan, S., Jiang, X.: Effect of free carboxylic acid groups in UiO-66 analogues on the adsorption of dyes from water: plausible mechanisms for adsorption and gate-opening behavior. J. Mol. Liq. **283**, 160–166 (2019). https://doi.org/10.1016/j.molliq.2019.03.100
39. Sun, W., Li, H., Li, H., Li, S., Cao, X.: Adsorption mechanisms of ibuprofen and naproxen to UiO-66 and UiO-66-NH_2: batch experiment and DFT calculation. Chem. Eng. J. **360**, 645–653 (2019). https://doi.org/10.1016/j.cej.2018.12.021
40. Noor, T., Raffi, U., Iqbal, N., Yaqoob, L., Zaman, N.: Kinetic evaluation and comparative study of cationic and anionic dyes adsorption on Zeolitic Imidazolate frameworks based metal organic frameworks. Mater. Res. Express. **6**(12) (2019)
41. Sun, S., Yang, Z., Cao, J., Wang, Y., Xiong, W.: Copper-doped ZIF-8 with high adsorption performance for removal of tetracycline from aqueous solution. J. Solid State Chem. **285**, p. 121219 (2020). https://doi.org/10.1016/j.jssc.2020.121219

The Simplex Lattice Design of Experiments: Formulation of Low-IFT Foam Surfactant for EOR Application

Nur Fatin Syazwanie Zahari, Anita Ramli, Norhidayah Ahmad Wazir, Wasan Saphanuchart, and Khaled Abdalla Elraies

Abstract The main objective of this work was to determine the optimum surfactant formulation to get low interfacial tension (IFT) and foam half-life value by using simplex lattice mixture Design of Experiment (DOE) method. A quadratic model was fitted to the first response, that is foam half-life, whereas special quartic model was fitted to IFT. Both models have been experimentally validated based on the ANOVA table. Two diagnostic tools were also used to help in investigate in diagnosing the models, which are the Normal Plot of Residuals and Box-Cox Plot. Besides that, the impacts of the surfactant component fractions on foam half-life and IFT variation were also studied followed by optimizing the models to get an optimum formulation that will produce high stability and minimum IFT value. The predicted and experimental value IFT were found to be in good agreement with value of 0.05 mN/, which close to the predicted value of 0.04 mN/m, whereas for foam half-life, the experimental value is 286 s which is not in good agreement to the predicted value of 338 s.

Keywords IFT · Foam half-life · Design of experiment

N. F. S. Zahari · A. Ramli (✉)
Fundamental and Applied Sciences Department and Centre for Biofuel and Biochemical Research, Institute of Self- Sustainable Building, Universiti Teknologi PETRONAS, 32610 Seri Iskandar, Perak, Malaysia
e-mail: anita_ramli@utp.edu.my

N. F. S. Zahari
e-mail: nur_17007653@utp.edu.my

N. A. Wazir · W. Saphanuchart
PETRONAS Research Sdn. Bhd, 43000 Bangi, Malaysia
e-mail: hidayah_wazir@petronas.com.my

K. A. Elraies
Petroleum Engineering Department, Universiti Teknologi PETRONAS, 32610 Seri Iskandar, Perak, Malaysia
e-mail: khaled.elraies@utp.edu.my

S. A. Abdul Karim et al. (eds.), *Proceedings of the 6th International Conference on Fundamental and Applied Sciences*, Springer Proceedings in Complexity,
https://doi.org/10.1007/978-981-16-4513-6_22

1 Introduction

Due to our world market demand, oil as non-renewable energy resource which possesses great utilization has urged the oil and gas industry to come out with more advanced technology to overcome the shortage of petroleum supply. There is still a large part of residual oil which are still arrested in the low permeability area in the reservoir, and the recent technology has been a great helpful to increase oil recovery. To overcome on this issue, the tertiary oil recovery or also known as enhanced oil recovery has attracted great attention in research and development on every aspect to increase the efficiency of oil recovery to fully utilize of oil resources [1]

The uses of chemicals such as IFT and foam surfactants have been used for decades in enhancing oil recovery especially in a great challenging reservoir condition [2]. Unique characteristic chemical structure of surfactant itself, makes it a very interesting chemical in oil recovery, where it has both hydrophilic and hydrophobic part in one chemical structure. Surfactants are being used to promote the imbibition of displacing fluid by reducing the interfacial tension between crude oil and displacing fluid which is inhibited by the high capillary force in the porous rock [3]. However, reducing the interfacial tension is not enough to get a good oil recovery, but the IFT need to be reduced at a very low value ($<10^{-2}$ mN/m) for an emulsion to form.

IFT surfactant only does not enough to sweep out the fluids and recently foam has grabbed a lot of researcher's attention in mobility part. Foams have been reported in few studies to help in improving sweeping efficiency of the displacing fluids and crude oil in gas flooding [4]. Few studies have also highlighted the advantage of foam where it is able to flow far to the low permeabilities region [5] and are bearable with harsh reservoir condition, such as high temperatures and salinities compared to when polymer is utilized [6]. Before current advances in EOR which is foam, they only use natural gas for mobility control, but major problem occurred when it is injected to EOR. Natural gas possesses low density and viscosity, thus it tends ascend to the top of the reservoir [7]. Major issue that always been mentioned in most case studies, related to its stability. Foam is well-known with its thermodynamically unstable system but with the use of surfactants, the foam stability can be improved [8].

The integration IFT and foam surfactants have been addressed in few researches specifically in fractured carbonate reservoirs and also proven to have higher oil-recovery [9–11] but this combination has some issues raised regarding to it. Lachaise described the challenge in combining IFT and foam surfactant has made the foam stability decreased as the IFT properties dominate to the ultra-low IFT value [12]. Another research papers has also mentioned this issue [9, 10]. This challenge is arising due to the different chemical structure for foams and low-IFT surfactants, thus affecting both IFT and foam properties contradictory. This is actually a clear evidence that for a formulation to have a dual-function ability may compromise the performances between low-IFT and foam properties. Besides that, it is also a big challenge, especially for the oil and gas industry because some chemical supplier, they do not state the actual composition for a commercial surfactant and will cost a large amount of time to do testing and characterization. Despite all the hurdles, the idea to have a dual-function surfactant have successfully proven to have incremental in oil recovery in carbonate type of reservoir however, it hasn't been discussed for sandstone type reservoir.

All these studies which attempted to formulate and design an effective a dual-function formulation is by selecting new surfactant that fall into these two categories, IFT and foam surfactant. However, conventional optimization for formulation is time consuming whilst also being expensive as this method would require a large amount of samples [13]. The optimum point for conventional optimization usually impossible to precisely find and therefore, a well-designed data collection method would help the experiments to get valid results statistically [14]. DoE is one of the systematic approaches to solve problem that can be applied to data collection and statistical analysis. This approach is satisfactory and compelling than other method, for example, classical one-at-a-time or mathematical methods since it can study numerous variables simultaneously with a smaller number of observations, costs and time saving [13]. In this study, an attempt was made to formulate an optimal dual-function surfactant formulation by getting specific target by minimizing the IFT value and has high foam stability according to a three-component mixture design. Design Expert® software was used to obtain an optimal formulation. The parameters in this experiment design would be the surfactant compositions. This paper concentrates in detail the simultaneous effect of these parameters on IFT and foam stability utilizing DoE.

2 Materials and Methodology

2.1 Materials

Three surfactants were used in this formulation which focusing on anionic and amphoteric surfactant because this study concentrates on sandstone type reservoir. Both type of surfactants are most widely used type of surfactant because it shows low adsorption on the rock's surface because of the negatively charged surface. Table 1 lists all the surfactants that were used in this study without any more treatment aside

Table 1 Type of surfactants that were used in this study

Surfactant	Supplier	Type
Cocamidopropyl betaine (CAB)	Stepan	Amphoteric
Amidopropyl betaine (AB)	Oil Chem Technologies	Amphoteric
Alpha olefin sulfonate (AOS)	AkzoNobel	Anionic

Table 2 Brine compositions that used in this study

Ions	SW (ppm)
Sodium, Na^+	9498
Potassium, K^+	379
Calcium, Ca^{2+}	294
Magnesium, Mg^{2+}	1140
Strontium, Sr^{2+}	4.695
Sulfate, $SO_4{}^{2-}$	2400
Cloride, Cl^-	17,023
Bicarbonate, $HCO_3{}^-$	82

for dilution with brine or sea water (SW). The brine compositions are shown in Table 2.

2.2 *Sample Preparation and Experimental Procedure*

The hydrocarbon phase or crude oil collected from east coast of Peninsular Malaysia oilfield was used in this investigation in all IFT and foam half-life tests and the aqueous phase contained seawater along with three surfactants that were listed in Table 1. The aqueous phase was mixed by using magnetic stirrer for one hour because some solution took a long time to homogenized. Since this research work is focusing on finding a right composition for the aqueous phase in which the IFT value between the two phases is at minimum value and the foam half-life value at maximum value. Both IFT and foam half-life value will be set as responses and to be measured based on the different compositions of aqueous phases assigned by the Design Expert software.

2.3 *Experimental Procedure of IFT and Foam Stability Measurements*

For IFT measurements, both hydrocarbon and aqueous phases density should be determined. The density value was carried out by using a handheld density meter, DMA 35 Basic purchased from Anton Paar. The sample was taken directly from a specific solution such as hydrocarbon and aqueous phases and the density data was collected directly without the need for further calculations at room temperature. Each test was replicated for three times.

Next, the IFT value between the hydrocarbon and aqueous phases were measured by using a Spinning Drop Tensiometer from KRUSS, Germany at 96 °C. The aqueous phases were injected first to a capillary tube until it full and then half μl of the hydrocarbon phase was injected into the capillary tube. The capillary tube was then inserted into the instrument and the diameter of the oil droplet was measured at 2800–3200 rpm. Three replicates were used for each test. Vonnegut's expression was used to estimate the IFT:

$$\sigma = \Delta\rho\omega^2 4R^3 \quad (1)$$

where,

σ	IFT
Δρ	fluid densities difference
Ω	angular velocity
R	droplet's radius.

For foam stability measurement, the test was carried out using Teclis foam scan (Teclis, France) at 96 °C and 1 bar pressure. Total volume to be injected in the foam scan using 9:1, 8 mL of crude oil was injected first followed by 72 mL surfactant solution. 100 mL/min flow rate of nitrogen gas was kept constant as the gas was purged to the surfactant solution from the bottom of the tube until the desired volume of foam (200 mL) was produced. The time taken for the foam to reach 200 mL was recorded as an indication for its ability to foam. Once it reached 200 mL, the foam will be started to collapse until it is at 100 mL. The interval time it took to reach 100 mL was recorded as its half-life and the flow of nitrogen gas was then stopped. Figure 1 shows a conceptual diagram for this foam test.

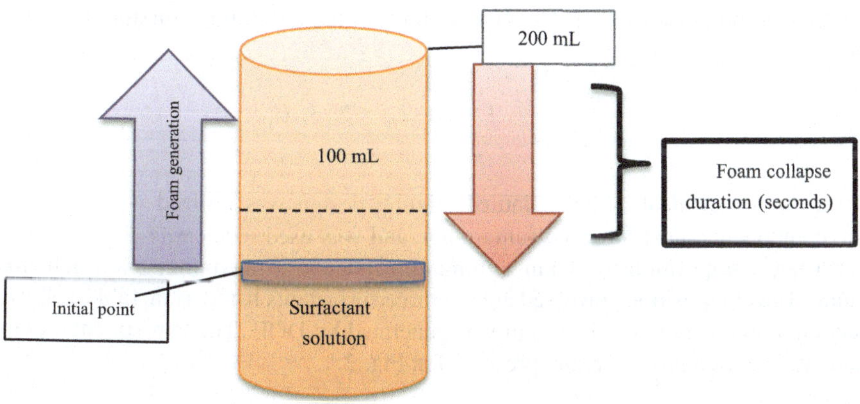

Fig. 1 General diagram for foam test

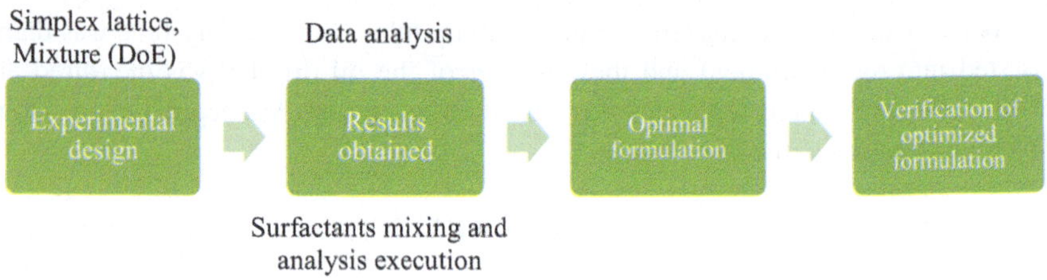

Fig. 2 Experimental optimization process

2.4 *Simplex Lattice Mixture Design*

Experimental designed by using Design of Experiment (DOE) is an appropriate tool which provides a simple strategy in optimizing a process and to come out with an optimal formulation of a specific mixture [15]. It is also a great tool for experimental conception relationship evaluations dependent on three sections including design, analysis and optimization [16]. A mixture experiment is a unique type of response surface experiment design in which the components of a mixture are the main variables and the response is dependently to the proportions of each component and not on the amount of the mixture. The key benefit of this approach is the minimization of the quantity of speculative try and error with more effective resolution on formulation mixing. Most mixture designs, there are restrictions specifically on component proportions (Xj) which limit the user to explore the entire simplex region. These restrictions here are define as the lower (Lj) and upper (Uj) constraints on the component proportions [17]. The general form of the constrained mixture problem are;

$$\sum_j X_j = 1 \quad \text{and} \quad \mathrm{L_j} \le \mathrm{X_j} \le \mathrm{U_j} \tag{2}$$

In this experiment, simplex lattice mixture design was utilized where are all the components must have the same range and was used to identify the optimum surfactant composition to obtain a minimum IFT and a maximum foam half-life value. The compositions involved are the three surfactants listed as in Table 1. Each random composition in this design was generated by DOE. The experimental plan and optimization procedure are presented in Fig. 2.

2.5 *Data Analysis*

To find a right approximation for the actual functional relationship between the independent variables and the response surface is necessary. The response was utilized to build up an empirical model that correlated both responses in this experiment, IFT and foam half-life values with the different component compositions which are the variables. The model was tested the adequacy of the fitted model, the shape of the three-dimensional response surface, and the optimal formulation was determined. Regression analysis and analysis of variance (ANOVA) were carried out by using Design Expert® version 11 software, Stat-Ease, Inc., Minneapolis, USA).

3 Results and Discussion

3.1 Data Analysis and Model Validation

Table 3 shows the empirical outcomes from the designed experiment. For single surfactant, the IFT value varies from 0.0348378 to 0.650383 mN/m. Foam half-life value for single surfactant varies from 171 to 391 s. For binary mixture, the IFT value show better performance which decreased from 0.0188537 to 0.294219 mN/m. For foam half-life, the value increased from 198 to 493 s. It is also observed that for the mixtures with fraction of all the three components, the IFT value increases from 0.04114535 mN/m from to 0.307773 mN/m whereas for foam half-life the

Table 3 Results of designed experiment

Run	Components (%)			Responses	
	A: Cocamidopropyl betaine	B. Amidopropyl betaine	C. Alpha olefin sufonate	Foam half-life (s)	IFT (mN/m)
1	0	0	0.5	203	0.58026
2	0	0	0.5	171	0.650383
3	0	0.5	0	320	0.0348378
13	0	0.5	0	385	0.05429
9	0.5	0	0	391	0.499275
15	0.5	0	0	358	0.565519
2	0	0.333333	0.166667	198	0.0188537
11	0	0.166667	0.333333	220	0.151302
7	0.333333	0	0.166667	493	0.235088
10	0.166667	0	0.333333	477	0.274669
5	0.166667	0.333333	0	413	0.11307
12	0.333333	0.166667	0	419	0.294219
4	0.166667	0.166667	0.166667	315	0.103698
8	0.0833333	0.0833333	0.333333	379	0.307773
14	0.0833333	0.333333	0.0833333	340	0.0414535
16	0.333333	0.0833333	0.0833333	453	0.157069

value varies from 315 to 340 s. These observations have shown that the presence of two or more components improves the physicochemical properties of the individual surfactant for the IFT and the foam half-life.

Using Design Expert software, a quadratic model was fitted to the foam half-life experimental results, whereas special quartic was fitted to the IFT. For foam half-life, power transformation with lambda of 1.42 was found to have best fit to the model. To predict the foam half-life between the aqueous phase and crude oil regarding the actual factors of mixture components, the final obtained model is expressed as in Eq. 3. As for IFT, special quartic was the suggested model to fit. In order to have the best model to fit, transformation of square root was found to have better fit in need to improve the model. Final equation obtained is expressed as in Eq. 4.

The predicted foam half-life and IFT that derived from the model versus actual foam half-life/IFT that was obtained from experiments are shown in Fig. 3 and this figure shows that the model was successful in portraying the correlation between the mixture components with R-squared value of 0.8947 and 0.9923, respectively. The adequate precision value that measures the signal-to-noise ratio, which is 13.237 for foam half-life, whereas 29.764 is for IFT, which is greater than 4 and indicates an adequate signal. Therefore, these models can be used to navigate the design space.

Foam half-life

$$+373.33A + 348.92B + 206.53C + 185.95AB + 862.25AC - 326.32BC \quad (3)$$

Sqrt (IFT)

$$\begin{aligned} &+ 0.73A + 0.20B + 0.79C - 0.12AB - 1.15AC - 1.05BC \\ &- 9.48A^2BC - 2.72AB^2C + 12.74ABC^2 \end{aligned} \quad (4)$$

where A, B, and C represent of each fraction of cocamidopropyl betaine, amidopropyl betaine, and alpha olefin sulfonate in weight percent (wt.%), respectively.

Analysis of variance was performed to investigate the fitness and significance of the model. From the ANOVA result (Table 4), it can show how individual, and interaction of variables influence on both responses, IFT and foam half-life of the mixture.

F value in ANOVA defines a ratio of two variances of responses. This value tells if the means between two populations are significantly different. A significant result means that the obtained results likely did not happen by chance. If a result is not statistically significant, the null hypothesis cannot be rejected. As for foam half-life response, the F tabulated is 9.01 and the F calculated is 16.99. F tabulated for IFT response is 8.85 and the F calculated is 113.46. Both F calculated values have bigger value than the tabulated, therefore, null hypothesis are rejected, and the models are statistically agreed with the predicted value at 95% confidence level.

However, F value alone is not enough, and p value should also be considered. *P* value is determined by the F statistic and is the probability that the results could have happened by chance. A value less than 0.05 demonstrates that there is a significant

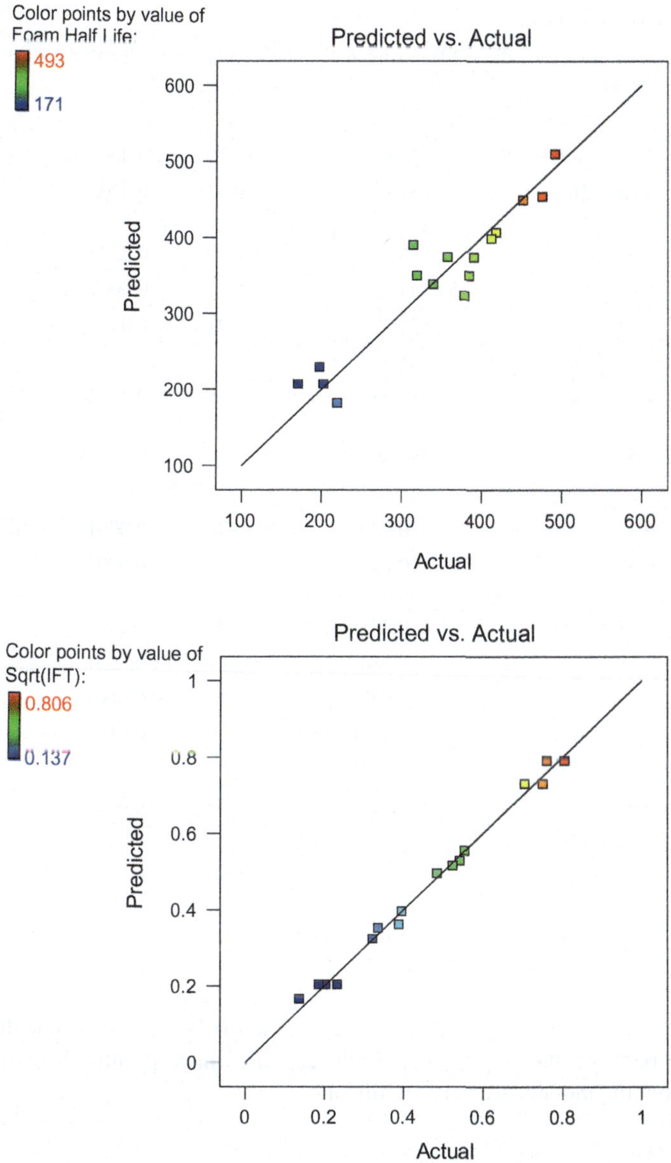

Fig. 3 Both predicted IFT and foam half-life values against actual IFT and foam half-life values

Table 4 ANOVA result table

Source	Sum of squares	df	Mean square	F value	p-value Prob > F	Remarks
Foam half-life						
Model	1.390E + 005	5	27,796.86	16.99	0.0001	Significant
Linear mixture	67,503.56	2	33,751.78	20.63	0.0003	
AB	2766.91	1	2766.91	1.69	0.2226	
A	59,494.52	1	59,494.52	36.37	0.0001	
BC	8520.93	1	8520.93	5.21	0.0456	
Residual	16,358.64	10	1635.86			
Lack of Fit	13,189.64	7	1884.23	1.78	0.3422	Not significant
Pure Error	3169.00	3	1056.33			
IFT						
Model	0.71	8	0.089	113.46	< 0.0001	Significant
Linear mixture	0.51	2	0.26	326.94	< 0.0001	
AB	9.855E−004	1	9.855E−004	1.26	0.2992	
AC	0.088	1	0.088	112.60	< 0.0001	
BC	0.073	1	0.073	93.69	< 0.0001	
A^2BC	9.470E−003	1	9.470E−003	12.08	0.0103	
AB^2C	7.779E−004	1	7.779E−004	0.99	0.3524	
ABC^2	0.017	1	0.017	21.78	0.0023	
Residual	5.489E−003	7	7.841E−004			
Lack of fit	2.383E−003	4	5.958E−004	0.58	0.7028	Not significant
Pure Error	3.105E−003	3	1.035E−003			

difference between the mean. A value larger than 0.10 indicates that there is no difference between the means [18]. Both responses have p value less or equal to 0.0001 show the models are very significant.

3.2 Diagnostic

Diagnostic test is done, primarily to study the model graphically. Normal plot of the Residuals is the first diagnostic tool and are presented in Fig. 4 for both IFT and foam half-life. This plot shows whether the residuals follow a normal distribution and a change in transformation (if required) can improve the points of the distribution.

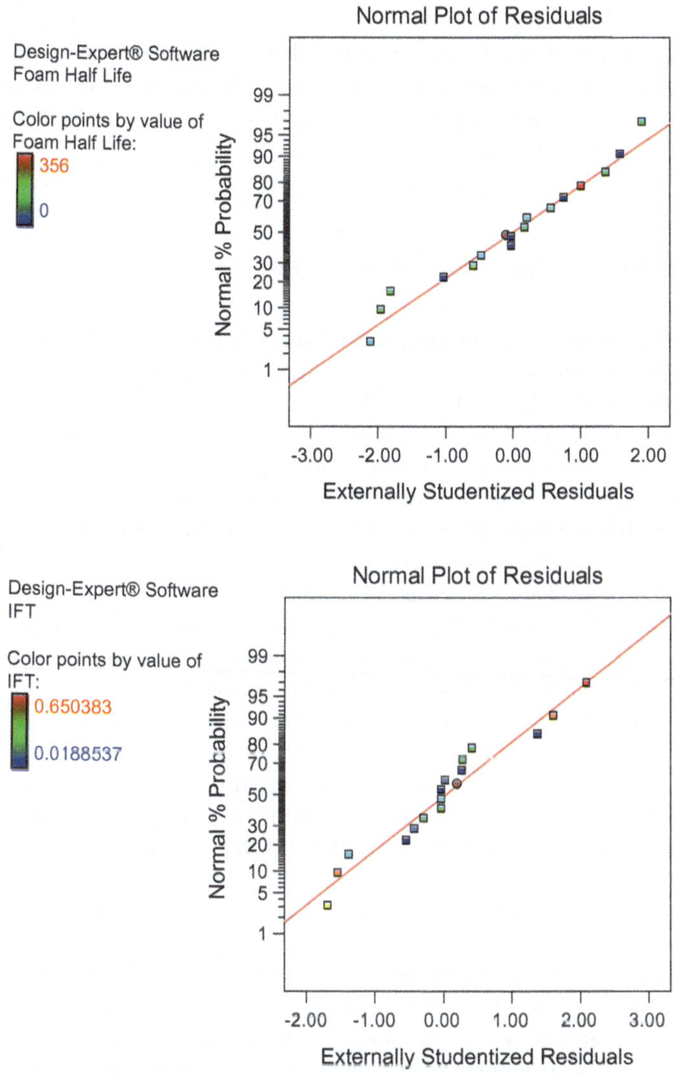

Fig. 4 Normal plot of residuals

Another diagnostic tool that is also consider important in confirming the model is Box-Cox Plot as presented in Fig. 4. When the error (residuals) is related to the magnitude of the response, then it is necessary to transform the response. The transformation is called as power law transformation that is applied to data using power function, meanwhile the plot of natural log of the sum of the squares of the residuals against lambda is the Box-Cox plot. The lambda value indicates the power to which all data should be raised. So, the right lambda value can be chosen based on the Box-Cox plot guideline. The minimum point found in the Box-Cox plot is the transformation based on the lambda value that is recommended by the Design Expert software. For foam half-life response, there is no any recommendation on the transformation, the current lambda of 1 is chosen as the best possible transformation. As for IFT, the software recommended to make square root transformation as the best possible transformation.

3.3 *Mixture Components Impact Study*

Based on the ANOVA table in Table 4, for foam half-life, AC and BC parameter are considered to be important and can have a huge effect on the foam half-life value. As for AB parameter, the *p*-value is high, however after removing the parameter, the R-squared value becomes lower compared to when it is still in the equation, hence the term is not removed. According to the quadratic model of foam half-life in Eq. 3, the linear terms of A, B and C which are the fractions of cocamidopropyl betaine, amidopropyl betaine and alpha olefin sulfonate, respectively, have positive effect on the foam half- life stability. Because their coefficient are positives, the increase of these parameters leads to an increase in foam stability.

As for the IFT, AB term shows a high *p*-value, but the term cannot be removed from the model as the term is not hierarchical and it will affect the R-squared, adjusted R-squared and predicted R-squared to be lower. The *p*-value for AB^2C is also high which is more than 0.05 but it is also not removed from the model because it causes the R-squared and predicted R-squared values to be lower. According to the special quartic model of IFT in Eq. 4, the linear terms of A, B and C have negative effects on the IFT reduction. Because their coefficients are positive, the decrease of these parameters leads to an increase in IFT. On the other hand, binary mixture of the components showed positive effects on IFT reduction because of their negative coefficients. In addition, the fitted models (Eqs. (3)-(4)) can also be presented in the form of two- dimensional contour plots and three-dimensional response surface plots as seen in Fig. 5. The response surface plots demonstrate that a wide region for different proportions of surfactant can be selected when maintaining the desired response. High proportions of amidopropyl betaine can reduce the IFT value, while low proportions of cocamidopropyl betaine and alpha olefin sulfonate can help to improve on the foam half-life value.

3.4 *Optimization*

For this hybrid surfactant formulation to function well at desired condition, it need to have low IFT and high foam half-life value. In order to achieve the ideal composition shape to satisfy both low IFT and high foam half-life value, the target goal for each component and the numerical optimization response were chosen. As shown in Fig. 6, the foam half-life reaches a maximum at a certain fraction of the components, and

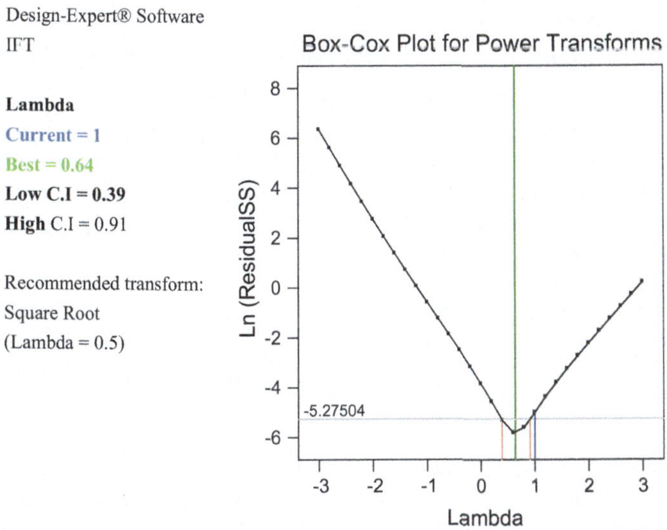

Fig. 5 Box-Cox plot for power transformation

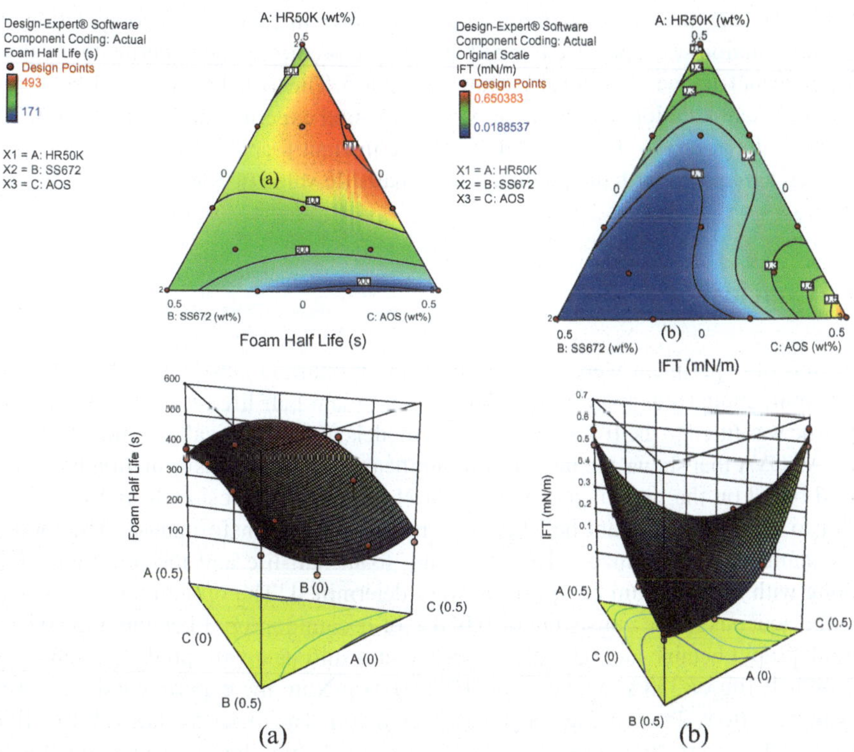

Fig. 6 Contour and response surface plots for foam half-life (**a**) and IFT (**b**)

Table 5 Numerical optimization

Parameter	Goal	Experimental region		Predicted	Experimental
		Lower	Upper		
Cocamidopropyl betaine	In range	0	0.5	0.058	-
Amidopropyl betaine	In range	0	0.5	0.391	-
Alpha olefin sulfonate	In range	0	0.5	0.051	-
Foam half-life (seconds)	Maximize	171	493	338	286
IFT (mN/m)	In range	0.01	0.04	0.04	0.05

same goes with IFT, the IFT value reaches a minimum at a certain fraction of the components. The numerical optimization can provide the minimum and maximum level of responses, which are IFT and foam half-life, within the range of factors, which are fractions of components, by setting the factor goals to 'maximize' and 'in range'. Both parameters and responses with the upper and lower limits that are required to run the numerical optimization are shown in Table 5. The composition of the formulation was optimized, together with the predicted and experimental value for foam half-life and IFT are also shown in Table 5. The foam half-life / IFT between the surfactant solution and crude oil were measured experimentally under 96 °C to evaluate the accuracy of the model. The experimental foam half-life is 286 s is not in good agreement with the predicted foam half-life value of 338 s.

4 Conclusion

16 runs of experiment were designed by using mixture in Design Expert software. Quadratic model was fitted to the experimental foam half-life response and special quartic was fitted to the IFT response. Both models were statistically verified through the ANOVA that shows the models are significant. The best transformation has been used based on the two diagnostic plots of Normal Plot of Residuals and Box-Cox Plot to IFT whereas foam half-life does not need any transformation. The model was numerically optimized. The maximum foam half-life and the minimum IFT along with the optimum composition were determined. The optimum composition of the aqueous phase consisting of 0.058% of cocamidopropyl betaine, 0.391% of amidopropyl betaine and 0.051% of alpha olefin sulfonate with predicted value for foam half-life of 338 s, whereas for IFT is 0.04 mN/m. the experimental value for foam half-life is 286 s and as for IFT is 0.05 mN/m. The experimental value of IFT is quite close to the predicted value but for foam half-life, the experimental value is not close to the predicted value. Simplex lattice mixture DOE can therefore be used

as an optimization tool and method for optimization of surfactants in EOR but it is not accurate for foam test by using the Teclis foam scan.

Acknowledgements The authors are grateful to PETRONAS Group Research and Technology (GR&T) and YUTP 015LC0-145 research grant for funding this research and for financial assistance

References

1. Janiga, D., Czarnota, R., Stopa, J., Wojnarowski, P.: Performance of nature inspired optimization algorithms for polymer enhanced oil recovery process. J. Petrol. Sci. Eng. **154**, 354–366 (2017)
2. Mumtaz, M., Tan, I.M., Mushtaq, M.: Synergistic effects of surfactants mixture for foam stability measurements for enhanced oil recovery applications. In: SPE Saudi Arabia Section Annual Technical Symposium and Exhibition, Society of Petroleum Engineers (2015)
3. Massarweh, O., Abushaikha, A.S.: The use of surfactants in enhanced oil recovery: a review of recent advances. Energy Rep. **6**, 3150–3178 (2020)
4. Jones, S., van der Bent, V., Farajzadeh, R., Rossen, W.R., Vincent-Bonnieu, S.: Surfactant screening for foam EOR: Correlation between bulk and core-flood experiments. Colloids Surf. A: Physicochem. Eng. Aspects **500**, 166–176 (2016)
5. Zhang, Y., Xiangan, Y., Jianxia, D., Yu, L.: New and effective foam flooding to recover oil in heterogeneous reservoir. In: SPE/DOE Improved Oil Recovery Symposium. Society of Petroleum Engineers (2000)
6. Quadri, S.M.R., Shoaib, M., AlSumaiti A.M., Alhassan, S.M.: Screening of polymers for EOR in high temperature, high salinity and carbonate reservoir conditions. In: International Petroleum Technology Conference, International Petroleum Technology Conference (2015)
7. Agneta, M., Zhaomin, Li., Chao, Z., Gerald, G.: Investigating synergism and antagonism of binary mixed surfactants for foam efficiency optimization in high salinity. J. Petroleum Sci. Eng. (2018)
8. Myers, D.: Surfactant Science and Technology, 3rd edn, p. 380. Wiley, New Jersey (2006)
9. Chevallier, E., Tchamba, O., Chambert, M., Bekri, S., Martin, F., Gautier, S.: Foams with ultra-low interfacial tensions for an efficient EOR process in fractured reservoirs. In: SPE Asia Pacific Enhanced Oil Recovery Conference, Society of Petroleum Engineers (2015)
10. Chevallier, E., Bouquet, S., Gland, N., Douarche, F., Batot, G.: Advanced EOR foam in naturally fractured carbonates reservoirs: optimal balance between foam and interfacial tension properties. In: SPE Middle East Oil and Gas Show and Conference, Society of Petroleum Engineers (2019)
11. Dong, P., Puerto, M., Ma, K., Mateen, K., Ren, G., Bourdarot, G., Morel, D., Biswal, S.L., Hirasaki, G.: Ultralow-interfacial-tension foam injection strategy investigation in high temperature ultra-high salinity fractured carbonate reservoirs. In: SPE improved oil recovery conference, Society of Petroleum Engineers (2018)
12. Lachaise, J., Breul, T., Graciaa, A., Marion, G., Monsalve, A., Salager, J.L.: Foaming properties of surfactant-oil-water systems in the neighbourhood of optimum formulation. J. Dispers. Sci. Technol. **11**(5), 443–453 (1990)
13. Lee, K.-M., Gilmore, D.F.: Formulation and process modeling of biopolymer (polyhydroxyalkanoates: PHAs) production from industrial wastes by novel crossed experimental design. Process Biochem. **40**(1), 229–246 (2005)
14. Jeirani, Z., Jan, B.M., Ali, B.S., Noor, I.M., Hwa, S.C., Saphanuchart, W.: The optimal mixture design of experiments: alternative method in optimizing the aqueous phase composition of a microemulsion. Chemom. Intell. Lab. Syst. **112**, 1–7 (2012)

15. Jaroenkhasemmeesuk, C., Prasertpong, P., Thankmongkhon, Y., Tippayawong, N.: Simplex lattice approach to optimize yields of light oil products from catalytic cracking of bio-oil with mixed catalysts. Chem. Eng. Commun. **204**(6), 677–688 (2017)
16. Zhu, L., Li, K., ding, H., Zhu, X.: Studying on properties of bio-oil by adding blended additive during aging. Fuel, **211**, 704–711 (2018)
17. Kamoun, A., Chaabouni, M., Sergent, M., Phan-Tan-Luu, R.: Mixture design applied to the formulation of hydrotropes for liquid detergents. Chemom. Intell. Lab. Syst. **63**(1), 69–79 (2002)
18. Baroutian, S., Arous, M.K., Raman, A.A.A., Sulaiman, N.M.N.: Potassium hydroxide catalyst supported on palm shell activated carbon for transesterification of palm oil. Fuel Process. Technol. **91**(11), 1378–1385 (2010)

Effects of Metal Ratio and Promoter on Physicochemical Properties of CuFe Catalyst

Nor Hafizah Berahim, Nabilah Saafie, Onn Chiew Mey, Nur Amirah Suhaimi, Nur Insyirah Zulkifli, and Noor Asmawati Mohd Zabidi

Abstract A series of CuFe catalysts have been synthesized on Al_2O_3 support using the impregnation method. Effects of metal ratio and potassium promoter on the physicochemical properties of these catalysts were investigated by FESEM/EDX, N_2 adsorption, H_2-TPR and CO_2-TPD techniques. Results showed that Cu:Fe ratio and potassium promoter did not change the morphological and textural properties of the CuFe catalyst. Potassium promoter improved the reducibility and doubled the number of the medium basic sites in the CuFe catalyst. The potassium-promoted catalyst increased the CO_2 conversion by a factor of four compared to that of the un-promoted CuFe catalyst.

Keywords CuFe · Catalyst · CO_2 conversion

N. H. Berahim · N. Saafie · O. C. Mey · N. A. Suhaimi · N. I. Zulkifli · N. A. Mohd Zabidi (✉)
Universiti Teknologi PETRONAS, 32610 Seri Iskandar, Perak, Malaysia
e-mail: noorasmawati_mzabidi@utp.edu.my

N. H. Berahim
e-mail: nor_19001638@utp.edu.my

N. Saafie
e-mail: nabilah.saafiie@utp.edu.my

O. C. Mey
e-mail: onn.chiew_25050@utp.edu.my

N. A. Suhaimi
e-mail: amirah.suhaimi@utp.edu.my

N. I. Zulkifli
e-mail: nur_19000323@utp.edu.my

N. H. Berahim
PETRONAS Research Sdn Bhd, 43000 Kajang, Selangor, Malaysia

S. A. Abdul Karim et al. (eds.), *Proceedings of the 6th International Conference on Fundamental and Applied Sciences*, Springer Proceedings in Complexity,
https://doi.org/10.1007/978-981-16-4513-6_23

1 Introduction

The constant rising of carbon dioxide (CO_2) emission have been identified among the most serious environmental impacts because of its contribution to climate change. Even so, because of its abundance and renewable nature, CO_2 is also a potential source of carbon which can be applied in the CO_2 hydrogenation process [1, 2]. Converting the very stable CO_2 molecules is very challenging, thus development of highly active catalyst is critical for the CO_2 conversion step [3].

The catalytic CO_2 hydrogenation reaction to C1 alcohol has received a lot of attention and has made significant progress [4]. Methanol production via CO_2 hydrogenation process has been commercialized in Reykjavik, Iceland [5]. Significant progress has recently been made in heterogeneous catalytic CO_2 hydrogenation to a variety of high-value and easily marketable chemicals and fuels comprising two or more carbons (C2 + species), such as dimethyl ether (DME) [6], olefins [7], liquid fuels [8], and higher alcohols [9]. The C2 alcohol synthesis is more difficult compared to that of C1 alcohol because of the severe inertness of CO_2 molecules, the high barrier of C–C coupling and the multiple component-based responses that contribute to the production of C2 alcohol.

The C1 alcohol synthesis using CuZnO-based catalysts have been investigated extensively. If the catalysts contain alkaline metals, small amounts of higher alcohols are produced. The ability to promote higher alcohols synthesis in CO hydrogenation increases with the alkaline atomic size, Li < Na < K < Rb < Cs [10]. In addition, for each type of support and synthesis technique, optimal loading of the alkali metals must be adjusted. Even though the presence of alkali metals can lead to higher selectivity at the given load, further additions may block the Cu and ZnO active sites, thereby reducing the reactivity [11]. The use of K_2CO_3 at 0.5 wt.% loading in promoting $Cu/ZnO/Al_2O_3$ catalyst resulted in highest selectivity to higher alcohols [12]. However, the maximum selectivity to higher alcohols in the case of Co/Cu/ZnO was achieved at 4.1 wt.% potassium promoter loading [13]. Nonetheless, using modified methanol-based catalysts, methanol was still the major product, and selectivity to higher alcohols was low, due to the limitation in reaction mechanism [14]. Because of its high selectivity to higher alcohols, as well as its low cost, Fischer Tropsch (FT)-based catalysts, particularly Cu–Co, Cu-Fe catalysts, have received a lot of attention. However, there are still several issues that need to be addressed with these modified FT catalysts, including hydrocarbon selectivity, low alcohol productivity, and a wide range of reaction products [15]. Appropriate synthesis technique, addition of promoters, and suitable supports are critical for obtaining evenly dispersed and well-structured bi or multi metallic catalysts and improving the performance of FT-based catalysts for alcohols formation [16].

In this work, we report the synthesis of bimetallic CuFe catalysts on alumina support in the presence of K promoter. The physicochemical properties of the synthesized catalysts have been investigated for the CO_2 hydrogenation reaction.

2 Materials and Methods

2.1 Materials

Copper (II) nitrate trihydrate ($Cu(NO_3)_2 \cdot 3H_2O$) and iron(III) nitrate nonahydrate ($Fe(NO_3)_3 \cdot 9H_2O$) were used as precursors and supplied by Sigma-Aldrich. Potassium nitrate (KNO_3) and aluminium oxide (γ-Al_2O_3) were obtained from Merck.

2.2 Catalyst Synthesis

Bimetallic catalysts (CuFe) at 15 wt.% loading was synthesized on Al_2O_3 support at 50:50 (CFA-5050) and 85:15 (CFA-8515) weight ratios via impregnation method. Potassium was added as a promoter at 1% weight loading. The K-promoted CuFe catalyst was denoted as CFAK-5050 and CFAK-8515, respectively. The aqueous solutions of the precursors were impregnated into the treated Al_2O_3 support. The mixture was stirred for 24 h, then filtered and washed with deionized water. The paste formed was dried in an oven at 120 °C for 12 h then calcined in air at 350 °C for 4 h.

2.3 Catalyst Characterizations

The physicochemical properties of the prepared bimetallic CuFe catalysts were investigated via several characterization techniques. The morphology and elemental composition of the synthesized catalysts were analyzed on Zeiss Supra 55 VP equipment equipped with EDX. The textural properties were measured on a Micromeritics ASAP 2020 via nitrogen adsorption–desorption analyses at −196 °C. The reduction behavior of the catalysts was studied using a Thermo Finnigan TPD/R/O 110 CE equipped with a thermal conductivity detector (TCD). Typically, 40 mgcatalyst was placed in the quartz tube and was pre-treated under the flow of pure N_2 at 250 °C for 1 h to remove moisture and impurities. The analysis was performed using 5% H_2/Ar (20 mL/min) up to 950 °C at ramping rate of 10 °C/min. The CO_2-TPD experiments were conducted using the same instrument as that of TPR. Typically, 40 mg of the synthesized catalyst was placed in the quartz tube and pre-treated at 250 °C under He flow (20 mL/min) for 50 min. The CO_2 sorption then continued by flowing CO_2 at 10 mL/min for 30 min at temperature of 75 °C. The desorption of CO_2 was conducted by purging He gas with a flow rate 20 mL/min through the sorbent bed and ramping the temperature from 40 to 950 °C at 10 °C/min.

2.4 Catalytic Performance

The catalytic performance was evaluated for the CO_2 hydrogenation reaction using a fixed-bed reactor (Microactivity Reference, PID Eng&Tech). The catalyst was reduced in-situ prior to the reaction in H_2 at 250 °C for 2 h. The reaction was performed at 250 °C, 20 bar with H_2:CO_2 = 3:1. Products were analyzed using the online gas chromatograph at 30 min intervals.

3 Results and Discussions

3.1 Morphology and Composition

Figure 1 shows the FESEM images and the corresponding EDX elemental mappings of the synthesized CuFe catalysts. These FESEM images show that all samples exhibit irregular morphology and metal-oxide particles were distributed uniformly onto the Al_2O_3 support. The addition of 1 wt.% potassium promoter did not produce significant difference on both morphologies and elemental distribution on the surface of the synthesized samples (Table 1). However, results of EDX mapping show that K promoter was not uniformly distributed on the catalyst surface since it was not detected in some of the EDX spots analyses.

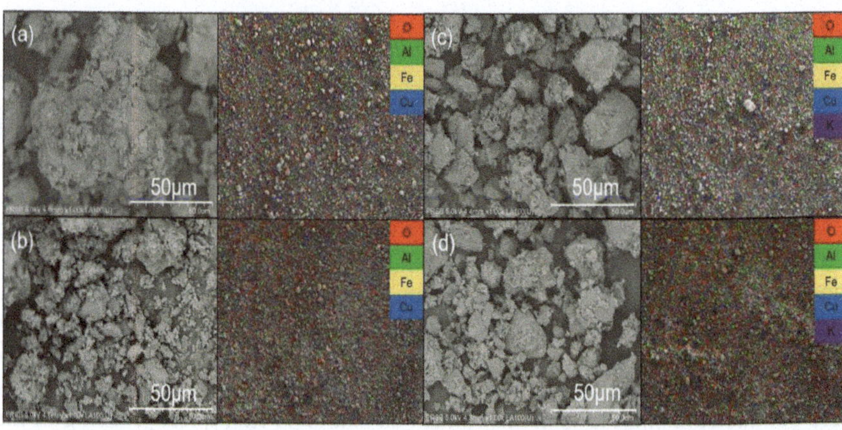

Fig. 1 FESEM micrographs at 1.00 K magnification and EDX mapping of **a** CuFe (50:50) **b** CuFe (85:15) **c** K-promoted CuFe (50:50) and **d** K-promoted CuFe (85:15)

Table 1 Elemental composition by EDX (average of 5 spots)

No	Catalyst	Elements (wt.%)				
		O	Al	Cu	Fe	K
1	CFA-5050	41.66	34.31	9.24	14.79	-
2	CFA-8515	39.98	29.25	25.98	4.79	-
3	CFAK-5050	40.68	34.08	9.22	15.99	0.11[a]
4	CFAK-8515	40.30	32.24	22.38	4.96	0.19[b]

[a]K was only detected on two spots (0.14 wt.%, 0.07 wt.%)
[b]K was only detected on three spots (0.18 wt.%, 0.16 wt.%, 0.24 wt.%)

Table 2 Textural properties

No	Catalyst	BET surface area (m^2/g)	Pore Volume (cm^3/g)	Pore size (Å)
1	γ-Al_2O_3 (Support)	202	0.5	75.4
2	CFA-5050	123	0.2	72.5
3	CFA-8515	113	0.2	84.7
4	CFAK-5050	138	0.3	71.2
5	CFAK-8515	114	0.3	86.6

3.2 *Textural Properties*

The textural properties of the samples are listed in Table 2. The catalyst support (γ-Al_2O_3) had the highest BET surface area of 202 m^2/g and pore volume of 0.5 cm^3/g. The values of the BET surface areas and pore volumes decreased on the impregnated samples. These results indicated that the metal-oxide particles have filled up the pores in the catalyst support after the impregnation process.

3.3 *Reducibility and Basicity Studies*

The H_2-TPR results of the four catalysts are shown in Fig. 2, where two distinct reduction temperature ranges were observed at low (120–300 °C) and high (> 340 °C) temperature regions. A stepwise reduction behavior of CuO can be observed on the first shoulder peak in which the α and β peaks are assigned to the reduction of highly dispersed CuO and bulk CuO, respectively. These peaks corresponded to the reduction of CuO to Cu metallic [17]. The broad γ peaks (>340 °C) for all samples are related to the reduction of Fe oxides ($Fe_2O_3 \rightarrow Fe_3O_4 \rightarrow FeO \rightarrow Fe$) [18]. Furthermore, the addition of potassium to CuFe (50:50) catalysts improved the reducibility of bulk CuO as the α peak showed up at lower temperature compared to that of the un-promoted sample. However, the effect is reversed for the sample containing higher Cu content in the catalyst.

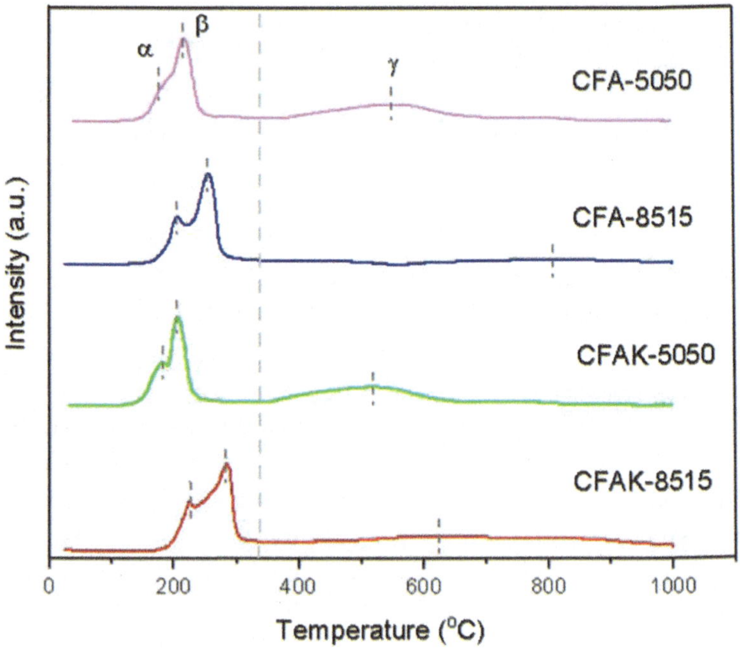

Fig. 2 H_2-TPR profiles for the catalysts

The basic sites on the catalyst surface were measured by CO_2-TPD as shown in Fig. 3. The CO_2 desorption profiles display three different strength of basic sites, assigned as weak (α, T < 300 °C), moderate (β, T = 300–700 °C) and strong (γ, T > 700 °C). In this work, alumina which is slightly acidic, was used as the support. Since CO_2 is acidic gas, the adsorption and dissociation of CO_2 can be promoted by the basic sites of the catalyst. The weak basic sites are related to the surface OH– groups, the moderately basic sites are attributed to the metal–oxygen pairs (Cu–O, Fe–O, Al–O) and the strong basic sites are associated with the coordinately unsaturated O^{2-} ions (low coordination oxygen atoms) existing from partial breakup of metal–oxygen pairs [19]. The amount of β sites in the un-promoted 50:50 CuFe catalyst was 7 μmol/g. The presence of K promoter increased the amount of the β sites in the 50:50 CuFe catalyst to 12 μmol/g, thus could enhance the CO_2 adsorption.

3.4 Catalytic Performance

Table 3 shows the pre-liminary results of the catalytic performance of CuFe catalysts in a CO_2 hydrogenation reaction conducted at 250 °C and 20 bar. The products detected from the CO_2 hydrogenation reaction include methanol, ethanol, methane, methyl formate (MF) and CO. The equal ratio CuFe catalyst resulted in higher CO_2

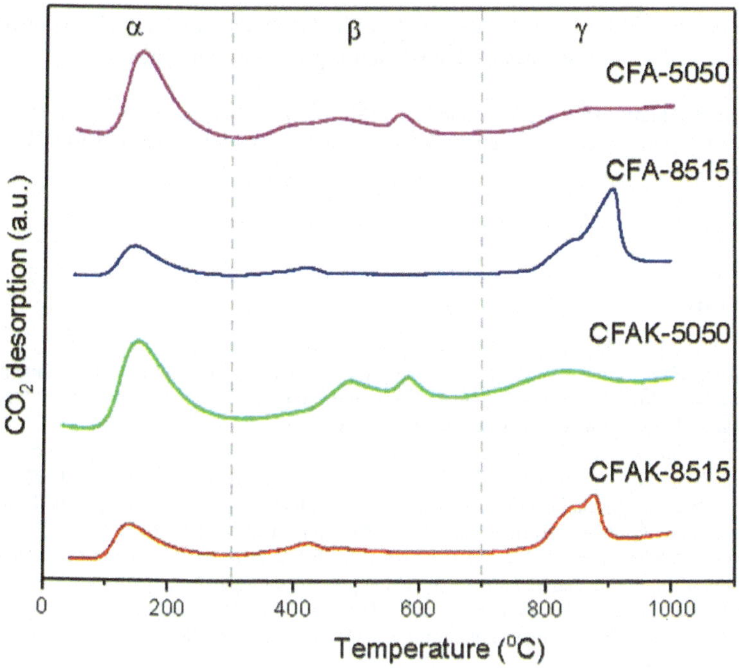

Fig. 3 CO_2-TPD profiles for the samples

Table 3 Catalytic performance

Catalysts	CO_2 conv (%)	Selectivity (%)				
		MeOH	EtOH	CH_4	MF	CO
CFA-5050	11	1.8	1.2	–	3.4	93.6
CFA-8515	6	2.9	0.3	–	1.4	95.4
CFAK-5050	45	4.6	4.1	18.9	3.6	68.9

conversion compared to the catalyst containing higher amount of Cu. The presence of K promoter enhanced both the CO_2 conversion as well as the alcohols selectivity. Nevertheless, CO remained as a major product over these catalysts.

4 Conclusion

Effects of Cu:Fe ratio and potassium promoter on the physicochemical properties and performance of $CuFe/Al_2O_3$ catalysts were investigated. Increasing the Cu content in the catalyst formulation did not change the catalyst morphology but shifted the reduction peak to higher temperature. However, the presence of K promoter improved the reducibility of the CuFe catalyst and resulted in an increase in the CO_2 conversion by a factor of four compared to that of the un-promoted catalyst.

Acknowledgements The authors would like to express their appreciation to the YUTP research grant (cost center: 015LC0-111) and the MRA Research grant (cost center 015MDO-031) for the financial support.

References

1. Garbo, M.D., Usman, M., Khan, S., Shehzad, F., Galadima, A., Ehsan, M.F., Ghanem, A.S., Humayun, M.: CO_2 towards fuel: a review of catalytic conversion of carbon dioxide to hydrocarbons. J. Environ. Chem. Eng. **9**(104756), 1–19 (2021)
2. He, M., Sun, Y., Han, B.: Green carbon science: scientific basis for integrating carbon resource processing, utilization, and recycling. Angew. Chem. Int. Ed. **52**, 9620–9633 (2013)
3. Deng, K., Hu, B., Lu, Q., Hong, X.: Cu/g-C_3N_4 modified ZnO/Al_2O_3 catalyst: methanol yield improvement of CO_2 hydrogenation. Catal. Commun. **100**, 81–84 (2017)
4. Roy, S., Cherevotan, A., Peter, S.C.: Thermochemical CO_2 hydrogenation to single carbon products: Scientific and technological challenges. ACS Energy Lett. **3**, 1938–1966 (2018)
5. Kothandaraman, J., Goeppert, A., Czaun, M., Olah, G.A., Prakash, G.K.: Conversion of CO_2 from air into methanol using a polyamine and a homogeneous ruthenium catalyst. J. Am. Chem. Soc. **138**, 778–781 (2016)
6. Saravanan, K., Ham, H., Tsubaki, N., Bae, J.W.: Recent progress for direct synthesis of dimethyl ether from syngas on the heterogeneous bifunctional hybrid catalysts. Appl. Catal. B **217**, 494–522 (2017)
7. Li, Z., Qu, Y., Wang, J., Liu, H., Li, M., Miao, S., Li, C.: Highly selective conversion of carbon dioxide to lower olefins. ACS Catal. **7**, 8544–8548 (2017)
8. Wei, J., Ge, Q., Yao, R., Wen, Z., Fang, C., Guo, L., Xu, H., Sun, J.: Directly converting CO_2 into a gasoline fuel. Nat. Commun. **8**, 15174–15181 (2017)
9. Yang, C., Mu, R., Wang, G., Song, J., Tian, H., Zhao, Z., Gong, J.: Hydroxyl-mediated ethanol selectivity of CO_2 hydrogenation. Chem. Sci. **10**, 3161–3167 (2019)
10. Spivey, J.J., Egbebi, A.: Heterogeneous catalytic synthesis of ethanol from biomass derived syngas. Chem. Soc. Rev. **36**, 1514–1528 (2007)
11. Slaa, J.C., Vanommen, J.G., Ross, J.R.H.: The synthesis of higher alcohols using modified Cu/ZnO/Al_2O_3 catalysts. Catal. Today **15**, 129–148 (1992)
12. Smith, K.J., Anderson, R.B.: The Higher Alcohol synthesis over promoted Cu/ZnO catalysts. Can. J. Chem. Eng. **61**, 40–45 (1983)
13. Boz, I.: Higher alcohol synthesis over a K promoted Co_2O_3/CuO/ZnO/Al_2O_3 catalyst. Catal. Lett. **87**, 187–194 (2003)
14. Sá, J.: Fuel Production with Heterogeneous Catalysis. CRC Press, Boca Raton (2014)
15. Gao, W., Zhao, Y.F., Liu, J.M., Huang, Q.W., He, S., Li, C.M., Zhao, J.W., Wei, M.: Catalytic conversion of syngas to mixed alcohols over CuFe based catalysts derived from layered double hydroxides. Catal. Sci. Technol. **3**, 1324–1332 (2013)
16. Lu, R.L., Mao, D.S., Yu, J., Guo, Q.S.: Enhanced activity of CuFe/SiO_2 catalyst for CO hydrogenation to higher alcohols by pretreating the support with ammonia. J. Ind. Eng. Chem. **25**, 338–343 (2015)
17. Ding, M., Tu, J., Qiu, M., Wang, T., Ma, L., Li, Y.: Impact of potassium promoter on Cu-Fe based mixed alcohols synthesis catalyst. Appl. Energy **138**, 584–589 (2015)
18. Sun, C., Mao, D., Han, L., Yu, J.: Effect of impregnation sequence on performace of SiO_2 supported Cu-Fe catalysts for higher alcohols synthesis from syngas. Catalyst Commun. **84**, 175–178 (2016)

19. Ayodele, O.B., Tasfy, S.F.H., Zabidi, N.A.M., Uemura, Y.: Co-synthesis of methanol and methyl formate from CO_2 hydrogenation over oxalate ligand functionalized ZSM-5 supported Cu/ZnO catalyst. J. CO_2 Utilization **17**, 273–283 (2017)

Ultrasonication Assisted Extraction of Lipids from *Chlorella vulgaris* with [Bmim][$MeSO_4$] as an Additive

Ninna Sakina Binti Azman, Noraini Abd Ghani, and Lam Man Kee

Abstract Extraction of lipids using combination of methods with ultrasonic probe and ionic liquid were done towards the cultivated *Chlorella vulgaris*. In this research, [Bmim][$MeSO_4$] were synthesized by using alkylation and metathesis methods. Effect of methanol, time reaction and IL loading in extracting the lipids were investigated by using ultrasonication. Results indicate that, 40 min gives the highest lipids extraction with 1.25% of [Bmim][$MeSO_4$] and methanol. Effect of temperature towards stirring methods were also studied in 40 min duration. 23.81% lipids were extracted at 40 °C as the optimum temperature. At room temperature, ultrasonication methods gave three times higher lipids extracted compared to stirring methods which are 26.78% and 9.25%, respectively.

Keywords Ionic liquids · *Chlorella vulgaris* · Lipid extraction · Ultrasonication

1 Introduction

Microalgae is one of the most widely used groups of material in metabolic products. The advantages of microalgae such as easy to cultivate and rapid growth period make them dominant in various applications including food supplements and biofuel [1]. With the depletion of non-renewable fuels and climate change had triggered the

N. S. B. Azman · N. A. Ghani (✉)
Department of Fundamental and Applied Sciences, Universiti Teknologi PETRONAS, 32610 Seri Iskandar, Perak, Malaysia
e-mail: noraini.ghani@utp.edu.my

Centre of Research in Ionic Liquids, Universiti Teknologi PETRONAS, 32610 Seri Iskandar, Perak, Malaysia

N. S. B. Azman
e-mail: ninna_17007415@utp.edu.my

L. M. Kee
Department of Chemical Engineering, Universiti Teknologi PETRONAS, 32610 Seri Iskandar, Perak, Malaysia
e-mail: lam.mankee@utp.edu.my

S. A. Abdul Karim et al. (eds.), *Proceedings of the 6th International Conference on Fundamental and Applied Sciences*, Springer Proceedings in Complexity,
https://doi.org/10.1007/978-981-16-4513-6_24

search for sustainable and renewable fuels. Basically, biodiesel that is generated from edible oil is known as first generation while second generation is biodiesel that produced from non-edible oil. Biodiesel that comes from microalgal biomass is categorized as third generation biodiesel and it shows the most promising species as it has the ability to accumulate large amounts of lipids which around 20–50% of dry weight [1–3]. Subsequently, previous study reported it was found that in some microalgae's species about 70% of its mass comprised of unsaturated fats [4].

Mainly, there are four steps involved in the production of biodiesel that is derived from microalgae: cultivation of biomass, separation of biomass from the culture media, extraction of the lipids and transesterification of lipids to final products [5]. However, due to the thick cell walls of microalgae, disruption process is important for the lipid's recovery. Here, disruption methods can be classified into two which are mechanical and non-mechanical methods. Mechanical methods include oil expeller or press, ultrasound assisted, and microwave assisted, while, non-mechanical methods or known as chemical methods consist of solvent extraction, supercritical CO_2 and ionic liquid extraction [6].

Organic solvents that are used in this method are methanol and chloroform show a great effect towards the lipid's recovery. However, due to the nature of the solvents itself which mostly toxic, flammable and give the bad impact towards health and environment are the major drawbacks [7]. To minimize these drawbacks, ionic liquids (ILs) which known as green solvents shows the great ability as a substitute solvent in the extraction of lipids. It is known as green solvents due to the features it's having such as low vapor pressure and toxicity along with the properties can be modified regarding to solubility, electrical conductivity, hydrophobicity and polarity [8]. Fundamentally, IL are salts consist of comparatively large asymmetric organic cations joined with smaller organic or inorganic anions which at temperature between 0 to 140 °C it remains in liquid state. Generally, the cations are comprised of a nitrogen containing ring structure such as pyridine or imidazolium where a broad range of functional side group can be attached. At this point, the polarity of the IL can be varied by depending on the structure of the side group added [8].

In fact, extraction of lipids from microalgae is much tougher compared with the extraction of lipids from oil seeds. This is because, microalgae are single-cell organisms with immensely strong cell walls that possibly hard to break [7]. Hence, instead of solvent extraction alone, lipid's recovery can be enhanced with mechanical pretreatment by imposing shear stress forces that help breakdown of algal cell walls consequently discharge the cell contents including lipids [9]. According to the Ellison et al. [9] from the research done by Prabakaran and Ravindran it was found that ultrasound was the most efficient method in lipids recovery from *Chlorella* sp. compared to the other techniques investigated including autoclaving, microwave, bead beating, and a 10% sodium chloride (NaCl) solution [9, 10].

In this work, comparison on the efficiency of lipids extraction techniques were done towards *Chlorella vulgaris* (*C. vulgaris*) with the present of [Bmim][$MeSO_4$] as additive. By using the technique ultrasonication, effect of solvent, time and IL loading were investigated. On the other hand, effect of temperature was investigated by using stirring method. Lastly, fatty acids extracted were calculated in order to determine the efficiency of different technique used.

2 Experimental

2.1 Materials

1-methylimidazole, sodium methyl sulfate ($NaMeSO_4$) for IL synthesized were purchased from Sigma, Malaysia. The chemicals were used without further purification. Methanol (analytical grade), n-hexane (analytical grade) were also purchased from Sigma, Malaysia and used without further purification. *C. vulgaris* were cultured using BG-11 medium.

2.2 Synthesis of 1-Butyl-3-Methylimidazolium Chloride and 1-Butyl-3-Methylimidazolium Methyl Sulfate

1-butyl-3-imidazolium chloride [Bmim][Cl] were synthesized by using 1-methylimidazole and chlorobutane without presence of solvent. 1-methylimidazole was mixed with chlorobutane and refluxed at 60 °C for 3 days. Then, the solutions were washed using acetic acid 3 times and once with ethyl acetate. The mixture was then going for rotary evaporator to purify it. The purified IL were analyzed using proton nuclear magnetic resonance (^{1}HNMR) by using deuterium oxide as a solvent.

1-butyl-3-methylimidazolium methyl sulfate, [Bmim][$MeSO_4$] were synthesized using the synthesized [Bmim][Cl] and sodium methyl sulfate ($NaMeSO_4$) with metathesis method. The ratio between [Bmim][Cl] and $NaMeSO_4$ was 1:1. First, about 0.7669 g of $NaMeSO_4$ were dissolved in 100 mL of methanol. Then, 10 g of [Bmim][Cl] was added into the solutions. The mixture was shaken vigorously to make sure its homogeneous. Then the solutions were stirred for 24 h. After 24 h, the IL went for rotary evaporator in order to remove the solvent. Then, the IL were washed using acetone and filtered by using filter paper to remove of NaCl precipitate. Solvent will be removed from the mixture using rotary evaporator to obtain pure IL. The IL was characterized by proton NMR.

2.3 Cultivation of C. vulgaris

C. vulgaris was cultured in 5 L volume of BG-11 medium. The medium was sterilized using autoclave for 15 min at 120 °C to eliminate the impurities and cool at room temperature for 1 day. 500 mL of *C. vulgaris* were added in the 5 L prepared medium and supplied with sufficient oxygen and lights. Algae sample were taken from the culture and analyse using optical density measurement every day for 21 days. Triplicate samples were tested with UV–Vis Spectrophotometer at wavelength 668 nm. 1 mL of the algae sample were diluted with 3 mL deionized water before tested with the UV–Vis.

2.4 Ultrasonication Irradiation

Effect of solvents

0.2 g of *C. vulgaris* was mixed with different percentage of [Bmim][$MeSO_4$] (0.5–2.5%) in the 50 mL jacketed tube. After that, 50 mL of methanol were added. The solution was subjected to ultrasonication (Vibracell 20 kHz, Sonics, USA, 13 mm titanium probe) for 20 min with default pulse and amplitude 2:1 and 50% respectively at 25 °C. After the reaction completed, 1:1 ratio of hexane and water were added. Centrifugations take place to separate the layers. The lipids layer with hexane were taken leaving the water and IL. The solution was washed again using water to make sure only lipids and hexane were left. Then proceed with rotary evaporation to remove the solvent and the residue was weighed to measure the yield. Based on the lipids weighted, the total lipid yield (wt.%) was calculated using Eq. 1. The procedure was repeated with water as a solvent to determine the effect of solvents towards the lipid extraction efficiency. All the data were triplicate.

$$Total\ lipid\ yield(\text{wt.\%}) = \frac{mass\ of\ lipids\ extracted}{mass\ of\ C.\ vulgaris} \times 100\% \qquad (1)$$

Effect of IL Loading and Time Reaction

Lipid extraction process was repeated using the same methods for 30 and 40 min. This is to determine the best IL loading percentage. The range of IL loading were reduced from 0.5 to 1.5% with 0.25% interval as the previous results shows plateau extraction yield after 1.5%. The data were triplicate, and control were run without the presence of IL.

After the best IL loading achieved, the same protocol was applied to various duration (10, 20, 30, 40, 50 and 60 min) in optimizing the time reaction for ultrasonication irradiation. All the reactions were triplicate to validate the accuracy of the data obtained.

2.5 Stirring Method

Triplicate sample of 20 mg of *C. vulgaris* was mixed with [Bmim][$MeSO_4$] (0.5–1.5%) and 5 mL of methanol under 400 rpm magnetic stirring for 40 min at room temperature. 1:1 ratio of hexane and water were added once the reaction ended. The sample were centrifuge to separate IL and lipids layers. Hexane containing the lipids were taken and washed with water to remove the polar compounds. The crude lipid was obtained by evaporation of the hexane phase using rotary evaporator, and the residue was weighed to measure the yield [8]. This procedure was repeated by increasing the temperature to 30, 40, and 50 °C.

3 Results and Discussion

3.1 *^{1}HNMR of the Synthesized IL*

The synthesized IL were characterized using ^{1}HNMR. Solvent used were deuterium oxide were used as a solvent which its peak appeared at 4.7 nm. In this process there is no addition of solvent, where the dissociation of ions occurred when heat was introduced. The purity of the [Bmim][Cl] synthesized are proven by the presence of the peaks as follows δ0.83 (3H, t), δ1.24 (2H, m), δ1.76 (2H, m), δ3.81 (3H, s), δ4.12 (2H, t), δ7.41 (2H, d), δ8.65 (1H, s).

Metathesis method was used in synthesizing [Bmim][$MeSO_4$]. Metathesis is a process where the addition of first generation of IL with salt anion. The presence of methanol in this synthesis act as a medium for the metathesis to occur. Dissociation of sodium methyl sulfate in methanol lead to the formation of methyl sulfate ion (negative charge) and sodium ion (positive charge). The addition of [Bmim][Cl] in the solution had caused the attraction between chloride ion and sodium ion which known as coulombic attraction. Basically, coulombic attraction is a mutual attraction between ions of opposite charge [11]. Sodium chloride salt had formed with ionic bond leaving 1-butyl-3-methylimidazolium methyl sulfate. The purity of synthesized [Bmim][$MeSO_4$] are proven by the presence of the peaks as follows, δ0.81 (3H, t), δ1.20 (2H, s), δ1.73 (2H, p), δ3.60 (3H, s), δ3.77 (3H, s), δ4.08 (2H, t), δ7.36 (2H, d) and δ8.59 (1H, s). This results similar with the study done by Bagno et al. [11].

3.2 *Cultivation of* C. vulgaris

In this cultivation of *C. vulgaris*, there are several stages of the growth curve of cultured *C. vulgaris*. By using UV–Vis equipment, the optical density of the microalgae was taken. Increases in optical density value indicate the increases of chlorophyll number and growth of the algae. Based on Fig. 1, day 1 until day 7 is the exponential phase, 8–12 linear phase and 13–18 stationary phase [4]. Day 19 showed the highest growth rate where there is stagnant reading until day 21. Thus, at day 19, *C. vulgaris* were harvested.

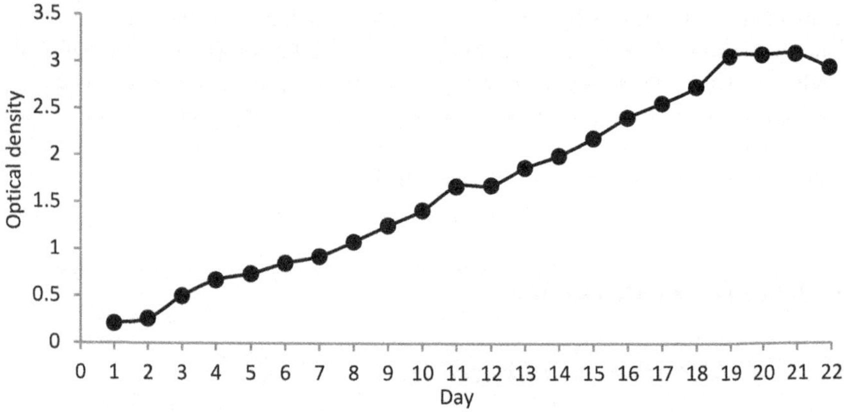

Fig. 1 Growth curve of *Chlorella vulgaris*

3.3 Ultrasonication Irradiation

Mixture of IL and methanol can penetrate the cells or break down the cell wall to release lipids. Water and methanol were chosen to test their effects on the extraction of lipids. Figure 2 shows the extraction of lipids using 2 different solvents under ultrasonication for 20 min at room temperature. Results shows methanol give higher percentage of lipids extracted which are 8 times higher compared to water. This shows the presence polar covalent molecule methanol helps in disruption of microalgae cytomembrane. Besides, the presence of IL and organic solvent help the extraction of a complex components with polar lipids (i.e., phospholipids, glycolipids, sterols,

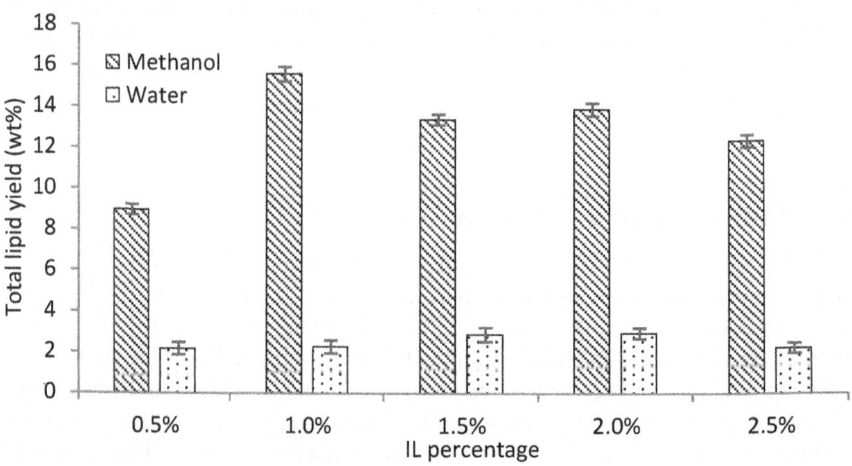

Fig. 2 Comparison between methanol and water in extraction of lipids by using ultrasonication

carotenoids) presence in the cytoplasm. In the cell membrane this complex is strongly linked via hydrogens bonds to protein. IL break the hydrogen bonds in the microalgal cell walls, subsequently enhanced the microalgal lipid extraction [3]. This is because, the van der Waals interactions formed between non-polar organic solvents and neutral lipids are not capable to disrupt these membrane-based lipid-protein associations [12].

However, when the IL loading increases the percentage of lipids extracted become plateau. As in the graph the highest extracted lipids were at 1.0%, when the IL loading increases the lipids extracted were constant at 14% and decreases with 2.5% IL loading. This is due to high concentration of IL was not conducive for the extraction of microalgal lipids. The role of IL in this system is to disrupt the structure of the fiber bindle in the cell wall which made it easy to release the lipid from inside the microalgae cells.

From these, the IL loading intervals were decreased to 0.25 which were from 0.5 until 1.5%. Figure 3 shows the effect of time in ultrasonic irradiation towards extraction of lipids. Here graph shows, as the time increase the lipids extraction also increase. This is because when the time of the cell exposed towards the irradiation longer, more cell-wall was broken hence more lipids can be extracted. Furthermore, the graph also shows the best IL loading is at 1.25% for 3 different temperatures.

In this study, highest lipids extracted were achieved at 40 min with 26.78% lipids extracted. However, when the time duration increases the percentage of lipids extracted start to decrease at 50 and 60 min as in Fig. 4. Basically, ultrasonication probe produce implosion of bubbles that emits shockwaves, generating chemical and mechanical energy breaking the cell walls and releasing the desired intracellular compounds into the solutions [13]. Thus, the long exposure of ultrasonication will cause microalgae denatured because ultrasonication produce heat, as a result less lipid extracted. In previous study, extraction of lipids using ultrasonic bath are able

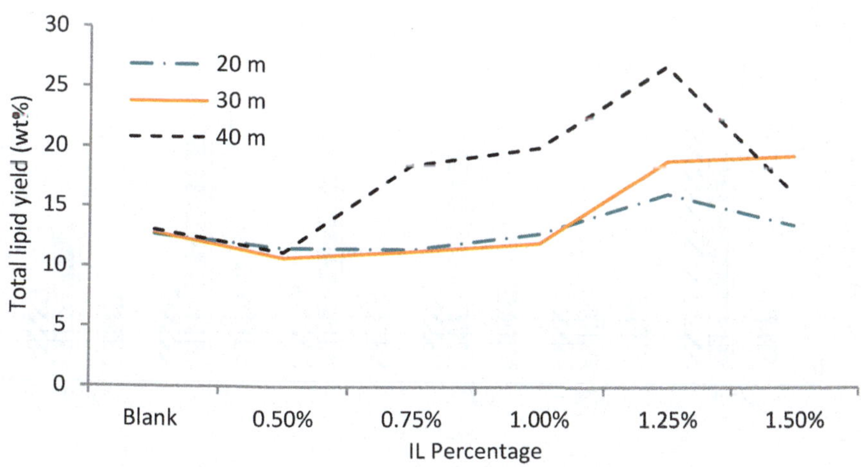

Fig. 3 Extraction of lipids by using ultrasonication with [Bmim][$MeSO_4$] at different time

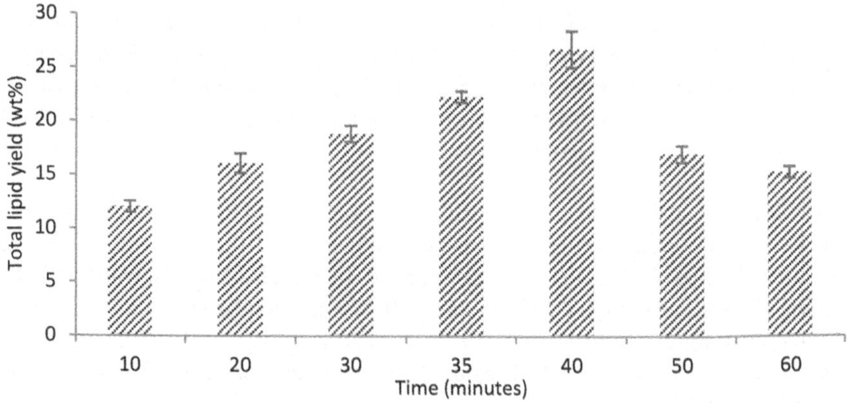

Fig. 4 Effect of time towards ultrasonication irradiation with 1.25% [Bmim][MeSO$_4$]

to extract lipids 7.4% at 2 h' time of reaction [14]. In this experiment, three times lipids are able to be extracted using ultrasonic probe in 40 min.

3.4 Stirring Effect

Reaction temperature and time were the main factors that affected hydrogen bond network damage of cellulose and the efficiency of extraction of total lipids. Figure 5 shows the effect of temperature using stirring method. Four different temperatures were tested with 40 min reaction time at 400 rpm. Highest lipids extracted were

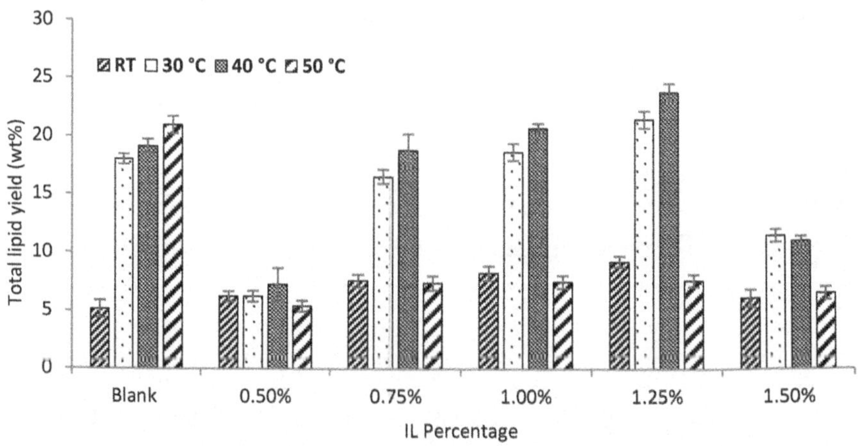

Fig. 5 Effect of temperature in extraction of lipids by using stirring method

achieved at 1.25% IL loading for all temperature. This result in agreement with Zhou et al., which are increasing the temperature will decrease the percentage of lipids extracted. Longer duration and higher temperature will not lead to more lipids obtained may be due to denatured of lipids at high temperature.

By comparing the effect of ultrasonication irradiation and stirring method, it shows that ultrasonication are able to extract more lipids at room temperature. With 1.25% of [Bmim][$MeSO_4$], 26.78% of lipids were successfully extracted by using ultrasonic probe. However, only 9.25% of lipids able to extract when using stirring at the same condition. This prove that while using ultrasonication methods, the lipids can be extract twice amount compared to stirring methods. This is because, a physical device that supplies high energy can simply increase the diffusion rate of the reaction system [15] which suggest that the ultrasonication as cellular disruption method will extract large lipids yield.

4 Conclusion

In this work, extraction of lipids by using ultrasonic probe with [Bmim][$MeSO_4$] were done at various time durations. With 1.25% IL loading, highest lipids were extracted at 40 min by 26.78%. Results also indicate that high temperature was not conducive for lipids extraction as it will cause lipids denatured. Study on the effect of various temperature by using stirring methods were done which gave high lipids extracted at 40 °C. Conclusively, the best lipids extraction was achieved with ultrasonication method compared to stirring methods by 3 times higher at room temperature.

Acknowledgements As a way of gratitude, the author would like to thank to various people who were involved directly or indirectly in order to make this project done. The author also would like to thank FRGS (cost centre: 0153AB-K80) for the financial support, Fundamental Applied Sciences Department and Centre of Research in Ionic Liquids (CORIL) of Universiti Teknologi PETRONAS, UTP to complete this project.

References

1. Araujo, G.S., Matos, L.J.B.L., Fernandes, J.O., Cartaxo, S.J.M., Gonçalves, L.R.B., Fernandes, F.A.N., Farias, W.R.L.: Extraction of lipids from microalgae by ultrasound application: prospection of the optimal extraction method. Ultrason. Sonochem. **20**(1), 95–98 (2013)
2. Mofijur, M., Rasul, M.G., Hassan, N.M.S., Nabi, M.N.: Recent development in the production of third generation biodiesel from microalgae. Energy Procedia **156**, 53–58 (2019)
3. Zhu, B., Chen, G., Cao, X., Wei, D.: Molecular characterization of CO_2 sequestration and assimilation in microalgae and its biotechnological applications. Biores. Technol. **224**, 1207–1215 (2017)
4. Yatirajula, S.K., Shrivastava, A., Saxena, V.K., Kodavaty, J.: Flow behavior analysis of *Chlorella vulgaris* microalgal biomass. Heliyon **5**(6), e01845 (2019)

5. Lee, A.K., Lewis, D.M., Ashman, P.J.: Disruption of microalgal cells for the extraction of lipids for biofuels: processes and specific energy requirements. Biomass Bioenerg. **46**, 89–101 (2012)
6. Mubarak, M., Shaija, A., Suchithra, T.V.: A review on the extraction of lipid from microalgae for biodiesel production. Algal Res. **7**, 117–123 (2015)
7. Chen, X., Hu, L., Xing, R., Liu, S., Yu, H., Qin, Y., Li, K., Li, R., Li, P.: Ionic liquid-assisted subcritical water promotes the extraction of lipids from wet microalgae *Scenedesmus* sp. Eur. J. Lipid Sci. Technol. **117**(8), 1192–1198 (2015)
8. Young, G., Nippgen, F., Titterbrandt, S., Cooney, M.J.: Lipid extraction from biomass using co-solvent mixtures of ionic liquids and polar covalent molecules. Sep. Purif. Technol. **72**(1), 118–121 (2010)
9. Ellison, C.R., Overa, S., Boldor, D.: Central composite design parameterization of microalgae/cyanobacteria co-culture pretreatment for enhanced lipid extraction using an external clamp-on ultrasonic transducer. Ultrason. Sonochem. **51**, 496–503 (2019)
10. Prabakaran, P., Ravindran, A.D.: A comparative study on effective cell disruption methods for lipid extraction from microalgae. Lett. Appl. Microbiol. **53**, 150–154 (2011)
11. Bagno, A., D'Amico, F., Saielli, G.: Computing the NMR spectrum of a bulk ionic liquid phase by QM/MM methods. J. Phys. Chem. B **110**(46), 23004–23006 (2006)
12. Santos, R.R.D., Moreira, D.M., Kunigami, C.N., Aranda, D.A.G., Teixeira, C.M.L.L.: Comparison between several methods of total lipid extraction from *Chlorella vulgaris* biomass. Ultrason. Sonochem. **22**, 95–99 (2015)
13. Soto-Sierra, L., Stoykova, P., Nikolov, Z.L.: Extraction and fractionation of microalgae-based protein products. Algal Res. **36**, 175–192 (2018)
14. Kim, Y., Choi, Y., Park, J., Lee, S., Yang, Y., Kim, H.J., Park, T., Kim, Y.H., Lee, S.H.: Ionic liquid-mediated extraction of lipids from algal biomass. Biores. Technol. **109**, 312–315 (2012)
15. Kim, Y., Park, S., Kim, M.H., Choi, Y., Yang, Y., Kim, H.J., Kim, H., Kim, H., Song, K., Lee, S.H.: Ultrasound-assisted extraction of lipids from *Chlorella vulgaris* using [Bmim][$MeSO_4$]. Biomas and Bioenergy **56**, 99–103 (2013)

Nano and Sustainable Technology

Role of Surface Modification in Synthesis of Structurally Well-Defined Silica Nanoparticles for Oil and Gas Applications

Hasnah Mohd Zaid, Muhammad Adil, and Mohd Arif Agam

Abstract Silica nanoparticles, as an improved oil recovery agent, have shown the ability to stimulate oil mobility. However, typical oil reservoir conditions with high temperature, salinity and pH modifies the way that nanoparticles interact with the porous medium, by causing the agglomeration leading to the particles' retention. Spherical, monodisperse silica nanoparticles (SiO_2 NPs) were synthesized via a one-step sol–gel method assisted by *in-situ* surface modification and characterized using various techniques. In this work, Triton X-100, as a surface-modification agent, was added during the hydrolysis of tetraethoxysilane (TEOS) in methanol to study their effect on morphology and size of the nanoparticles under varying ratio. The particle shape and size was determined using a field-emission scanning electron microscopy (FESEM), exhibiting the well-defined spherical particles with particle diameter between 13 and 27 nm. The obtained results also showed that an increase in the ratio of methanol to the surfactant [Methanol]/[Triton TX-100] produces a decrease in particle size. Furthermore, an energy dispersive x-ray spectroscopy (EDX) pattern was used to analyse the compositional ratio of silicon and oxygen to verify the surface-modification of SiO_2 nanoparticles. By X-ray diffraction (XRD), a broad peak of pure amorphous nature of SiO_2 is observed. While, the chemical binding of Triton X-100 on to the SiO_2 nanoparticles was studied by infrared spectroscopy (FTIR). Lastly, surface area of nanoparticles showed an increment as the molar ratio [Methanol]/[Triton TX-100] increased. All these results indicate that the surface modification plays an important role in effectively control the size and morphology of the SiO_2 nanoparticles, without the presence of agglomerations.

Keywords Silica · In-situ surface modification · Spherical nanoparticles · Nonionic surfactant

H. M. Zaid (✉) · M. Adil
Department of Fundamental and Applied Sciences, Universiti Teknologi PETRONAS, 32610 Seri Iskandar, Perak Darul Ridzuan, Malaysia

M. A. Agam
Faculty of Applied Science and Technology, Universiti Tun Hussein Onn Malaysia, UTHM Pagoh Campus, 84600 Muar, Johor, Malaysia
e-mail: arif@uthm.edu.my

S. A. Abdul Karim et al. (eds.), *Proceedings of the 6th International Conference on Fundamental and Applied Sciences*, Springer Proceedings in Complexity,
https://doi.org/10.1007/978-981-16-4513-6_25

1 Introduction

Nanoparticles have emerged as potential materials in the light of the decline in production from existing oil fields and the increasing difficulty of extracting hydrocarbons. Among them, SiO_2 NPs are widely considered [1–4]. They are environmentally friendly particularly in high reservoir temperature and pressure [5–7]. Once nanoparticles are incorporated into the subsurface, they can build or split emulsions or change the wettability of porous media [8, 9], thereby improving oil recovery. On the other hand, surfactant flooding has constraints as surfactants cannot be stabilized over long periods, particularly in the presence of oil, extreme temperature, and salinity. The surfactant's robustness can also be considerably improved by introducing surface-modified nanoparticles [7, 10].

The nanometer-scale synthesis of silica particles has been conducted using techniques such as vapor phase and liquid phase, the latter being called sol–gel process which has produced better results [11, 12]. Sol–gel is commonly used due to its ability to manipulate the particle size, morphology, and size distribution through careful monitoring of experimental parameters that influence the hydrolysis rate and condensation reactions. Stöber, who originally reported the SiO_2 NPs prepared using the sol–gel method [11, 12], based his research on the TEOS polymerization reactions in dilute ethanol catalyzed by ammonium hydroxide (NH_4OH) solution. Silica particles were produced in sizes ranging between 0.05 and 2 μm in diameter [11, 13]. The formation of silica particles from TEOS reactions is generally accepted to be presented as following [14]:

Hydrolysis

$$Si(OC_2H_5)_4 + H_2O \rightarrow Si(OC_2H_5)_3OH + C_2H_5OH \quad (1)$$

Water condensation

$$\equiv S{-}O{-}H + H{-}O{-}Si \rightarrow \equiv Si{-}O{-}Si\equiv + H_2O \quad (2)$$

Alcohol condensation

$$\equiv S{-}O{-}C_2H_5 + H{-}O{-}Si\equiv \rightarrow \equiv Si{-}O{-}Si + C_2E_5OH \quad (3)$$

Additives including, surfactants, electrolytes, and organic acids can also control the size and morphology of SiO_2 NPs using Stöber synthesis as a platform. Kim et al. have generated silica nanoparticles using sodium iodide as an electrolyte, while Rehman et. al. employed ammonium bromide to regulate the size of SiO_2 NPs [15, 16]. Wang and colleagues reported SiO_2 NPs in cubic form using tartaric acid as an organic acid [17]. Ikari et. al. used a binary surfactant method to inhibit the growth of particles with a nonionic amphiphilic agent by producing SiO_2 NPs with hexagonally arranged mesopores [18]. Möller and colleagues produced mesoporous silica particles of nanometer size by using triethanolamine and cationic surfactant as a base and template, respectively [19]. Venkatathri used polyvinyl pyrrolidine and various types of cationic surfactants to synthesize mesoporous silica nanospheres with uniform size and morphology [20]. In addition, Arrigada and Osseo-Asare produced nanoscale silica particles by employing controlled TEOS hydrolysis with reverse microemulsion [21], in the presence of NP-5 as a surfactant [22]. The resulting particles were noticeably uniform and monodispersed, with a diameter ranging between 30 and 70 nm. The primary advantage of the reverse microemulsion approach over Stöber's method is the generation of an ideal environment for particle size control.

In this paper, nonionic surfactant Triton X-100 is used to form sol–gel assisted by reverse micelle microemulsion to synthesize the spherical silica nanoparticles with a diameter ranging from 13 to 145 nm by varying the methanol-surfactant molar ratio as well as pH of the solution. The result is a simple, single-step procedure for self-assembling structurally well-defined silica nanoparticles of predetermined size. The silica nanoparticles' structure, morphology, and size were characterized through field-emission scanning electron microscopy (FESEM), energy dispersive x-ray spectroscopy (EDX), X-ray diffraction (XRD), and infrared spectroscopy (FT-IR).

2 Methodology

Silica nanoparticles were synthesized using the sol–gel process aided by reverse micelle microemulsion, employing tetraethylorthosilicate (TEOS, Merck) as a silica precursor, Triton X-100 (Sigma-Aldrich) as a non-ionic surfactant, methanol (98%, Merck) as a co-surfactant, Cyclohexane (99.5%, Merck) as an oil phase, ammonia (25%, Merck) as a pH adjuster and deionized water ($\geq$18 MΩ). The chemicals were used without further purification.

Surfactant, co-surfactant, and oil phase are initially combined at room temperature to form a microemulsion at 700 rpm for 15 min. Upon the generation of microemulsion, the medium is modified to a basic pH with a solution of ammonium hydroxide to catalyze the polymerization reactions needed to grow the silica nanoparticles that are produced in the next step. Finally, TEOS is applied to facilitate the hydrolysis and condensation of the silica molecules inside the micelle. The synthesis was fixed at 2 h [23] in order to obtain the small size particles by reducing the nucleation period. The microemulsion then was broken with the introduction of ethanol (VWR chemicals). The mixture was centrifuged (3 $\times$ 10 min. 6000 rpm) and washed with ethanol and water to recover the nanoparticles by removing the residual amount of surfactant and co-surfactant still present on the nanoparticles. After that, the particles were oven dried at 80 °C for overnight to obtain the dry nano-silica particles. Although the co-surfactant charges are the source of interaction with the surfactant charges to establish the micelle, the molar ratio [Methanol]/[Triton X-100] provides an appropriate environment within the micelle for the precursor's polymerization reactions.

Table 1 Variation in the molar ratio of [Methanol]/[Triton X-100] for nano-silica synthesis

pH	Synthesis parameters		
	[Methanol]/[Triton X-100]	$[H_2O]$/[Triton X-100]	$[H_2O]$/[TEOS]
10	4.5	9.2	59.1
	7.6	9.2	59.1
12	4.5	9.2	59.1
	7.6	9.2	59.1

Therefore, the influence of molar ratio of [Methanol]/[Triton X-100] on the size and particle shape was studied, keeping other reagents constant. The variation of NH_4OH solution also induces a change in the solution's pH, which affects the particle size directly. Table 1 shows the variation of the molar ratio of [Methanol]/[Triton X-100] and pH.

The size and shape of the as-synthesized particles were determined with a field emission scanning electron microscope (FESEM, Zeiss Supra 55VP), equipped with energy dispersive X-ray spectroscopy (EDX) for elemental analysis. The term particle size used in this article refers to the mean diameter of the silica nanoparticles. The average diameter was estimated for near-spherical particles. The XRD pattern of SiO_2 nanoparticles was obtained using X-ray diffraction (XRD, Bruker D8) with CuKα ($\lambda = 1.5406$ Å) radiation source. The measurements were conducted between a 2θ of 10°–90° ± 0.1 at an accelerated voltage of 40 kV and current 40 mA. The composition of the silica samples were analyzed using Fourier Transform Infrared (FT-IR) spectrometer (Perkin Elmer) in the frequency range of 4000–400 cm^{-1}, where the measured transmittance was within ± 0.1%T. On the other hand, surface area and porosity were determined using nitrogen adsorption–desorption (77 K) and the Brunauer– Emmett–Teller (BET) algorithm used by ASAP 2020 adsorption analyzer (Micromeritics). The samples were degassed overnight at 250 °C under vacuum (10–3 mmHg). Assuming that the silica particles are homogeneous, non-porous, and spherical, the average particle size D (nm) was determined using spherical model equation [24, 25]:

$$D = 6/\rho S_{sp} \quad (4)$$

where ρ is the density of silica particles (2.0×10^6 gm^{-3} for silica particles synthesized using wet-synthesis method [26]) and S_{sp} is the surface area of the spherical nanoparticles.

3 Results and Discussion

Nanoparticles are formed within the microemulsion in four stages: (i) TEOS interaction with the microemulsion; (ii) TEOS hydrolysis and monomers formation; (iii) nucleation; and (iv) particles growth. Reaction time is significant in the last two phases due to the constant collision and coalescence of water droplets in the microemulsion. Contact reactions between TEOS and ammonia force $Si(OH)_4$ to condense within the microemulsion's aqueous core. The generation of a large number of cores causes a reduction in their size, which is reflected in small particle size, so that all the cores are formed at the same time. Table 2 shows the impact of the change of [Methanol]/[Triton TX-100] on the particle size at a basic pH of around 10, while the other synthesis parameters were kept constant.

For the particles produced by the direct addition of TEOS and NH_4OH, the particle size and polydispersity are directly dependent on the stage of nucleation and growth. However, methanol as a co-surfactant plays the role of micelle stabilizer in reverse micelle microemulsion synthesis and the means to provide the chemical reaction within it. Micelles serve as reservoirs in the silica cores formation. As a result, particles show a decrement in diameter as the amount of methanol increases in the microemulsion. Figure 1 shows that the SiO_2 NPs have a particle size of (a) 27.9 ± 3.6 nm and (b) 13.1 ± 3.2 nm, when the molar ratio of [Methanol]/[Triton TX-100] changes from 4.5 to 7.6. As the methanol volume increased, the particle size decreases, and they also become more dispersed. This is due to the presence of a smaller amount of Triton X-100 in the micelles, compared to the methanol which results in a drop size being smaller for the availability of polar groups of Triton X-100 to bind methanol groups. On the other hand, the particle size decreases with the increase in the volume of methanol as more nuclei develop within the silicon micelle, resulting in smaller particle sizes.

Table 2 Change in particle size related to variation in [Methanol]/[Triton TX-100]

[Methanol]/[Triton X-100][a]	Particle Size (nm)
4.5	27.9 ± 3.6
7.6	13.1 ± 3.2

Fig. 1 Photomicrographs of FESEM of silica nanoparticles synthesized via sol–gel at varying [Methanol]/ [Triton X-100] ratio of **a** 4.5 and **b** 7.6 and pH = 10

Fig. 2 FESEM images of silica nanoparticles synthesized by varying [Methanol]/ [Triton X-100] **a** 4.5 and **b** 7.6 at pH = 12 having a corresponding particle diameter of **a** 128.8 ± 11.8 nm and **b** 119.3 ± 8.1 nm

The change in pH of the system causes a change in the molar ratio of water-surfactant, which influences not only the particle size but also their morphology (as shown in Fig. 2). In this case, pH value was increased from 10 to 12 by the addition of ammonium hydroxide. An increase in pH causes a change not only in particle shape, but also in size. At higher pH values, larger particle sizes (in the range of nanometers) with well-defined spherical morphology and monodispersity are obtained. The explanation behind the change in particle size, when the pH of the system varies, can be the speed of polymerization reaction. To change the pH value from 7.8 (pH_i) to 12, a greater amount of ammonium hydroxide is required, producing a large amount of OH and a higher rate of hydrolysis. This contributes to a little number of nuclei produced that form larger particles [27].

For compositional analysis of SiO_2 nanoparticles, EDX spectra of the surfactant-modified silicas were presented in Fig. 3, and compared to study the influence of molar ratio of [Methanol]/[Triton TX-100] in the silica matrix. The EDX spectra of surfactant-modified SiO_2 NPs clearly shows the presence of carbon (C). However, the concentration of C noticeably increases to 38% at [Methanol]/[Triton TX-100]

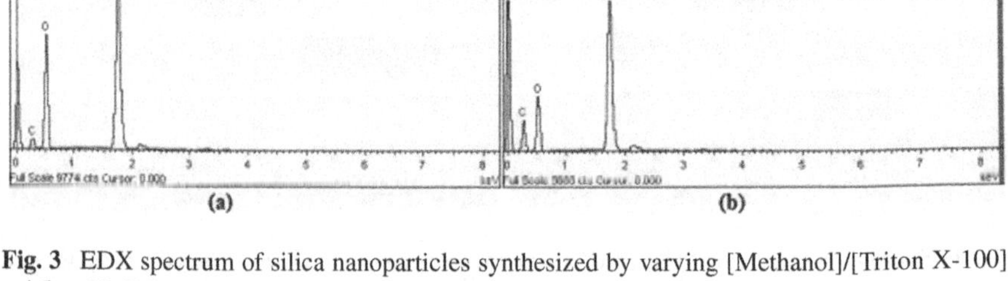

Fig. 3 EDX spectrum of silica nanoparticles synthesized by varying [Methanol]/[Triton X-100]: **a** 4.5 and **b** 7.6 at pH = 10

= 7.6, which confirmed that Triton TX-100 was more thoroughly incorporated into silica via sol–gel synthesis.

XRD patterns of silica nanoparticles shown a typical broad halo in Fig. 4, which indicate the amorphous nature of as-synthesized nano-SiO_2 by sol–gel technique. These results are in accordance with the previous studies [28, 29].

A representative FTIR spectrum of SiO_2 nanoparticles synthesized using different molar ratio of [Methanol]/[Triton TX-100] at pH 10 and 12 is shown in Fig. 5. The very intense and broad band appearing at ~ 1102 and 474 cm^{-1} is assigned to extension and flexural vibrations of Si–O–Si bonds [30], which reveals that a dense silica network was formed. On the other hand, ~ 800 cm^{-1} absorption band comes

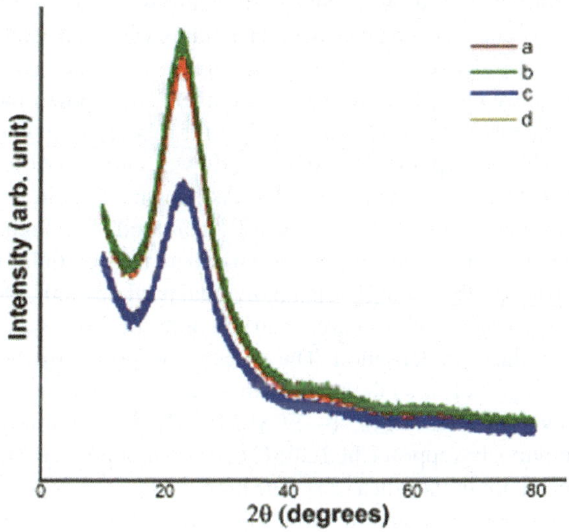

Fig. 4 XRD pattern of SiO_2 nanoparticles at varying molar ratio [Methanol]/[Triton X-100] of **a** 4.5, **b** 7.6 at pH = 10 and **c** 4.5 and **d** 7.6 at pH = 12

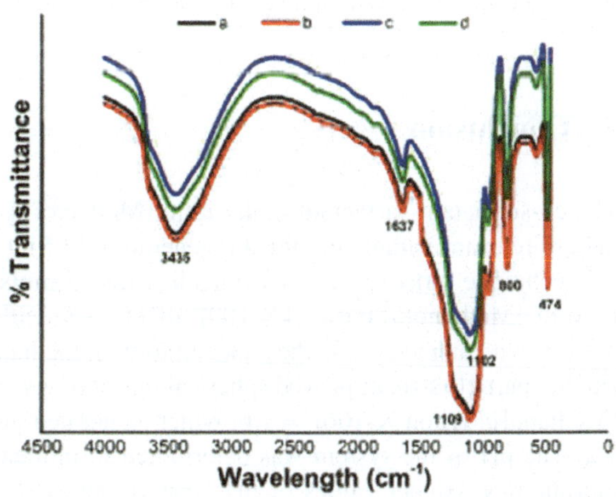

Fig. 5 FTIR spectra of nano silica particles synthesized using molar ratio of [Methanol]/ [Triton X-100]: **a** 4.5, **b** 7.6 at pH = 10 and **c** 4.5 and **d** 7.6 at pH = 12

Table 3 Structural and surface properties of nano-silica prepared via a variation in [methanol]/[triton x-100]

Properties	[Methanol]/[Triton X-100][b]	
	4.5	7.6
BET surface area (m^2 g^{-1})	104.9	236.3
Micropore area (m^2 g^{-1})	5.39	12.6
Micropore volume (cm^3 g^{-1})	0.00255	0.00503
Average pore diameter (nm)	22.55	14.26

from the vibration of (SiO_4) tetrahedrons, while the absorption bands at ~ 3435 and 1637 cm^{-1} originate from O–H bonding vibration of adsorbed molecular water. FTIR spectra also revealed that, as the particle size is decreased, the band at 1109 cm^{-1} was slightly moved to lower wave number. This finding indicates that the local bonding structures of Si and O atoms shift for relatively small particles [31, 32].

N_2 adsorption–desorption study was carried out to analyze the solid state characteristics of the nanopowder. As shown in Table 3, S_{BET} increased as the molar ratio [Methanol]/[Triton TX-100] increased. Although, the surface area can decrease readily due to the agglomeration, where nanoparticles join together to create a bulky structure that is larger than individual particles [20]. On the other hand, pore volume (V_p) was found to increase in the order of 7.6 < 4.5, correlating to the increase in capillary condensation. The pores were formed by the voids present between i) the nanoparticles, ii) the nanoparticles and agglomerates and iii) the agglomerates and also within (see Fig. 1—FESEM) [24]. In addition, pore size diameter in Table 3 obviously supports the IUPAC definition of pore size i.e. 2–50 nm, for mesoporosity behavior in the silica [25]. The high volume and pore size of nano-samples reveal the presence of open structures associated with agglomeration [24]. These open structures result in nanoparticles being loosely packed. Both samples revealed micropores within the structure, which might represent the fine pores within the agglomerates.

4 Conclusion

The co-surfactant-surfactant molar ratio [Methanol]/[Triton TX-100], and solution pH are relevant parameters for the generation of SiO_2 NPs synthesized using sol–gel technique aided by reverse micelles microemulsion. A change in the molar ratio of [Methanol]/[Triton TX-100] affects the synthesis, where the particle size reduces with an increase in the concentration of methanol. The condition resulting in smaller particle size, improved spherical morphology, and monodispersity was when [Methanol]/[Triton X-100] $= 7.6$, which generated an approximate size of 13.1 ± 3.2. The pH of the system was determined to influence both the particle size and morphology. Higher values of pH generate an increase in the OH, which in turn increase monodispersity, but also generate larger particles due to the formation of a smaller number of nuclei. In XRD spectra, the amorphous structure of the prepared nanoparticles has been confirmed by a single wide peak. While, EDX and FTIR results revealed that Triton X-100 was successfully incorporated with the modified nanoparticles. The structural and physical properties of the as-synthesized nanoparticles were adjustable by the quantity of surfactant added. The incorporation of a relatively large amount of Triton X-100 into silica results in an increase in particle size, which correlates to a decrease in surface area.

Acknowledgements The authors would like to acknowledge Universiti Teknologi PETRONAS (YUTP Grant No. 0153AA-H37) for funding the research.

References

1. Ebrahim, T., Mohsen, V.S., Mahdi, S.M., Esmaeel, K.T., Saeb, A.: Performance of low-salinity water flooding for enhanced oil recovery improved by SiO_2 nanoparticles. Pet. Sci. **16**(2), 357–365 (2019)
2. Evdokimov, I.N., Eliseev, N.Y., Losev, A.P., Novikov, M.A.: Emerging petroleum-oriented nanotechnologies for reservoir engineering. In: SPE Russian Oil and Gas Technical Conference and Exhibition (2006)
3. Hendraningrat, L., Li, S., Torsæter, O.: A coreflood investigation of nanofluid enhanced oil recovery. J. Pet. Sci. Eng. **111**, 128–138 (2013)
4. Roustaei, A., Saffarzadeh, S., Mohammadi, M.: An evaluation of modified silica nanoparticles' efficiency in enhancing oil recovery of light and intermediate oil reservoirs. Egypt. J. Pet. **22**(3), 427–433 (2013)
5. Lau, H.C., Yu, M., Nguyen, Q.P.: Nanotechnology for oilfield applications: challenges and impact. J. Pet. Sci. Eng. **157**, 1160–1169 (2017)
6. Olajire, A.A.: Review of ASP EOR (alkaline surfactant polymer enhanced oil recovery) technology in the petroleum industry: prospects and challenges. Energy **77**, 963–982 (2014)
7. Singh, R., Mohanty, K.K.: Synergistic stabilization of foams by a mixture of nanoparticles and surfactants. In: SPE Improved Oil Recovery Symposium (2014)
8. Romero-Zerón, L.: Introduction to enhanced oil recovery (EOR) processes and bioremediation of oil-contaminated sites. BoD Books on Demand (2012)
9. ShamsiJazeyi, H., Miller, C.A., Wong, M.S., Tour, J.M., Verduzco, R.: Polymer-coated nanoparticles for enhanced oil recovery. J. Appl. Polym. Sci. **131**(15) (2014).
10. Emrani, A.S., Ibrahim, A.F., Nasr-El-Din, H.A.: Mobility Control using nanoparticle-stabilized CO_2 foam as a hydraulic fracturing fluid. In: SPE EUROPEC Featured at 79th EAGE Conference and Exhibition (2017)
11. Stöber, W., Fink, A., Bohn, E.: Controlled growth of monodisperse silica spheres in the micron size range. J. Colloid Interface Sci. **26**(1), 62–69 (1968)
12. Canton, G., Ricco, R., Marinello, F., Carmignato, S., Enrichi, F.: Modified Stöber synthesis of highly luminescent dye-doped silica nanoparticles. J. Nanoparticle Res. **13**(9), 4349–4356 (2011)
13. Abarkan, I., Doussineau, T., Smaihi, M.: Tailored macro/microstructural properties of colloidal silica nanoparticles via microemulsion preparation. Polyhedron **25**(8), 1763–1770 (2006)
14. Rahman, I.A., Vejayakumaran, P., Sipaut, C.S., Ismail, J., Bakar, M.A., Adnan, R., Chee, C.K.: An optimized sol–gel synthesis of stable primary equivalent silica particles. Colloids Surf. A Physicochem. Eng. Asp. **294**(1–3), 102–110 (2007)

15. Rahman, I.A., Vejayakumaran, P., Sipaut, C.S., Ismail, J., Bakar, M.A., Adnan, R., Chee, C.K.: Effect of anion electrolytes on the formation of silica nanoparticles via the sol–gel process. Ceram. Int **32**(6), 691–699 (2006)
16. Kim, S.-S., Kim, H.-S., Kim, S.G., Kim, W.-S.: Effect of electrolyte additives on sol-precipitated nano silica particles. Ceram. Int. **30**(2), 171–175 (2004)
17. Yu, K., Guo, Y., Ding, X., Zhao, J., Wang, Z.: Synthesis of silica nanocubes by sol–gel method. Mater. Lett. **59**(29–30), 4013–4015 (2005)
18. Ikari, K., Suzuki, K., Imai, H.: Structural control of mesoporous silica nanoparticles in a binary surfactant system. Langmuir **22**(2), 802–806 (2006)
19. Moeller, K., Kobler, J., Bein, T.: Colloidal suspensions of nanometer-sized mesoporous silica. Adv. Funct. Mater. **17**(4), 605–612 (2007)
20. Venkatathri, N.: Synthesis of mesoporous silica nanosphere using different templates. Solid State Commun. **143**(10), 493–497 (2007)
21. Yong, C.P., Gan, L.M.: Microemulsion polymerizations and reactions, pp. 257–298. Polymer Particles, Springer (2005)
22. Zou, H., Wu, S., Shen, J.: Polymer/silica nanocomposites: preparation, characterization, properties, and applications. Chem. Rev. **108**(9), 3893–3957 (2008)
23. Jaramillo, N., Paucar, C., García, C.: Influence of the reaction time and the Triton X-100/cyclohexane/methanol/H_2O ratio on the morphology and size of silica nanoparticles synthesized via sol–gel assisted by reverse micelle microemulsion. J. Mater. Sci. **49**(9), 3400–3406 (2014)
24. Gregg, S.J.: Adsorption, Surface Area and Porosity, 2nd edn. Academic Press, London (1982)
25. Vansant, E.F., Van Der Voort, P., Vrancken, K.C.: Characterization and Chemical Modification of the Silica Surface. Elsevier, Ansterdam (1995)
26. Michael, G., Ferch, H.: Schriftenreihe Pigmente. Degussa Ag. **11**, 63 (1991)
27. Arriagada, F.J., Osseo-Asare, K.: Synthesis of nanosize silica in a nonionic water-in-oil microemulsion: effects of the water/surfactant molar ratio and ammonia concentration. J. Colloid Interface Sci. **211**(2), 210–220 (1999)
28. Gorji, B., Ghasri, M.R.A., Fazaeli, R., Niksirat, N.: Synthesis and characterizations of silica nanoparticles by a new sol-gel method. J. Appl. Chem. Res. **6**(3), 22–26 (2012)
29. Shen, X., Zhai, Y., Sun, Y., Gu, H.: Preparation of monodisperse spherical SiO_2 by microwave hydrothermal method and kinetics of dehydrated hydroxyl. J. Mater. Sci. Technol. **26**(8), 711–714 (2010)
30. Duran, A., Serna, C., Fornes, V., Navarro, J.M.F.: Structural considerations about SiO_2 glasses prepared by sol-gel. J. Non. Cryst. Solids **82**(1–3), 69–77 (1986)
31. Li, X., Cao, Z., Zhang, Z., Dang, H.: Surface-modification in situ of nano-SiO_2 and its structure and tribological properties. Appl. Surf. Sci. **252**(22), 7856–7861 (2006)
32. Rahman, I.A., Vejayakumaran, P., Sipaut, C.S., Ismail, J., Chee, C.K.: Size-dependent physicochemical and optical properties of silica nanoparticles. Mater. Chem. Phys. **114**(1), 328–332 (2009)

Sensitivity of Nickel Oxide Nanoflakes Layer on Extend Gate Field Effect Transistor for pH Sensor

Dauda Abubakar, Naser M. Ahmed, and Zakariyya Uba Zango

Abstract Nickel oxide (NiO) nanoflakes films for Extended gate field effect transistor (EGFET) pH sensors were synthesized and deposited by chemical bath deposition (CBD) method on indium titanium oxide (ITO) with glass substrate. The properties of NiO nanoflakes layer and pH sensing behaviour of the device were studied. Field emission scanning electron microscopy (FESEM) result shows that the NiO/Ni$(OH)_2$ films consists of porous net-like structure made up of interconnected nanoflake wall, with wall thickness of about 300 nm for the annealed NiO layer sample at 300 °C. XRD (X-ray diffraction) analysis demonstrated that the NiO nanoflakes has a cubic structure with the preferential orientation of the films at (100), (200) and (220). The devices exhibited superior sensing characteristic, due to the small size of the crystal and wide pore/flakes wall size of the NiO layer. The results confirmed that the NiO nanoflakes membrane has high sensitivity (72 μA/pH and 57 mV/pH) as compared with previews reports on NiO as sensing membrane for pH sensor.

Keywords Nickel oxide · Nanoflakes · pH sensor · EGFET · Chemical bath deposition

1 Introduction

The role Hydrogen ion concentration in the field of physical, biological and chemical sciences is important. A precise pH concentration measurement for a particular

D. Abubakar (✉)
Department of Physics, Bauchi State University Gadau, 65 Itas/Gadau, Bauchi, Nigeria

N. M. Ahmed
School of Physics, Universiti Sains Malaysia, 11800 Pulau Pinang, Malaysia
e-mail: naser@usm.my

Z. U. Zango
Department of Chemistry, College of Natural and Applied Sciences, Al-Qalam University, Katsina, Nigeria

S. A. Abdul Karim et al. (eds.), *Proceedings of the 6th International Conference on Fundamental and Applied Sciences*, Springer Proceedings in Complexity,
https://doi.org/10.1007/978-981-16-4513-6_26

system is needed in various experimental analysis for research and experimental applications. Glass electrodes have been used in the measurement of pH due to its higher sensitivity. But the high instability in corrosive solutions, high impedance of glass membrane and miniaturization difficulty fragility, limit the glass electrode application for pH sensor devices [1].

Therefore, to minimize the effect related to glass electrode, the Ion sensitive field effect transistor (ISFET) has been employed based metal oxide field effect transistor (MOSFET). However, as time goes on, the ISFET was observed to be facing some imperfection due to its highly instability in solution of saline water with sensitivity of light [2].

The Extended gate field effect transistor (EGFET) was advanced with the FET isolation structure from the metal oxide, that separates part of the chemical solution from the FET. EGFET structure is made of two separate parts; (1) commercial standard MOSFET and (2) sensing membrane cell. It comes along with so many advantages, which include easy maintenance, low cost, simple structure, free from chemical environment as well as light and temperature sensitivity. Nowadays, EGFETs has got great attention due to its vast advantages over others [3].

However, different metal oxides have been developed as the sensing membrane for EGFETs, such as PtO_2, ZnO [4], SnO_2 [5], Ta_2O_5, TiO_2 [6], WO_3 [7], etc. [8]. But the performance of some of these metal oxides appears to be average for pH sensing. Currently, the development of device having easy fabrication with low cast sensor membrane is the main focus in this area. Therefore, in this report, an effort was made in obtaining high pH sensitivity with low cost and easy fabrication of NiO nanostructure sensing membrane as a novel pH sensor based EGFET.

2 Experimental

2.1 Preparation of NiO Nanoflakes Sensing Membrane

The Nickel oxide (NiO) layers were deposited using chemical bath deposition (CBD) technique. All chemicals were of analytical reagent grade and used without further purification. De-ionized water was used in this synthesis and in washing steps of NiO layer deposition. The precursor solution for Nanoflakes Nickel oxide layer was obtained by dissolving 2.63 g (1 mol) of nickel sulphate hexahydrate ($NiSO_4.6H_2O$) in 10 ml and 2.24 g (1 mol) of potassium chloride (KCl) in 30 ml of deionized water. Further of it; 40 ml of 35% ammonia solution was added as complex agent. Colour changes were observed in the solution upon addition of ammonia, demonstrating the precipitation of nickel hydroxides particles from pale-green to blue-violet colour [9]. The complex agents used slow down the precipitation action and enables the formation of $Ni(OH)_2$ [10]. The total volume of the solution was made up to 100 ml by adding 20 ml of de-ionized water solution into 100 ml beaker. The mixture was maintained at 300 rpm stirring of approximately 30 min. The pH of the resultant solution was measuring to be 13 due to excess ammonia solution. ITO/glass was used as the substrates in this study. The substrates layers were chemically and ultrasonically cleaned before deposition, in acetone and ethanol for 900 s at 40 °C to eliminating any greasy track then rinsed with de-ionized water. The clean substrates were immersed vertically in the solution and then heated. The deposition of Nanoflakes Nickel hydroxide was made at a solution temperature of 90 ^{0}C. The sample was annealed at 300 °C temperature for 90 min.

2.2 *Characterization Techniques*

X-ray diffraction analyses were obtained with an X-ray diffraction high resolution system (Model: Panalytical X'Pert PRO MRD PW3040) consisted of CuKα (1.5406 Å) radiation. The nanostructures and surface morphology of the films were analysed with a Field Emission Scanning Electron Microscopy (FESEM Model: FEI Nova NanoSEM 450).

2.3 *Measurement of pH Sensor*

The analysis of NiO nanoflakes membranes was made using two (2) different source meter (Keithley Series 2400) through personal computer (PC) system cable connection. Measurement was carried out using LabTracer software that aids in initiating the pH sensor measurements by input data and receiving output data signal which were saved and analysis as well. The setup measurement for the pH sensor is shown in Fig. 1. The NiO nanoflakes sensing membrane was placed at the bottom of electrochemical cell that contains the pH buffer solution. The connection of the cell was made through the gate terminal of a commercially standard MOSFET (CD4007UB). Ag/AgCl electrode was employed as the reference electrode due to its provision of stable reference voltage to the sensing system [11]. The reference electrode was inserted into the pH buffer solution contained inside the cell. The measurement was carried out at room temperature (RT) (25 $^{°C}$) and wait for 60 s before the begin ning of the pH measurements. The reference electrode was connected to the one of the source meter for inputting reference voltage (V_{REF}) to the electrode while the second source meter was used to input the drain-source voltage (V_{DS}) to the drain and source terminals of the MOSFET for drain-source current (I_{DS}) output readingand then measured the current–voltage (I–V) characteristics in pH = 4, 6, 8, 10 and 12 pH buffer solutions at RT. The V_{DS} was set at 0.3 V, which is the voltage between the drain and the source of the MOSFET and the V_{REF} was set at 3 V, which is the voltage that supply to the reference electrode.

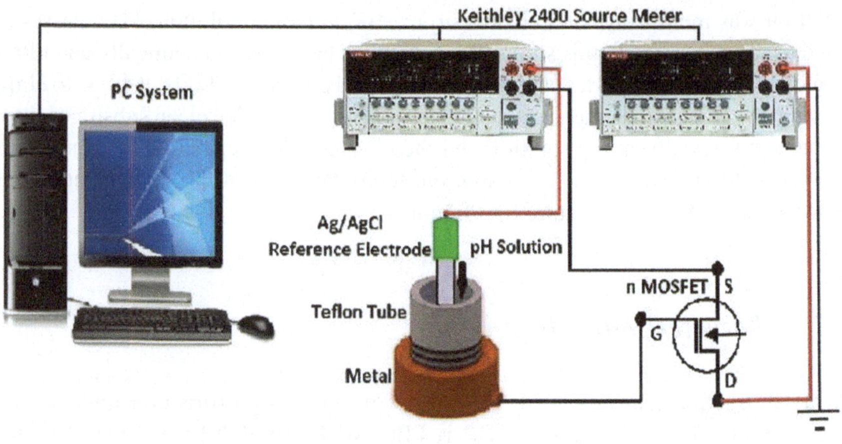

Fig. 1 Setup diagram of pH sensor measurement system

3 Results and Discussion

Figure 2 shows the X-ray diffraction pattern and the values related to peaks of chemically grown, nanoporous $Ni(OH)_2$ with the subsequent annealed sample at 300 °C. From the pattern of the as-grown films sample, the $Ni(OH)_2$ phase was identified. The peaks at 33.15°, 38.62°, 59.18°, 62.73° and 72.68° that were labeled according to the (ICSD 00-014-0117) with Hexagonal unit cell parameters of a = 3.1260 (Å), b = 3.1260 (Å) and c = 4.6050 (Å) with angles Alpha = 90°, Beta = 90° and Gamma

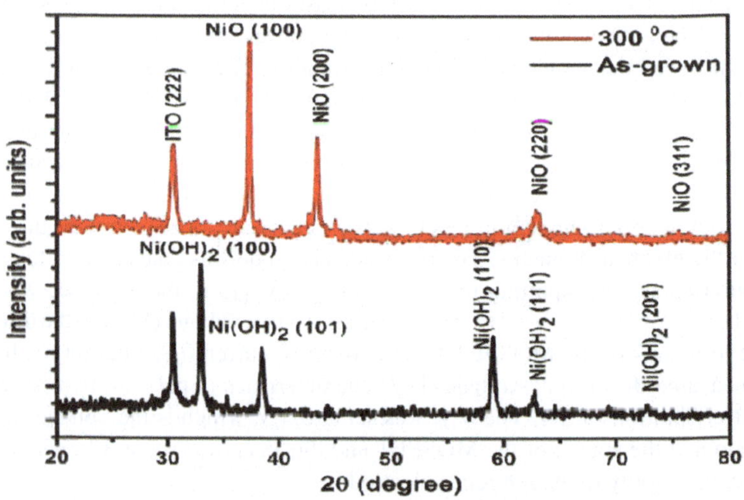

Fig. 2 XRD pattern of the $NiO/Ni(OH)_2$ nanoflakes as-grown and at 300 °C

= 120°. Also, the peak for cubic ITO structure pattern has been identified for both as-grown and annealed sample with peak orientation of (222) at 2θ = 30.62°.

Meanwhile, X-ray diffraction annealed samples that attribute to cubic NiO was detected for the films on ITO/Glass substrate (00-047-1049) crystals with the lattice parameter of a, b and c = 4.1771 Å, and angle of α = β = γ = 90. The peaks obtained were at 2θ = 37.25°, 43.28°, and 62.88° with the preferential orientation of the films being at (100), (200) and (220) respectively, that indicates high crystallinity with the well defined intense peak at (100) indices. This indicates the formation of polycrystalline NiO films after annealing, with no trace of $Ni(OH)_2$ phase on the treated films. Which shows a complete $Ni(OH)_2$ to NiO transformation. The mean size of the crystallite (D) was estimated based on the full width at the half maximum (FWHM) peak position and the corresponding angle values was obtained, using Scherer's equation of Eq. (1). A small crystal size of 19.84 nm was obtained.

$$D = \frac{0.9\lambda}{\beta \cdot \cos\theta} \quad (1)$$

where: D is the mean size of the crystalline, K is a dimensionless shape factor, λ is the X-ray wavelength; β is the line broadening at half the maximum intensity (FWHM) in radian and θ is the Bragg angle.

FESEM measurements were carried out in order to determine the micro/nanostructural properties of the chemically deposited NiO films; as shown in Fig. 3. It can be seen that the $NiO/Ni(OH)_2$ films consists of grown porous net-like structure made up of interconnected nanoflakes' wall with a thickness of 30–70 nm of the annealed samples. The diameter of pores ranges from 50 to 300 nm with an observable difference on the dimension of thin films micro/nanostructure as the 300 °C, that shows high surface to volume ratio. The flakes wall appears to have a

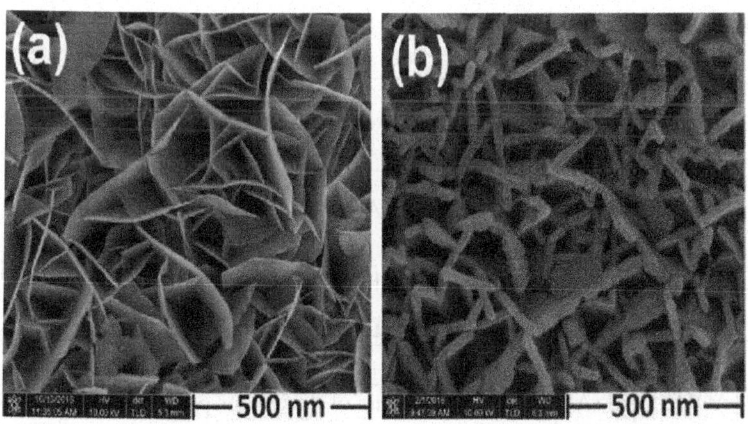

Fig. 3 FESEM image and EDS spectrum nanoflakes for **a** as-grown $Ni(OH)_2$ and **b** 300 °C NiO annealed sample

Table 1 Atomic and the weightpercent of the as-grown and annealed sample obtained from EDS EDX spectrum

Sample	As grown		300 °C	
Element	O	Ni	O	Ni
Atomic %	60.23	38.77	50.16	49.84
Weight %	29.51	70.49	21.35	78.65

rough surface with annealing temperature. The porous nanostructure, surface roughness and surface to volume ratio can be an important factor that provides easy access of ions surface reaction in sensor application [12] (Table 1).

The atomic ratio O: Ni of the 300 ^{0}C sample appears with non-stoichiometric value, which makes it behave as a p-type material due Nickel vacancy.

3.1 Sensitivity and Linearity of pH Sensor

At the first contact of nanostructures NiO with an electrolyte, primary adsorption of $[OH^-]$ and $[H^+]$ occurs. This procedure is made at a low degree of the surface coverage. Furthermore, the surface reaction (NiO–OH) group occurs with $[OH^-]$ ions in basic solution and $[H^+]$ ions in acidic solution. The rate of this reaction is associated with the free electrons or holes concentration at the surface. The process involves charge transfer to the solid, which is distributed by transport of electron over the near surface region of the solid to create a depletion layer in solution of acid and then to improved or generate an accumulation layer at the base. The site-dissociation model was established to describe the oxide layer charging mechanism in the pH sensor [8]. The oxide charging as the result of equilibrium between the AOH and H^+ ions surface sites in the solution bulk was given. The surface reactions were as follows [13]

$$AOH \overset{K_A}{\leftrightarrow} AO^- + H_S^+ \tag{2}$$

$$AOH_2^+ \overset{K_B}{\leftrightarrow} AOH + H_S^+ \tag{3}$$

where H_S^+ is the H^+ ions located close to the solid surface and $K_A = \frac{[AO^-][H^+]_S}{[AOH]}$ and $K_B = \frac{[AOH][H^+]_S}{[AOH_2^+]}$ are the equilibrium constants of Eqs. (2) and (3) respectively AO^-, AOH_2^+ and AOH represent negative, positive and neutral surface sites, respectively. $[AO^-][AOH_2^+]$ and [AOH] are the numbers of these sites per surface area. $[H^+]_S$ is the H^+ ions surface activity.

In view of the pH electrode, the surface potential is obtained by the H^+ ions exchange between the binding sites on the surface of the sensing membrane and the solution. Also, the pH electrode selectivity and chemical sensitivity are controlled by the properties of electrolyte/insulator interface. The pH electrode response obeys the Nernstian equation, which represent straight line equation as;

$$E = E_O - \frac{RT}{nF} In([H^+]) \tag{4}$$

$$E = E_O + Slope(T) \cdot \text{pH} \tag{5}$$

where E is the measured potential, E_0 is a characteristic constant for the ion-sensitive/external electrode system, and the slope factor term "*RT/F*" is known as the Nernst factor, which provides the amount of change in total potential (mV) for all change in pH unit. T the absolute temperature (in Kelvin), n is the signed ionic charge, R is the gas constant, F the Faraday constant and $[H^+]$ the molar concentration of the free ionic species. Nernst equation describes electrochemical cell potential as a function of ions concentrations taking part in the reaction. The ideal response was assumed to be 59.2 mV/pH at 25 °C [14].

Based on the site binding theory, the number of binding sites residing on the sensing membrane could result to changes in the surface potential voltage between the electrolyte and the interface of the sensing layer which depend on pH at point of zero charge (pH_{pzc}). The actual change in surface potential voltage would depend on the pH value of the electrolytic solution. Based on the above mentioned site binding model, the surface potential voltage of NiO films between the electrolytic pH solution interface and the sensing layer can be expressed as [15];

$$\psi = 2.303 \frac{kT}{q} \frac{\beta}{\beta + 1} (\text{pH}_{pzc} - \text{pH}) \tag{6}$$

where, $\beta = \frac{2q^2 N_S \left(\frac{K_B}{K_A}\right)^{1/2}}{kTC_{DL}}$ is the sensitivity parameter, which is dependent on the surface density of hydroxyl groups and reflects the chemical sensitivity of the gate insulator, $\text{pH}_{pzc} = -\log_{10}\left(\frac{K_A}{K_B}\right)^{1/2}$ is the pH value at the zero charge point of the sensing membrane, K_A and K_B are acid and base equilibrium constants, respectively. N_S is the total number of surface sites per unit area, C_{DL} is the electrical double layer capacitance derived by the Gouy–Chapman–Stern model, k is Boltzmann's constant, T is the absolute temperature, q is the charge of an electron [3].

Similar to the MOSFET, the linear relationship between the gate voltage (V_G) and the drain current (I_D) for an EGFET is expressed as [16];

$$I_D = \mu_n C_{ox} \frac{W}{2L} \left[2V_{DS}(V_{GS} - V_{TH}) - V_{DS}^2\right] \tag{7}$$

where L is the gate length, μ_n is the charge-carrier effective mobility, C_{ox} is the gate oxide capacitance per unit area and W is the gate width, V_{DS} is the drain-source voltage and V_{TH} is the threshold voltage of MOSFET relating to the surface potential[17]. Base on this sensing model, the pH sensitivity is therefore defined as [18]:

$$Sensitivity = \frac{\Delta\psi}{\Delta \text{pH}} = \frac{\Delta V_{TH}}{\Delta \text{pH}} = \frac{\Delta V_{GS}}{\Delta \text{pH}} \tag{8}$$

In the acidic region, more positive surface potential of the NiO films leads to a smaller applied gate voltage. As the pH value raises, the V_G value increased as well. Accordingly, the pH sensitivity measurement can be obtained from the slope of gate voltage V_G versus pH solution.

The commercial MOSFET (CD 4007UB) has been confirmed to be a eligible EGFET transducer due to its good linearity of over 99%. For operation of EGFET at $V_{GS} = 3$ V, it is within the operational region in pH values conversion to I_{DS} with remarkable sensitivity. The MOSFET controls the electron-holes flow from source (S) to drain (D) by the control of size and shape of the conductive channel. The channel size controls the gate-source voltage (Vgs) positivity ($[H^+]$) or negativity ($[OH^-]$)). The I_{DS} dependency on the pH value is because of the accumulation of $[H^+]$ ions in the acidic solution on the sensing surface, which is similar to applying an additional positive voltage at the MOSFET gate [14]. This would result to increase of the conductive channel with increase of the concomitant current. On the contrary, the conductive channel constricted or closed for the basic solution with extra negative voltage application, hence, current decreases as shown in Fig. 4 [17].

Figure 4 shows the (I–V) characteristics for NiO membranes using the Fig. 1 setup with different pH value of 4, 6, 8, 10 and 12 buffer solutions. Figure 4a, b indicates the (I_{DS}–V_{GS}) and (I_{DS}–V_{DS}) curves for linear regimes and saturation, respectively. In the saturation regime, V_{GS} was set fixed at 3 V, and in linear regime, V_{DS} was set fixed at 0.3 V. The (I–V) curves were then plotted to get the sensitivity for both membranes from the slope of Fig. 4c, d.

The current was decrease with the increase in pH value in the saturation regime while the threshold voltage shifted to the right with increase in the pH value. In the saturation regime, where the low pH value is similar to high accumulation of H^+ ions in the acidic solution on the sensing surface. The high amounts of ions result in additional positive voltage to be applied to the gate, in which conductive channel increases but depletion region decreases, hence current increases. Unlike for a basic solution, where the high pH value denoting high OH^- ions and low H^+ ions is equivalent to applying negative voltage to the gate, the conductive channel decreases through increasing the depletion region, thereby decreasing the current flow. Such voltage is inversely proportional to the current, the threshold voltage shifted to the right with the increase in the pH value for the linear regime. As a result, voltage is directly proportional to pH value.

The site-binding model was used in this research work to show the chemical processes that take place at the interface between the available $[H^+]$ ions in the electrolyte and membrane. The membrane material elements enclose water shells, which hinders any contact with the interface and inhibit chemical reactions, thus the small sized $[H^+]$ ions will neither be hydrated nor shelled. This feature facilitates

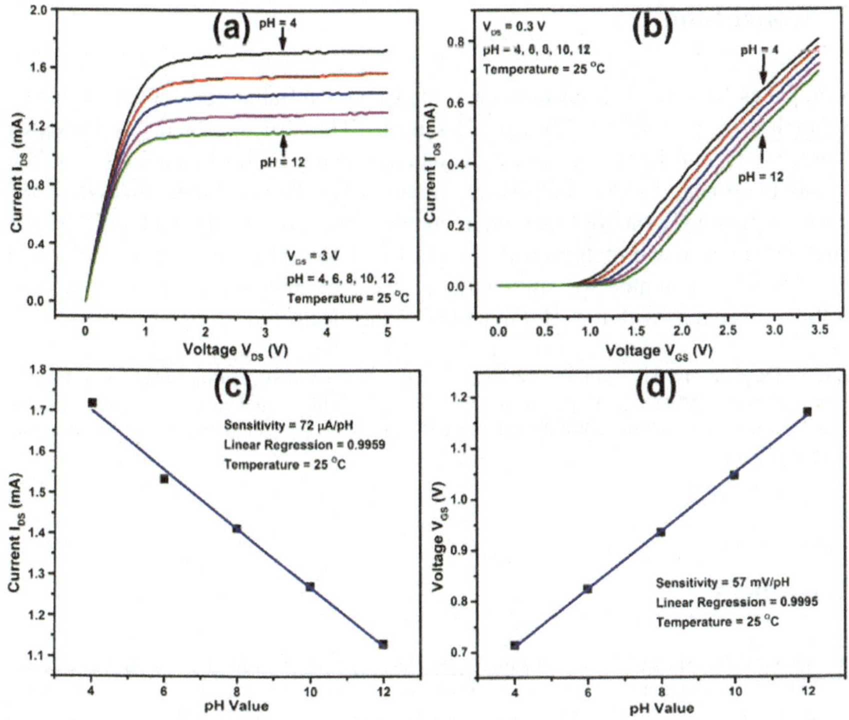

Fig. 4 I-V curve characteristics in **a** saturation and **b** linear regime with corresponding sensitivity and linearity **c** and **d** NiO membrane

chemical reactions at the interface. On the contrary, $[OH^-]$ ions are the only ions that can bond to the available sites on the interface surface.

Linearity and Sensitivity can be determined by linear fitting between I_{DS} and V_{GS} values against pH values obtained at fixed value of V_{DS} (4 V) and fixed value of I_{DS} (0.4 mA) of the I-V curves of Fig. 4 respectively using Eq. (8). From the linear fitting, the voltage sensitivity obtained as slope and linearity as linear regression of NiO membrane were found to be 57 mV/pH; 0.9995 and 72 μA/pH; 0.9959, for both linear (I_{DS}–V_{GS} characteristics) and saturation (I_{DS}–V_{DS} characteristics) regimes, respectively as shown in Fig. 4d, c. The result obtained is much better than the earlier report works using NiO as membrane for pH sensing [17, 19]. This apparent improvement in sensitivity can be associated to the high surface-to-volume ratio with the small sized NiO nanoporous crystals and its high crystallinity, which provides more surface binding sites and larger effective sensing areas. The NiO nanostructures have a larger surface area for adsorbing additional OH^- and H^+ ions and have increased oxygen-related binding that could sense H + ions effectively in the acid solution region. The result suggests that the Nanoflakes NiO membrane synthesis based on Chemical Bath Deposition is potentially an essential materials for EGFET based pH sensors.

4 Conclusion

Nanoflakes NiO membrane was made using Chemical Bath Deposition to determine its applicability as EGFET based pH sensors. The structural and morphological characteristics of the membranes were analyzed using FESEM and XRD. The NiO membrane sensitivity (57 mV/pH) was found to be relatively higher than the works reported previewly on NiO membrane as pH sensor, due to the high crystallinity, small crystal size and high porosity properties of NiO Nanostructured layer, which provided higher number of sites with surface area and increased the number of exchangeable ions with the $[H^+]$ that exist in the electrolyte.

Acknowledgements The authors wish to convey their sincere appreciation to School of Physics, Universiti Sains Malaysia, Department of Physics, Bauchi State University, Gadau together with GOT Incentive Grant 1001/PFIZIK/822055 for the wonderful supports of this successful accomplishment.

References

1. Sabah, F.A., Ahmed, N.M., Hassan, Z., Al-Hardan, N.H.: Sensitivity of CuS and CuS/ITO EGFETs implemented as pH sensors. Appl. Phys. A Mater. Sci. Process. **122**(9), 1–6 (2016)
2. Yossawat, R., et al.: High sensitive nanocrystal titanium nitride EG-FET pH sensor. Adv. Mater. Res. **802**(September), 232–236 (2013)
3. Tsai, C., Chou, J., Sun, T., Hsiung, S.: Study on the sensing characteristics and hysteresis effect of the tin oxide pH electrode. Sens. Actuators B Chem. **108**, 877–882 (2005)
4. Chang, S., et al.: Investigation of ZnO-nanowire-based extended-gate field-effect-transistor pH sensors prepared on glass substrate. Sci. Adv. Mater. **4**(xx), 1–5 (2012)
5. Fernandes, J.C., Nascimento, R.A.S., Mulato, M.: Effects of measurements conditions on an extended-gate FET used as pH sensor. Mater. Res. **19** (2016)
6. Rosdan, M.A., Herman, S.H., Abdullah, W.F.H., Kamarozaman, N.S., Syono, M.I.: Sputtered titanium dioxide thin film for extended-gate FET sensor application, pp. 219–222 (2013)
7. De Castro, R., Tadeu, D., Mulato, M., Maria, E.: Comparative sensibility study of WO 3 ph sensor using EGFET and ciclic voltammetry. Mater. Res. **18**(1), 15–19 (2015)
8. Wu, S., Wu, Y., Lin, C.: High performance ISFET-based pH sensor utilizing low-cost industrial-grade touch panel film as the gate structure, pp. 2131–2133 (2014)
9. Adekunle, A.S., et al.: Comparative catalytic properties of $Ni(OH)_2$ and NiO nanoparticles towards the degradation of nitrite (NO_2\n-) and nitric oxide (NO). Int. J. Electrochem. Sci. **9**(6), 3008–3021 (2014)
10. Ezema, F.I., Ekwealor, A.B.C., Osuji, R.U.: Optical properties of chemical bath deposited CdS thin films. J. Optoelectron. Adv. Mater. **9**(6), 1898–1903 (2007)
11. Khanna, V.K.K.V.: Fabrication of ISFET microsensor by diffusion-based Al gate NMOS process and determination of its pH sensitivity from transfer characteristics. Indian J. Pure Appl. Phys. **50**(3), 199–207 (2012)
12. Inamdar, A.I., Kim, Y., Pawar, S.M., Kim, J.H., Im, H., Kim, H.: Chemically grown, porous, nickel oxide thin-film for electrochemical supercapacitors. J. Power Sources **196**(4), 2393–2397 (2011)
13. Ali, G.M., Dhaher, R.H., Abdullateef, A.A.: pH sensing characteristics of EGFET based on Pd- Doped ZnO thin films synthesized by sol-gel method. **4**(2), 234–238 (2015)

14. Wu, Y.-L., Chang, S.-P., Chang, S.-J., Weng, W.Y., Lin, Y.-H.: A novel ph sensor using extended-gate field-effect transistors with Ga_2O_3 nanowires fabricated on a novel pH sensor using extended-gate. Sci. Adv. Mater. **6**, 1–4 (2014)
15. Chiu, Y., Tseng, C., Lee, C.: Nanostructured EGFET pH sensors with surface-passivated ZnO thin-film and nanorod array. IEEE Sens. J. **12**(5), 930–934 (2012)
16. Huang, B., Lin, J., Yang, Y.: ZnO/silicon nanowire hybrids extended-gate field-effect. **160**(6), 78–82 (2013)
17. Lin, F., Chang, H., Hsiao, S.: Preparation and characterization of nickel oxide-based EGFET pH sensors. In: Ninth International Conference on Sensing Technology Preparation, pp. 402–405 (2015)
18. Vieira, N.C.S., Figueiredo, A., Faceto, A.D., De Queiroz, A.A.A., Zucolotto, V., Guimarães, F.E.G.: Sensors and actuators B: chemical dendrimers/TiO_2 nanoparticles layer-by-layer films as extended gate FET for pH detection. Sens. Actuators B. Chem. **169**, 397–400 (2012)
19. Ibupoto, Z.H., Khun, K., Willander, M.: Development of a pH sensor using nanoporous nanostructures of NiO. J. Nanosci. Nanotechnol. **14**(9), 6699–6703 (2014)

Synthesis and Characterization of Nickel–Cobalt Oxide ($NiCo_2O_4$) for Promising Supercapacitor Material

Poppy Puspitasari, Mahdalena Julia, Avita Ayu Permanasari, Maizatul Shima Shaharun, Timotius Pasang, and Muhamad Fatikul Arif

Abstract Supercapacitor takes advantage of the broad electrode surface and thin dielectric material to reach higher capacity compared to the conventional capacitor. These factors make it possible for the supercapacitor to have larger energy density compared to the conventional and larger power density compared to a battery. Thus, supercapacitor would become a right solution as an energy storage device. $NiCo_2O_4$ is a good oxide material and exciting to learn due to its promising material and good electronic conductivity. $NiCo_2O_4$ was synthesised using self-combustion method then sintered at 400, 500, 600 °C in time variations of 60, 120, 180 min before testing with the XRD, SEM-EDAX, and VSM tests. The XRD test results showed a single-phase because of the appearance of crystallite in all samples in all temperature variations. The crystallite size, then, was calculated using the Scherrer equation. The results were: sample with 60 min sintering duration had 6.699 nm crystallite size, the sample with 120 min sintering duration had 4.678 nm crystallite size, and sample with 180 min sintering duration had 5.883 nm crystallite size. The morphology in the SEM test displayed the particles in small and nano size. The magnetic property test showed that the samples were superparamagnetic with the highest coercivity and retentivity was found in the sample of 500 °C temperature and 60 min sintering process..

P. Puspitasari (✉) · M. Julia · A. A. Permanasari
Universitas Negeri Malang, East Java, Indonesia
e-mail: poppy@um.ac.id

A. A. Permanasari
e-mail: avita.ayu.ft@um.ac.id

M. S. Shaharun
Universiti Teknologi PETRONAS, Perak, Malaysia
e-mail: maizats@utp.edu.my

T. Pasang
Oregon Institute of Technology, Klamath Falls, USA
e-mail: tim.pasang@oit.edu

M. F. Arif
Institut Teknologi Sumatera, Lampung, Indonesia
e-mail: mf.arif@mt.itera.ac.id

S. A. Abdul Karim et al. (eds.), *Proceedings of the 6th International Conference on Fundamental and Applied Sciences*, Springer Proceedings in Complexity,
https://doi.org/10.1007/978-981-16-4513-6_27

Keywords Self-combustion · Magnetic saturation · Retentivity · Coercivity

1 Introduction

In current modern life, electrical energy is a significant necessity that cannot be avoided. Various technology nowadays most required electricity storage device. There is a huge interest in researchers to develop and perfect more efficient energy storage such as supercapacitor. A supercapacitor, also known as ultracapacitor or electrochemical capacitor, utilises the electrode surface and thin dielectric electrolyte solution to achieve capacitance several times greater than conventional capacitors [1, 3].

Supercapacitors are used in applications that require energy sources that have a faster charge/wear cycle than long-lasting energy sources, supercapacitors are widely used in cars, buses, trains, cranes and elevators [4]. Electrode material, as one of the key components of super capacitors, is closely related to energy density and very high power density in the case of high-cycle supercapacitors. Electrode material, as one of the key components of supercapacitors, is closely related to energy density and very high power density in the case of high-cycle supercapacitors [5, 6].

$NiCo_2O_4$ generally assumed as a mix valence of oxide that adopts pure spinel structure where nickel occupies the octahedral site and cobalt is distributed in octahedral and tetrahedral sites [7]. Nickel–cobalt oxide ($NiCo_2O_4$) receives interests to be used in supercapacitor because of its good electron conductivity, stable structure, environmentally friendly, low cost, easy preparation, and higher electrical conductivity compared to NiO and CoO [8–10].

2 Experimental Setup

2.1 Material Preparation

The $NiCo_2O_4$ samples were made with the self-combustion method using Ni $(NO_3)_2.6\ H_2O$ and Co $(NO_3)_2.9\ H_2O$ was mixed using nitric acid as the fuel. The samples were dissolved using distilled water and stirred using magnetic stirrer in continuous heating and stirring. Slow evaporation from a homogeneous solution produced a very thick gel and further heating resulted in spontaneous combustion. The samples then oven-dried at 110 °C for 8 h, sintered at 400, 500, and 600 °C for 60, 120, and 180 min respectively.

2.2 *Characterization*

Nickel Cobalt ($NiCo_2O_4$) nanopowder sample were then tested with XRD (X-Ray Diffraction Model PanAnalytical, Type: E'xpert Pro) with long-range angle 10°–90° 2θ to find the crystallinity properties, SEM–EDX (Scanning Electron Microscopy with EDAX feature Model FEI, Type: Inspect-S50) in 100.000 magnification to get the morphologies and nano-elements produced, FTIR (Fourier Transform Infra-Red Model Shimadzu, Type: IRPrestige21) start from 500 until 4000/cm wavelength to find the functional groups and bonds, and VSM (Vibrating Sample Magnetometer Model OXFORD, Type VSM 1.2H/CF/HT) up to 1 T of Magnetic force to find the magnetic coercivity, retentivity, or saturation from all samples.

3 Result and Discussion

3.1 *Phase Identification*

The XRD chart in Fig. 1 shows that all samples are single-phase $NiCo_2O_4$ with peak crystallites of [311], [400], and [440] in samples of 400 °C variation; the highest peak is [400]. The sample with 60 min variation has crystallite of 8.977 nm, the sample with 120 min duration has crystallite of 6.279 nm, and the sample with 180 min duration has crystallite of 7.828 nm as displayed in Table 1.

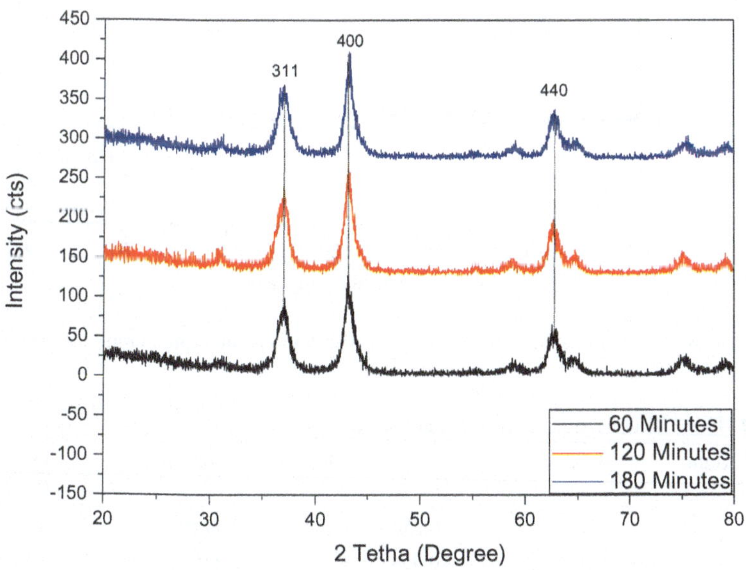

Fig. 1 XRD Graph for $NiCo_2O_4$ at Sintering Duration at 400 °C

Table 1 Phase Identification of NiCo2O4 at 400 °C

Sample (min)	Height (cts)	FWHM (2Th.)	d-spacing (Å)	D (nm)
60	88.35	0.5510	2.0970	6.69
120	101.60	0.7872	2.0943	4.68
180	93.02	0.6298	2.0916	5.88

Figure 2 presents that all samples with 500 °C temperature variation form single-phase with each has five peaks: [220], [311], [400], [511], and [440] with the highest intensity is [400]. The sample with 60 min duration has crystallite size of 24.712 nm, the sample with 120 min duration has crystallite of 19.151 nm, and the sample with 180°C minutes variation has 28.874 nm crystallite as shown in Table 2. Figure 3 presents the similar matter to Fig. 2 where samples have single-phase $NiCo_2O_4$ with five peaks: [220], [311], [400], [511], and [440] and [400] as the highest peak. The

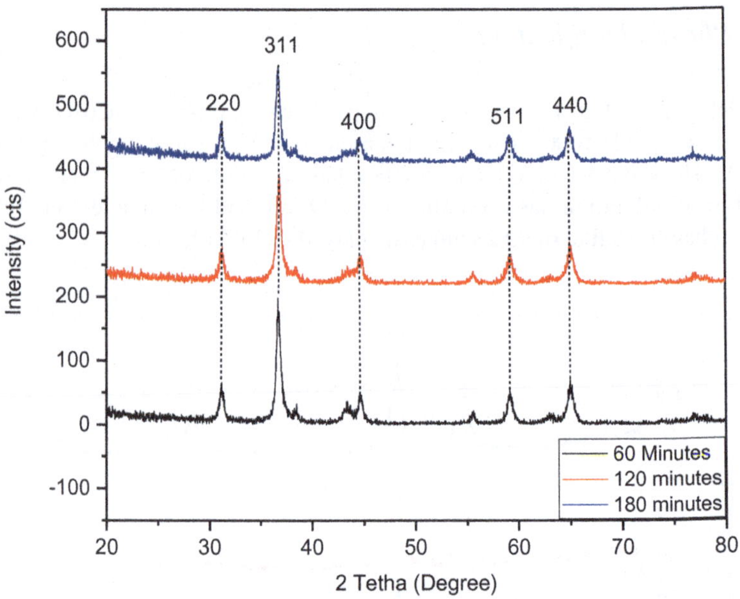

Fig. 2 The X-Ray diffraction results of $NiCo_2O_4$ with Various Sintering duration at 500 °C

Table 2 Phase identification of $NiCo_2O_4$ at 500 °C

Sample (min)	Height (cts)	FWHM (2Th.)	d-spacing (Å)	D (nm)
60	173.7	0.2755	2.4496	24.71
120	150.9	0.3542	2.4453	19.15
180	141.1	0.2362	2.4518	28.87

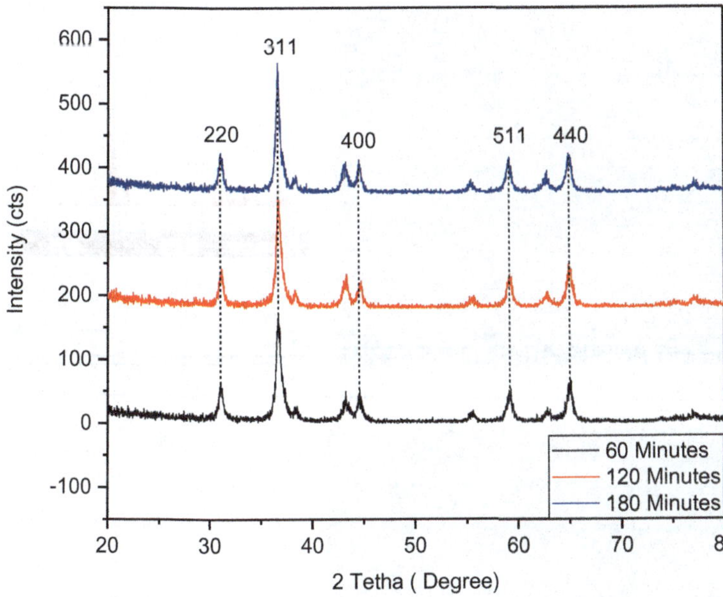

Fig. 3 The X-Ray Diffraction Results of NiCo2O4 with Various Sintering Duration at 600 °C

Table 3 Phase identification of $NiCo_2O_4$ at 600 °C

Sample (min)	Height (cts)	FWHM (2Th.)	d spacing (Å)	D (nm)
60	148.46	0.1968	2.44907	34.58
120	156.19	0.3542	2.44670	19.17
180	188.68	0.1968	2.44586	34.48

sample with 60 min duration has crystallite of 34.577 nm, the sample with 120 min duration has crystallite of 19.172 nm, and the sample with 180 °C minutes variation has 34.485 nm crystallite as shown in Table 3. Based on these three charts, the $NiCo_2O_4$ nanoparticle has a cubic structure and centered on the surface [11–15], supported with the images from the SEM test. From all charts, the high and low-intensity peaks display the size of the crystallite, and the crystallite particle can be calculated using the Scherrer formula [16, 17].

3.2 *Morphological Analysis*

Based on the Fig. 4, generally, all nine samples have similarities in their morphologies. The nanopowder morphology was close and adhered to each other and form nano bulks. The content from the EDAX test was closer to the theory, in which Nickel

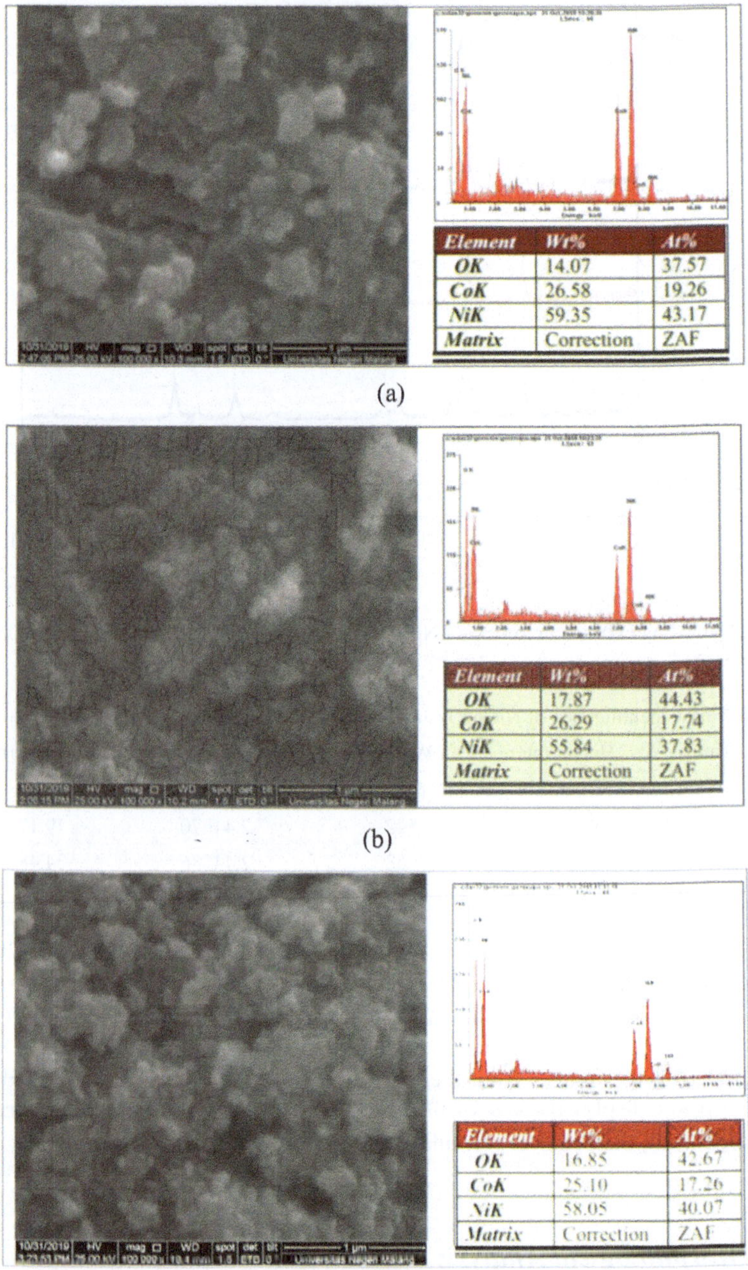

(a)

Element	Wt%	At%
OK	14.07	37.57
CoK	26.58	19.26
NiK	59.35	43.17
Matrix	Correction	ZAF

(b)

Element	Wt%	At%
OK	17.87	44.43
CoK	26.29	17.74
NiK	55.84	37.83
Matrix	Correction	ZAF

(c)

Element	Wt%	At%
OK	16.85	42.67
CoK	25.10	17.26
NiK	58.05	40.07
Matrix	Correction	ZAF

Fig. 4 Morphologies of $NiCo_2O_4$ at 400 °C with Sintering Duration of **a** 60 min, **b** 120 min, **c** 180 min

(Ni) was higher than Cobalt (Co) with the ratio of 2:1. However, differed from the theory, the oxygen level here was high due to impurities that causes an increase in the oxide level. Moreover, there were large-size items as the result of particle unification during the sintering process [18, 19]. All sample's images show that the particle sizes are nano, relatively consistent with the XRD method that applies the Scherrer formula [16, 17]. The increase in sintering temperature was also followed with the increase of particle size; which meant that the surface of other particles as impurities disappear [19–23].

3.3 Magnetic Properties

The VSM test results showed that at 400 °C, the highest saturation was found in the sample at 60 min duration with the value of 1.343 emu/g, whereas the highest retentivity was found in the sample at 120 min duration with the value of 0.140 emu/g and the highest coercivity was found in the sample at 180 min duration with the value of 0.0308 T. The samples at 500 °C had the highest magnetic saturation and coercivity in the sample with 180 min duration with the value of 4.781 emu/g and 0.0240 T and highest retentivity in the sample with 60 min duration with the value of 1.235 emu/g. The samples at 600 °C had the all the highest magnetic saturation, retentivity, and coercivity in the sample with 60 min duration with the value of 1.986 emu/g, 0.215 emu/g, and 0.0240 T. It can be observed that from all results that the highest magnetic saturation and retentivity occurred in the sample with 500 °C temperature variant and 180 min and 60 min duration with the value of 4.781 emu/g and 1.235 emu/g respectively while the highest coercivity occurred in the sample with 400 °C temperature and 180 min duration that was 0.0308 T (Figs. 5, 6 and 7; Tables 4, 5 and 6).

Figure 8 displays the curve characteristic where variants with 400 °C temperature has superparamagnetic property, supported by the small particles between 7 ± 2 nm and the highest coercivity value of 0.0308 T [24]. Meanwhile, variants with 500 °C have higher magnetic saturation and retentivity with the value of 4.781 emu/g and 0.0240 T, but lower coercivity with the value of 0.0240 emu/g. These phenomena was caused by the half of NiO phase that turned into antiferromagnetic [25, 26]. Variants at 600 °C also experienced a decrease of coercivity, magnetic saturation, and retentivity. This is because the increasing of sintering temperature caused the changing of NiO phase from ferromagnetic into antiferromagnetic and reduced the NiO phase [25, 27–29] (Table 7).

4 Conclusion

Synthesis and characterisation using the self-combustion method were successfully conducted. The phase identifications showed that the samples had single-phase as

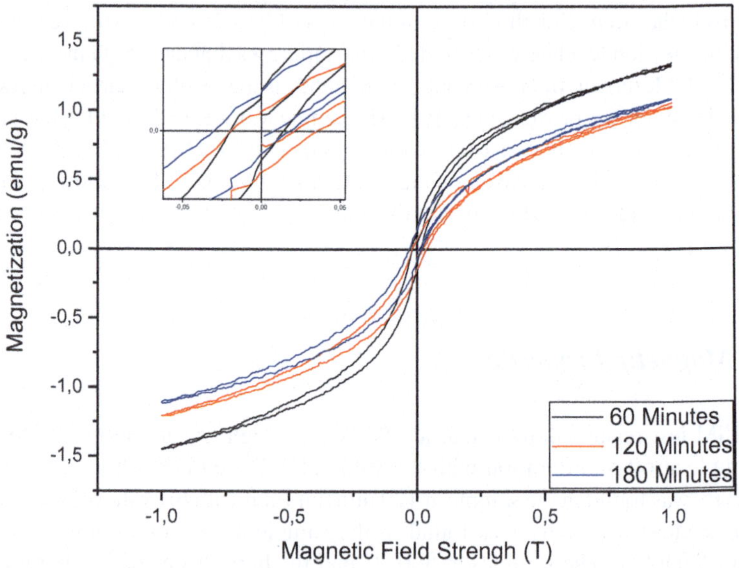

Fig. 5 Curves comparison of $NiCo_2O_4$ with different sintering time at 400 °C temperature

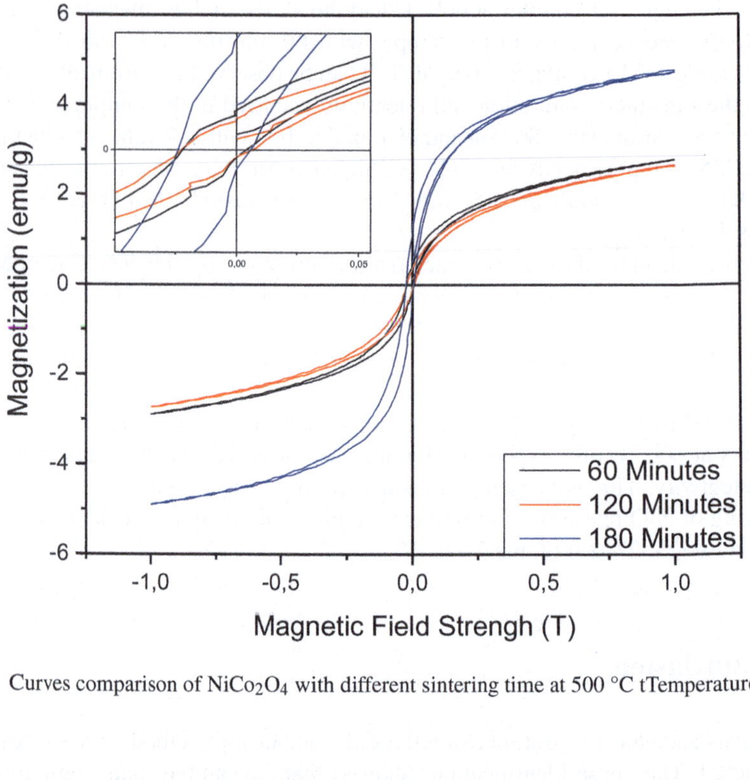

Fig. 6 Curves comparison of $NiCo_2O_4$ with different sintering time at 500 °C tTemperature

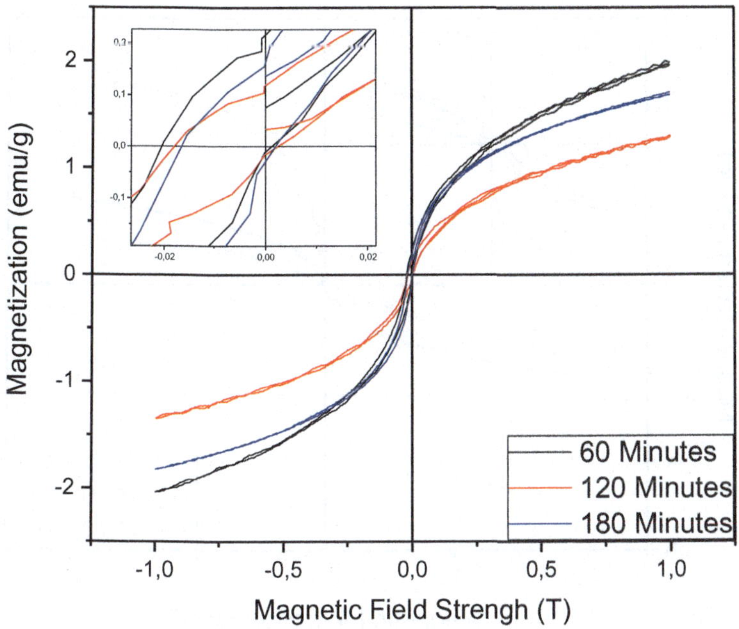

Fig. 7 Curves comparison of $NiCo_2O_4$ with different sintering time at 600 °C temperature

Table 4 Magnetic property analysis of $NiCo_2O_4$ based on hysteresis curve at 400 °C temperature

Sample (min)	Hc (T)	Mr (emu/g)	Ms (emu/g)
60	0.0202	0.128	1.343
120	0.0201	0.140	1.056
180	0.0308	0.072	1.086

Table 5 Magnetic property analysis of $NiCo_2O_4$ based on hysteresis curve at 500 °C temperature

Sample (min)	Hc (T)	Mr (emu/g)	Ms (emu/g)
60	0.0219	1.235	2.781
120	0.0227	0.438	2.669
180	0.0240	0.352	4.781

Table 6 Magnetic property analysis of $NiCo_2O_4$ based on hysteresis curve at 600 °C temperature

Sample (min)	Hc (T)	Mr (emu/g)	Ms (emu/g)
60	0.0204	0.215	1.986
120	0.0180	0.119	1.301
180	0.0166	0.162	1.706

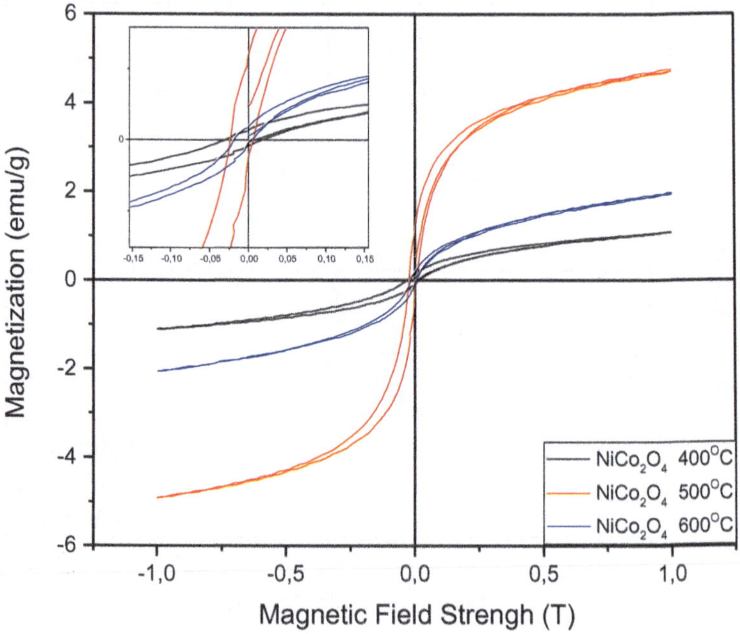

Fig. 8 Hysteresis Curves comparison of $NiCo_2O_4$ with different sintering temperature at 400, 500, 600 °C temperature

Table 7 Magnetic property analysis of $NiCo_2O_4$ based on hysteresis curve at 400, 500, 600 °C temperature

Sample (min)	Hc (T)	Mr (emu/g)	Ms (emu/g)
400 °C	0.0308	0.072	1.086
500 °C	0.0240	0.352	4.781
600 °C	0.0204	0.215	1.986

proved by the matching peak of $NiCo_2O_4$. All charts with sintering temperature variations displayed the $NiCo_2O_4$ nanoparticle in cubic structure centered on the surface, supported by the morphological analysis. The morphological analysis of $NiCo_2O_4$ presented the conversion from small particle become larger particle as the sintering temperature increases. The shape of $NiCo_2O_4$ shows as nanosphere powder, while the EDAX test resulted in the domination of Ni elements. The magnetic properties from the VSM test showed the highest coercivity at 400°C temperature and the highest retention as 500 °C temperature, while samples at 600 °C had a declination of coercivity, retentivity, and magnetic saturation. Samples with higher temperature had lower coercivity because of the decrease or disappearance of NiO phase from ferromagnetic into antiferromagnetic. Therefore, the best material for supercapacitor is $NiCo_2O_4$ sintered at 500 °C for 60 min.

Acknowledgements The author would like to thank the Ministry of Education, Culture, Research, and Technology, for Hibah Penelitian Dasar Unggulan Perguruan Tinggi (PDUPT) 2021 for PP. Also thanks to The Mineral and Advanced Material Central Laboratory of Auckland University of Technology, Dr. Yuan Tao.

References

1. Ozoemena, K.I., Chen, S. (eds.): Nanomaterials in Advanced Batteries and Supercapacitors. Springer International Publishing, Cham (2016)
2. Drew, J.: Supercapacitors Can Replace a Backup Battery for Power Ride-Through Applications
3. Di Ventra, M., Evoy, S., Heflin, J.R. (eds.): Introduction to Nanoscale Science and Technology. Kluwer Academic Publishers, Boston (2004)
4. Tehrani, Z., et al.: Large-area printed supercapacitor technology for low-cost domestic green energy storage. Energy **118**, 1313–1321 (2017). https://doi.org/10.1016/j.energy.2016.11.019
5. Levitt, A.S., Alhabeb, M., Hatter, C.B., Sarycheva, A., Dion, G., Gogotsi, Y.: Electrospun MXene/carbon nanofibers as supercapacitor electrodes. J. Mater. Chem. A **7**(1), 269–277 (2019). https://doi.org/10.1039/C8TA09810G
6. Sevilla, M., Diez, N., Ferrero, G.A., Fuertes, A.B.: Sustainable supercapacitor electrodes produced by the activation of biomass with sodium thiosulfate. Energy Storage Mater. **18**, 356–365 (2019). https://doi.org/10.1016/j.ensm.2019.01.023
7. Wu, Y.Q., Chen, X.Y., Ji, P.T., Zhou, Q.Q.: Sol–gel approach for controllable synthesis and electrochemical properties of $NiCo_2O_4$ crystals as electrode materials for application in supercapacitors. Electrochim. Acta **56**(22), 7517–7522 (2011). https://doi.org/10.1016/j.electacta.2011.06.101
8. Mondal, A.K., et al.: A microwave synthesis of mesoporous $NiCo_2O_4$ nanosheets as electrode materials for lithium-ion batteries and supercapacitors. ChemPhysChem **16**(1), 169–175 (2015). https://doi.org/10.1002/cphc.201402654
9. Bashir, S., Hanumandla, P., Huang, H.-Y., Liu, J.L.: Nanostructured materials for advanced energy conversion and storage devices: safety implications at end-of-life disposal. In: Li F, Bashir S, Liu L (eds.) Nanostructured Materials for Next-Generation Energy Storage and Conversion, pp. 517–542. Springer, Berlin (2018)
10. Wang, C., Sun, P., Qu, G., Yin, J., Xu, X.: Nickel/cobalt based materials for supercapacitors. Chin. Chem. Lett. **29**(12), 1731–1740 (2018). https://doi.org/10.1016/j.cclet.2018.12.005
11. de Fısica, D., de Microsco, S.: Mesoporous $NiCo_2O_4$ Spinel: Influence of Calcination Temperature over Phase Purity and Thermal Stability, p. 8. https://doi.org/10.1021/cg900648q
12. Kim, T.: Synthesis and characterization of $NiCo_2O_4$ nanoplates as efficient electrode materials for electrochemical supercapacitors, p. 29
13. Analysis of oxygen reduction reaction pathways on Co_3O_4, $NiCo_2O_4$, Co_3O_4-Li_2O, NiO, NiO-Li_2O, Pt, and Au electrodes in alkaline medium, p. 31
14. Liu, S.: Mesoporous $NiCo_2O_4$ nanoneedles grown on three dimensional graphene networks as binder-free electrode for high-performance lithium-ion batteries and supercapacitors, p. 30
15. Jiang, W., Hu, F., Yao, S., Sun, Z., Wu, X.: Hierarchical NiCo2O4 nanowalls composed of ultrathin nanosheets as electrode materials for supercapacitor and Li ion battery applications. Mater. Res. Bull. **93**, 303–309 (2017). https://doi.org/10.1016/j.materresbull.2017.05.036
16. Razak, J.A., Sufian, S., Ku Shaari, K.Z., Puspitasari, P., Hoe, T.K., Yahya, N.: Synthesis, characterization and application of $Y_3Fe_5O_{12}$ nanocatalyst for green production of NH_3 using magnetic induction method (MIM). In: Presented at the International Conference on Fundamental and Applied Sciences 2012: (ICFAS2012), Kuala Lumpur Convention Centre, Kuala Lumpur, Malaysia, pp. 633–638 (2012). https://doi.org/10.1063/1.4757548

17. Li, Q., Wang, L.-S., Hu, B.-Y., Yang, C., Zhou, L., Zhang, L.: Preparation and characterization of NiO nanoparticles through calcination of malate gel. Mater. Lett. **4** (2007)
18. Zhao, Y., Zhang, X., He, J., Zhang, L., Xia, M., Gao, F.: Morphology controlled synthesis of nickel cobalt oxide for supercapacitor application with enhanced cycling stability. Electrochim. Acta **174**, 51–56 (2015). https://doi.org/10.1016/j.electacta.2015.05.162
19. Naseri, M., et al.: Structure and physical properties of NiO/Co_3O_4 nanoparticles. Metals **6**(8), 181 (2016). https://doi.org/10.3390/met6080181
20. Hsu, C.-T.: Synthesis and characterization of mesoporous spinel $NiCo_2O_4$ using surfactant-assembled dispersion for asymmetric supercapacitors. J. Power Sources 10 (2013)
21. Kong, L.-B., Lu, C., Liu, M.-C., Luo, Y.-C., Kang, L.: Effect of surfactant on the morphology and capacitive performance of porous $NiCo_2O_4$. J. Solid State Electrochem 9 (2013)
22. Zhou, Q., et al.: Ordered assembly of $NiCo_2O_4$ Multiple hierarchical structures for high-performance pseudocapacitors. ACS Appl. Mater. Interfaces 9 (2014)
23. Xu, K., Yang, J., Li, S., Liu, Q., Hu, J.: Facile synthesis of hierarchical mesoporous $NiCo_2O_4$ nanoflowers with large specific surface area for high-performance supercapacitors. Mater. Lett. **187**, 129–132 (2017). https://doi.org/10.1016/j.matlet.2016.10.083
24. Verma, S., Joshi, H.M., Jagadale, T., Chawla, A., Chandra, R., Ogale, S.: Nearly monodispersed multifunctional $NiCo_2O_4$ spinel nanoparticles: magnetism, infrared transparency, and radiofrequency absorption. J. Phys. Chem. C **112**(39), 15106–15112 (2008). https://doi.org/10.1021/jp804923t
25. Zhu, C.M., Wang, L.G., Liu, F.C., Kong, W.J.: Exchange bias behaviors up to room temperature in $NiCo_2O_4$/NiO nanoparticle system. Ceram. Int. **45**(8), 9878–9883 (2019). https://doi.org/10.1016/j.ceramint.2019.02.028
26. Cabo, M., Pellicer, E., Rossinyol, E., Castell, O., Suriñach, S., Baró, M.D.: Mesoporous $NiCo_2O_4$ spinel: influence of calcination temperature over phase purity and thermal stability. Cryst. Growth Des. **9**(11), 4814–4821 (2009). https://doi.org/10.1021/cg900648q
27. Hao, C., Zhou, S., Wang, J., Wang, X., Gao, H., Ge, C.: Preparation of hierarchical spinel $NiCo_2O_4$ nanowires for high performance supercapacitors. Ind. Eng. Chem. Res. **57**(7), 2517–2525 (2018). https://doi.org/10.1021/acs.iecr.7b04412
28. Yang, X., et al.: Controllable synthesis and magnetic properties of hydrothermally synthesized $NiCo_2O_4$ nano-spheres. Ceram. Int. **43**(12), 8585–8589 (2017). https://doi.org/10.1016/j.ceramint.2017.03.121
29. Nakate, U.T., Kale, S.N.: Microwave assisted synthesis and characterizations of $NiCo_2O_4$ nanoplates and Electrical, magnetic properties. Mater. Today Proc. **3**(6), 1992–1998 (2016). https://doi.org/10.1016/j.matpr.2016.04.101

Microwave Absorption of Coconut Wasted Derived Activated Carbon

Hassan Soleimani, Jemilat Yetunde Yusuf, Noorhana Yahya, Amir Reza Sadrolhosseini, Maziyar Sabet, and Lawal Adebayo Lanre

Abstract Recently, electromagnetic (EM) interference as an adverse effect of the proliferation of technologies utilizing EM waves has become a serious problem. EM wave absorbers have been extensively utilized as an effective means of eliminating EMI. Activated carbon derived from agricultural waste has been identified as potential material to fabricate an ideal EM wave absorber. This study investigates activated carbon prepared from coconut fiber as an EM wave absorber at X-band frequency. Coconut fiber Activated carbon at 600 °C exhibited the highest surface area of 228.689 m^2/g and 5.9 nm number of pores which is beneficial for microwave absorption. COMSOL Multiphysics software was used to simulate the EM scattering parameters of the coconut fiber activated carbon. A good microwave absorption performance was observed minimum reflection loss value of −41.78 dB at 10.8 GHz at a thickness of 4.0 mm. This result shows that activated carbon prepared from coconut fiber can function as an EM wave absorber at X band frequency.

Keywords Microwave absorber · Activated carbon · Coconut fiber

H. Soleimani (✉) · J. Y. Yusuf · N. Yahya · L. A. Lanre
Fundamental and Applied Science Department, Universiti Teknologi PETRONAS, 32610 Seri Iskandar, Perak, Malaysia
e-mail: hassan.soleimani@utp.edu.my

J. Y. Yusuf
e-mail: jemilat_19001626@utp.edu.my

N. Yahya
e-mail: Noorhana.yahya@utp.edu.my

L. A. Lanre
e-mail: lawal_17010034@utp.edu.my

A. R. Sadrolhosseini
Department of Physics, Faculty of Science, University Putra Malaysia, 43400UPM Serdang, Selangor, Malaysia

M. Sabet
Department of Petroleum and Chemical Engineering, Jalan Tungku Link, Gadong, Universiti Teknologi Brunei (UTB), 1410 Darussalam, Brunei
e-mail: maziyar.sabet@utb.edu.bn

S. A. Abdul Karim et al. (eds.), *Proceedings of the 6th International Conference on Fundamental and Applied Sciences*, Springer Proceedings in Complexity,
https://doi.org/10.1007/978-981-16-4513-6_28

1 Introduction

The use of electronic gadgets and wireless communication devices has been of benefit to the whole world. However, their excessive usage has increased the menace of electromagnetic (EM) pollution and EM interference (EMI). EM pollution threatens human health whereas EMI disturbs the adequate function of electronic devices [1–4]. To tackle this problem, fabricating microwave absorbing materials (MAMs) has attracted a lot of research attention [5, 6]. MAMs can dissipate the EM wave by transforming the EM wave into heat and other forms of energy [7]. The mechanism of loss in EM wave absorption depends greatly on the dielectric loss and magnetic loss ability of the MAMs. The microwave absorption performance (MAP) of an absorber can be evaluated by the relative permittivity (ε_r) and relative permeability (μ_r) [8, 9].

$$\varepsilon_r = \varepsilon' - j\varepsilon'' \tag{1}$$

$$\mu_r = \mu' - j\mu'' \tag{2}$$

ε' is the real permittivity and μ' is the real permeability which determines the amount of energy storage. The ε'' is the imaginary part of the permittivity and μ'' is the imaginary permeability which indicates energy dissipation. The dissipation of energy arises from various loss mechanisms such as conduction, dipolar, resonance, and relaxation [10]. The degree of energy loss can be evaluated by the dielectric loss ($\tan\delta_\varepsilon$) and magnetic loss tangent ($\tan\delta_m$) [4, 11].

$$\tan\delta_m = \frac{\mu''}{\mu'} \tag{3}$$

$$\tan\delta_\varepsilon = \frac{\varepsilon''}{\varepsilon'} \tag{4}$$

Hence, for a material to effectively absorb EM waves it must possess good dielectric and/or magnetic loss ability. Carbon materials have been reported in several studies as good EM absorptive materials due to their unique properties including lightweight, large surface area, and good dielectric properties. Graphene [12–15] and carbon nanotubes (CNTs) [16, 17] have been reported by several researchers as good MAMs. However, their expensive and time-consuming method of preparation limits their wide application. Recently, agricultural wastes products such as rice husk, sugarcane bagasse, cotton, and walnut shell have been employed as carbon precursors to produce activated carbon. Activated carbons are prepared by simple carbonization of carbon precursors at high temperatures in the presence of an inert gas followed by the activation with chemical activating agents. Activated carbons have been widely utilized in different applications including MAMs due to their lightweight, low-density large surface area, low cost of preparation, easy processability, well-defined porosity, and good dielectric loss properties [18, 19]. Coconut waste is one of the most abundant agricultural waste in the world with Malaysia producing about 5280 kg of coconut wastes annually per hectare of land [20]. Coconut husk generally contains about 30% (by weight) of coconut peats and 70% coconut fiber. Coconut fiber contains about 38–50% of carbon and high content of lignin which makes them stable and durable[1]. Salleh et al. [21] reported the MAP of a single-layer coconut shell base absorber. The activated carbon absorber was prepared at different temperatures. A minimum RL peak of 22 dB was reached at 9.6 GHz at 6.4 mm thickness. Chen et al. [22] also, studied the MAP of rice husk ash calcined at different temp. Yew et al. [23] also reported the MAP of raw coconut peat. A minimum RL of -38.58 dB was achieved at 12.0 GHz. Seok et al. [24] also studied the dielectric properties of coconut shell activated carbon and coconut shell powder. The coconut shell activated carbon displayed a high dielectric constant of 7.24 compared to the powdered coconut shell with 3.77. This makes the coconut shell activated carbon a good EM absorber. In this work, the measured permeability and permittivity was used in simulating the MAP of the activated carbon derived from Coconut fiber (CF) at X-band frequency.

2 Methodology

2.1 Experimental Procedure

Sample Collection

Coconut husk was collected from a local coconut store at Sri Iskandar, Malaysia.

Sample Preparation

The CF was separated from the husk and cleaned with deionized water. The CF sample was oven-dried and grounded to a fine powder after drying.

Preparation of Activated carbon

Chemical Activation

The grounded CF was activated using KOH as an activating agent at a 1:1 impregnation ratio. 20 g of KOH was dissolved in 100 ml of deionized water. 20 g of CF was impregnated in the dissolved KOH solution and stirred for 6 h and dried overnight. The impregnation ratio is the dry weight of the powdered sample divided by the weight of the dry activating agent.

Carbonization of chemically Activated Sample

The impregnated sample was carbonized in a tubular furnace under steady flow of nitrogen gas (N_2) at 5 °C/min different temperatures (400, 500, and 600 °C for

1 h). The carbonized sample was collected when the furnace cooled down to room temperature (20 °C).

Neutralization

The CF activated carbon was neutralized by washing several times with 0.1 M HCL solution, then with deionized water several times until the pH is neutral. The sample was dried at 60 °C for 2 h. This is to ensure the sample is free from chemical traces.

Characterization

The crystallinity of the samples was studied using the XRD to identify the phase structure of the sample in the diffraction range of 10°–80°. The Renishaw Raman spectroscopy operated at room temperature with a wavelength excitation of 514 nm was used to study the state of carbon present in the sample. The surface area and pore size of the sample, the nitrogen isothermal desorption adsorption was analyzed using the Brunauer Emmett- Teller equation and BJH calculations. The EM characterization was also carried out to measure the permeability and permittivity of the samples.

Composite Preparation and Fabrication process

Paraffin wax was used as the matrix due to its transparency to the EM wave. It is well known that the MAM particles should be smaller than the skin depth [17–19]. Therefore, a 1:1 weight ratio of CF-activated carbon to paraffin was used for fabrication. A Rectangular mould with an inner dimension of 14 cm length, 2.3 cm width, and 1 cm height was used to fabricate the CF activated carbon into a rectangular-shaped microwave absorber. CF activated carbon composite of different thickness was fixed into the WR-90 waveguide adapter and the EM parameters were measured at the X-band frequency (8.2–12.4 GHz) using the Vector Network Analyzer (VNA) (Fig. 1).

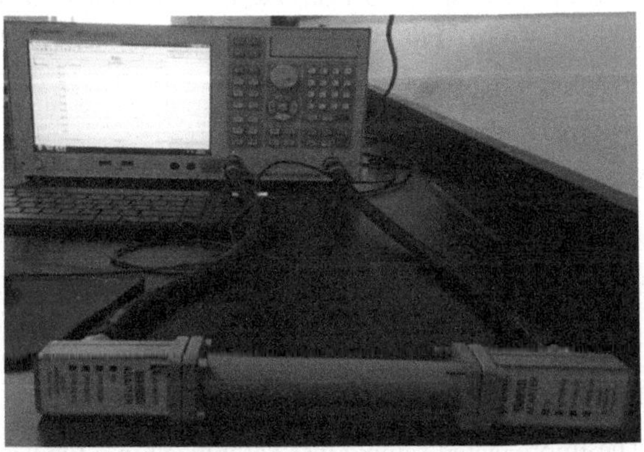

Fig. 1 A Vector Network Analyzer

COMSOL Multiphysics Simulation

COMSOL Multiphysics software based on the finite element method (FEM) was used to simulate rectangular waveguide with three dimensions (3D) geometry. The finite element method (FEM) is the numerical method used in the electromagnetic wave frequency domain and is governed by the equation below:

$$\nabla \times \mu_r^{-1}(\nabla \times \cup E) - k_0^2\left(\varepsilon_{r-}\frac{j\sigma}{\omega\epsilon_0}\right)E \tag{5}$$

where, μ_r and ε_r are the permeability and permittivity, σ is the conductivity, ω is the angular frequency, ε_0 is the permittivity of free space and k corresponds to wave vector. FEM is based on a spatial discretization, in a rectangular waveguide, the electric field was discretized using tetrahedron elements. The FEM analysis was used for simulating the EM properties of CF-activated carbon composite. The dimension of the rectangular waveguide was 2.3, 1 and 14 cm for width, height, and length respectively and the geometry set up was modeled into 3D geometry. Firstly, the cross-section of the waveguide was drawn on a work plane and then extruded to 3D. The 3D Rectangular waveguide was partitioned into three domains. Air was filled in domains 1 and 3. The CF activated carbon composite was filled in domain 2. The measured permittivity and permeability value of the CF activated carbon was used to estimate the refractive index which was input as the property of the material. Aluminum was selected and applied to all boundaries except for two boundaries. These two boundaries were labeled port 1 and port 2. Electromagnetic wave frequency was selected as the physics. Transverse electric TE_{10} mode was used as the propagating mode through the waveguide in this study. The transverse electric indicates that the electric field is transverse to the direction of the propagating wave. TE10 mode is known as the mode with the lowest cut-off frequency that can transmit through the rectangular waveguide. This model applies the RF module's port boundary conditions (BC) to solve the problem associated with wave propagation. The mesh (Fig. 2) was created for the rectangular waveguide and the electromagnetic properties were studied in the X band frequency. COMSOL Multiphysics software utilizes the BC to automatically generate the reflection loss (RL), reflection, and transmission coefficient values (S_{11} and S_{21}) [25].

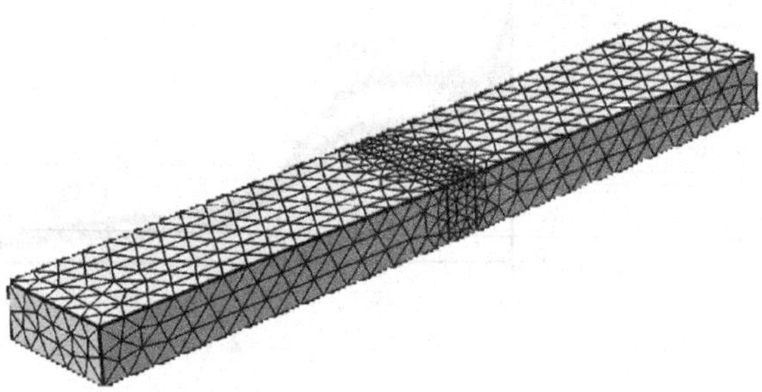

Fig. 2 Mesh of the CF activated carbon absorber

3 Results and Discussion

Figure 3 shows the XRD diffraction pattern of CF samples at different carbonization temperatures. The samples all exhibit one broad peak at (002) at 24° and one shoulder peak (100) at 44° plane reflection of graphite. The presence of these peaks could be ascribed to the amorphous nature of the samples. The broad peak observed in CF 400 is more obvious compared to that of CF 500 and CF 600. This may be due to the amorphous nature of CF 400.

RAMAN SPECTRA

To further clarify the state of carbon atoms in the CF samples, the Raman spectrum was studied. Figure 4 shows that all the samples exhibit two strong peaks at 1368 and 1590 cm^{-1} which can be attributed to the D and G band of carbon materials, respectively. The D band observed from the result agrees with the sp^3 hybridized carbon atom of distorted or amorphous carbon whereas the G band represents the sp2 hybridized carbon or graphitized carbon [18]. The I_D/I_G ratio represents the degree of graphitization in each sample. The I_D/I_G ratio of CF-400, CF-500, CF 600 are 1.15, 1.16, 1.17, respectively. It was observed that the degree of graphitization increases with an increase in carbonization temperature. More defects are introduced to the carbon as the temperature increases which is beneficial for EM absorption.

The Brunauer-Emmert-Teller (BET) SAP analysis was used to identify the pore distributions of CF and characterize their specific surface area. Liu et al. [22] reported that the surface area and pore size can influence the scattering parameter in EM absorption. Table 1 shows the BET calculation models and BJH calculation of surface

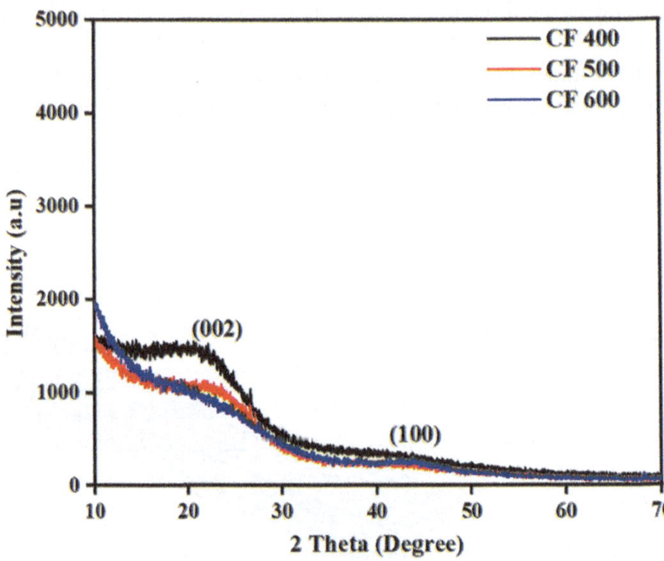

Fig. 3 XRD pattern for CF samples

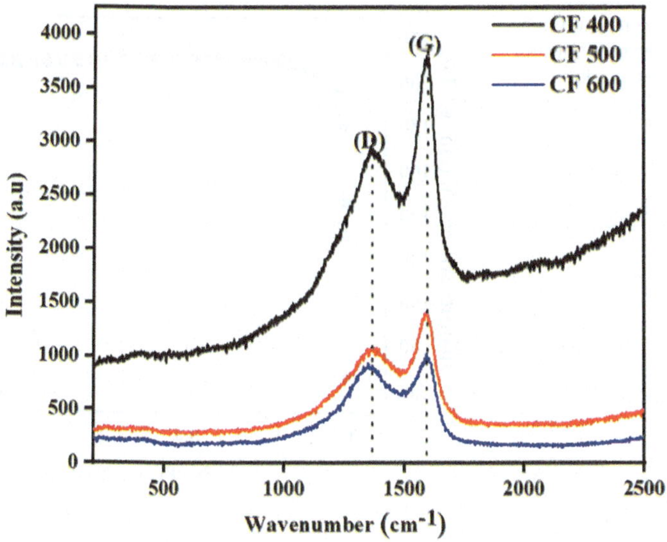

Fig. 4 Raman spectra of CF samples

Table 1 Nitrogen adsorption and desorption isotherms for CF at 600 °C

Samples	BET surface area (m^2/g)	Total pore volume (cm^3/g)	Pore size (nm)
CF 400 °C	139.395	0.07144	4.3
CF 500 °C	197.336	0.05541	5.4
CF 600 °C	228.689	0.15726	5.9

area and pore distribution at different temperatures. The BET surface area increases with an increase in temperature likewise the pore sizes also increase. This means the higher the temperature the more the surface area and pore size. CF 600 exhibits the highest surface area and number of pores. Figure 5 shows a type I isotherm Nitrogen adsorption and desorption with a hysteresis loop for CF activated carbon at 600 °C. The adsorption shows that a strong interaction exists between the sample surface and adsorbate. The rapid rise in Nitrogen uptake at low a relative pressure (0.0–0.9) indicates micropore adsorption [18]. The pore size was 5.9 nm which would extend the EM wave routes of propagation. The high surface area of CF 600 would effectively trap microwaves and form junctions which could increase interfacial polarization. Thus, enhancing microwave absorption.

Fig. 5 Nitrogen adsorption and desorption isotherms for CF carbonized at 600 °C

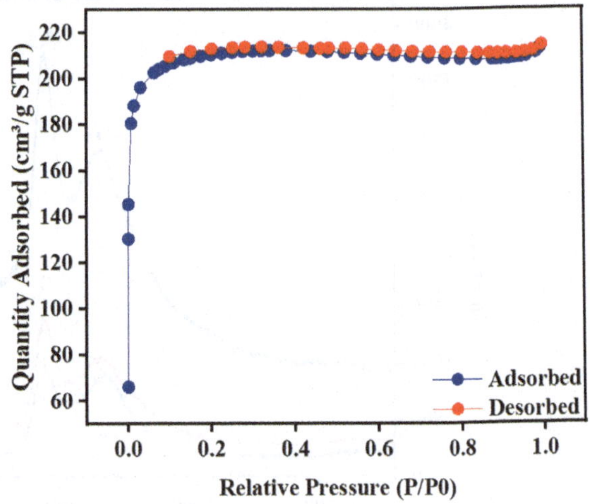

4 Microwave Aborption Characterization

Computational Simulation (COMSOL) Result

The computational simulation result obtained for CF 600 °C shows that the reflection coefficient increases at thin thickness as the thickness increases, this implies that the reflection coefficient gradually decreases to zero with a decrease in frequency. Whereas the transmission coefficient increases with increasing thickness of the sample. The reflection loss (RL) of the simulated CF 600 at a thickness of 2.0–5.0 mm is as shown in Fig. 6. At thickness of 2.0 and 3.0 mm, the RL loss value is less than −10 dB, this implies that CF 600 exhibit weak absorption intensity. At

Fig. 6 Frequency dependence of reflection loss of simulated CF 600

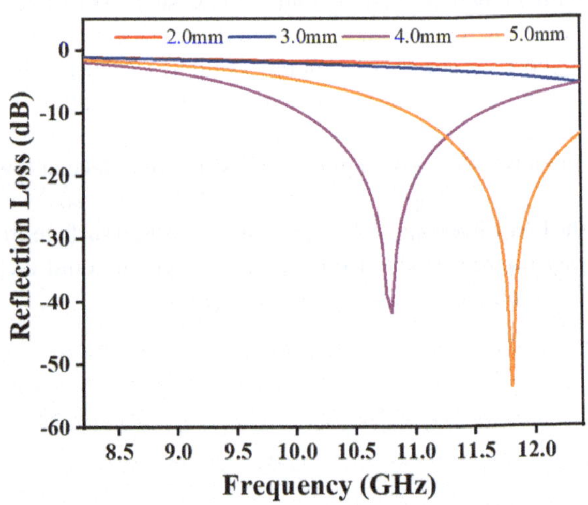

4.0 mm thickness, a minimum RL value of −41.78 dB was achieved at 10.8 GHz with EAB of 2.11 GHz. Whereas at 5.0 mm thickness, a minimum RL value of −53.5 dB was achieved at 11.8 GHz with an EAB of 2.0 GHz. This result shows that there is a good impedance matching between the CF 600 and free space and thus results in enhanced MAP.

5 Conclusion

Coconut Fiber activated carbon was prepared using a chemical activation method. The XRD and Raman spectra show that the obtained CF activated carbon is a good source of dielectric material. The presence of pores in the CF-activated carbon absorber may induce multiple reflection and scattering of EM waves. The MAP was simulated using the measured permittivity and permeability. The RL curve reveals that CF activated carbon possesses good microwave absorption ability with minimum RL values less than −20 dB. This means about 99.9% of the EM wave was successfully dissipated. Therefore, the use of agricultural waste as a microwave absorber due to their surface area and porosity provides an alternative way for waste management and opens a new route for the fabrication of green MAMs.

Acknowledgements The authors wish to acknowledge the support of the Yayasan Universiti Teknologi PETRONAS (YUTP) research grant through cost center 015LC0-143, and Universiti Teknologi PETRONAS.

References

1. Zhao, H., et al.: Biomass-derived porous carbon-based nanostructures for microwave absorption. Nano-Micro Lett. **11**(1) (2019). https://doi.org/10.1007/s40820-019-0255-3
2. Adebayo, L.L., et al.: Recent advances in the development OF Fe_3O_4-based microwave absorbing materials. Ceram. Int. **46**(2), 1249–1268 (2019). https://doi.org/10.1016/j.ceramint.2019.09.209
3. Soleimani, H., Abbas, Z., Yahya, N., Soleimani, H., Ghotbi, M.Y.: Microwave characterization of lanthanum iron garnet-filled PVDF-polymer composite using rectangular waveguide at X-band frequency. J. Compos. Mater. **46**(12), 1497–1501 (2012). https://doi.org/10.1177/0021998311423857
4. Panwar, R., Puthucheri, S., Singh, D., Agarwala, V., Lee, J.-R.: Microwave absorption properties of FSS-impacted composites as a broadband microwave absorber. Adv. Compos. Mater **26**(2), 99–113 (2016). https://doi.org/10.1080/09243046.2016.1232009
5. Singh, K., Ohlan, A., Bakhshi, A.K., Dhawan, S.K.: Synthesis of conducting ferromagnetic nanocomposite with improved microwave absorption properties. Mater. Chem. Phys. **119**(1–2), 201–207 (2010). https://doi.org/10.1016/j.matchemphys.2009.08.060
6. Melvin, G.J.H., Wang, Z., Ni, Q.-Q., Siambun, N.J., Rahman, M.M.: Electromagnetic wave absorption properties of rice husks carbonized at 2500 °C (2017)

7. Jian, X., et al.: Facile synthesis of Fe_3O_4/GCs composites and their enhanced microwave absorption properties. ACS Appl. Mater. Interfaces **8**(9), 6101–6109 (2016). https://doi.org/10.1021/acsami.6b00388
8. Wang, Y., Du, Y., Xu, P., Rong, Q., Han, X.: Recent advances in conjugated polymer-based microwave absorbing materials. Polymers (Basel) **9**(1) (2017). https://doi.org/10.3390/polym9010029
9. Wu, Z., et al.: Hierarchically porous carbons derived from biomasses with excellent microwave absorption performance. ACS Appl. Mater. Interfaces **10**(13), 11108–11115 (2018). https://doi.org/10.1021/acsami.7b17264
10. Wang, C.C., et al.: Overview of carbon nanostructures and nanocomposites for electromagnetic wave shielding. Carbon **140**, 696–733 (2018). https://doi.org/10.1016/j.carbon.2018.09.006
11. Adebayo, L.L., Soleimani, H., Yahya, N., Abbas, Z., Ridwan, A.T., Wahaab, F.A.: Investigation of the broadband microwave absorption of citric acid coated Fe_3O_4/PVDF composite using finite element method. Appl. Sci. **9**(18) (2019). https://doi.org/10.3390/app9183877
12. Meng, F., et al.: Graphene-based microwave absorbing composites: a review and prospective. Compos. B Eng. **137**, 260–277 (2018). https://doi.org/10.1016/j.compositesb.2017.11.023
13. Li, J.S., Xie, Y.Z., Lu, W.B., Chou, T.W.: Flexible electromagnetic wave absorbing composite based on 3D rGO-CNT-Fe_3O_4 ternary films. Carbon **129**, 76–84 (2018)
14. Park, S., et al.: The effect of concentration of graphene nanoplatelets on mechanical and electrical properties of reduced graphene oxide papers. Carbon **50**(12), 4573–4578 (2012). https://doi.org/10.1016/j.carbon.2012.05.042
15. Zhao, X., Li, J., Liu, J., Zhang, L., Yang, C.: Graphene decorated by $Co_2(OH)_3Cl$: a promising lightweight wideband microwave absorber. J. Alloys Compounds **811** (2019). https://doi.org/10.1016/j.jallcom.2019.151905
16. Shu, R., et al.: Facile preparation and microwave absorption properties of RGO/MWCNTs/$ZnFe_2O_4$ hybrid nanocomposites. J. Alloy. Compd. **743**, 163–174 (2018). https://doi.org/10.1016/j.jallcom.2018.02.016
17. Yu, T., et al.: One-pot hydrothermal synthesis of Co_3O_4/MWCNTs/graphene composites with enhanced microwave absorption in low frequency band. ChemNanoMat **5**(6), 847–857 (2019). https://doi.org/10.1002/cnma.201900173
18. Qiu, X., Wang, L., Zhu, H., Guan, Y., Zhang, Q.: Lightweight and efficient microwave absorbing materials based on walnut shell-derived nano-porous carbon. Nanoscale **9**(22), 7408–7418 (2017). https://doi.org/10.1039/c7nr02628e
19. Zhou, P., et al.: Walnut shell-derived nanoporous carbon@Fe_3O_4 composites for outstanding microwave absorption performance. J. Alloy. Compd. **805**, 1071–1080 (2019). https://doi.org/10.1016/j.jallcom.2019.07.177
20. Tan, I.A., Ahmad, A.L., Hameed, B.H.: Preparation of activated carbon from coconut husk: optimization study on removal of 2,4,6-trichlorophenol using response surface methodology. J. Hazard Mater. **153**(1–2), 709–717 (2008). https://doi.org/10.1016/j.jhazmat.2007.09.014
21. Salleh, M.K.M., Yahya, M., Awang, Z., Muhammas, W.N.W., Mozi, A.M., Yaacob, N.: Single layer coconut shell-based microwave absorbers. In: IEEE TENCON (2011)
22. Liu, S.-T., Chen, X.-G., Zhang, A.-B., Yan, K.-K., Ye, Y.: Electromagnetic performance of rice husk ash. BioResources **9**(2) (2014). https://doi.org/10.15376/biores.9.2.2328-2340
23. Yew, B.S., Wee, F.H.: Agricultural waste based-coco peat microwave absorber. Int. J. Eng. Sci. Emerg. Technol. **7**(2) (2014)
24. Ab Jabal, S.N., Seok, Y.B., Hoon, W.F.: Carbon composition, surface porosities and dielectric properties of coconut shell powder and coconut shell activated carbon composites. ARPN J. Eng. Appl. Sci. **11**(6) (2016)
25. Soleimani, H., Yahya, N., Abbas, Z., Soleimani, H., Zaid, H.M.: Determination reflection and transmission coefficients of lanthanum iron garnet filled PVDF-polymer nanocomposite using finite element method modeling at microwave frequencies. J. Nano Res. **21**, 151–157 (2012). https://doi.org/10.4028/www.scientific.net/JNanoR.21.151

Assessment of In-Depth Transport and Retention of Zinc Oxide Nanoparticles Using a Coreflood Approach

Mohammed Falalu Hamza, Hassan Soleimani, Abdelazim Abbas Ahmed, and Hassan Ali

Abstract The study is aimed at investigating the in-depth migration and retention of ZnO nanoparticles (NPs) used in enhanced oil recovery. The nanofluid (0.03%) was prepared according to a standard procedure of dispersing the NPs in brine. The hydrodynamic diameter and zeta potential (ζ) were investigated to ascertain the aggregation and stability of the NPs in brine, respectively. The flooding scenarios were maintained throughout at 0.2 mL min^{-1} flow rate and started by injecting 2 pore volume (*PV*) of brine followed by 3 PV of nanofluid and lastly another 3 PV of brine post-flush. The effluents were collected at around every 0.2 PV to ascertain the NPs breakthrough, while the pressure drop (Δp) across the core was recorded to assess the in-depth migration and retention of the NPs. The rock's permeability (K) and porosity (ϕ) before and after the flooding was investigated to further ascertain the impairment of the rock properties due to possible retention of the NPs. The results of the Δp show that nanofluid could generate Δp of 109 ± 19 mbar slightly higher than the waterflood baseline (107 ± 12 mbar) due to structural disjoining pressure resulting from spreading of the NPs in the core. However, large Δp (210 ± 29 mbar) was noticeable after brine post-flush signifying trapping of the NPs in the porous media. The NPs effluent concentration profile shows that a breakthrough had occurred after injecting 1.53 *PV* having a lower concentration than the initial injection. Further analysis of K and ϕ after the NPs injection revealed that the K and ϕ had reduced by 52 and 23.3%, respectively due to the NPs trapping. These observations could be related to high aggregation (415 ± 34 nm) of the ZnO in brine with moderate stability of − 35 ± 3 mV as described by the ζ.

M. F. Hamza (✉)
Department of Pure and Industrial Chemistry, Bayero University Kano, Kano 3011, Nigeria
e-mail: mfhamza.chm@buk.edu.ng

H. Soleimani · A. A. Ahmed
Institute of Hydrocarbon Recovery, Universiti Teknologi PETRONAS, 32610 Seri Iskandar, Perak Darul Ridzuan, Malaysia
e-mail: hassan.soleimani@utp.edu.my

H. Soleimani · H. Ali
Fundamental and Applied Science Department, Universiti Teknologi PETRONAS, 32610 Seri Iskandar, Perak Darul Ridzuan, Malaysia

S. A. Abdul Karim et al. (eds.), *Proceedings of the 6th International Conference on Fundamental and Applied Sciences*, Springer Proceedings in Complexity,
https://doi.org/10.1007/978-981-16-4513-6_29

Keywords Nanoparticles · Zinc oxide · Retention · Migration

1 Introduction

The rate of oil production by secondary water flooding could not meet up with the high oil demand because its operation leaves behind about 60–70% of oil in the reservoir [1, 2]. Enhanced oil recovery (EOR) was purposely introduced to extract additional oil left after water flood [3]. This technology has been widely accepted by oil companies and has attracted widespread interests for further investigations into practical application in the oil fields [4]. In this aspect, nanoparticles (NPs) have been researched and have recorded success as ideal candidates for improving oil recovery [5]. A number of scientific research studies shows that NPs account to about 4–51% incremental factor in the EOR operations [6–9]. Their ultra-small sizes typically in a nano range (1–100 nm) enabled them to have penetration power through the pore throats to alter the petrophysical properties such as wettability and interfacial force in order to mobilize the trapped oil [5]. However, it becomes paramount important to understand their transport behaviors in the reservoir and the degree of retention to predict their practical applications in the oil fields [10]. The mobility of the NPs dispersion and retention are the two important components constituting the transport of the NPs in the reservoirs. They give information about the fractions of the NPs that can survive and reach the oil zones [10]. Similarly, because of their relative small sizes, the transport is mostly governed by a dominant mechanism called 'random Brownian diffusion' [11]. Zinc oxide (ZnO) NPs is among the widely investigated particles because it demonstrates promising oil recovery at laboratory scale. For that reason, the present study is an attempt to make investigation into the migration and retention of the ZnO NPs that has been limitedly reported. The approach employed in this work is the concept of pressure gradient (Δp) before and after the NPs injection as well as changes in permeability (K) and porosity (ϕ) of the porous media to investigate the retention parameters. The research could be an avenue in providing with the scientific information guiding the selection of the ZnO NPs and may also be compared with other NPs for potential field applications.

2 Experimental

2.1 *Material and Methods*

The ZnO NPs with an average particle diameter of 18 nm was purchased from US Research Nanomaterials, Inc., USA and used as received without further purifications. The Sodium chloride used in the preparation of synthetic brine was purchased from Sigma Aldrich Malaysia. The 3 inches Berea-gray sandstone was purchased from Polygon Scientific Resources, Malaysia.

2.2 Nanofluid Preparation

Exactly 3% of brine was prepared by dissolving certain mass of NaCl in deionized water until completely dissolved. Thereafter, the predetermined mass of the NPs was dispersed in certain volumes of brine to make up 0.03% of nanofluid. Thereafter, the mixture was stirred using a magnetic stirrer for few hours until a homogenous solution has been ensured. Subsequently, the nanofluid was agitated using an Ultrasonic device (Ultrasonics Corporation, USA) at 25 °C for 1 h to minimize particle aggregation and *pH* was recorded as 7.47. The hydrodynamic diameter and zeta potential (ζ) of the NPs dispersion was monitored by particle size distribution (PSD)-Zeta potential equipment to ascertain the aggregation and dispersion stability of the NPs.

The calibration curves of different NPs concentrations (0.01–0.05%) in brine was established showing linear trend relationship with the refractive index to enable the determination of the NPs concentrations in the effluents received during flooding experiments.

2.3 Core Characterization

The core sample was firstly cleaned and dried in oven at 80 °C for 24 h, and subsequently the weight, length and diameter were recorded as shown in Table 1. The core was then saturated with the 3% brine using a desiccator. The saturation was achieved in 24 h under vacuum until no air bubbles from the core and constant weight was obtained [3]. Liquid pore volume (*PV*) was determined by mass difference before and after the saturation.

The POROPERM (CORVAL 30) Vinci technology instrument shown in Fig. 1 is a fully automated multi-samples permeameter and porosimeter dedicated to measure the porosity and permeability to helium gas under multiple confining pressures ranging from 400 to 10,000 psi. The reservoir, manifold and the core sample were filled with a Helium gas. After thermal equilibrium, the outlet valve was opened to initiate the pressure transient. When the upstream pressure is reduced to about 85% of the filled pressure, during which a smooth pressure profile is established throughout the core. The rock compressibility factors, fracture volume and real pore volume are subsequently computed in triplicate. The porosity and permeability measurements to helium gas were conducted before and after the coreflood displacement to determine the rock permeability and porosity impairments.

Similarly, the coreflood equipment (Vinci Technology, FES 350) shown in Fig. 2

Table 1 Sandstone core plug properties

Sandstone type	Length (mm)	Diameter (mm)	Dry weight (g)	Wet weight (g)	PV (mL)	$K_{(brine)}$ (mD)
Berea gray	76.04	38.04	182.19	199.75	17.56	101.7

Fig. 1 The POROPERM (CORVAL 30) Vinci technology instrument

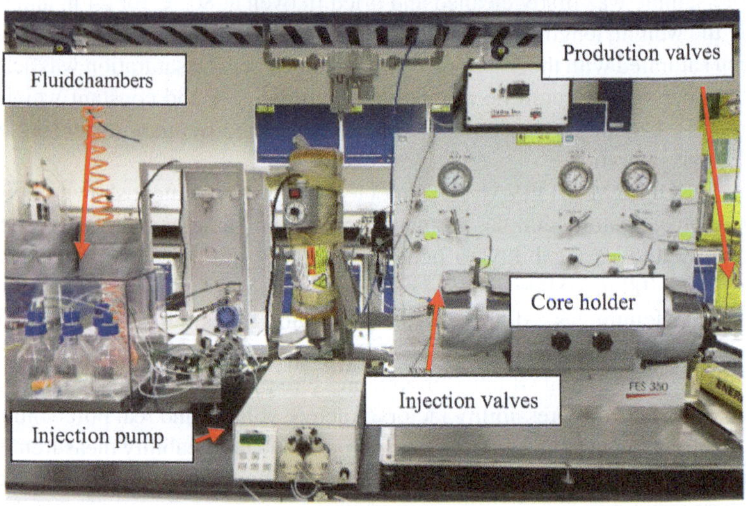

Fig. 2 The coreflood displacement equipment used to ascertain the migration of NPs in the porous media

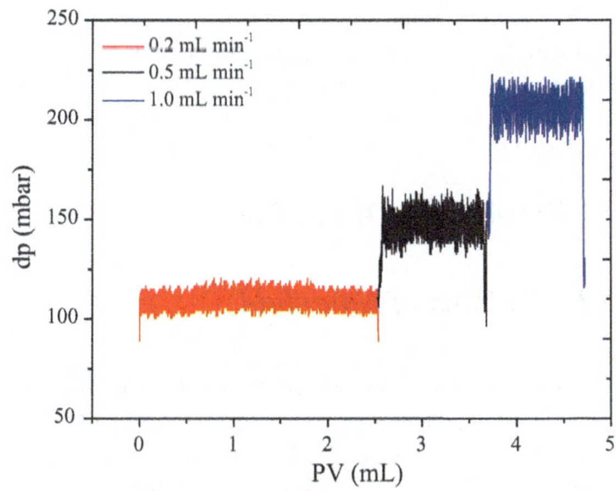

Fig. 3 The dp profile due to injection of brine at different flow rates

was used to determine the liquid permeability by flooding experiments at room temperature. Initially, several *PV* of brine was injected into the core at different flow rates of 0.2, 0.5, and 1.0 mL min^{-1} until the pressure drops (Δp) are stabilized.

The recorded Δp pressures shown in Fig. 3 were used in the Darcy equation given in Eq. 1 to calculate the core's permeability (K) due to brine [12].

$$q = \frac{KA}{\mu L} * \Delta p \tag{1}$$

where; K is permeability in (mD), q is the flow rate (ml s^{-1}), A is the core area (cm^2), L is the core length (cm), Δp is the differential pressure across the core and μ is the brine viscosity (cp).

2.4 *Assessment of NPs Migration and Retention*

The same core flood equipment was used to assess the NPs migration and retention in the porous body. It started by injecting about 2 *PV* of brine at 0.2 mL min^{-1} to establish a baseline Δp prior to nanofluid injection. Thereafter, 3 *PV* of nanofluid was injected at the same flow rate and the Δp was observed and recorded. Approximately, for each 0.2 *PV* nanofluid injection, the sample effluent was collected and analysed by the refractive index to determine the concentration of the NPs. Lastly, about 3 *PV* of brine post-flush was then re-injected by maintaining the flow rate until the NPs is undetected in the effluents at this stage. The Δp behaviour due to possible NPs retention in the cores was observed.

At the end of the experiment, the core was removed from the equipment, dried and weighed again to re-analyze the K and ϕ to investigate the core impairment due retention of the NPs in the cores.

3 Results and Discussion

3.1 Stability of Nanofluid

The hydrodynamic measurement of the nanofluid shown in Fig. 4 exhibit a wide particles size distribution between 250 and 1750 nm suggesting that high aggregation of the NPs had occurred in the brine despite sonication procedure. This means that, approximately, about 23 particles each having 18 nm size agglomerate together to form a cluster of particles with average diameters of 415 ± 34 nm due to interfacial interaction [11]. This may suggest that the stability of this nanofluid may not be sufficiently achieved due to a large aggregation observed, to achieve efficient NPs transport in the porous media the typical size must be less than 100 nm otherwise it is predicted to encounter transport difficulties [10].

This was further confirmed after conducting the ζ measurement which describes the potential difference between the mobile dispersion medium and the stationary layer of the dispersion medium attached to the dispersed particle [13]. Excellent nanofluid stability should demonstrate good ζ greater than ± 60 mV [14], However, the dispersion of the ZnO NPs in this work reports the average ζ of – 35 ± 3 mV as shown in Fig. 5, indicating moderate stability. The study shows that among the factors affecting the nanofluid stability are likely the pH and ionic strength [15] which could be the reason of the moderate stability observed in this study because of high brine ionic strength (3%) used. The use of this ionic strength is inevitable

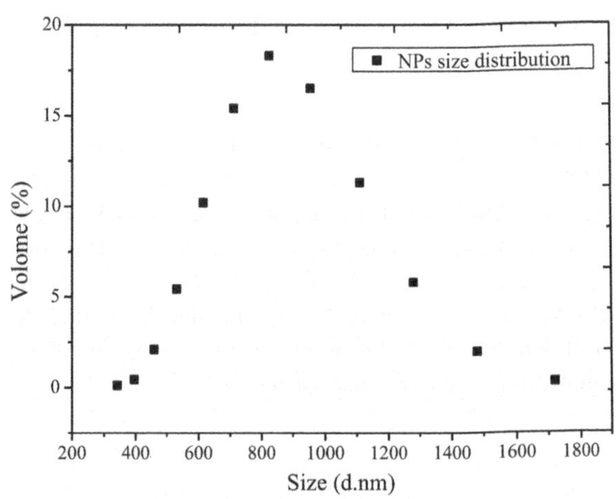

Fig. 4 Hydrodynamic aggregation of the NPs

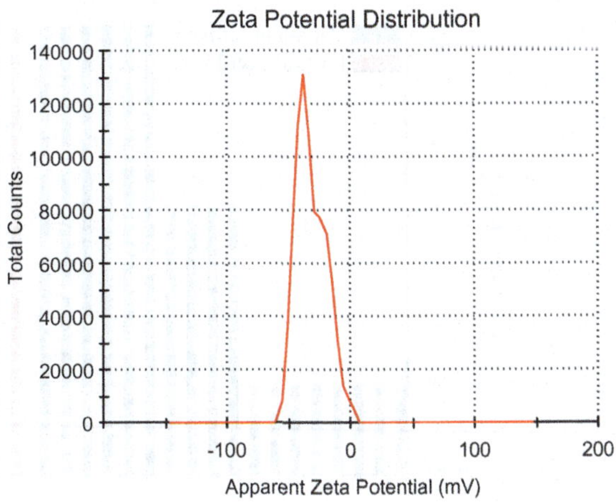

Fig. 5 The distribution of the ζ recorded for the dispersion of the ZnO NPs in brine

as most reservoirs are associated with high salinity, consequently, it is an ideal for investigation in this work.

3.2 *Migration and Retention*

Figure 6 shows the coreflood scenarios performed which enable the intriguing observations on the migration and retention of the NPs behaviors. The baseline Δp

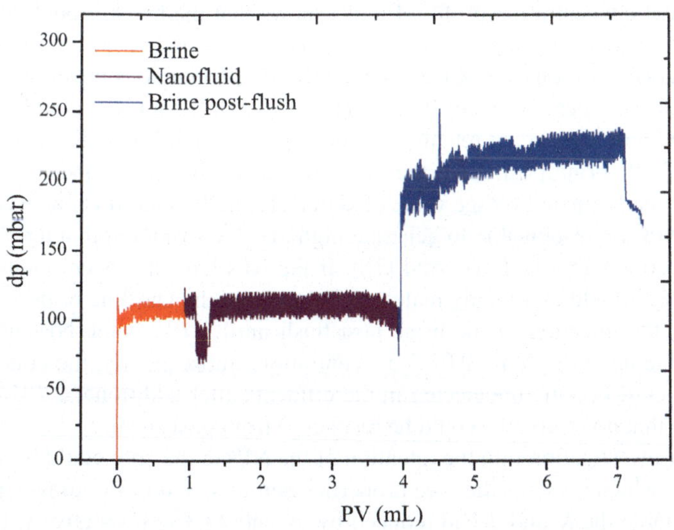

Fig. 6 The Δp showing migration and retention of the NPs in the core

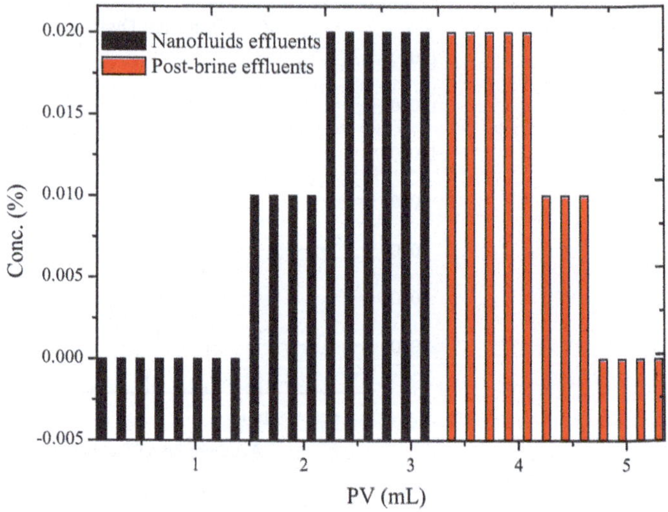

Fig. 7 The NPs effluents profile

observed due to 1 *PV* initial water injection established that waterflood could only maintain the Δp below 125 mbar with an average value of 107 ± 12 mbar. However, after the 3 *PV* nanofluid injections, the Δp increased slightly with an average of 109 ± 19 mbar resulting from the structural disjoining pressure due to partial migration and spreading of the NPs across the core [16]. Hendraningrat et al. [17] reports that nanofluid has a tendency to build appreciable Δp than the baseline waterflood which is attributed to the activity of the NPs at pore throats. Correspondingly, the NPs effluent concentration profile (Fig. 7) shows that the breakthrough of the NPs had occurred after injecting 1.53 *PV* but having 0.01% lower concentration than the injection concentration (0.03%). Subsequently, after continuous injection until 3 *PV* it produces at concentrations of 0.02%. This shows that ZnO NPs could slightly be transported with significant retention in the porous media. It is noticeable that after injecting 3 *PV* brine post-flush, the Δp significantly increased and maintained at 150–250 mbar with an average value of about 210 ± 29 mbar. It is believed that the retained NPs are responsible to generate high Δp due suspicion that a reduction in the core permeability had occurred [18]. If the NPs have not been retained in the core, the Δp should expectedly match with the Δp of the baseline brine.

During the injection of the brine post-flush until 1 *PV*, some NPs in the core were producing at the same 0.02% concentration. Subsequently, the concentration decreases until became undetected in the effluents after additional 2 *PV* injection, indicating that no more NPs could be recovered from post-flush.

Further investigation into the retention of the NPs in the core could be supported by Table 2 which presents the core properties before and after flooding experiment. From the table the K and ϕ had reduced by 52 and 23.3%, respectively, indicating that NPs had been retained [18]. Consequently, the mass of ZnO NPs retained in the

Table 2 Sandstone core plug properties after NPs injection

Parameters	Results
Dry weight after flooding (g)	183.84
$K_{(brine)}$ reduction (%)	52.00
$K_{(air)}$ reduction (%)	19.24
ϕ reduction (%)	23.25
NPs retention (g)	1.65

core was found to be 1.65 g. The weak migration and high retention of the NPs could be attributed to the high aggregation and moderate stability of the ZnO in brine as described by the PSD and ζ in this work, respectively.

4 Conclusion

This study is an attempt to make investigations into the migration and retention of the ZnO NPs commonly used in EOR. For economic reasons, the research is necessary to provide scientific information guiding the selection of the NPs for field applications. The work shows that ZnO NPs at 0.03% has moderate stability with high aggregation in brine (3%). Its transport in porous media is weak and demonstrates high retention. Thus, for further investigation, the pH and brine ionic strength should be controlled and its surface can be possibly modified to improve the dispersion stability and ultimately the migration distance. Furthermore, investigation into kinetic mechanisms such as 'random Brownian motion' under the thermal influence could help to ascertain the distance migration in the reservoirs.

Acknowledgements The authors acknowledged the support of Yayasan Universiti Teknologi PETRONAS (YUTP grant: 015LC0143).

References

1. Hamza, M.F., Sinnathambi, C.M., Merican, Z.M.A.: Recent advancement of hybrid materials used in chemical enhanced oil recovery (CEOR): a review. Mater. Sci. Eng. **206**(1), 012007 (2017)
2. Li, S.: An Experimental Investigation of Enhanced Oil Recovery Mechanisms in Nanofluid Injection Process. Ph.D. Thesis, Norwegian University of Science and Technology, Trond heim, Norway (2016)
3. Bila, A., Stensen, J.Å., Torsæter, O.: Experimental investigation of polymer-coated silica nanoparticles for enhanced oil recovery. Nanomaterials **9**(6), 822 (2019)
4. Hamza, M.F., Soleimani, H., Merican, Z.M.A., Sinnathambi, C.M., Stephen, K.D., Ahmad, A.A.: Nano-fluid viscosity screening and study of in situ foam pressure buildup at high-temperature high-pressure conditions. J. Petrol. Explor. Prod. Technol. 1–12 (2019)

5. Agista, M.N., Guo, K., Yu, Z.: A state-of-the-art review of nanoparticles application in petroleum with a focus on enhanced oil recovery. Appl. Sci. **8**(6), 871 (2018)
6. Cheraghian, G.: Effects of titanium dioxide nanoparticles on the efficiency of surfactant flooding of heavy oil in a glass micromodel. Petrol. Sci. Technol. **34**, 260–267 (2016)
7. Joonaki, E., Ghanaatian, S.: The application of nanofluids for enhanced oil recovery: effects on interfacial tension and coreflooding process. Petrol. Sci. Technol. **32**, 2599–2607 (2014)
8. Onyekonwu, M.O., Ogolo, N.A.: Investigting the use of nanoparticles in enhancing oil recovery. In: Proceedings of the Nigeria Annual International Conference and Exhibition, Tinapa-Calabar. Nigeria, Society of Petroleum Engineers, Richardson, TX, USA (2010)
9. Ehtesabi, H., Ahadian, M.M., Taghikhani, V.: Enhanced heavy oil recovery using TiO2 nanoparticles: investigation of deposition during transport in core plug. Energy Fuels **29**, 1–8 (2014)
10. Rodriguez Pin, E., Roberts, M., Yu, H., Huh, C., Bryant, S.L.: Enhanced migration of surface-treated nanoparticles in sedimentary rocks. In: SPE Annual Technical Conference and Exhibition. Society of Petroleum Engineers (2009)
11. Zhang, W., Crittenden, J., Li, K., Chen, Y.: Attachment efficiency of nanoparticle aggregation in aqueous dispersions: modeling and experimental validation. Environ. Sci. Technol. **46**(13), 7054–7062 (2012)
12. Lock, E., Ghasemi, M., Mostofi, M., Vamegh. R.: An experimental study of permeability determination in the lab. WIT Trans. Eng. Sci. 80, 221–230 (2012)
13. Lu, G.W., Gao, P.: Emulsions and microemulsions for topical and transdermal drug delivery. In: Handbook of Non-invasive Drug Delivery Systems, pp. 59–94. William Andrew Publishing (2010)
14. Kumar, A., Dixit, C.K.: Methods for characterization of nanoparticles, pp. 43–58. Advances in Nanomedicine for the Delivery of Therapeutic Nucleic Acids, Woodhead Publishing (2017)
15. Choudhary, R., Khurana, D., Kumar, A., Subudhi, S.: Stability analysis of Al2O3/water nanofluids. J. Exp. Nanosci. **12**(1), 140–151 (2017)
16. Li, S., Torsaeter, O., Lau, H.C., Hadia, N.J., Stubbs, L.P.: The impact of nanoparticle adsorption on transport and wettability alteration in water-wet berea sandstone: an experimental study. Front. Phys. **7**, 74 (2019)
17. Hendraningrat, L., Li, S., Torsæter, O.: A coreflood investigation of nanofluid enhanced oil recovery. J. Petrol. Sci. Eng. **111**, 128–138 (2013)
18. Aadland, R.C., Dziuba, C.J., Heggset, E.B., Syverud, K., Torsæter, O., Holt, T., Gates, I.D., Bryant, S.L.: Identification of nanocellulose retention characteristics in porous media. Nanomaterials **8**(7), 547 (2018)

Application of ANN Model in Sandstone Reservoir Using Electromagnetic Parameters for Predicting Recovery Factor

Surajudeen Sikiru, Hassan Soleimani, and Noorhana Yahya

Abstract Reservoir fluid is one of the major parameters that play a significant role in oil mobility, the fluid flow through the porous of the reservoir, the mechanism at which this fluid has been migrated are some of the major concerns in enhanced oil recovery industries. So, it of great significance to use an appropriate technique to calculate or to predict the oil recovery from the reservoir. The study aims to use an artificial neural network (ANN) model with electromagnetic parameters such as real and imaginary permittivity, real and imaginary permeability, and reflection loos at different concentrations of NaCl electrolyte. Deep neural network (DNN) approach with 280 nodes in each hidden layer and one output, it was revealed from the obtained result the correlation coefficient given by the DNN is R2 of 0.994 for estimated recovery factor and R2 of 0.894 for predicted recovery factor. This research result shows a good prediction of RF with the reservoir rock and fluid properties in terms of cost and production effectiveness.

Keywords Reservoir fluid · Electromagnetic parameters · Deep neural network (DNN) · Enhanced oil recovery

1 Introduction

The oil prediction of reservoir has been a major constrain for oil and gas industries. It is important factor to petroleum reservoir management [1, 2]. Primary recovery process is the use of natural energy source to lift the oil from the reservoir zone, there is a decline in the reservoir pressure due to long process of the oil from the same field.

S. Sikiru (✉) · H. Soleimani · N. Yahya
Department of Fundamental and Applied Sciences, Universiti Tecknologi PETRONAS, Seri Iskandar, PETRONAS, Malaysia

H. Soleimani
e-mail: hassan.soleimani@utp.edu.my

N. Yahya
e-mail: noorhana_yahya@utp.edu.my

S. A. Abdul Karim et al. (eds.), *Proceedings of the 6th International Conference on Fundamental and Applied Sciences*, Springer Proceedings in Complexity,
https://doi.org/10.1007/978-981-16-4513-6_30

At this stage oil production is not economically feasible, to diminish the deterioration of reservoir pressure, and drive the oil in the reservoir to the production well, the used of water flooding will be required known as secondary recovery mechanisms which entail the injection of water into the reservoir well. The declination in all filed in most of the world oil fields make it difficult to discovered new oil filed, in the process of primary recovery and secondary recovery techniques only 20–50% of the original oil-in place can be recovered [3–6]. There is still about 60% of the oil trapped in the reservoir zone after due process of primary and secondary recovery method due to high mobility of the injected water induced by low viscosity [7–11]. The mobility can be explained as the effective permeability of the reservoir divided by the viscosity of the water [12]. After primary recovery and secondary recovery, tertiary recovery will take place. Few methods in tertiary recovery or EOR are thermal, gas injection, chemical, and others. Thermal recovery methods accelerate hydrocarbon recovery by raising the temperature of the formation and reducing hydrocarbon viscosities [13–15].

The injection of chemicals in chemical methods is one of the famous methods in EOR. It is involving the use of long-chained molecules called polymers to increase the effectiveness of waterflood. The use of detergent-like surfactants also helps lower interfacial tension (IFT) that often prevents oil droplets from moving through a reservoir. Other's method is approaching new technology, which is a nanomaterial, microbial, acoustic, and electromagnetic. Few kinds of research and fewer pilot-scale application has been carried out, yet the technique still requires further research to prove their technical and economic viability. The electromagnetic (EM) technique utilizes radiofrequency and microwave regions to improve oil mobility in very deep formations, thin pay-zones, and low permeability formation [16–22]. Furthermore, EM is employed in nanotechnology EOR, which involves transferring of electrical energy from EM waves to the dielectric and resistive nanomaterials to reduce the oil interfacial tension and viscosity, thereby charge potential generated by EM waves causes separation of negative and positive ion in the reservoir environments e.g. rock surface cations positive charges [19, 23–26].

The artificial Neural network (ANN) is a function of the human brain. Naturally, the brain can learn new things and adapt to new environments. The brain has the most remarkable ability to analyze incomplete and unclear, fuzzy evidence, and make its own decision out of it. The human brain is the most primitive animal which has more capability than most advanced computers. The brain has different functions in the body system just like controlling the physical part of the body and some complex activities such as imagining, learning, thinking, visualizing, etc., also the activities that cannot be described in physical terms. An artificial thinking machine is still beyond the capacity of the most advanced supercomputers. The brain is made of cells called neurons. The structure of the brain is illustrated in Fig. 1. The interconnection of such cells (neurons) makes up the neural network or the brain. There are about 1011 neurons in the human brain and about 10,000 connections with each other. ANN is an imitation of the natural neural network where the artificial neurons are connected in a similar way as the brain network [27–29].

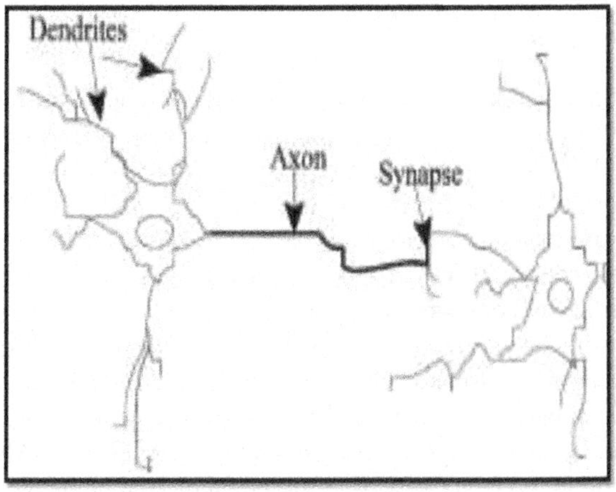

Fig. 1 Brain neuron change

Artificial Neural networks (ANN) is an information-processing system that has certain performance features in common with biological neural networks. A distinctive neural network is a multilayered system comprising of a single input layer, a single or double hidden layer, and a single output layer. Each layer is composed of basic processing elements called neurons. Each neuron is connected to the neurons of the adjacent layer with the connection eights between 0 and 1. The signals between the neurons are multiplied by the associated connection weights and added up together as Eq. 1, and then used as the net input of the neuron [30].

$$NET = \sum_{k=1}^{n} 1_k W^k \tag{1}$$

NET represents the net input of the neuron, l is the input variable, W is the connection weight, k is the index and n are the numbers of input variables. The activation function was applied to each neuron to its net input to determine its output signal and the signal transmitted to the next neuron. The sigmoid faction in Eq. 2. Is the activation function commonly used [30]

$$f(NET) = \frac{1}{1 + eNET^2} \tag{2}$$

This study present electromagnetic parameters in prediction of enhanced oil recovery using ANN model using real and imaginary permittivity, real and imaginary permeability, and X-ray diffraction (XRD) for the comparison between two electromagnetic data at different concentration of electrolyte from 3000 pp, 5000 ppm, 7000 ppm, 9000 ppm and 11,000 ppm, respectively. XPS characterization was used for the oil group present and the type of combination between carbon and oxygen, then XRF give the illustration of the elemental composition present on the sandstone after the infused of NaCl electrolyte.

1.1 Methodology

Saturated Berea sandstone was used to replicate the reservoir sandstone, prepared at different five concentration of NaCl electrolyte using 3000, 5000, 7000, 9000, and 11,000 ppm concentration, respectively [7, 8, 14]. The Berea sandstone were aged in brine solution for 15 days to create a spontaneous imbibition. We then proceeded in saturated in crude oil for another seven (7) days, the rock sample was put in autoclave vessels at 70 °C at 1000 psia pressure to replicate the reservoir condition. This was repeated for each sample at a specific concentration of electrolyte, each rock sample were dehydrated in oven at 50 °C for 12 h. The rock sample analysis using Network analyzer of Keysight N5071C for electromagnetic characterization. Also, we crushed in an agate mortar to increase the exposed surface area for the XRF analysis for the elemental composition. Artificial Neural network (ANN) with deep neural network module (DNN) were used for the prediction of the recovery factor. The total number of the data set involved in this analysis were 2456 for real and imaginary permittivity, real and imaginary permeability, reflection loss at each concentration of electrolyte with seven hidden layers. Then also the 60% of the data were used for the train the ANN model, 20% was used for the validation of the model and the remaining 20% were used to test the trained model for the better accuracy. Figure 2 show the ANN deep neural model used in the study.

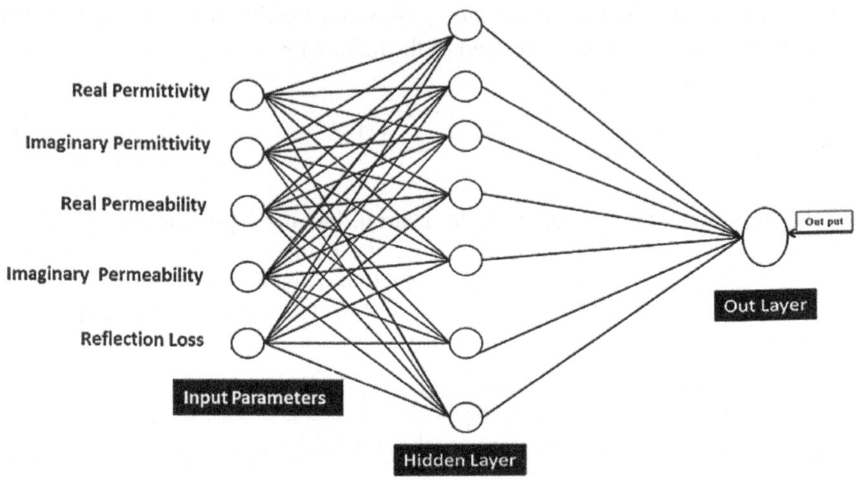

Fig. 2 Schematic diagram of deep neural networks used with 5 input parameters of electromagnetic properties obtained from different influx of NaCl electrolyte concentration

1.2 Result and Discussion

Minerology and elemental composition Analysis

The reservoir comprises of clays minerals which are majorly in the form of quartz, K-feldpar, albite, Kaolinite, chlorite, siderite, clionchlore, and muscovite. The peak from the XRD analysis after the influx of the electrolyte indicate quartz, clinochlore and kaolinite (Fig. 3) while the XRF shows the major composition of the reservoir sandstone which are Si, Al, Mg, fe, K, Na, Cl, and Ca (Table 1), then FTIR anal-

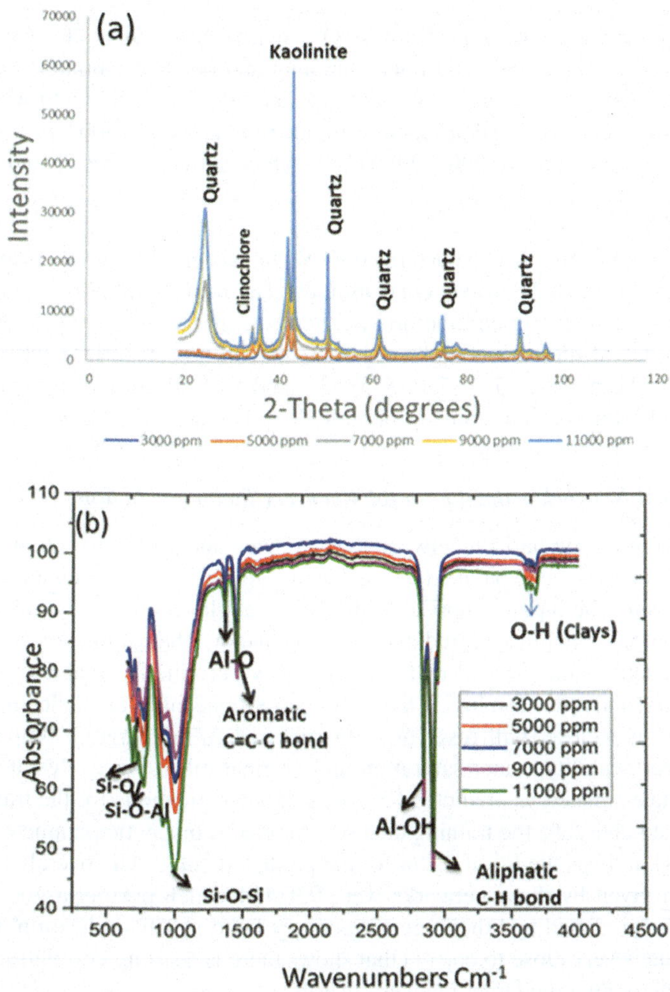

Fig. 3 **a** XRD spectral of the reservoir rock sample at different NaCl electrolyte concentration, **b** Fourier transform infrared spectroscopy (FTIR) of the rock sample for five different concentration of NaCl electrolyte

Table 1 XRF elemental composition

Concentration (ppm)	Elements (%)							
	Si	Na	Cl	Ca	Al	Mg	Fe	K
3000	69.3	3.72	2.10	1.20	6.61	0.22	6.05	7.80
5000	69.7	3.70	3.10	1.13	6.92	0.20	5.47	8.50
7000	68.4	2.10	3.20	0.99	6.62	0.36	6.98	7.40
9000	62.9	2.80	2.19	1.54	7.74	0.22	12.70	7.13
11,000	63.2	2.50	2.29	1.22	8.38	0.27	9.54	8.38

ysis shows the characteristic peaks of Si–O = quartz, Si–O–Si = siloxane function group, Si–O–A, O–H, and C–H bond starching and bending vibration were identified at 600 Cm^{-1} for 3000, 5000, 9000 pp, and 800 Cm^{-1}, Si–O–Si were found at 1000 Cm^{-1}, C=C–C bond aromatic were identified around 1500 Cm^{-1}, the C–H bond was identified around 2900–2980 Cm^{-1} which shown in (Fig. 3b and Table 1).

XPS analysis

Figure 4 show the result of characterized XPS analysis of C1s and O1s peaks for the saturated at different electrolyte concentration. The peak at 288.80 eV and 290 eV represented the aliphatic and the aromatic carbon merge with oxygen atoms, the linking modes of the carbon and oxygen atoms can be categorized into three (3) categories. The peaks at 534, 536, 535, 536, and 540 eV correspond wit carbons combine with the oxygen atoms through hydroxyls, carbonyl, and carboxyl group [31]. The XPS result analysis were summarized in Table 2 for better clarity.

Deep Neural Network propagation for Recovery factor prediction

Figure 5 shows a comparison between the estimated and predicated recovery factor from the electromagnetic parameters used in the analysis for both training data point test data point. The input comprised of five (5) input layer and seven hidden layers with one output which is the predicate recovery factor. The parameters comprise of electromagnetic parameters of real and imaginary permittivity, real and imaginary permeability, and reflection loss at five consecutive concentration of electrolyte. The hidden layer is enclosed with reservoir properties such as Temperature, fluid viscosity, porosity, permeability, water saturation and original oil in place. Result shows in Fig. 5 is the estimated vs predicted RF based on DNN methods for the training and the predicted data sets, the training data sets for electromagnetic parameters has the R^2 of 0.994 and has the R^2 of 0.894 for the predicted value. The overall coefficient correlation given by these networks were 0.92434 which portray a good relation between the predicted and the targeted recovery factors. This indication it because the R^2 value is very close to one (1) that shows there is a strong correlation, and the network can predict the target recovery factors.

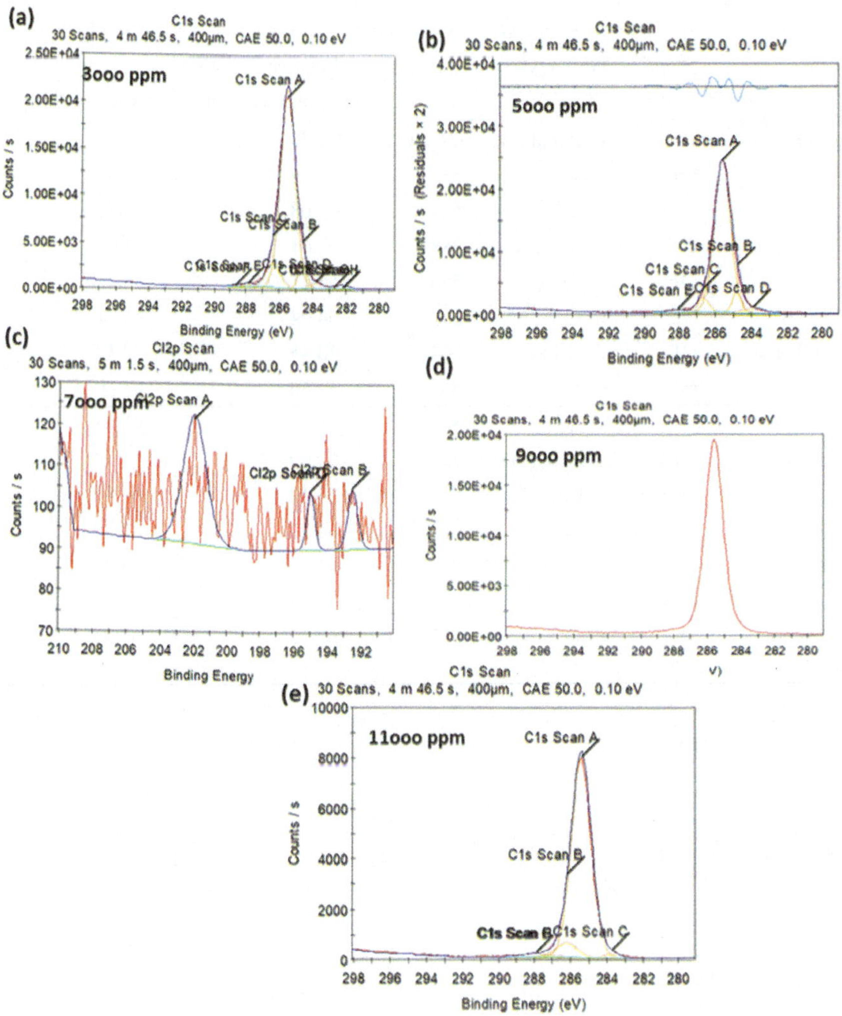

Fig. 4 XPS analysis of C1s and O1s for the saturated sandstone at five consecutive NaCl electrolyte concentration

2 Conclusion

This research work present DNN model for prediction of recovery factor using electromagnetic parameters based on reservoir rock and fluid properties. The study make used of five input parameter, seven (7) hidden layers and one (1) output, the input layers in generated from NaCl electrolyte concentration for the electromagnetic field parameters. It was revealed from the result obtained that the model successfully predicted with 70% accuracy. The estimated R^2 of 0.994 and predicted R^2 of 0.894 show that ANN techniques can be used to predict the performance of oil recovery.

Table 2 XPS analysis for the saturated sandstone at different NaCl electrolyte concentration

Con.	Name	Group type	B.E. (eV)	Relative Peak (%)	
				Before reaction	After reaction
3000 ppm	C1s	C(al and ar)	288.80	61.02	62.11
	O1s	C=O	534	32.06	32.09
		C–O	532	34.08	36.36
5000 ppm	C1s	C(al and ar)	288.50	62.61	63.01
	O1s	C=O	536	65.11	64.22
		C–O	530	64.04	65.11
7000 ppm	C1s	C(al and ar)	204	50.45	52.12
	O1s	C=O	535	62.56	63.80
		C–O	531	61.22	61.67
9000 ppm	C1s	C(al and ar)	284	63.09	63.78
	O1s	C=O	536	64.12	64.45
		C–O	528	64.23	63.89
11,000 ppm	C1s	C(al and ar)	290	64.23	64.69
	O1s	C=O	540	64.88	64.89
		C–O	535	62.12	62.48

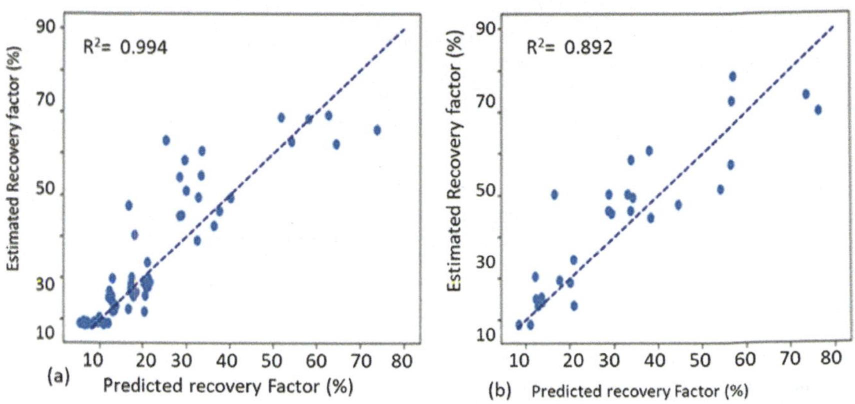

Fig. 5 Estimated versus predicted recovery factor by the DNN correlation using electromagnetic parameters

Acknowledgements The authors express their appreciation to PETRONAS research fund (PRF) for providing scholarship under Center for graduate studies.

References

1. Olalekan, S., Abdullahi, M., Olabisi, A.: Modeling of solar radiation using artificial neural network for renewable energy application. J. Appl. Phys. **10**, 6–12 (2018)
2. Kang, P.-S., Lim, J.-S., Huh, C.: Artificial neural network model to estimate the viscosity of polymer solutions for enhanced oil recovery. Appl. Sci. **6**, 188 (2016)
3. Kang, P.-S., Lim, J.-S., Huh, C.: Screening criteria and considerations of offshore enhanced oil recovery. Energies **9**, 44 (2016)
4. Ali, A.M., Yahya, N., Mijinyawa, A., Kwaya, M.Y., Sikiru, S.: Molecular simulation and microtextural characterization of quartz dissolution in sodium hydroxide. J. Petrol. Explor. Prod. Technol. 1–16 (2020)
5. Afeez, Y., Yahya, N., Nyuk, C.M., Al-qaseem, B., Qureshi, S., et al.: Investigation on Nanoparticles Effect on Interfacial Forces for Enhanced Oil Recovery (2019)
6. Yahya, N., Ali, A.M., Wahaab, F.A., Sikiru, S.: Spectroscopic analysis of the adsorption of carbon based nanoparticles on reservoir sandstones. J. Mater. Res. Technol. (2020)
7. Sikiru, S., Yahya, N., Soleimani, H.: Photon–phonon interaction of surface ionic adsorption within electric double layer in reservoir sandstone. J. Market. Res. **9**, 10957–10969 (2020)
8. Sikiru, S., Yahya, N., Soleimani, H., Ali, A.M., Afeez, Y.: Impact of ionic-electromagnetic field interaction on Maxwell-Wagner polarization in porous medium. J. Mol. Liquids 114039 (2020)
9. Surajudeen, S., Rostami, A., Soleimani, H., Yahya, N., Afeez, Y., Oluwaseyi, A., et al.: Graphene: outlook in the enhance oil recovery (EOR). J. Mol. Liquids 114519 (2020)
10. Yusuff, A.O., Yahya, N., Zakariya, M.A., Sikiru, S.: Investigations of graphene impact on oil mobility and physicochemical interaction with sandstone surface. J. Petrol. Sci. Eng. 108250 (2020)
11. Aliu, O., Sakidin, H., Foroozesh, J., Sikiru, S.: Determination of CO2 diffusivity in subsurface systems using molecular dynamics simulation. In: AIP Conference Proceedings, pp. 050014 (2020)
12. Sheng, J.J.: Modern Chemical Enhanced Oil Recovery: Theory and Practice. Gulf Professional Publishing (2010)
13. Siavashi, M., Doranehgard, M.H.J.A.T.E.: Particle Swarm Optimization of Thermal Enhanced Oil Recovery from Oilfields with Temperature Control, vol. 123, pp. 658–669 (2017)
14. Surajudeen, S., Yahya, N., Soleimani, H., Musa, A.A., Afeez, Y., Rostami, A.: Effect of adsorption on saturated sandstone within electric double layer on solid/liquid inter-phase. Petrol Coal **61** (2019)
15. Aliu, T., Sakidin, H., Yahya, N., Sikiru, S., Ali, A.M.: Dynamics of Nanoparticles Propagation in Porous Media (2019)
16. Sahni, A., Kumar, M., Knapp, R.B.: Electromagnetic heating methods for heavy oil reservoirs. In: SPE/AAPG Western Regional Meeting (2000)
17. Gunal, O.G.: Alteration of Asphaltic Crude Rheology with Electromagnetic and Ultrasonic Irradiation, vol. 26, pp. 263–272 (2000)
18. Yin, X., Kong, L., Zhang, L., Cheng, L., Travitzky, N., Greil, P.J.I.M.R.: Electromagnetic Properties of Si–C–N based Ceramics and Composites, vol. 59, pp. 326–355 (2014)
19. Zaid, H.M., Latiff, N.R.A., Yahya, N., Soleimani, H., Shafie, A.: Application of electromagnetic waves and dielectric nanoparticles in enhanced oil recovery. J. Nano Res. 135–142 (2014)
20. Latiff, N.R.A., Yahya, N., Zaid, H.M., Demiral, B.: Novel enhanced oil recovery method using dielectric zinc oxide nanoparticles activated by electromagnetic waves. In: 2011 National Postgraduate Conference, pp. 1–7 (2011)
21. Chakma, A., Jha, K.: Heavy-oil recovery from thin pay zones by electromagnetic heating. In: SPE Annual Technical Conference and Exhibition (1992)
22. Davletbaev, A., Kovaleva, L., Babadagli, T.: Mathematical modeling and field application of heavy oil recovery by radio-frequency electromagnetic stimulation. J. Petrol. Sci. Eng. **78**, 646–653 (2011)

23. Ross, S., Morrison, E.: Colloidal Systems and Interfaces (1988)
24. Zaid, H.M., Yahya, N., Latiff, N.R.A.: The effect of nanoparticles crystallite size on the recovery efficiency in dielectric nanofluid flooding. J. Nano Res. 103–108 (2013)
25. Soleimani, H., Yahya, N., Latiff, N.R.A., Zaid, H.M., Demiral, B., Amighian, J.: Novel enhanced oil recovery method using Co2+ xFe2+ 1-xFe3+ 2O4 as magnetic nanoparticles activated by electromagnetic waves. J. Nano Res. 111–116 (2014)
26. Alqasem, B., Sikiru, S., Ali, E.M., Rostami, A., Ganeson, M., Nyuk, C.M., et al.: Effect of electromagnetic energy on net spin orientation of nanocatalyst for enhanced green urea synthesis. J. Mater. Res. Technol. (2020)
27. Kukreja, H., Bharath, N., Siddesh, C., Kuldeep, S.J.I.J.A.R.I.I.E.: An Introduction to Artificial Neural Network, vol. 1, pp. 27–30 (2016)
28. Wang, S.-C.: Artificial neural network. In: Interdisciplinary Computing in Java Programming, pp. 81–100. Springer, Berlin (2003)
29. Olalekan, S., Abdullahi, M., Olabisi, A.J.J.o.A.P.: Modeling of Solar Radiation Using Artificial Neural Network for Renewable Energy Application, vol. 10, pp. 6–12 (2018)
30. Steffones, K., Tyagi, A.J.: Enhanced Oil Recovery Screening using Artificial Neural Network, vol. 15, p. 4 (2014)
31. Ma, Q., Yang, Z., Zhang, L., Lin, R., Wang, X.: Generation of hydrogen sulfide during the thermal enhanced oil recovery process under superheated steam conditions. RSC Adv. **9**, 33990–33996 (2019)

Design and FEA Analysis of MEMS Acoustic Sensors for High Voltage Applications

Farooq Ahmad, Ayesha Rafique, and Syed Muhammad Arslan Gilani

Abstract MEMS based acoustic sensor are playing vital role in science and technology for many years because of their miniaturization, low cost, low power consumption and high sensitivity. In this research article, MEMS acoustic sensor with optical readout is presented by optimizing and designing of membrane in such a way that it captures very low strength signal up to 0.2 Pa in resonance mode and relatively higher strength signal in static mode showing deflection in μm ~ pm range. The sensor membrane is designed using CMOS technology that is why it is more sensitive and easily mounted on high voltage power equipment, on multiple locations without electromagnetic interference. Fabrication of sensor performed in simulation environment using Intellisuite software with the help of deposition and etching technology that is more sensitive and less time consuming. Using ANSYS Workbench testing of proposed solution of MEMS based sensor in power cable gives the sensitivity of 500 nm/Pa at the resonance frequency of 2.04 MHz with quality factor of 33.124 and amplitude of 11.627 pm. In power transformer, sensitivity of 1700 pm/Pa is obtained at resonance frequency of 265 kHz with quality factor of 33.124 and amplitude of 92.39 pm.

Keywords MEMS · CMOS · UHF · AE · PD · UV · MPW

1 Introduction

Regular health mentoring of electrical apparatus is one of the important factors for reliable and continuous operation of power system. When power apparatus like overhead power cables, power transformers and switchgears are exposed to worst environment under pollution and high humidity, their insulation degradation occurs that reduces the insulation capacity of gas, liquid and solid insulation materials [1, 2]. Insolation degradation and poor design initially leads to partial discharge and if it persist for long time can cause flashover that is full line ground discharge.

F. Ahmad (✉) · A. Rafique · S. M. A. Gilani
Department of Electrical Engineering, University of Engineering and Technology, Lahore, Pakistan

S. A. Abdul Karim et al. (eds.), *Proceedings of the 6th International Conference on Fundamental and Applied Sciences*, Springer Proceedings in Complexity,
https://doi.org/10.1007/978-981-16-4513-6_31

Partial discharge does not fully bridge the insolation therefore online monitoring of PD phenomenon is important parameter for safety of power components because it reduces the risk of complete failure of power system. Long term partial discharge releases continuous energy that produces dangerous harmonics in the system [3, 4].

Corona, internal and surface discharges are the three main types of partial discharge and all of them have different information data and pattern [5]. Partial discharge is a random signal that bursts into different kind of energy signals like chemical signals, acoustic signals, electrical signals, electromagnetic signals and structural and physical changes around the affected insolation material. Partial discharge identifies with the help of many established conventional and unconventional methods. Electrical method is considered as conventional method and unconventional methods are surrounded with UHF, chemical, optical (interferometry fiber optic sensors and fiber Bragg gratings) and acoustic methods [6]. Each method is opted according to different apparatus being monitored. Acoustic method has certain advantages over other methods: quicker and easier instrumentation setup, very effective for location of PD source and immune to electromagnetic interference and noise. Conventional acoustic emission (AE) sensors have cost and size issues that are resolved by introducing micro-electro-mechanical-system (MEMS) technology with AE sensing technology. A research on related work has been reviewed and it is found that large number of MEMS AE sensor also used for PD detection having three basic transduction methods like piezoelectric, piezoresistive and capacitive [7]. Magnetic sensors can also use for PD detection [8].

Different sensors that use different design and working principle methodology such that Chemical detection, electrical detection, UHF detection and optical detection are bulky in size, consume large quantity of power, susceptible of electromagnetic noise and difficult integration with ICS.

Optical fiber-based sensors are not suitable for PD detection due to its complex setup. In this research article MEMS based acoustic sensor with optical readout designed, optimized, and FEA analysis performed for the power cables and power transformers.

2 Sensor Design

3D model of proposed acoustic sensor is shown in Fig. 1. The sensor structure includes three materials that are silicon substrate, SiO_2 layer at backside.

There is an Aluminum (Al) layer for making shining surface like mirror at front side. Sensor membrane consists of two layers SiO_2 and Al. When external stress applied on that membrane, it gets displaced.

In this research work, sensor was designed using certain materials such as silicon, silicon dioxide and aluminum. Selected material properties given in Table 1.

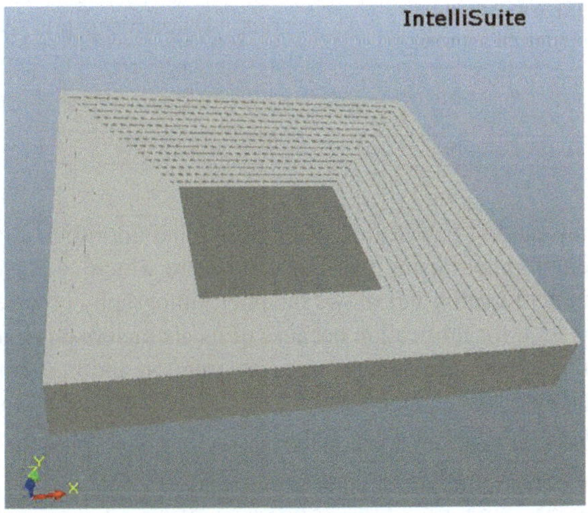

Fig. 1 3D model of design sensor

Table 1 Properties of selected materials

Sr.#	Materials/properties	Symbol	Young's modulus (GPa)	Poisson's ratio	Density (kg/m^3)
1	Silicon	Si	140	0.265	2.28
2	Silicon dioxide	SiO_2	67	0.15	2.65
3	Aluminum	Al	69	0.32	2.56

3 Optimization and Virtual Fabrication

In power cables frequency and pressure, range of acoustic signal produced by partial discharge is KHz to MHz and 0.5 Pa to 10 kPa respectively. As partial discharge is random signal so a certain frequency that is 2 MHz is selected for simulation and testing purposes. Similarly, pressure and frequency range of acoustic signal in power transformer is 0.2 Pa to 20 kPa and 20 kHz to 1 MHz respectively. Because of acoustic wave propagation signal strength becomes weak so average frequency is almost 260 kHz and that frequency is picked as resonance frequency for testing and simulation purposes of power transformer. Optimization is performed according to parameters and location of membrane in power cables. Parametric optimization of membrane is performed by varying dimension that is area and thickness of membrane. Choice of area and thickness is in such a way that it meets resonance frequency requirement and gives relatively maximum deflection. Optimized parameter of design sensor membrane for cable and transformer is given in Table 2.

The proposed sensor is compatible with MPW (Multi Project Wafer) approach [9]. Fast modification and development of packaging and fabrication technologies allowed the production of MEMS devices such that MEMS acoustic sensors and

Table 2 Membrane optimized parameters

Sr.#	Dimensions (μm)	Power cable	Power transformer
1	Length	250	350
2	Width	250	350
3	Thickness	2.5	3

resonators. Fabrication of proposed and optimized sensor membrane in simulation environment using Intelisuite software which is considered the best tool that have many features and shows the photolithography process in steps. CMOS technology is used for fabrication because of its characteristics like low propagation delay, high noise immunity, low power consumption and high input impedance [10, 11]. Side and cross-sectional view of fabrication is shown in Fig. 2. Photolithography process contains the following steps.

- The definition of silicon substrate.
- Thermal deposition of SiO_2 on silicon substrate.
- Deposition of photoresist by spin coating method on opposite side of Si substrate.
- Exposure of UV on specific part of photoresist that is the required area of etching.
- Wet etching of Si substrate for the development of model and sensor membrane
- Wet etching of extra photoresist that is not required.
- Deposition of Al on the front side by sputtering method to make the surface shining like mirror so that light source easily reflect making 90° angle.

4 Simulation Results and Discussion

Simulated and theoretical results in power cables and power transformers are explained using ANSYS and MATLAB software. Optimization of sensor membrane is performed by changing dimensions that is area and thickness or location of sensor in power cables and power transformers. While relationship between thickness and deflection is reverse as compared to direct relation between area and deflection that is shown in Figs. 3 and 4.

Area and resonance frequency of sensor membrane is inversely proportional to each other on the other hand thickness is directly proportional to the resonance frequency that is observed from Figs. 5 and 6.

Optimized location of membrane for the transformer is at the center of the transformer inside the transformer tank and for power cable perpendicular placement of membrane is exact location where membrane gives the maximum deflection and meets the resonance frequency criteria.

Optimized and fabricated sensor membrane is placed at the optimized location as shown in Fig. 7. Real environment in power transformer is created using ACT acoustic extension to make transformer tank as acoustic body. Acoustic pressure strength of partial discharge is 0.2 Pa to 20 kPa approximately or higher so we assign transformer tank to the acoustic body and provide that pressure to the membrane through acoustic

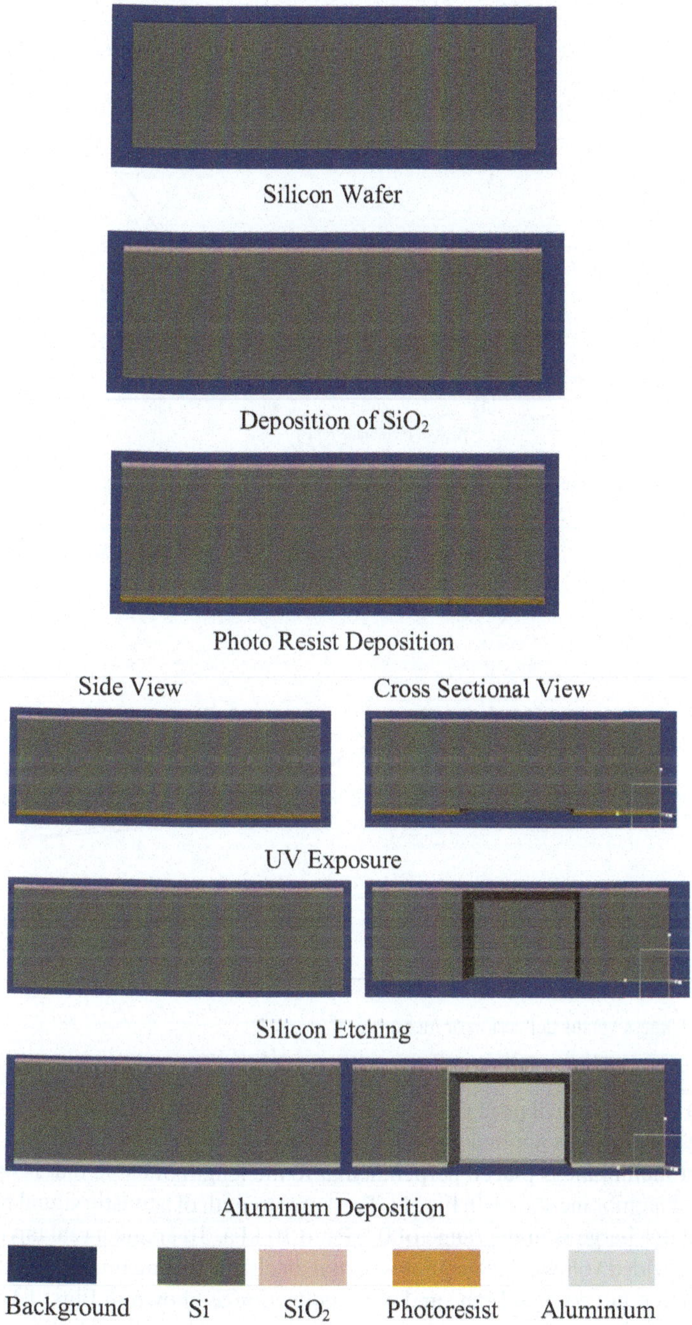

Fig. 2 Fabrication steps in Intellisuite software

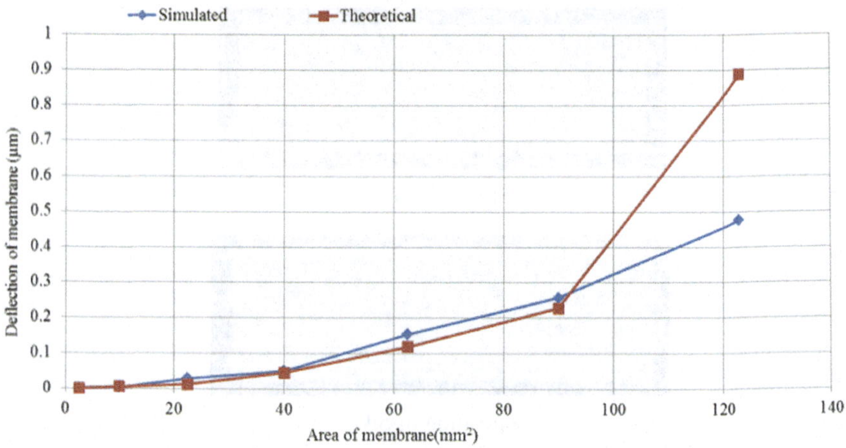

Fig. 3 Area versus deflection of sensor membrane

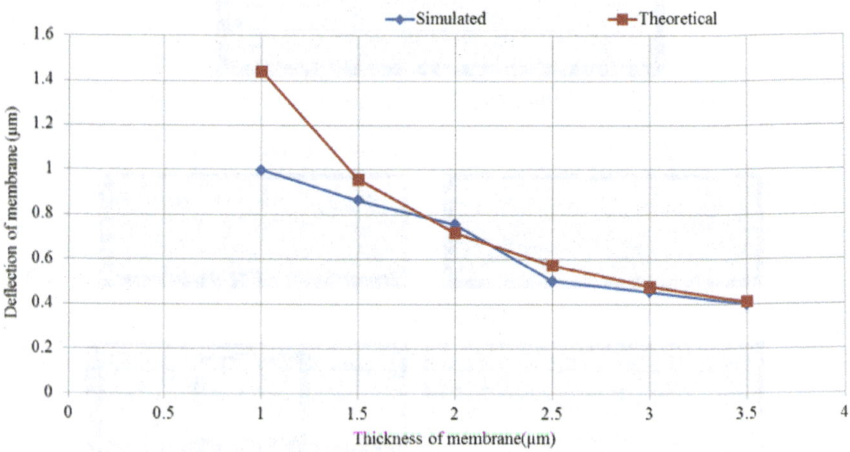

Fig. 4 Thickness versus deflection of membrane

body as a consequence it produce deflection shown in Fig. 8. Deflection of membrane gradually increases as applied pressure increases shown in Figs. 9 and 10.

Sensor membrane is placed perpendicular to the length of cable that is optimized location of membrane shown in Fig. 11. Pressure strength of acoustic signal generated by partial discharge is in the range of 0.5 Pa to 10 kPa. To create a real environment cable is provided voltage that generates partial discharge having pressure in the 0.5 Pa to 10 kPa that fall on membrane and gives deflection as shown in Figs. 12 and 13.

Voltage in electrical system and pressure in mechanical system they are analogous to each other. Calibration curve between voltage and pressure is obtained to identify how much voltage is needed to generate PD signal in the range of 0.5 Pa to 10 kPa

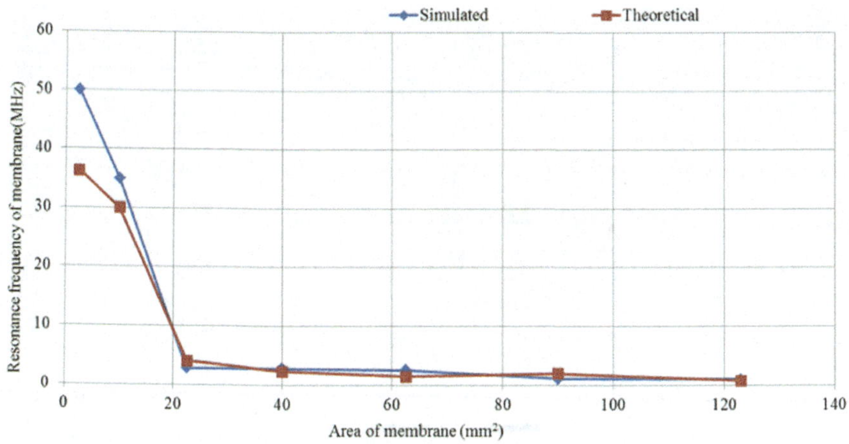

Fig. 5 Area versus resonance frequency of sensor membrane

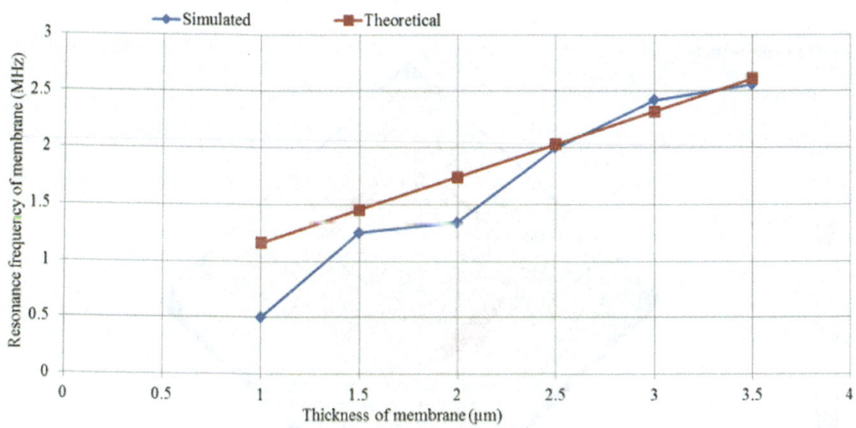

Fig. 6 Thickness versus resonance frequency of sensor membrane

through the help of ANSYS software. Calibration curve is represented in Fig. 14. Figure 15 shows that according to calibration curve pressure obtained respective of the pressure range in literature review that pressure is applied on sensor membrane and pressure verses deflection curve of sensor membrane is achieved.

Theoretical and simulated results obtained from ANSYS and MATLAB respectively compared that shows minimal difference between the two values. Tables 3 and 4 shows the comparison between the data obtained from ANSYS and MATLAB.

Theoretical and simulated results obtained from ANSYS and MATLAB respectively compared that shows minimal difference between the two values. Tables 3 and 4 shows the comparison between the data obtained from ANSYS and MATLAB.

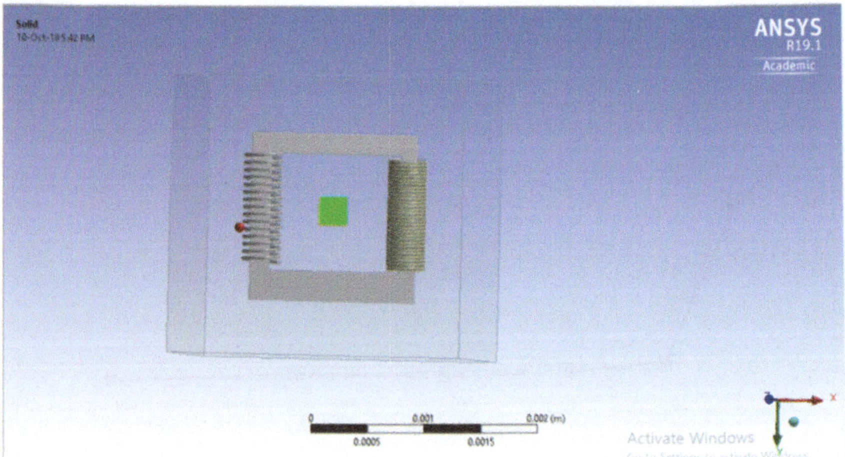

Fig. 7 Optimized sensor membrane at optimized location

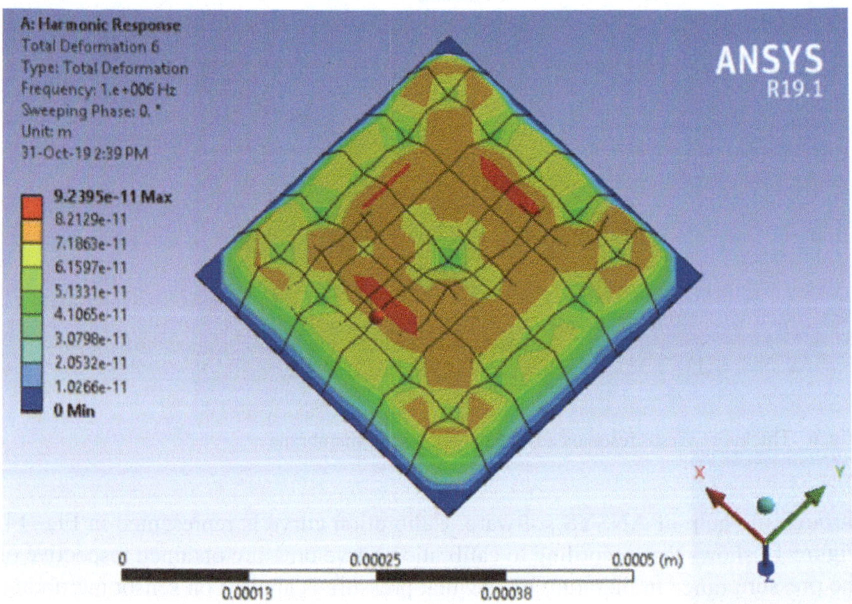

Fig. 8 Deflection of membrane due to applied pressure

5 Conclusions

In this research article, MEMS based acoustic sensor designed and explained for the power cables and power transformers in the simulation. Design, micro fabrication

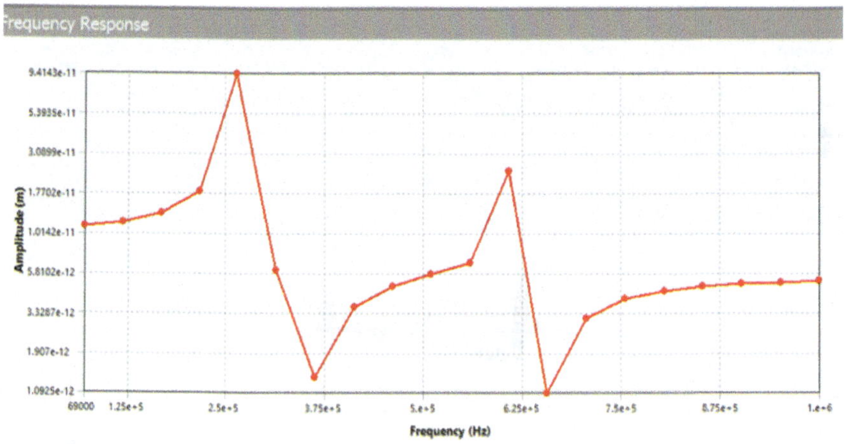

Fig. 9 Resonance frequency versus amplitude of deflection

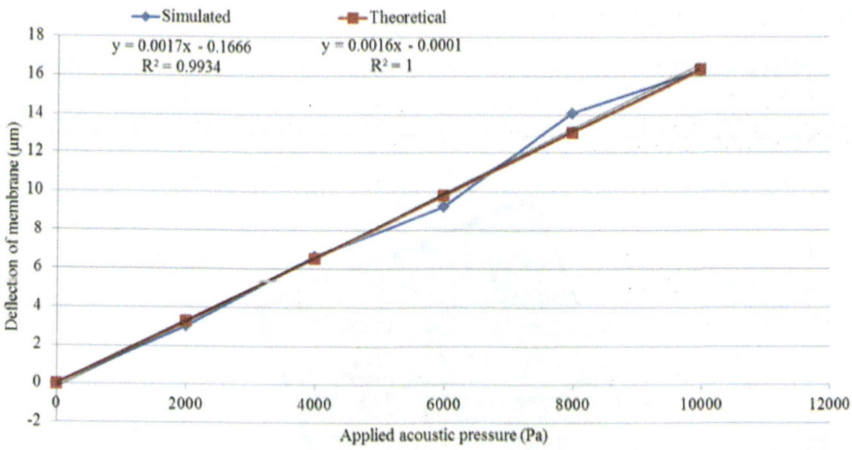

Fig. 10 Applied acoustic pressure versus deflection of membrane

and testing of MEMS based acoustic sensor with optical readout is explained for high voltage application such that power cables and power transformers in the simulation environment by the help of ANSYS, Intellisuite and MATLAB software. Sensor is tested for power cable and simulation is performed for 0.5 Pa acoustic pressure that is lower limit of acoustic pressure. 0.5 Pa pressure is applied on the membrane that gives deflection of 92.39 pm with quality factor of 33.124 which is compared by theoretical value of amplitude that is 90.02 pm with quality factor 32. From the Table 4 it is shows that sensitivity difference is 5.8% that clearly verified the design of sensor.

Testing of sensor is than performed for power transformer at lower limit of acoustic pressure signal that is 0.2 Pa that is applied on the membrane and gives deflection with

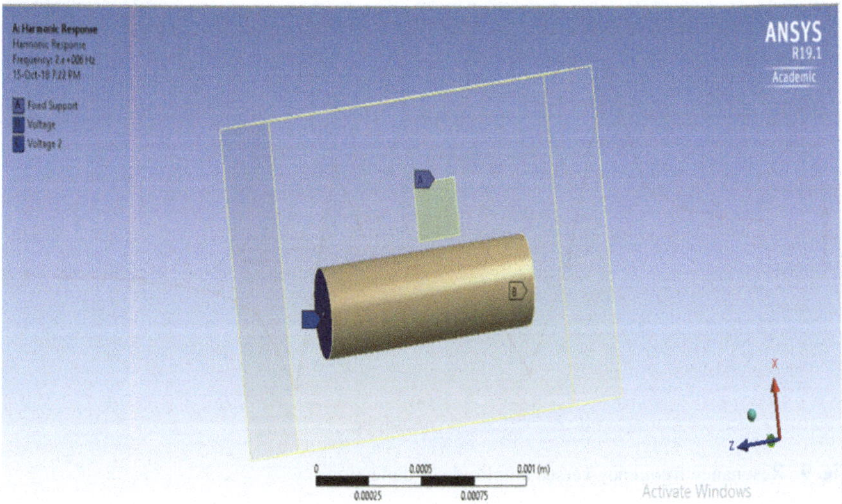

Fig. 11 Optimized membrane at optimized location

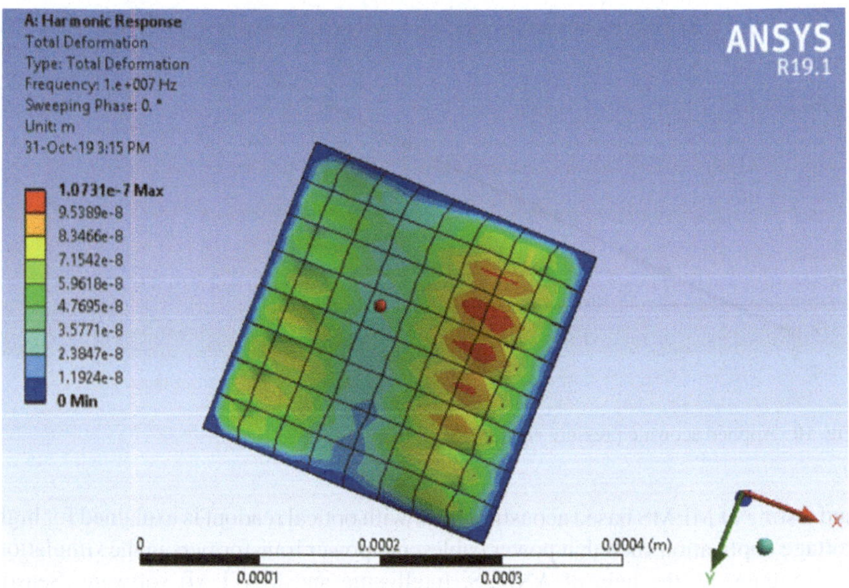

Fig. 12 Deflection of membrane due to applied acoustic pressure

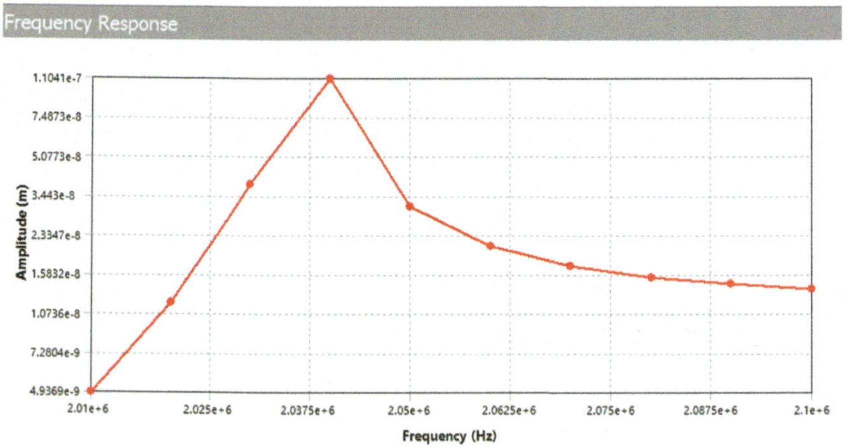

Fig. 13 Resonance frequency versus amplitude of deflection

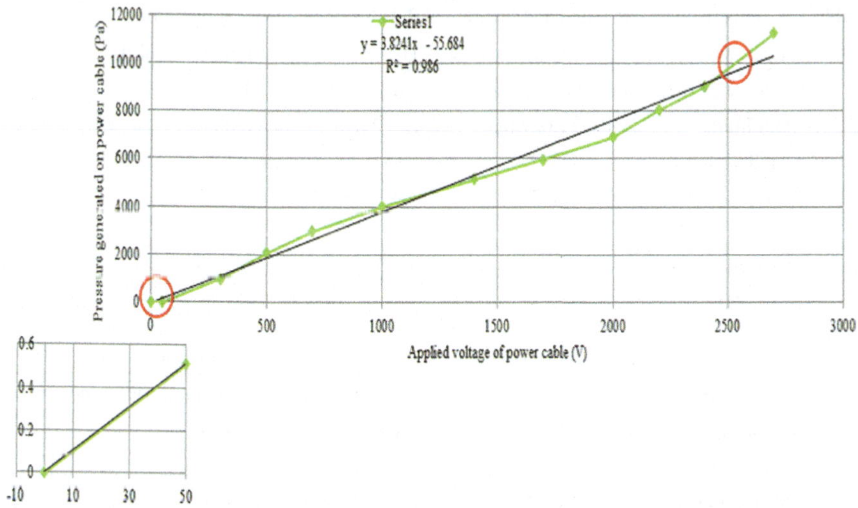

Fig. 14 Calibration curve of voltage versus acoustic pressure

amplitude 11.627 μm with quality factor of 40.803 and compared with theoretical results that is 12.70 μm with quality factor 39.5354 in power cable. Requirements, that sensor sense the lower limit of acoustic pressure with relatively higher and measurable amplitude, higher sensitivity and quality factor is achieved.

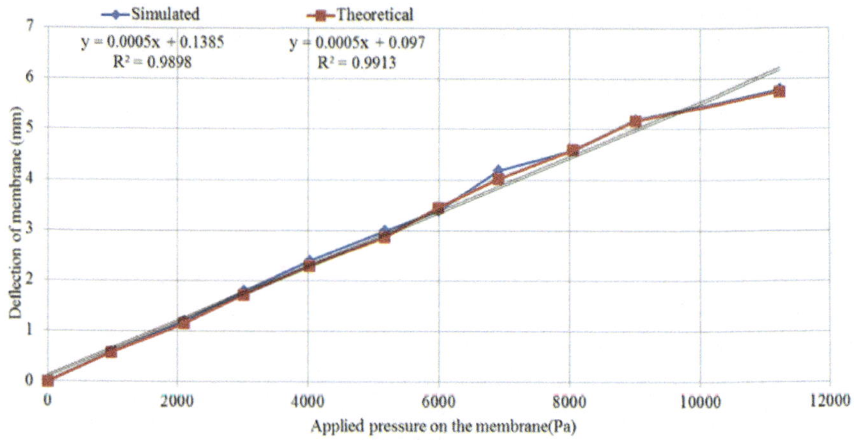

Fig. 15 Applied pressure versus deflection of membrane

Table 3 Comparison of simulated and theoretical data for power cable application

Sr.#	Parameters	Simulated data	Theoretical data	Percentage difference (%)
1	Resonance frequency	2.04 MHz	2 MHz	1.9
2	Amplitude	11.627 μm	12.70 μm	8.6
3	Quality factor	40.803	39.5354	3.1
4	Damping ratio	0.01224	0.0126468	3.2
5	Sensitivity	500 nm/Pa	500 nm/Pa	0
6	Linearity	0.9898	0.9913	0.1

Table 4 Comparison of simulated and theoretical data for power transformer application

Sr.#	Parameters	Simulated data	Theoretical data	Percentage difference (%)
1	Resonance frequency	265 kHz	260 kHz	1.8
2	Amplitude	92.39 pm	90.02 pm	2.5
3	Quality factor	33.124	32	3.3
4	Damping ratio	0.015094	0.015625	3.3
5	Sensitivity	1700 pm/Pa	1600 pm/Pa	5.8
6	Linearity	0.9934	1	0.6

Acknowledgements The authors would like to thank Higher Education Commission, Pakistan to support this research in terms of simulation tool provision under the project No: 9138/Punjab/NRPU/R&D/HEC/2017, and special thanks to Al-Khwarizmi Institute of Computer Science, University of Engineering and Technology, Lahore, Pakistan for their support in the successful completion of this work.

References

1. Tamus, Z.A., Cselko, R., Berta, I.: Partial discharge measurement in low voltage cable diagnostics. In: Annual Report Conference on Electrical Insulation Dielectric Phenomena (2008)
2. Liu, J., Yang, Y., Wang, J., Wang, Q.: Novel sensor system for online partial discharge (PD) detecting on insulator of transmission lines. In: 2nd IEEE Conference on Industrial Electronics and Applications, pp. 614–617 (2007)
3. Suwarno: Partial discharges in high voltage insulations. In: IEEE International Conference on Electrical Engineering and Computer Science (2014)
4. Hapeez, M.S., Abidin, A.F., Hashim, H., Hamzah, M.K., Hamzah, N.R.: Analysis and classification on different types of partial discharges by harmonic orders. Electron. Eng. **19**(9) (2013). ISSN 1392-1215
5. Huang, K., Ma, N., Pi, H., Yan, X., Du, Z., Li, H., Hu, H., Zhang, W., Li, H.: A technique to identify the types of partial discharges. In: Asia-Pacific Power and Energy Engineering Conference (2012)
6. Suwanasri, C., Thawonsukanan, P., Ruankon, S., Suwanasri, T.: Partial discharge detection in high voltage equipment using high frequency current transducer. GMSARN Int. J. **7**, 53–58 (2013)
7. Shanmuganantham, T., Gogoi, U.J.: A study scheme of energy harvesting process of MEMS piezoelectric pressure sensor. In: International Conference on Circuit, Power and Computing Technologies [ICCPCT], pp. 1–5 (2016)
8. Dennis, J., et al.: Optical characterization of Lorentz force based CMOS-MEMS magnetic field sensor. Sensors **15**(8), 18256–18269 (2015)
9. Dennis, J.O., et al.: Post micromachining of MPW based CMOS–MEMS comb resonator and its mechanical and thermal characterization. Microsyst. Technol. **22**(12), 2909–2919 (2016)
10. Dixit, P., Suhasaria, P., Singh, A., Patil, A.: VLSI using CMOS fabrication. Int. J. Sci. Res. Publ. **7**(4) (2017)
11. Santos, P.M., Casimiro, A.P., Lanca, M., Castro Simas, M.I.: High-voltage solutions in CMOS technology. Microelectron. J. **33**, 609–617 (2002)

Hubbard's Modified Density Functional Theory Calculations for the Electronic Structure and Optical Properties of Carbon Doped Anatase TiO_2

Norani Muti Mohamed, Farman Ullah, Robabeh Bashiri, Chong Fai Kait, Mohamed Shuaib Mohamed Saheed, and Muhammad Umair Shahid

Abstract Reducing the bandgap of TiO_2 to an optimal range by doping with other atoms is a very effective method having great potential in solar energy applications. The fundamental effect of structural changes upon the electronic structure of doped semiconducting TiO_2 is very much important to explore an effective doping configuration. A detailed computational study is therefore, required, to better understand the effect of different doped materials upon the structural, electronic and optical properties. We systematically study the carbon doped anatase TiO_2, using first-principle density functional theory (DFT) calculations to determine the effect of carbon concentration on the structural, electronic and optical properties of C doped TiO_2. We optimize the geometric structures of carbon doped anatase TiO_2 using generalized gradient approximation (GGA) with Perdew–Burke–Ernzerhof (PBE) potential (GGA+PBE). Furthermore, to study optical and electronic properties, we perform the calculation with GGA+Hubbard potential (GGA+U) exchange correlational functional. The results confirm that GGA+PBE produce more accurate results for the geometric structure of undoped and carbon doped TiO_2, closer to the experimental results. Moreover, GGA+U functional presents the bandgap energies of doped and undoped systems that are close to the actual values at lower computational cost.

N. M. Mohamed · F. Ullah (✉) · R. Bashiri · C. F. Kait · M. S. M. Saheed · M. U. Shahid
Centre of Innovative Nanostructure and Nanodevices (COINN), Universiti Teknologi PETRONAS, 32610 Bandar Seri Iskandar, Perak, Malaysia
e-mail: farman_19001014@utp.edu.my

Department of Fundamental & Applied Sciences, Universiti Teknologi PETRONAS, 32610 Bandar Seri Iskandar, Perak, Malaysia

N. M. Mohamed
e-mail: noranimuti_mohamed@utp.edu.my

R. Bashiri
e-mail: robabeh.bashiri@utp.edu.my

C. F. Kait
e-mail: chongfaikait@utp.edu.my

M. S. M. Saheed
e-mail: shuaib.saheed@utp.edu.my

S. A. Abdul Karim et al. (eds.), *Proceedings of the 6th International Conference on Fundamental and Applied Sciences*, Springer Proceedings in Complexity,
https://doi.org/10.1007/978-981-16-4513-6_32

As a result of carbon doping, new impurity levels have been introduced into the bandgap region of TiO_2 that leads to the decrease of bandgap energy. Narrowing the bandgap resulting in shift of the optical absorption edge to the visible region that might enhance the photocatalytic activity.

Keywords Anatase TiO_2 · Photocatalyst · Doping · Bandgap reduction · Density functional theory (DFT) · Optical absorption

1 Introduction

Energy crises and climate changes are the main reasons for researchers to direct their attention towards green renewable energy resources. Hydrogen production through solar photocatalytic water-splitting technology is assumed to be a clean renewable energy technology, in order to deal with energy crisis and environmental challenges in future [1]. Semiconductor photocatalysts has an important role as they are providing an easy way for utilization of solar energy. Since Fujishima's report in 1972, Titanium dioxide (TiO_2), has received intense attention as a promising photocatalytic material due to their excellent physical and chemical properties [1, 2] for photocatalyst applications for photocatalyst applications. But the wide bandgap of TiO_2 (valued 3.2 eV for anatase) restricts their usefulness to ultraviolet radiation (< 387 nm) mainly, which is only about 3–5% of the total sunlight. The utilization of sun light is therefore playing a very minor role, to activate this metal oxide to promote oxidation and reduction surface reactions [3]. To improve the photocatalysis of TiO_2, it is necessary to know that how to increase optical absorptivity coefficient of TiO_2 into the visible range (43% of solar spectrum), having wavelength range (380–700 nm) therefore, it is one of the most important subjects for researchers now a days. To modify their light absorptivity and electronic structure, incorporation of the impurities into the crystal structure, is the most feasible techniques. Considerable research work is going on in order to minimize their band gap to an optimal position. Many approaches such as doping of metal cations and non-metal anions have been attempted. TiO_2 doping with nonmetals including N, C, F, B and S etc. [4, 5] modify the electronic structure and hence properties by creating new impurity energy state in the host crystal structure [4, 6].

Carbon is considered an appropriate dopant element for TiO_2 in order to minimize their band gap, and therefore, improving the performance of the photocatalyst by increasing the visible light absorption. Some theoretical and experimental studies have been performed for carbon doped TiO_2 photocatalyst [6, 7]. Recently, Computational simulation methods, like density functional theory (DFT), is attracting more attention as a designing tool, as it is predicting many characteristics of the photocatalytic materials and to choose the suitable one for the practical application [8]. It is because, these methods provide a possibility to conduct research on complicated structures without expensive and/or dangerous experiments. Previously, most of the researchers conducted DFT calculations for the band gap of carbon doped

TiO_2 conventional DFT methods. These methods underestimated the value of the bandgap for strongly corelated systems like transition metal oxides. To get more effective results for the C-doping of TiO_2 systems, we need a detailed information as the mechanism is not clear too much yet [6, 9].

The goal of this work is to understand the effect of carbon concentration on the structural, optical and electronic properties of C doped TiO_2 by using DFT calculations. For structural properties GGA+PBE and for electronic and optical properties, GGA+Hubbard potential (GGA+U) approach were applied, respectively. It is notable to highlight that GGA+U approach produces better result for band gap energy calculation in comparison with conventional DFT methods. We investigated different structural, electronic and optical properties like lattice constants, cell parameters, band structure, density of states, energy band gap and absorption of the undoped and doped model using the concept of substitutional doped model. Geometry optimization of the undoped and the different doped TiO_2 model (2.08 and 4.17 at.% concentration of C), have been done with plane-wave pseudopotential approach. Geometries of the different undoped and doped structures calculated accurately and, the band gaps energies and density of states calculation also agree very well with the experimental ones.

2 Computational Methodology

In this study, all the theoretical calculations and designing of computational models have been performed with Cambridge sequential total energy package (CASTEP), one of the modules in Materials Studio program. It is based on the total energy plane-wave pseudopotential DFT approach [10]. Firstly, geometries of the structures have been optimized within the framework of GGA+PBE exchange correlational functional. The structural properties were calculated by applying GGA+PBE potential. For a system like transition metal oxides, having strongly localized orbitals (d- and f-orbitals), the bandgap energy may not be calculated too much accurately with GGA+PBE functionals [11]. Generally, the band gaps of transition metal oxides calculated with GGA+PBE functional are usually underestimated (approximately 50% smaller than the experimental value), because of the reason that it describes the strong Coulomb interactions less accurately between electrons of the d- and/or f-orbitals [12]. Therefore, in the present research work, we prefer the GGA+U scheme, for electronic and optical properties calculations, to produce more accurate results. GGA+U method produces more accurate results because of the reason that it considers the electronic interactions in d-shell, as strongly correlated interaction by introducing an on-site Coulomb's correction term and hence describe the systems with localized d and f electrons also in a good way [12, 13]. In this method, Hubbard-potential is used in order to obtain the value of bandgap energy, which is much closer to experimental value. Different values of the Hubbard parameter (U) may be used for the valence Ti 3d electrons and the energy gap values calculated accordingly. In our present case, we choose the value of $U = 8.47$ eV for localized

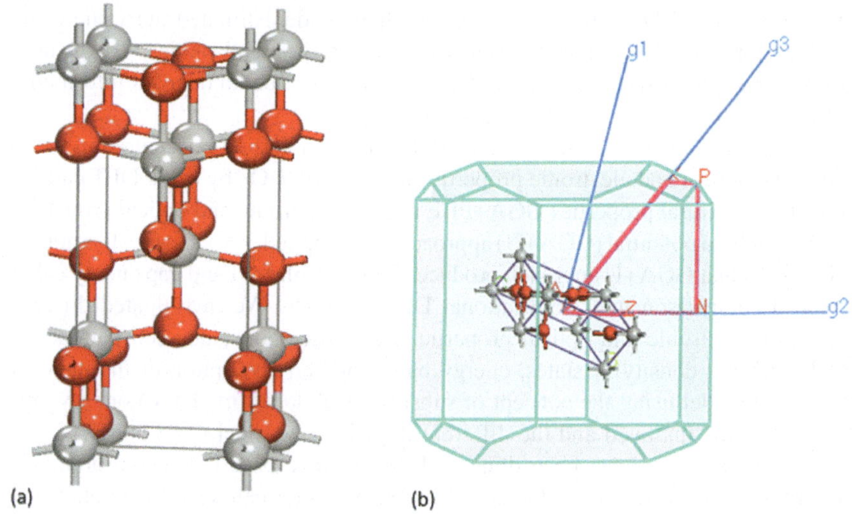

Fig. 1 DFT simulation images for **a** anatase TiO_2 unit cell in real apace and **b** in reciprocal space

electrons and the result obtained for band gap energy is much closer to the experimental ones. This method also provides a good balance between the accuracy of the calculated results and the available computational resources.

To construct C-doped systems, we built the 48-atoms 2×2×1 anatase TiO_2 supercell models. Simulation of the doped systems was conducted for the said 2×2×1 optimized supercell, containing different carbon atoms at the substituted position of the titanium atoms, as shown in Fig. 1b, c. Optimization of the atomic positions and cell parameters, were performed with residual forces below 0.03 eV/Å. To accelerate the computational runs, ultra-soft pseudopotential functional were applied to the electronic configuration of Ti; $3s^2$, $3p^6$, $3d^2$, $4s^2$, C; $2s^2$ $2p^2$ and O; $2s^2$, $2p^4$. A 3× 3×2 k-point mesh was used for numerical sampling integration, over the Brillouin zone, with Monkhorst–Pack method. The cutoff energy for the valence electrons wavefunction was 300 eV. The self-consistent field (SCF) iterations convergence threshold equal to 2×10^{-6} eV/atom. The convergence parameters chosen with the energy values 2×10^{-5} eV/atom and the stress on the atom limited below 0.1 GPa. The force was set as 0.03 eV/Å and the atomic displacement less than 2×10^{-3} Å.

3 Result and Discussion

3.1 *Structural Optimization*

Anatase TiO_2 has a tetragonal structure with lattice parameters a = 3.776 Å, b = 3.776 Å and c = 9.486 Å. Each Titanium atom is bonded to its four nearest oxygen neighbors. Figure 1a, b shows the computationally generated DFT simulation images for anatase TiO_2 unit cell in real apace and in reciprocal space.

Figure 2a shows a 2×2×1 supercell for pure anatase TiO_2 including 16 titanium and 32 oxygen atoms. Figure 2b, c represents carbon-doped TiO_2 supercell with 2.08 at.% (1 Ti replaced by C) and 4.17 at.% (2 Ti replaced by C), respectively. The concept of a supercell is introduced in order to effectively represent a bulk solid even if we simulate this very small block (supercell). Also, if the system contains a non-periodic entity such as a vacancy, we can apply the supercell approach by including the vacancy into the supercell [11].

To analyze the structural modifications as a result of carbon dopants incorporation, the optimized lattice parameters of the undoped and C-doped TiO_2 structures is then compared. For undoped system, the optimized bulk lattice parameters are $a = b =$

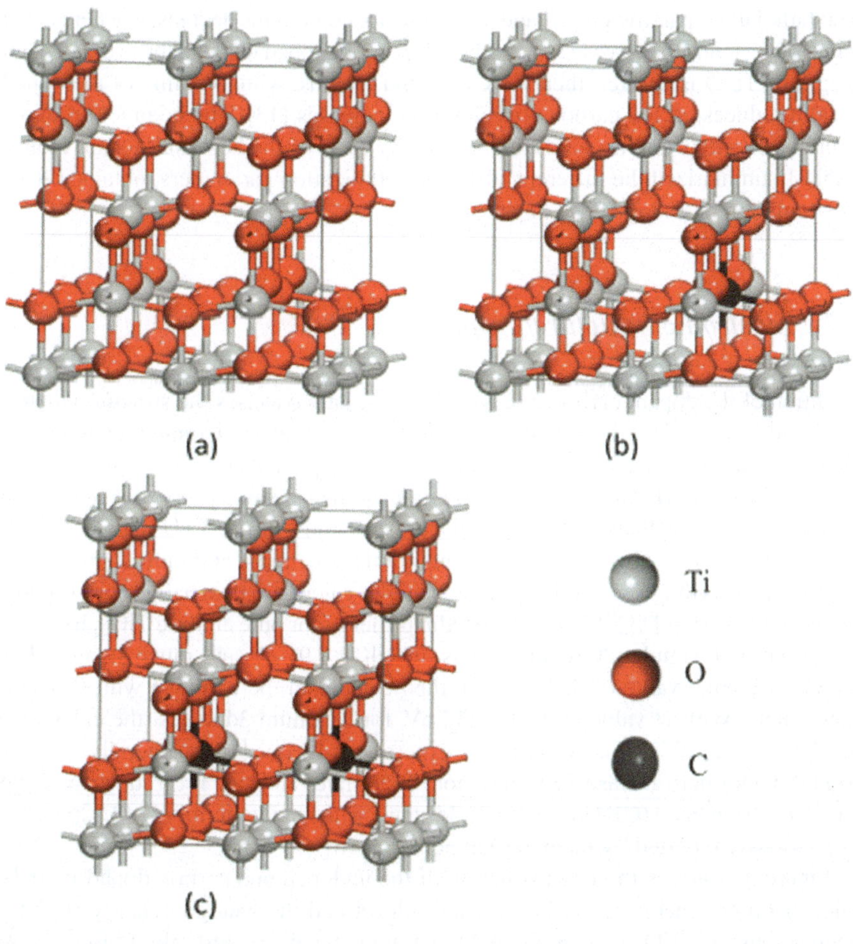

Fig. 2 (2×2×1) anatase TiO_2 supercell models with different concentrations of C **a** 0 at.%, **b** 2.08 at.%, **c** 4.17 at.%

Table 1 Comparison of the optimized lattice parameters

C (at.%)	Lattice parameters (Å)			Bond lengths (Å)		ΔV (%)
	a	*b*	*c*	Ti–O	C–O	
0.00	3.797	3.797	9.556	1.947	–	–
2.08	3.786	3.786	9.497	1.932	1.827	0.94
4.17	3.772	3.772	9.585	1.924	1.841	1.17

3.797 A° and $c = 9.549$ Å, that agrees well with the experimental values of the anatase having values, $a = b = 3.785$ Å and $c = 9.502$ Å [12]. It also indicating that our calculation method is more reliable. For carbon doped system having 2.08% doping concentration, the optimized bulk lattice parameters are $a = b = 3.786$ Å and $c = 9.497$ Å, while for 4.17% doped system, $a = b = 3.772$ Å, and $c = 9.585$ Å. The bulk lattice parameters '*a*' and '*b*' of the doped systems decreases, whereas the value of 'c' tends to increase after carbon doping. In doped models, the average bond length of Ti–O is shorter, then the C–O bond lengths, while volume of the doped system reduces, as the carbon concentration increases [13]. The contraction in the overall volume is due to the differences in the radii of the host and dopant atoms. Table 1 summarized the different structure optimization parameters, obtained from of simulation of undoped and doped systems.

3.2 Electronic Structure/Properties

Addition of the dopant carbon atoms to the TiO_2 lattice causes the structural distortions and as well as modifying the electronic structures of semiconducting materials also. New impurity states introduced in the bandgap causing changes in the electronic structure [14]. The electronic structure and their related properties have great impact on the performance of TiO_2. In the present study, the DFT+U (Hubbard coefficient) approach has been adopted to find out the electronic and optical properties. DFT+U approach improves the bandgap energy calculation partially more reliable as discussed earlier [11]. Firstly, the bandgap energy for pure anatase TiO_2, has been found out with simple GGA functionals, which is 1.91 eV and much smaller than the experimental value of 3.20 eV. We then calculated the bandgap with GGA+U functionals. With U value equal to 8.47 eV, for Titanium 3d states, the calculated bandgap is 3.13 eV and is closer to the experimental value [16]. Generally, valence band (VB) for pure anatase TiO_2, has more O 2p states and less Ti 3d states, whereas the conduction band (CB) has more Ti 3d states with smaller number of O 2p states, as previously reported by many researchers [15, 16].

Figure 3 indicates the band structure of the undoped and carbon doped models. The zero-point energy is the Fermi energy level and the bandgap energy (Eg) for undoped anatase TiO_2 is 3.13 eV, as shown in the band structure plot in Fig. 3a. At 2.08% C concentration, the bandgap decreases to 3.02 eV, due to the carbon-doped

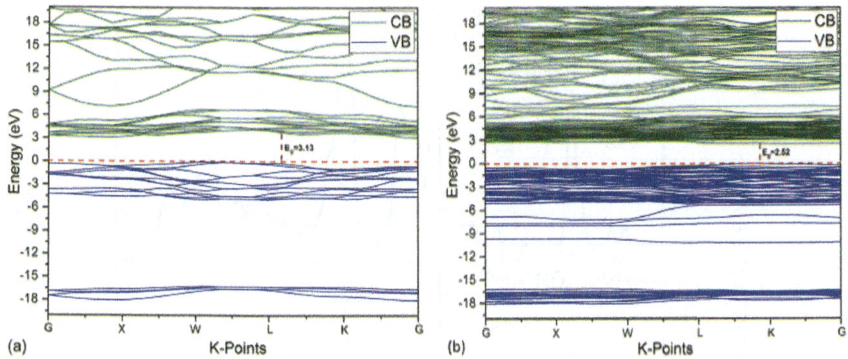

Fig. 3 Electronic band structure of anatase TiO_2 **a** 0 at.% (undoped) and **b** 4.17 at.%. Red-dashed line corresponds to the Fermi energy level

impurity states in the bandgap region. At 4.17 at.% as shown in Fig. 3b, the bandgap of C-doped TiO_2 narrows to 2.52 eV as the conduction band minima (CBM) shifts toward the Fermi level continuously, which results in an effective reduction of the bandgap of TiO_2. The result also indicates that the decrease in bandgap energy from 2.08 at.% to 4.17 at.% is more obvious than that from 0 to 2.08 at.%. By narrowing the bandgap, the excitation energy of the electrons in the valence band decreases as the electrons in the valence band excited to impurity states easily, and subsequently to the conduction band also [17]. As electron transition energy decreases in C-doped TiO_2, the material will absorb more visible light as the redshift of the optical absorption edge occurs.

The density of states (DOS) graphs also confirming the reduction in the bandgap energy as shown in Fig. 4. The DOS graphs also showing small peaks appearing in the valence band (VB) region for carbon-doped models, that were not present in the undoped TiO_2, and is due to the induction of carbon impurity states. Thus, the valence band was found to be more broaden as the concentration of incorporated carbon increases into TiO_2. An increase in the bandwidth of the valence band (WVB) might benefit the hole mobility in VB [18]. In this way, the effect of both i.e. narrowing the bandgap and increase in the hole mobility can improve the photocatalytic performance of the system.

3.3 Optical Properties

Pure anatase TiO_2 only absorbs ultraviolet light radiation due to wide bandgap energy, due to which, their application range is very limited. To shift the absorption edge to the longer wavelength range, red shift could be attributed to the bandgap reduction and is due to the introduction of impurity states in the bandgap of TiO_2 [19, 20]. We plotted the optical absorption (UV–visible) spectra for undoped and C-doped

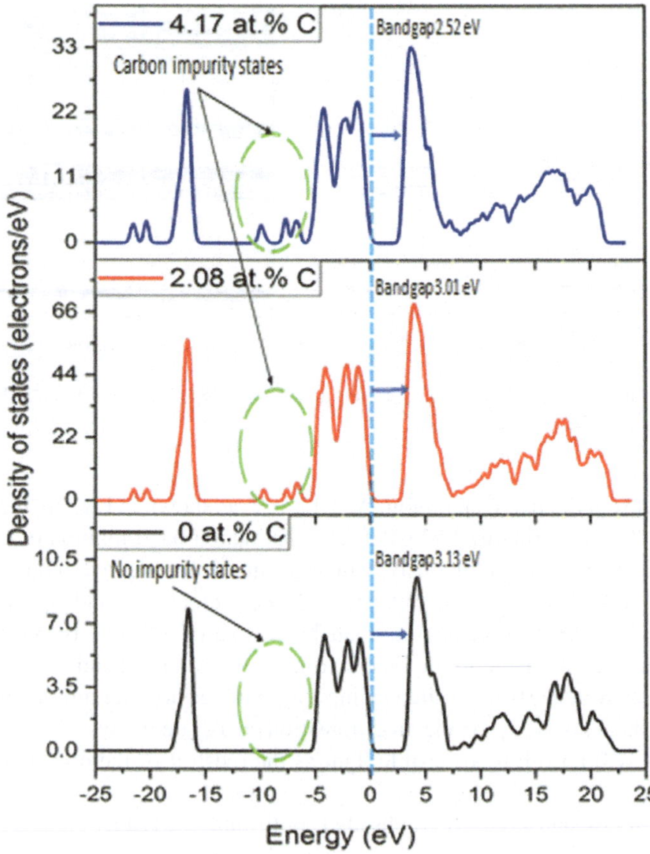

Fig. 4 Density of states of TiO_2 doped with 0 at.%, 2.08 at.% and 4.17 at.% of C

models, as shown in Fig. 5. The absorption spectra show that undoped TiO_2 mainly absorbs the UV radiation in the wavelength region between 200 and 360 nm, which agrees very well with previous theoretical calculations [21, 22]. After the C-doping of TiO_2, the changes in electronic structure extend the solar absorption edge from the UV towards the longer wavelength visible region, resulting in the red-shift of the optical absorption edge. Compared with undoped TiO_2, the doped model with 4.17 at.% C-doping concentration, the absorption band edge shifted up to 450 nm and thus resulting in red-shift of the absorption band edge. This redshift of the absorption edge is due to the bandgap narrowing, which reduces the energy required for electron excitation. It is to be noted that the C-doping not only resulting in red-shift of the optical spectra, but also improving the absorption visible light significantly.

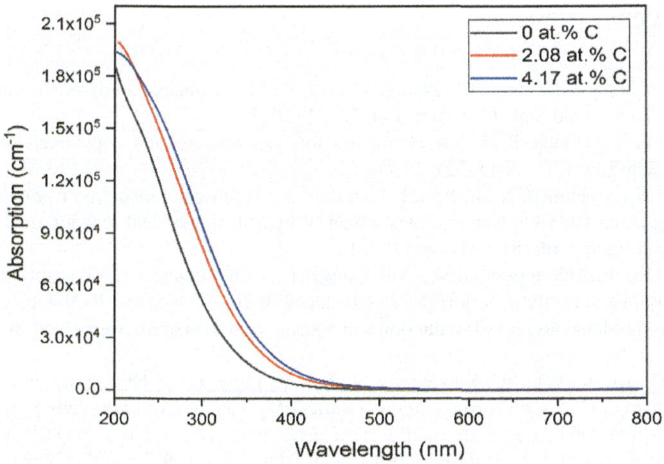

Fig. 5 The absorption plot for undoped and carbon doped anatase TiO_2

4 Conclusions

In this study, we investigate the structural properties of undoped and carbon doped TiO_2, through DFT calculations using GGA+PBE exchange correlational functional. A simple and efficient GGA+U technique is also used to find out the electronic and optical properties of anatase TiO_2 by investigating the influence of different carbon doping concentrations. The calculated results imply that carbon concentration induces new impurity energy states in the band structure of TiO_2 and higher concentrations of carbon effectively narrowing the bandgap of anatase TiO_2 to absorb more visible light. The bandgap energy value for pure anatase TiO_2, with the GGA+U technique ($U = 8.47$ eV), is around 3.13 eV and is in very close agreement with experimental value (3.20 eV). An increase in carbon impurities decreases the bandgap energy from 3.13 eV, for undoped TiO_2 (0 at.% C), to 3.01 and 2.52 eV for 2.08 and 4.17% of carbon doping respectively. As a result of band gap reduction, the redshift in the optical absorption edge occurs, that improving the visible light absorption also. Therefore, we conclude that the carbon doping of TiO_2 with a proper mechanism is a feasible method to improve the photocatalytic performance of TiO_2 under solar light absorption.

Acknowledgements This work is supported financially by Universiti Teknologi PETRONAS (UTP) and ministry of higher education (MOHE) Malaysia, Fundamental Research Grant Scheme (FRGS) (Grant No. FRGS/1/2019/STG07/UTP/01/1).

References

1. Liu, Y., Liang, W., Zhang, W., Zhang, J., Han, P.: First principle study of Cu and N-doped anatase TiO_2. Solid State Commun. **164**, 27–31 (2013)
2. Fujishima, A., Honda, K.: Electrochemical photolysis of water at a semiconductor electrode. Nature **238**(5358), 37–38 (1972)
3. Bashiri, R., Mohamed, N.M., Kait, C.F., Sufian, S.: Hydrogen production from water photo-splitting using Cu/TiO_2 nanoparticles: effect of hydrolysis rate and reaction medium. Int. J. Hydrogen Energy **40**(18), 6021–6037 (2015)
4. Giannakas, A., Antonopoulou, M., Daikopoulos, C., Deligiannakis, Y., Konstantinou, I.: Characterization and catalytic performance of B-doped, B-N co-doped and B–N–F tri-doped TiO_2 towards simultaneous Cr (VI) reduction and benzoic acid oxidation. Appl. Catal. B **184**, 44–54 (2016)
5. Chen, H., Li, X., Wan, R., Kao-Walter, S., Lei, Y., Leng, C.: A DFT study on modification mechanism of (N, S) interstitial co-doped rutile TiO_2. Chem. Phys. Lett. **695**, 8–18 (2018)
6. Li, X., et al.: A DFT study on the modification mechanism of (Cr, C) co-doping for the electronic and optical properties of anatase TiO_2. Comput. Mater. Sci. **129**, 295–303 (2017)
7. Mohamed, M.A., et al.: Revealing the role of kapok fibre as bio-template for In-situ construction of C-doped g-C3N4@ C, N co-doped TiO_2 core-shell heterojunction photocatalyst and its photocatalytic hydrogen production performance. Appl. Surf. Sci. **476**, 205–220 (2019)
8. Würger, T., et al.: Adsorption of acetone on rutile TiO_2: a DFT and FTIRS study. J. Phys. Chem. C **122**(34), 19481–19490 (2018)
9. Shao, J., et al.: In situ synthesis of carbon-doped TiO_2 single-crystal nanorods with a remarkably photocatalytic efficiency. Appl. Catal. B **209**, 311–319 (2017)
10. Clark, S.J., et al.: First principles methods using CASTEP. Zeitschrift für Kristallographie-Crystalline Mater. **220**(5/6), 567–570 (2005)
11. Lee, J.G.: Computational Materials Science: An Introduction. CRC Press (2016)
12. Li, X., Gao, H., Liu, G.: A LDA+ U study of the hybrid graphene/anatase TiO_2 nanocomposites: Interfacial properties and visible light response. Comput. Theor. Chem. **1025**, 30–34 (2013)
13. Zhang, H.-X., Zhu, Y.F., Zhao, M.: Interface charge transfer and enhanced visible light response of graphene/anatase TiO_2 (110) systems with and without oxygen vacancy: a DFT+ U calculation. Appl. Surf. Sci. **420**, 105–109 (2017)
14. Zou, Y., Shi, J.-W., Ma, D., Fan, Z., Lu, L., Niu, C.: In situ synthesis of C-doped TiO_2@ g-C3N4 core-shell hollow nanospheres with enhanced visible-light photocatalytic activity for H_2 evolution. Chem. Eng. J. **322**, 435–444 (2017)
15. German, E., Faccio, R., Mombru, A.W.: Comparison of standard DFT and Hubbard-DFT methods in structural and electronic properties of TiO_2 polymorphs and H-titanate ultrathin sheets for DSSC application. Appl. Surf. Sci. **428**, 118–123 (2018)
16. Long, R., English, N.J.: Density functional theory studies of doping in titania. Mol. Simul. **36**(7–8), 618–632 (2010)
17. Zhu, Z., Chen, J.-Y., Su, K.-Y., Wu, R.-J.: Efficient hydrogen production by water-splitting over Pt-deposited C-HS–TiO_2 hollow spheres under visible light. J. Taiwan Inst. Chem. Eng. **60**, 222–228 (2016)
18. Wu, H.-C., Li, S.-H., Lin, S.-W.: Effect of Fe concentration on Fe-doped anatase TiO_2 from GGA. Int. J. Photoenergy **2012** (2012)
19. Zhang, J., et al.: Engineering the absorption and field enhancement properties of Au–TiO_2 nanohybrids via whispering gallery mode resonances for photocatalytic water splitting. ACS Nano **10**(4), 4496–4503 (2016)

20. Basu, K., et al.: 4.3 Highly Stable Photoelectrochemical Cells for Hydrogen Production Using, I dedicate this thesis to my family, p. 111 (2018)
21. Yan, J., et al.: Heterostructured CoO/3D-TiO_2 nanorod arrays for photoelectrochemical water splitting hydrogen production. J. Solid-State Electrochem. **21**(2), 455–461 (2017)
22. Kandiel, T.A., Takanabe, K.: Solvent-induced deposition of Cu–Ga–In–S nanocrystals onto a titanium dioxide surface for visible-light-driven photocatalytic hydrogen production. Appl. Catal. B **184**, 264–269 (2016)

Temperature-Conductivity Relationship of Solid Polymer Electrolyte Membrane Based on PVA-K_2CO_3

Bashir Abubakar Abdulkadir, John Ojur Dennis, Mohd Fadhlullah Abd Shukur, Mohamed Mahmoud Elsayed Nasef, and Fahad Usman

Abstract Solid polymer electrolyte for energy storage devices draws extensive considerations nowadays. In this study, we prepared an eco-friendly, flexible and transparent solid polymer electrolyte (SPE) from biodegradable polymer and conducting salt of the poly vinyl alcohol and potassium carbonate (PVA-K_2CO_3) by solution casting method. The prepared electrolyte was characterized using Fourier transform infrared (FTIR) for their functional groups and electrochemical impedance spectroscopy (EIS) for their electrochemical properties. The result for characterization shows that, there is formation of composite between polymer and the salt that designates the complete composites formation between polymer and the salts, which formed PVA-K_2CO_3. The obtained PVA-K_2CO_3 electrolyte with good flexibility exhibits highest conductivity of 9.4×10^{-5} mScm^{-1} at 373.15 K. The outcomes equally show that, the electrolyte in this study is dependent on temperature as ionic conductivity increases with increase in temperature. The good results found in this study showed that, the PVA-K_2CO_3 as a newly developed SPE has a prospective applications in solid-state and flexible energy storage devices.

Keywords Solid polymer electrolyte · Poly vinyl alcohol · K_2CO_3 · Electrochemical impedance spectroscopy

---

B. A. Abdulkadir · J. O. Dennis · M. F. Abd Shukur (✉) · F. Usman
Fundamental and Applied Sciences Department, Universiti Teknologi PETRONAS, 32610 Seri Iskandar, Perak Darul Ridzuan, Malaysia
e-mail: mfadhlullah.ashukur@utp.edu.my

B. A. Abdulkadir
e-mail: abubakar_g03619@utp.edu.my

F. Usman
e-mail: fahad_17004459@utp.edu.my

M. M. E. Nasef
Malaysia-Japan International Institute of Technology, Universiti Teknologi Malaysia, Jalan Sultan Yahya Petra, 54100 Kuala Lumpur, Malaysia
e-mail: mahmoudeithar@cheme.utm.my

S. A. Abdul Karim et al. (eds.), *Proceedings of the 6th International Conference on Fundamental and Applied Sciences*, Springer Proceedings in Complexity,
https://doi.org/10.1007/978-981-16-4513-6_33

1 Introduction

Electrolyte is an essential part of energy storage devices like supercapacitors, which control device electrochemistry [1]. Conventionally, liquid electrolyte is consider a suitable candidate, which exhibit high ionic conductivity. However, it has many issues, like leakage problem, which may lead to an explosion owing to the use of flammable solvents. In order to solve these concerns, scientists produced solid polymer electrolyte (SPEs) that can fundamentally settle the difficulties of using liquid electrolytes. SPE is prepared from a mixture of host polymer such as PVA and a conducting salt such as K_2CO_3 in a present of a solvent, such as water [2]. Fundamentally, good energy storage device like supercapacitors depend on the invention of electrolyte such as SPEs that have high ionic conductivity and wide potential window. The choices for polymer matrices are versatile. Polymers were reported to provide a substitute way and draw a lot of consideration for application in supercapacitors as either electrolyte material or electrolyte membrane. Polymer is known to be very fast in redox reaction that result in pseudo-capacitance stored in a supercapacitor in addition to the EDL capacitance. Furthermore, polymers are electrically conducting materials [3]. Polymers have a π-conjugated backbone which contains a single (C–C) and double (C=C) carbon bonds. Different polymers such as poly(pyrrole) (PPy), poly (methylmethacrylate) (PMMA) and polyaniline (PANI) have been studied in supercapacitor owing to their conductivity, simple way of preparation, effectiveness, and environmental benign [4].

However, the major disadvantages of using SPEs is crystallinity that may lead to low ionic conductivity and poor potential window and ion transport occurs primarily through amorphous region. Therefore, to overcome these disadvantages and improve SPEs conductivity, there is need to reduce the crystalline nature of the polymer by incorporating a conducting salt that can affect polymer segmental motion. The conjugated backbones of these polymers can easily be modified by incorporating conducting salt to reduce the crystallinity and enhance electrochemical performance of the electrolyte. Among these polymers, PVA has attracted particular interest because its semi-crystalline, can trap liquid electrolytes, and provide mechanical support [5]. Therefore, PVA can offer not only high ionic conductivity, but also good mechanical properties. As a favourable conductive polymer electrolyte material, PVA have been extensively studied and reported due to its simple way of preparation, environmental benign, good mechanical and temperature resistance, dielectric constant, non-toxicity, and the capability to absorbed salts easily owing to the existence of polar groups (–OH) [6]. This study is aim to report the development and characterization of SPE based on PVA-K_2CO_3 by solution casting technique for application in energy storage devices.

2 Material and Methods

2.1 Materials

Materials used in this study; K_2CO_3 (anhydrous) and PVA (hydrolyzed 99%) were purchased from Sigma-Aldrich (002008792-H), WI, USA. The materials were all of analytical grade and used without any further treatment. For all experiments preparation, deionized water were used.

2.2 Development and Synthesis of SPE (PVA-K_2CO_3)

The SPE of K_2CO_3 doped PVA (PVA-K_2CO_3) were developed by mixing appropriate quantities of salts and the polymer into 20 mL of deionized water in a distinct glass beaker until the content were dissolved completely [2]. The two solutions were mix together and continuously heated at 80 °C with a constant stirring until homogeneous, viscous and clear solution were obtained. The solution was casted onto plastic Petri dish to dry at room temperature. The dried film was kept in a desiccator packed with silica gel for more drying. Figure 1a shows the picture of the synthesized SPE. Clearly, the electrolyte film is freestanding and appropriate for the fabrication of any energy storage device. Mechanical properties of the prepared SPE in this study is found to have excellent bending and stretching properties as can be observed in Fig. 1b. The electrolyte film is easy to bend into different and desired shapes. Similar types of solid electrolyte were reported previously [7, 8].

(a) (b)

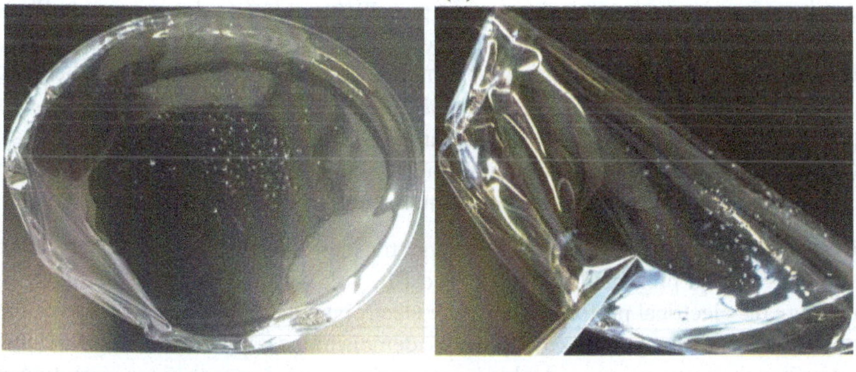

Fig. 1 The view of **a** free-standing film and **b** flexible and foldable SPE based on PVA-K_2CO_3 (30 wt% K_2CO_3) composite

2.3 Characterization

The Fourier transform infrared (FTIR) (Bruker Instruments, model Aquinox 55, Germany) spectroscopy was carried out on an FTIR spectrometer in the 4000–400 cm^{-1} range using KBr pellets with the scanning resolution of 4 cm^{-1}. The Impedance Spectroscopy of the prepared SPE was studied using AUTOLAB/AUT51018 (potential/galvanostat) electrochemical impedance spectroscopy (EIS). A blocking electrode cell packed with SPE sandwiched between two stainless steel was used for the study [9]. The analysis was conducted over a frequency and temperature ranges of 0.1–10^5 Hz and 303.15–383.15 K, respectively. The conductivity of the sample was calculated using Eq. (1) [10].

$$\sigma = \frac{t}{R_b A} \quad (1)$$

where t is the thickness of the film (cm), R_b (Ω) is the bulk resistance and A is the electrode/electrolyte contact area (cm^2).

3 Result and Discussions

3.1 Characterization

Chemical Structure. FTIR spectroscopy was used to examine the interactions amid PVA and conducting salt in PVA-K_2CO_3 electrolyte. By blending polymer and salt, there would be extra sites for exchange and jumping for ions that make the ionic conductivity to improve. The spectrum of pure PVA and K_2CO_3 incorporated SPE is shown in Fig. 2. The result shows that, band of hydroxyl of PVA is seen at 3294 cm^{-1}. However, after K_2CO_3 successfully incorporated into the polymer system, the peak moved to a lower band of 3266 cm^{-1}, replicating the physical connections between polymer matrix and the ions [11]. The changes, shifting and enlargement in the peak intensities observed after incorporation of the salts obviously displaying effective realization of SPEs from PVA-K_2CO_3 composites [12].

Resistance. EIS used to examine the charge transport mechanism of the electrolyte and the result is presented in Fig. 3. The EIS is a versatile procedure employ to measure the electrical performance of any electrolyte material. The responses of the impedance from the synthesized electrolyte demonstrated a spike at both high and low frequency range representing double layer capacitance owing to the ions migration at the electrode–electrolyte interface [13]. The high-frequency intercept on the axis of real impedance in the plot signifies the cell resistance known as bulk resistance (R_b) (shown in Table 1) which reveals the bulk properties of the prepared sample. Based on R_b value, ionic conductivity was computed following Eq. (1). The result show that,

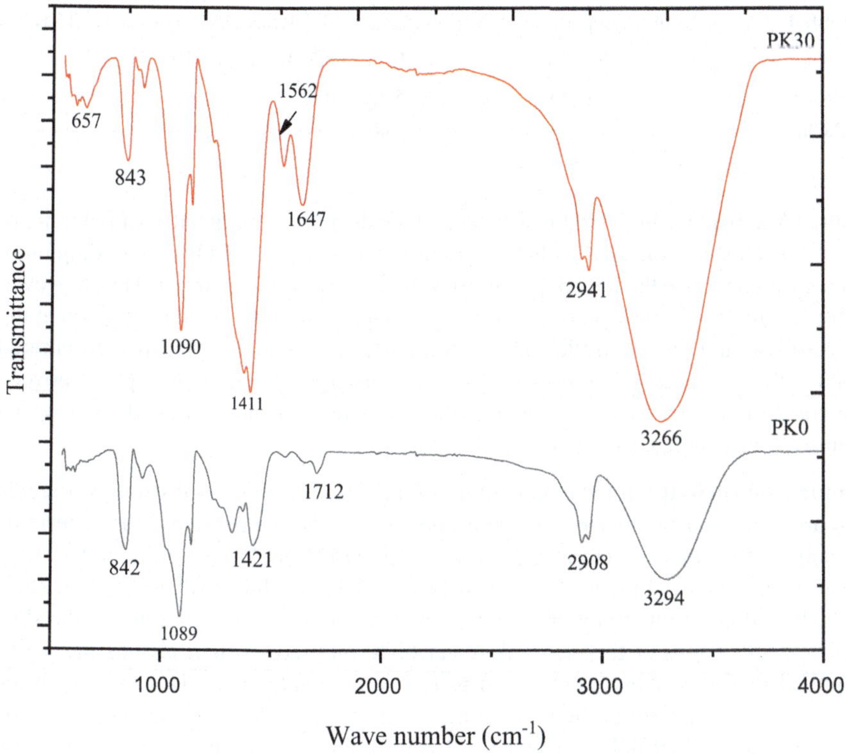

Fig. 2 FTIR spectra of PVA (PK0) and PK30 (30 wt% K_2CO_3) solid polymer electrolytes

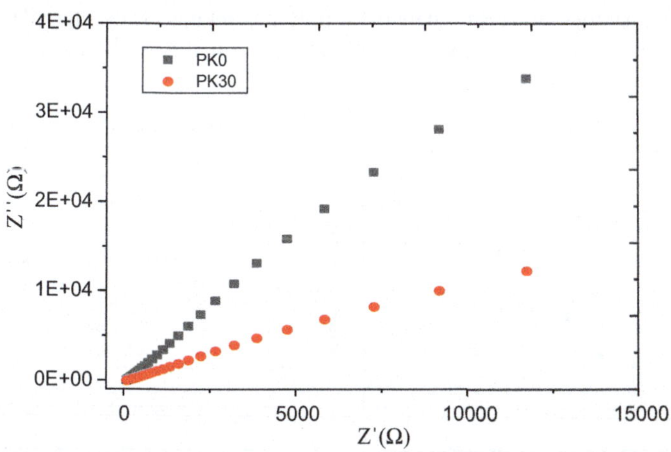

Fig. 3 **a** Cole–Cole plot for PVA (PK0) and PVA-K_2CO_3 (30 wt% K_2CO_3) (PK30) composites at 373.15 K

Table 1 Electrochemical properties of the synthesized solid polymer electrolytes at 373.15 K

Electrolyte	R_b (Ω)	Ionic conductivity, σ ($mScm^{-1}$)
PK0	182.33	5.50×10^{-7}
PK30	117.15	9.40×10^{-5}

pure PVA displays high bulk resistance, which decrease significantly with doping of K_2CO_3. This is associated with the presence and effect of K_2CO_3 in providing more ions, which reduce the R_b value of the developed electrolyte material. This, however, showed and confirmed the role of the salt in increasing interfacial interaction at the electrolyte and the electrode surface. Similarly, as conducting salt is incorporated and with the successful composite formation observed in Fig. 2, the space between anode and cathode is reduce; hence, the diffusion resistance is condensed which subsequently increase the ionic conductivity [8].

Ionic Conductivity. Ionic conductivity of an electrolyte material strongly affected by mobility and number of ions that depends on the temperature [14]. The ionic conductivity with respect to temperature of the SPE from PVA-K_2CO_3 in this study was computed using the impedance analysis and the result is displayed in Fig. 4a, b. The low ionic conductivity observed at low temperature can be credited to the slow movement of ions at the electrode–electrolyte interface, which lessens the active applied field that subsequently lead to low ionic conductivity. However, with the increase in temperature, there is strong ion hopping within the electrodes and the time of the applied field is small for the accumulation of charge, hence the ionic conductivity improved [15]. Similarly, an increase in conductivity with temperature increase can be related to the rise in movement and number of ions, and this is because with the rise in temperature, the chain in the polymer obtains faster internal modes that expedite inter and intra chain ion jumping in SPE. From the result, ionic conductivity

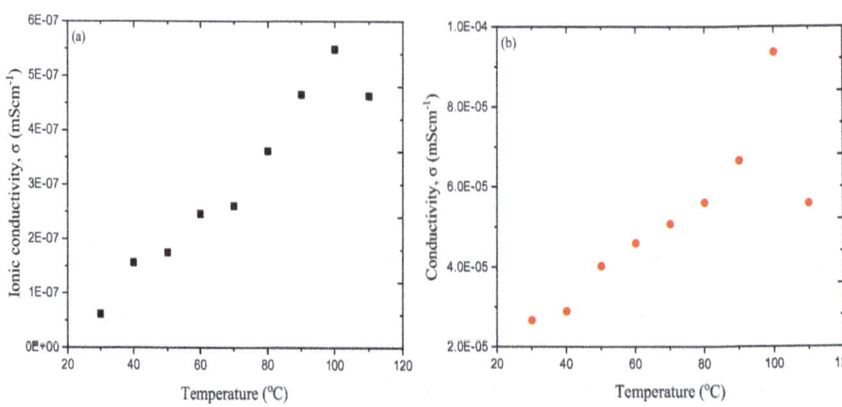

Fig. 4 Effect of temperature on the ionic conductivity of **a** pure PVA and **b** PVA-K_2CO_3 (30 wt% K_2CO_3) at 373 K

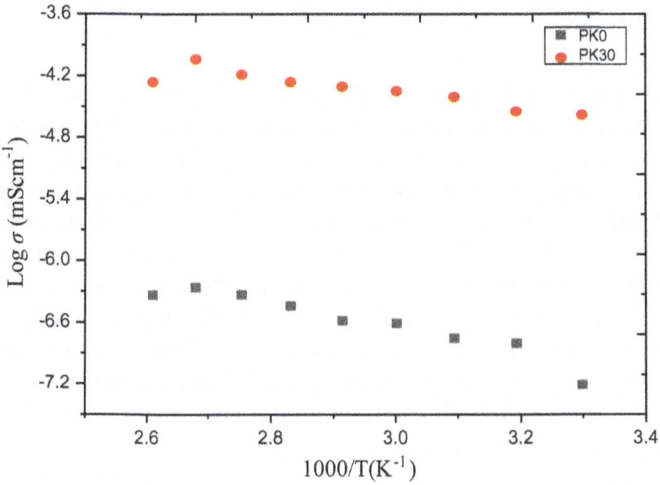

Fig. 5 Conductivity against temperature of pure PVA (PK0) and PVA-K_2CO_3 electrolytes

increased with the rise in in temperature where the highest ionic conductivity of 9.4 $\times$ 10^{-5} $mScm^{-1}$ was achieved at 100 °C, which suggested that the conductivity is temperature dependent. However, beyond 100 °C, a decrease in conductivity was observed. This decrease can be credited to the conversion of active K^+ in K_2CO_3 to inactive K^0, which decrease the number of charge carriers that contribute in the conduction process [16].

Similarly, the linear variation of temperature-dependence conductivity pure PVA and PVA-K_2CO_3 electrolyte is depicted in Fig. 5 suggesting an Arrhenius-type thermal activated process of the prepared SPEs; hence, the conductivity can be expressed as:

$$\sigma = \sigma_o \exp\left(\frac{-Ea}{k_B T}\right) \tag{2}$$

where σ is the ionic conductivity ($mScm^{-1}$), Ea is activation energy (eV), k_B is a Boltzmann constant and T stand for the absolute temperature (K). An increase in conductivity with a steady rise in temperature is observed for both electrolytes. This is mainly because of the increase in the modes of molecules vibration in the polymer-salt complex at higher temperature [1]. It was reported that, fast mode of vibration of the molecules in the complex helps in bond rotation in the polymer matrix. Consequently, ionic transportation is enhance with increasing temperature. Similar result study of increase with increase in temperature were reported previously [17, 18].

4 Conclusion

Solid polymer electrolyte film based on PVA-K_2CO_3 composite that are freestanding and flexible were prepared and characterized in this study. The structural results confirm that, K_2CO_3 has successfully incorporated into the polymer matrix of PVA, which confirm composite formation. The addition of salt into polymer increase the ionic conductivity significantly, which might be link to the rise in number of ions on the surface of the polymer. The electrolyte show that, ionic conductivity is temperature dependent where an increase in temperature results in increase ionic conductivity where the highest conductivity of 9.4×10^{-5} mScm^{-1} was achieved at 373.15 K which might be credited to the growth in number and movement of ions from the K^+. Therefore, this study indicate that the solid polymer electrolyte based on PVA-K_2CO_3 is a capable electrolyte for the fabrication of any energy storage device owing to their high conductivity and flexibility.

Acknowledgements The authors are grateful to the YUTP (Grant no: 015LC0-317) for the financial support and to the Universiti Teknologi PETRONAS for providing research facilities and graduate assistantship.

References

1. Aziz, S.B., Woo, T.J., Kadir, M.F.Z., Ahmed, H.M.: A conceptual review on polymer electrolytes and ion transport models. J. Sci. Adv. Mater. Devices **3**, 1–17 (2018)
2. Polu, A.R., Kumar, R.: Preparation and characterization of PVA based solid polymer electrolytes for electrochemical cell applications. Chin. J. Polym. Sci. **31**(4), 641–648 (2013)
3. Wang, K., Zhang, X., Li, C., Sun, X., Meng, Q., Ma, Y.: Chemically crosslinked hydrogel film leads to integrated flexible supercapacitors with superior performance. Adv. Mater. **27**, 7451–7457 (2015)
4. Gao, Y.: Graphene and polymer composites for supercapacitor applications: a review. Nanoscale Res. Lett. **12**(387), 1–17 (2017)
5. He, T., Jia, R., Lang, X., Wu, X., Wang, Y.: Preparation and electrochemical performance of PVdF ultrafine porous fiber separator-cum-electrolyte for supercapacitor. J. Electrochem. Soc. **164**(13), 379–384 (2017)
6. Miao, Y., Yan, J., Huang, Y., Fan, W., Liu, T.: Electrospun polymer nanofiber membrane electrodes and an electrolyte for highly flexible and foldable all-solid-state supercapacitors. RSC Adv. **5**, 26189–26196 (2015)
7. Muchakayala, R., Song, S., Wang, J., Fan, Y., Bengeppagari, M., Chen, J., Tan, M.: Development and supercapacitor application of ionic liquid-incorporated gel polymer electrolyte films. J. Ind. Eng. Chem. **59**, 79–89 (2018)
8. Hu, X., Fan, L., Qin, G., Shen, Z., Chen, J., Wang, M., Yang, J., Qiang, C.: Flexible and low temperature resistant double network alkaline gel polymer electrolyte with dual-role KOH for supercapacitor. J. Power Sources **414**, 201–209 (2019)
9. Shukur, M.F., Yusof, Y.M., Zawawi, S.M.M., Illias, H.A., Kadir, M.F.Z.: Conductivity and transport studies of plasticized chitosan-based proton conducting biopolymer electrolytes. Phys. Scr. **T157**, 1–6 (2013)
10. Yusof, Y.M., Illias, H.A., Shukur, M.F., Kadir, M.F.Z.: Characterization of starch-chitosan blend-based electrolyte doped with ammonium iodide for application in proton batteries. Ionics **23**, 681–697 (2017)

11. Wang, J., Zhao, Z., Song, S., Ma, Q., Liu, R.: High performance poly (vinyl alcohol)-based Li-Ion conducting gel polymer electrolyte films for electric double-layer capacitors. Polymers **10**, 1–15 (2018)
12. Nishihara, M., Terayama,Y., Haji, T., Lyth, S. M., Satokawa, S., Matsumoto, H.: Proton-conductive nano zeolite-PVA composite film as a new water-absorbing electrolyte for water electrolysis. eXPRESS Polym. Lett. **12**, 256–264 (2018)
13. Shujahadeen, B.A., Brza, M.A., Pshko, A.M., Kadir, M.F.Z., Hamsan, M.H., Abdulwahid, R.T., Woo, H.J.: Increase of metallic silver nanoparticles in Chitosan: AgNt based polymer electrolytes incorporated with alumina filler. Res. Phys. **13**, 1–10 (2019)
14. Yadav, N., Mishra, K., Hasmi, S.A.: Optimization of porous polymer electrolyte for quasi-solid-state electrical double layer supercapacitor. Electrochim. Acta **235**, 570–582 (2017)
15. Salman, Y.A.K., Abdullah, O.G., Hanna, R.R., Aziz, S.B.: Conductivity and electrical properties of chitosan-methylcellulose blend biopolymer electrolyte incorporated with lithium tetrafluoroborate. Int. J. Electrochem. Sci. **13**, 3185–3199 (2018)
16. Aziz, S.B., Abidim, Z.H.Z., Arof, A.K.: Effect of silver nanoparticles on the DC conductivity in chitosan–silver triflate polymer electrolyte. Phys. B **405**, 4429–4433 (2010)
17. Ji, Y., Liang, N., Xu, J., Qu, R., Chen, D., Zhang, H.: Solid polymer electrolyte membranes based on quaternized polysulfone and solvent-free fluid as separators for electrical double-layer capacitors. Electrochim. Acta **283**, 97–103 (2018)
18. Abdulkadir, B.A., Dennis, J.O., Shukur, M.F.B.A., Nasef, M.M.E., Usman, F.: Preparation and characterization of gel polymer electrolyte based on PVA-K_2CO_3. Polym. Technol. Mater. **59**, 679–1697 (2020)

Simulation and Experimental Study for Electromagnetic Absorption in Sandstone with $SrFeO_3$ Nanofluid

Hassan Ali, Hassan Soleimani, Noorhana Yahya, and Mohammed Falalu Hamza

Abstract Electromagnetic absorption properties of sandstone saturated with Strontium ferrite ($SrFeO_3$) nanofluid are discussed in this paper. The synthesis of nanoparticles was carried out using the sol–gel combustion method and calcinated at a temperature of 300–400 °C in a furnace. X-ray diffraction pattern verified the crystalline structure of $SrFeO_3$, and FESEM images confirmed the morphology of nanoparticles. The dielectric and magnetic studies were carried out using the vector network analyzer (VNA) to measure the dielectric losses in dry sandstone and the sandstone saturated with $SrFeO_3$ nanofluids. Comparative analysis of the electric permittivity, complex permittivity, and the dielectric losses of sandstone with $SrFeO_3$ provided an insight about the microwave absorbing properties. Simulations of these materials with electromagnetic field verified experimental results, as $SrFeO_3$ has the greater capacity to absorb the electromagnetic waves and can be used as a potential material for enhanced oil recovery.

Keywords Enhanced oil recovery · Nanofluid · Dielectric permittivity · Dielectric loss

---

H. Ali (✉) · H. Soleimani · N. Yahya
Department of Applied Sciences, Universiti Teknologi PETRONAS, Seri Iskandar, Perak, Malaysia
e-mail: hassan_16005513@utp.edu.my

H. Soleimani
e-mail: hassan.soleimani@utp.edu.my

N. Yahya
e-mail: noorhana_yahya@utp.edu.my

H. Ali
Department of Physics, University of Narowal, Narowal, Pakistan

M. F. Hamza
Department of Pure and Industrial Chemistry, Bayero University Kano, Kano, Nigeria

S. A. Abdul Karim et al. (eds.), *Proceedings of the 6th International Conference on Fundamental and Applied Sciences*, Springer Proceedings in Complexity,
https://doi.org/10.1007/978-981-16-4513-6_34

1 Introduction

Nanoparticles and nanofluids have attained the attraction of scientists and researchers in the field of oil and gas in the past decade [1]. The unique properties of dielectric and magnetic particles because of their small size and interaction with electromagnetic (EM) field has made such particles suitable for enhanced oil recovery (EOR) [2, 3]. These nanoparticles manipulate the petrophysical properties such as viscosity, wettability, and interfacial tension in a reservoir on the application of the EM field [4, 5]. This phenomenon has been adopted in a variety of other applications such as damping devices, micro-robotics, memory devices, and cancer treatment [6].

Multiferroic materials exhibit that they can spontaneously switch the internal order because of their magnetoelectric coupling [7]. Among such ferrites, strontium ferrite ($SrFeO_3$) has an advantage over others because of the applications of EM absorption [8]. These materials can absorb the EM waves which is incident on them, and later on, dissipate the absorbed energy in form of heat [9]. The factors affecting the absorption of EM waves in nanomaterials are the complex permittivity (ε''), complex permeability (μ''), and dielectric loss factor (δ) [10–12].

In this work, a simulation technique based on the finite element method is proposed for the absorption of EM waves in sandstone saturated with $SrFeO_3$ nanofluid and dry sandstone. The experimental work is conducted to synthesize nanofluid and to find out the dielectric properties of sandstone and sandstone saturated with $SrFeO_3$ nanofluid.

2 Methodology

2.1 Simulation of EM Waves in the Porous Medium

To describe the scattering and absorption of EM waves in a porous medium saturated with nanofluid, a simulation technique based on the finite element method (FEM) has been adopted [13]. A system of equations comprising of Darcy's law for two-phase flow, continuity equation, and Maxwell's equations have been used to simulate the propagation of EM waves in the porous medium. The setup for the EM wave transmitter is described in Sect. 2.2.

2.2 Electromagnetic Transmitter

The halfwave dipole antenna operating at 8–12.5 GHz frequency was designed. The half waves were driven by a discrete face port or lumped port between the two rods of the dipole antenna. The same dipole with different impedance was used as a receiver on the other side of the porous medium near the outlet to calculate the EM losses

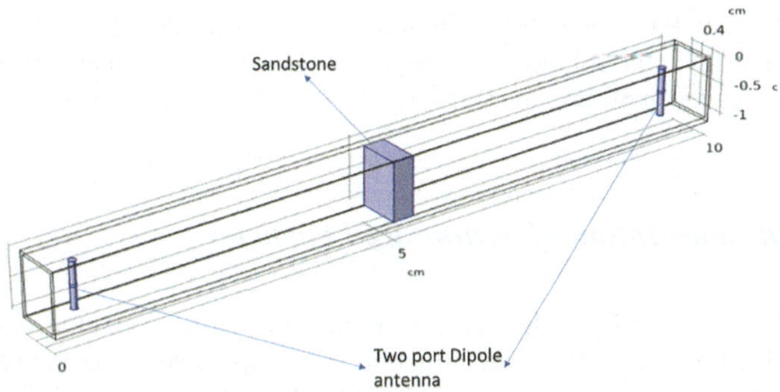

Fig. 1 Simulation model of porous medium placed between two dipole antennas

Table 1 Input parameters for dipole antenna

Description	Name	Expression
Permeability	μ	$4\pi \times 10^{-7}$ H/m
Relative permittivity of $SrFeO_3$	ε_{sr}	10.15
Relative permittivity of oil	ε_o	2.5
Relative permittivity of rock	ε_r	3.966
Relative permeability	μ_o	1
Electric conductivity	σ	0.05 [S/m]
Applied frequency	f_o	8–12.5 GHz

in the porous medium as shown in Fig. 1. The parameters and constants used for simulation are given in Table 1.

2.3 *Materials Used*

Strontium ferrite was prepared using Strontium nitrate pentahydrate $(Sr(NO_3)_2 \cdot 5H_2O$, iron nitrate nonahydrate $(Fe(NO_3)_2 \cdot 9H_2O)$, along with oxalic acid and ethanol was used. All the chemicals used were of analytical grade from Sigma Aldrich.

2.4 *Synthesis of Nanoparticles*

Strontium ferrite $SrFeO_3$ nanoparticles were synthesized using the sol–gel combustion route by making separate solutions of strontium nitrate in ethanol and iron nitrate in water. 11.05 g of strontium nitrate was mixed in 100 mL of ethanol, distilled water was added dropwise to get a clear solution because of the low solubility of strontium nitrate in ethanol. Later, both the nitrate solutions were mixed for 8 h and the pH was maintained as 0.19. the solution was filtered and dried in an oven at 160 °C for 2 h and was calcinated at 500 °C for 5 h.

2.5 *Characterization of Synthesized Particles*

X-ray diffraction (XRD) technique was used with Cu-Ka radiation operated at 45 kV and 40 mA to analyze the structure of nanoparticles. The morphological and elemental analysis were carried out by using FESEM to obtain EDX spectra of particles. Dielectric properties of the sandstone with nanofluid and dry sandstone were studied using a vector network analyzer.

2.6 *Measurement of Dielectric Properties by VNA*

The dielectric properties of lossy materials are calculated using the relative permittivity, relative permeability, and tangent loss given by the following equations [14].

$$\varepsilon = \varepsilon' - j\varepsilon'' \tag{1}$$

$$\tan\delta = \frac{\varepsilon''}{\varepsilon'} \tag{2}$$

where ε' is the real part of permittivity and is defined as the dielectric constant to store energy in the material. ε'' is referred to as dielectric loss factor, is a complex quantity and influences the energy absorption and attenuation in the material. Another important factor tan (δ) is attributed as tangent loss and it contributes to the dielectric loss in heterogeneous media [15]. The sandstone samples from the Angsi-E2 oil field were cut in specific dimensions and saturated with $SrFeO_3$ nanofluids. The dielectric measurements at a frequency of 8–12.5 GHz have been carried out using a vector network analyzer (VNA).

3 Results and Discussion

3.1 *XRD Analysis*

The X-ray diffraction patterns of $SrFeO_3$ are shown in Fig. 2. Some impurity peaks were obtained while the peaks at 33.327, and 44.653 confirmed the formation

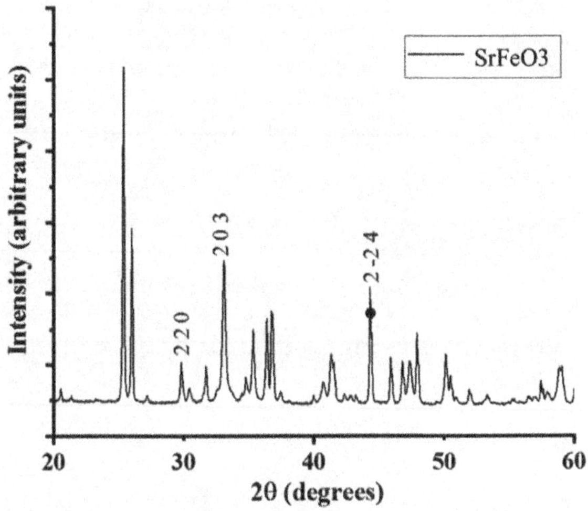

Fig. 2 X-ray diffraction pattern of $SrFeO_3$ nanoparticles calcinated at 500 °C

of strontium ferrite corresponding to hkl parameters of (220), (203), and (2–24) respectively.

3.2 *FESEM Analysis*

FESEM analysis of $SrFeO_3$ in Fig. 3 shows that strontium ferrites obtained from the gels of the solution with oxalic acid and thermal treatment at 500 °C consist of mostly large agglomerates. We can observe the porous morphology of the agglomerated particles with very few particles that are loosely bound. This porosity is due to the combustion of oxalic acid and the partial sintering of the particles.

3.3 *EDX Analysis*

Energy dispersive X-ray (EDX) of $SrFeO_3$ nanoparticles was carried out for the elemental analysis as shown in Fig. 4. The EDX results of strontium ferrite confirmed the elemental presentation of strontium, iron, and oxygen atoms with an atomic percentage of 9.66%, 11.16%, and 79.18% respectively.

3.4 *Dielectric Permittivity* (ε')

The real part of dielectric permittivity (ε'), dielectric loss (ε''), and dielectric loss factor (tan δ) is measured for the saturation of sandstone with $SrFeO_3$ nanofluid (S_{Sr})

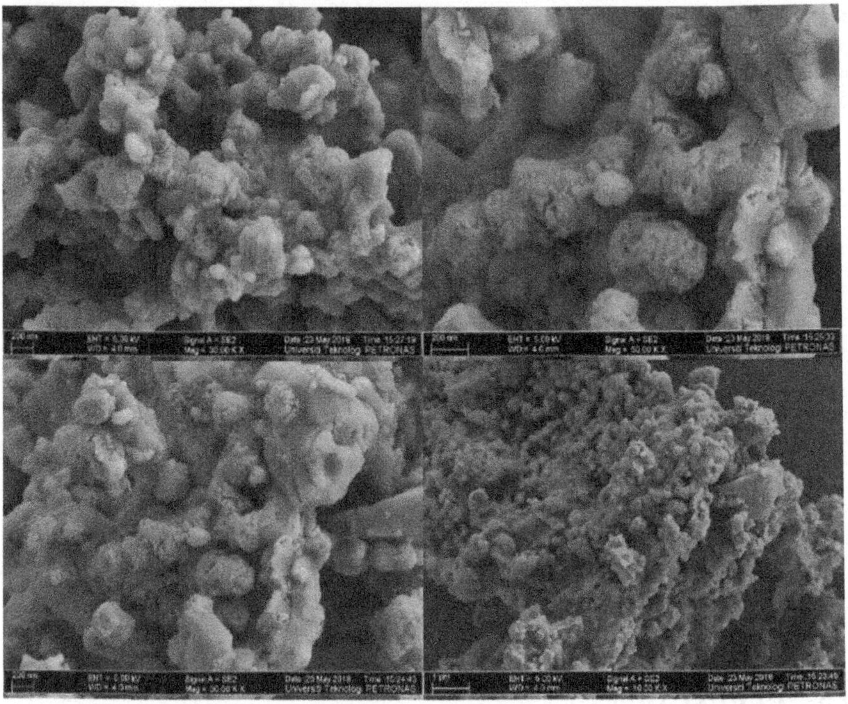

Fig. 3 SEM analysis of $SrFeO_3$ calcinated at 500 °C

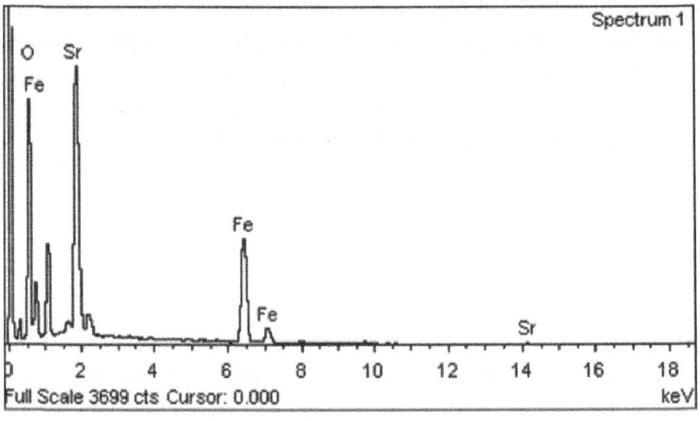

Fig. 4 EDX spectra of $SrFeO_3$

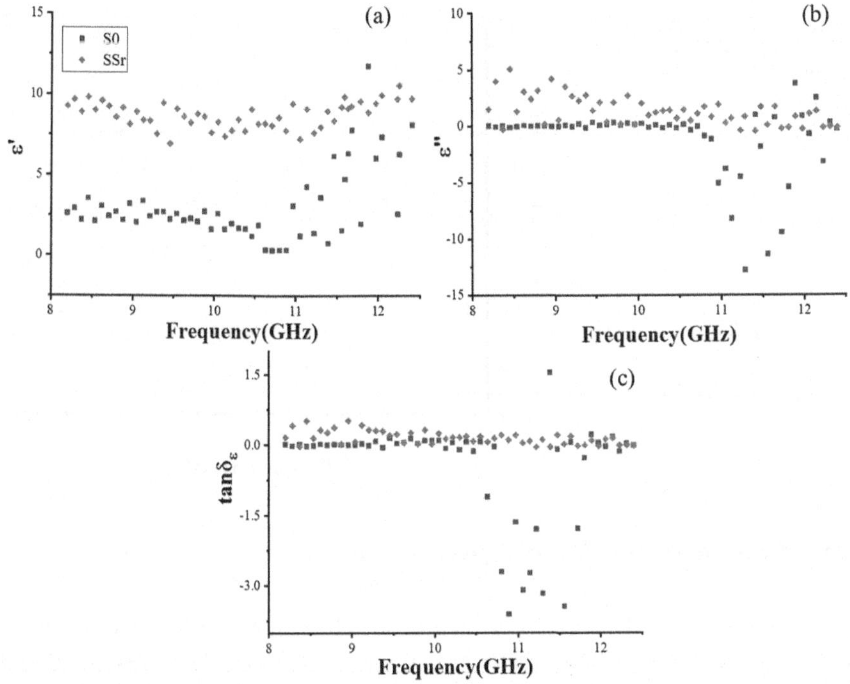

Fig. 5 Behavior of dielectric permittivity (ε'), dielectric loss (ε''), and dielectric loss factor (tan δ) as a function of frequency for dry rock (S_0), and saturation with $SrFeO_3$ nanofluid (S_{Sr})

and dry sandstone (S_0) at a range of frequencies 8.5–1.25 GHz (Fig. 5a–c). The analysis of the dielectric permittivity confirmed that at a relatively low frequency ε' is higher for S_{Sr} as compared to S_0 while at a higher frequency of 1.25 GHz there is an abrupt increase in ε' for S_0. At this sharp change of curves, behavior occurs in that frequency range due to the effective conductivity dependence. This behavior of S_{Sr} for greater dielectric permittivity is due to the presence of additional relaxation due to the polarization of $SrFeO_3$ nanoparticles. The dielectric loss with respect to frequency for the sandstone saturated in $SrFeO_3$ nanofluid and dry sandstone is shown in Fig. 5b. It is observed that the dielectric loss is higher for S_{Sr} samples as compared to the sandstone. The lower value of ε'' for S_0 shows that the conduction has increased and at a higher frequency the dielectric losses become much less. The graph shows that the dielectric loss increases with the increasing frequency in S_0. This happens because of the space charge polarization and reduced ionic conductivity in the dry sandstone. The dielectric loss factor (tan δ) also varies with increasing frequency for rock saturated with $SrFeO_3$ nanofluid and dry rock is shown in Fig. 5c. The maximum loss factor is for $SrFeO_3$ at 8.35 GHz frequency which is due to the relaxation of ions which happens because of the ions present in the material.

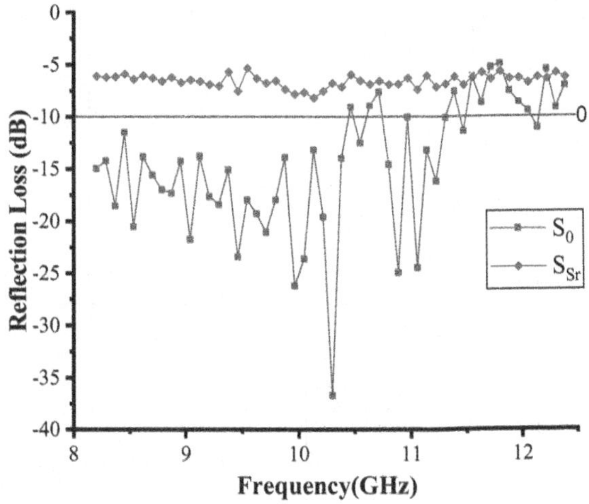

Fig. 6 Reflection loss (RL) curves for S_0 and S_{Sr} at X-band (8.2–12.4 GHz) frequency

3.5 Simulated Result for Reflection Loss

The reflection loss (RL) curve is defined by the absorption of EM waves having a frequency less than − 10 dB, where 90% of EM wave is absorbed. Figure 6 shows the variation of the RL curve for S_0 and S_{Sr} at a frequency of 8–12.5 GHz, the same as in the vector network analyzer. For S_{Sr} the RL value was more than − 10 dB, while for S_0 the value is less for most frequencies. The reason for high absorption for S_0 is the factors such as for dry sandstone the pores were not filled with nanofluid and the surface area was more. Also, the porous structure was responsible for greater energy loss and greater microwave absorption.

4 Conclusion

Simulation results show that the reflection loss in dry rock increases with an increase in frequency while it decreases for the rock with $SrFeO_3$ nanofluid because of the presence of ions the permittivity value increases, and more waves are absorbed. The results from the network analyzer show that with the inclusion of nanoparticles the dielectric permittivity ε' of the system increases and hence the dielectric loss ε'' decreases which is beneficial for the effective activation of particles and can be helpful for enhanced oil recovery. Also, the dielectric loss factor is maximum for $SrFeO_3$ as compared to dry rock, and high absorption of EM waves. So, $SrFeO_3$ can be used as a potential material for enhanced oil recovery.

Acknowledgements The authors wish to acknowledge the project funding; YUTP research grant 015LCO-143 and Universiti Teknologi PETRONAS Malaysia for providing financial support through the graduate assistant (GA) scheme.

References

1. Almahfood, M., Bai, B.: The synergistic effects of nanoparticle-surfactant nanofluids in EOR applications. J. Petrol. Sci. Eng. Rev. **171**, 196–210 (2018). https://doi.org/10.1016/j.petrol.2018.07.030
2. Hamza, M.F., Merican, Z.M.A., Soleimani, H., Abghari, S.Z., Sinnathambi, C.M., Stephen, K.D.: A Laboratory Study of Chemical Enhanced Oil Recovery (CEOR) in Compartmentalized Sandstone Reservoir: A Case Study of a 2-D Phase Macro-model Reservoir (2006)
3. Soleimani, H., et al.: Synthesis of ZnO nanoparticles for oil–water interfacial tension reduction in enhanced oil recovery. Appl. Phys. A **124**(2), 128 (2018)
4. Wahaab, F.A., Yahya, N., Shafie, A., Soleimani, H., Rostami, A., Ganeson, M.: Determination of optimum frequency for electromagnetic-assisted nanofluid core flooding. Appl. Sci. **9**(21), 4608 (2019)
5. Adil, M., Zaid, H.M., Chuan, L.K., Alta'ee, A.F., Latiff, N.R.A.: Microscopic evolution of dielectric nanoparticles at different calcination temperatures synthesized via sol-gel auto-combustion. In: AIP Conference Proceedings, vol. 1669, no. 1, p. 020016. AIP Publishing (2015)
6. Anantharaman, M., et al.: Tailoring magnetic and dielectric properties of rubber ferrite composites containing mixed ferrites. Bull. Mater. Sci. **24**(6), 623–631 (2001)
7. Latiff, N.R.A., Soleimani, H., Zaid, H.M., Adil, M.: Magnetoviscous effect of ferrite-based magnetic fluid for EOR application. In: AIP Conference Proceedings, vol. 1787, no. 1, p. 050021. AIP Publishing (2016)
8. Luo, J., Shen, P., Yao, W., Jiang, C., Xu, J.: Synthesis, characterization, and microwave absorption properties of reduced graphene oxide/strontium ferrite/polyaniline nanocomposites. Nanoscale Res. Lett. **11**(141) (2016). https://doi.org/10.1186/s11671-016-1340-x
9. Adebayo, L.L., Soleimani, H., Yahya, N., Abbas, Z., Ridwan, A.T., Wahaab, F.A.: Investigation of the broadband microwave absorption of citric acid coated Fe3O4/PVDF composite using finite element method. Appl. Sci. **9**(18), 3877 (2019)
10. Zhang, X., et al.: Microwave absorption properties of the carbon-coated nickel nanocapsules. Appl. Phys. Lett. **89**(5), 053115 (2006)
11. Ali, H., et al.: Absorption of electromagnetic waves in sandstone saturated with brine and nanofluids for application in enhanced oil recovery. J. Taibah Univ. Sci. **14**(1), 217–226 (2020)
12. Ali, J.A., Kolo, K., Manshad, A.K., Mohammadi, A.H.: Recent advances in application of nanotechnology in chemical enhanced oil recovery: effects of nanoparticles on wettability alteration, interfacial tension reduction, and flooding. Egypt. J. Petrol. (2018)
13. Ali, H., Soleimani, H., Yahya, N., Baig, M., Rostami, A.: Finite element method for modelling of two phase fluid flow in porous media. J. Phys. Conf. Series **1123**(1), 012002 (2018)
14. Jamwal, G., Prakash, J., Chandran, A., Gangwar, J., Srivastava, A., Biradar, A.: Effect of nickel oxide nanoparticles on dielectric and optical properties of nematic liquid crystal. In: AIP Conference Proceedings, vol. 1675, no. 1, pp. 030065. AIP Publishing (2015)
15. Ansari, S.A., et al.: Temperature dependence anomalous dielectric relaxation in Co doped ZnO nanoparticles. Mater. Res. Bull. **47**(12), 4161–4168 (2012)

Flexible Photoelectrode Based on Binder-Free Paste for Dye-Sensitized Solar Cell

Norani Muti Mohamed, Muhammad Umair Shahid, Kamilah Ramly, Balbir Singh Mahinder Singh, Chong Fai Kait, Siti Noor Azella Zaine, and Farman Ullah

Abstract The flexible Dye-sensitized solar cell (DSC) has a huge potential in portable devices. However, the fabrication of flexible DSC is a challenge as it is tough to sinter TiO_2 film at low temperature. Therefore, in this study, a binder-free TiO_2 paste is presented, which has been sintered on a flexible conductive substrate using UV exposure. The UV exposure time is optimized based on the XRD, FTIR, FESEM and UV–Vis spectrophotometer. The XRD has shown that TiO_2 is perfectly crystalline and UV exposure transforms anatase and brookite completely into the rutile phase. This transformation is investigated by sintering the same type of samples in an oven. It is found that oven sintering does not support anatase transformation into rutile. Moreover, FTIR, FESEM and UV–Vis transmittance has revealed that a 5 h of UV exposure is the optimum time to degrade the organic matter of the paste and properly sinter the film on a flexible substrate.

N. M. Mohamed · M. U. Shahid · K. Ramly · B. S. M. Singh · C. F. Kait · F. Ullah (✉)
Fundamental and Applied Sciences Department, Centre of Innovative Nanostructure and Nanodevices (COINN) , Universiti Teknologi PETRONAS, Seri Iskandar, Perak, Malaysia
e-mail: farman_19001014@utp.edu.my

N. M. Mohamed
e-mail: noranimuti_mohamed@utp.edu.my

K. Ramly
e-mail: kamilah_16001900@utp.edu.my

B. S. M. Singh
e-mail: balbir@utp.edu.my

C. F. Kait
e-mail: chongfaikait@utp.edu.my

M. U. Shahid
Department of Materials Science and Engineering, Institute of Space Technology, Islamabad, Pakistan

S. N. A. Zaine
Department of Chemical Engineering, Universiti Teknologi PETRONAS, Seri Iskandar, Malaysia
e-mail: nurazella.zaine@utp.edu.my

S. A. Abdul Karim et al. (eds.), *Proceedings of the 6th International Conference on Fundamental and Applied Sciences*, Springer Proceedings in Complexity,
https://doi.org/10.1007/978-981-16-4513-6_35

Keywords Flexible photoelectrode · Dye-sensitized solar cell · TiO_2 · UV exposure

1 Introduction

Photovoltaic (PV) is a promising future source of sustainable production of electricity. PV had been installed up to 30 GW producing about 25 TWh of electricity [1]. It has been predicted that 15–30% of all electricity could generate from solar energy by 2050 [2]. Currently, most of the PV systems are based on mono or multi-crystalline silicon (c-Si), having a market share of 90%. By creating more cost-effective equipment and easier manufacturing of solar cells, reducing production costs has become an immediate need to make them an inexpensive way to generate renewable energy [1, 3].

Nowadays, the demand for portable power sources has been increased due to mobile electronic devices bring convenience to us, owing to the short serving time [4]. Amongst the available flexible devices, the flexible dye-sensitized solar cells show high potential for the portable power due to ease of manufacturing, low production cost, relatively potential conversion efficiency and excellent flexibility [5].

There are few challenges to translate DSC from glass to a flexible substrate, amount these, preparation of highly disperse binder-free paste and its sintering on a flexible substrate are most important [6, 7]. Typically, high-temperature sintering is required for semiconductor TiO_2 film, which is not possible with a flexible substrate [8, 9]. Low-temperature sintering cannot degrade binders properly, thus poor electron transport, consequently, low photoconversion efficiency (PCE). There are different methods to prepare or sinter TiO_2 films such as electrophoretic deposition, electrochemical deposition, and laser sintering [10–12]. Moreover, UV light irradiation has also been used to sinter TiO_2 film on a flexible substrate [13]. However, the UV exposure time for binder-free paste has not been explored well. Therefore, in current work, binder-free TiO_2 highly dispersed printable paste has been presented to employe in the fabrication of flexible photoelectrode. The UV irradiation time is optimized based on the various characterization techniques such as XRD, FTIR, FESEM and UV–Vis spectroscopy.

2 Methodology

2.1 Synthesis of TiO_2 Nanoparticles and Binder-Free Paste

A 0.7 M solution of TTIP in water was prepared, immediately white preceptees were formed [14]. Before adding water, TTIP was modified with 5 mL of 2-propanol under stirring. The solution was sonicated with probe type sonicator for 20 min to disperse the white precipitates. Later concentrated nitric acid was added so that its molarity becomes 0.41 M and the temperature of the solution was raised to 80 °C and kept for 45 min under stirring. Later the solution was transferred to the Teflon lined autoclave with 24 mL DI water and treated at 240 °C for 12 h. The white precipitates were collected after cooling and washed with water and centrifuged three times.

A 6 g of TiO_2 was dispersed in 150 mL of ethanol with the help of probe sonicator for 30 min, then 2 mL of acetic acid was added and again sonicate for 10 min. Afterwards, an amount of 0.6 mL Triton-X 100 surfactant was added under stirring, after 5 min of stirring sonicated for 10 min. Later, 24 mL of terpineol was added into TiO_2 dispersion under stirring, after 5 min of stirring sonicated for 10 min. Finally, TiO_2 paste was obtained by removing ethanol.

2.2 *Fabrication of Flexible Photoelectrode*

Photoelectrode was fabricated with a simple tape casting process. First patterns were prepared on a flexible conducting substrate using laser scriber; then the tape was applied following the marking. The TiO_2 paste was then fill in the cavity between the tape and cast with the help of sharp scale. Later, samples were first dried at room temperature for 5 min then transferred to the UV light chamber. Samples were irradiated for 0.5–5 h after each 15 min lamp was switched off then switched on again after resting of 10 min. Figure 1 shows the fabricated flexibility of the photoelectrode after UV exposure (5 h) without cracking.

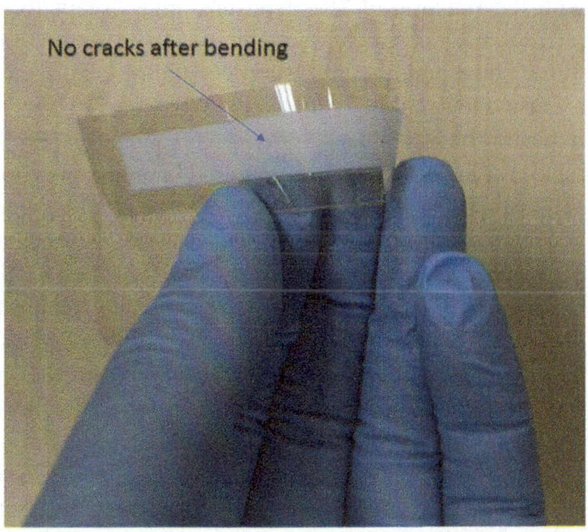

Fig. 1 Illustrating flexibility of the film without cracking

3 Characterization

The X-ray Diffraction (XRD) of the Bruker diffractometer with Cu Kα (λ = 0.1540 nm) is used to identify the crystalline structure of prepared TiO2 nanoparticles. Grazing angle (GXRD) of Bruker has been employed to identify phase changes in TiO2 film under due to UV exposure. Zeiss Supara55 VP Field Emission Scanning Electron Microscope (FESEM) is used to acquire the morphology of the photoelectrodes. UV–Vis Spectrophotometer (Cary series) of Agilent technologies is used to measure the optical transmittance of flexible photoelectrodes.

4 Results and Discussion

XRD is employed to identify the crystalline phases of prepared TiO_2 nanoparticles and determine the effect of UV exposure on the crystalline phases. Figure 2a, b presents the XRD pattern of TiO_2 nanoparticles. The peaks are matched with the three phases of TiO_2; anatase, rutile, and brookite. The intense peak at 25.28° corresponds to (101) plane of the anatase, while the minor peaks at 27.32° and 30.80 show the rutile phase at (110) and brookite at (121) orientation respectively [15]. The phase composition analysis of TiO_2 revealed that mainly anatase is present with ~ 83% while rutile is 8%, and rest is brookite.

Moreover, Fig. 2a also shows the XRD patterns of the TiO_2 film with different UV exposure time. There is not a significant difference in XRD patterns of TiO_2 with the function of UV exposure time, except for the peak intensity. The peak intensity has been increased with the UV exposure time, that ascribed to the amount of the corresponding phase and crystallinity. Further, it can be seen that the whole anatase and brookite completely converted into the rutile phase surprisingly. That indicates the rutile phase is thermodynamically more stable with UV exposure. The minor rutile peaks which were found in prepared nanoparticles at ~ 37.9°, 41.2° and 68.9° are also entirely vanished, only the previous intense peak (27.32°) has grown further. It shows that TiO_2 crystallites are transformed into the rutile (110) orientation completely.

To confirm this effect of UV, the same TiO_2 binder-free paste was tape cast on the glass and sintered in the oven at 120 °C; Fig. 2b shows the XRD patterns. It shows that there is no significant difference between the oven treated film and primary TiO_2 particles. Therefore, this result confirms that the transformation of anatase and brookite into rutile is due to UV exposure.

The degradation of organic solvent, stabilizer and surfactant in TiO_2 paste depends upon the UV exposure time and intensity. Here intensity is fixed, only exposure time is varied. FTIR spectrum was taken at different UV exposure time, such as; 0 h, 0.5 h, 1 h, and 4 h in order to evaluate the effect of UV exposure on degrading the solvent and surfactant. Figure 3 shows that there are several absorption peaks in 0 h UV exposed sample (the sample without UV exposure), with an increase in UV exposure time

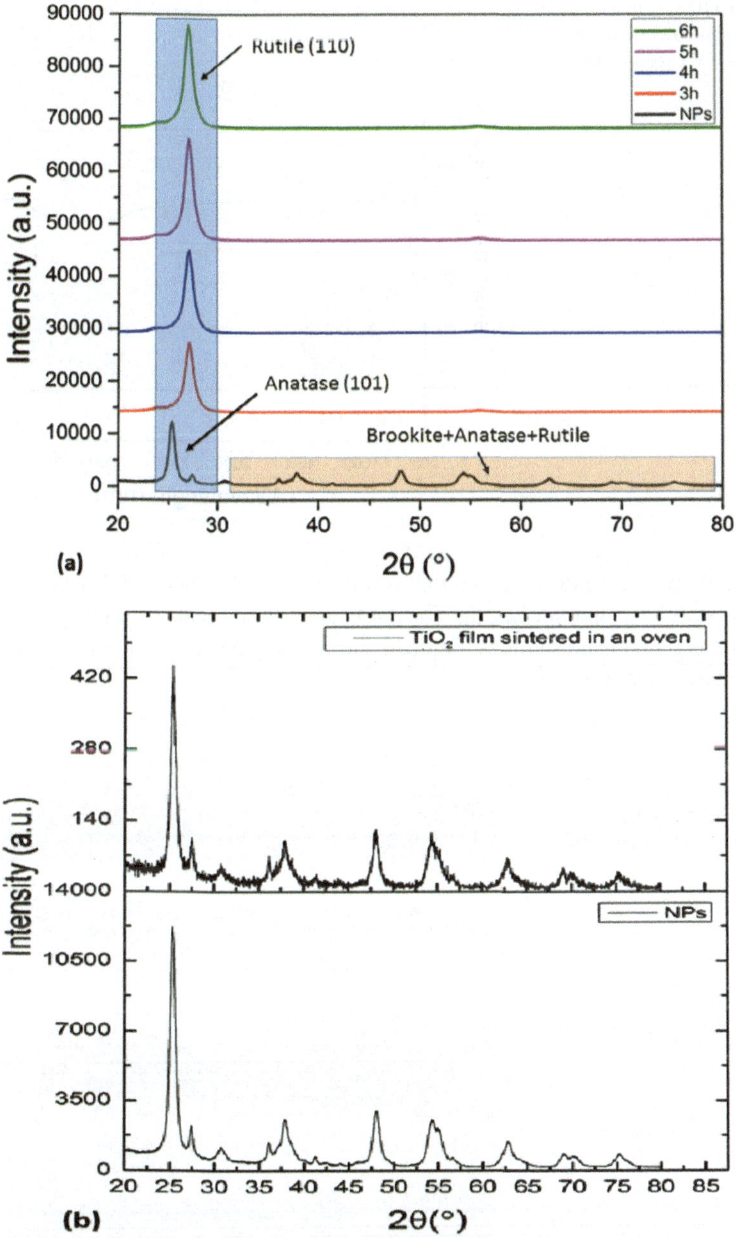

Fig. 2 **a** XRD patterns of TiO_2 NPs and film with different UV exposure time, **b** XRD pattern of TiO_2 sintered in an oven

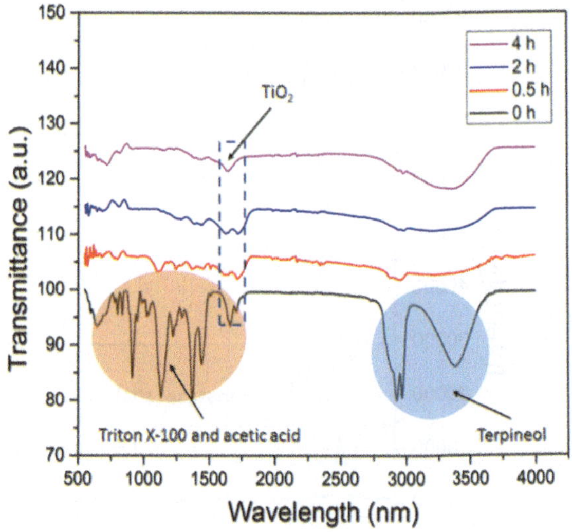

Fig. 3 FTIR spectra of flexible photoelectrodes with different UV exposure time

functional groups are disappeared that is attributed to the successful degradation of solvent, stabilizer and surfactant. The TiO_2 absorption peak as highlighted in the spectrum is found in all samples. Figure 4 shows the appearance of real samples with different exposure time.

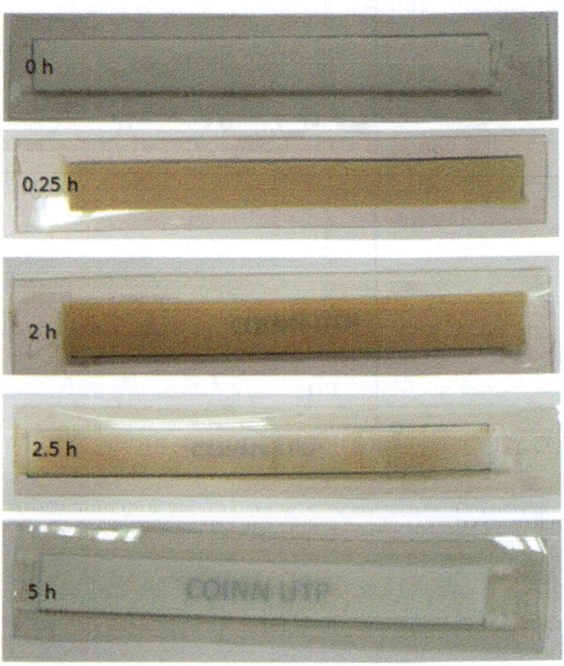

Fig. 4 Displaying the appearance of flexible photoelectrode under different UV exposure time

Figure 5a–c show the morphology of flexible TiO_2 photoelectrodes with UV exposure times 3 h, 4 h and 5 h, respectively. On comparison of the morphology of these photoelectrode film, it is observed that film treated at higher time 4 and 5 h are relatively denser as compared to the 3 h one. However, there is no significant difference found the morphology of 4 h, and 5 h UV treated films. Moreover, using EDX the composition of the film was determined in order to trace carbon contents which indicate the organic residue of stabilizers and surfactant. Figure 5d shows that carbon contents dropped significantly from ~ 11 to ~ 5.4%. However, there is no significant difference in carbon contents found between 4 and 5 h exposure time.

Figure 6 shows the optical transmittance of flexible photoelectrodes sintered with different UV exposure time. The photoelectrodes have shown higher optical transmittance in higher wavelength region. The transmittance decreased with a decrease in wavelength. The transmittance of 2 and 3 h photoelectrodes is low than 4 and 5 h and behaves differently. The curve for these 2 and 3 h photoelectrode shows abrupt fall in transmittance if go less than 550 nm. It shows the absorption of which attributed to the residue organic components and improper degradation. Moreover, the photoelectrode

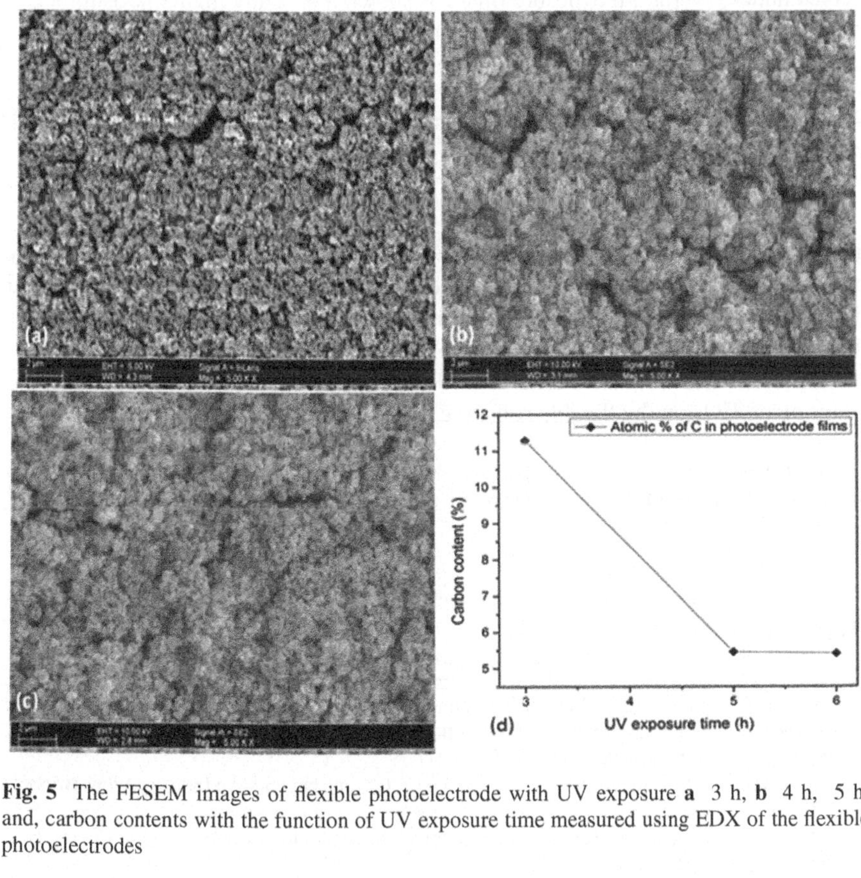

Fig. 5 The FESEM images of flexible photoelectrode with UV exposure **a** 3 h, **b** 4 h, 5 h, and, carbon contents with the function of UV exposure time measured using EDX of the flexible photoelectrodes

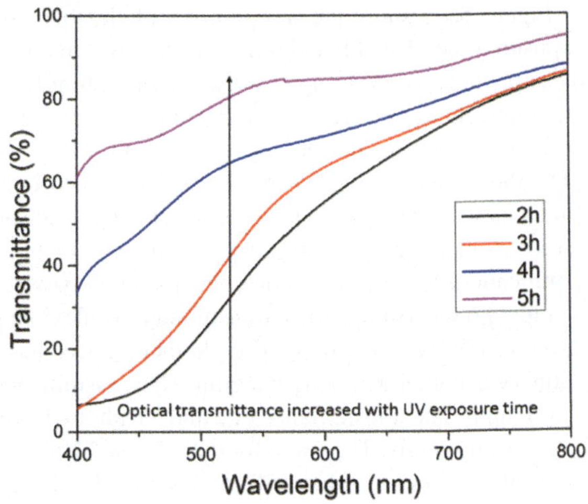

Fig. 6 Optical transmittance of flexible photoelectrodes with different UV exposure time

treated under UV for 5 h exposure time has shown drastically higher transmittance, which is attributed to the successful sintering and proper degradation of organic matters.

5 Conclusion

The flexible photoelectrode based on the binder-free TiO_2 paste is successfully presented. The UV exposure time to sinter TiO_2 film on the flexible substrate and degrade organic matter of the paste is optimized. It is found that UV exposure immediately transforms anatase and brookites phases of TiO_2 into rutile phases with (110) plane preferred orientation. FESEM, FTIR and UV–Vis have shown that for proper sintering of TiO_2 on the flexible substrate the optimum is 5 h.

Acknowledgements This work is supported by Ministry of Higher Education (MOHE), Malaysia under the research grant LRGS Nanomite (0153-AB k50).

References

1. Hashmi, G., Miettunen, K., Peltola, T., Halme, J., Asghar, I., Aitola, K., Toivola, M., Lund, P.: Review of materials and manufacturing options for large area flexible dye solar cells. Renew. Sustain. Energy Rev. **15**(8), 3717–3732 (2011)
2. Hoffmann, W.: A Vision for PV Technology up to 2030 and beyond–An Industry View. Brussels (2004)
3. El Chaar, L., El Zein, N.: Review of photovoltaic technologies. Renew. Sustain. Energy Rev. **15**(5), 2165–2175 (2011)

4. Feng, H., Jiao, X., Chen, R., Zhu, X., Liao, Q., Ye, D., Zhang, B., Zhang, W.: A microfluidic all-vanadium photoelectrochemical cell with the N-doped TiO2 photoanode for enhancing the solar energy storage. J. Power Sources **419**, 162–170 (2019)
5. Li, G., Sheng, L., Li, T., Hu, J., Li, P., Wang, K.: Engineering flexible dye-sensitized solar cells for portable electronics. Sol. Energy **177**, 80–98 (2019)
6. Chiu, W.-H., Lee, K.-M., Hsieh, W.-F.: High efficiency flexible dye-sensitized solar cells by multiple electrophoretic depositions. J. Power Sources **196**(7), 3683–3687 (2011)
7. Weerasinghe, H., Sirimanne, P., Franks, G., Simon, G., Cheng, Y.: Low temperature chemically sintered nano-crystalline TiO2 electrodes for flexible dye-sensitized solar cells. J. Photochem. Photobiol., A **213**(1), 30–36 (2010)
8. Pichot, F., Pitts, J.R., Gregg, B.A.: Low-temperature sintering of TiO2 colloids: application to flexible dye-sensitized solar cells. Langmuir **16**(13), 5626–5630 (2000)
9. Li, X., Lin, H., Li, J., Wang, N., Lin, C., Zhang, L.: Chemical sintering of graded TiO2 film at low-temperature for flexible dye-sensitized solar cells. J. Photochem. Photobiol., A **195**(2–3), 247–253 (2008)
10. Ming, L., Yang, H., Zhang, W., Zeng, X., Xiong, D., Xu, Z., Wang, H., Chen, W., Xu, X., Wang, M.: Selective laser sintering of TiO2 nanoparticle film on plastic conductive substrate for highly efficient flexible dye-sensitized solar cell application. J. Mater. Chem. A **2**(13), 4566–4573 (2014)
11. Chen, Z., Tian, Y., Li, S., Zheng, H., Zhang, W.: Electrodeposition of arborous structure nanocrystalline SnO2 and application in flexible dye-sensitized solar cells. J. Alloy. Compd. **515**, 57–62 (2012)
12. Chen, H.-W., Huang, K.-C., Hsu, C.-Y., Lin, C.-Y., Chen, J.-G., Lee, C.-P., Lin, L.-Y., Vittal, R., Ho, K.-C.: Electrophoretic deposition of TiO2 film on titanium foil for a flexible dye-sensitized solar cell. Electrochim. Acta **56**(23), 7991–7998 (2011)
13. Oh, Y., Lee, S.-N., Kim, H.-K., Kim, J.: UV-assisted chemical sintering of inkjet-printed TiO2 photoelectrodes for low-temperature flexible dye-sensitized solar cells. J. Electrochem. Soc. **159**(10), H777–H781 (2012)
14. Shahid, M.U., et al.: Solvent exfoliated graphene incorporated mixed phase TiO2 transparent photoelectrode for the efficient and color transparent dye-sensitized solar cell. Solar Energy **206**, 317–329 (2020)
15. Thamaphat, K., Limsuwan, P., Ngotawornchai, B.: Phase characterization of TiO2 powder by XRD and TEM. Kasetsart J. (Nat. Sci.) **42**(5), 357–361 (2008)

Ni-Doped TiO_2 Nanotubes for Low Concentration H_2 Sensing

Summaira Abbasi, Norani Muti Mohamed, Balbir Singh Mahinder Singh, and Mohamed Shuaib Mohamed Saheed

Abstract The Growing importance of H_2 in different fields such as in industries, fuel cell technology and medicine has increased the need for rapid detection of H_2. In this work, H_2-gas sensitive n-type material based on undoped and Ni-doped TiO_2 nanotubes were synthesized and their H_2 sensing properties were investigated. For the first time, Ni was doped into anodic nanotubular TiO_2 by hydrothermal reaction. Morphological and crystallographic analysis by electron microscope and X-ray diffraction suggested that Ni doping has no effect on surface morphology of the tube structure. Whereas, X-ray photoemission spectroscopy revealed that Ni has been incorporated interstitially in the TiO_2 network, which in turn induced effective reduction of bandgap, thereby improving the gas sensing properties towards low concentration of H_2 gas. The H_2 sensing properties of the nanotubes exposed to 20, 70 and 100 ppm of H_2 were investigated at operating temperature of 50, 100 and 200 °C. Ni-doped TiO_2 nanotube effectively improved the hydrogen sensing of the sensor compared to undoped TiO_2 sensor.

Keywords Hydrogen gas sensor · TiO_2 · Ni doping

S. Abbasi (✉) · N. M. Mohamed · M. S. M. Saheed
Centre of Innovative Nanostructures and Nanodevices (COINN), Universiti Teknologi PETRONAS, Seri Iskandar, Perak Darul Ridzuan, Malaysia

N. M. Mohamed
e-mail: noranimuti_mohamed@utp.edu.my

M. S. M. Saheed
e-mail: shuaib.saheed@utp.edu.my

B. S. M. Singh
Centre for Foundation Studies, Universiti Teknologi PETRONAS, Seri Iskandar, Malaysia
e-mail: balbir@utp.edu.my

S. A. Abdul Karim et al. (eds.), *Proceedings of the 6th International Conference on Fundamental and Applied Sciences*, Springer Proceedings in Complexity,
https://doi.org/10.1007/978-981-16-4513-6_36

1 Introduction

H_2 is a promising and capable nontoxic energy source. Due to which, the rapid industrialization deems it necessary to constantly monitor this inflammable gas (4%in air). The application of H_2 gas varies across a wide range of industrial uses. Not only that, but it also has a medical relevance to it. The presence of low concentration of H_2 in breath (< 25 ppm) is found to be a biomarker for certain diseases such as bacterial growth, lactose intolerance and NEC. Over the years, a wide range of different detection principles, methods, and materials have been adopted to develop the hydrogen sensors. Conventional sensing material and mechanism include semiconductor metal oxides [1, 2], Schottky diodes [3, 4], catalytic sensors[5], electrochemical sensors, optical and acoustic etc. [6].

Semiconductor metal oxides (SMO) with the notable features of low cost, simplicity and long term stability, has proved as potential candidates for gas sensing technologies [1]. In this kind of sensor, H_2 gas diffuses, reacting with the adsorbed oxygen on the surface, resulting changes in its electrical resistance. A number of researches have shown that performance of a SMO gas sensors is greatly related to surface reactions [7].

Commonly and dominantly used SMOs for gas sensors are ZnO, SnO_2 etc. However, with the advent of nanotechnology, various other nanostructures of metal oxides have been studied, and advancement in this technology to modify the morphology as well as surface properties of the nanomaterial has been achieved. Among a number of studies on numerous metal oxides, a rapid growing interest in TiO_2 has been observed due to its wide band gap (~ 3.2 eV) p rendering its usage in catalytic, sensing and energy related applications [8, 9].

The application of TiO_2 nanotubes for hydrogen gas sensing illustrates the remarkable and readily measurable properties. At room temperature, in response to 1000 ppm hydrogen TiO_2 nanotube arrays ~ 1 μm in length demonstrated an unprecedented change in the electrical resistance of 8.7 orders of magnitude [10]. Due to its notable ability, it has proven to be an ideal candidate for hydrogen sensing [11]. However, despite of being sensitive enough against resistance, TiO_2 nanotubes carries a drawback in terms of band gap. The crystalline forms of TiO_2 nanotubes contain anatase and rutile structures. The anatase is of great interest especially in the fields of sensors as well as photoelectrochemical [12]. However, the band gap of anatase is around 3.2 eV, resulting in relatively poor electrical conductivity and increase in device resistance. This prevents it from further improvement in the sensitivity.

The sensitivity can be enhanced by increasing the conductivity of the sensing element. The existing literature suggests that the doping is an efficient way to improve the properties of TiO_2 nanotubes [13]. Extensive efforts have been made to extend the functionalities of titania by modifying the band structure with a few dopant materials. Only few reports with doped or alloyed TiO_2 nanotube arrays are published. Li et al. [14, 15] used a NiTi alloy (50.8 at.% Ni). The doped nanotubes were found to be sensitive to hydrogen atmospheres in the temperature range from room temperature to 200 °C. A wide-range of sensing has been carried out from 50 ppm to 2%. Since

Nickel doping has shown remarkable improvement, therefore a wide space for the efficiency improvement in H_2 sensing properties is expected. Hence the focus dopant in the work would be nickel. However, one drawback encountered is that by using NiTi alloy, the doping level of Ni is fixed, which renders further improvement in H_2 gas sensing.

In this work as opposed to using NiTi alloy, the nickel doped TiO_2 nanotubes are synthesized using a two-step route as first synthesized using anodization in viscous electrolyte consisting of ethylene glycol of NH_4F. Then the nanotubes are annealed and then doped with nickel using hydrothermal reaction. The hydrogen gas sensing properties of the prepared samples were investigated to low concentrations of H_2 gas. With the integration and development of the synthesized nanotubes as stand along sensor, a dedicated gas chamber is further used to investigate thoroughly the gas sensing abilities. The remarkable sensing results has open new frontiers in the H_2 gas sensing with high sensitivity and quick response in detection at room and higher working temperatures.

2 Methodology

2.1 Synthesis and Characterization of the Sensing Element

Titanium foil 0.127 mm thickness, with the purity of 99.8% and platinum foil 0.25 mm thickness were used. Chemicals used were ethylene glycol, NH_4F, NH_4OH, $NiSO_4$, acetone and ethanol. All the chemicals were of reagent grade.

Prior to the experiment, the titanium foil (99.8% purity) was first mechanically polished using sandpaper, followed by ultrasonication in a solution of acetone and ethanol. Titanium foil was then rinsed with deionized water and dried in air. The as-prepared TiO_2 nanotube samples are amorphous. The samples are then annealed in a furnace at 500 °C for three hours to improve crystallinity. This improves the efficiency of nanotubes (Nanostructure). The samples were then washed with distilled water and dried. The anodization was carried out with titanium plate as the anode and platinum as cathode in an electrochemical cell consisting of two-electrodes. The potential was supplied by a DC power supply connected to current sensor interfaced to a PC using PASCO 750 interface. The experiments were performed at room temperature. The electrolyte consisted of ethylene glycol and 0.27 M NH_4F and 3 wt.% H2O. The anodization was done at potentials 40 V for 2 h. Hydrothermal was carried out using the as-prepared samples for 24 h, in a mixture of NH_4OH and $NiSO_4$.

2.2 Sensor Configuration and Gas Detection Measurements

Like any semiconductor resistive type sensor, the electrodes are pasted on the top of the TiO_2 and Ni-TiO_2 nanotubes for measuring the changes in the electrical characteristics. In this work, gold electrodes are prepared by using silver epoxy directly onto the nanotubes. The sensing element is fixed on a glass substrate and thin gold wires were wire bonded for creating electrical contacts from the Au electrodes to by using silver paste.

2.3 Hydrogen Gas Sensing Setup

The test chamber for H_2 gas-sensing consists of a dedicated flow type stainless steel chamber (4 L). Mass Flow Controllers (MFC) is used to control the gas flow through the tube. The variation in sample resistance in air containing certain concentration of hydrogen is observed by computer-controlled digital multimeter (Agilent 34410A). All testing are carried out in laboratory environment under room temperature, unless specified. The schematic diagram of testing setup is shown in Fig. 1.

In order to control the operation temperature, the sensor was placed on an aluminum block attached to a thermocouple. The gas tube has an outer diameter of 8 mm and the distance between the tube mouth and the sensor was 40 mm.

The gas chamber is first flushed using an inert gas e.g. Ar. Then air is introduced, in order to provide an ambient environment. The H_2 gas of desired concentration (20 ppm) is inserted and contained. The changes in electrical characteristics are then measured. The desired concentration of H_2 has been created by inserting an exact volume of H_2 to the chamber and the concentration was verified with a calibrated commercial H_2 sensor (Linde). The measurement is repeated for each H_2 concentration at different operating temperatures. The response of the sensor to hydrogen is defined as:

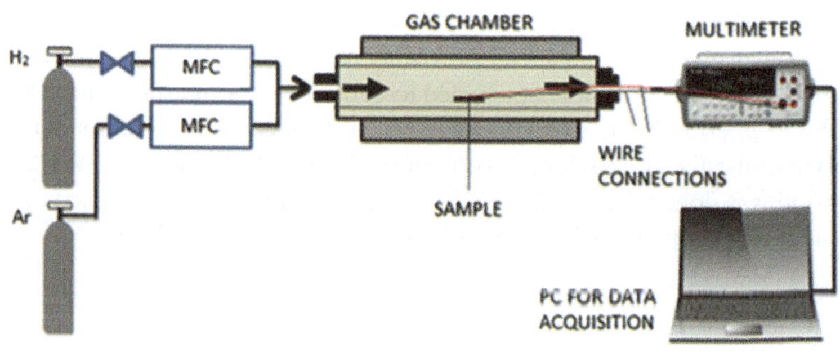

Fig. 1 Schematic of gas sensing setup

$$S = \frac{R_0 - R_H}{R_H} \tag{1}$$

where R_o is the resistance in the absence of hydrogen, and R_H is the resistance in the presence of hydrogen.

3 Results and Discussion

3.1 *Characterization of Ni Doped TiO_2 Nanotube*

After the synthesis, the structure of nanotubes were characterized using FESEM and EDX measurements were performed to analyze the elemental composition of the nanotubes. Figure 2 presents the morphology of top and cross-sectional images of the nanotubes annealed at 500 °C. The result showed that after anodization of Ti for 2 h at 40 V, well organized nanotubes were successfully achieved with average diameter of 90 nm and tube length of about 6 μm (Fig. 2a). In Fig. 2a, the EDX result revealed that the nanotubes consisted of three elements: Ti, O, and F.

The TiO_2 nanotubes are doped with nickel, in order to modify their properties, and to improve the gas sensing characteristics. Figure 2b shows the morphological images of TiO_2 nanotubes after doping with nickel. The cross-sectional image shows

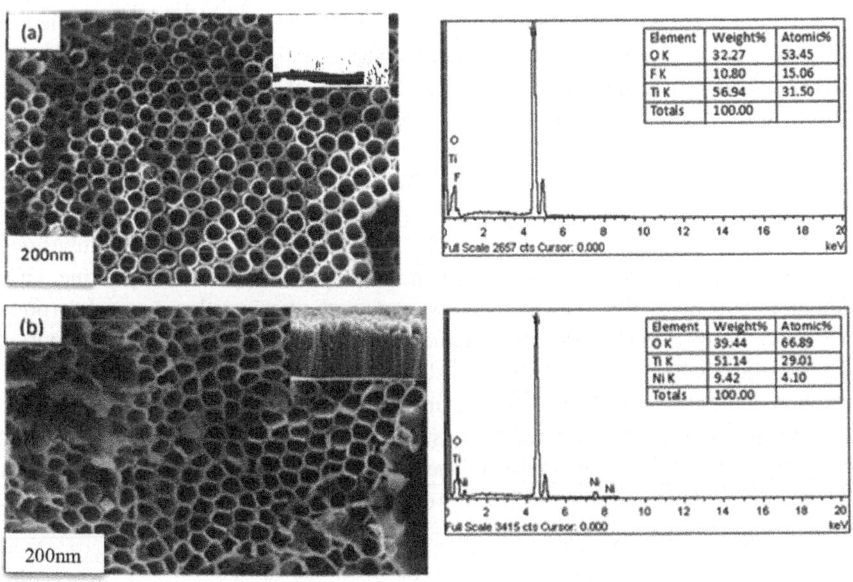

Fig. 2 **a** FESEM images and EDX result of TiO_2 nanotubes before doping with nickel, **b** FESEM images and EDX result of TiO_2 nanotubes after doping with nickel

Fig. 3 XRD patterns of pristine TiO_2 nanotubes and Ni doped TiO_2 nanotubes, showing absence of nickel peak

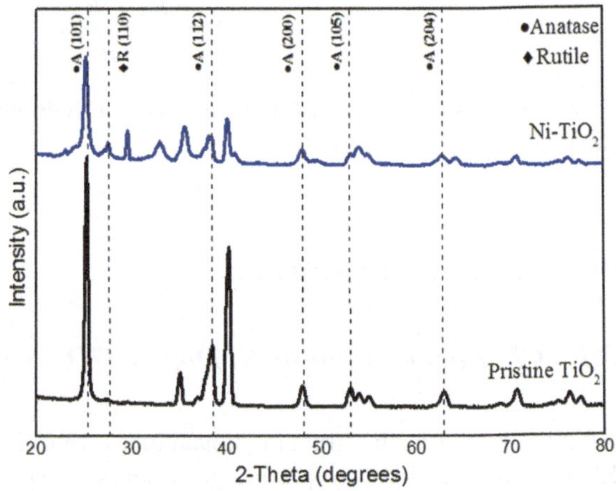

slightly collapsed nanotubes at the top, overall tubular structure is retained. The EDX result confirms the presence of Ti, O and Ni elements. The atomic percentage of Ti, O and Ni was 29.01%, 66.89%, and 4.10% respectively.

XRD analysis is conducted to examine and study the crystallinity of TiO_2 and Ni-TiO_2 nanotubes. Crystal structure and stability of the TiO_2 nanotubes need to be taken into consideration for different purposes such as DSCs, photocatalytic applications and as a sensing element. As mentioned above, the anatase crystal phase of TiO_2 is preferred, since it has a higher charge carrier mobility compared to the rutile phase [16]. Figure 3 shows XRD pattern of pure and Ni-doped TiO2 films annealed at 500 °C in order to convert the amorphous material into anatase. Typical peaks in XRD pattern of TiO2 films are observed at 2θ values of 25.28°, 38.08°, 48.04°, and 55.02°, which are assigned to 101, 112, 200, and 211 planes respectively. Studies show that the observed crystalline peaks correspond to anatase TiO_2 phase [17]. In both samples, the peaks corresponding to the formation of the anatase phase can be observed, showing that incorporation of Ni preserves the anatase structure.

Figure 4 presents the comparison of absorbance of pristine TiO_2 nanotubes and Ni-TiO_2 nanotubes. It can be seen that the absorption intensity of pristine TiO_2 nanotubes cuts off at 396 nm which corresponds to the intrinsic band gap absorption of anatase TiO_2 nanotube, indicating a bandgap of 3.2 eV, consistent with the reported values [18]. For Ni-TiO_2 nanotubes red shift was observed and showed absorption, which is centered on 480–590 nm. A substantial enhancement in the visible light absorption is observed compared to pristine TiO_2 nanotube, which is considered due to the doping of Ni ions in TiO_2. The optical band gap energy is calculated by extrapolating the linear portion of the (ahυ)2 versus hυ. Based on this, the band gap is calculated to be 2.5 eV.

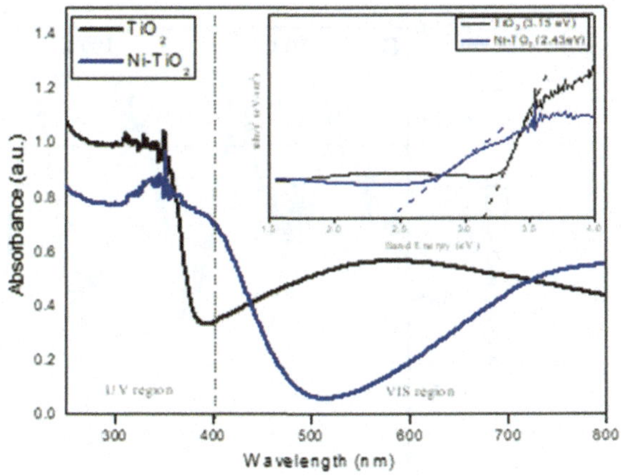

Fig. 4 UV–Vis spectrum of pristine TiO_2 and Ni-doped TiO_2 nanotubes. The inset is a plot of $(ah\upsilon)^2$ versus energy

3.2 *Hydrogen Detection Properties*

Electrical characterization of TiO_2 nanotubes based sensor is performed using the setup described previously. The chamber is first flushed with flowing air through it, for half an hour, after which H_2 gas of specific concentration, mixed with 95% of N_2 gas is introduced at a fixed flow rate of 200 sccm and contained to study the response. Figure 5a describes the resistance versus time saturation and maximal response of TiO2 sensor to 20, 70 and 100 ppm respectively of H_2 at operating temperature of 50, 100 and 200 °C for 2 cycles of the response and recovery. Baseline resistance for TiO_2 nanotubes is found to be 6.3 MΩ.

In order to investigate the behavior of the sensor on repeated hydrogen exposure for 20 ppm, the temperature was varied in discrete steps from 0 to 200 °C while keeping the hydrogen concentration constant at 20 ppm; the chamber is flushed with air after each exposure to hydrogen. When H_2 gas of 20 ppm is introduced, there is decrease in resistance by a factor of 0.5 MΩ at 50 °C, and when H_2 is flushed out, the sensor recovers almost back to its baseline resistance. Change in the temperature results in the reduction of resistance by a factor of 1 MΩ at 100 °C and 1.3 MΩ 200 °C respectively. To obtain the certainty analysis, the process is repeated by flowing in 70 and 100 ppm of H_2 gas concentration.

For Ni-TiO_2 nanotube sensor, TiO_2 nanotube with nickel doping level of 4% is employed. For Ni-TiO_2 the baseline resistance is found to be 5.8 MΩ. The change in baseline resistance signifies the increase in conductivity on the surface of the doped nanotubes. Following the same procedure of gas testing as TiO_2 nanotube sensor, the Ni-TiO_2 nanotubes sensor is also tested with different concentration of H_2 (20 ppm, 70 ppm, 100 ppm) at various operating temperatures. Figure 5b shows the resistance versus time plot for 20, 70 and 100 ppm respectively at different temperatures.

Overall, the increase in temperature resulted in the decrease in baseline resistance and also substantial reduction in the hysteresis. There is a linear increase in sensitivity

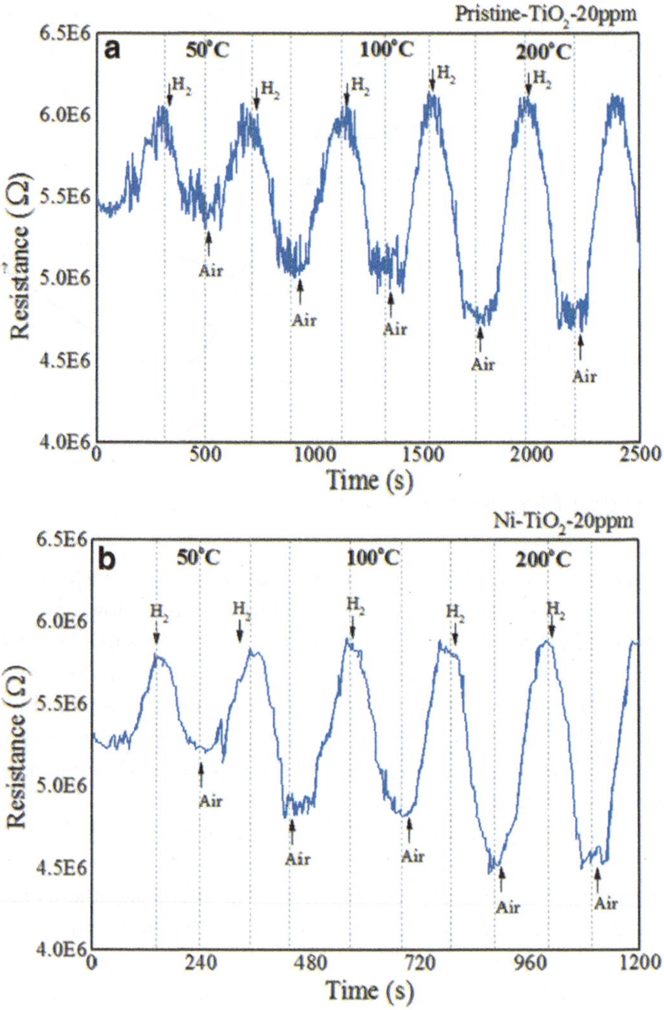

Fig. 5 **a** 20 ppm H_2 sensing of pristine TiO_2 at 50, 100, and 200 °C temperature. **b** 20 ppm H_2 sensing of Ni–TiO_2 at 50, 100, and 200 °C temperature. Sensitivity versus H_2 gas concentrations at different operating temperature for TiO_2 nanotubes

with respect to temperature. The same behavior can be observed for the increase in H_2 concentration.

The lowest sensor sensitivity (14%) for the pristine TiO_2 nanotubes is at 50 °C for 20 ppm H_2 concentration. It increases as H_2 concentration increases, achieving up to 27% at 100 ppm. At 100 °C, the sensitivity at 20 ppm is 22% for 20 ppm, 34% for 70 ppm and 42% for 100 ppm, the improvement in the sensitivities is associated with the increase in temperature. The similar trend of increment is also seen at 200 °C, the response range from 29% (20 ppm) to 50% (100 ppm) is obtained.

4 Conclusion

In the present work, for the first time highly-organized anatase-phases TiO_2 nanotubes doped with nickel is successfully synthesized for the detection of low concentration of contained H_2 gas. The main objective of this research is to develop sensing element based on nickel doped TiO_2 nanotubes with improved sensing properties by anodization method and integrating it into sensor configuration. The Ni–TiO_2 nanotubes compared to pristine TiO_2 showed enhanced sensitivity toward low concentration of H_2 surrounding. Sensor showed fast response and recovery time, and showed excellent sensitivity at 200 °C. Hence, the results revealed that the sensor is not only able to detect 20 ppm of H_2 gas in N_2 surrounding, but also utilization of this sensor in various application specially breath analyzer has been realized. Furthermore, the new method of doping can open new doors towards enhancing the sensing properties even more by changing the doping levels of Ni.

References

1. Phanichphant, S.: Semiconductor metal oxides as hydrogen gas sensors. Procedia Eng. **87**, 795–802 (2014)
2. Gu, H., Wang, Z., Hu, Y.: Hydrogen gas sensors based on semiconductor oxide nanostructures. Sensors **12**, 5517–5550 (2012)
3. Iwanaga, T., Hyodo, T., Shimizu, Y., Egashira, M.: H 2 sensing properties and mechanism of anodically oxidized TiO2 film contacted with Pd electrode. Sens. Actuators, B Chem. **93**, 519–525 (2003)
4. Skucha, K., Fan, Z., Jeon, K., Javey, A., Boser, B.: Palladium/silicon nanowire Schottky barrier-based hydrogen sensors. Sens. Actuators, B Chem. **145**, 232–238 (2010)
5. Han, C.-H., Hong, D.-W., Kim, I.-J., Gwak, J., Han, S.-D., Singh, K.C.: Synthesis of Pd or Pt/titanate nanotube and its application to catalytic type hydrogen gas sensor. Sens. Actuators, B Chem. **128**, 320–325 (2007)
6. Hübert, T., Boon-Brett, L., Black, G., Banach, U.: Hydrogen sensors–a review. Sens. Actuators, B Chem. **157**, 329–352 (2011)
7. Wang, C., Yin, L., Zhang, L., Xiang, D., Gao, R.: Metal oxide gas sensors: sensitivity and influencing factors. Sensors **10**, 2088–2106 (2010)
8. Sreekantan, S., Saharudin, K.A., Lockman, Z., Tzu, T.W.: Fast-rate formation of TiO2 nanotube arrays in an organic bath and their applications in photocatalysis. Nanotechnology **21**, 365603 (2010)
9. Xie, Z.-B., Adams, S.: A reliable TiO 2 nanotube membrane transfer method and its application in photovoltaic devices. Electrochim. Acta **62**, 116–123 (2012)
10. Paulose, M., Varghese, O.K., Mor, G.K., Grimes, C.A., Ong, K.G.: Unprecedented ultra-high hydrogen gas sensitivity in undoped titania nanotubes. Nanotechnology **17**, 398 (2005)
11. Fomekong, R.L., Kelm, K., Saruhan, B.: High-temperature hydrogen sensing performance of Ni-doped TiO2 prepared by co-precipitation method. Sensors **20**, 5992 (2020)
12. Luttrell, T., Halpegamage, S., Tao, J., Kramer, A., Sutter, E., Batzill, M.: Why is anatase a better photocatalyst than rutile?-Model studies on epitaxial TiO2 films. Sci. Rep. **4** (2014)
13. Zaleska, A.: Doped-TiO2: a review. Recent Patents Eng. **2**, 157–164 (2008)
14. Li, Z., Ding, D., Liu, Q., Ning, C., Wang, X.: Ni-doped TiO(2) nanotubes for wide-range hydrogen sensing. Nanoscale Res. Lett. **9**, 118–118 (2014)

15. Simonetti, E.A.N., de Oliveira, T.C., do Carmo Machado, A.E., Silva, A.A.C., dos Santos, A.S., de Simone Cividanes, L.: TiO2 as a gas sensor: the novel carbon structures and noble metals as new elements for enhancing sensitivity–a review. Ceramics Int. (2021)
16. Baraton, M.-I., Merhari, L.: Surface chemistry of TiO 2 nanoparticles: influence on electrical and gas sensing properties. J. Eur. Ceram. Soc. **24**, 1399–1404 (2004)
17. Sharma, S.D., Singh, D., Saini, K.K., Kant, C., Sharma, V., Jain, S.C., et al.: Sol–gel-derived super-hydrophilic nickel doped TiO2 film as active photo-catalyst. Appl. Catal. A: General **314**, 40–46 (2006)
18. Lv, L., Bai, X., Ye, Z.: Construction of N-doped TiO 2/SnO 2 heterostructured microspheres with dominant 001 facets for enhanced photocatalytic properties. CrystEngComm **18**, 7580–7589 (2016)

Effect of Nickel Doping on the Optical and Morphological Properties of Titanium Dioxide Nanotubes

Summaira Abbasi, Norani Muti Mohamed, Balbir Singh Mahinder Singh, Asad Mumtaz, and Mohamed Shuaib Mohamed Saheed

Abstract Doping is considered to be one of the effective approaches to modify the properties of TiO_2 nanomaterials. The work presented is the investigation on the effect of nickel doping using successive ion layer adsorption-reaction (SILAR) method on the optical and morphological properties of titanium dioxide nanotubes, synthesized by anodizing Ti foil electrochemically. Morphology of the nanotubes is studied under field emission scanning electron microscope (FE-SEM) with EDX result revealed the incorporation of nickel in TiO_2 nanotube. The anatase crystal structure of titania was identified by X-ray diffraction (XRD). Furthermore, optical properties of nickel doped TiO_2 nanotube are also investigated using UV–VIS spectroscopy, in which red shift is observed, indicating change in the band gap. The promising results of the reduction in the bandgap of the doped TiO_2 opens new frontier in the development of sensors and photovoltaic cell, where the bandgap tailoring is essential for an improved performance.

Keywords TiO_2 Nickle doping · SILAR XRD · Properties of Ni-TiO_2

S. Abbasi (✉) · N. M. Mohamed · A. Mumtaz · M. S. M. Saheed
Centre of Innovative Nanostructures and Nanodevices (COINN), Universiti Teknologi PETRONAS, Seri Iskandar, Perak Darul Ridzuan, Malaysia

N. M. Mohamed
e-mail: noranimuti_mohamed@utp.edu.my

M. S. M. Saheed
e-mail: shuaib.saheed@utp.edu.my

B. S. M. Singh
Centre for Foundation Studies, Universiti Teknologi PETRONAS, Seri Iskandar, Malaysia
e-mail: balbir@utp.edu.my

S. A. Abdul Karim et al. (eds.), *Proceedings of the 6th International Conference on Fundamental and Applied Sciences*, Springer Proceedings in Complexity,
https://doi.org/10.1007/978-981-16-4513-6_37

1 Introduction

Nanotechnology has achieved immense attention owing to its investigation in novel nanostructured materials and is also believed to hold the key to enhance contemporary and advanced technologies. Among such materials, TiO_2 has been found to be one of the most investigated compounds [1]. This recognition can be mainly attributed to its perspective utilization in photoelectrochemical applications such as Solar cells [2, 3] and photocatalysis [4, 5]. It is a very functional material consisting of features like wide band gap, excellent ion intercalation properties, environmental friendly, excellent biocompatibility and electronic properties. It has also demonstrated outstanding performance in other fields e.g. hydrogen generation [6], sensors [7–9], and biological applications [10].

Various approaches have been employed to synthesize TiO_2 nanotubes which include sol–gel method, template-assisted, hydrothermal and electrochemical. Since properties of TiO_2 nanotube change significantly when its morphology is altered therefore, electrochemical anodization synthesis method is the most preferred choice as it facilitate in controlling the nanotube's morphology [10]. By changing the anodizing parameters and electrolyte parameters e.g. viscosity, composition, the morphological features are controlled.

The crystalline form of TiO_2 nanotube contains anatase and rutile structures. The anatase is of great interest especially in the fields of photoelectrochemical and sensors. The band gap of anatase is around 3.2 eV resulting in relatively poor electrical conductivity and increase in device resistance. In addition to this, the reactivity and chemical stability of TiO_2 nanotube is only high under UV light ($\lambda < 387$ nm), and is low under visible light ($\lambda > 400$ nm), the major part of solar spectrum. The existing literature suggests that the above problems can effectively be approached by doping, which is an efficient way to improve the properties of TiO_2 nanotubes. Generally compared with pristine TiO_2, TiO_2 doped with transition metal-ion usually has a narrower band gap, larger specific surface area and smaller grain size, and its optical absorption expands and a red shift can be observed. This leads to improved photocatalytic properties and improved sensing properties compared to undoped TiO_2 [11–13]. The effect of TiO_2 doped with some amount of Ni has been investigated by various groups. A report on the improved sensing of the nanostructured 2.5% Ni-doped TiO_2 thin film to liquified petroleum gas has been presented [14]. A Study on photocatalyst properties of Ni-TiO_2 film for splitting $Na_2S + Na_2SO_3$ solution into H_2 Ni-doping resulted in a broader optical absorption range of the nanotubes and a red shift with increase of doping degree [15, 16].

It is envisaged that with the fabrication of a doped-TiO_2 nanotubular structure, improved photoelectrochemical and sensing properties can be realized. To date, very rare work has been done on doping TiO_2 nanotubes after synthesizing the nanotubes. In this work, the doping of TiO_2 nanotubes after its synthesis has been successfully been achieved, whereby, giving the flexibility to change the doping level as per requirement. TiO_2 nanotubes have been synthesized by anodization method, and have been doped with Ni via successive ionic layer adsorption and reaction (SILAR) method. In order to demonstrate the effects of Ni doping on the semiconducting properties of TiO_2, energy band gap of Ni-TiO_2 was calculated. The band gap is calculated by using the equation $EG = 1240/\lambda_e$, where EG is bandgap, and λ_e is the absorption edge spectrum.

2 Experimental

2.1 Methods and Material

Titanium foil 0.127 mm thickness, with the purity of 99.7% and Platinum foil 0.25 mm thickness were used. Chemicals used were all reagent grade: Ethylene glycol (99%Ammonium), Fluoride, Nickel Nitrate, Ethanol, Methanol and Acetone.

2.2 Synthesis of Titanium Dioxide Nanotubes (TiO_2)

Prior to the experiment, the titanium sheet (99.8% purity) was first mechanically polished using sandpaper, followed by ultra-sonication in a solution of acetone and ethanol. Titanium foil was washed with deionized water and it was air dried. The anodization was then performed with titanium plate as the anode and platinum as cathode in a two-electrode system electrochemical. The potential was supplied by a DC power supply connected to current sensor interfaced to a PC using PASCO 750 interface. The experiments were performed at room temperature. The electrolyte consisted ethylene glycol and 0.30 M NH_4F and 3 wt% H_2O. The anodization was done at potentials 40 V for 2 h.

2.3 Annealing

The as-prepared TiO_2 samples are amorphous. The samples are then annealed in a furnace at 500 °C for three hours to improve crystallinity. This improves the efficiency of nanotubes. After that the samples were cleansed with distilled H_2O and dried.

2.4 Doping with Nickel

Doping of TiO_2 was carried out using a subsequent ionic layer absorption reaction (SILAR) technique as previously reported [10, 11]. One SILAR cycle, consisted of submerging a TiO_2 sample in 0.1 M $Ni(NO_3)_2$ solution for one minute, followed by washing with methanol and then again submerging in in methanol/water solution (50/50 vol.) for one minute. It was done under magnetic stirring, ending with washing in methanol/water. Eight to twelve SILAR cycles, were completed, after which the samples were vacuum dried overnight at 80 °C.

2.5 *Characterization*

After annealing, the morphological structures of the sample was observed under field effect scanning electron microscope (FE-SEM), and the crystal structure of the samples was examined X-ray Diffraction (XRD) and optical properties were studied using Ultraviolet–visible spectroscopy (UV–VIS).

3 Results and Discussion

Titanium foil of 0.127 mm thickness was anodized in a bath containing ethylene glycol, 0.30 M Ammonium fluoride was used as the source of fluoride, mixed in 3 wt% H_2O. The anodizing voltages used were 40 V, whereas the anodization time observed was 2 h and the growth of the nanotubes was analyzed. The water content is kept to the minimum since the presence of H_2O assists in dissolution of nanotubes, resulting in shorter length.

Figure 1 shows the FE-SEM images of pristine and doped nanotubes synthesized at 40 V for 2 h. During the anodization bubbles can be seen evolving at the cathode, suggesting the formation of H_2 as described by previous works [4]. The distribution of the nanotubes is very uniform having diameter around 80 nm (Fig. 1a) with smooth

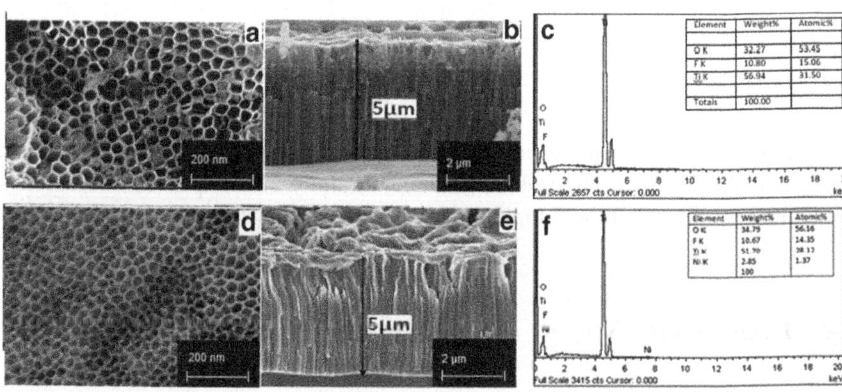

Fig. 1 The surface morphology of TiO_2 nanotubes anodized at 40 V for 2 h. **a** Top view, **b** cross-section of nanotubes, **c** EDX patterns of the TiO_2 nanotubes, **d** top view of Ni doped TiO_2, **e** cross-section of Ni doped TiO_2 nanotubes, **f** EDX patterns of Ni doped TiO_2

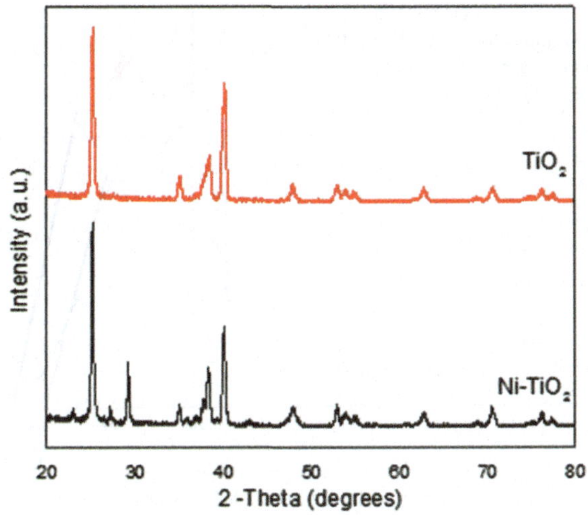

Fig. 2 XRD pattern of pristine and doped TiO_2 nanotube

walls. The cross-sectional thickness of the nanotube was found to be 5 μm as shown in Fig. 1b. The EDX analysis of the pristine nanotubes (Fig. 1c) revealed that the nanotubes consisted of three elements namely Ti, O and F. The weight percentage of the Ti, O and F element at the top surface of TiO_2 nanotubes were 56.94%, 32.27% and 10.80%, respectively. Figure 1d and e shows the top and cross sectional view of doped TiO_2 nanotubes. It is observed that the structure of the nanotube is maintained even after doping, and there isn't any surface dissolution taking place, hence the length of the nanotubes is also same. The EDX analysis of Ni-TiO_2 in Fig. 1f indicated the presence of Ni along with Ti, O and F elements. The weight percentages of Ni, Ti, O and f were found to be 2.85%, 51.70%, 34.79% and 10.67% respectively.

The XRD pattern of TiO_2 and Ni-TiO_2 is shown in Fig. 2. From the analysis, TiO_2 exists in two main crystallographic forms i.e. anatase and rutile, with anatase being the dominant phase. Compared to rutile, anatase phase considered more reactive, sensitive and also it has higher charge carrier mobility compare to the rutile phase. The phase structure of nanotubes (Fig. 2) shows various peaks indicating the presence of anatase and rutile. It can be found that the nanotubes could still retain their crystallographic form after doping. In contrast to the EDX analysis, the XRD does not detect any crystalline form of Ni. Therefore, there isn't any differences found between the two patterns. It is believed that Ni may have dissolved into the matrix, or exists in the amorphous form.

Optical properties of TiO_2 nanotube and Ni-TiO_2 nanotube are measured using UV–Vis spectroscopy with a wavelength range of 200–800 nm. The energy band gap of TiO_2 has been known to be 3.2 eV [12, 13] as evident in Fig. 3. However, red shift of around 50 nm is observed for Ni-TiO_2 and the band-gap is calculated to be 2.8 eV. The reduction in the band-gap is attributed to the creation of sub-states below the conduction band and above valence band, resulting in the narrowing of the band

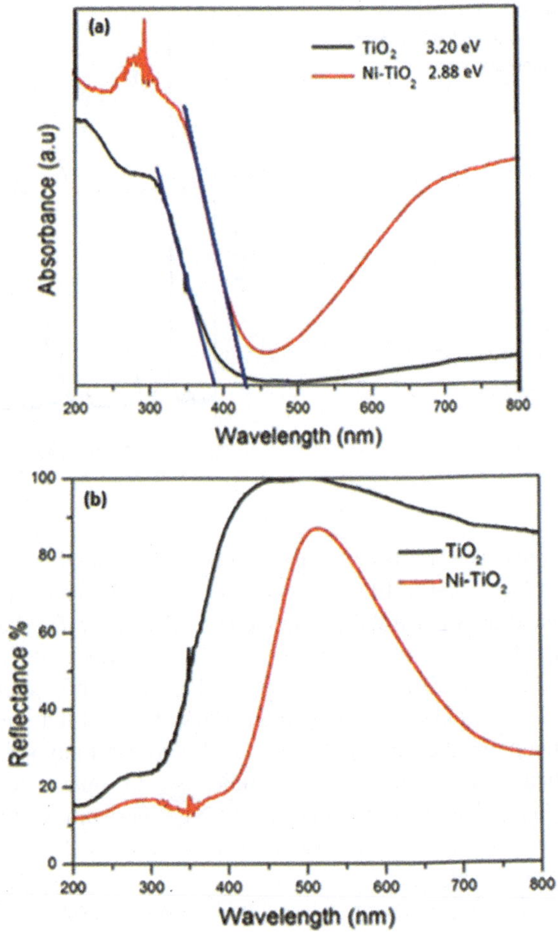

Fig. 3 UV–Vis spectra (**a**) and reflectance (**b**) of the two samples TiO_2 and Ni-TiO_2 nanotube

gap. The Ni-TiO_2 nanotube also shows the absorption in the range > 550 nm region. This is because the oxygen defects have been created which allows the sample to absorb in this region too. The reflectance graph (Fig. 3) shows strong reflectance at > 320 nm for the TiO_2 nanotubes, whereas for Ni-TiO_2, the reflectance starts from around 400 nm, supporting the results in the absorption. The change in band gap also indicates the presence of nickel doping.

4 Conclusion

In summary, simple method to dope TiO_2 nanotube with nickel has been demonstrated using SILAR technique. FE-SEM result showed the nanotubes are highly oriented with diameter around 80 nm and length approximately 5 μm. EDX conducted confirmed the doping of nickel in TiO_2 nanotube, which is found to be 2.85%. Compared with TiO_2 nanotube, about 50 nm red shift in the spectrum of UV–Vis is observed as well. The interstitial doping is expected to improve photoelectrochemical as well as sensing performance compared to undoped TiO_2 nanotube and warrants further studies. The new preparation method provides great advantage of flexibility by altering the doping levels according to conditions as per required and reduced preparation cost.

Acknowledgements This research was supported by the Top Down Nano Fund (0153AB-B47) funded by the Ministry of Science, Technology and Innovation. We would like to show utmost gratitude to the Universiti Teknologi PETRONAS for providing lab facility and financial support throughout the project.

References

1. Roy, P., Berger, S., Schmuki, P.: TiO_2 nanotubes: synthesis and applications. Angew. Chem. Int. Ed. **50**, 2904–2939 (2011)
2. Mor, G.K., Varghese, O.K., Paulose, M., Shankar, K., Grimes, C.A.: A review on highly ordered, vertically oriented TiO_2 nanotube arrays: fabrication, material properties, and solar energy applications. Sol. Energy Mater. Sol. Cells **90**, 2011–2075 (2006)
3. Nikolay, T., Larina, L., Shevaleevskiy, O., Ahn, B.T.: Electronic structure study of lightly Nb-doped TiO_2 electrode for dye-sensitized solar cells. Energy Environ. Sci. **4**, 1480–1486 (2011)
4. Sun, M., Ma, X., Chen, X., Sun, Y., Cui, X., Lin, Y.: A nanocomposite of carbon quantum dots and TiO_2 nanotube arrays: enhancing photoelectrochemical and photocatalytic properties. RSC Adv. **4**, 1120–1127 (2014)
5. Leong, K.H., Chu, H.Y., Ibrahim, S., Saravanan, P.: "Palladium nanoparticles anchored to anatase TiO_2 for enhanced surface plasmon resonance-stimulated, visible-light-driven photocatalytic activity. Beilstein J. Nanotechnol. **6**, 428–437 (2015)
6. Ye, M., Gong, J., Lai, Y., Lin, C., Lin, Z.: High-efficiency photoelectrocatalytic hydrogen generation enabled by palladium quantum dots-sensitized TiO_2 nanotube arrays. J. Am. Chem. Soc. **134**, 15720–15723 (2012)
7. Perillo, P., Rodríguez, D.: A room temperature chloroform sensor using TiO_2 nanotubes. Sens. Actuators B: Chem. **193**, 263–266 (2014)
8. Kılınç, N., Şennik, E., Işık, M., Ahsen, A.Ş., Öztürk, O., Öztürk, Z.Z.: Fabrication and gas sensing properties of C-doped and un-doped TiO_2 nanotubes. Ceram. Int. **40**, 109–115 (2014)
9. Bindra, P., Hazra, A.: Selective detection of organic vapors using TiO_2 nanotubes based single sensor at room temperature. Sens. Actuators B: Chem. **290**, 684–690 (2019)
10. Grimes, C.A., Mor, G.K.: TiO_2 Nanotube Arrays: Synthesis, Properties, and Applications. Springer (2009)
11. Smith, Y.R., Gakhar, R., Merwin, A., Mohanty, S.K., Chidambaram, D., Misra, M.: Anodic titania nanotube arrays sensitized with Mn-or Co-doped CdS nanocrystals. Electrochim. Acta **135**, 503–512 (2014)
12. Seong, M., Kim, S., Yoo, H., Choi, J.: Doping of anodic nanotubular TiO_2 electrodes with MnO_2 for use as catalysts in water oxidation. Catal. Today **260**, 135–139 (2016)
13. Seo, H., Baker, L.R., Hervier, A., Kim, J., Whitten, J., Somorjai, G.A.: Generation of highly n-type titanium oxide using plasma fluorine insertion. Nano Lett. **11**, 751–756 (2011)

14. Patil, L., Suryawanshi, D., Pathan, I., Patil, D.: Nickel doped spray pyrolyzed nanostructured TiO_2 thin films for LPG gas sensing. Sens. Actuators B Chem. **176**, 514–521 (2013)
15. Alijani, M., Ilkhechi, N.N.: Effect of Ni doping on the structural and optical properties of TiO_2 nanoparticles at various concentration and temperature. SILICON **10**, 2569–2575 (2018)
16. Cahyaningsih, D., Taufik, A., Saleh, R.: Effect of Ni doping on the structural and optical properties of TiO_2 nanoparticles prepared by co-precipitation method. J. Phys.: Conf. Ser. 012017 (2020)

Performance of Organic Polymer Electrolyte Based on Extracted Aloe Vera Polysaccharide Compared with Mannose, Agarose and Carboxymethyl-Cellulose (CMC) for DSCs Application

N. A. S. Yuharmon, Norani Muti Mohamed, Chong Fai Kait, and K. Y. Cheong

Abstract In this work, the performance of integrated DSC based on AVG/CMC-rGO-I^-/I_3^- gel polymer electrolyte has been compared with the mannose/CMC-DMSO-rGO-I^-/I_3^-, agarose/CMC-DMSO-rGO-I^-/I_3^- and CMC/PEG-DMF-rGO-I^-/I_3^- gel polymer electrolyte. The gel polymer electrolyte were characterized using field-emission scanning electron microscopy (FESEM) in order to examine the cross-section images of the gel polymer electrolyte thickness, current–voltage (I-V) measurement and electrochemical impedance spectroscopy (EIS). The fabricated test cells with an active area of 0.40 cm^2 incorporated with AVG/CMC-rGO-I^-/I_3^- gel polymer electrolyte, possess the highest conversion efficiency (η) of 0.84%. However, it still considered lower compared to mannose (2.35%), agarose (3.55%) and CMC (5.76%) gel polymer electrolyte, due to the lower percentage of the extracted polysaccharides.

Keywords Gel polymer electrolyte · Efficiency · Polysaccharides

N. A. S. Yuharmon (✉) · C. F. Kait
Fundamental and Applied Science Department, Universiti Teknologi PETRONAS, 32610 Seri Iskandar, Perak, Malaysia
e-mail: nur_16001694@utp.edu.my

C. F. Kait
e-mail: chongfaikait@utp.edu.my

N. M. Mohamed
Centre of Innovative Nanostructure and Nanodevices, Universiti Teknologi PETRONAS, 32610 Seri Iskandar, Perak, Malaysia
e-mail: noranimuti_mohamed@utp.edu.my

K. Y. Cheong
School of Material and Mineral Resources, Universiti Sains Malaysia, Engineering Campus, 14300 Nibong Tebal, Pulau Pinang, Malaysia
e-mail: srcheong@usm.my

S. A. Abdul Karim et al. (eds.), *Proceedings of the 6th International Conference on Fundamental and Applied Sciences*, Springer Proceedings in Complexity,
https://doi.org/10.1007/978-981-16-4513-6_38

1 Introduction

Greenhouse emission can be reduced by a developing of renewable and green energy. Solar technology is one type of renewable energy that have an advantages and potential for power generation [1]. Solar cells, or also known as photovoltaic cells able to convert photon with specific wavelength to electricity. The physicals of solar cells is depend on the photovoltaic influenced namely the generation, trapping, recombination and transport of electron–hole pairs throughout the semiconducting materials and within the contact electrodes [2].

Photovoltaic technologies based on inorganic materials have been commercially used until now. However, the price is costly and it also need more energy consuming for preparation methods. The materials that has been used also are toxicity and contains lower natural abundance. Therefore, organic materials has been introduced in order to avoid those problems. Unfortunately, the efficiency of organic-based photovoltaic cells cannot be compared with the purely inorganic-based photovoltaic technologies [3, 4].

The most efficient dye solar cells (DSCs) is reported to possess high conversion efficiency greater than 11%, which required low-cost renewable energy source. However, liquid electrolytes have a major problem where the electrolyte loss caused by the leakage and volatilization, which lead to the limitation of the long-term usage of DSCs. Solidifying liquid electrolyte by forming a gel or quasi-solid-state electrolytes is the primary solution in order to overcome this problem. It is because gel or quasi-solid-state electrolytes able to make the sealing process easier and directly capable to minimize the loss of electrolytes for enhanced durability [5].

Nowadays, an organic material such as k-carrageenan and agarose that contain in seaweed, cellulose and chitosan has been choose by researcher to act as an electrolytes for solar cells application. An organic materials has been used because it has an advantages such as environmentally friendly, production cost for large-scale fabrication is lower, easier processing technique compared to liquid electrolytes and non-toxic [6]. Besides, it also able to act as a dielectric layer with a good dielectric performance and exhibit charge trapping capabilities [5].

However, electrolyte based on an organic itself will cause several problem namely lower ionic conductivity and the bonding within the layer become weaker [7]. Furthermore, this electrolytes will suffer from solvent leakage and also demand careful sealing treatment at high temperature environments due to their instability under high temperature [8].

In this research, bioactive polysaccharide of Aloe Vera in the form of manually extracted gel while comparing with the commercial mannose (also known as polysaccharide) from Aloe Vera gel, agarose from bioactive compound seaweed and plant extracted carboxymethyl-cellulose (CMC) has been proposed when all of this organic materials act as an bio- or organic polymer electrolytes for DSCs. All of this organic materials has been choose because it have a potential organic dielectric for viable electronic application [9].

2 Materials and Methods

2.1 Materials

Extracted Aloe Vera gel using Soxhlet extractor. Commercialize mannose, agarose, carboxymethyl-cellulose (CMC), polyethylene glycol (PEG) and graphene oxide (GO) were obtained from Sigma. Iodide/tri-iodide organic solvent based electrolyte (EL-HSE), N719 (ruthenium 535-bis TBA), dimethylformamide (DMF) and dimethyl sulfate (DMSO) as a solvent to dilute polymer salt and TiO_2 were used as received without purification.

2.2 Preparation of Organic Gel Polymer Electrolytes

The extracted Aloe gel polymer electrolyte were prepared by mixing the extracted Aloe gel, which is in gelling form with CMC and reduce graphene oxide (rGO). Then, all of this mixture were dissolved in iodide/tri-iodide organic solvent-based electrolyte and being stirred at 20 °C. As for commercial mannose and agarose gel polymer electrolytes, fixed amount of mannose and agarose (4.00 g) mixed with CMC, rGO and iodide/tri-iodide organic solvent-based electrolyte were dissolved in DMSO solvents (4.00 ml) in a beaker with continuous stirring, also at constant heating of 20 °C. While for CMC gel polymer electrolyte, PEG (3.50 g) has been mixed with CMC, rGO and iodide/tri-iodide organic solvent-based electrolyte. It has been dissolved in DMF and left for continuous stirring (70 °C).

2.3 Dye Solar Cells (DSCs) Fabrication

A double-layer TiO_2 was coated on fluorine-doped tin oxide (FTO) working glass electrode while platinum (Pt) was coated on the FTO conducting glass electrode by screen printing. After that, the FTO/TiO_2 was annealed and then sensitized in an N719 dye solution at room temperature for 24 h. The organic gel polymer electrolyte then being screen printed on FTO/TiO_2/N719 dye working glass by using doctor blade method and left it for a minute so that the electrolyte will diffuse homogenize on it. DSCs were fabricated by glued both of TiO_2 photo-anode and Pt-sputtered cathode in sandwiched form. A DSCs test cell (active area 0.400 cm^2) were completed for each type of organic gel polymer electrolytes.

3 Results and Discussion

3.1 *Field-Emission Scanning Electron Microscopy (FESEM-Thickness)*

The cross-section images of the printed gel polymer electrolytes film taken using FESEM at 100× of magnification are shown in Fig. 1. There have two layer that been detected for all samples which are the upper layer attributed to the organic gel polymer electrolyte layer while the bottom layer correspond to the ITO glass substrates. As for AVG/CMC-rGO-I^-/I_3^- gel polymer electrolyte film exhibit the thickness around 115.8 μm. After that, mannose/CMC-DMSO-rGO-I^-/I_3^-, agarose/CMC-DMSO-rGO-I^-/I_3^- and CMC/PEG-DMF-rGO-I^-/I_3^- gel polymer electrolyte films have a thickness around 183.1 μm, 90.3 μm and 562.1 μm, respectively.

However, CMC/PEG-DMF-rGO-I^-/I_3^- gel polymer electrolyte films was take over the thickness. It acquire the highest thickness followed with mannose/CMC-DMSO-rGO-I^-/I_3^-, AVG-Ethanol-Hexane (7:3)/CMC-rGO-I^-/I_3^- and agarose/CMC-DMSO-rGO-I^-/I_3^- gel polymer electrolytes. Therefore, the performance of all electrolytes can be relate with the thickness of each gel polymer electrolytes that has been observed from the cross-sectional images.

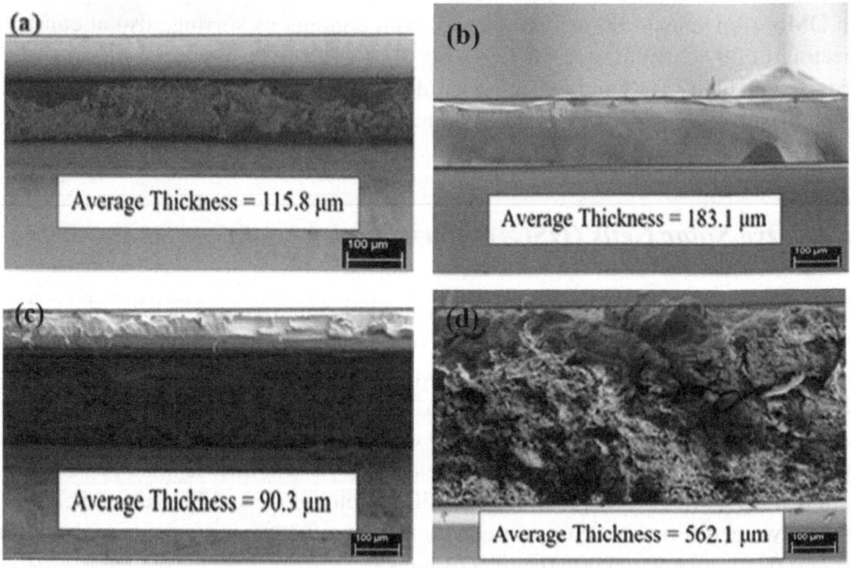

Fig. 1 Typical cross-sectional FESEM images of (a) AVG-Ethanol-Hexane (7:3)/CMC-rGO-I^-/I_3^-, (b) mannose/CMC-DMSO-rGO-I^-/I_3^-, (c) agarose/CMC-DMSO-rGO-I^-/I_3^- and (d) carboxymethyl cellulose/PEG-DMF-rGO-I^-/I_3^- gel polymer electrolytes deposited on the ITO substrates

3.2 *Current–Voltage (I-V) Performance*

The performance of integrated DSC based on AVG/CMC-rGO-I^-/I_3^- gel polymer electrolyte has been compared with the mannose/CMC-DMSO-rGO-I^-/I_3^-, agarose/CMC-DMSO-rGO-I^-/I_3^- and CMC/PEG-DMF-rGO-I^-/I_3^- gel polymer electrolyte. The photoelectrochemical measurement of all electrolytes were determined by the current–voltage (I-V) curves, as shown in Fig. 2. The overall light-to-electricity conversion efficiency (η), fill factor (FF), open-circuit voltage (V_{OC}) and short-circuit current (I_{SC}) has been tested under illumination of 100 Mw/cm^2 intensity of simulated light. The summarized data of all photovoltaic parameter has been listed in the Table 1.

As can be seen, the test cells with active area 0.400 cm^2 that has been fabricated based on AVG/CMC-rGO-I^-/I_3^- gel polymer electrolyte exhibits I_{SC} of 0.50 mA, V_{OC} of 0.80 V and the FF of 2.11. The energy conversion efficiency (η) obtained is 0.84%. After being compared with the commercial main product

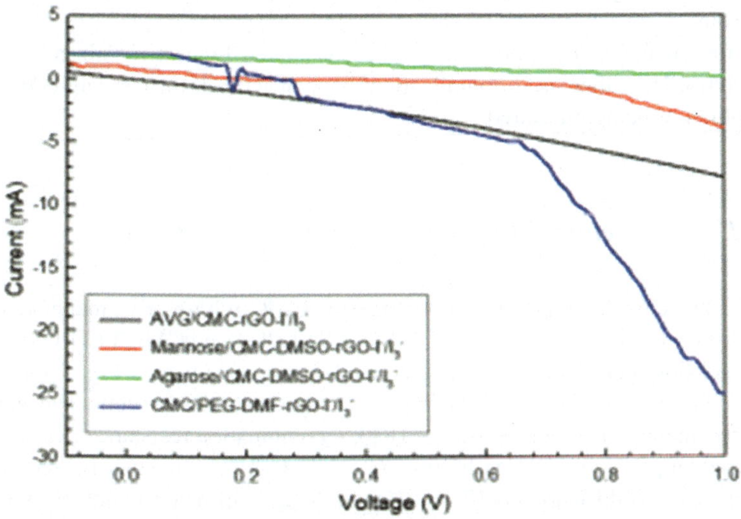

Fig. 2 I-V curve of gel polymer electrolytes for solar cells application

Table 1 The photoelectrochemical performance parameters of the test cells for gel polymer electrolytes with active area of 0.400 cm^2

Gel polymer electrolytes	V_{OC}, V	I_{SC}, mA	FF	η, %
AVG/CMC-rGO-I^-/I_3^-	0.80	0.50	2.11	0.84
Mannose/CMC-DMSO-rGO-I^-/I_3^-	0.86	0.99	2.77	2.35
Agarose/CMC-DMSO-rGO-I^-/I_3^-	0.80	1.85	2.78	4.12
CMC/PEG-DMF-rGO-I^-/I_3^-	0.90	1.99	3.22	5.76

such as mannose, agarose and CMC, it can be conclude that CMC/PEG-DMF-rGO-I^-/I_3^- gel polymer electrolyte possess highest values of efficiency which is 5.76% followed by agarose/CMC-DMSO-rGO-I^-/I_3^- gel polymer electrolyte (4.12%) and mannose/CMC-DMSO-rGO-I^-/I_3^- gel polymer electrolyte (2.35%).

From the table also, it shows that the value of V_{OC} is different to each other except for AVG/CMC-rGO-I^-/I_3^- gel polymer electrolyte and agarose/CMC-DMSO-rGO-I^-/I_3^- gel polymer electrolyte, 0.80 V. From the previous study, similar V_{OC} indicates that there is no difference in the value of Fermi level energy of TiO_2 and redox couples potential.

After that, AVG/CMC-rGO-I^-/I_3^- gel polymer electrolyte and CMC/PEG-DMF-rGO-I^-/I_3^- gel polymer electrolyte shows the large difference in terms of I_{SC}, where the values is too gap in between of 0.50 mA to 1.99 mA. This situation can be related to the injection of the efficiency of electron from dye molecules to the semiconductor and the recombination rate between the injected electrons and redox couples in the gel polymer electrolytes [6]. However, the dye molecules and TiO_2 photoelectrode in this study does not being modified. Therefore, it can be conclude that the electron injection efficiency which has been controlled by a photon energy can be considered as constant.

As for fill factor, FF values, it can be considered highest for all gel polymer electrolytes. Thus, there is no complications happened inside the test cells component during the I-V being measured.

3.3 Electrochemical Impedance Spectroscopy (EIS)

Figure 3 presents the impedance plots for AVG/CMC-rGO-I^-/I_3^-, mannose/CMC-DMSO-rGO-I^-/I_3^-, agarose/CMC-DMSO-rGO-I^-/I_3^- and CMC/PEG-DMF-rGO-I^-/I_3^- gel polymer electrolytes. The equivalent circuit given in Fig. 4 have been employed for the curve fitting of the impedance spectra of the test cells. The analyzed properties namely transport resistance (R_T), recombination resistance (R_{BR}), chemical capacitance ($C\mu$), electron lifetime (τ_n), reaction rate constant for recombination (k), electron diffusion coefficient (D_n), effective diffusion length (L_n) and the steady-state electron density in the conduction band (n_s) were summarized in Table 2.

From the table above, it presents that the value of R_{BR} for AVG/CMC-rGO-I^-/I_3^-, agarose/CMC-DMSO-rGO-I^-/I_3^- and CMC/PEG-DMF-rGO-I^-/I_3^- gel polymer electrolyte higher compared to R_T, where the arc form in Nyquist plot showed a true circle, whilst mannose/CMC-DMSO-rGO-I^-/I_3^- gel polymer electrolyte possess lower R_{BR} ($R_{BR} < R_T$). When the D_n values increased, there is no effect to the Nyquist plot. However, decreasing the values of D_n will significantly changes the shapes of Nyquist plot. It can be seen through the Fig. 3 (c). Mannose/CMC-DMSO-rGO-I^-/I_3^- gel polymer electrolyte shows that when the values of R_T decreased with an increase in D_n, there is no drastic effect was observed [10].

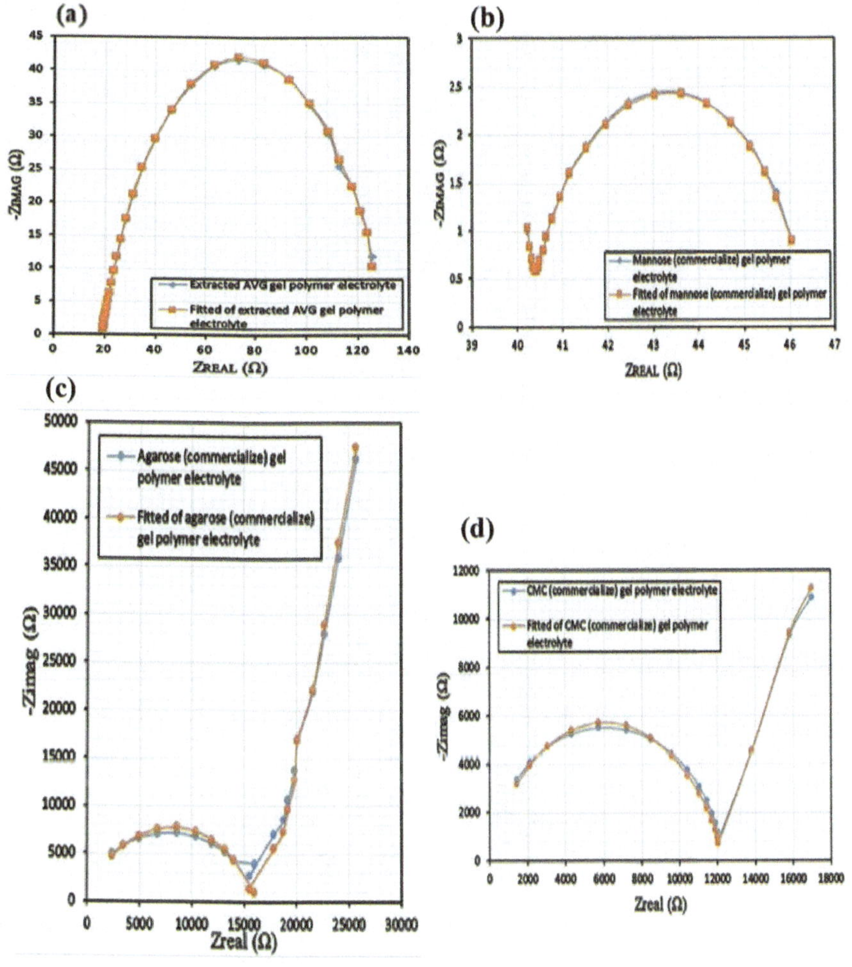

Fig. 3 The EIS analysis plots for (a) AVG/CMC-rGO-I^-/I_3^-, (b) mannose/CMC-DMSO-rGO-I^-/I_3^-, (c) agarose/CMC-DMSO-rGO-I^-/I_3^- and (d) CMC/PEG-DMF-rGO-I^-/I_3^- gel polymer electrolyte with 0.400 cm^2 active area

Fig. 4 Equivalent circuit model of the EIS of gel polymer electrolytes test cells

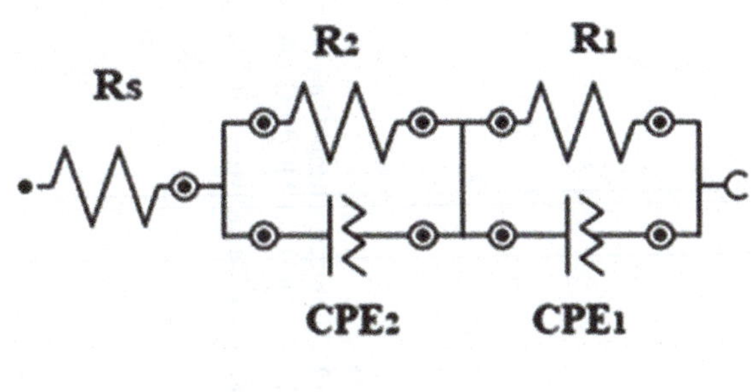

Table 2 The electrochemical properties of DSC integrated with (a) AVG/CMC-rGO-I^-/I_3^-, (b) mannose/CMC-DMSO-rGO-I^-/I_3^-, (c) agarose/CMC-DMSO-rGO-I^-/I_3^- and (d) CMC/PEG-DMF-rGO-I^-/I_3^- gel polymer electrolyte

Gel Polymer Electrolytes	R_t (Ω)	R_{br} (Ω)	C_μ (μF)	τ_n (s)	k (s^{-1})	D_n (cm^2/s)	L_n (μm)	Con (Ωcm/s)	n_s (cm^{-3})
a	30.62	81.19	0.226	1.83×10^{-3}	5.45×10^2	5.310×10^{-5}	163	5.310×10^{-5}	7.57×10^{17}
b	9.6838	6.11	0.021	1.26×10^{-7}	7.95×10^6	0.050	0.79	0.058	6.90×10^{14}
c	1.57×10^5	15,522	7.370	0.114	8.74	8.624×10^{-9}	0.31	1.628×10^{-4}	2.47×10^{17}
d	11,489	27,028	0.000457	1.24×10^{-5}	8.10×10^4	0.0019	150	2.626	1.53×10^{13}

4 Conclusion

The performance of AVG/CMC-rGO-I^-/I_3^- gel polymer electrolyte has been compared with the mannose/CMC-DMSO-rGO-I^-/I_3^-, agarose/CMC-DMSO-rGO-I^-/I_3^- and CMC/PEG-DMF-rGO-I^-/I_3^- gel polymer electrolytes. The cross-sectional images taken from FESEM shows that CMC/PEG-DMF-rGO-I^-/I_3^- gel polymer electrolyte film (562.1 μm) is the thicker compared to AVG/CMC-rGO-I^-/I_3^- (115.8 μm), mannose/CMC-DMSO-rGO-I^-/I_3^- (183.1 μm) and agarose/CMC-DMSO-rGO-I^-/I_3^- (90.3 μm) gel polymer electrolytes film. As for the I-V results, it representing that CMC/PEG-DMF-rGO-I^-/I_3^- gel polymer electrolyte possess the highest efficiency which is 5.76% followed by agarose/CMC-DMSO-rGO-I^-/I_3^- gel polymer electrolyte (4.12%), mannose/CMC-DMSO-rGO-I^-/I_3^- gel polymer electrolyte (2.35%) and AVG/CMC-rGO-I^-/I_3^- gel polymer electrolyte (0.84%). After that, the EIS analysis shows when the electron density in the conduction band higher, the recombination rate become lower, thus resulting in high efficiency performance.

Acknowledgements The authors would like to express their appreciation to the Centre of Innovative, Nanostructures & Nanodevices (COINN), University of Technology PETRONAS for all support of the project and also to the colleagues that always give support and encouragement in this meantime.

References

1. Gong, J., Liang, J., Sumathy, K.: Review on dye-sensitized solar cells (DSSCs): fundamental concepts and novel materials. Renew. Sustain. Energy Rev. **16**(8), 5848–5860 (2012)
2. Archer, M.D., Green, M.A.: Clean Electricity from Photovoltaics. World Scientific Publishing Company (2014)
3. Nazeeruddin, M.K., Baranoff, E., Grätzel, M.: Dye-sensitized solar cells: a brief overview. Sol. Energy **85**(6), 1172–1178 (2011)
4. Yu, Z.: Liquid redox electrolytes for dye-sensitized solar cells. In: KTH Chemical Science and Engineering (2012)
5. Hsu, H.-L., Hsu, W.-T., Leu, J.: Effects of environmentally benign solvents in the agarose gel electrolytes on dye-sensitized solar cells. Electrochim. Acta **56**(17), 5904–5909 (2011)
6. Bella, F., et al.: From seaweeds to biopolymeric electrolytes for third generation solar cells: an intriguing approach. Electrochim. Acta **151**, 306–311 (2015)
7. Su'ait, M., Rahman, M., Ahmad, A.: Review on polymer electrolyte in dye-sensitized solar cells (DSSCs). Sol. Energy **115**, 452–470 (2015)
8. Ye, M., et al.: Recent advances in dye-sensitized solar cells: from photoanodes, sensitizers and electrolytes to counter electrodes. Mater. Today **18**(3), 155–162 (2015)
9. Lim, Z.X., Cheong, K.Y.: Effects of drying temperature and ethanol concentration on bipolar switching characteristics of natural Aloe vera-based memory devices. Phys. Chem. Chem. Phys. **17**(40), 26833–26853 (2015)
10. Adachi, M., et al.: Determination of parameters of electron transport in dye-sensitized solar cells using electrochemical impedance spectroscopy. J. Phys. Chem. B **110**(28), 13872–13880 (2006)

Effect on Addition of Low Loading of Graphene Oxide to the Physical Characterization of Electrospun Polystyrene Fiber Mat

Kamilah Ramly, Norani Muti Mohamed, and Chong Fai Kait

Abstract Electrospun fiber made from electrospinning process is popular in the development of flexible devices. For polystyrene to be applied in various application that will experience physical endurance such as bending, pulling and folding; graphene oxide was added as filler to enhance the mechanical strength of the fiber mat. Wet electrospinning of polystyrene and polystyrene embedded with low content of graphene oxide were carried out and the change of the physical characteristic was observed. Based on the morphology, the non-bead fiber diameter was decreased after the addition of the graphene oxide due to high conductivity of the polymer solution. In addition, the chemical reaction between polymer matrix is poorly resolved as shown in FTIR spectra of PS + GO, however in RAMAN spectra the G band of the GO peaks was observed in the PS + GO peak. There was also some interaction of GO polystyrene polymer chain as based on the XRD; the PS + GO peak were becoming less intense as compared to PS fiber. Finally, the mechanical strength of the PS + GO fiber was also increased upon the addition of the GO despite of the addition of GO is considered small, ranging around 0.01%, proving the dispersion of the low loading of nanofiller inside the fiber mat.

Keywords Electrospinning · Low loading filler · Graphene oxide

K. Ramly (✉) · N. M. Mohamed · C. F. Kait
Fundamental and Applied Sciences Department, Universiti Teknologi PETRONAS, Seri Iskandar 32610, Perak, Malaysia

N. M. Mohamed
e-mail: noranimuti_mohamed@utp.edu.my

C. F. Kait
e-mail: chongfaikait@utp.edu.my

N. M. Mohamed · C. F. Kait
Center of Innovative Nanostructures and Nanodevices (COINN), Universiti Teknologi PETRONAS, Seri Iskandar 32610, Perak, Malaysia

S. A. Abdul Karim et al. (eds.), *Proceedings of the 6th International Conference on Fundamental and Applied Sciences*, Springer Proceedings in Complexity,
https://doi.org/10.1007/978-981-16-4513-6_39

1 Introduction

The production of flexible substrate nanofiber has increased since the last decade proportional to its demand increase. As the flexible substrate is lightweight and foldable to a certain degree, it is suitable to be attached to consumer end-product that emphasize in wearability, portability and rollability without sacrificing its performance. In electrospinning process, nanoweb parameters such as diameter, porosity and alignment can be finely tuned, as compared to other methods such as drawing, phase separation and self-assembly. Electrospun nanofiber produced by electrospinning has a large surface area and high porosity with very small pore [1, 2]. As the result, it can be widely applied in energy harvesting, energy storage, oil–water separation, drug delivery and bone or skin tissue engineering [3, 4].

Typical electrospinning unit consist of two electrodes (one is the needle while the other is the collector) and a DC voltage supply. Initially, the drop of polymer solution injected by the syringe at the tip of the needle were spherical in shape, due to its surface tension. When voltage is applied, electrostatic force will overcome the surface tension, transforming the spherical shape into an oval shape and finally deformed into a conical shape called "Taylor cone". Further increment of electrostatic force will pull the polymer solution drop and produce jet solution which will be grounded onto the surface of the collector. Based on the fundamental of electrospinning process, it could be foreseen that there are many parameters involved in developing a reproducible fiber. The morphology of the electrospun fiber is highly depending on the process parameter such as the conductivity of the polymer solvent, the density of polymer, surrounding humidity, feeding rate and applied voltage. Apart from having high thermal and chemical resistance, electrospinning of polystyrene alone is prone to low mechanical strength and challenging to be processed when it is subjected to the specific application that require high durability of the membrane [5]. For this reason, fillers were added to the wet electrospinning in order to counter the drawback of low mechanical strength of single electrospun fiber. Wet electrospinning is facile compared to melt electrospinning because it is difficult to maintain a high melting point polymer in liquid form at room temperature. Additionally, wet electrospinning also can reduce the potential of the polymer to undergo chemical reaction under high temperature. Aside from that, physical blending of filler either miscible or immicible can also be electrospun by wet electrospinning [6]. This research focuses on the impact of low loading of graphene oxide embedded within the polystyrene fiber, with the objective to forsee the effect on the physical properties of the final product. It is proven that the addition of small amount of graphene into polymers can enhance the reinforcement, damage tolerance, electrical and thermal conductivity [7].

Graphene is a well-known material to be used for electrochemical energy storage due to its unique structure and properties, for instance, ultra-large specific surface area, decent chemical stability, excellent electrical conductivity, and high mechanical properties [8]. Due to its wide potential application such as supercapacitor, optical element in fiber laser, lithium ion battery, sensor and solar cell, various method have been developed to produce graphene since its early discovery in 2004 [9, 10]. However, major drawback of graphene is that it is insoluble in many solvent, contributing to its poor interaction when dispersed in polymer matrix. Unlike pristine graphene which consisted of pure carbon that is highly conductive in nature, graphene oxide is bound with several functional groups that contains oxygen namely epoxy, hydroxyl, carboxyl and ester. As for the case of graphene oxide, the material would still have a good connectivity with the ability to be dissolved in many organic solvents because of the functional group attached at the edge of the GO plane.

2 Methodology

2.1 *Materials*

Polystyrene (Mw = 280,000 g/mol), N-N,dimethylformamide (DMF) and graphene oxide in water dispersion (Sigma Aldrich) were used without further purification.

2.2 *Production of Nonwoven Polystyrene Based Fiber Mat*

For a start, Polystyrene solution were prepared in 20% (w/v) in DMF. It is stirred for 12 h to promote homogenous mixture. The solution is then inserted into plastic syringe and the nonwoven polystyrene (PS) fiber were produced with the electrospinning unit with applied voltage of 15 kV, distance from tip of the needle to the collector is 10 cm and the feeding rate of 1 ml/h is used. Height of needle is to be fix therefore the needle should be perpendicular to rotating collector covered in aluminium foil.

Graphene oxide (GO) in water dispersion were dried at 80 °C for 12 h to eliminate water content in graphene prior to the electrospinning process. 0.01% Graphene oxide were added to PS solution and sonicated for 1 h to promote homogenous dispersion. The PS + GO solution were then electrospun using the earlier mentioned parameter while maintaining the surrounding humidity.

2.3 *Characterizations*

Fourier Transform Infrared Spectroscopy (FTIR) of graphene oxide dispersed in KBr pellet was recorded using ATR technique with Pelkin Elmer, Spectrum One. The wavenumber was ranged from 400 to 4000 cm^{-1}. As for the PS and PS + GO membrane, the membrane was carefully peeled from the aluminium foil and tested as a thin film. Morphology characterization of graphene oxide were carried out using

Fig. 1 Tensile test on fiber mat

FESEM (VPFESEM, Zeiss Supra55 VP). All samples were coated with gold before testing. An area of (0.5 × 0.5) cm^{-2} nonwoven mat as spun fiber attached with the collector substrate (aluminium foil) from the center were used.

Raman Spectroscopy was recorded using Horiba JobinYvon HR800 and the wavenumber spanning from 1000 to 2000 cm^{-1} using 514 nm source. Raman Spectra is then fitted using gaussian function using OriginLab. As for XRD measurement, the scan was carried out from 5° to 40° using the X'Pert3 Powder & Empyrean, PANalytical model. Further analysis on mechanical strength of the nanofiber mat were carried out using a 100 N Universal Tensile Machine. The sample were prepared according to ASTM D882. The tensile rate was set at 500 (in. min^{-1}), the gage length of the sample was 50 mm and the thickness for both PS fiber and PS + GO fiber was comparable at about 0.3 mm. The tensile test arrangement on the fiber mat is shown in Fig. 1.

3 Discussion

It was observed that the non-beads fiber of polystyrene fiber mat diameter reduced after the GO is embedded into the fiber upon electrospinning process, as depicted in Fig. 2a. The diameter of the initial PS fiber mat was ranging from 2 to 5 μm, while after the GO addition, the size is decreased in diameter to around 1–3.5 μm as shown in Fig. 2b. This morphology most likely was affected by the increased in solution conductivity causing the low surface tension at the tip of the electrospinning needle, decreasing the flow rate and eventually causing a thinner solution jet captured by the collector [11].

Both Raman and Fourier Transform Infrared Spectroscopy (FTIR) were used as characterization tool to determine the functional group in graphene oxide due to different information can be extracted, such as Raman is active when polarization is

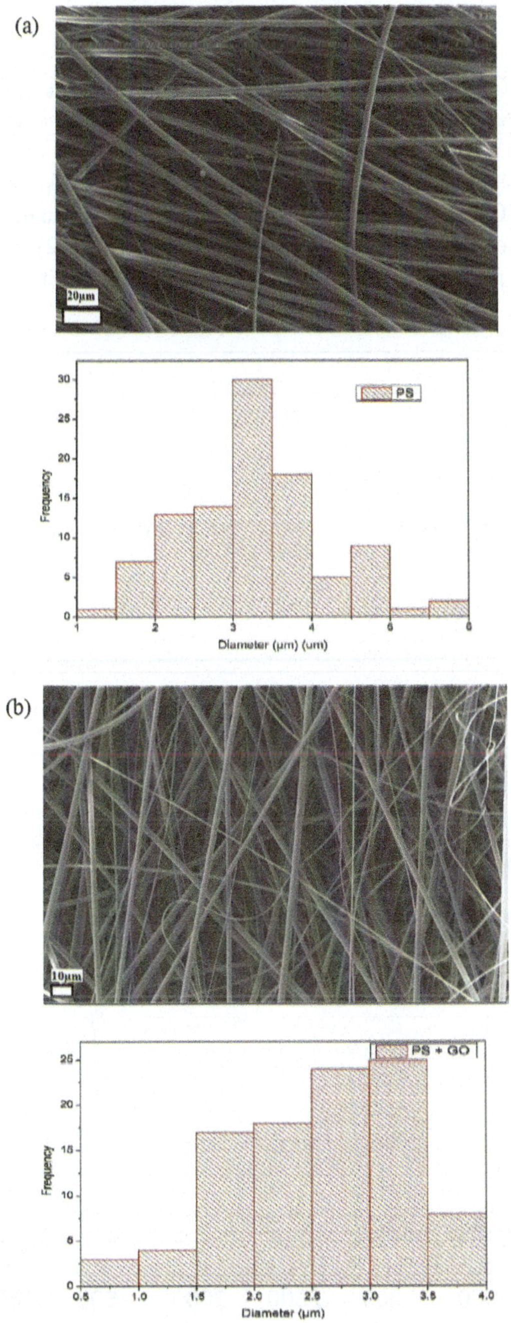

Fig. 2 Electrospun fiber for polystyrene and **b** polystyrene and GO

changed during the vibration. On the other hand, IR is active when dipole moment is changed during the vibration.

IR is excellent for carbonyl (C=O) species while Raman is quite variable. This band is easily identified in the IR spectrum because of its intensity and lack of interference from most other group frequencies. According to Fig. 3 the carbonyl in Graphene oxide can be seen at 1713 cm^{-1}. As carbonyl peak found in this region fell between 1750 and 1700 (middle of carbonyl range) it can be considered as aldehydes, ester, carbamate, ketones and carboxylic acids [12]. In addition to that, the characteristic peaks of hydroxyl (OH) stretching at 3166 cm^{-1} and the C=C also

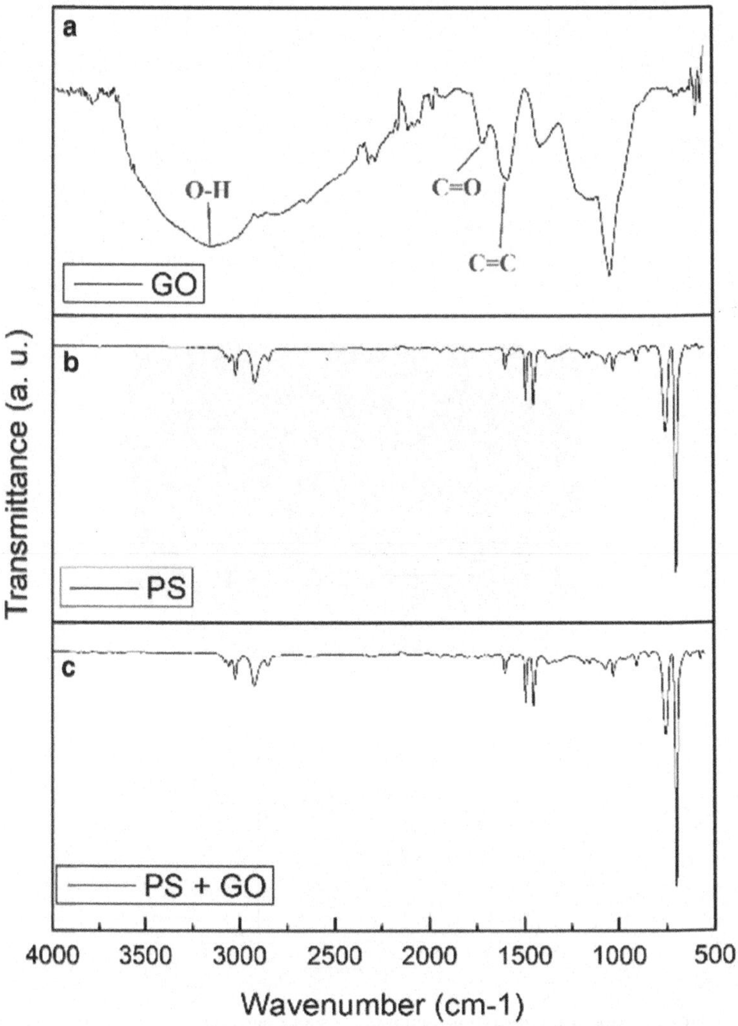

Fig. 3 FTIR Spectra of **a** Graphene oxide **b** polystyrene membrane and **c** PS + GO membrane

can be seen at 1586 cm^{-1}. Based on the FTIR spectra alone, the PS + GO absorption bands were poorly resolved since it is similar to PS spectra, possibly because of randomly dispersed graphene oxide and very small loading of graphene oxide in the fiber mat. Additionally, it is also attributed to the physical attachment of graphene oxide that stucked in between the polystyrene chains.

In contrast to FTIR spectra, Raman spectra in Fig. 4 shows the characteristic of graphene oxide in polystyrene embedded with GO fiber mat. In GO spectrum, The D band (the first peak of Raman Spectra) represents the defect concentration or measure of disorder in the C–C bonds within the graphitic materials. The G-band (Second peak of Raman Spectra) is associated with in-phase vibration of C–C bonds and is a measure of graphitization [13]. The profound C=C showing the graphitizition (G

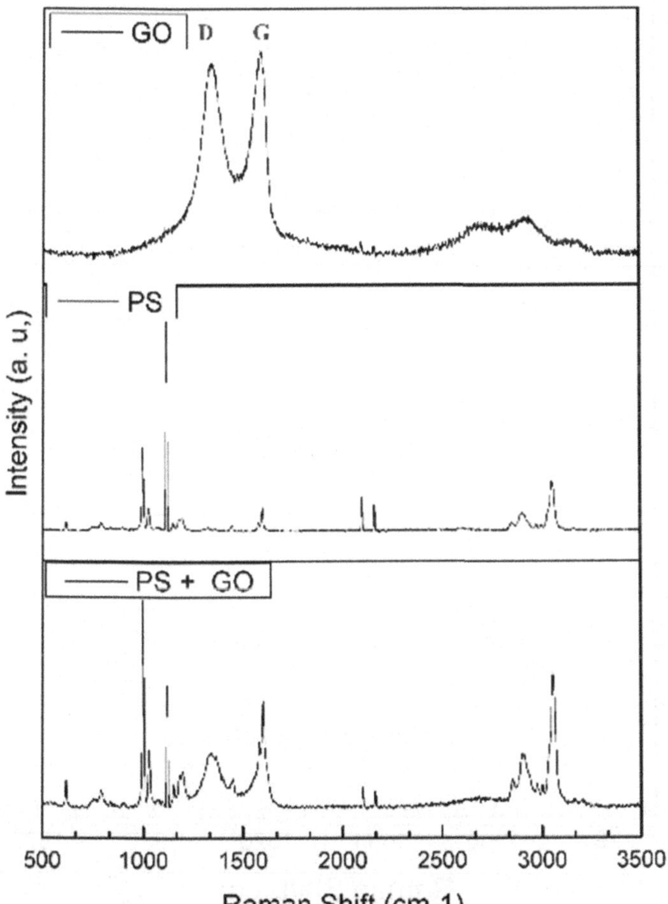

Fig. 4 Distribution of Raman spectra for GO, PS and PS + GO

band) of carbon-based material at 1595 cm^{-1} also can be seen at PS + GO spectrum as shown in Fig. 5.

Figure 6 shows XRD pattern for GO, Polystyrene fiber (PS), composite of polystyrene and graphene oxide fiber (PS + GO). The GO sample having prominent peak at 10.5° showing the graphene oxide characteristic as studied by (ref). PS which is having two broad peak at 9.5 and 18.8 were decreased upon addition of GO as shown in PS + GO XRD pattern, indicates there are dispersion of GO in Polystyrene sheet [14].

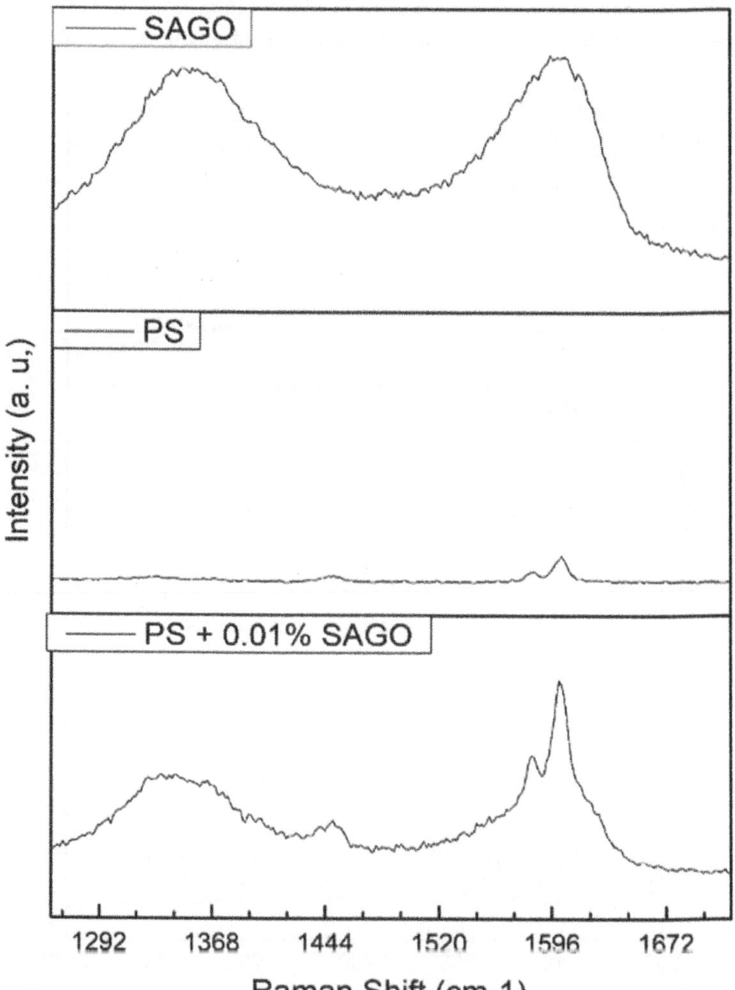

Fig. 5 D and G band of GO, PS and PS + GO

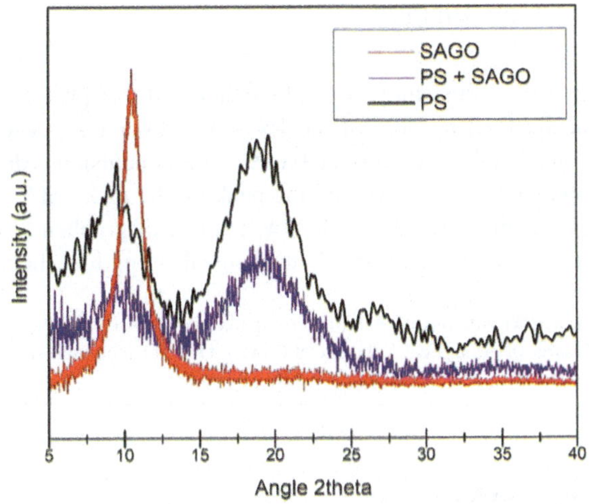

Fig. 6 Distribution of XRD peak for GO, PS + GO and PS

The improvement in tensile strength and Elastic's modulus as shown in Table 1 indicates a good dispersion of GO in the PS polymer. It also may be due to the elongated porous after the addition of filler as observed in Fig. 7 as compared to the initial Polystyrene fiber. The graphene oxide is suspected to increase the conductivity of the PS solution, causing it to inject faster towards the collector drum.

Sample	Elastic modulus	Tensile strength	Max load	Max elongation	Elongation @ break
PS	3.31	0.12	0.745	3.54	9.97
PS + GO	19.74	0.82	3.426	4.74	6.76

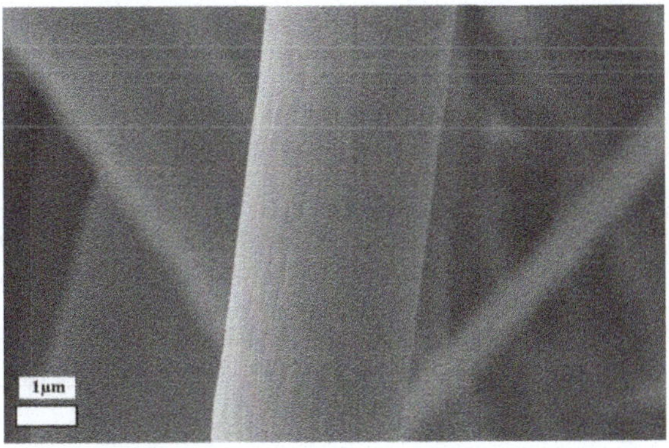

Fig. 7 Porous structure of individual fiber for **a** PS fiber and **b** PS + GO fiber

4 Conclusions

Based on this study, it can be concluded that profound changes could be observed in the final product of the PS + GO fiber mat, despite of the addition of GO is considered minute around 0.01%. This conclusion is drawn based on the morphology obtained by FESEM, and the peak of GO in RAMAN spectra. Additionally, it was also demonstrated that the low loading of graphene oxide dispersed in polystyrene fiber mat can increase the mechanical strength of the PS + GO as high as 6%.

Acknowledgements Author would like to thank LRGS-NanoMITe for the grant given and the Center of Analytical Laboratory (CAL), UTP for technical support.

References

1. Feng, S., Shen, X.: Electrospinning and mechanical properties of polystyrene and styrene-isoprene-styrene block copolymer blend nanofibres. J. Macromol. Sci. Part B Phys. **49**(2), 345–354 (2010)
2. Huang, J., Deng, H., Song, D., Xu, H.: Electrospun polystyrene/graphene nanofiber film as a novel adsorbent of thin film microextraction for extraction of aldehydes in human exhaled breath condensates. Anal. Chim. Acta **878**, 102–108 (2015)
3. Lee, M.W., An, S., Latthe, S.S., Lee, C., Hong, S., Yoon, S.S.: Electrospun polystyrene nanofiber membrane with superhydrophobicity and superoleophilicity for selective separation of water and low viscous oil. ACS Appl. Mater. Interfaces **5**(21), 10597–10604 (2013)
4. Yoshimoto, H., Shin, Y.M., Terai, H., Vacanti, J.P.: A biodegradable nanofiber scaffold by electrospinning and its potential for bone tissue engineering. Biomaterials **24**(12), 2077–2082 (2003)
5. Yoon, J.W., Park, Y., Kim, J., Park, C.H.: Multi-jet electrospinning of polystyrene/polyamide 6 blend: thermal and mechanical properties. Fash. Text. **4**(1) (2017)
6. Greiner, A., Wendorff, J.H.: Electrospinning: a fascinating method for the preparation of ultrathin fibers. Angew. Chem. Int. Ed. **46**(30), 5670–5703 (2007)
7. Yeole, N., Kutcherlapati, S.N.R., Jana, T.: Polystyrene-graphene oxide (GO) nanocomposite synthesized by interfacial interactions between RAFT modified GO and core-shell polymeric nanoparticles. J. Colloid Interface Sci. **443**, 137–142 (2015)
8. Ma, X., Zhoua, X., Xua, L.: Preparation of nanoporous graphene sheets via free radicals oxidation of graphene oxide and their application in lithium ion battery. Mater. Res. Express **4**(7) (2017)
9. Cui, C., et al.: Highly electroconductive mesoporous graphene nanofibers and their capacitance performance at 4 V. J. Am. Chem. Soc. **136**(6), 2256–2259 (2014)
10. Bao, Q., et al.: Graphene-polymer nanofiber membrane for ultrafast photonics. Adv. Funct. Mater. **20**(5), 782–791 (2010)
11. Aboamera, N.M., Mohamed, A., Salama, A., Khattab, A.: Characterization and mechanical properties of electrospun cellulose acetate/graphene oxide composite nanofibers. Mech. Adv. Mater. Struct. **26**(9), 765–769 (2019)

12. Larkin, P.J.: IR and Raman Spectroscopy—Principles and Spectral Interpretation (2011)
13. Munir, K.S., et al.: Quantitative analyses of MWCNT-Ti powder mixtures using Raman spectroscopy: the influence of milling parameters on nanostructural evolution. Adv. Eng. Mater. **17**(11), 1660–1669 (2015)
14. Uddin, M.E., Layek, R.K., Kim, N.H., Hui, D., Lee, J.H.: Preparation and properties of reduced graphene oxide/polyacrylonitrile nanocomposites using polyvinyl phenol. Compos. Part B Eng. **80**, 238–245 (2015)

Cellulose Nanofibers from Palm Bio-waste as Separator Membrane for EDLC

Khairul Anuar Jamaluddin, John Ojur Dennis, Mohd Fadhlullah Abd Shukur, Hisyam Jusoh, Bashir Abubakar Abdulkadir, Irwan Othman, and Imtias Amir

Abstract Empty fruit bunch (EFB) is a kind of bio-waste formed during the production process of palm oil. EFB is what remains of the fresh fruits bunch after the fruit has been removed for oil pressing. Therefore, this bio-waste was suggested for developing separator as the electronic component for electric double layer capacitor (EDLC). Separator made from paper pulp causes deforestation contributing more carbon dioxide building up in the atmosphere. Hence, bio-waste could possibly alternative to existing material for the separator. The objectives of this study are to prepare separators (microfiber and nanofiber) from EFB and characterize it in terms of morphology, surface area, porosity, and ionic conductance. The fibers were prepared from bio-waste through laboratory synthesis. The preparation of EFB into a cellulose microfiber and nanofiber covers; washing process, alkaline treatment and

K. A. Jamaluddin · J. O. Dennis · M. F. Abd Shukur (✉) · B. A. Abdulkadir
Fundamental of Applied Science, Universiti Teknologi PETRONAS, Seri Iskandar, Perak, Malaysia
e-mail: mfadhlullah.ashukur@utp.edu.my

K. A. Jamaluddin
e-mail: khairulanuar_jamalud@utp.edu.my

J. O. Dennis
e-mail: johndennis@utp.edu.my

B. A. Abdulkadir
e-mail: abubakar_g03619@utp.edu.my

H. Jusoh
Civil and Environmental Engineering, Universiti Teknologi PETRONAS, Seri Iskandar, Perak, Malaysia

I. Othman · I. Amir
Centralized Analytical Laboratory, Universiti Teknologi PETRONAS, Seri Iskandar, Perak, Malaysia
e-mail: irwan_othman@utp.edu.my

I. Amir
e-mail: imtias@utp.edu.my

S. A. Abdul Karim et al. (eds.), *Proceedings of the 6th International Conference on Fundamental and Applied Sciences*, Springer Proceedings in Complexity,
https://doi.org/10.1007/978-981-16-4513-6_40

digestion, grinding, bleaching, dissolution, and electrospinning. The characterization and performance test such as morphology, ionic conductivity, surface area and porosity was performed. This project successfully produced cellulose microfiber membrane (separator) made from 100% EFB and cellulose nanofiber membrane (separator) by electrospinning technique. The measured cellulose microfiber diameter was found to be 7 μm. The ionic conductivity of microfiber separator was measured to have 1.600×10^{-3} S cm^{-1} while the surface area and porosity tends to have the value of 1.638 m^2/g and 48.73%. The diameter of cellulose nanofiber from EFB was found to be 42 nm. The measured ionic conductivity indicated the value was 1.552×10^{-3} S cm^{-1} where the value of surface area and porosity were 7.298 m^2/g and 51.15%, respectively.

Keywords Empty fruit bunch (EFB) · Separator · Cellulose nanofiber · Electric double layer capacitor (EDLC)

1 Introduction

The electrochemical double-layer capacitor (EDLC) is composed of two electrodes, electrolyte and separator. An emerging technology plays a major role to fulfill the demands in electronic device and system industries for present and future. It is able to store larger amount of energy than conventional capacitors [1]. The separator prevents the occurrence of electrical contact between two electrodes, yet it is ion-permeable whereby allowing ionic charge transfer to take place [2]. Generally, polymer and paper separators are used with organic electrolytes while ceramic and glass fiber separators are often used with aqueous electrolytes. The used of conventional separator from papers comprises macro/microscopic cellulose fibers. Although these materials are found produce a better performance yet consumes a high cost of operation and production compared using abundance bio-waste material. Polymer based separator encounter poor thermal shrinkage and weak mechanical properties; it is difficult to fully ensure electrical isolation between electrodes [3]. Moreover, their intrinsically hydrophobic character and low porosity raised serious concerns over insufficient electrolyte wettability, which could directly impair ionic transport through the separators. In addition, the high processing cost of polyolefin separators remains a critical challenge. Cellulose fibers with porous structure and electrolyte absorption properties are considered to be a good potential substrate for the separator. Separator is usually a porous membrane or mat sandwiched between anode and cathode. The main function of separator is to physically isolate the anode and cathode in order to avoid electrical short circuits while at the same time provide an ionic charge pathway in liquid electrolyte throughout the interconnected porous structure [4]. The selection on the bio-waste materials determines the performance of the separator is depending on the pore size distribution, surface area and specific capacitance. The nano-sphere of bio-waste obtained from oil palm tree was found its application as separator for supercapacitors due to its porous nature, good surface area, high electrical resistance, high ionic conductance, and low thickness [5].

2 Palm Bio-waste

2.1 *Empty Fruit Bunch (EFB)*

Empty fruit bunch (EFB) waste become liability to the country because of the industries are not fully utilize the usage of this waste. EFB is classified to have high lignin (34.37%) and low cellulose content (39.13%) fiber compared to high cellulose fibers. [6]. Hence, the EFB could possibly be utilized to produce separator from bio-waste product. Thus, high porosity and surface area of separator material is required for investigation. The scope of this study is to explore the potential empty fruit bunch (EFB) application to a flexible separator membrane by converting waste of oil palm-based into added value products. The preparation of EFB into a cellulose fiber and cellulose nanofiber covers seven processes; washing process, alkaline treatment and digestion, grinding, bleaching, dissolution, and electrospinning. The separators performance from developed cellulose microfiber separator and cellulose nanofiber separator were evaluated based on the standard properties such as surface area, porosity and ionic conductivity.

2.2 *Electrospinning*

Electrospinning mechanism is used for production of nanoscale fibers on various applications including medication, filtration, material and nano-electronics. This electrospinning has become a satisfactory method for creating nanofibers since the last decade [7]. It is a most favored method for producing a fiber which used electric force to draw charged threads of polymer solutions or polymer melts up to fiber diameters at size up to hundred nanometers. Electrospinning shares qualities of both electro-spraying and conventional solution dry spinning of the fibers. This process does not involve the use of coagulation chemistry and high temperatures to produce fiber from the solution [8]. Hence, makes the practice principally suitable for fiber production using large and complex molecules.

3 Material and Method

This study primarily focuses on investigating the characteristic of empty fruit bunch (EFB) as alternative separator for electric double layer capacitor. Prior to EFB analyses, the samples were obtained from local area palm oil estate at Kilang Sawit Felcra Nasaruddin, Bota, Perak. This material will go through the processes in the making of the separator. The determination of EFB characteristics were conducted based on specified experiment namely morphology, surface area, porosity, and conductivity. Scanning Electron Microscopes (SEM) and Field Emission Scanning Electron Microscopy (FESEM) was used in this project to analyze the morphology of the cellulose fiber. The chemical microanalysis data was obtained from Energy Dispersive X-Ray Spectroscopy (EDX) analysis (not explain in this paper). For measuring the surface area and porosity, the Brunauer, Emmett and Teller (BET) machine was used. Then, the usage of Electrochemical Impedance Spectroscopy (EIS) machine gave the result of ionic conductivity. By using the Nyquist Graph, the value of bulk resistance (Rb) used to calculate the conductivity as shown in Eq. 1.

$$\sigma = (L/R_b A) \tag{1}$$

where, σ, L, Rb and A is denoted as conductivity, thickness of the electrode together with separator, bulk resistance and area of the electrode.

Nanofiber Electrospinning Machine is used to produce cellulose nanofiber in this study. In this process, the fibers were forced by electrostatic and mechanical to spin from the tip of a fine spinneret. The electrostatic repelling overcomes the surface tension force of the polymer solution, the liquid spills out of the spinneret and forms an extremely fine continuous filament. The solvents used to dissolve this natural polymer (cellulose) are H20, N,N-dimethylformamide (DMF), Acetone, Trifluoroacetic Acid (TFA) and Dichloroethylene (DCE). The schematic diagram of the electrospinning was illustrated in Fig. 1.

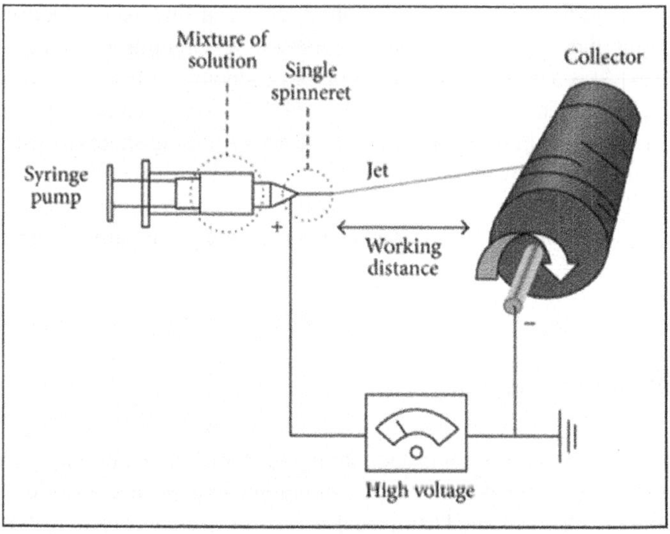

Fig. 1 Schematic diagram of the electrospinning setup as illustrated in [8]

4 Results and Discussion

Scanning Electron Microscope (SEM) analysis was conducted on raw empty fruit bunch (EFB) and paper form cellulose microfiber as shown in Fig. 2. The produced imaged were compared on fiber thickness with cellulose nanofiber. Figure 2a shows the image of the raw EFB while Fig. 2b depicted the image of the cellulose fiber which is more compact compared to the raw EFB where it has more surface area

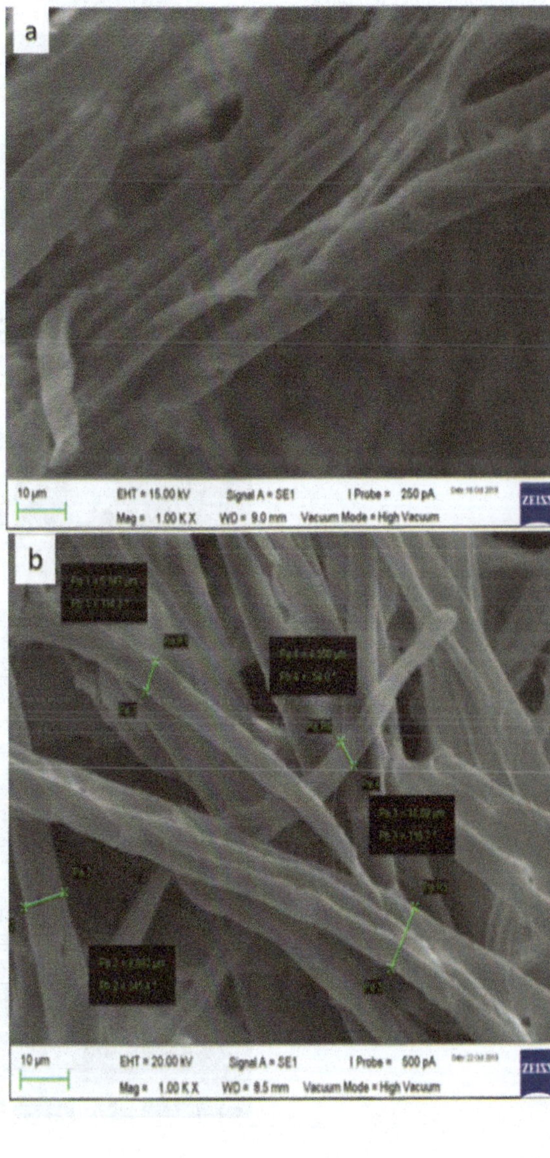

Fig. 2 Image from scanning electron microscopy (SEM) for; **a** raw EFB and **b** cellulose fiber

which can store more electrolytes. The thickness of the measured fiber is between 4 and 10 μm.

Field Emission Scanning Electron Microscope (FESEM) was carried out to determine the cellulose nanofiber morphology and thickness. Two cellulose nanofiber separators were prepared based on different duration of dissolution. Figure 3a and b shows the morphology of the cellulose nanofiber comparatively was differentiate by

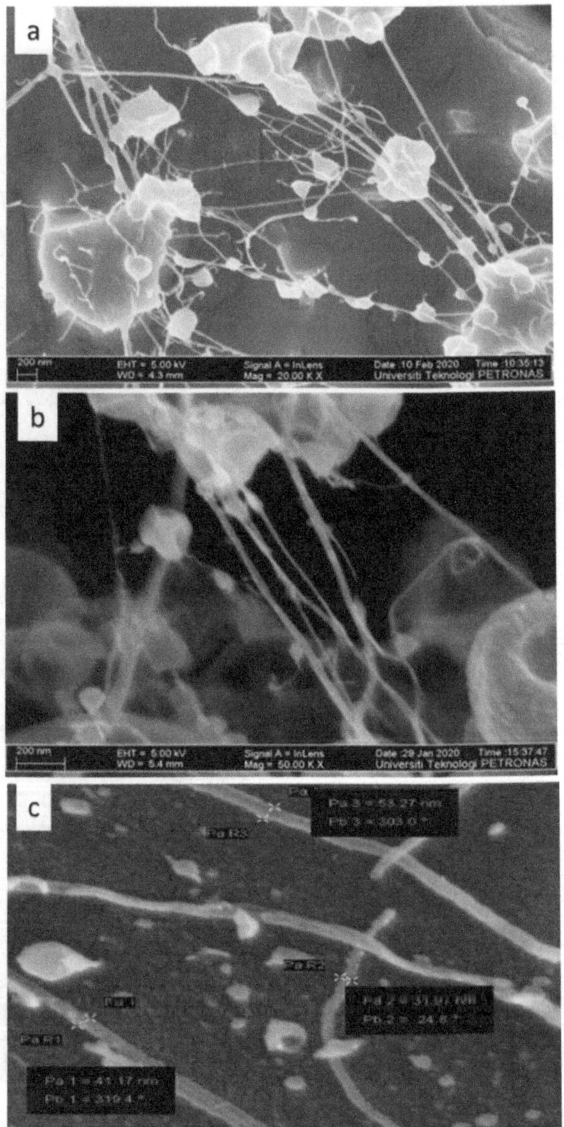

Fig. 3 FESEM image of cellulose nanofiber **a** 3 days dissolution, **b** 5 days dissolution, **c** cellulose nanofiber size

Table 1 Characterization and performance test for cellulose fiber separator and cellulose nanofiber separator

Analysis	Cellulose fiber	Cellulose microfiber separator	Cellulose nanofiber separator (3 days)	Cellulose nanofiber separator (4 days)	Cellulose nanofiber separator (5 days)
Morphology (fiber dia.)	7 μm	7 μm	42 nm	NA	46.5 nm
Thickness of the separator	NA	129 μm	29 μm	46 μm	14 μm
Conductivity	NA	1.600×10^{-3} S cm^{-1}	1.552×10^{-3} S cm^{-1}	NA	0.738×10^{-3} S cm^{-1}
Surface area	2.949 m^2/g	1.638 m^2/g	7.298 m^2/g	NA	5.642 m^2/g
Pore size	5.037 nm	2.941 nm	2.135 nm	NA	2.758 nm
Porosity	38.37%	48.73%	51.15%	NA	39.56%
Dissolution viscosity	NA	NA	206 cP	191 cP	164 cP
Dissolution conc.	NA	NA	1.5 wt%	1.5 wt%	1.5 wt%

NA Not applicable

the dissolution duration time at 3 days and 5 days, respectively. Figure 3a image is the most desirable with a lot of nanofibers compared to Fig. 3b. This is due that at 3 days dissolution the produced fiber having higher conductivity and thinner layer than dissolution at 5 days. The average diameter size measured for the cellulose nanofiber was 42 nm for 3 days dissolution (Fig. 3c) while 46.5 nm for 5 days dissolution. This proved that the obtained cellulose nanofiber by using electrospinning method is effected by the variables such as dissolution, applied voltage and collector distance [7, 9].

The summary of the analysis is described in Table 1 indicating the performance of EFB as a separator for both micro and nano size separators. The separator thickness measured for cellulose microfiber separator was 129 μm, while the thickness for cellulose nanofiber separator was 29 μm, 46 μm and 14 μm. The cellulose nanofiber separator is thinner than the cellulose microfiber separator. The dissolution viscosity recorded at 25 °C was 206 cP for 3 days dissolution, 191 cP for 4 days dissolution, and 164 cP for 5 days dissolution by using the same concentration 1.5 wt%. The surface area of cellulose nanofiber separator was found to be 7.298 m^2/g and the porosity value was 51.15%. This proves that higher porosity separator ensures the membrane has excellent electrolyte retention and enhanced ionic conductivity.

5 Conclusion

Empty fruit bunch (EFB) is a greener and sustainable alternative material for separators in electric double layer capacitor (EDLC). Various technologies have been applied to convert palm oil waste to renewable energy and value-added products. The huge amount of waste generated has to be utilized efficiently to promote sustainability. The analyzed results using EFB shows the separator can possibly be commercialized as an option to current material's separator which made from pulp. Hence, cellulose from EFB waste can be suggested as alternative material in separator production. This can prevent global warming from becoming worse. Thus, cellulose nanofiber separator is disclosed, which are electrically isolated, ionic conductive, and high surface area. For this project, the diameter of the fiber produced was at nanoscale where the diameter was found to be 42 nm. The measured ionic conductivity indicated the value was 1.552×10^{-3} S cm^{-1}. The value of surface area and porosity were 7.298 m^2/g and 51.15%, respectively. Throughout this study, it can be concluded that the production of cellulose nanofiber separator from empty fruit bunch shall be further optimized in order to achieve a uniform distribution of the fiber during electrospinning process.

Acknowledgements The authors would like to thank Universiti Teknologi PETRONAS for providing the financial support through YUTP-FRG (Grant no: 015LC0-317). The authors would like to acknowledge team of Fundamental Applied Science Department especially the technologists of Applied Physics Laboratory, Universiti Teknologi PETRONAS for their technical supports.

References

1. Sharma, P., Bhatti, T.: A review on electrochemical double-layer capacitors. Energy Convers. Manag. **51**(12), 2901–2912 (2010)
2. Alecrim, V., et al.: Exfoliated MoS2 for paper based supercapacitors and photodetectors. In: ICFPE, 5th International Conference of Flexible and Printed Electronics, pp. 437–438 (2014)
3. Chun, S.-J., Choi, E.-S., Lee, E.-H., Kim, J.H., Lee, S.-Y., Lee, S.-Y.: Eco-friendly cellulose nanofiber paper-derived separator membranes featuring tunable nanoporous network channels for lithium-ion batteries. J. Mater. Chem. **22**(32), 16618–16626 (2012)
4. Sun, X.-Z., Zhang, X., Huang, B., Ma, Y.-W.: Effects of separator on the electrochemical performance of electrical double-layer capacitor and hybrid battery-supercapacitor. Acta Physico-Chim. Sin. **30**(3), 485–491 (2014)
5. Divyashree, A., Manaf, S.A.B.A., Yallappa, S., Chaitra, K., Kathyayini, N., Hegde, G.: Low cost, high performance supercapacitor electrode using coconut wastes: eco-friendly approach. J. Energy Chem. **25**(5), 880–887 (2016)
6. Sgriccia, N., Hawley, M., Misra, M.: Characterization of natural fiber surfaces and natural fiber composites. Compos. A Appl. Sci. Manuf. **39**(10), 1632–1637 (2008)
7. Mirjalili, M., Zohoori, S.: Review for application of electrospinning and electrospun nanofibers technology in textile industry. J. Nanostruct. Chem. **6**(3), 207–213 (2016)

8. Mori, C., et al.: Nanostructured polylactic acid/candeia essential oil mats obtained by electrospinning. J. Nanomaterials **2015** (2015)
9. Fathilah, W.F.W., Othaman, R.: Electrospun cellulose fibres and applications. SainsMalaysiana **48**(7), 1459–1472 (2019)

Study the Impact of Shading on Series-Connected PV Modules

Easter Joseph and Balbir Singh Mahinder Singh

Abstract This research paper investigates the consequences of shading by varying the solar irradiances falling on PV modules using MATLAB/ Simulink software. Besides, the effect of different solar irradiances and temperature on PV arrays were also evaluated. Partial shading and full shading are inevitable in a PV system and have a severe impact on the power generation. The shading of the PV modules represents a critical variable in the optimization of solar power generation and energy prediction. Shading has a paramount influence on the current and voltage and could even destroy some or all the cells in the PV module in the event of a hot spot. In this paper, a bypass diode is applied in PV modules to mitigate shading issues. Output power is retained at a higher level when PV modules are connected with bypass diodes. Conversely, environmental conditions such as solar irradiance appear to have a greater impact on short circuit current meanwhile temperature has a prominent effect on voltage open circuit.

Keywords Shading effect · Solar irradiance · Temperature · MATLAB/Simulink · Bypass diode

1 Introduction

Over the past decades, the part of electricity production from PV has seen huge growth, especially in tropical countries due to the rapid development in PV technologies as well as various incentives offered by the government. In Malaysia, a small-scale solar electricity generating system (SEGS) is picking up, especially for rural electrification projects [1]. The SEGS operates independently of the national grid and is a competitive and effective way of providing electricity to those who do not

E. Joseph (✉) · B. S. M. Singh
Department of Fundamental and Applied Sciences (FASD), Universiti Teknologi PETRONAS, 32610 Seri Iskandar, Perak, Malaysia

B. S. M. Singh
e-mail: balbir@utp.edu.my

S. A. Abdul Karim et al. (eds.), *Proceedings of the 6th International Conference on Fundamental and Applied Sciences*, Springer Proceedings in Complexity,
https://doi.org/10.1007/978-981-16-4513-6_41

have access to the national grid, particularly for the indigenous community in remote areas. Despite being more affordable for rural electrification thanks to government subsidization, the investment for the entire system dramatically relies on the overall electricity output of the PV module. Thus, energy prediction in a PV system (PVS) is essential particularly for the standalone SEGS to monitor and detect the presence of faults through the performance of PV panels [2]. The major factors that influence the performance of PV modules are solar radiation, module temperature, dirt/soiling, shading, and PV modules' tilt angle and orientation [3–5]. The occurrence of such events and faults may induce a high amount of energy loss that is difficult to observe and encompass, leading to critical financial losses.

The solar radiation level is greatly affected by the weather conditions especially during the cloudy or rainy day which in turn has a serious impact on the performance of the PV modules. Even though the solar rays can penetrate the Earth's atmosphere, they are not entirely managed to reach the Earth's surface. This is due to the solar rays being absorbed, scattered, and reflected by air molecules, water vapor, clouds, dust, and pollutants [6]. Semiconductor devices such as PV cell is extremely responsive towards temperature. The temperature of the PV module increases with solar radiation and air temperature which has impacts on its current short circuit (I_{SC}), voltage open circuit (V_{OC}), power and fill factor (FF).

Shading normally occurs when several cells or modules are under the shade caused by clouds, buildings, trees, etc. which is difficult to avoid. Shading resulted in mismatches in the generated photocurrents of the individual cells of a module. Besides, soiling on PV could also lead to partial shading caused by soil patches such as leaves, bird dropping, and dirt patches [7]. Since the shaded cell produces a much lower current than the unshaded cell, there will be a discrepancy of current flow in each cell connected in series. Likewise, the maximum power point (MPP) current of the shaded cell is also lower than the rest of the cells around it. According to a previous study on hot-spotting, the temperature of the shaded cell can reach up to 10–20 °C higher than the neighboring cells under even illumination conditions and about 10–20% energy loss yearly [8]. The common solution to the shading problem is connecting Bypass Diodes (BD) across a sub-string of cells in the PV module. It acts as an open switch when no shading occurs and acts as a closed switch whenever there are current mismatches between the cells by allowing the generated current to pass through the BD instead of the shaded sub-string [9].

This paper investigates the impact of different shading conditions and how the BD improves the performance of the PV string under non-uniform conditions. Besides, the effect of various solar radiations and temperatures on the performance of the PV array was also studied. This research is simulated using Matlab/Simulink.

2 Methodology

This study is divided into two parts. The first part is the analysis of various solar radiations and temperature on PV array. The second part is the evaluation of the shading effect on PV string with and without the presence of BD.

2.1 *PV Array for Solar Radiation and Temperature*

Commercial PV modules were connected in series to form a string. Three strings were then connected in parallel to form an array as shown in Fig. 1. Each PV module consists of 20 cells where each cell has 0.6 V. The datasheets of the reference PV module for this simulation are given in Table 1. The PV datasheets values are derived at STC (*Standard Test Conditions*) which represents irradiance at 1000 W/m^2, module temp. at 25 °C, wind speed of 0 m/s, air mass (AM 1.5), and light incidence angle at 0°. The equivalent circuit of a PV cell is shown in Fig. 2 which is comprised of a photo-current source (I_{ph}) in parallel with a single diode (D), a shunt resistor (R_{sh}), and a series resistor (R_s) [10]. Normally, the value of R_{sh} is extremely large

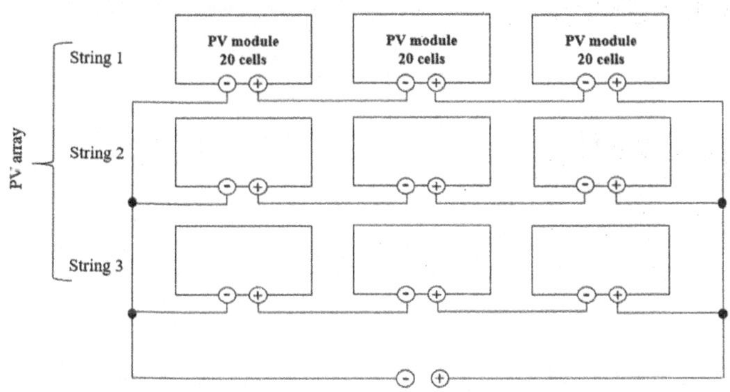

Fig. 1 An array made up of 3 strings

Table 1 Electrical characteristics data of reference PV module

Symbol	Name	Value
I_{sc}	Short circuit current (A)	8.62
V_{oc}	Open circuit voltage (V)	37.92
V_{max}	Voltage at max power (V)	30.96
I_{max}	Current at max power (A)	8.07
P_{max}	Rated power (W)	250
N_s	Number of cells connected in series	60
N_p	Number of PV strings connected in parallel	3

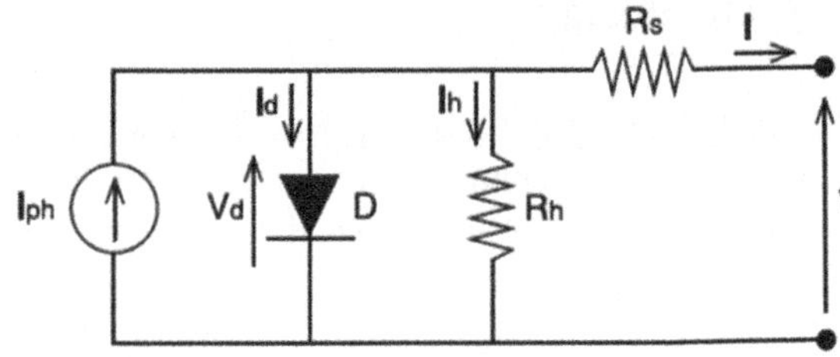

Fig. 2 One diode model of single PV cell

meanwhile R_s is very small. The PV modeling equations of I_{ph}, saturation current (I_0), reverse saturation current (I_{rs}), shunt resistance (I_{sh}), and current output (I) can be obtained from [11].

The PV array is used to study the effect of different solar radiations (G) and temperatures (T) on the output electrical parameters. Five (5) subsystems were created according to the PV modeling equations to calculate $I_{ph,}$ I_0, $I_{sh,}$ $I_{r,}$ and output I. These subsystems were interconnected to produce a PV array model as illustrated in Fig. 3. The workspace is added in this model to calculate the output I, power (P), and voltage (V) based on the variable T and G. Simulink is used to demonstrate the behavior of PV cells under different solar irradiance and temperature as given in Table 2.

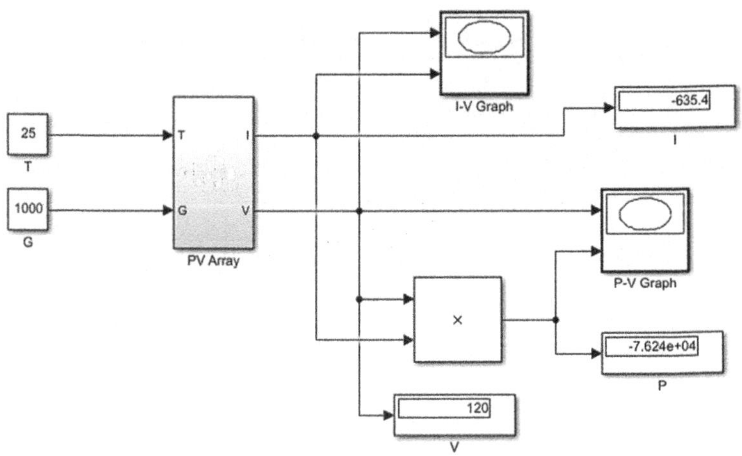

Fig. 3 PV array model

Table 2 Various solar irradiance and temperature of PV array

Solar irradiance, G (W/m^2)			Temperature, T (°C)		
1000	700	500	25		
1000			30	40	50

2.2 PV String for Shading

A PV string with similar electrical characteristics is used to investigate the effect of partial and full shading. The I–V and P–V characteristics of PV string will be analyzed with and without the presence of BD. The characteristics curves of the PV string with BD exhibit non-linearity with only one MPP. These complexities can be evaluated with computing environments, such as MATLAB [12]. The model containing a BD is illustrated in Fig. 4. The temperature difference between the shaded and non-shaded modules and the reverse breakdown effect in the shaded module are not taken into account in this study. Initially, the three modules are simulated in a circuit devoid of the BD. Afterward, each module is connected to BD as shown in Fig. 5. Three successive simulations were conducted with modules illuminated with different solar irradiances as portrayed in Table 3 [13].

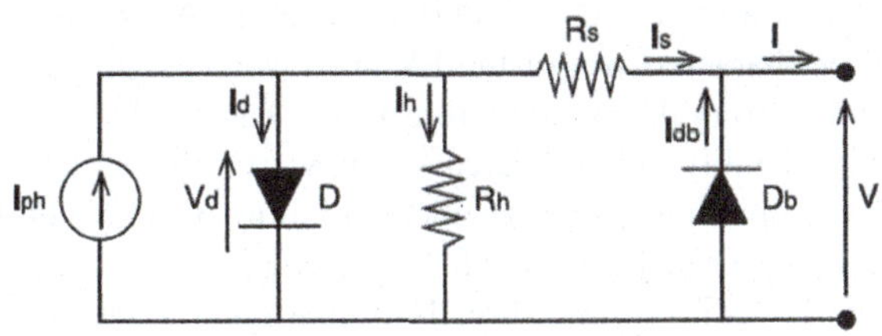

Fig. 4 One diode model with bypass diodes

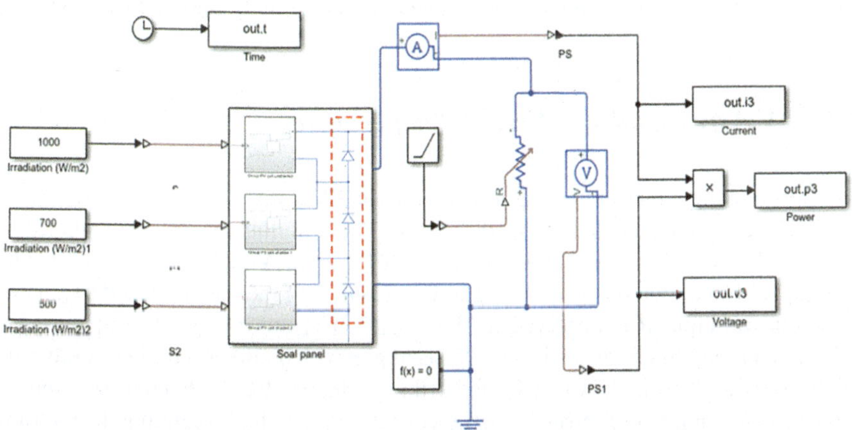

Fig. 5 The PV modules connected in series with bypass diode highlighted in a red box

Table 3 Various solar irradiance and temperature of PV array

Case	Solar irradiance (W/m^2)		
	Module 1	Module 2	Module 3
1	1000	1000	1000
2	1000	1000	500
3	1000	700	500
4	1000	500	0

3 Results and Discussion

3.1 PV Array Under Different Solar Irradiances

The effect of solar radiation on PV array can be measured by increasing the value of G from 500, 700 to 1000 W/m^2 at a constant temperature at 25°. The I–V and P–V characteristics of the PV array with varying irradiation are shown in Fig. 6a and b. From the simulation results, it can be observed that both I_{sc} and V_{oc} reduces considerably as G is decreasing, however, the reduction in V_{oc} is much less than the corresponding I_{sc}. The incident G falling on a PV cell has a prominent effect on the photogenerated current which in turn affects the output I as well. It can be deduced that I_{sc} varies almost directly proportional to the incident G. Likewise, since both parameters I_{sc} and V_{oc} increases, this contributes to an increase in P output of the PV array. Therefore, solar irradiance has a huge dominance on I_{sc} and P at a constant T.

3.2 PV Array Under Different Temperatures

To determine the effect of varying temperature on PV array, T was varied from 25°, 35° to 45° at constant irradiation of 1000 W/m^2. The Simulink results are displayed in Fig. 7a and b. Based on I–V characteristics of the PV array, V_{oc} decreases significantly as compared to I_{oc} which shows a slight dropped as T is increasing. Similarly, a net reduction in output P is observed as the cell's temperature rises. The increase of PV cell's temperature reduces its energy band gap which allows a higher number of electrons to excite from the valence band to the conduction band. This will contributes to the growing number of intrinsic carrier concentrations, thus escalating the internal carries recombination rates [3]. An increase in dark saturation current lowers the V_{oc} more than it increases the I_{sc} as saturation current of the diode of PV cell is highly temperature-dependent [14]. Hence, under a constant illumination of 1000 W/m^2, module current increases slightly while voltage drops at a higher rate, resulting in a larger drop in power output.

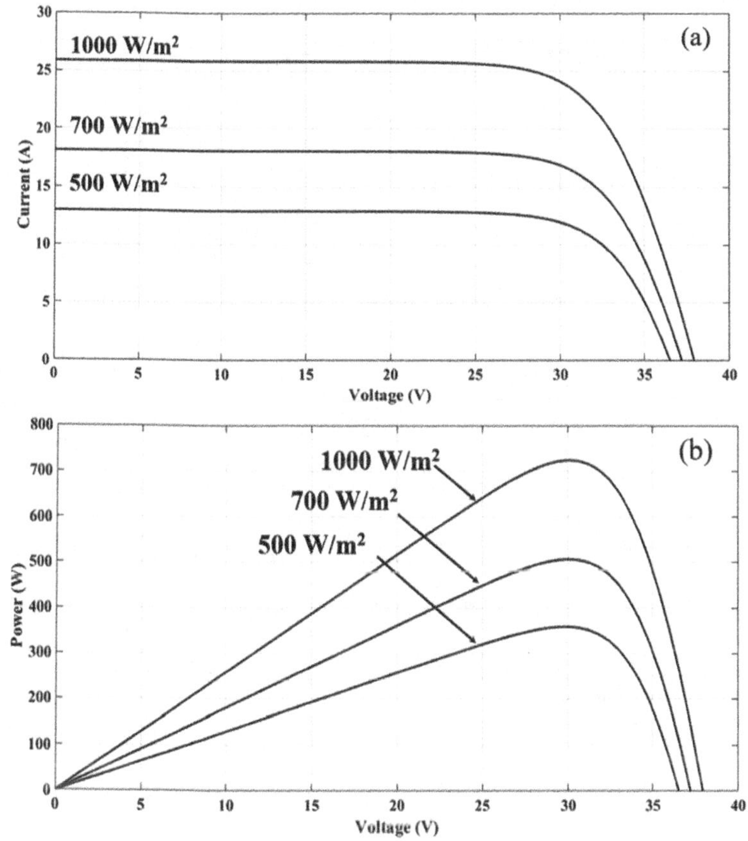

Fig. 6 Impacts of variations in irradiance on the **a** I–V and **b** P–V characteristics of a PV array

3.3 *PV String Under Different Shading*

The I–V and P–V characteristics of the three series-connected PV modules with constant irradiation (Case 1) and temperature at 1000 W/m^2 and 25° respectively are shown in Fig. 8. The maximum power is recorded at 249.83 W. It can be observed that the results are comparable for both simulations with and without the presence of a BD. Similar curves are produced as in Fig. 6 for both characteristics when the constant irradiation for all three modules is decreased to 700 and 500 W/m^2. Based on the results, we can conclude that as long as the three modules are equally illuminated, the output I and V depend solely on the solar irradiation, regardless of the presence of a BD. Subsequently, the output P also relies on the values of both parameters.

In contrast, when PV modules are simulated at different illuminations without the BD for case 2, a significant degradation to the output P from 249.83 to 125.53 W can be seen due to voltage reverse biased and hot spotting as displayed in Fig. 9. The curves obtained are equivalent to the power of the least lighted module which

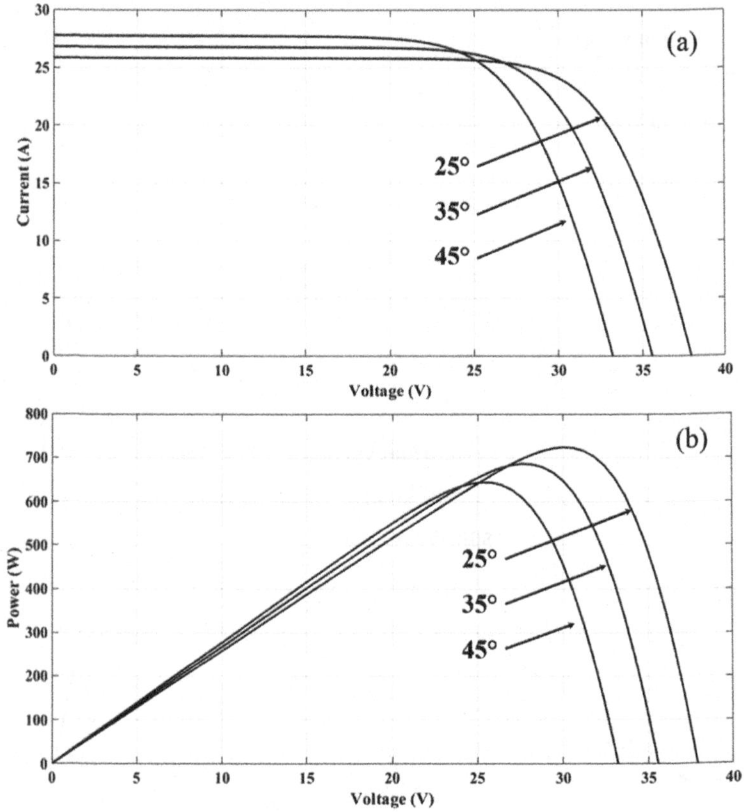

Fig. 7 Impacts of variations of cell temperature on the **a** I–V and **b** V–P characteristics of a PV array

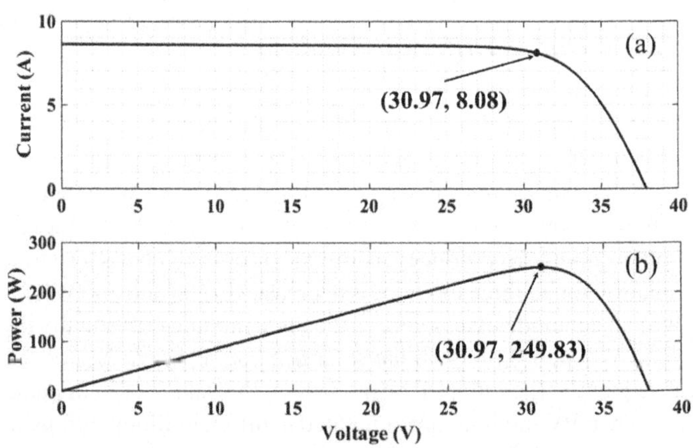

Fig. 8 **a** I–V characteristic and **b** P–V characteristics of three PV modules illuminated at 1000 W/m^2 with and without bypass diode

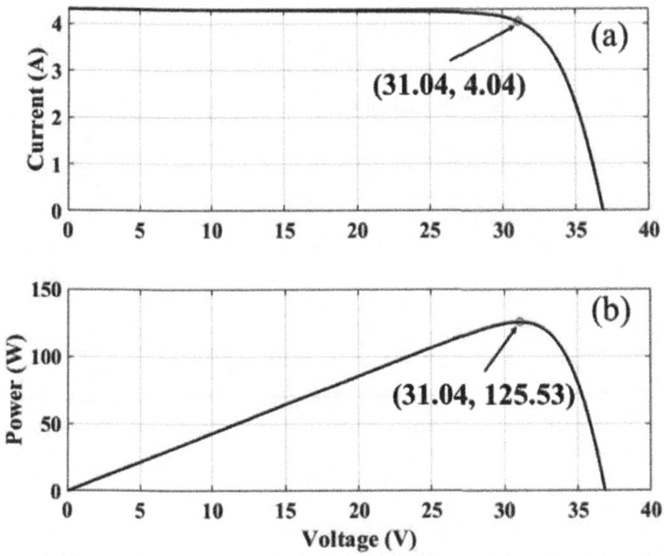

Fig. 9 **a** I–V characteristic and **b** P–V characteristics of three PV modules illuminated at 1000, 1000, and 500 W/m^2 without bypass diode

is 500 W/m^2 [13]. In the case of series-connected PV modules, the performance is critically affected by the non-uniform solar radiation. During the power generation process, the shaded module generates less current than the other two modules. This causes the "good" modules to reduce their generated currents and match to the current drop by increasing the V_{OC}. Consequently, the shaded module becomes reversed biased which allow the current to flow in the opposite direction, causing a large fraction of power generated by the unshaded modules being dissipated due to high resistance in the diode. Ultimately, when one or few cells or modules in series string experience voltage reverse biased, the energy production will drop considerably due to overheating or hot spotting. In a situation where the difference of illumination levels is high between cells, there is an immense possibility of PV cells to get damaged.

The structure of the I–V and P–V curves becomes more complex when the PV modules are connected with BD at different irradiances (Case 2) as shown in Fig. 10. The effect of shading has created several local MPPs with only one global maximum which correspond to the lowest lighted module. Note that, the existence of two local MPPs on the graph are corresponding to the two different illuminations on the PV modules. The circuit with the BD generates output P at 160.85 W, which is 21.96% greater than the output P without BD. Clearly, a BD maintains the output P at a higher level. For Case 3, the PV modules are illuminated under three different solar irradiances. The power generated is similar as in Case 2 for PV modules without BD. Conversely, for the PV modules with the presence of BD, the I–V and P–V characteristics show three local MPPs that are consistent with three different

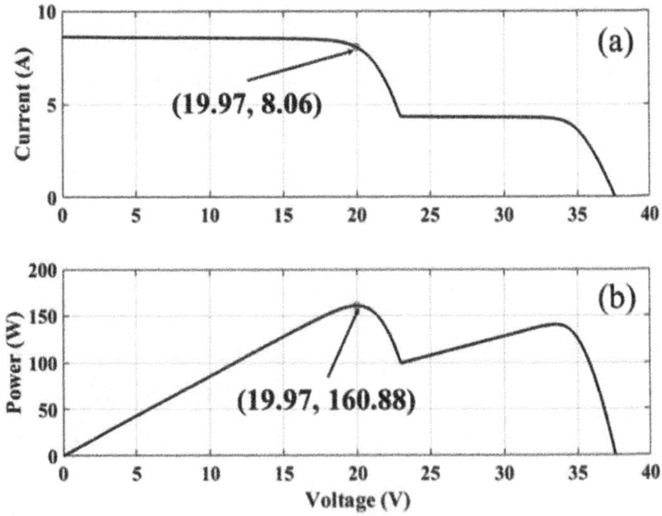

Fig. 10 **a** I–V characteristic and **b** P–V characteristics of three PV modules illuminated at 1000, 1000, and 500 W/m^2 with bypass diode

illuminations (Fig. 11). The output P decreases to 138.43 W, but the V_{OC} rises to 33.07 V due to the current adjustment of the unshaded module. In Case 4, each of the PV modules is put under full irradiance, partial, and full shading (Fig. 12). Based on the curves obtained, only two local MPPs are present as it considers a full

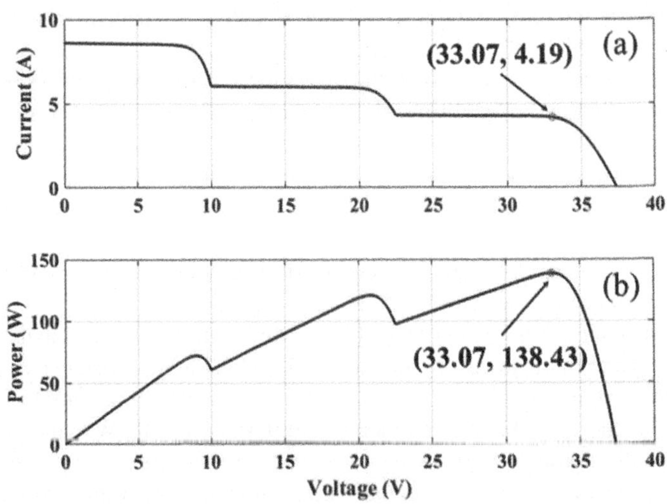

Fig. 11 **a** I–V characteristic and **b** P–V characteristics of three PV modules illuminated at 1000, 1000, and 500 W/m^2 without bypass diode

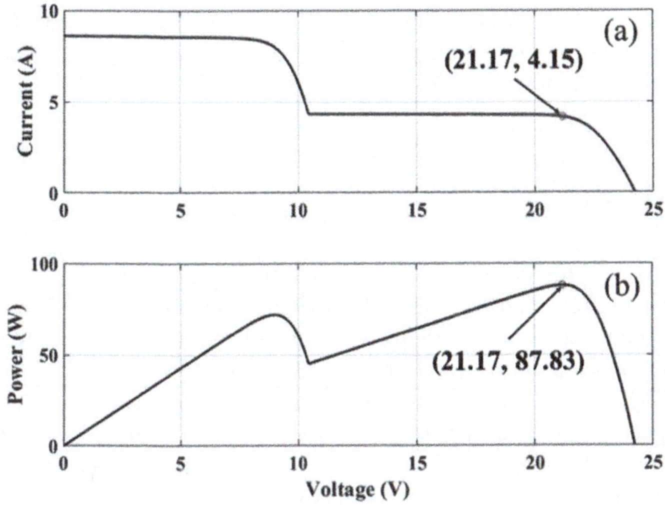

Fig. 12 **a** I–V characteristic and **b** P–V characteristics of three PV modules illuminated at 1000, 1000, and 500 W/m^2 without bypass diode

shaded module is not generating any current. With the presence of BD, an output P of 87.83 W is still attainable in comparison to when the BD is removed which brought the output P to almost zero. The over-heating of the full shaded module contributes to the low output P as a consequence of a large variation in terms of current generated between modules that may terminate the entire system.

To understand the effect of BD on the shading conditions, the path taken by the current is illustrated in Fig. 13 when one of the modules is put under the shading effect. In this state, the BD becomes active and demonstrates a lower impedance as compared to the shaded module. At this point, the BD creates a short circuit around the shaded module which allows the current to flow through it, thus avoiding the shaded module. With regards to this, we can conclude that BD protects the solar cells by preventing the current to flow through the shaded cells, hence halting the formation of hot spots and reducing the voltage reverse biased. Partial and full shading causes a

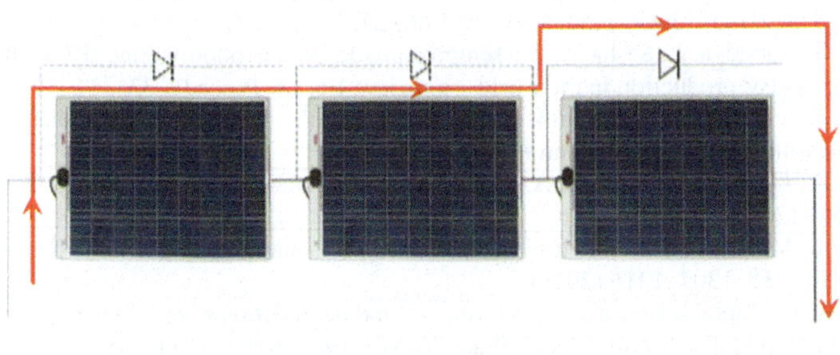

Fig. 13 The path took by the current when shading occurs

large reduction in the power generation of PV modules, and their impact towards the output P is not linear with a BD, which is why it is crucial to understand the process to reduce the power losses in a PV system [15].

4 Conclusion

The objective of this paper is to investigate the impact of various shading, solar irradiances, and temperature on PV modules by using MATLAB/ Simulink. The PV array model was developed with subsystem blocks and user-friendly icons and dialogs to produce a meaningful outcome. The shading effect on PV modules was conducted by changing the illuminations level from full irradiance to partial and full shading. There is a significant drop in the performance of the PV modules when one or more modules were put under partial and full shading due to hot spot and voltage reverse biased. In the absence of a BD, the power generated is according to the lowest illumination received by the modules. In contrast, employing a BD in the PV connection prevents a vast drop of output P with the emergence of multiple MPPs. On the other hand, output I has a positive correlation with the amount of sunlight since it is directly related to the number of photons absorbed by the PV cell while output V has a negative correlation with cell temperature since it increases the internal recombination rates.

Acknowledgements This research work is supported by the Yayasan Universiti Teknologi PETRONAS (YUTP) (015LC0-294).

References

1. Muhammad-Sukki, F., et al.: Solar photovoltaic in Malaysia: the way forward. Renew. Sustain. Energy Rev. **16**(7), 5232–5244 (2012)
2. Rahman, M.M., et al.: Global modern monitoring systems for PV based power generation: a review. Renew. Sustain. Energy Rev. **82**, 4142–4158 (2018)
3. Vidyanandan, K.V.: An overview of factors affecting the performance of solar PV systems. Energy Scan (House J. Corp. Plan. NTPC Ltd.) **27**, 2–8 (2017)
4. Dubey, S., Sarvaiya, J., Seshadri, B.: Temperature dependent photovoltaic (PV) efficiency and its effect on PV production in the world—a review. Energy Procedia **33**, 311–321 (2013)
5. Matter, K., et al.: Matlab/Simulink modeling to study the effect of partially shaded condition on Photovoltaic array's Maximum Power Point (2018)
6. Foster, R., Ghassemi, M., Cota, A.: Solar Energy: Renewable Energy and the Environment, pp. 1–337 (2009)
7. Maghami, M.R., et al.: Power loss due to soiling on solar panel: a review. Renew. Sustain. Energy Rev. **59**, 1307–1316 (2016)
8. Kaplanis, S., Kaplani, E.: Energy performance and degradation over 20 years performance of BP c-Si PV modules. Simul. Model. Pract. Theory **19**(4), 1201–1211 (2011)
9. Hu, Y., et al.: Identifying PV module mismatch faults by a thermography-based temperature distribution analysis. IEEE Trans. Device Mater. Reliab. **14**(4), 951–960 (2014)

10. Singh, G.K.: Solar power generation by PV (photovoltaic) technology: a review. Energy **53**, 1–13 (2013)
11. Gray, J.: Phys. Solar Cell 82–129 (2011)
12. Ramabadran, R., Mathur, D.: Impact of partial shading on solar PV module containing series connected cells. SHORT PAPER Int. J. Recent Trends Eng. **2** (2009)
13. Boukdir, A., et al.: Modeling and simulation of the shading effect on the performance of a photovoltaic module in the presence of the bypass diode. E3S Web Conf. **37**, 06002 (2018)
14. Yaprak, D., et al.: A roadmap to uranium ionic liquids: anti-crystal engineering. Chemistry **20**, 6482–6493 (2014)
15. Mermoud, A., Lejeune, T.: Partial Shadings on PV Arrays: By-Pass Diode Benefits Analysis, pp. 6–10 (2010)

Mathematical and Analytical Methods in Science and Technology

CFD Simulation of Petcoke Gasification in an Updraft Gasifier

Dennis Ling Chuan Ching and Iman Eslami Afrooz

Abstract Gasification is an environmentally benign solution for producing energy from industrial wastes like petcoke. This can be done by partial combustion of petcoke within a fluidized bed gasifier (FBG). Since the fabrication of gasifier reactor is costly and time-consuming, the use of a reliable numerical technique reduces these extra costs significantly. Computational fluid dynamics (CFD) is a powerful tool to accurately validate the reactor design. In this study, CFD is used to predict the syngas composition of the petcoke gasification. It is found out that CFD model can predict well the syngas composition.

Keywords CFD · Simulation · Gasification · Fluidized bed gasifier

1 Introduction

Every year, millions of metric tons petroleum coke (petcoke) are produced worldwide. However, according to the United Nations Statistics Division (UN Data) in 2013, Malaysia is 29th in the world ranking with 181,000 metric tons of petcoke production from its refineries. This amount of petcoke can be used as feedstock for fluidized bed gasifier (FBG) reactors to produce synthesis gas (syngas). Therefore, there is a need to improve the efficiency of petcoke gasification process. This can be done through simulation modeling using Computational Fluid Dynamics (CFD).

CFD has been used widely to simulate, predict and optimize reactor equipment designs. The feedstock gasification efficiency can also be examined and enhanced using CFD modelling and simulation. Discrete Phase Model (DPM) is an accurate method in modelling of solid fuel particles. In this technique, solid fuel particles injected into the gasification chamber are treated as dispersed phase and being solved

D. L. C. Ching · I. E. Afrooz (✉)
Fundamental and Applied Sciences Department, Universiti Teknologi PETRONAS, 32610 Seri Iskandar, Perak, Malaysia

D. L. C. Ching
e-mail: dennis.ling@utp.edu.my

S. A. Abdul Karim et al. (eds.), *Proceedings of the 6th International Conference on Fundamental and Applied Sciences*, Springer Proceedings in Complexity,
https://doi.org/10.1007/978-981-16-4513-6_42

by Lagrangian trajectory calculations including coupling with the continuous phase (fluid phase). The heat and mass transfer between gas and solid phases are then calculated. However, the DPM is only applicable if the solid phase volume fraction is less than 10% [1]. Since the loading mas of solid-to-gas phase is low in gasification process, DPM can be utilized as a powerful tool in thermal and hydrodynamic characteristics estimation of gasification.

Gasification simulation has been widely discussed in the literature [2–6]. There is however a lack of reports on the capability of CFD as a method for predicting the thermal characteristics of petcoke gasification. Recently, Runstedtler et al. [7] used CFD model to investigate the performance of a high-pressure entrained-flow gasifier for petcoke gasification over the range of operating conditions. They compared the CFD results with the experimental values and the good agreements were observed. Bockelie et al. [8] used CFD for modeling the entrained flow gasifiers to validate its ability in producing the reasonable results and the good agreement was observed between their CFD results and those from literature. Chyou et al. [9, 10] simulated the coal gasification within a cross-type gasifier using the commercial CFD solver ANSYS FLUENT. They also successfully developed their model to the more realistic model with DPM to include heterogeneous finite-rate reactions and coal particles tracking.

The above discussion attests that CFD is an efficient, accurate and cost-effective tool in investigating and understanding the gasification process. This paper is an attempt to demonstrate the application of CFD in an updraft gasifier simulation.

2 Methodology

In this study, ANSYS Fluent® (v.15) is used for the simulation of FBG reactor. The geometry and boundary conditions used in this study is given in Fig. 1. The FBG consists of a bed, a feeder and a riser. The geometry is a cylindrical vessel with diameter and height of 0.0762 m and 1 m, respectively. Air is blown from the inlet at the bottom of the bed and the feedstock is injected from the feeder located 10 cm from the bed inlet. Petcoke was used as feedstock in the present study. The petcoke characteristic is shown in Table 1.

The transient pressure based solver with K-epsilon (k-ε) turbulence model [11] was employed.

$$\frac{\partial}{\partial t}(\rho k) + \frac{\partial}{\partial x_i}(\rho k u_i) = \frac{\partial}{\partial x_j}\left[\left(\mu + \frac{\mu_t}{\sigma_k}\right)\frac{\partial k}{\partial x_j}\right] + P_k + P_b - \rho\varepsilon - Y_M + S_k \tag{1}$$

$$\frac{\partial}{\partial t}(\rho \varepsilon) + \frac{\partial}{\partial x_i}(\rho \varepsilon u_i) = \frac{\partial}{\partial x_j}\left[\left(\mu + \frac{\mu_t}{\sigma_\varepsilon}\right)\frac{\partial \varepsilon}{\partial x_j}\right]$$

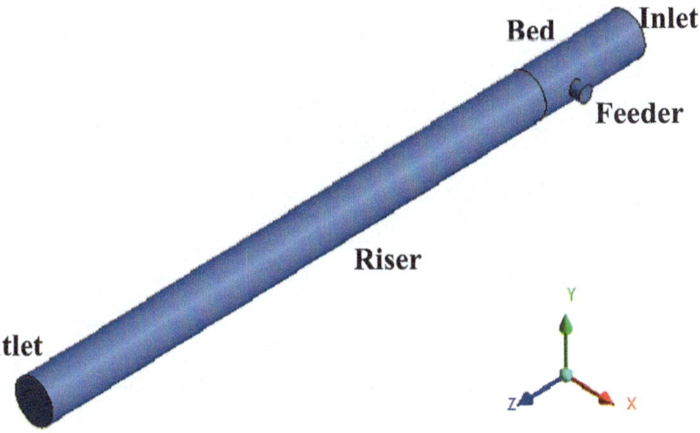

Fig. 1 Geometry of fluidized bed gasifier

$$+ C_{1\varepsilon}\frac{\varepsilon}{k}(P_k + C_{3\varepsilon}P_b) - C_{2\varepsilon}\rho\frac{\varepsilon^2}{k} + S_\varepsilon \tag{2}$$

The 3-D Navier–Stokes equations for conservation of mass (Eq. 3), momentum (Eq. 4) and energy (Eq. 5) and Discrete Ordinates (DO) for radiation model (Eq. 6) were also used.

$$\frac{\partial\rho}{\partial t} + \frac{\partial}{\partial x}(\rho u) + \frac{\partial}{\partial y}(\rho v) + \frac{\partial}{\partial z}(\rho w) = 0 \tag{3}$$

$$\rho\left[\frac{\partial\vec{V}}{\partial t} + \vec{V}.\nabla\vec{V}\right] = -\nabla p + \mu\nabla^2\vec{V} + \frac{1}{3}\mu\nabla\left(\nabla.\vec{V}\right) + \overrightarrow{F_b} \tag{4}$$

$$\frac{\partial}{\partial t}(\rho e_t) + \nabla.\left[\vec{V}(\rho e_t + p)\right] = \nabla.\left[k\nabla T + \left(.\vec{V}\right)\right] + \dot{S}_g \tag{5}$$

$$\nabla.(I(\vec{r},\vec{s})\vec{s}) + (a+\sigma_s)I(\vec{r},\vec{s}) = an^2\frac{\sigma T^4}{\pi} + \frac{\sigma_s}{4\pi}\int_0^{4\pi} I(\vec{r},\vec{s}')\Phi(\vec{r},\vec{s}')d\Omega' \tag{6}$$

Moreover, species transport equations with the Finite-Rate/Eddy-Dissipation reaction model were incorporated. Table 2 shows chemical reactions applied to the model.

The simple scheme together with least squares cell based theorem was used to compute the gradient. The Second Order schemes were also used for all the discretization terms.

Table 1 Petcoke characteristics

Proximate analysis (wt. %)				Ultimate analysis (wt. %)					Calorific value (MJ/g)
Moisture	Volatiles	Ash	Char	C	H	N	S	O	3.6211
6.000	14.800	0.140	79.060	83.290	3.576	1.680	5.528	5.926	

Table 2 Chemical reactions included in the simulation

$C + O_2 = CO_2$	$CH_4 + 2O_2 = CO_2 + H_2O$
$C + 0.5O_2 = CO$	$CH_4 + H_2O = CO + 3H_2$
$C + CO_2 = 2CO$	$CO + 0.5O_2 = CO_2$
$C + H_2O = CO + H_2$	$CO + H_2O = CO_2 + H_2$
$C + 2H_2 = CH_4$	$CO_2 + H_2 = CO + H_2O$

3 Results and Discussion

The simulations were carried out for 4500 number of time steps with the time step size of 0.01 s. Figure 2 shows the temperature contours for the entrained-flow gasifier and petcoke particles at selected time intervals for 45 s of simulation. As can be clearly seen from the photo, the temperature increases along the reactor due to the combustion of the volatiles matters. This provides the energy required for petcoke gasification. The petcoke injection was stopped after 10 s. Almost all the petcoke particles left the furnace chamber after 30 s where the gasification reactions stopped.

Figure 3 shows the mole fractions of the species at the outlet. Compared to typical synthesis gas composition for petcoke and coal gasification [12, 13], it can be asserted that the CFD model can predict well the gasification syngas composition. However, the simulation was conducted on the basis of random reactor design. Therefore, the results can be varied upon reactor size and operating conditions.

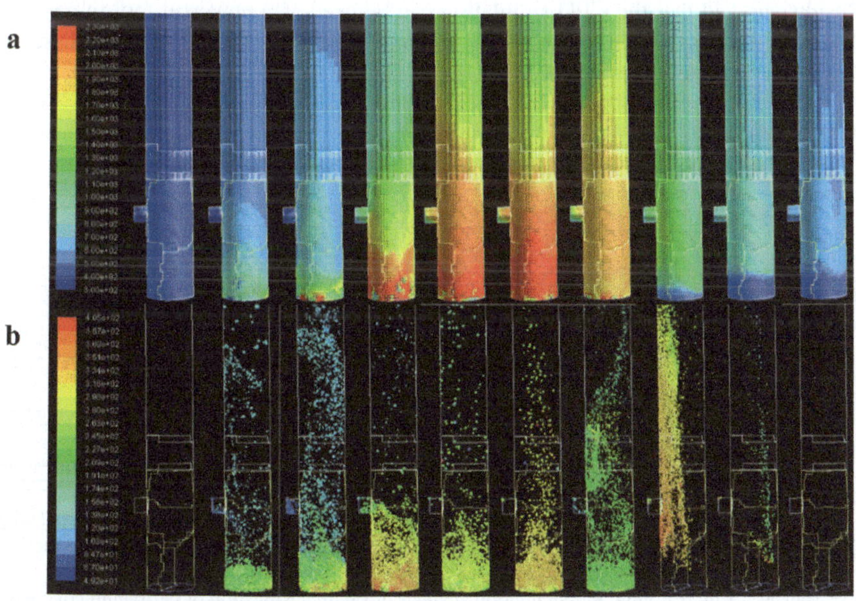

Fig. 2 Temperature contours of **a** FBG, **b** petcoke particles during 45 s

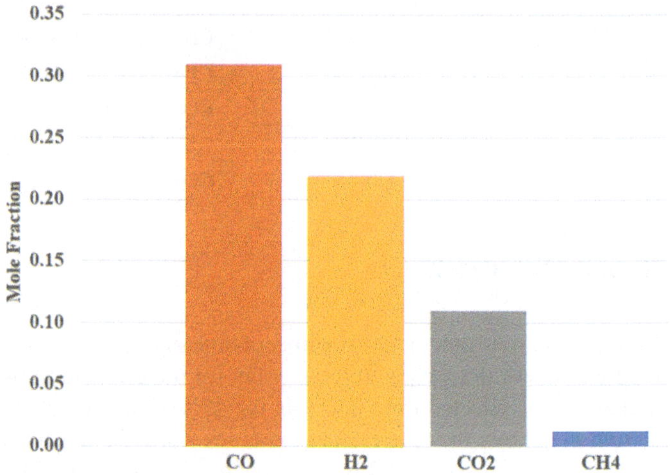

Fig. 3 Components mole fraction at exit

4 Conclusion

In this paper, a CFD model was applied to predict the singas composition of an upward fluidized bed gasifier for petcoke gasification. The temperature gradient of gasifier furnace and fuel particles and syngas composition were presented and discussed. Results showed that the CFD simulation can be used as a powerful tool in gasifier design and analysis. This simple model can also be used to help design full-scale reactors to address sizing, performance, and safety.

Acknowledgements The authors express their sincere gratitude to the University Technology PETRONAS for YUTP financial support (grant number 015LC0-078).

References

1. ANSYS Fluent Theory Guide, ANSYS, Inc., 275 Technology Drive Canonsburg, PA 15317 (2013)
2. Nagpal, S., Sarkar, T.K., Sen, P.K.: Simulation of petcoke gasification in slagging moving bed reactors. Fuel Process. Technol. **86**, 617–640 (2005)
3. Wan, W., Dai, Z., Yu, G., Wang, F.: System simulation of two kinds of petcoke gasification process for hydrogen. Int. J. Energy Res. **38**, 1162–1170 (2014)
4. Lazzaroni, E., Elsholkami, M., Martelli, E., Elkamel, A.: Design and simulation of a petcoke gasification polygeneration plant integrated with a bitumen extraction and upgrading facility and net energy analysis. Energy **141**, 880–891 (2017)
5. Tian, W., Yan, F., Liang, R.: Simulation analysis of steam gasification of petroleum coke with CaO. Pet. Sci. Technol. **36**, 2170–2183 (2018)
6. Lee, H., Choi, S., Paek, M.: Interpretation of coal gasification modeling in commercial process analysis simulation codes. J. Mech. Sci. Technol. **24**, 1515–1521 (2010)

7. Runstedtler, A., Yandon, R., Duchesne, M., Hughes, R., Boisvert, P.: Conversion of petroleum coke in a high-pressure entrained-flow gasifier: comparison of computational fluid dynamics model and experiment. Energy Fuels **31**, 5561–5570 (2017)
8. Bockelie, M., Denison, M., Chen, Z., Linjewile, T., Senior, C., Sarofim, A., et al.: CFD modeling for entrained flow gasifiers. In: Gasification Technologies Conference (2002)
9. Chyou, Y.-P.D., Huang, C.-B., Luan, Y.-T., Wang, T. Z.: Numerical simulation of gasification process in a cross-type two-stage gasifier. In: 27th International Pittsburgh Coal Conference, Istanbul, Turkey (2010)
10. Luan, Y.-T., Chyou, Y.-P., Wang, T.: Numerical analysis of gasification performance via finite-rate model in a cross-type two-stage gasifier. Int. J. Heat Mass Transf. **57**, 558–566 (2013)
11. Launder, B. E., Sharma, B. I.: Application of the energy dissipation model of turbulence to the calculation of flow near a spinning disc. Lett. Heat Mass Transfer **1**(2), 131–138 (1974)
12. Murthy, B.N., Sawarkar, A.N., Deshmukh, N.A., Mathew, T., Joshi, J.B.: Petroleum coke gasification: a review. Can. J. Chem. Eng. **92**, 441–468 (2014)
13. Available: https://www.netl.doe.gov/research/coal/energysystems/gasification/gasifipedia/syngas-composition

Multivariate Gaussian Process Regression for Evaluating Electromagnetic Profile in Screening Process of Seabed Logging Application

Muhammad Naeim Mohd Aris, Hanita Daud,
Khairul Arifin Mohd Noh, and Sarat Chandra Dass

Abstract Computer simulation is an important task for reservoir modeling in screening process of seabed logging. Information acquired from the computer simulation could provide reliable information of electromagnetic (EM) profile and subsurface underneath the seabed. However, the computer simulation could be a time-consuming task in the screening process due to its intricate mathematical equations. In this paper, a predictive model based on Gaussian process regression (GPR) is used to provide information of EM profile at various observations with low time consumption. Multivariate GPR model is developed based on computer simulation outputs. Normalized magnitude versus offset plots are analyzed to eliminate data from any undesired wave interaction. Root mean square error and coefficient of variation between the GPR model and the computer simulation outputs at untried observations are computed. On average, the resulting error was 0.0352 and the coefficient of variation was less than 0.5%. This indicates the multivariate GPR model is well-fitted and capable of evaluating EM profile with low processing-time.

Keywords Multivariate regression · Gaussian process · Seabed logging

M. N. Mohd Aris (✉) · H. Daud · K. A. Mohd Noh
Universiti Teknologi PETRONAS, 32610 Seri Iskandar, Perak, Malaysia
e-mail: muhammad_16000991@utp.edu.my

H. Daud
e-mail: hanita_daud@utp.edu.my

K. A. Mohd Noh
e-mail: khairula.nmoh@utp.edu.my

S. C. Dass
Heriot-Watt University Malaysia, 62200 Putrajaya, Malaysia
e-mail: s.dass@hw.ac.uk

S. A. Abdul Karim et al. (eds.), *Proceedings of the 6th International Conference on Fundamental and Applied Sciences*, Springer Proceedings in Complexity,
https://doi.org/10.1007/978-981-16-4513-6_43

1 Introduction

Seabed logging (SBL) is an application of controlled-source electromagnetic (CSEM) surveying technique in marine environment to remotely detect and characterize hydrocarbon-filled reservoirs beneath the seafloor. This promising application exploits the fact that hydrocarbon reservoirs have higher electrical resistivity (approximately 30–500 Ωm) than its surroundings such as seawater (0.5–2.0 Ωm) and sedimentary rocks (1.0–2.0 Ωm) [1]. SBL survey is based on the use of an array seabed receivers as the EM signal recorders and a moving source called as horizontal electric dipole (HED) transmitter. The source is normally elevated at 30–50 m from the seafloor and it continuously emits low frequency, typically from 0.1 to 10.0 Hz, from different positions through the seawater and the subsurface [2]. Generally, offshore hydrocarbon-saturated reservoirs are embedded within conductive seabed geological formations [3]. Based on the property of electrical resistivity, electromagnetic (EM)-based surveying techniques can provide crucial information that complement seismic surveying technique to characterize hydrocarbon-filled reservoirs beneath the seafloor. In the SBL, screening process prior could identify whether the exploration regions are suitable for marine CSEM mapping or not. In other words, it can determine the applicability of the SBL given the potential target areas. Screening process is done based on synthetic forward modelling codes by computing and simulating synthetic EM datasets. Although this procedure could be disputed due to the CSEM synthetic modelling can be poorly constrained, but according to the [4], the study could indicate the limitation of the SBL application towards the "EM-friendly" regions.

Computer simulation for screening process uses powerful numerical modeling techniques to compute the synthetic EM profile. In geophysics applications, finite-element method (FEM) is commonly employed in the marine CSEM forward modeling due to its powerful capability of modeling offshore reservoirs compared to the other numerical methods. However, solving the linear equations system using FEM algorithms can become a time-consuming task [5]. Researchers in [6] mentioned that evaluating the integrals and solving the linear equations in the FEM require high computational time. In addition, only a few realizations of tried observations of the computer simulation are capable of being acquired after the very lengthy computational time [7]. Addressing these problems, calibration of Gaussian process regression (GPR) and computer simulation is proposed in this research work to evaluate the EM profile at all possible observations in the screening process of the SBL application with low computational time. Multivariate GPR models are developed based on prior information generated through computer simulation technology (CST) outputs. In the context of this research work, the source-receiver separation distance (i.e., offset), depth of hydrocarbon layer from the seabed surface (i.e., thickness of overburden) and the transmission frequency are the multivariate independent variables used corresponding to the magnitude of E-field as the dependent variable. The main goal of this paper is to extend the use of GPR in the SBL and determine its viability as a predictive intelligent computational method to evaluate the EM profile in multivariate distribution. The literature of GPR methodology is briefly discussed in next section.

2 Statistical Background—Gaussian Process Regression (GPR)

Gaussian process regression has successfully been applied in numerous applications especially in computer experiments and machine learning (ML). This intelligent computational method becomes very famous due to its flexibility and its nature of being a probabilistic and non-parametric model [8]. According to [9], Gaussian process regression (GPR) has convenient properties for many tasks of modeling especially for the use of analyzing data in statistics application and ML. This method provides predictions through Bayesian inference [10]. It utilizes the prior knowledge of a condition associated to an event to describe the probability of the event. Bayesian approach gives an advantage by providing significant and interpretable information on the subject and group levels [11]. As in the GPR, the inference begins with a prior distribution and knowledge is updated as data points are being observed, then a posterior distribution over function is yielded as the output.

Studies related to calibration of statistical methodology and computer experiments can be found in [12–16]. The details of the use of GPR in computer experiments also was thoroughly explained in [7]. Learning regression functions from given datasets is an essential task in ML and computer vision in many applications. According to [17], kernel machines such as Support Vector Machines (SVM) and GPR have shown an outstanding performance in learning non-linear mappings for big-dimensional data and low-dimensional as well. However, the regularization of parameters in SVM needs cross validation [18], while GPR learns the parameters without the process. The other advantage is, GPR is a probabilistic approach where it provides fully predictive distribution. Ideally, GPR also is capable of measuring confidence of the predictions. It can provide the predictive variance and show error bars along the predictions. If the predictive distribution spreads widely over the values range, it indicates that the model is uncertain [10].

In other application, GPR also has been utilized in optimizing the received-signal-strength (RSS) threshold to generate useful proximity reports [19]. The complexity of computation from GPR-based RSS models was compared with the other RSS models. From the results, low-cost and low-complex proximity report was able to be generated through the use of GPR algorithms. Next, the viability of GPR in geophysics application has been proven by [20] in determining the reservoir porosity and permeability of the southern basin of the South Yellow Sea. From the findings, estimation of reservoir porosity and permeability using GPR consumed lower computational time than using radial-basis-function-neural-network (RBFNN), back-propagation-neural-network (BPNN) and generalized-regression-neural-network (GRNN). Researchers in [21] and [22] applied GPR as the alternative forward modeling technique in inversion methodology to estimate hydrocarbon depth and resistivity in the SBL, respectively.

3 Data and Method

Subsections 3.1, 3.2 and 3.3 explain synthetic data acquisition using the computer simulation technology (CST) software, methodology of the multivariate Gaussian process regression (GPR) and validation of the GPR model, respectively.

3.1 Synthetic Data Acquisition

As mentioned earlier, the CST software is used to replicate the SBL models and acquire the synthetic EM datasets (i.e., SBL data). The CST software offers CST EM Studio (EMS) module as low-frequency solver to discretize Maxwell's equations. In this research work, two types of SBL models were replicated and modeled which are SBL model with presence of hydrocarbon layer (i.e., target model) and SBL model with all sediment half-space beneath the seabed surface (i.e., reference model). The target model consists of four horizontal layers which are air, seawater, hydrocarbon and sediment, while the reference model has no hydrocarbon layer. These synthetic SBL models are parameterized based on the literature. Illustration of these two models is depicted in Fig. 1.

From Fig. 1, air and seawater layers for both target and reference models have similar thicknesses which are 300 m and 1000 m, respectively. The thickness of the overburden layer indicates the depth of hydrocarbon layer from the seafloor. Note that the CST software designs the simulation models in three-dimensional (3D) (length × width × height: 20 km × 20 km × 5 km). In order to understand the EM profile with various possibilities of marine environment, various depths of hydrocarbon are tested (1000–2000 m with gap of 200 m each) and thickness of hydrocarbon layer replicated in the target models is fixed at 200 m with volume of 80km^3 (length × width × height: 20 km × 20 km × 0.2 km). A 270 m-long transmitter is elevated at 30 m from the seabed and positioned at the center of both the simulation models with multiple low-frequencies emitted from the source (0.125 Hz, 0.25 Hz, 0.375 Hz and 0.5 Hz). The current of the source is kept at 1250 A. Receivers are positioned on the seabed and set to have gap of approximately 100 m each along the length of the SBL models. Note that static source-receiver combination is exercised in this research work along 20 km of offset distance.

The property of electrical resistivity is the key parameter in the SBL application. Relevant electrical conductivity (inverse of electrical resistivity) is parameterized for each layer of the simulation models. As of the prior knowledge, 24 EM datasets of the target models for 20 km-range source-receiver separation distances, four different transmission frequencies (0.125 Hz, 0.25 Hz, 0.375 Hz, 0.5 Hz) and six different

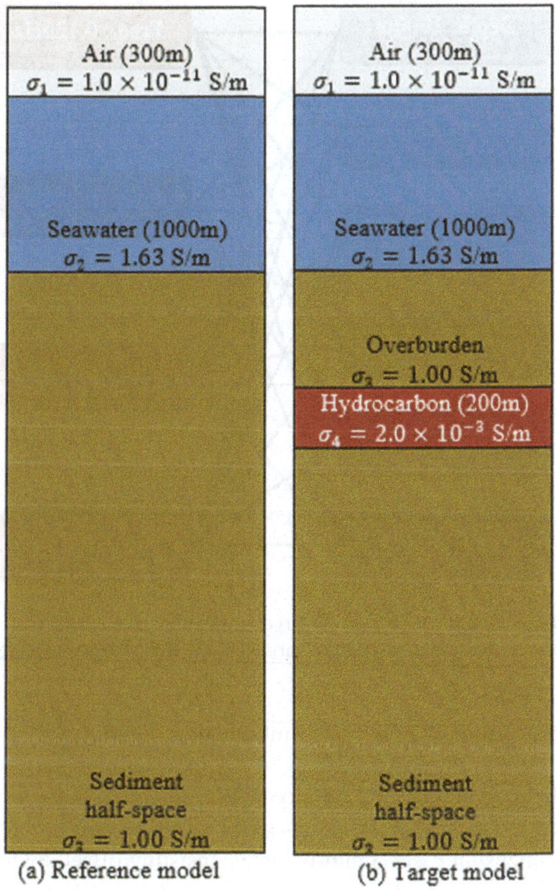

Fig. 1 **a** SBL model without hydrocarbon layer, **b** SBL model with hydrocarbon layer. Sediment above the hydrocarbon layer is known as overburden

tested hydrocarbon depths (1000–2000 m with an increment of 200 m) are generated using the CST software, while four EM datasets of the reference models (for 20 km-range offsets and four different frequencies) are generated as well. The multivariable of the target models is portrayed in Fig. 2.

3.2 Developing Multivariate Gaussian Process Regression (GPR) Model

Gaussian process (GP) can be defined as an infinite collection of random variables where any of the finite subsets is normally distributed as well [23–25]. In the context of this research work, zero-mean function $m(x) = 0$ is assumed as the prior since it is a linear combination of random variables that is treated as Gaussian distribution (i.e., normal distribution). 24 training sets of n_{train} different input specifications

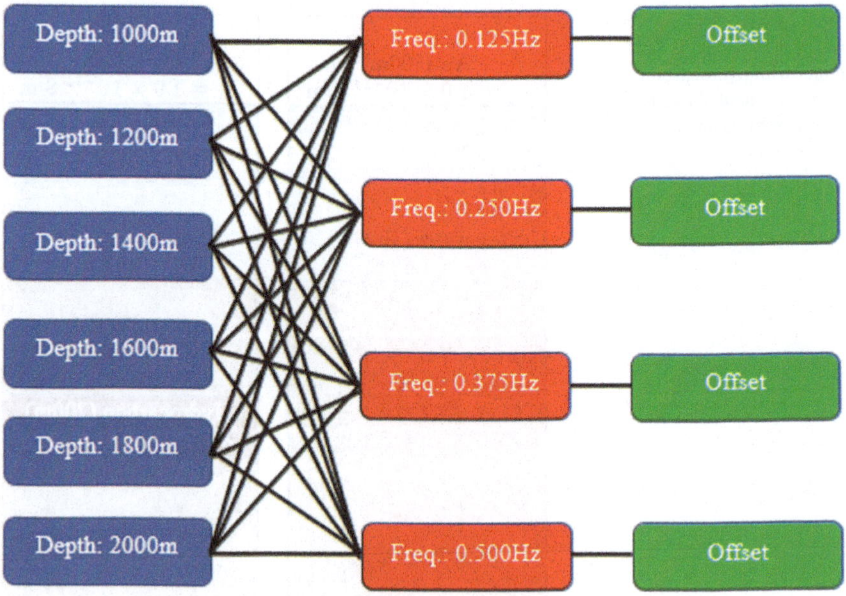

Fig. 2 Multivariable of the target model (hydrocarbon depth × frequency × offset: 6 × 4 × 1). For the reference model, the variable of hydrocarbon depth is ignored

are assumed to have the multivariate input, x, at spatial locations $x \in \mathbb{R}^{n_{train} \times n_d}$ and the output variable, $y \in \mathbb{R}$, represents as the magnitude of electric field (E-field). There are 62 data points for every EM dataset. In this research work, 1488 input–output specifications of the target model (i.e., n_{train}) are observed with spatial dimension $n_d = 3$ multivariate consisting offset, hydrocarbon depth and frequency. Thus, $x = (p, q, r)$ where $p_a | a = 1, 2, \ldots, 62$, $q_b | b = 1, 2, \ldots, 6$, $r_c | c = 1, \ldots, 4$ denote the offset, depth and frequency, respectively. The GPR on function f is then expressed in (1).

$$f(x) \sim GP\big(0, k\big(x, x'\big)\big) \quad (1)$$

Mean function in the GPR represents the expected value of the function f at the input point, while covariance function defines the correlation between the points. Squared exponential (SE) covariance function, as shown in (2), is exercised in this research work.

$$k\big(x, x'\big) = \sigma_f^2 \exp\left(-\frac{d^2}{2}\right) \quad (2)$$

where

$$d = \frac{|x - x'|}{\ell} \tag{3}$$

There are two hyper-parameters involved in (3) which are signal variance, σ_f and characteristic-length scale, ℓ. In this research work, these hyper-parameters are optimized by minimizing negative marginal log-likelihood as expressed in (4).

$$\log p(y|X, \theta) = \frac{-\left[n \log(2\pi) + \log|K| + y^T K^{-1} y\right]}{2} \tag{4}$$

where $\theta = \{\sigma_f, \ell\}$ and K is a covariance matrix of the training datasets obtained by using (2). Logarithmic identifier is used to simplify the integral expression involved in marginal likelihood equation. In order to get the expected function value f^* which is represented by m^*, prior joint Gaussian distribution given the test input point $x^* \in \mathbb{R}^{n_{test} \times n_d}$ (untried observation) is expressed in (5).

$$\begin{bmatrix} m \\ m^* \end{bmatrix} \sim G\left(0, \begin{bmatrix} K(X, X) + \sigma_n^2 I & K(X, X^*) \\ K(X, X^*)^T & K(X^*, X^*) \end{bmatrix}\right) \tag{5}$$

where σ_n denotes the signal noise that is normally distributed, $K(X, X)$ is K, $K(X, X^*)$ represents the covariance matrix of the training–testing data and $K(X^*, X^*)$ is the covariance matrix of the testing-testing data. For the test input points, GPR model is used to evaluate EM profile for $n_{test} = 62 \times 15 \times 3 = 2790$ different test input specifications (offset × depth × frequency). The untried observations of the hydrocarbon depth are from 1050 to 1950 m with a gap of 50 m (the prior information is excluded) and for the transmission frequencies are 0.1875 Hz, 0.3125 Hz and 0.4375 Hz.

Here, the posterior conditional Gaussian distribution given X, y and x^* is expressed in (6).

$$p(m^*|X, y, x^*) \sim G(m^*|\mu^*, \kappa^*) \tag{6}$$

where

$$\mu^* = K(X, X^*)^T \left[K(X, X) + \sigma_n^2 I\right]^{-1} y \tag{7}$$

and

$$\kappa^* = K(X^*, X^*) - K(X, X^*)^T \left[K + \sigma_n^2 I\right]^{-1} K(X, X^*) \tag{8}$$

The GPR outputs are defined in terms of predictive mean μ^* and predictive variance κ^*. The predictive mean is the EM profile evaluated by the GPR at the untried observations, while the predictive variance represents the uncertainty quantification of the predictive distribution in terms of ± two standard deviations (95% confidence interval).

3.3 Electromagnetic Profile Validation

This subsection briefly discusses the GPR model validation using root mean square error (RMSE) and coefficient of variation (CV) in percentage. The purpose of this subsection is to quantify the differences between the true EM values (generated through the CST) and the estimates (the multivariate GPR model). The RMSE and CV are calculated based on (9) and (10), respectively.

$$RMSE = \sqrt{MSE} = \sqrt{\frac{1}{z}\sum\left(y_i - y_i^{\cdot}\right)^2} \tag{9}$$

and

$$CV = \frac{RMSE}{\left|\mu_{y^{\cdot}}\right|} \times 100\% \tag{10}$$

where z is number of data, y is the true value (i.e., CST output), $y^{\cdot}$ is the estimate evaluated by the multivariate GPR and $\mu_{y^{\cdot}}$ is the average of the estimates.

4 Results and Discussion

Normalized magnitude versus offset (MVO) was plotted for all the EM datasets of the simulation models for the purpose of identifying EM data acquired from insignificant waves contribution in the computer simulation. Note that the normalized MVO was calculated by dividing the EM profile from the target model with the EM profile from the reference model. The normalized MVO plot is capable of determining the anomalies underneath the seafloor. If the calculated value is approaching to one, it means that the EM profile of the target model apparently looks like the EM profile of the reference model. This indicate that the surveyed area may absolutely have low certainty of the presence of hydrocarbon-saturated reservoir underneath the seafloor.

Note that all the simulation models were modeled with 20 km source-receiver separation distance and the source was located at the center of the simulation models. Thus, all the CST outputs were symmetry. The EM signal traveled at equal separation distances from the source location which was at coordinate of (x = 10,000, y = 10,000) to the left (0 km) and right (20 km) of the model boundaries. Therefore, only half of the CST outputs (10–20 km) were considered for the visualization and GPR modeling purposes in order to make it more interpretable. The normalized MVO plots for all the EM datasets are depicted in Fig. 3.

Based on Fig. 4, there are six different plots where the x-axis represents the source-receiver separation distance (i.e., offset) and the y-axis is the log10 of magnitude of E-field. These plots were grouped based on the depth of hydrocarbon layer. For every hydrocarbon depth, four prior multi-frequencies were exercised in the

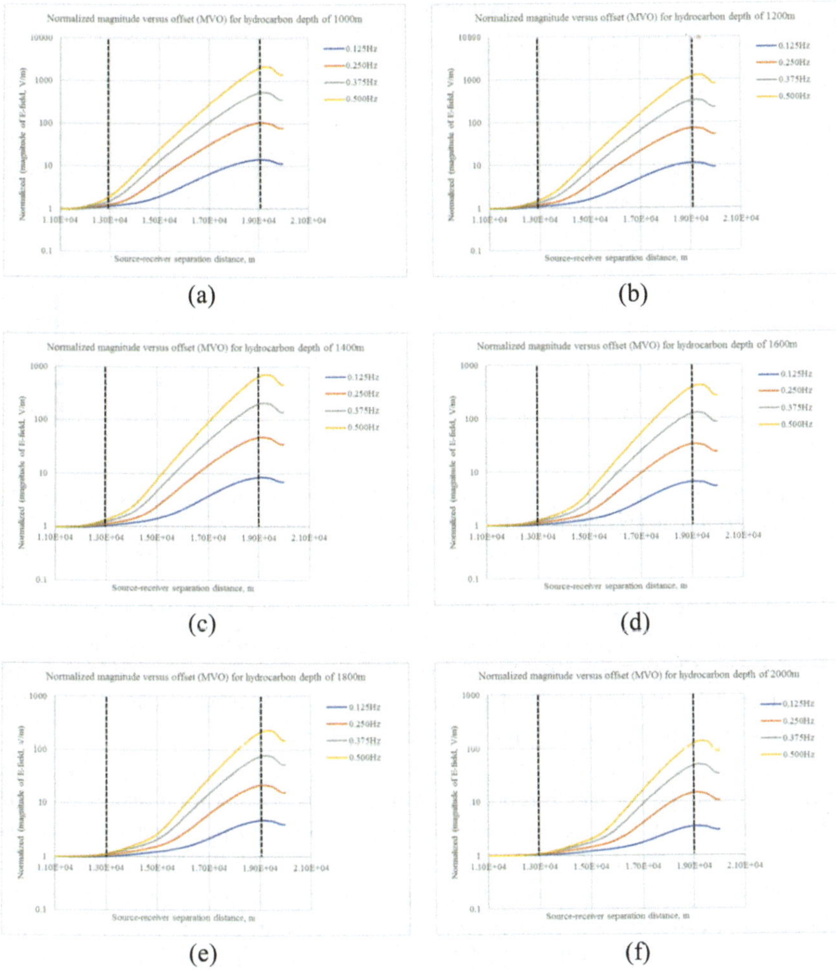

Fig. 3 Normalized magnitude versus offset (MVO); **a** 1000 m, **b** 1200 m, **c** 1400 m, **d** 1600 m, **e** 1800 m, **f** 2000 m

computer simulation models. Every curve defines normalized magnitude at different transmission frequency (0.125 Hz, 0.25 Hz, 0.375 Hz and 0.5 Hz). It is known that this research work replicated simulation target models with hydrocarbon layer positioned at different depths. Thus, from the normalized MVO plots, the anomalies are present where at certain offsets. The normalized EM profiles become irrelevant at short and long source-receiver separation distances. At short offset which is approximately less than 13 km (< ~ 3 km from the source), the normalized magnitude values are less than one. Theoretically, The EM responses of the target model should be higher than responses from the reference model. Thus, this research work assumed insignificant wave contribution (e.g., direct wave) occurred at the short offset distances. Referring

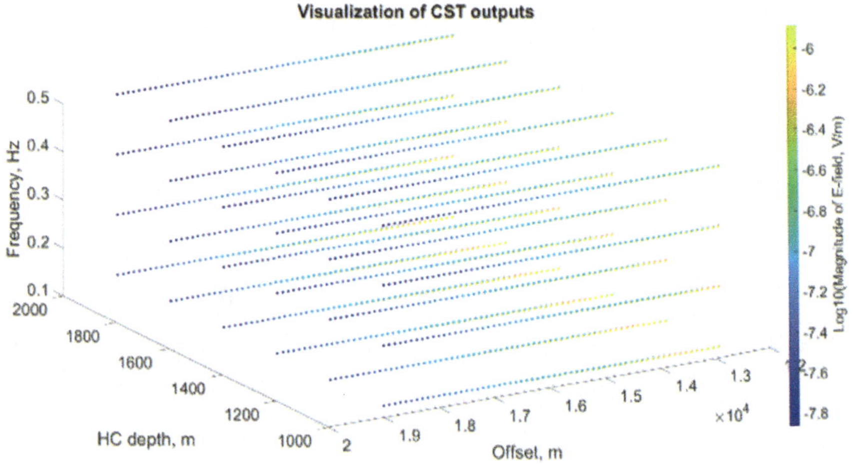

Fig. 4 EM profiles used as the prior knowledge of the GPR model. The colors represent the magnitude of E-field in logarithmic scale

to Fig. 3, at long offset (> ~ 9 km from the source), the normalized magnitudes suddenly decrease after gradually increase from the center of the models due to presence of noise at long source-receiver separation distances. Therefore, this research work inferred to only process the EM data points at offsets of approximately 13 km to 19 km (~3–9 km from the source), as shown in Fig. 4.

Figure 4 shows the visualization of all the CST outputs utilized for the GPR. There are three different axes which represent source-receiver separation distance, depth of hydrocarbon (HC) and transmission frequency. Legend provided in the figure indicates the magnitude of E-field in logarithmic scale with base 10. Based on the color codes, the EM profile exponentially decreases from the smallest offset distance to the biggest distance. This proves that the CST software is capable of providing relevant EM profile to the GPR. The CST consumed approximately 18 min to generate four different EM datasets with different frequencies for each of the hydrocarbon depth. Thus, in total, the time for the CST software to generate the 24 EM profiles of the target models was approximately 108 min (~1 h 48 min). For the comparison purposes, the multivariate GPR model is depicted in Fig. 5.

Note that GPR algorithms used in this research work can be found in study presented by [9]. The multivariate GPR model consisted prior knowledge from the CST outputs and the posterior distribution (i.e., estimates) at the untried observations. In total, Fig. 5 presents $62 \times 21 \times 7 = 9114$ different EM data points. From the figure, there are seven tested frequencies, 21 hydrocarbon depths and 62 source receiver separation distance points. GPR took about approximately less than five minutes to evaluate 9114 EM data points. The difference between the time taken of the computer simulation and GPR to generate the EM profiles is very significant. Computer simulation could consume approximately 378 min (~6 h 18 min) in

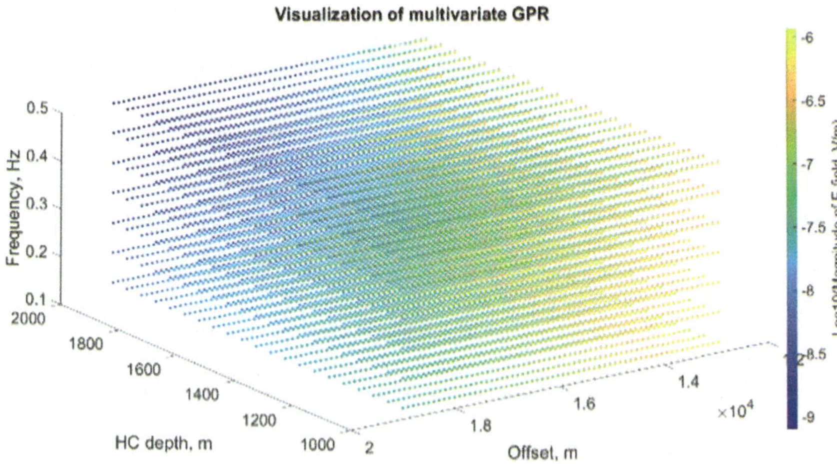

Fig. 5 Multivariate GPR model. The colors represent the magnitude of E-field in logarithmic scale

order to generate EM profiles with the 21 depths of hydrocarbon and four frequencies. Furthermore, the capability of the GPR to provide such multivariable input points with huge number of outputs (big input–output pair combination) with low time consumption should be highlighted as well. Besides, this GPR modeling also provides the uncertainty quantification along the predictions. The resulting predictive variances revealed very small values. This means that the multivariate GPR is able to fit the synthetic EM datasets very well in order to produce new information.

The EM profiles estimated by the multivariate GPR were validated and compared with the CST outputs at the untried observations. Simulations for target models at the untried observations were modeled and run separately for the purpose of the GPR model validation. Root mean square error (RMSE) and coefficient of variation in percentage (CV) were computed. Low RMSE and CV indicate the GPR is able to evaluate reliable EM profile in the screening process of the SBL application. Figure 6 shows the comparison of MVO plots between the CST outputs as the true values and the estimates (i.e., GPR model) at untried observations.

The random untried hydrocarbon depths were 1150 m, 1700 m and 1950 m, while 0.1875 Hz and 0.3125 Hz for the frequency. Based on Fig. 6, the x-axis represents the offset and the y-axis represents the log10 of the magnitude of E-field. The red curve denotes the estimates, while the blue curve represents the CST outputs. From the figure, the comparisons look like no significant differences. This inference was strongly strengthened by the numerical proof of the computed RMSE and CV. All the RMSE and CV values for the comparisons shown in Fig. 6 are tabulated in Table 1.

Based on Table 1, all the resulting RMSE and CV values are very small. In average, the RMSE was 0.0352 which was less than 0.04, while the average of the

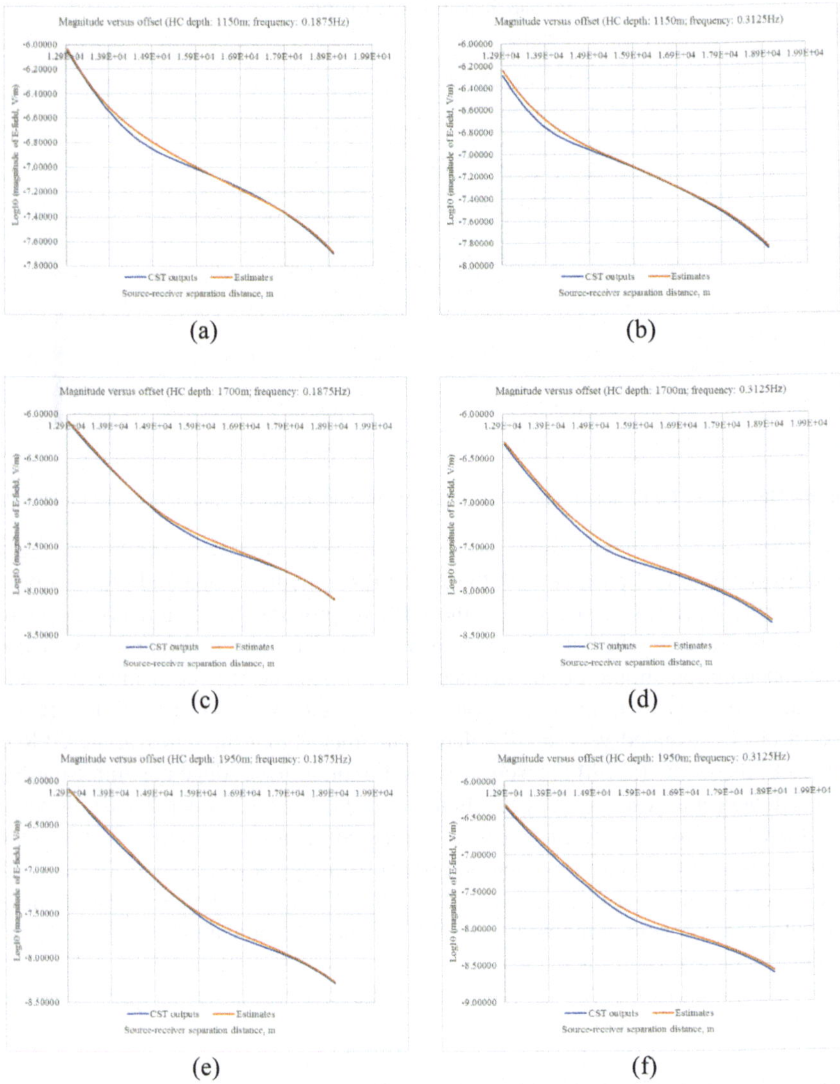

Fig. 6 Comparison of magnitude versus offset (MVO) plots between CST outputs and GPR at random untried observation; **a** depth: 1150 m, frequency: 0.1875 Hz, **b** depth: 1150 m, frequency: 0.3125 Hz, **c** depth: 1700 m, frequency: 0.1875 Hz, **d** depth: 1700 m, frequency: 0.3125 Hz, **e** depth: 1950 m, frequency: 0.1875 Hz, **f** depth: 1950 m, frequency: 0.3125 Hz

CV was 0.48% which was less than 0.5%. This indicates that the GPR model is well-fitted and capable of providing such beneficial information of EM profile at untried observations in the screening process of the SBL application.

Table 1 The resulting RMSE and CV at the untried observations

HC depth (m)	Frequency (Hz)	RMSE	CV (%)
1150	0.1875	0.0289	0.4152
	0.3125	0.0333	0.4686
1700	0.1875	0.0268	0.3699
	0.3125	0.0450	0.5985
1950	0.1875	0.0277	0.3769
	0.3125	0.0493	0.6430

5 Conclusion

Multivariate Gaussian process regression (GPR) methodology was proposed in this research work to evaluate electromagnetic (EM) profile in the screening process of the SBL application. The GPR has proved that this methodology is able to estimate the huge number of input–output pair specifications of EM profile with low time consumption compared to the computer simulation. The error measurements (i.e., root mean square error and coefficient of variation) also revealed very small values and percentages which were less than 0.04 and 0.5%, respectively. This means that the multivariate GPR model for the untried observations was successfully developed with deviation from the computer simulation outputs. This attempt could help geophysicists to understand and pre-determine the effect of the setup of the seabed logging (SBL) application such as the transmission multi-frequency used during the survey and the geometry configurations such as hydrocarbon depth towards the EM profile behaviors before in-depth analysis. Besides de-risking the hydrocarbon exploration in the SBL, this attempt also will be very useful for the higher-dimensional modeling that faces insufficient information. Therefore, GPR could be very helpful processing tool for the screening process of the SBL to evaluate EM profile at all possible observations with low time consumption.

Acknowledgements The authors would like to thank to those who have contributed to this research work. This research work is sponsored by Yayasan Universiti Teknologi PETRONAS-Fundamental Research Grant (YUTP-FRG) (cost center: 015LC0-055).

References

1. Gelius, L.J.: Multi-component processing of sea bed logging data. PIERS ONLINE **2**, 589–593 (2006)
2. Chiadikobi, K.C., Chiaghanam, O.I., Omoboriowo, A.O., Etukudoh, M.V., Okafor, N.A.: Detection of hydrocarbon reservoirs using the controlled-source electromagnetic (CSEM) method in the 'Beta' field deep water offshore Niger Delta, Nigeria. Int. J. Sci. Emerg. Technol. **3**(1), 7–18 (2012)

3. Persova, M.G., Soloveichik, Y.G., Domnikov, P.A., Vagin, D.V., Koshkina, Y.I.: Electromagnetic field analysis in the marine CSEM detection of homogeneous and inhomogeneous hydrocarbon 3D reservoirs. J. Appl. Geophys. **119**, 147–155 (2015)
4. Stefatos, A., Boulaenko, M., Hesthammer, J.: Marine CSEM technology performance in hydrocarbon exploration—limitations or opportunities. First Break **27**, 65–71 (2009)
5. Li, Y., Key, K.: 2D marine controlled-source electromagnetic modeling: Part 1—An adaptive finite-element algorithm. Geophysics **75**, WA51–WA62 (2007)
6. Bakr, S.A., Pardo, D., Mannseth, T.: Domain decomposition Fourier finite element method for the simulation of 3D marine CSEM measurements. J. Comput. Phys. **255**, 456–470 (2013)
7. Mohd Aris, M.N., Daud, H., Mohd Noh, K.A., Dass, S.C.: Model calibration of stochastic process and computer experiment for MVO analysis of multi-low-frequency electromagnetic data. Processes **8**(5), 605 (2020)
8. Liu, D., Pang, J., Zhou, J., Peng, Y., Pecht, M.: Prognostics for state of health estimation of lithium-ion batteries based on combination Gaussian process functional regression. Microelectron. Reliab. **53**, 832–839 (2013)
9. Rasmussen, C.E., Nickisch, H.: Gaussian processes for machine learning (GPML) toolbox. J. Mach. Learn. Res. **11**, 3011–3015 (2010)
10. Chan, L.L.T., Liu, Y., Chen, J.: Nonlinear system identification with selective recursive gaussian process model. Ind. Eng. Chem. Res. **52**, 18276–18286 (2013)
11. Turner, B.M., Forstmann, B.U., Wagenmakers, E.J., Brown, S.D., Sederberg, P.B., Steyvers, M.: A Bayesian framework for simultaneously modeling neural and behavioral data. Neuroimage **72**, 193–206 (2013)
12. Sacks, J., Welch, W., Mitchell, T., Wyan, H.P.: Design analysis of computer experiments. Stat. Sci. **4**, 409–423 (1989)
13. Santner, T.J., Williams, B.J., Nolts, W.I.: The design and Analysis of Computer Experiments. Springer, Berlin Heidelberg (2003)
14. Higdon, D., Gattiker, J., Williams, B., Rightley, M.: Computer model calibration using high dimensional output. J. Amer. Statist. Assoc. **103**, 570–583 (2008)
15. Hills, R.G.: Model validation: Model parameter and measurement uncertainty. J. Heat Transfer **128**, 339–351 (2006)
16. Wang, S., Chen, W., Tsui, K.L.: Bayesian validation of computer models. Technometrics **51**, 439–451 (2009)
17. Ranganathan, A., Yang, M.H.: Online sparse matrix Gaussian process regression and vision applications. In: Forsyth, D., Torr, P., Zisserman, A. (eds.) European Conference on Computer Vision (ECCV 2008), vol. 5302, pp. 468–482. Springer, Berlin Heidelberg (2008)
18. Chu, W., Ghahramani, Z.: Gaussian processes for ordinal regression. J. Mach. Learn. Res. **6**, 1019–1041 (2005)
19. Yin, F., Zhao, Y., Gunnarsson, F., Gustafsson, F.: Received-signal-strength threshold optimization using Gaussian processes. IEEE Trans. Signal Process. **65**, 2164–2177 (2017)
20. Asante-Okyere, S., Shen, C., Ziggah, Y.Y., Rulegeya, M.M., Zhu, X.: Investigating the predictive performance of Gaussian process regression in evaluating reservoir porosity and permeability. Energies **11**, 3261 (2018)
21. Daud, H., Mohd Aris, M.N., Mohd Noh, K.A., Dass, S.C.: A novel methodology for hydrocarbon depth prediction in seabed logging: Gaussian process-based inverse modeling of electromagnetic data. Appl. Sci. **11**(4), 1492 (2021)
22. Mohd Aris, M.N., Daud, H., Mohd Noh, K.A., Dass, S.C.: Stochastic process-based inversion of electromagnetic data for hydrocarbon resistivity estimation in seabed logging. Mathematics **9**(9), 935 (2021)
23. Fang, D., Zhang, X., Yu, Q., Jin, T.C., Tian, L.: A novel method for carbon dioxide emission forecasting based on improved Gaussian processes regression. J. Clean. Prod. **173**, 143–150 (2018)

24. Kong, D., Chen, Y., Li, N.: Gaussian process regression for tool wear prediction. Mech. Syst. Signal Process **104**, 556–574 (2018)
25. Rasmussen, C.E., Williams, C.K.I.: Gaussian Processes for Machine Learning. The MIT Press, Cambridge (2006)

On the Optimal Control of the Thermal Exchange on a Two-Phase Boundary Layer

Zh. Zhanabekov, G. Abduakhitova, and U. Kusherbayeva

Abstract This paper considers a problem of optimal control of an injection into an incompressible laminar boundary layer with boundary surface. Based on the necessary condition for the extremum of the auxiliary functional, a system of equations is obtained for determining the Lagrange multipliers. The dependence of the viscosity and thermal diffusivity coefficients on temperature is taken into account. A special case is considered. A method for solving the problem is given.

Keywords Optimal control of the boundary layer · Injected fluid · Injection into the boundary layer · Hydrodynamic problem · Laminar boundary layer · Minimization of heat flux

1 Introduction

In modern technology there are many devices that work under conditions of a gas stream flowing around their surfaces at high speeds and high temperatures (gas flow in nozzles and combustion chambers of jet engines, in interscapular channels of gas turbines, etc.). An increase in surface temperature due to large heat fluxes from the streamlined gas to the wall and aerodynamic heating can exceed the maximum permissible values that ensure the strength of the device.

One of the possible ways to protect heated surfaces is the use of porous cooling [1, 2].

The problem of optimal control of the boundary layer was first considered in [3], where it was assumed that the physical properties of the injected fluid and the incoming flow are similar.

In some practical problems, such an assumption significantly limits the control possibilities; therefore, problems of injection into the boundary layer of a medium that differs in its physical properties from the incident flow may be of interest.

Zh. Zhanabekov · G. Abduakhitova (✉) · U. Kusherbayeva
Al-Farabi Kazakh National University, Almaty, Kazakhstan

S. A. Abdul Karim et al. (eds.), *Proceedings of the 6th International Conference on Fundamental and Applied Sciences*, Springer Proceedings in Complexity,
https://doi.org/10.1007/978-981-16-4513-6_44

In [4], under the mentioned assumption, the hydrodynamic problem of the optimal control of friction in a laminar boundary layer with a boundary surfaces considered.

In this paper, we study the question of minimizing the heat flux from a streamlined gas to a wall.

2 Problem Statement

A stream, having a speed $U(x)$, temperature T_∞, density ρ_2 and viscosity ν_2, outside the boundary layer, flows around an asymptotically thin body, from the surface of which a foreign liquid with a density of $\rho_{1,}$, viscosity ν_1, is supplied, which forms the inner layer Fig. 1.

It is assumed that the amount of injected fluid is small, so that the effect of injection only affects the fluid flow in a thin parietal layer 1 and does not introduce disturbances in potential flow 2. We assume that there is a stable border between the media. In this case, the system of equations of a steady incompressible two-phase boundary layer in a rectangular coordinate system takes the form

$$F_{i1} \equiv U_i \frac{\partial U_i}{\partial x} + V_i \frac{\partial U_i}{\partial y} - \frac{\rho_2}{\rho_1} U \frac{dU}{dx} - \frac{\partial}{\partial y}(\nu_i \tau_i) = 0, \quad (1)$$

$$F_{i2} \equiv \frac{\partial U_i}{\partial x} + \frac{\partial V_i}{\partial y} = 0, \quad (2)$$

$$F_{i3} \equiv U_i \frac{\partial T_i}{\partial x} + V_i \frac{\partial T_i}{\partial y} - \frac{1}{\sigma_i} \frac{\partial}{\partial y}(\nu_i q_i) = 0, \quad (3)$$

$$F_{i4} \equiv \tau_i - \frac{\partial U_i}{\partial y} = 0, \quad (4)$$

$$F_{i5} = q_i - \frac{\partial T_i}{\partial y} = 0, \; F_{i6} \equiv \nu_i - f_i(T_i) = 0. \quad (5)$$

The boundary conditions are of the form:

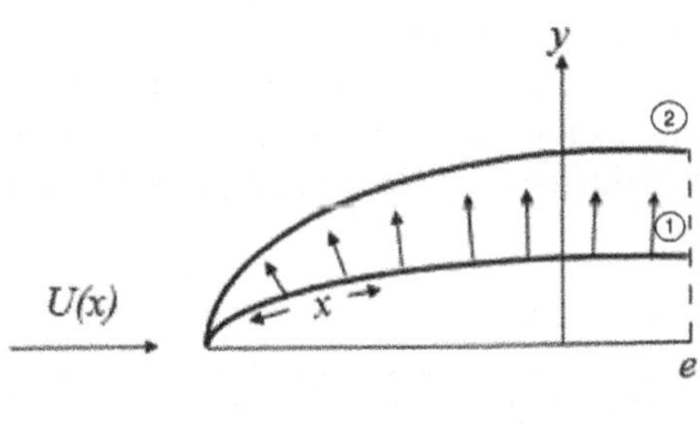

Fig. 1 Asymptotically thin body from the surface of which a foreign liquid is supplied that forms parietal layer

$$U_1 = 0, V_1 = V_w(x), T_1 = T_w \text{ if } y = 0, \tag{6}$$

$$\begin{aligned} &U_2 = U(x),\ T_2 = T_\infty \text{ if } x \to \infty, \\ &U_2 = U^*(y),\ T_2 = T^*(y) \text{ if } x = 0. \end{aligned} \tag{7}$$

The conditions on the boundary surface:

$$\begin{aligned} &U_1 \dot{y}_0 - V_1 = U_2 \dot{y}_0 - V_2 = 0, \\ &\mu_1 \frac{\partial U_1}{\partial y} = \mu_2 \frac{\partial U_2}{\partial y},\ U_1 = U_2 = U_0 \end{aligned} \tag{8}$$

$$k_1 \frac{\partial T_1}{\partial y} = k_2 \frac{\partial T_2}{\partial y},\ T_1 = T_2 = T_0, \tag{9}$$

where σ_i are Prandtl numbers ν_i and k_i are coefficients of viscosity and thermal conductivity which depend on temperature. The following variational problem is posed: among all admissible controls $V_\omega(x)$ determine one that would result in a minimum

$$Q = -\int_0^l k_1 \left(\frac{\partial T_1}{\partial y} \right)_{y=0} dx, \tag{10}$$

the amount of heat transferred from the hot gas stream to the plate at a given power of the cooling system

$$P = \int_0^l a(x) V_\omega^2(x) dx, \tag{11}$$

where $a(x)$ is the thickness of the porous wall. We believe that given $V_\omega(x)$, solution of problem (1)–(5) with conditions (6)–(9) exists and is unique.

3 Main Body

Guided by the general approach to solving variational problems on a conditional extremum [5], we compose the functional:

$$I = \int_0^l (-k_1 q_1(x) + \alpha a V_\omega^2(x)) dx + \sum_{i=1}^{2} \iint_{D_i} \left(\sum_{j=1}^{6} \lambda_{ij} F_{ij} \right) dx dy, \tag{12}$$

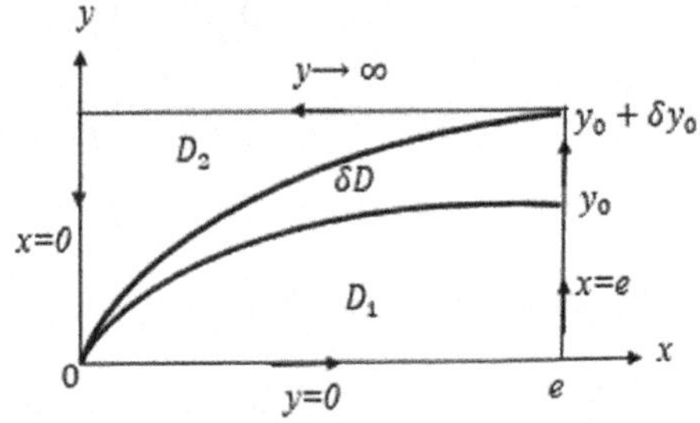

Fig. 2 Line bounding regions D_1 and D_2

where D_1 is the area bounded by lines $y = 0,\ x = l,\ y = y_0(x)$; D_2 is the area bounded by lines $y = y_0(x), x = l, y \to \infty, x = 0$ Fig. 2. Parameters $\alpha,\ \lambda_{i1},\ \lambda_{i2},\ \lambda_{i3},\ \lambda_{i4},\ \lambda_{i5},\ \lambda_{i6}$ are Lagrange multipliers.

The first variation of functional (12) will be:

$$\begin{aligned}\delta I = & \int_0^l \left[-k_1 \delta q_1 - q_1 \frac{\partial k_1}{\partial T_1} \delta T_1 + 2\alpha a\,(x) V_w \delta V_w \right] dx \\ & + \sum_{i=1}^{2} \iint_{D_i} \Big(L_{i1} \tilde{\delta} V_i + L_{i2} \tilde{\delta} \tau_i + L_{i3} \tilde{\delta} q_i \\ & - L_{i4} \tilde{\delta} U_i - \ L_{i5} \tilde{\delta} T_i + L_{i6} \tilde{\delta} v_i \Big) dxdy \\ & + \sum_{i=1}^{2} \iint_{D_i} \Big[\frac{\partial}{\partial x} \Big(M_{i1} \tilde{\delta} U_i + \lambda_{i3} U_i \tilde{\delta} T_i \\ & + \frac{\partial}{\partial y} \Big(M_{i2} \tilde{\delta} U_i + \lambda_{i2} \tilde{\delta} V_i + \ M_{i3} \tilde{\delta} T_i - M_{i4} \tilde{\delta} v_i \\ & - \lambda_{i1} v_i \tilde{\delta} \tau_i - \frac{v_i}{\sigma_i} \lambda_{i3} \tilde{\delta} q_i \Big) \Big] dxdy. \end{aligned} \tag{13}$$

Consider the necessary condition of the extremum $\delta I = 0$.

In expression (13), we leave only the control variation δV_w, which is independent, and exclude the remaining variations. For this, we select the factors $\lambda_{ij} (i = 1, 2;\ j = 1, 2, \ldots, 6)$ so that they satisfy the system of equations:

$$\begin{aligned} & \lambda_{i1} \frac{\partial U_i}{\partial y} - \frac{\partial \lambda_{i2}}{\partial y} + \lambda_{i3} \frac{\partial T_i}{\partial y} = 0, \\ & U_i \frac{\partial \lambda_{i1}}{\partial x} + V_i \frac{\partial \lambda_{i1}}{\partial y} + \lambda_{i1} \frac{\partial V_i}{\partial y} + \frac{\partial \lambda_{i2}}{\partial x} - \lambda_{i3} \frac{\partial T_i}{\partial x} + \frac{\partial}{\partial y} \left(v_i \frac{\partial \lambda_{i2}}{\partial y} \right) = 0, \\ & U_i \frac{\partial \lambda_{i3}}{\partial x} + V_i \frac{\partial \lambda_{i3}}{\partial y} + \frac{1}{\sigma_i} \frac{\partial}{\partial y} \left(v_i \frac{\partial \lambda_{i3}}{\partial y} \right) - \left(\tau_i \frac{\partial \lambda_{i1}}{\partial y} + \frac{q_i}{\sigma_i} \frac{\partial \lambda_{i3}}{\partial y} \right) \frac{\partial f_i}{\partial T_i} = 0. \end{aligned} \tag{14}$$

We determine the boundary conditions for system (14) from the condition that the coefficients are equal to zero for variations of those functions whose values at the boundary are not specified. Considering the contour integrals in expression (13) along the lines $x = 0$, $y = 0$, $x = l$, $y \to \infty$ and satisfying the necessary condition for the extremum, we find the boundary conditions for the solution of (14):

$$\begin{aligned}&\lambda_{11} = 0, \lambda_{13} = \rho_1 \left(c_p\right)_1 \; if \; y = 0,\\&\lambda_{21} \to 0, \lambda_{23} \to 0, \lambda_{22} \to 0, \; if \; y \to \infty,\\&\lambda_{11}U_1 + \lambda_{12} = \lambda_{21}U_2 + \lambda_{22} = 0, \; \lambda_{13}U_1 = \lambda_{23}U = 0, \; if \; x = l.\end{aligned} \tag{15}$$

We now consider the integral over the boundary surface $y = y_0(x)$ in the expression for the first variation of functional (13):

$$\begin{aligned}\delta l_2 = \sum_{i=1}^{2}(-1)^i \int_0^l \Big[&(\lambda_{i2}\dot{y}_0 - \lambda_{i4})\tilde{\delta} U_i - \lambda_{i5}\tilde{\delta} T_i - \lambda_{i2}\tilde{\delta} V_i\\&+ \nu_i\lambda_{i1}\tilde{\delta}\tau_i + \frac{\nu_i}{\sigma_i}\lambda_{i3}\tilde{\delta} q_i \left(\tau_i\lambda_{i1} + \frac{q_i}{\sigma_i}\lambda_{i3}\right)\tilde{\delta}\nu_i\Big]dx\end{aligned} \tag{16}$$

Given Eqs. (4), (5) and relations (8), (9), we obtain a relation between the variations of functions $U_i, V_i, T_i, \tau_i, q_i, \nu_i$:

$$\begin{aligned}&U_1 = \delta U_2 = \delta u_0, \; \delta V_i = \delta U_i \dot{y}_0 + U_i \delta \dot{y}_0,\\&\delta\tau_2 = \frac{\mu_1}{\mu_2}\delta\tau_1, \delta T_1 = \delta T_2 = \delta T_0,\\&\delta q_2 = \frac{k_1}{k_2}\delta q_1, \delta\nu_1 = \delta\nu_2 = \delta\nu_0.\end{aligned} \tag{17}$$

The relation between the total variations and the variations for fixed integration regions has the form [6]

$$\begin{aligned}&\tilde{\delta} U_i = \delta U_i - \frac{\partial U_i}{\partial y}\delta y_0, \tilde{\delta} V_i = \delta V_i - \frac{\partial V_i}{\partial y}\delta y_0,\\&\tilde{\delta}\tau_i = \delta\tau_i - \frac{\partial \tau_i}{\partial y}\delta y_0, \tilde{\delta} T_i = \delta T_i - \frac{\partial T_i}{\partial y}\delta y_0,\\&\tilde{\delta} q_i = \delta q_i - \frac{\partial q_i}{\partial y}\delta y_0, \tilde{\delta}\nu_i = \delta\nu_i - \frac{\partial \nu_i}{\partial y}\delta y_0,\end{aligned} \tag{18}$$

Variation of the derivative function $y_0(x)$ can be eliminated by setting

$$\lambda_{i2}U_i\delta\dot{y}_0 = \frac{d}{dx}(\lambda_{i2}U_i\delta y_0) - \frac{d\lambda_{i2}}{dx}U_i\delta y_0 - \lambda_{i2}\frac{dU_i}{dx}\delta y_0, \tag{19}$$

where

$$\frac{d\lambda_{i2}}{dx} = \frac{\partial\lambda_{i2}}{\partial x} + \dot{y}_0\frac{\partial\lambda_{i2}}{\partial y}, \frac{dU_i}{dx} = \frac{\partial U_i}{\partial x} + \dot{y}_0\frac{\partial U_i}{\partial y}$$

are total derivatives with respect to x along the curve $y_0(x)$. If we integrate, taking into account expressions (17)–(19), then for δI_2 we will have:

$$\begin{aligned}\delta I_2 = \int_0^l \Bigg\{ &(\lambda_{14} - \lambda_{24})\delta U_0 + (\lambda_{15} - \lambda_{25})\delta T_0 + \left(\nu_2\frac{\mu_1}{\mu_2}\lambda_{21} - \nu_1\lambda_{11}\right)\delta\tau_1 \\ &+ \left(\frac{\nu_2}{\sigma_2}\cdot\frac{k_1}{k_2}\lambda_{23} - \frac{\nu_1}{\sigma_1}\lambda_{13}\right)\delta q_1 + \left(\tau_2\lambda_{21} - \tau_1\lambda_{11} + \frac{q_2}{\sigma_2}\lambda_{23} - \frac{q_1}{\sigma_1}\lambda_{13}\right)\delta\nu_0 \\ &+ \sum_{i=1}^{2}(-1)^i\Bigg[\lambda_{i4}\frac{\partial U_i}{\partial y} - \nu_i\lambda_{i1}\frac{\partial\tau_i}{\partial y} \\ &+ \frac{\partial\lambda_{i2}}{\partial y}U_i\dot{y}_0 + \frac{\partial\lambda_{i2}}{\partial x}U_i + \lambda_{i5}\frac{\partial T_i}{\partial y} - \frac{\nu_i}{\sigma_i}\cdot\frac{\partial q_i}{\partial y}\lambda_{i3} \\ &- \left(\tau_i\lambda_{i1} + \frac{q_i}{\sigma_i}\lambda_{i3}\right)\frac{\partial\nu_i}{\partial y}\Bigg]\delta y_0\Bigg\}dx + [(\lambda_{22} - \lambda_{12})U_0\delta y_0]_{x=l}\end{aligned} \tag{20}$$

When integrating, we assume that $\delta y_0(0) = 0$. Equating the coefficients of free variations to zero in (20), we obtain the conditions on the boundary surface:

$$\begin{aligned}&\lambda_{14} = \lambda_{24}, \lambda_{15} = \lambda_{25}\frac{\lambda_{11}}{\rho_1} = \frac{\lambda_{21}}{\rho_2}, \frac{\lambda_{13}}{p_1c_{\rho 1}} = \frac{\lambda_{23}}{p_2c_{\rho 2}}, \\ &\sum_{i=1}^{2}(-1)^i\Bigg[\lambda_{i4}\frac{\partial U_i}{\partial y} - \nu_i\lambda_{i1}\frac{\partial\tau_i}{\partial y} + \frac{\partial\lambda_{i2}}{\partial y}U_i\dot{y}_0 + \frac{\partial\lambda_{i2}}{\partial x}U_i + \lambda_{i5}\frac{\partial T_i}{\partial y} \\ &-\frac{\nu_i}{\sigma_i}\lambda_{i3}\frac{\partial q_i}{\partial y} - \left(\tau_i\lambda_{i1} + \frac{q_i}{\sigma_i}\lambda_{i3}\right)\frac{\partial\nu_i}{\partial y}\Bigg] = 0,\end{aligned} \tag{21}$$

$$\lambda_{12} = \lambda_{22}\ if\ x = l.$$

Now the variation of functional (12) is written as

$$\delta I = \int_0^l \left[2\alpha a(\mathrm{x})V_\omega - \lambda_{12}(x,0)\right]\delta V_\omega dx. \tag{22}$$

Since only a factor λ_{12} is included in (22), we resolve (14) with respect to λ_{i2}. Eliminating λ_{i1} from the first two equations in a system (14) and using (1)–(3), we obtain the equation for λ_{i2}, which admits the first integral [7]:

$$\nu_i \frac{\partial^2 \lambda_{i2}}{\partial y^2} - \left(\frac{2\nu_i U_{iyy}}{U_{iy}} + \nu_{iy} - V_i \right) \frac{\partial \lambda_{i2}}{\partial y} + U_i \frac{\partial \lambda_{i2}}{\partial x} = f_i(x, y) + \varphi_i(x),$$

where

$$f_i(x, y) = \left(\nu_i - \frac{\nu_i}{\sigma_i} \right) T_{iy} \frac{\partial \lambda_{i3}}{\partial y} + \left[\left(\nu_i + \frac{\nu_i}{\sigma_i} \right) T_{iyy} - \frac{2\nu_i U_{iyy} T_{iy}}{U_{iy}} + \left(\frac{1}{\sigma_i} - 1 \right) \nu_{iy} T_{iy} \right] \lambda_{i3}, \tag{23}$$

and $\varphi_i(x)$ is some integration function determined from the boundary conditions.

Consider the special case when $\nu_i = const, k_i = const$. Then, the function λ_{i3} is found from the equation

$$\frac{\nu_i}{\sigma_i} \frac{\partial^2 \lambda_{i3}}{\partial y^2} + U_i \frac{\partial \lambda_{i3}}{\partial x} + V_i \frac{\partial \lambda_{i3}}{\partial y} = 0 \tag{24}$$

with boundary conditions

$$\begin{gathered} \lambda_{13} = \rho_1 c_{p1} \text{ if } y = 0, \\ \lambda_{23} \to 0 \text{ if } y \to \infty, \\ \lambda_{13} U_1 = \lambda_{23} U_2 = 0 \text{ if } x = l \end{gathered} \tag{25}$$

and conditions on the boundary surface

$$\frac{\lambda_{13}}{\rho_1 c_{\rho 1}} = \frac{\lambda_{23}}{\rho_2 c_{\rho 2}}, \frac{\kappa_1}{p_1 c_{p1}} \frac{\partial \lambda_{13}}{\partial y} = \frac{\kappa_2}{p_2 c_{p2}} \frac{\partial \lambda_{23}}{\partial y}. \tag{26}$$

Solving the boundary conditions (15) and (21) relative to λ_{i2}, we find:

$$\begin{aligned} &\frac{\partial \lambda_{12}}{\partial y} = p c_{p1} \cdot q_1(x, 0) \text{ if y } = 0, \\ &\lambda_{12} = 0, \lambda_{22} = 0 \text{ if } x = l, \\ &\frac{\partial \lambda_{22}}{\partial y} \to 0, \lambda_{22} \to 0 \text{ if } x \to \infty. \end{aligned} \tag{27}$$

On the boundary surface, we will have:

$$\sum_{i=1}^{2} (-1)^i \mu_i \left(\frac{\nu_i U_{iyy}}{U_{iy}} - V_i \right) \frac{\partial \lambda_{i2}}{\partial y} - U_i \frac{\partial \lambda_{i2}}{\partial x} - \frac{\nu_i}{\sigma_i} T_{iy} \frac{\partial \lambda_{i3}}{\partial y}$$

$$+\left(\frac{\nu_i}{\sigma_i}T_{iyy} - \nu_i\frac{T_{iy}U_{iyy}}{U_{iv}}\right)\lambda_{i3} + \varphi_i(x) = 0,$$

$$\nu_1\left[\frac{\partial\lambda_{12}}{\partial y} - T_{1y}\lambda_{13}\right] = \nu_2\left[\frac{\partial\lambda_{22}}{\partial y} - T_{2y}\lambda_{23}\right], \quad (28)$$

$$\varphi_1(x) = \varphi_2(x).$$

Thus, to construct a control V_w, that delivers the minimum to functional (10), it is necessary to solve the system of Eqs. (1)–(5) and (23) with conditions (6)–(9) and (27), (28). The relationship between optimal control and function λ_{12} is given by the relation

$$2\alpha a\cdot(x)\ V_\omega - \lambda_{12}(x,0) = 0,$$

following from (22).

The sequential descent method [8] allows one to approximately construct the optimal control. First, we set the first approximation of control $V_\omega^{(1)}(x)$ by which we construct a solution to system (1)–(5), and then to Eq. (23). Knowing the first approximation $U^{(1)}$, $V^{(1)}$, $T^{(1)}$ $u\lambda_{12}^{(1)}$, we calculate the value of functionals (10) and (11). It should be expected that the control $V_\omega^{(1)}(x)$ will not immediately satisfy the isoperimetric condition (11) and deliver a minimum to the functional (10), therefore, it is necessary to change the control in the direction of better correspondence.

According to the steepest descent method, we set:

$$\delta V_w^{(1)} = -\varepsilon\left[2\alpha a(\mathrm{x})V_w^{(1)} - \lambda_{12}^{(1)}(x,0)\right]. \quad (29)$$

Here the constant ε determines the step of approximation. Assuming $\delta P^{(1)} = 0$, we get the equation for determining α:

$$\int_0^l 2a(x)V_\omega^{(1)}\left[2\alpha a(x)V_w^{(1)} - \lambda_{12}^{(1)}\right]dx = 0. \quad (30)$$

An unknown quantity ε is found from the condition $P^{(1)} = C_1 - P^{(1)}$, where $P = C_1$. Now the variation $\delta V_w^{(1)}$ is completely determined.

Setting a small value to ε, we make several approximations before reaching $P = C_1$. In each approximation, for the obtained control $V_w^{(s)} + \delta V_w^{(s)}$ (s is the approximation number), the system (1)–(5) and (23) is solved anew, its own factor is determined, and its own step ε is set. Satisfying the isoperimetric condition (11), we proceed to constructing a control delivering min Q. The construction procedure remains the same as when constructing a control satisfying the isoperimetric condition (11), but in each step we assume that $\delta P^{(s)} = 0$. We assume that the control is constructed after the condition $\overline{Q} = \min Q$ is met precisely.

4 Conclusion

In the formulation and solution of many problems of aerodynamics and hydrodynamics, an important role is played by the theory of the boundary layer. This theory allows us to characterize the ability of a transport object (ship, plane) to withstand the resistance of a gaseous and liquid medium during movement. The development of methods for controlling the boundary layer is associated with the goal of weakening or preventing flow separation on the streamlined surface of the ship and aircraft skin, reducing heat transfer at high flow rates [9, 10]. One of the promising control methods is the possibility of laminarization of the boundary layer using suction. The next promising method for controlling the boundary layer is to inject a gaseous jet along the streamlined surface of a moving object [11, 12]. Today, increasing the pace in the study of methods for managing the boundary layer is a promising direction. The modern concept of fluid dynamics and aerodynamics of continuous facilities reveals a variety of theoretical and experimental studies in the technical needs of different countries of the world. In this regard, the fundamental assertion is that the wide possibilities for the development of aviation and the navy are characterized by the development of the theory of the boundary layer.

References

1. Morduchow, M.: On heat transfer over a sweat cooled surface in laminar compressible flow. J. AeroSpace Sci. **19**(10) (1952)
2. Loitsyansky, L.G.: Laminar boundary layer. Fizmatgiz (1962) [In Russian]
3. Divakov, O.G., Sirazetdinov, T.K.: On optimal control of the boundary layer. IWUZ, Aeronautical engineering, issue 3 (1969) [In Russian]
4. Afanasyev, A.A.: Optimal blowing of foreign liquid into the boundary layer. Tr. KAI, issue 147 (1972) [In Russian]
5. Sirazetdinov, T.K.: Optimal problems of gas dynamics. IVUZ, Aviation equipment, issue 2 (1963) [In Russian]
6. Gelfand, I.M., Fomin, S.V.: The calculus of variations. Fizmatgiz (1961) [In Russian]
7. Garaev, K.G.: On invariant variational problems. In: Collection: Materials of the First Volga Conference on Automatic Control, vol. 1, Kazan (1971) [In Russian]
8. Sirazetdinov, T.K.: The method of sequential descent to optimize systems with distributed parameters. In: Collection: Materials of the First Volga Conference on Automatic Control, vol. 1, Kazan (1971) [In Russian]
9. Lushchik, V.G., Yakutenko, A.B.: Friction and heat transfer in a boundary layer on a permeable surface by injection. Thermophys. High Temp. **43**(6), 880–887 (2005) [In Russian]
10. Plotkina, V.A., Starodubtsev, P.A., Shevchenko, A.P.: Promising Methods and Techniques for Controlling the Border Layer of Vehicles. Vladivostok, Pacific Naval Institute S.O. Makarova (2015) [In Russian]
11. Kuznetsov, V.K.: Mathematical modeling in the problems of optimal control of a laminar boundary layer in supersonic flows. Abstract of Dissertation for the Degree of Candidate of Physical and Mathematical Sciences. Kazan (2010)
12. Garaev, K.G., Kuznetsov, V.K.: On optimal injection into the boundary layer in a supersonic flow. Vestn. of KSTU named after A.N. Tupolev **4**, 143–151 (2012) [In Russian]

Surface Reconstruction Using Rational Quartic Triangular Spline

Nur Nabilah Che Draman, Samsul Ariffin Abdul Karim, Ishak Hashim, and Yeo Wee Ping

Abstract Reconstruction a smooth surface from non-uniform points is a major challenge in many areas. This paper discusses C^1 surface scattered data interpolation using rational quartic triangular patches with two different existing convex combination. In order to achieve C^1 continuity, we apply a rational corrected that obtained from convex combination between three local schemes. To validate our proposed method, we tested the scheme by using two test functions and two real applications from established dataset. We measured the performance by comparing errors which is maximum error, root mean square (RMSE), coefficient of determination (R^2) and central processing unit (CPU). All numerical and graphical results are obtained using MATLAB R2019a. From the result, we found that the proposed scheme is better than existing scheme.

Keywords C^1 continuity · Local scheme · Scattered data interpolation · Convex combination · Rational quartic triangular patches

N. N. C. Draman
Fundamental and Applied Sciences Department, Universiti Teknologi PETRONAS (UTP), 32610 Seri Iskandar, Perak, Malaysia
e-mail: nur_19001020@utp.edu.my

S. A. Abdul Karim (✉)
Fundamental and Applied Sciences Department and Centre for Systems Engineering (CSE), Institute of Autonomous System, Universiti Teknologi PETRONAS (UTP), 32610 Seri Iskandar, Perak, Malaysia
e-mail: samsul_ariffin@utp.edu.my

I. Hashim
Department of Mathematical Sciences, Faculty of Science and Technology, Universiti Kebangsaan Malaysia, 43600 UKM, Bangi, Selangor, Malaysia
e-mail: ishak_h@ukm.edu.my

Y. W. Ping
Faculty of Science, Universiti Brunei Darussalam, Bandar Seri Begawan BE1410, Brunei Darussalam
e-mail: weeping.yeo@ubd.edu.bn

S. A. Abdul Karim et al. (eds.), *Proceedings of the 6th International Conference on Fundamental and Applied Sciences*, Springer Proceedings in Complexity,
https://doi.org/10.1007/978-981-16-4513-6_45

1 Introduction

Computer Aided Geometric Design (CAGD) has been widely used in scattered data interpolation. Scattered data interpolation scheme is aimed to reconstruct smooth surface from non-uniform data sets where it is sparse everywhere. This problem usually arises in many areas such as seamount, rainfall, geochemical, oil and gas and others. There are several methods that we can use to solve scattered data interpolation such as Delaunay triangulation [1], moving last square (MLS) [2] and radial basis function (RBF) [3]. In this paper, we focus on Delaunay triangulation to reconstruct the surface. Many researchers have investigated scattered data interpolation using Delaunay triangulation. Karim et al. [4] proposed spatial interpolation to estimate the unknown data at certain point. Besides, shape preserving interpolation is considered especially positivity preserving interpolation where the end of the result for the surface is always positive. They used Delunay triangulation based method to estimate the unknown points of rainfall at spatial localization using cubic Bezier triangular patches. Ali et al. [5] discussed the scattered data interpolation to visualize the rainfall data and digital elevation in Malaysia by using cubic Timmer triangular patches. In [5], they calculated the inner ordinates with two existing schemes which are Goodman and Said [6] and Foley and Opitz [7]. Hussain and Hussain [8] proposed C^1 positive scattered data interpolation using rational cubic function defined on triangular grid. Each boundary of the triangle is constructed by rational cubic function with two shape parameters. The final scheme comprising the convex combination of three side vertex interpolations. To preserve the positivity, the simple sufficient data independent is derived. Safraz et al. [9] presented positivity preserving scheme using the side vertex method and applied to scattered data interpolation. A C^1 piecewise rational cubic function with four shape parameters is used for interpolation along edge of a triangle. This scheme applied Delaunay triangulation to triangulate the sparse data. The sufficient data dependent constrains is derived on two free parameters while the other two are left for user to enhanced the shape of data.

The objective of this paper is to reconstruct a smooth surface using rational quartic triangular patches with three parameters. The rest of this paper is organized as follow. Description of rational quartic triangular patches and the properties is shows in Sect. 2. Based on it, Sect. 3 construct derivation of sufficient condition of C^1 scattered data interpolation. The numerical and graphical results are given in Sect. 4. Conclusions are given in Sect. 5.

2 Rational Quartic Triangular Patches

Quartic triangular patch with three shape parameters is an extension over a triangular domain of the basis from two shape parameter [10]. Let the barycentric coordinate α, β, γ on triangle P with vertices V_1, V_2 and V_3 such that $x + y + z = 1$ and $x, y, z \geq$

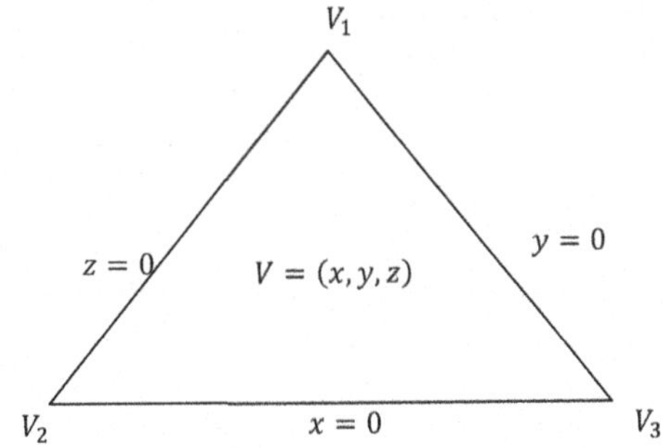

Fig. 1 Triangle T

0. Any points $V(x, y) \in R^2$ inside the triangle (including at the vertices) can be expressed as

$$V = xV_1 + yV_2 + zV_3$$

As shown in Fig. 1.

Rational quartic triangular patches are defined as follow: Definition 1. Let $\alpha, \beta, \gamma \in [0, +\infty)$, given control points $T_{p,q,r} \in R^3 (p, q, r \in N, p + q + r = 3)$ and a triangular domain $D = (x, y, z)|x + y + z = 1, x \geq 0, y \geq 0, z \geq 0$. It is defined as

$$R(x, y, z) = \sum_{p+q+r=3} B^3_{p,q,r}(x, y, z) T_{p,q,r}, \quad (x, y, z) \in D \tag{1}$$

the quartic rational triangular patch with three shape parameters α, β, γ with ten basis functions. From (1), it can be written as

$$\begin{aligned}
R(x, y, z) = {} & \frac{x^2}{1 + \alpha(1 - x)} B_{3,0,0} + \frac{y^2}{(1 + \beta(1 - y))} B_{0,3,0} + \frac{z^2}{(1 + \gamma(1 - z))} B_{0,0,3} \\
& + \frac{(x^2 y[2 + \alpha + 2\alpha(1 - x)])}{(1 + \alpha(1 - x))} B_{2,1,0} + \frac{(x^2 z[2 + \alpha + 2\alpha(1 - x)])}{(1 + \alpha(1 - x))} B_{2,0,1} \\
& + \frac{(xy^2[2 + \beta + 2\beta(1 - y)])}{(1 + \beta(1 - y))} B_{1,2,0} + \frac{(x^2 y[2 + \beta + 2\beta(1 - y)])}{(1 + \beta(1 - y))} B_{0,2,1} \\
& + \frac{(xz^2[2 + \gamma + 2\gamma(1 - z)])}{(1 + \gamma(1 - z))} B_{1,0,2} + \frac{(yz^2[2 + \gamma + 2\gamma(1 - z)])}{(1 + \gamma(1 - z))} B_{0,1,2} \\
& + 6xyzB_{1,1,1}
\end{aligned} \tag{2}$$

Figure 2 shows barycentric coordinate quartic rational triangular patch that lies on triangle patch.

Fig. 2 Rational quartic triangular patch basis

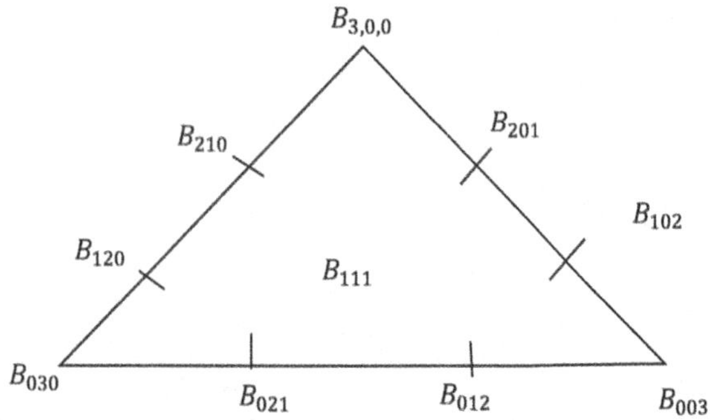

3 Derivation of C^1 Scattered Data Interpolation

The problem statement for scattered data interpolation can be described as follows: Given the functional scattered data

$$(a_i, b_i, z_i), \quad i = 1, 2, \ldots, N \tag{3}$$

We wish to construct a smooth C^1 surface $z = P(a_i, b_i)$ such that

$$z_i = F(a_i, b_i), \quad i = 1, 2, \ldots, N \tag{4}$$

3.1 *Local Scheme*

This scheme comprising convex combination of three local schemes P_1, P_2 and P_3 are defined as below.

1. Saaban et al. [11]

$$R(x, y, z) = \frac{(yzP_1 + xzP_2 + xyP_3)}{(yz + xz + xy)}, \tag{5}$$

2. Hussain and Hussain [8]

$$R(x, y, z) = \frac{(x^2y^2P_1 + y^2z^2P_2 + x^2z^2P_3)}{(x^2y^2 + y^2z^2 + x^2z^2)}. \tag{6}$$

where the local scheme P_i, $i = 1, 2, 3$ is obtained by replacing $P^i_{1,1,1}$ in [4] as shown in Fig. 3.

The inner triangular points for the local schemes P_i, $i = 1, 2, 3$ is calculated by using cubic precision method that of Foley and Opitz [7].

Fig. 3 The directional P_1, P_2, P_3

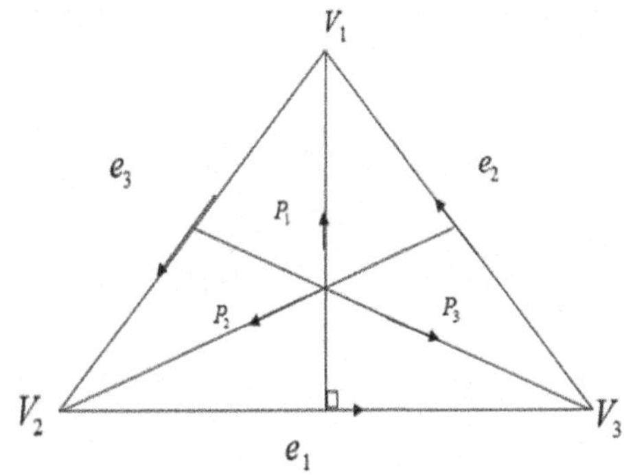

Fig. 4 Notations on a triangle T

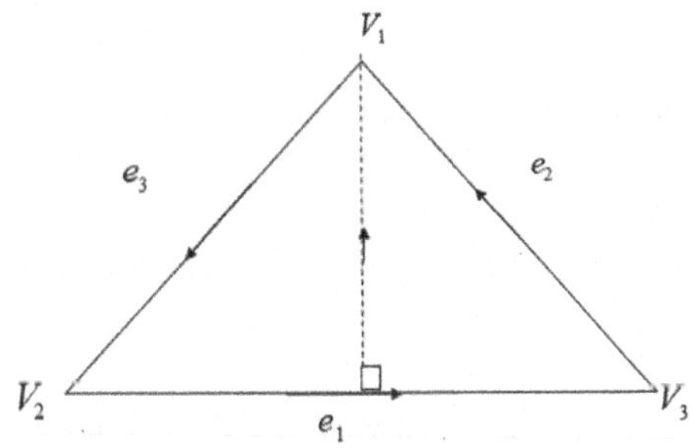

Goodman and Said [6] scheme is used to calculate the edge ordinates for each triangle. Figure 4 shows a directional triangle T with three vertices (V_1, V_2, V_3), three edges (e_1, e_2, e_3) and one face. Note that,e_3, e_2, e_1 are opposite to V_3, V_2, V_1 respectively shown in Draman et al. [12].

To achieve C^1 continuity along all edges as shown in Fig. 5, the following equations must be satisfied.

$$c_{2,0,1} = r^2 b_{2,1,0} + 2st b_{0,2,1} + 2rs b_{1,2,0} + s^2 b_{0,3,0} + 2rt b^1_{1,1,1} + t^2 b_{0,1,2} \quad (7)$$

$$c_{2,1,0} = r^2 b_{2,0,1} + 2st b_{0,1,2} + 2rt b_{1,0,2} + s^2 b_{0,2,1} + 2rs b^1_{1,1,1} + t^2 b_{0,0,3} \quad (8)$$

$$b_{2,1,0} = u^2 c_{2,0,1} + 2vw c_{0,1,2} + 2uw c_{1,0,2} + v^2 c_{0,2,1} + 2uv c^1_{1,1,1} + w^2 c_{0,0,3} \quad (9)$$

$$b_{2,0,1} = u^2 c_{2,1,0} + 2vw c_{0,2,1} + 2uw c_{1,2,0} + v^2 c_{0,3,0} + 2uv c^1_{1,1,1} + w^2 c_{0,1,2} \quad (10)$$

To find $b^1_{1,1,1}$ in (7) and (8), we need to add these equations together. Thus, we obtain

$$b^1_{1,1,1} = \frac{1}{2r(s+t)} \left(c_{2,0,1} + c_{2,1,0} - r^2(b_{2,1,0} + b_{2,0,1}) - s^2(b_{0,3,0} + b_{0,2,1})\right)$$
$$+t^2(b_{0,1,2} + b_{0,0,3}) - 2st(b_{0,2,1} + b_{0,1,2} - rs b_{1,2,0} - 2rt b_{1,0,2})$$

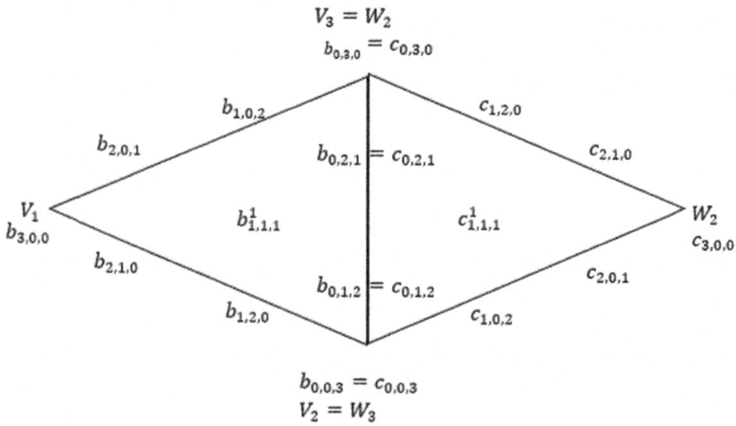

Fig. 5 Two adjacent rational quartic triangular patches

Similarly, with (9) and (10) we obtain the following

$$\begin{aligned} c^1_{1,1,1} &= 1/2u(v+w)\left(b_{2,0,1} + b_{2,1,0} - u^2(c_{2,1,0} + c_{2,0,1}) - v^2(c_{0,3,0} + c_{0,2,1})\right) \\ &\quad + w^2(c_{0,1,2} + c_{0,0,3}) - 2vw(c_{0,2,1} + c_{0,1,2}) - uvc_{1,2,0} - 2uwc_{1,0,2} \end{aligned}$$

So, the final scheme can be written as

$$R(x, y, z) = \sum_{p+q+r=3, p.q.r\neq 1} b_{pqr} B^3_{p,q,r}(x, y, z) + 6xyz(c_1 b^1_{1,1,1} + c_2 b^2_{1,1,1} + c_3 b^3_{1,1,1}) \tag{11}$$

with

$$c_1 = \frac{yz}{(yz + xz + xy)}, \quad c_2 = \frac{xz}{(yz + xz + xy)}, \quad c_3 = \frac{xy}{(yz + xz + xy)} \tag{12}$$

and the other version of convex combination is used such that

$$\begin{aligned} c_1 &= \frac{(y^2z^2)}{(y^2z^2 + x^2z^2 + x^2y^2)}, \quad c_2 = \frac{(x^2z^2)}{(y^2z^2 + x^2z^2 + x^2y^2)}, \\ c_3 &= \frac{(x^2y^2)}{(y^2z^2 + x^2z^2 + x^2y^2)} \end{aligned} \tag{13}$$

Both form (12) and (13) have difference final degree of the rational scheme. Rational patches obtained from (12) is rational sextic with quartic denominator (6/4), while (13) will give a rational octic with quartic denominator (8/4). But both schemes require only 10 data points. This is the main advantages of the proposed scheme.

4 Result and Discussion

To validate our scheme, we measure the performance by comparing errors which is maximum error, root mean square error (RMSE), coefficient of determination (R^2) and CPU time. We generated the graphical and numerical result using MATLAB R2019a.

Noted that, CPU A presented intel COREi5-2410M CPU @ 2.30 GHz, and CPU B representing to intel COREi3-5005U CPU @ 2.00 GHz. The data we sample from two real application were obtained from Gilat [13] and shows in Tables 1 and 3 respectively.

Maxwells Equation

Molecules of a gas in a container are moving around at different speeds. Maxwell's speed distribution law gives the probability distribution $P(v)$ as a function of temperature and speed:

$$P(v) = 4\pi\left(\frac{M}{2\pi RT}\right)^{3/2} v^2 e^{\frac{-Mv^2}{2RT}}$$

Table 1 49 data of Maxwells equation

x	y	$P(v)$
0	70	0
0	120	0
0	150	0
0	220	0
0	270	0
0	300	0
0	320	0
100	70	0.0025
100	120	0.0012
100	150	0.0009
100	220	0.0005
100	270	0.0004
100	300	0.0003
100	320	0.0003
200	70	0.0043
200	120	0.0031
200	150	0.0025
200	220	0.0016
200	270	0.0013
200	300	0.0011
200	320	0.0010

(continued)

Table 1 (continued)

x	y	$P(v)$
300	70	0.0025
300	120	0.0031
300	150	0.0029
300	220	0.0024
300	270	0.0020
300	300	0.0019
300	320	0.0017
500	70	0.0001
500	120	0.0007
500	150	0.0010
500	220	0.0016
500	270	0.0018
500	300	0.0018
500	320	0.0019
750	70	0
750	120	0
750	150	0
750	220	0.0002
750	270	0.0004
750	300	0.0006
750	320	0.0006
1000	70	0
1000	120	0
1000	150	0
1000	220	0
1000	270	0
1000	300	0.0001
1000	320	0.0001

where M is the molar mass of the gas in kg/mol, $R = 8.31$ J/(mol K), is the gas constant, T is the temperature in kelvins, and v is the molecule's speed in m/s. Noted that, $0 \leq v \leq 1000$ m/s and $70 \leq T \leq 320$ K for oxygen (molar mass 0.032 kg/mol). Table 1 shows 49 data points that we used in Maxwells equations.

Figure 6 shows the true surface for MaxWells equation. Figure 7 shows Delaunay triangulation where all the data points are connected. The total number of triangle is 72 triangles. Figure 8 shows 3D interpolant for MaxWells. Meanwhile, Fig. 9 shows surface interpolant using our proposed method for $\alpha = 6, \beta = 1, \gamma = 0.5$

Table 2 shows the error analysis for two different version convex combination from Saaban et al. [11] and Hussain and Hussain [8]. From the result, we found

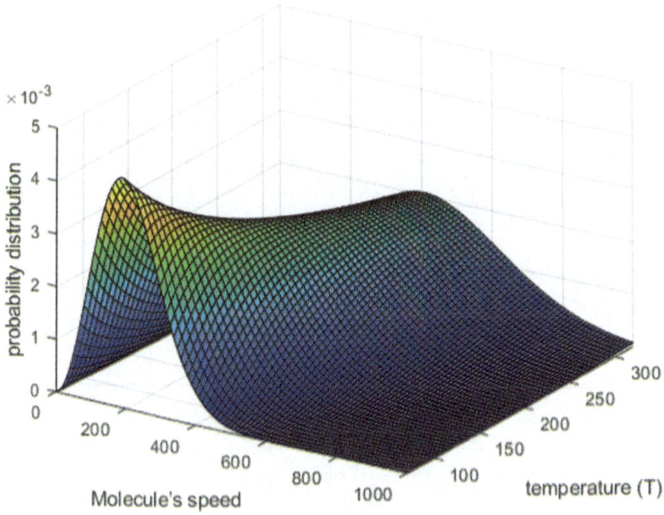

Fig. 6 True surface of Maxwells equation

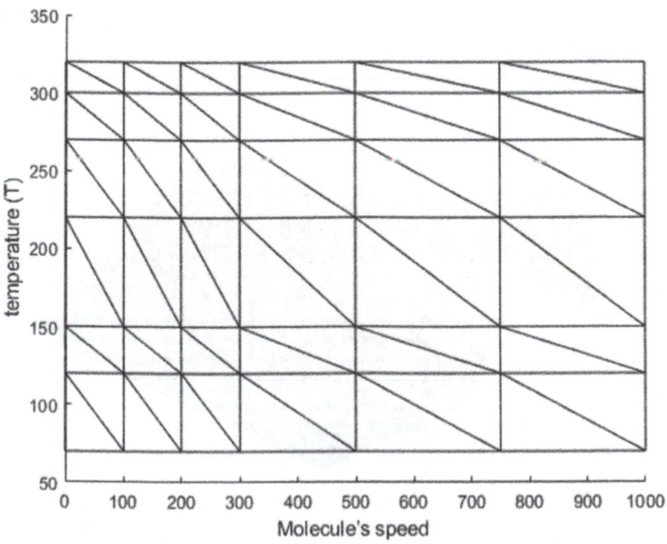

Fig. 7 Delaunay triangulations of Maxwells equation

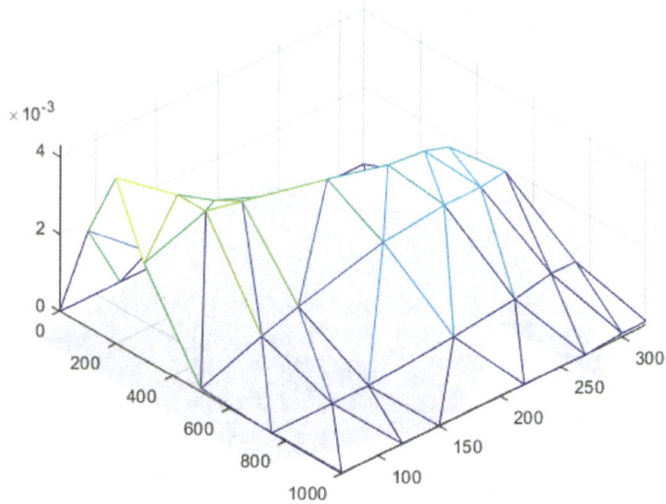

Fig. 8 3D interpolant of Maxwells equation

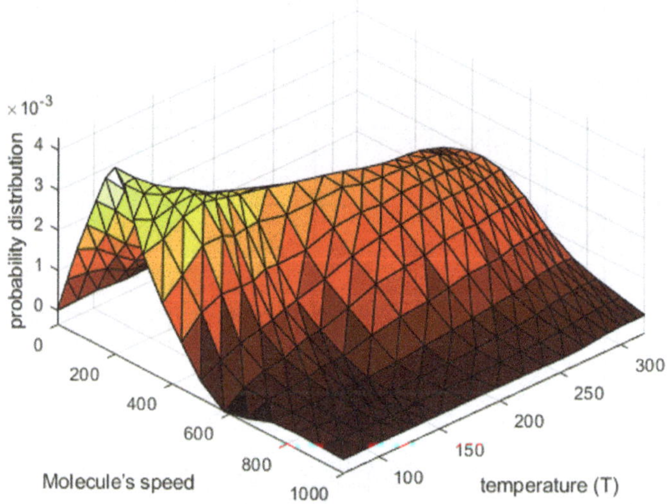

Fig. 9 Surface interpolant of Maxwells equation using our proposed method when ($\alpha = 6, \beta = 1, \gamma = 0.5$)

Table 2 Error measurements

Local scheme	Method	α	β	γ	Max error	RMSE	R^2	A	B	Average
Saaban et al. [11]	Cubic Ball	0	0	0	0.0007	0.0001	0.9841	0.4757	0.3587	0.4172
	Cubic Bézier				0.0007	0.0001	0.9836	0.4529	0.3978	0.4254
	Rational quartic	2	1	0.5	0.0007	0.0001	0.9841	0.4696	0.3814	0.4255
		6	1	0.5	0.0007	0.0001	0.9843	0.4573	0.3604	0.4089
		2	1	1	0.0006	0.0001	0.9837	0.5407	0.3707	0.4557
Hussain and Hussain [8]	Cubic Ball	0	0	0	0.0007	0.0001	0.9832	0.4969	0.3513	0.4241
	Cubic Bézier				0.0007	0.0001	0.9835	0.5305	0.4653	0.4979
	Rational quartic	2	1	0.5	0.0007	0.0001	0.9839	0.4642	0.3913	0.4278
		6	1	0.5	0.0007	0.0001	0.9841	0.4793	0.2496	0.3645
		2	1	1	0.0006	0.0001	0.9837	0.4425	0.3250	0.3838

that Saaban et al. [11] scheme generated less maximum error, RMSE and higher R^2 compared to Hussain and Hussain [8] scheme. Besides that, Saaban et al. [11] gives less computation time compared with Hussain and Hussain [8] (Table 2).

Figure 10 shows true surface of U.S Customary units. Figure 11 shows Delaunay triangulation that contains 48 triangles. Figure 12 shows 3D interpolant surface and Fig. 13 shows surface interpolant using our proposed method. There are 48 triangular patches with C^1 continuity.

Table 4 shows the error analysis for U.S. Customary equations with two different version of convex combination. From the numerical result, we found that the resulting scattered data interpolation using convex combination of Saaban et al. [11] generated less maximum error, RMSE and higher R^2 compared to Hussain and Hussain [8] scheme. Besides, it gives less computation compare to Hussain and Hussain [8] scheme. This is significant because we the user wanted to visualize a large sets of scattered data.

5 Conclusion

This paper discusses the comparison between two existing convex combination using rational quartic triangular patches with three free parameters on triangular domain. We construct C^1 continuity to the triangular patch by deriving the sufficient conditions on each adjacent triangle. In addition, we measured the performance by

Table 3 35 data of US customary

x	y	T_{wc}
0	0	35.7400000000000
0	10	41.9550000000000
0	25	51.2775000000000
0	30	54.3850000000000
0	50	66.8150000000000
5	0	−9.95688897543972
5	10	−3.74188897543971
5	25	5.58061102456028
5	30	8.68811102456029
5	50	21.1181110245603
10	0	−15.3165463021730
10	10	−9.10154630217302
10	25	0.220953697826981
10	30	3.32845369782698
10	50	15.7584536978270
30	0	−25.1282526432565
30	10	−18.9132526432565
30	25	−9.59075264325648
30	30	−6.48325264325647
30	50	5.94674735674352
45	0	−29.2079482557028
45	10	−22.9929482557028
45	25	−13.6704482557028
45	30	−10.5629482557028
45	50	1.86705174429720
60	0	−32.2673157952409
60	10	−26.0523157952409
60	25	−16.7298157952409
60	30	−13.6223157952409
60	50	−1.19231579524087
70	0	−33.9655117885826
70	10	−27.7505117885825
70	25	−18.4280117885825
70	30	−15.3205117885825
70	50	−2.89051178858255

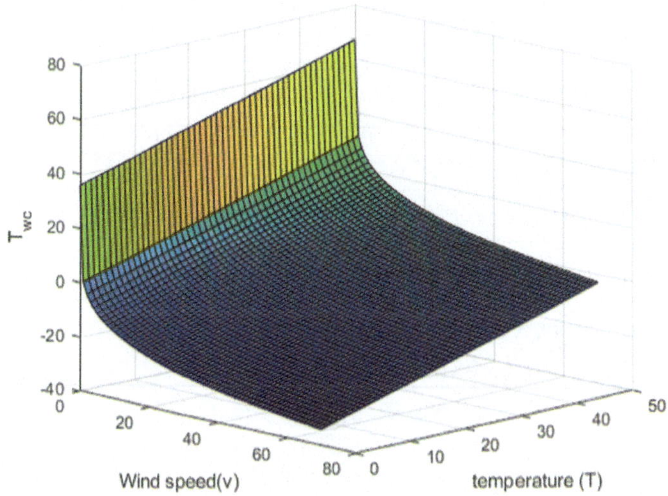

Fig. 10 True surface of U.S. customary unit

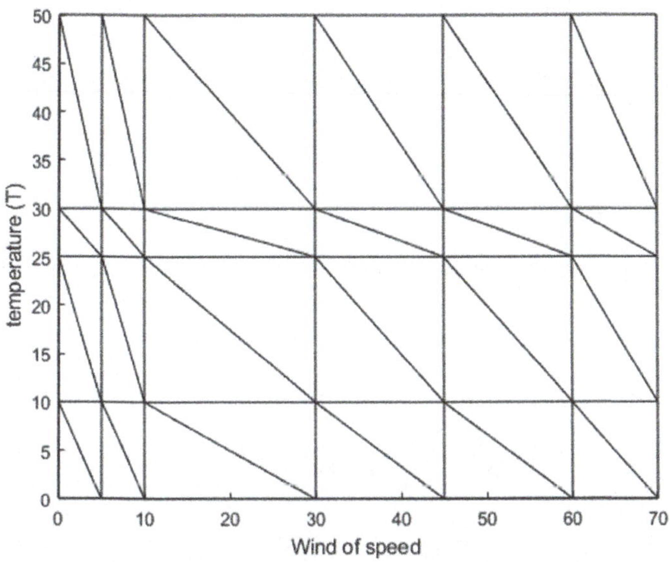

Fig. 11 Delaunay triangulations for U.S. customary

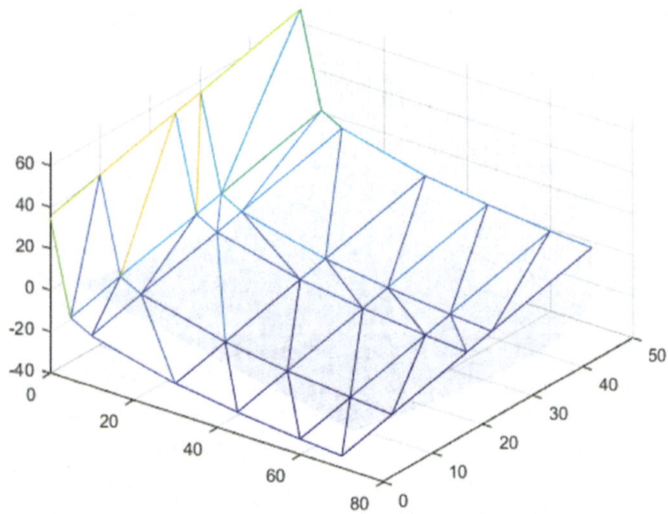

Fig. 12 3D interpolant for U.S. customary using our proposed method

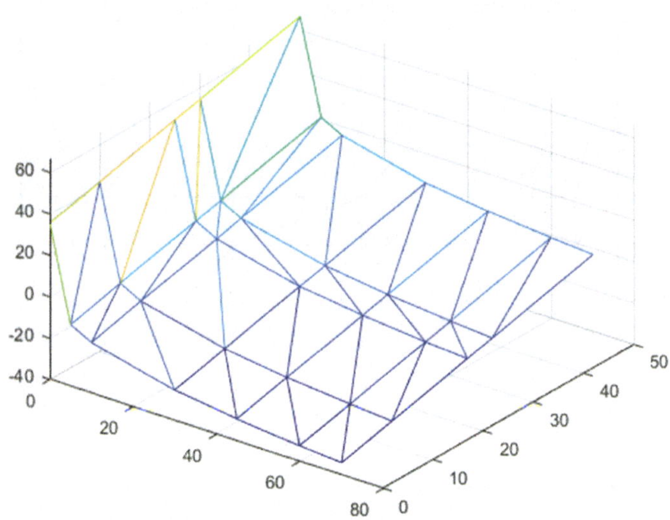

Fig. 13 Surface interpolant for U.S. customary using our proposed method ($\alpha = 1, \beta = 1, \gamma = 0$)

Table 4 Error analysis

Local scheme	Method	α	β	γ	Max error	RMSE	R^2	A	B	Average
Saaban et al. [11]	Cubic ball	0	0	0	18.5953	2.9592	0.9641	0.4867	0.3550	0.4209
	Cubic Bézier				18.5953	3.0370	0.9630	0.4821	0.4178	0.4500
	Rational quartic	1	0	0	18.3734	2.9317	0.9655	0.4764	0.4012	0.4388
		1	0	0.5	18.4057	2.9326	0.9655	0.4061	0.3246	0.3654
		1.5	0	1.5	18.8440	2.9288	0.9655	0.4484	0.3248	0.3866
Hussain and Hussain [8]	Cubic ball	0	0	0	18.5953	2.9903	0.9641	0.4346	0.2580	0.3463
	Cubic Bézier				18.5953	3.0419	0.9628	0.4362	0.3453	0.3908
	Rational quartic	1	0	0	18.3734	2.9556	0.9649	0.4312	0.3247	0.3780
		1	0	0.5	18.4057	2.9491	0.9651	0.4482	0.3194	0.3838
		1.5	0	1.5	18.8440	2.9442	0.9652	0.5066	0.4052	0.4559

calculating the Root Mean Square Error (RMSE), maximum error coefficient of determination (R^2) as well as CPU time (in seconds). From the result obtained, the proposed scheme using convex combination of Saaban et al. [11] give less error and less computation compared to the other convex combination form. For future research, we can apply shape preserving such as positivity by using the proposed rational quartic triangular patches.

Acknowledgements This research was fully supported by Universiti Teknologi PETRONAS (UTP) through a research grant YUTP: 0153AA-H24 (Spline Triangulation for Spatial Interpolation of Geophysical Data) and Ministry of Education, Malaysia through FRGS/1/2018/STG06/UTP/03/1/015MA0-020. The first author is supported through Graduate Research Assistant (GRA) Scheme.

References

1. Sulaiman, P.S., Jaafar, A.: Delaunay triangulation of a missing points. J. Adv. Sci. Eng. Res. **7**, 58–69 (2017)
2. Zhang, L., Gu, T., Zhao, J., Ji, S., Hu, M., Li, X.: An improved moving least squares method for curve and surface fitting. Math. Probl. Eng. (2013)
3. Cavoretto, R., De Rossi, A.: Adaptive meshless refinement schemes for RBF-PUM collocation. Appl. Math. Lett. **90**, 131–138 (2019)
4. Karim, S.A.A., Saaban, A., Hasan, M.K., Sulaiman, J., Hashim, I.: Interpolation using cubic Bézier triangular patches. Int. J. Adv. Sci. Eng. Inf. Technol. **8**(4–2), 2088–5334 (2018)

5. Ali, F.A.M., Karim, S.A.A., Saaban, A., Hasan, M.K., Ghaffar, A., Baleanu, D.: Construction of cubic timmer triangular patches and its application in scattered data interpolation. Mathematics **8**(2), 159 (2020)
6. Goodman, T.N.T., Said, H.B.: A C^1 triangular interpolant suitable for scattered data interpolation. Commun. Appl. Numer. Methods **7**(6), 479–485 (1991)
7. Foley, T.A., Opitz, K.: Hybrid cubic Bézier triangle patches. In: Mathematical Methods in Computer Aided Geometric Design II, pp. 275–286. Academic Press, USA (1992)
8. Hussain, M.Z., Hussain, M.: C^1 positive scattered data interpolation. Comput. Math. Appl. **59**(1), 457–467 (2010)
9. Sarfraz, M., Hussain, M.Z., Ali, M.A.: Positivity-preserving scattered data interpolation scheme using the side-vertex method. Appl. Math. Comput. **218**(15), 7898–7910 (2012)
10. Zhu, Y., Han, X., Liu, S.: Quartic rational said-ball-like basis with tension shape parameters and its application. J. Appl. Math. (2014)
11. Saaban, A., Majid, A.A., Piah, M., Rahni, A.: Visualization of rainfall data distribution using quintic triangular Bézier patches. Bull. Malays. Math. Sci. Soc. **32**(2), 2009 (2009)
12. Draman, N.N.C., Karim, S.A.A., Hashim, I.: Scattered data interpolation using rational quartic triangular patches with three parameters. IEEE Access **8**, 44239–44262 (2020). https://doi.org/10.1109/access.2020.2978173
13. Gilat, A.: MATLAB: An Introduction with Applications, 4th edn. Wiley, USA (2013)

Positivity Preserving Using C^2 Rational Quartic Spline Interpolation

Noor Adilla Harim and Samsul Ariffin Abdul Karim

Abstract This paper discusses the positivity preserving interpolation of C^2 rational quartic spline for positive data. The rational quartic spline has three different parameters which are α_i, β_i and γ_i. The proposed rational spline can achieve C^2 continuity without the need to solve any tridiagonal systems linear of equations, unlike some other splines that needed solving systems linear of the equation. The sufficient condition is derived on one parameter meanwhile the other two parameters are free parameters that can the user interpolate the final shape of the positive interpolating curve. These conditions will guarantee to produce a positive interpolating curve everywhere. Comparison with the existing schemes also is discussed in detail. From the graphical and numerical results, we found that the proposed scheme is better than existing schemes, since it has an extra free parameter to control the positive interpolating curve while maintaining C^2 continuity.

Keywords Interpolation · Rational quartic · Said-ball function · Monotonicity

1 Introduction

The modeling of shape preserving interpolation for curves and surfaces for selected given data has studied with various requirements such as smoothness of the interpolation curves and surfaces and the preservation of the final shape of the data. Positivity is one of shape preserving interpolation. The problem of positivity preserving interpolation occurs in visualizing the shape that cannot be negative which may arise if the data is taken from some scientific or social. Positivity preserving gain meaning when

N. A. Harim
Fundamental and Applied Sciences Department, Universiti Teknologi PETRONAS (UTP), 32610 Seri Iskandar, Perak, Malaysia

S. A. Abdul Karim (✉)
Fundamental and Applied Sciences Department and Centre for Systems Engineering (CSE), Institute of Autonomous System, Universiti Teknologi PETRONAS (UTP), Bandar Seri Iskandar, 32610 Seri Iskandar, Perak DR, Malaysia
e-mail: samsul_ariffin@utp.edu.my

S. A. Abdul Karim et al. (eds.), *Proceedings of the 6th International Conference on Fundamental and Applied Sciences*, Springer Proceedings in Complexity,
https://doi.org/10.1007/978-981-16-4513-6_46

theirs values are positive such as the amount of gas discharge during experiments by Brodlie and Butt [4], Brodlie et al. [5] and Sarfraz et al. [17].

A great deal of research on this topic has done, especially on the shape pre-serving interpolation. The cubic Hermite from Brodlie and Butt [6] and cubic polynomial spline from Fiorot and Tabka [8] preserve the positivity preserving by inserting the extra knots in the subinterval where the interpolation did not preserve the positivity. In order to achieve the final shape of interpolation curves, the user needs the first derivative modification. Duan et al. [7] and Bao et al. [3] have been introduced the rational cubic spline for control value control, inflection-point control, convexity control of the interpolation which were bases on the function values to achieve the C^2 continuous curve. Lamberti and Manni [15] investigated the approximation order of global C^2 shape preserving interpolation function using the parametric cubic curves. Abbas [1], Karim and Kong [14] and Abbas et al. [2] have discussed the positivity preserving interpolation by using C^2 rational cubic spline with free parameters.

The rational cubic spline with quadratic denominator in Hussain et al. [13] and Sarfraz et al. [17] have used for positivity, monotonicity and convexity preserving with C^2 continuity. Hussain et al. [10] have only one free parameter mean-while the Sarfraz et al. [17] have no free parameters.

Wang et al. [18] have introduced the rational quartic with linear denominator with two local control parameters for C^2 continuity meanwhile Harim et al. [11] have introduced the new rational quartic with quadratic denominator using three different values of parameters. Harim et al. [12] extended the Harim et al. [11] for positivity preserving interpolation for C^1 continuity. Harim et al. [10] extended for C^2 continuity for data interpolation. Han [9] gave an explicit representation of a C^2 continuity with a local control parameters which is can be preserve the shape preserving properties of the given data by choosing the suitable values of parameters. This scheme has some disadvantage such as cannot preserve the positivity properties as well as monotonicity properties in general. Zhu [19] introduced the rational quartic spline with cubic denominator with two local control parameters.

Numerical comparison between C^2 rational quartic spline and the works of Hussain et al. [13], Abbas et al. [2], and Karim and Kong [14] also has been made comprehensively. From all presented numerical result shape preserving interpolation by using new C^2 rational quartic spline gives more comparable results with existing scheme.

The remainder of the paper organized as follows. Section 2 introduced the new C^2 rational quartic spline with three parameters. In Sect. 3, we discuss positivity preserving interpolation using the proposed scheme, and also devoted for Results and Discussion. The summary is given in the final section.

2 Rational Quartic Spline Interpolation

This section will introduce C^2 rational quartic spline interpolation with three parameters. The rational quartic spline has been initiated by Harim et al. [11]. The main differences is that, in this paper the rational quartic spline has C^2 continuity while in work by Harim et al. [12] it has C^1 continuity. This section will introduce the C^2 rational quartic spline interpolant with three parameters. Given the scalar data (x_i, f_i), $i = 1, 2, ..n$ where $x_1 < x_2 < \cdots < x_n$ and the second derivative d_i, at the respective point x_i, $i = 0, 1, ..n$. Then the rational quartic spline with three parameters $\alpha_i, \beta_i > 0$ and $\gamma_i \geq 0$ on the interval $[x_i, x_{(i+1)}]$, $i = 0, 1, \ldots, n-1$ is given by:

$$S_i(x) = \frac{P_i(\theta)}{Q_i(\theta)} \tag{1}$$

where

$$P_i(\theta) = (1-\theta)^4\alpha_i f_i + (1-\theta)^3\theta A_i + (1-\theta)^2\theta^2 B_i + (1-\theta)\theta^3 C_i + \theta^4 \beta_i f_{(i+1)}$$

$$Q_i(\theta) = \alpha_i(1-\theta)^2 + \gamma_i(1-\theta)\theta + \beta_i\theta^2$$

$$S(x) = \frac{(1-\theta)^4\alpha_i f_i + (1-\theta)^3\theta A_i + (1-\theta)^2\theta^2 B_i + (1-\theta)\theta^3 C_i + \theta^4 \beta_i f_{(i+1)}}{\alpha_i(1-\theta)^2 + \gamma_i(1-\theta)\theta + \beta_i\theta^2} \tag{2}$$

with $h_i = x_{(i+1)} - x_i$, $\Delta_i = \frac{f_{(i+1)} - f_i}{h_i}$ and a local variable, $\theta = \frac{x - x_i}{h_i}$ where the, $\theta \in [0, 1]$. The following conditions, it will assure that the rational quartic spline interpolant in Eq. (1) has C^2 continuity.

$$\begin{aligned} S(x_i) &= f_i \\ S(x_{i+1}) &= f_{i+1} \\ S^1(x_i) &= d_i \\ S^1(x_{i+1}) &= d_{i+1} \\ S^2(x_{i+}) &= S^2(x_{i-}) \end{aligned} \tag{3}$$

where $S^2(x_i)$ denote second derivative respectively while for $S^2(x_i^-)$ shows the right and left hand side of the second derivative. By using condition in Eq. (3), the C^2 rational quartic spline interpolant with three parameters defined in Eq. (1) has the unknowns:

$$\begin{aligned} A_i &= (2\alpha_i + \gamma_i) f_i + \alpha_i h_i d_i \\ B_i &= (\alpha_i + \gamma_i) f_{i+1} + (\beta_i + \gamma_i) f_i \\ C_i &= (2\beta_i + \gamma_i) f_{i+1} - \beta_i h_i d_{i+1} \end{aligned} \tag{4}$$

The C^2 continuity,

$$S^2(x_{i+}) = S^2(x_{i-}), \quad i = 1, 2, 3, \ldots, n-1$$

will give that

$$\frac{2[(\beta_{i-1}+\gamma_{i-1})(d_i-\Delta_{i-1})]}{h_{i-1}\beta_{i-1}}=\frac{2[(\alpha_i+\gamma_i)(\Delta_i-d_i)]}{h_i\alpha_i} \tag{5}$$

3 Derivative Estimation

When the data interpolation is not from the true function, the second derivative need to be determined. There is a common method to estimate the second derivative which is the arithmetic mean method (AMM) by Delbourgo and Gregory [6]. Suppose a data (x_i, f_i), $i = 1, 2, ..n$ is considered that $x_1 < x_2 < \cdots < x_n$, The derivation of d_i at the point x_i, $i = 1, 2, \ldots, n$ are estimated as follow: Let assume $h_i = x_{i+1} - x_i$ and $\Delta_i = \frac{(f_{i+1}-f_i)}{h_i}$. For endpoint derivatives d_1 and d_n described as:

$$S^1(x_1) = d_i, \quad S^1(x_n) = d_n \tag{6}$$

which can be estimated by using the derivative estimation of Arithmetic Mean Method (AMM).

$$d_1 = \Delta_1 + (\Delta_1 - \Delta_2)\left(\frac{h_1}{(h_1+h_2)}\right) \tag{7}$$

$$d_n = \Delta_{n-1} + (\Delta_{n-1} - \Delta_{n-2})\left(\frac{h_{n-1}}{(h_{n-2}+h_{n-1})}\right) \tag{8}$$

At the other point, x_i, $i = 2, 3, \ldots, n-1$ the value d_i is calculated by using Eq. (5) after simplification

$$d_i = \left[\frac{\alpha_i\beta_{i-1}\Delta_{i-1}h_i + \alpha_i\beta_{i-1}\Delta_i h_{i-1} + \alpha_i\gamma_{i-1}\Delta_{i-1}h_i + \beta_{i-1}\gamma_i\Delta_i h_{i-1}}{\alpha_i\beta_{i-1}h_{i-1} + \alpha_i\beta_{i-1}h_i + \alpha_i\gamma_{i-1}h_i + \beta_{i-1}\gamma_i h_{i-1}}\right] \tag{9}$$

Equation (8) can be written as:

$$d_i = \left[\frac{(h_i\alpha_i(\beta_{i-1}\Delta_{i-1} + \gamma_{i-1}\Delta_{i-1}) + h_{i-1}\beta_{i-1}(\alpha_i\Delta_i + \gamma_i\Delta_i)}{h_i\alpha_i(\beta_{i-1}+\gamma_{i-1}) + h_{i-1}\beta_{i-1}(\alpha_i+\gamma_i)}\right], \quad i = 2, 3, \ldots, n-1 \tag{10}$$

4 Positivity Preserving Interpolation

In this section, the positivity preserving interpolation by using the proposed C^2 rational quartic spline interpolation defined by Eq. (1) will discussed in detail. We follow the same idea of Harim et al. [12] for positivity preserving interpolation. In this study, given that given that (x_i, f_i), $i = 1, 2, .., n$ with $x_1 < x_2 < \cdots < x_n$ is positive to avoid dividing by zero such that

$$f_i > 0, \quad i = 1, 2, ..n \tag{11}$$

We have adopted the method from Hussain et al. [13]. The following theorem gives sufficient conditions for the positivity preserving of the proposed scheme. It is data-dependent and has two free parameters to achieve the final interpolating curve of the positive result.

Theorem 1 *For strictly positive dataset, the rational quartic spline interpolant defined on the interval* $[x_1, x_n]$ *is positive in each subinterval* $[x_i, x_{(i+1)}]$, $i = 1, 2, \ldots, n-1$ *if the following sufficient conditions are satisfied:*

$$\gamma_i > \max\left\{0, \left[\frac{-2\alpha_i f_i - \alpha_i h_i d_i}{f_i}\right], \left[\frac{(-2\beta_i f_{i+1} + \beta_i h_i d_{i+1})}{f_{i+1}}\right]\right\} \tag{12}$$

Proof The sufficient conditions of the positivity are derived from the condition in Eq. (10) as follows: it can be observed that $\alpha_i f_i$, $\beta_i f_{i+1}$ and B_i is positive since $\alpha_i, \beta_i, \gamma_i > 0$ and $f_i, f_{i+1} > 0$. Therefore, $P_i(\theta)$ will be positive value when both of A_i and C_i are positive such that:

$$A_i > 0, \quad C_i > 0 \tag{13}$$

$A_i > 0$ if

$$\gamma_i > \frac{-2\alpha_i f_i - \alpha_i h_i d_i}{f_i} \tag{14}$$

$C_i > 0$ If

$$\gamma_i > \frac{(-2\beta_i f_{i+1} + \beta_i h_i d_{i+1})}{f_{i+1}} \tag{15}$$

The above constraints can combined and rewritten as:

$$\gamma_i > \max\left\{0, \left[\frac{-2\alpha_i f_i - \alpha_i h_i d_i}{f_i}\right], \left[\frac{(-2\beta_i f_{i+1} + \beta_i h_i d_{i+1})}{f_{i+1}}\right]\right\} \tag{16}$$

For the computer implementation, condition in Eq. (16) can rewritten as:

$$\gamma_i = \lambda_i + \max\left\{0, \left[\frac{-2\alpha_i f_i - \alpha_i h_i d_i}{f_i}\right], \left[\frac{(-2\beta_i f_{i+1} + \beta_i h_i d_{i+1})}{f_{i+1}}\right]\right\} \tag{17}$$

For $i = 1, 2, \ldots, n-1$

5 Numerical Example

This section discusses the numerical example of the data interpolation using the proposed scheme including comparison with some existing schemes for the comparison in C^2 continuity.

Example 1 Positive data from Hussain et al. [13] (Table 1).

Figure 1 shows the interpolating curve for positivity preserving interpolation curve by comparing the existing schemes and the proposed scheme of a rational quartic spline. Figure 1a is Hussain et al. [13] and that figure shows the negative result on the interval [0, 1] as shown in Fig. 2. Figure 1b, c show the scheme of Abbas et al. [2] and Karim and Kong [14]. Figure 1d default quartic polynomial interpolating curve when $\alpha_i = \beta_i = 1$ and $\gamma_i = 2$ and Fig. 1e, f shows the curve interpolation for positivity preserving by the proposed schemes with the values of parameters as shown in Table 2. Figure 1f gives a more pleasing visually curve compared with other proposed schemes.

Example 2 Positive data from Sarfraz et al. [16] (Table 3).

Figure 3 shows the interpolating curve for positivity preserving interpolation curve by comparing the existing schemes and the proposed scheme of a rational quartic spline. Figure 3a is Hussain et al. [13] and that figure shows the negative result as

Table 1 Data from Hussain et al. [13]

i	1	2	3	4
x_i	0	1	1.7	1.8
f_i	0.25	1	11.11	25
d_i	−7.2962	8.7962	123.4286	154.5714
d_i ([13])	−7.2962	2.7874	122.216	154.5714
d_i ([2])	−7.2962	3.079	69.242	154.5714
d_i ([14])	−7.2962	2.108	82.5421	154.5714
d_i (the proposed scheme)	−7.2962	7.2022	115.8239	154.5714

Table 2 Values of parameters

	i	1	2	3	4
	d_i	−7.2962	7.2022	115.8239	154.5714
Figure 1e	α_i	1.0	1.0	1.0	–
	β_i	0.1	0.1	1.0	–
	γ_i	27.283	5.40421	0.1	–
Figure 1f	α_i	5.0	0.1	0.1	–
	β_i	0.5	1.0	1.0	–
	γ_i	130.02	5.30543	0.1	–

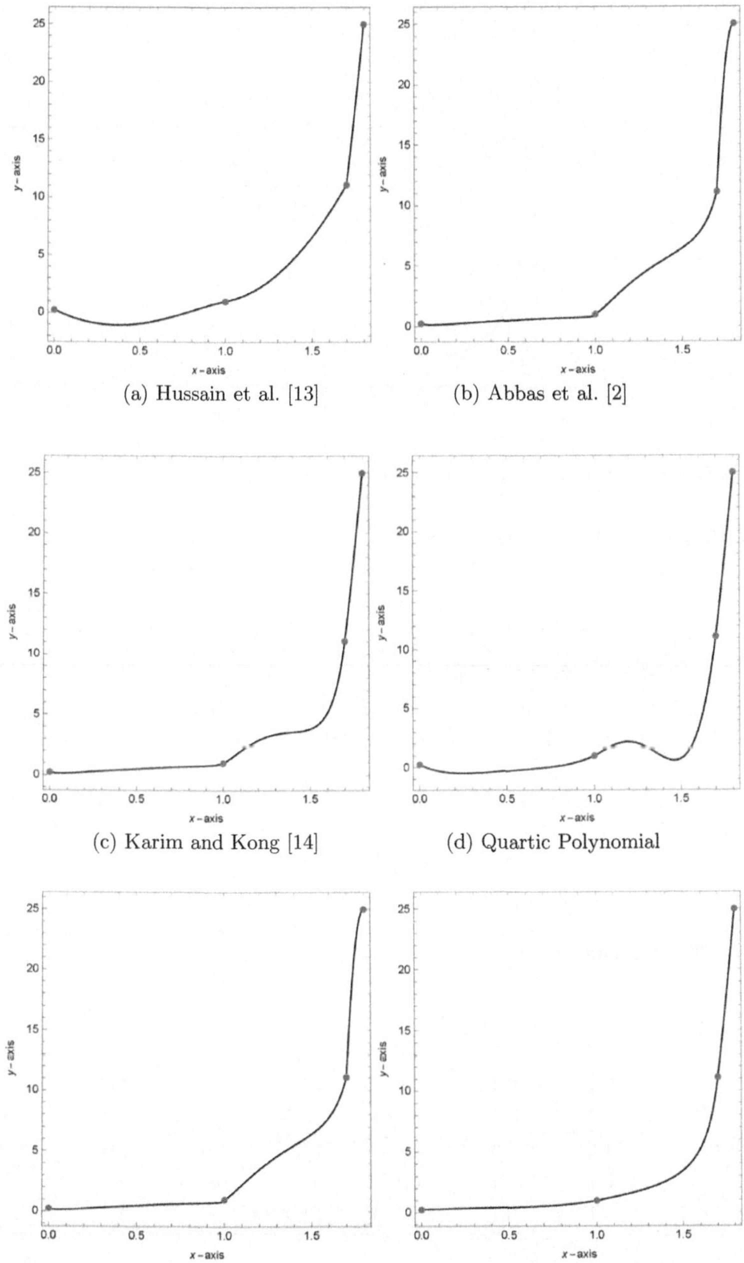

(a) Hussain et al. [13]

(b) Abbas et al. [2]

(c) Karim and Kong [14]

(d) Quartic Polynomial

(e) The proposed scheme

(f) The proposed scheme

Fig. 1 The interpolating curve

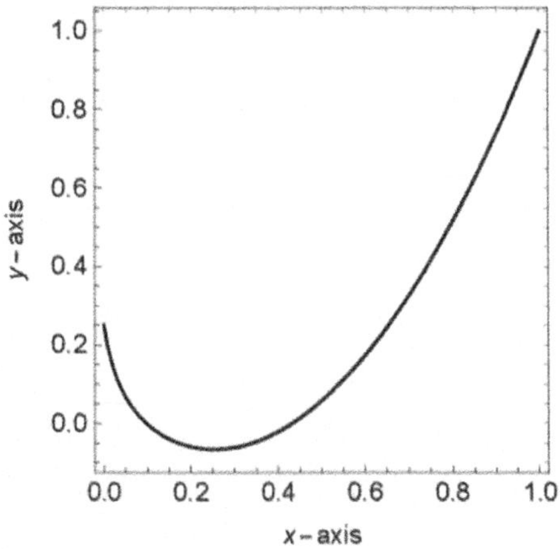

Fig. 2 The negative interpolating curve using Hussain et al. [13]

Table 3 Data from Sarfraz et al. [16]

i	1	2	3	4	5	6	7
x_i	2	3	7	8	9	13	14
f_i	10	2	3	7	2	3	10
d_i	−9.65	−6.35	3.25	0	−3.95	5.65	8.35
d_i ([13])	−9.65	−3.2782	4.6	2.579	−6.9076	4.678	8.35
d_i ([2])	−9.65	−2.7481	3.6429	1.7012	−5.7728	9.446	8.35
d_i ([14])	−9.65	−4.86	3.34	−0.48	−4.057	5.25	8.35
d_i (the proposed scheme)	−9.65	−6.6974	3.0625	0.14871	−4.1728	5.3125	8.35

Table 4 Values of parameters

	i	1	2	3	4	5	6	7
	d_i	−9.65	−6.6974	3.0625	0.14871	−4.1728	5.3125	8.35
Figure 3e	α_i	1.0	1.0	1.0	1.0	1.0	1.0	–
	β_i	0.1	0.1	1.0	1.0	1.0	1.0	–
	γ_i	0.1	0.1	0.7845	13.0472	1.0	1.0	–
Figure 3f	α_i	1.0	1.0	1.0	1.0	1.0	1.0	–
	β_i	0.1	0.1	0.1	0.1	0.1	0.1	–
	γ_i	0.1	0.1	0.7845	13.0074	0.1	0.1	–

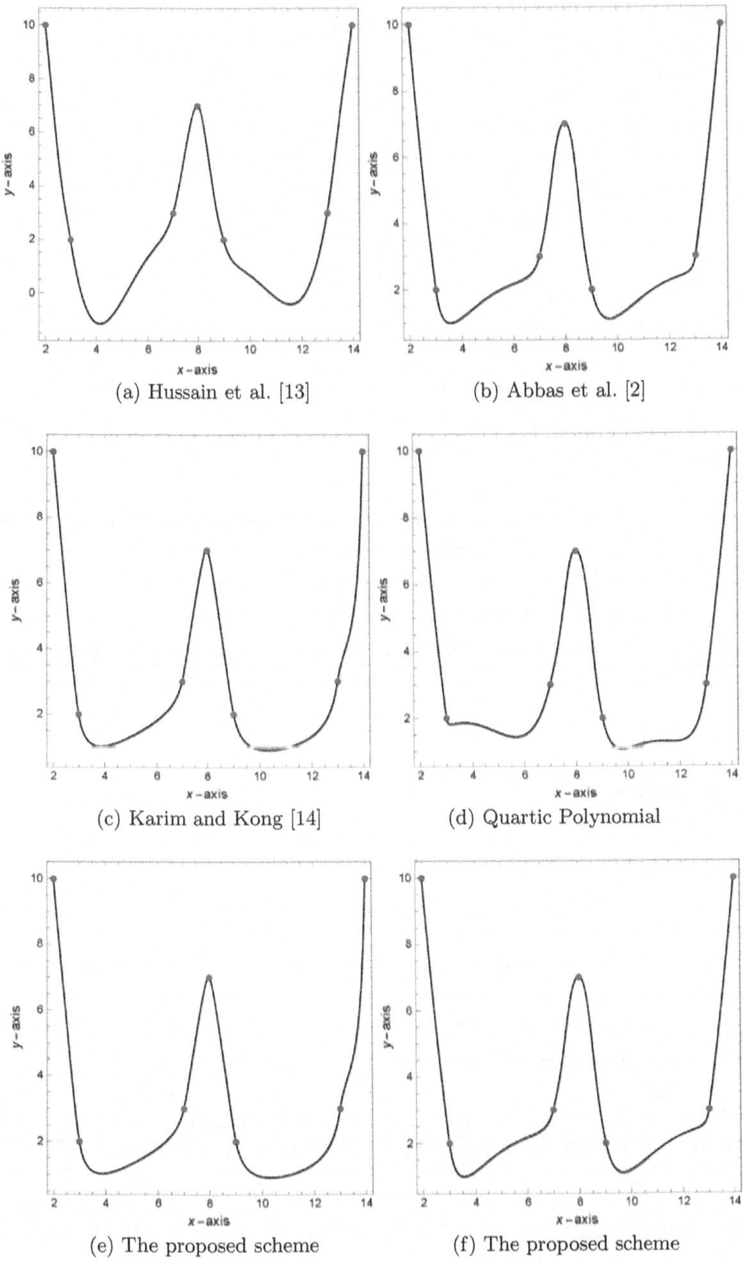

(a) Hussain et al. [13]
(b) Abbas et al. [2]
(c) Karim and Kong [14]
(d) Quartic Polynomial
(e) The proposed scheme
(f) The proposed scheme

Fig. 3 The interpolating curve

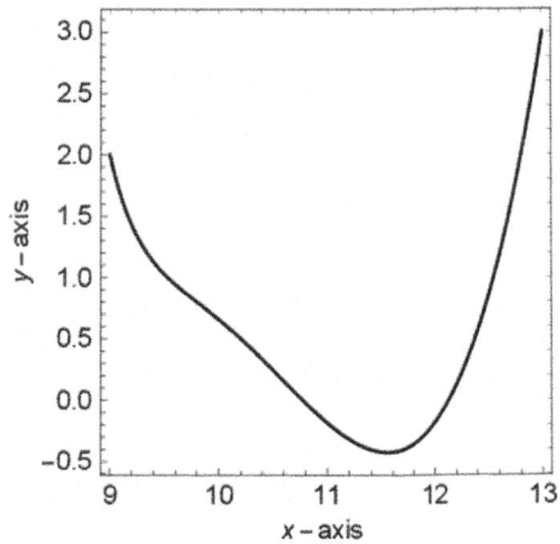

Fig. 4 The negative interpolating curve using Hussain et al. [13]

shown in Fig. 4. Figure 3b, c show the scheme of Abbas et al. [2] and Karim and Kong [14]. For Fig. 3c show the interpolating curve not very smooth on the interval [13, 14]. Figure 3d default quartic polynomial interpolating curve when $\alpha_i = \beta_i = 1$ and $\gamma_i = 2$. Figure 3e, f shows the curve interpolation for positivity preserving by the proposed schemes with the values of parameters as shown in Table 4. Figure 3e gives a more pleasing visually curve compared with other proposed schemes.

6 Conclusion

The new rational quartic spline scheme with three parameters with C^2 continuity can be used produce positive interpolating curves without solving a linear system of consistency equation for the derivative. The free parameters can be manipulated to preserve the positivity preserving interpolation curve. Based on our numerical results, we found that the proposed scheme preserves the positivity preserving interpolation everywhere compared with the work of Hussain et al. [13] that fails to produce positive interpolating curve everywhere.

Acknowledgements The author would like to acknowledge Universiti Teknologi PETRONAS (UTP) for the financial support received in the form of a research grant: FRGS/1/2018/STG06/UTP/03/1/015MA0-020.

References

1. Abbas, M.: Shape preserving data visualization for curves and surfaces using rational cubic functions. Doctoral dissertation. Universiti Sains Malaysia (2012)
2. Abbas, M., Majid, A.A., Ali, J.M.: Positivity-preserving C^2 rational cubic spline interpolation. ScienceAsia **39**(2), 208–213 (2013)
3. Bao, F., Sun, Q., Duan, Q.: Point control of the interpolating curve with a rational cubic spline. J. Vis. Commun. Image Represent. **20**(4), 275–280 (2009)
4. Brodlie, K.W., Butt, S.: Preserving convexity using piecewise cubic interpolation. Comput. Graph. **15**(1), 15–23 (1991)
5. Brodlie, K., Mashwama, P., Butt, S.: Visualization of surface data to preserve positivity and other simple constraints. Comput. Graph. **19**(4), 585–594 (1995)
6. Delbourgo, A., Gregory, J.A.: The determination of derivative parameters for a monotonic rational quadratic interpolant. Brunel University Mathematics Technical Papers Collection (1984)
7. Duan, Q., Wang, L., Twizell, E.H.: A new C^2 rational interpolation based on function values and constrained control of the interpolant curves. Appl. Math. Comput. **161**(1), 311–322 (2005)
8. Fiorot, J.C., Tabka, J.: Shape-preserving C^2 cubic polynomial interpolating splines. Math. Comput. **57**(195), 291–298 (1991)
9. Han, X.: Convexity-preserving piecewise rational quartic interpolation. SIAM J. Numer. Anal. **46**(2), 920–929 (2008)
10. Harim, A., Karim, S.A.A, Othman. M., Saaban, A.: Data interpolation using C^2 rational quartic spline. Int. J. Sci. Technol. Res., 1219–27433 (2019)
11. Harim, A., Karim, S.A.A., Othman. M., Saaban, A.: High accuracy data interpolation using rational quartic spline with three parameters. Int. J. Sci. Technol. Res., 1219–27432 (2019)
12. Harim, A., Karim, S.A.A, Othman. M., Saaban, A., Ghaffar, A., Nisar, K.S.: Positivity preserving interpolation using rational quartic spline. AIMS Math. (2019) (under revision)
13. Hussain, M.Z., Sarfraz, M., Shaikh, T.S.: Shape preserving rational cubic spline for positive and convex data. Egypt. Inf. J. **12**(3), 231–236 (2011)
14. Karim, A., Ariffin, S., Voon Pang, K.: Shape preserving interpolation using rational cubic spline. J. Appl. Math. **2016** (2016)
15. Lamberti, P., Manni, C.: Shape-preserving C^2 functional interpolation via parametric cubics. Numer. Alg. **28**(1–4), 229–254 (2001)
16. Sarfraz, M., Hussain, M.Z., Chaudary, F.S.: Shape preserving cubic spline for data visualization. Comput. Graph. CAD/CAM **1**(6), 185–193 (2005)
17. Sarfraz, M., Hussain, M.Z., Shaikh, T.S., Iqbal, R.: Data visualization using shape preserving C^2 rational spline. In: 2011 15th International Conference on Information Visualisation, pp. 528–533. IEEE (2011)
18. Wang, Y., Tan, J., Li, Z., Bai, T.: Weighted rational quartic spline interpolation. J. Inf. Comput. Sci. **10**(9), 2651–2658 (2013)
19. Zhu, Y.: C^2 positivity-preserving rational interpolation splines in one and two dimensions. Appl. Math. Comput. **316**, 186–204 (2018)

K-means Clustering Analysis for EEG Features of Situational Interest Detection in Classroom Learning

Ernee Sazlinayati Othman, Ibrahima Faye, Areej Babiker, and Aarij Mahmood Hussaan

Abstract This paper proposes a method to detect situational interest in classroom learning using k-means algorithms. The developed algorithm in this paper had been tested on features from ten students who experienced mathematics learning in a classroom. The subjects were given 21 min of Laplace lecture presentation with some interesting elements introduced. Electroencephalogram (EEG) signal was preprocessed and decomposed using Fast Fourier Transform. The mean power for each sub-frequency band was served as input to the k-means algorithm. Results showed that EEG features can be successfully clustered in the alpha frequency band at the frontal region when visual-auditory stimuli are introduced to the subjects.

Keywords Situational interest · Classroom learning · EEG · K-means clustering

1 Introduction

Learning in general and learning Mathematics specifically remains a challenge for some students. Studies from psychology have been demonstrated that interest is one of the factors that affected the learning process. Literature has categorized interest

E. S. Othman · I. Faye (✉)
Centre for Intelligent Signal and Imaging Research (CISIR), Universiti Teknologi PETRONAS, 32610 Seri Iskandar, Perak, Malaysia
e-mail: ibrahima_faye@utp.edu.my

Fundamental and Applied Science Department, Universiti Teknologi PETRONAS, 32610 Seri Iskandar, Perak, Malaysia

E. S. Othman
e-mail: ernee_17004938@utp.edu.my

A. Babiker
Electronic Engineering Department, Future University, Khartoum, Sudan

A. M. Hussaan
Computer Science Department, IQRA University, Karachi, Pakistan
e-mail: aarijhussaan@iqra.edu.pk

S. A. Abdul Karim et al. (eds.), *Proceedings of the 6th International Conference on Fundamental and Applied Sciences*, Springer Proceedings in Complexity,
https://doi.org/10.1007/978-981-16-4513-6_47

into two: situational and individual interest. The difference between these two is the activating and sustaining time. Rotgans [1] had observed that situational interest can be activated quickly, while individual interest can be improved by repetitive experiences and stimulating situational interest.

There are several ways to measure a student's interests. Previously, the occurrence of situational interest (SI) was evaluated through a personal perspective. Linnenbrink-Garcia [2] and Rotgans [3] reported that questionnaires and subjective self-reports are commonly used but unfortunately, it is not always accurate. One of the well-known SI questionnaires was developed by Mitchell [4] to measure situational interest in mathematics classrooms among secondary school students. However, a current study from Fuller [5] suffers from students imitating to take part, resulting in a doubted execution of the self-report evaluation method.

Here, physiology measures, for example, Pupil Diameter, Electrocardiogram (ECG) and Electroencephalography (EEG) give an alternative. Studies often used EEG in interest research related to favourite during web searching and advertisement [6, 7]. Unlike questionnaires, the signals obtained from EEG cannot be manipulated by participants and it will not cause interference to participant's attention, interest, or the session flow.

This paper focuses on SI instead of individual interest since it can have an impulsive impact on student's learning. It can also be effectively stimulated and estimated by a physiological detector. Hidi [8] reported that students who completed SI performed better at recalling thoughts and statements. Thus, uncovering this biomarker will be valuable for evaluating teaching techniques that will promote the improvement of classroom learning.

The objective of the present paper is to examine the ability of k-means cluster analysis to detect SI when a person is observing interesting visual-auditory stimuli. As the EEG signals were recorded when the subjects were learning in a classroom, the data are useful to check the real-time feedback of the students. The k-means clustering algorithm was used as an observation tool to determine to which class (interested or not interested) each sample belongs. This might be used as instant feedback to the teachers and educators to act immediately upon low or not interested students during the lesson.

The remaining sections of this paper are arranged as follows: Sect. 2 describes the related work about detecting SI, Sect. 3 discusses the details methodology, Sect. 4 presents the results and Sect. 5 summarizes the conclusion and future work.

2 Related Work

Unlike the psychological methods, usage of physiological sensors could qualify SI which is not leveled out of the subjects themselves. The study from Moldovan [9] has shown that learner's interest perceived using EEG have a significant consequence on learning in multimedia-based mobile learning. Even though there is literature on EEG associates with situational interest, to characterize and detect this type of interest in classroom learning is limited. Assessing attention is a way of measuring SI in a classroom. The EEG beta and theta power ratio were used by Ko et al. [10] to distinguish attentive and non-attentive students in a classroom setting. Besides, studies from Palva [11] reported that the attenuation or reduction of alpha frequency was used to indicates the attentional level.

3 Methodology

3.1 Participants

The research work was approved by Universiti Teknologi PETRONAS (UTP) Research Ethical Committee. A total of eleven healthy volunteers (9 males and 2 females) participated in this experiment. One subject's data was excluded because of high artifacts. They were undergraduate students of UTP. The subjects were explained about the experiment and signed consent forms before the start experiment is started. A presentation questionnaire was undertaken once the experiment was done. The score from the presentation's questionnaire will be used later as a result of subjective measurement.

3.2 EEG Recording

EEG signal was recorded with 500 samples per second using 8-channels Enobio headset. The channels used were: Fp1, Fp2, F3, F4, T7, T8, O1 and O2 with 2 reference electrodes according to the international 10–20 system. The EEG data were comprised of 4 min of eyes opened, 4 min of eyes closed and 21 min of Laplace Transform lecture. All the subjects were placed in a classroom setting and recording was obtained simultaneously with the assistant of researchers. Details of the experimental are described in [12]. To minimize the artifacts during the recording, the subjects were instructed to seat calmly, avoid unnecessary eye blinks and movement when possible.

3.3 Tasks

All the subjects have attended a lecture on Laplace Transform. The lecture was performed by UTP lecturer who teaches the same courses in the university. The lecture's presentation was designed with some interesting elements such as including

Table 1 Sub-points of the delivered lecture

Sub-points	Details	Length (s)
A	Introduction	33
B	Learning Objectives	21
C	Real-life problems	110
D	Definition of Laplace	124
E	Notation of Laplace Transform	26
F	History of Laplace Transform	142
G	Example	105
H	Unit step function	68
I	Challenge question	68
J	Table of Laplace Transform	81
K	Examples	94
L	Definition of inverse Laplace and linearity	149
M	Table of inverse Laplace Transform	79
N	Example	122
O	Summary	38

different examples while maintaining meaningful content throughout the presentation. The sub-points of the presentation are described in Table 1. The total length of EEG signal was 1260 s.

3.4 Methods

The raw EEG was preprocessed with a high pass filtering with a low cut-off frequency of 0.5 Hz was performed followed by a low pass filtering with a high cut-off frequency of 45 Hz. Using bandpass filtering, power line noise artifacts were removed. The artifacts from eye and muscle movement are also being corrected. Later, the preprocessed signal was divided into fifteen segments (15 subpoints) for each subject.

Reports from literature show that the EEG power bands can efficiently be used to characterize an individual's activity. To get a better understanding of situational interest effect on each student's brain, electrode map and power spectral density were plotted. The plots were generated using EEGLAB which is an open signal processing environment for electrophysiological signals. High or strong activated brain area is indicated by red colour and low or no activation is indicated by blue colour.

Based on Fig. 1, theta and alpha were more concentrated in the frontal lobe for low interested subjects compared to high interested subjects. Moreover, beta and gamma are more pronounce at frontal for high interested subjects compared to low interested subjects (Fig. 2). This might reflect the area affected when visual-auditory stimuli are exposed to the subjects. Hence, power density can be one of the features

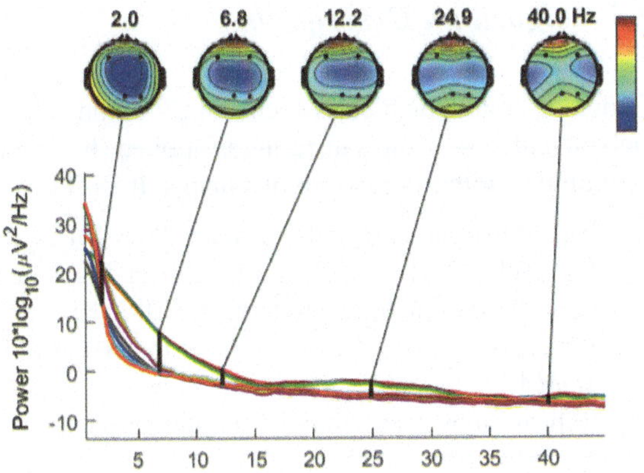

Fig. 1 An example of electrode map and power spectral density of low SI subject

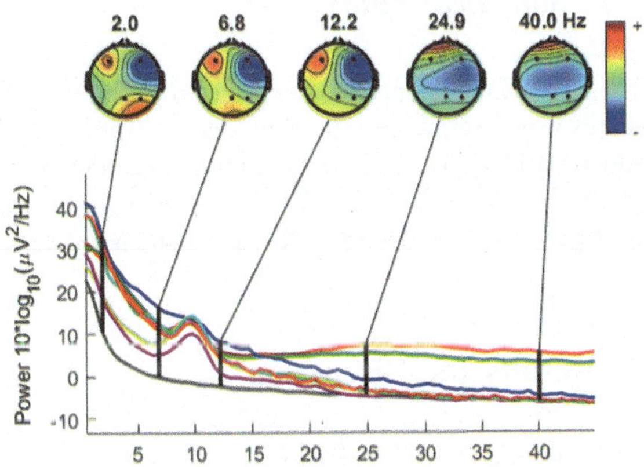

Fig. 2 An example of electrode map and power spectral density of high SI subject

to be future explored. The power of EEG signal was extracted and categorized into five frequencies. The frequency range is defined as follows: delta (0.5–4 Hz), theta (4–8 Hz), alpha (8–13 Hz), beta (13–30 Hz), and gamma (30–45 Hz). EEG power for all frequency bands was averaged within the subject.

A matrix M_{ij} was formed, where i is the number of rows representing the mean power for each band and j is the number of columns representing the number of channels. The mean power for each sub-frequency band was fed as input to the k-means algorithm to cluster the subject's SI level. Clustering analysis aims to classify n data distributions into k clusters in which each data belongs to the cluster with the nearest mean, serving as a sample of the cluster.

3.5 Algorithm Description

To demonstrate which class each subject belongs to, a quick and simple clustering method called k-means clustering is applied. It is an iterative procedure carried out to minimize within-class sum of squares. It works as follow [13]:

1. Randomly choose k initial centers $M = \{m_1, \ldots, m_k\}$.
2. For each $i \in \{1, \ldots, k\}$, set the cluster M_i to be the set of points in X that are closer to m_i than they are to m_j for all $j \neq i$.
3. For each $i \in \{1,\ldots,k\}$, set m_i to be the center of mass of all points in M_i: $m_i = 1/\,|M_i|\,\sum_{x \in Mi} x$.
4. Repeat Steps 2 and 3 until C no longer changes.

3.6 Statistical Analysis

The statistical evidence among the results obtained from k-means and presentation scores was determined. The Kolmogorov was used to test the normality of the data and Spearman to test the correlation between the k-means and presentation scores. The SPSS software version 24 was used to perform the statistical tests with a 95% confidence level and a 5% risk of claiming a difference.

4 Results and Discussion

The experiment was designed with a focus on assessing situational during mathematics classroom learning. Although eight electrodes were used during EEG recording, for analysis purposes the channels were paired according to their brain region. For example, for frontal polar region electrodes of Fp1-Fp2 were paired, the frontal region was F3-F4, the temporal region was T7-T8 and the occipital region was O1-O2 respectively.

After the k-means algorithm was applied through all frequency bands, the results show the two clusters are separated best in the alpha band. Examples of clustering results are shown in Fig. 3a–d. Based on an alpha inhibition hypothesis from Klimesch [14] and Pfurtscheller [15], who proposed that an indicator of active neuronal processing regions was reflected by a small amplitude of alpha oscillation, while large-amplitude alpha oscillations reflect the reduction and disconnection of task-irrelevant cortical areas. It could be seen that subjects in Cluster I is associated with high interest and subjects who in Cluster II is associated with a low-interested student.

Since alpha frequency shows a good biomarker to detect situational interest, we compare the accuracy of the k-means clustering through the brain region. The accuracy is calculated by how effective the k-means algorithm distinguishes the two clusters in all sub-points of the presentation. The frontal lobe has the highest accuracy

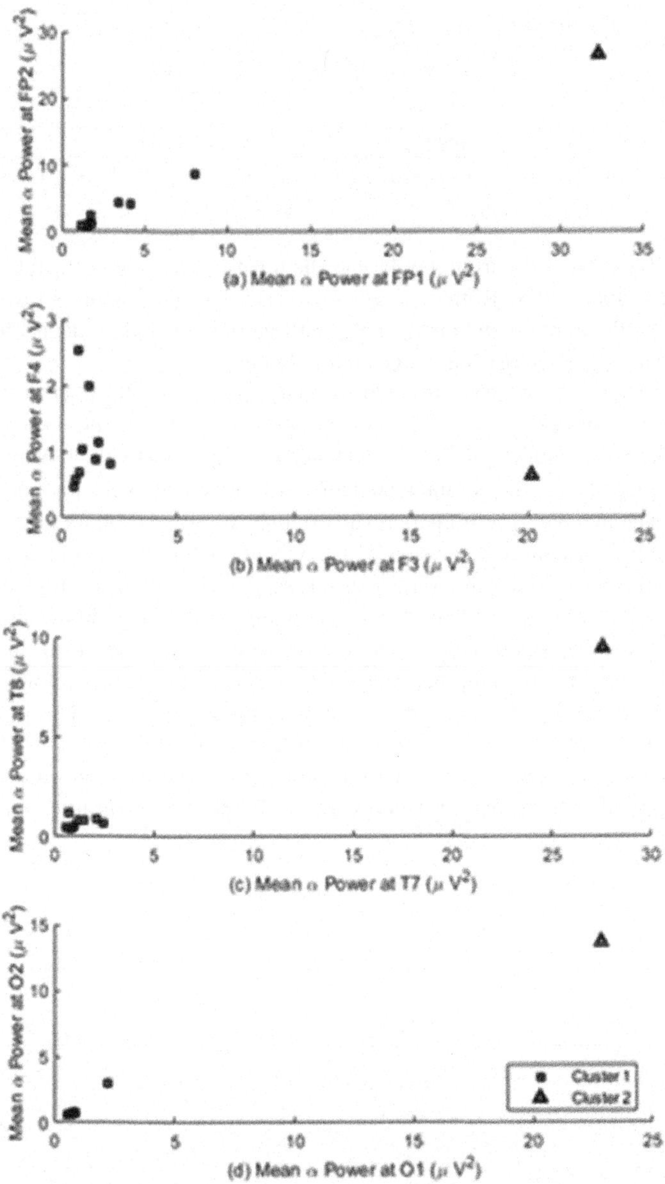

Fig. 3 The best k-means clustering in alpha frequency at one of the presentation's sub-point

Table 2 Clustering accuracy with alpha frequency band

Brain region	Accuracy (%)
Frontalpolar	53.33
Frontal	93.33
Temporal	80.00
Occipital	66.67

with 93.33% whereas the frontal polar has the lowest percentage (see Table 2). This finding was supported by Kothare [16] which describes the frontal cortex as mainly associated with attention and other important functions including decision-making, planning, working memory, language, and judgment.

Figure 4 shown a summarization of k-means clustering to all subjects. This figure shows the level of the subject's situational interest can be assessed through all sub-points of the presentation. The interested level was indicated by the low mean power of alpha frequency and the not interested level was indicated by the high mean power of alpha frequency. This preliminary result might be used as a primary idea to develop an instantaneous or automatic system to perceive the interest level of students while learning in the classroom. This could be used as a real-time indicator for teachers and educators to ensure level student's interests remain high during classroom learning.

Examining behavioural data which are presentation score and k-means (Fig. 5), showed a positive relationship between the high score in presentation with the high k-means score. Except for participant numbers three and eight. The presentation score will represent the SI score from the questionnaire. To check whether there is a correlation between the presentation score and k-means score, the Spearman test was applied. The correction value was $\rho = 0.388$ with a significant p-value (p

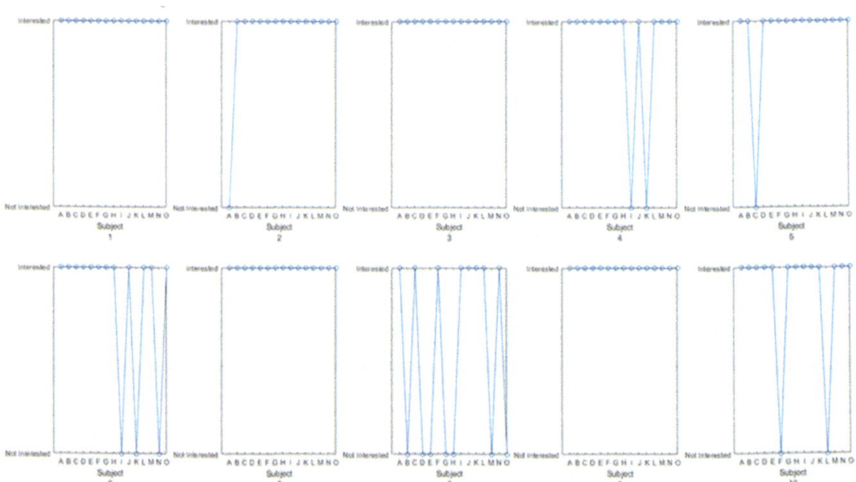

Fig. 4 Clustering of situational interest for each participant through all sub-points of presentation

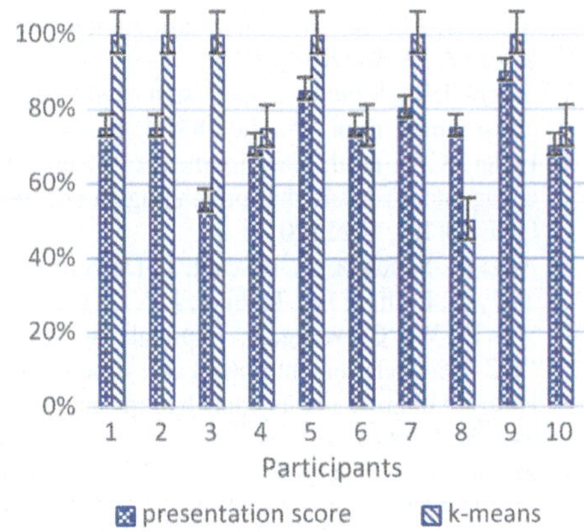

Fig. 5 Presentation score and k-means result

> 0.05) respectively indicating a weak correlation with no significant difference. This is because only four out of 15 sub-points were considered when calculating the presentation score.

5 Conclusion

This paper aims to investigate the ability of k-means clustering for situational interest identification compared to the presentation questionnaire score. The results show there is a weak correlation between them. However, this problem might be improved by increased sample size to increase the accuracy of the result. By synchronizing the length of the sub-points might also help the researcher to predict the optimum period of students could retain their interest level.

Instead of only mean power used, the first-order statistical features such as mod, median, variance, and standard deviation are recommended to serve as input. Further work will employ different clustering algorithms to improve cluster accuracy.

Acknowledgements This research was supported by Ministry of Education Malaysia under Higher Institutional Centre of Excellent (HICoE) Schemed awarded to Centre for Intelligent Signal and Image Research (CISIR) [015 MA0-050] and the international collaboration research grant between Universiti Teknologi PETRONAS and Iqra University [015ME0-074].

References

1. Rotgans, J.I., Schmidt, H.G.: Interest development: arousing situational interest affects the growth trajectory of individual interest. Contemp. Educ. Psychol. **49**, 175–184 (2017)
2. Linnenbrink-Garcia, L., et al.: Measuring situational interest in academic domains. Educ. Psychol. Meas. **70**(4), 647–671 (2010)

3. Rotgans, J.I., Schmidt, H.G.: Situational interest and learning: thirst for knowledge. Learn. Instr. **32**, 37–50 (2014)
4. Mitchell, M.: Situational interest: its multifaceted structure in the secondary school mathematics classroom. J. Educ. Psychol. **85**(3), 424–436 (1993)
5. Fuller, K.A., et al.: Development of a self-report instrument for measuring in-class student engagement reveals that pretending to engage is a significant unrecognized problem. PLoS ONE **13**(10), 1–22 (2018)
6. Ang, H., Sanchez, G., Pascual, J.: Detecting interest in video advertisements using EEG data analysis. Philipp. Inf. Technol. J. **7**(1), 4–12 (2014)
7. Wei, Z., Wu, C., Wang, X., Supratak, A., Wang, P., Guo, Y.: Using support vector machine on EEG for advertisement impact assessment. Front. Neurosci. **12**(March), 1–12 (2018)
8. Hidi, S.: Interest and its contribution as a mental resource for learning. Rev. Educ. Res. **60**(4), 549–571 (1990)
9. Moldovan, A.N., Ghergulescu, I., Muntean, C.H.: Analysis of learner interest, QoE and EEG-based affective states in multimedia mobile learning. In: Proceedings—IEEE 17th International Conference on Advance Learning Technologies, pp. 398–402. ICALT 2017 (2017)
10. Ko, L.W., Komarov, O., Hairston, W.D., Jung, T.P., Lin, C.T.: Sustained attention in real classroom settings: an EEG study. Front. Hum. Neurosci. **11**(July), 1–10 (2017)
11. Palva, S., Palva, J.M.: New vistas for α-frequency band oscillations. Trends Neurosci. **30**(4), 150–158 (2007)
12. Babiker, A., Faye, I., Mumtaz, W., Malik, A.S., Sato, H.: EEG in classroom: EMD features to detect situational interest of students during learning. Multimed. Tools Appl. **78**(12), 16261–16281 (2019)
13. Arthur, D., Vassilvitskii, S.: K-means++: the advantages of careful seeding. In: Proceedings of the Annual ACM-SIAM Symposium on Discrete Algorithms, pp. 1027–1035, vol. 07–09-Janu. (2007)
14. Klimesch, W.: Memory processes, brain oscillations and EEG synchronization. Int. J. Psychophysiol. **24**(1–2), 61–100 (1996)
15. Pfurtscheller, G.: Induced oscillations in the alpha band: functional meaning. Epilepsia **44**(s12), 2–8 (2003)
16. Kothare, S.V.: Atlas of EEG patterns, 2nd edn. In: Neurology, vol. 83, no. 7, pp. 668 LP –668, Aug. (2014)

Modelling the Disease Severity of Cocoa Black Pod Rot Disease

Albert Ling Sheng Chang, Suhaida Salleh, Ahmad Kamil Mohd Jaaffar, and Haya Ramba

Abstract Cocoa black pod rot disease, caused by *Phytophthora palmivora*, is the main disease that caused major losses in Malaysia. It is important for screening the cocoa genotypes available in Malaysia on its tolerance level against the cocoa black pod rot disease. Characterize the symptom of the cocoa black pod rot disease in determined the disease severity over time can assist the breeder in identified the tolerance level of the genotypes. In this study, we developed the nonlinear model on the disease severity for 50 cocoa genotypes available in Malaysia that later find the best fitted model for each genotype. Those selected nonlinear model was used to determine its area under disease progress curve (AUDPC) which used to identify the tolerance clones against cocoa black pod rot disease. Our results highlight the Gompertz model was well fitted to the disease severity of 50 cocoa genotypes and its estimation of AUDPC values showed few good tolerant clones against cocoa black pod rot disease with the lowest AUDPC values such as MCBC13, PBC221, BAL209, KKM19 and KKM17.

Keywords AUDPC · Nonlinear model · Cocoa · Black pod rot · Disease severity

---

A. L. S. Chang (✉) · A. K. M. Jaaffar
Malaysian Cocoa Board, Kota Kinabalu 88999, Sabah, Malaysia
e-mail: albert@koko.gov.my

A. K. M. Jaaffar
e-mail: kamil@koko.gov.my

S. Salleh
Cocoa Research and Development Center, Jengka 28000, Pahang, Malaysia
e-mail: suhaida@koko.gov.my

H. Ramba
Cocoa Research and Development Center, Kota Samarahan 93450, Sarawak, Malaysia
e-mail: hayaramba@koko.gov.my

S. A. Abdul Karim et al. (eds.), *Proceedings of the 6th International Conference on Fundamental and Applied Sciences*, Springer Proceedings in Complexity,
https://doi.org/10.1007/978-981-16-4513-6_48

1 Introduction

Cocoa black pod rot disease, caused by *Phytophthora palmivora*, is the main disease that caused major losses in Malaysia. The symptom of the disease began with pod lesions as small, hard, dark spots on any part of the pod, at any stage of pod development [1]. It is important to establish the effective screening method for selecting the tolerance genotypes in breeding program. However, selection of tolerance plant in the field would take many years as need to wait until a tree bears pods and the infection rates are not always sufficient under natural infection conditions may disrupt the reliability of assessment results [2]. Therefore, understanding the disease severity progress from the lesion growth in the laboratory is more effective in assisting the breeder in identify the tolerance level of cocoa genotypes available, especially in Malaysia. This could be done by modeling the disease severity of its lesion growth over the time. The nonlinear models are commonly used to model temporal changes in the proportion of plant disease [3]. Four nonlinear models that commonly used to describe the disease progress curve are the exponential model, monomolecular model, logistic model and Gompertz model [4]. The work of characterized the disease progress curve with nonlinear models and estimating the area under the disease progress curve (AUDPC) to related the estimated AUDPC to changes in the components of quantitative resistance has been carried out in many other plant diseases such as rice blast disease [5], fire blight disease in apple [6] and late blight disease in potato [7] but rarely used in cocoa screening test. Thus, this paper is to develop the nonlinear models on the severity of the cocoa black pod rot disease and estimating the area under the disease progress curve (AUDPC) of the best nonlinear model developed. These AUDPC values of tested genotypes will be used to identify the tolerance level of the genotypes against the cocoa black pod rot disease.

2 Materials and Methods

2.1 Samples Preparation

Fifty cocoa genotypes that were widely planted by cocoa smallholders in Malaysia are used in this study. The detached pod test described in [8] was used in preparing the pod sample for study the disease severity progress in each genotypes. Isolate of *Phytophthora* sp. from a naturally infected cocoa pod was used to inoculate the sample pods by a single point on the ridges of pod. Forty inoculated pods per genotype with mycelial plugs from seven days old *Phytophthora palmivora* culture grown on Potato Dextrose Agar (PDA) were incubated at room temperature (25 ± 2 °C) in the laboratory. The diameter of the established lesions were measured with a caliper meter throughout one to seven day after inoculation (DAI).

2.2 Measuring the Disease Severity

The assessment to the black pod disease severity was measured by the proportion of lesion size to the pod surface area. The pod surface was estimated as the prolate spheroid model [9, 10] and the lesion area was estimated based on the ellipse shape [11]. The disease severity of the black pod disease was expressed as;

$$y = \frac{The\ lesion\ area}{Pod\ surface\ area} = \frac{\pi r_1 r_2}{2\pi a^2 + 2\pi a\left(\frac{ac}{e}\right)\sin^{-1}\left(\sqrt{1-\frac{a^2}{c^2}}\right)} \quad (1)$$

where r_1 is a major radius of the lesion area, r_2 is a minor radius of the lesion area, a is a polar radius of the pod and c is an equatorial radius of the pod.

2.3 Fitting the Disease Progress Curve

Four nonlinear models were used to describe the disease progress curve or also known as growth-curve models such as exponential model, monomolecular model, logistic model and Gompertz model [4, 12]. The nonlinear models for disease progress curve are given as follows:

$$\text{Exponential model: } y = y_0 \exp(r_E t) \quad (2)$$

$$\text{Monomolecular model: } y = 1 - (1 - y_0)\exp(-r_M t) \quad (3)$$

$$\text{Logistic model: } y = \frac{1}{1 + \exp\left(-\ln\left[\frac{y_0}{1-y_0}\right] - r_L t\right)} \quad (4)$$

$$\text{Gompertz model: } y = \exp\left[\ln(y_0) \times \exp(-r_G t)\right] \quad (5)$$

where y_0 is the initial disease intensity and r_E, r_M, r_L and r_G are the rate parameter (constant).

2.4 The Goodness of Fit Test

Curve fitting on the black pod disease severity was done using PROC NLIN using numerical method of Levenburg-Marquardt to minimize the error sum of squares of fitted models [13].

In order to identify the best fitted model among the four nonlinear models, two goodness of fit tests were used, namely the Akaike Information Criterion (AIC) [14] and the Bayesian Information Criterion (BIC) [15]. The best model was selected based on the smallest value of AIC and BIC. The mathematical expression for both tests was given as follows;

$$AIC = 2p - 2\ln(L) \tag{6}$$

$$BIC = p\ln(n) - 2ln(L) \tag{7}$$

with p = number of parameters, n = sample size and $ln(L)$ = maximum log-likelihood of the estimated model and calculated as follows:

$$\ln(L) = 0.5 \times \left[-N \times \left(\ln(2\pi) + 1 - \ln(N) + ln\left(\sum_{i=1}^{n} x_i^2\right)\right)\right] \tag{8}$$

with $x_1, \ldots, x_n$ = the residuals from the nonlinear least squares fit and N = their number.

2.5 *Estimating the Area Under Disease Progress Curve*

The area under disease progress curve (AUDPC) was calculated for each nonlinear model fitted to the disease progress curve using the method of Yeh [16] based on the trapezoidal rule that approximate the area under a curve by dividing the area into a number of strips of equal width. Then, the sum of approximate area of each strip by the area of the trapezium formed will give the approximation of area under the curve. The trapezoidal rule can be presented as integral function given as follows:

$$\int_a^b f(x)dx = ((b-a)/n) \times (((f(a) + f(b))/2) + \Sigma f(a + i(b-a)/n)) \tag{9}$$

where the domain [a, b] of the integration function are subdivided into n strips with the points of $x_0, x_1, \ldots, x_n$. $x_0 = a$, $x_n = b$ and $x_r = x_0 + r\,(b-a)\,/\,n$.

In our case of study, $a = 1$, $b = 6$ and n is set to 50 subdivisions or strips while the $f(x)$ is given as nonlinear function model of each clone.

Table 1 ANOVA test on 50 cocoa genotypes

Source	DF	Pod length		Pod girth		Pod surface	
		Sum of squares	F value	Sum of squares	F value	Sum of squares	F value
Genotype	49	16.137	18.25*	6.224	8.69*	28.593	12.35*
Error	1950	35.192		28.501		92.025	
Corrected Total	1999	51.330		34.725		120.618	

Note: * Significant at 5% level

3 Results and Discussion

3.1 Samples Preparation

The cocoa genotypes used in this study varied in pod length, pod girth and pod surface area (see ANOVA Test in Table 1). The pod length from 50 genotypes was ranged between 13.38 cm to 22.08 cm, the girth pod were ranged between 20.13 and 26.59 cm and the pod surface area were ranged between 229.59 and 426.13 cm^2.

3.2 Disease Severity Curve Fitted Model

The overall *F*-value for 50 genotypes are significant at 5% level in all developed models except the Monomolecular model in the genotype KKM19, MCBC12 and PBC221 (see Table 2). Table 2 shows the values of AIC and BIC of four nonlinear models fitted on the disease severity of each genotype. The results of AIC test and BIC test show Gompertz model fitted well the data disease severity in 47 genotypes with the smallest value in AIC and BIC. Another three genotypes, QH326, MCBC6 and KKM1 were fitted well with the Logistic model. The results showed the disease severity progress curve fitted well to two nonlinear models which are Logistic model and Gompertz model. In order to select only one model to represent the cocoa black pod disease severity progress curve, the Gompertz model was chosen due to higher percentage of Gompertz model fitted well to the data compare to the Logistic model. The application of Gompertz model in describing the disease progress also supported by Plaut [17] and Berger [18] who reported that Gompertz model was better fit to other statistical models in plant diseases including estimation of epidemic rate, projection of future disease severity and determination of initial disease.

Table 2 Nonlinear models fitted on 50 cocoa genotypes

Genotype	Model	*F value*	AIC	BIC	Genotype	Model	*F value*	AIC	BIC
BAL209	Exp	533.17**	−95.49	−95.6	MCBC3	Exp	797.92**	−76.8	−76.91
	Gomp	1418.74**	−102.32	−102.43		Gomp	4296.27**	−88.56	−88.67
	Log	539.2**	−95.57	−95.68		Log	848.82**	−77.23	−77.34
	Mono	8.23*	−68.12	−68.23		Mono	7.28*	−45.96	−46.07
BAL244	Exp	622.55**	−63.39	−63.49	MCBC4	Exp	1079.99**	−85.49	−85.59
	Gomp	8568.55**	−81.71	−81.82		Gomp	3952.97**	−94.56	−94.66
	Log	728.28**	−64.48	−64.59		Log	1127.56**	−85.79	−85.89
	Mono	10.65*	−36.36	−36.46		Mono	10.82*	−54.7	−54.81
BR25	Exp	2139.36**	−89.26	−89.36	MCBC5	Exp	599.95**	−73.67	−73.78
	Gomp	20,054.60**	−104.91	−105.02		Gomp	4714.30**	−88.08	−88.19
	Log	2304.96**	−89.78	−89.89		Log	644.91**	−74.18	−74.29
	Mono	8.51*	−52.36	−52.47		Mono	12.07*	−47.62	−47.73
DESA1	Exp	1446.55**	−77.49	−77.6	MCBC6	Exp	4071.47**	−86.13	−86.24
	Gomp	12,750.4**	−92.71	−92.82		Gomp	2361.83**	−82.32	−82.43
	Log	1618.91**	−78.27	−78.38		Log	4491.77**	−86.82	−86.93
	Mono	8.16*	−43.1	−43.21		Mono	8.79*	−44.91	−45.02
KKM1	Exp	2509.74**	−83.36	−83.47	MCBC7	Exp	358.80**	−76.75	−76.86
	Gomp	1385.61**	−79.21	−79.31		Gomp	1184.81**	−85.08	−85.19
	Log	2623.40**	−83.67	−83.78		Log	371.64**	−76.99	−77.1
	Mono	9.92*	−46.19	−46.3		Mono	11.77*	−54.13	−54.24
KKM15	Exp	303.89**	−73.46	−73.57	MCBC8	Exp	430.42**	−82.41	−82.52

(continued)

Table 2 (continued)

Genotype	Model	*F value*	AIC	BIC	Genotype	Model	*F value*	AIC	BIC
	Gomp	838.88**	−80.54	−80.64		Gomp	1300.05**	−90.12	−90.23
	Log	314.51**	−73.7	−73.81		Log	441.13**	−82.58	−82.69
	Mono	9.22*	−50.62	−50.73		Mono	7.97*	−56.35	−56.46
KKM17	Exp	595.07**	−91.47	−91.57	MCBC9	Exp	643.07**	−66.85	−66.96
	Gomp	1917.90**	−99.64	−99.75		Gomp	8743.41**	−85.09	−85.2
	Log	605.14**	−91.58	−91.69		Log	732.12**	−67.75	−67.86
	Mono	7.05*	−62.51	−62.62		Mono	10.51*	−39.52	−39.62
KKM19	Exp	2807.70**	−99.06	−99.17	PBC112	Exp	596.98**	−88.75	−88.86
	Gomp	16,616.30**	−111.5	−111.61		Gomp	1340.05**	−94.39	−94.5
	Log	2905.69**	−99.3	−99.41		Log	607.01**	−88.87	−88.97
	Mono	5.57 ns	−58.09	−58.2		Mono	8.50*	−60.77	−60.87
KKM2	Exp	1405.10**	−86	−86.1	PBC123	Exp	512.01**	−83.99	−84.1
	Gomp	1873.72**	−88.01	−88.12		Gomp	1501.21**	−91.5	−91.61
	Log	1444.80**	−86.19	−86.3		Log	523.90**	−84.15	−84.26
	Mono	6.83*	−50.88	−50.99		Mono	6.13*	−55.37	−55.48
KKM22	Exp	379.71**	−85.72	−85.83	PBC130	Exp	746.97**	−70.69	−70.8
	Gomp	925.20**	−91.93	−92.04		Gomp	6001.21**	−85.26	−85.37
	Log	385.95**	−85.84	−85.94		Log	830.52**	−71.43	−71.54
	Mono	6.95*	−59.82	−59.93		Mono	15.58**	−44.62	−44.73
KKM25	Exp	699.04**	−81.22	−81.32	PBC131	Exp	1003.60**	−84.25	−84.36
	Gomp	3339.51**	−92.14	−92.25		Gomp	6951.23**	−97.79	−97.89

(continued)

Table 2 (continued)

Genotype	Model	*F value*	AIC	BIC	Genotype	Model	*F value*	AIC	BIC
	Log	731.77**	−81.53	−81.64		Log	1049.16**	−84.56	−84.67
	Mono	9.20*	−52.56	−52.67		Mono	6.42*	−51.18	−51.28
KKM26	Exp	339.92**	−69.62	−69.73	PBC137	Exp	813.15**	−81.81	−81.92
	Gomp	1136.45**	−78.03	−78.14		Gomp	3813.18**	−92.61	−92.72
	Log	359.21**	−70	−70.11		Log	853.42**	−82.15	−82.26
	Mono	11.63*	−47.3	−47.41		Mono	13.76**	−54.41	−54.51
KKM27	Exp	652.39**	−68.13	−68.24	PBC139	Exp	357.56**	−64.64	−64.75
	Gomp	9323.24**	−86.73	−86.83		Gomp	1330.79**	−73.8	−73.91
	Log	735.49**	−68.97	−69.08		Log	385.90**	−65.17	−65.28
	Mono	11.23*	−41.08	−41.19		Mono	14.52**	−43.28	−43.38
KKM28	Exp	323.75**	−77.78	−77.89	PBC140	Exp	288.07**	−67.71	−67.81
	Gomp	859.95**	−84.59	−84.7		Gomp	895.65**	−75.61	−75.71
	Log	332.46**	−77.97	−78.08		Log	304.61**	−68.09	−68.2
	Mono	9.11*	−54.43	−54.54		Mono	19.41**	−49.61	−49.72
KKM3	Exp	2527.07**	−88.23	−88.34	PBC159	Exp	432.41**	−69.41	−69.51
	Gomp	16,131.00**	−101.2	−101.31		Gomp	1367.50**	−77.44	−77.55
	Log	2767.09**	−88.87	−88.98		Log	461.14**	−69.85	−69.96
	Mono	7.60*	−49.57	−49.68		Mono	9.21*	−44.1	−44.21
KKM4	Exp	895.05**	−71.18	−71.29	PBC179	Exp	368.36**	−81.4	−81.5
	Gomp	10,115.80**	−88.14	−88.25		Gomp	1017.10**	−88.48	−88.58
	Log	1014.51**	−72.06	−72.16		Log	377.08**	−81.56	−81.67

(continued)

Table 2 (continued)

Genotype	Model	*F value*	AIC	BIC	Genotype	Model	*F value*	AIC	BIC
	Mono	14.47**	−43.4	−43.51		Mono	15.12**	−60.07	−60.18
KKM5	Exp	538.52**	−60.33	−60.44	PBC221	Exp	291.44**	−84.31	−84.42
	Gomp	10,652.80**	−81.19	−81.3		Gomp	517.02**	−88.29	−88.4
	Log	647.82**	−61.62	−61.73		Log	294.44**	−84.38	−84.49
	Mono	12.24*	−35.11	−35.21		Mono	4.85 ns	−58.49	−58.59
KKM6	Exp	398.08**	−70.59	−70.7	QH1003	Exp	450.65**	−73.03	−73.13
	Gomp	1774.35**	−81.02	−81.12		Gomp	1352.46**	−80.69	−80.8
	Log	422.62**	−71	−71.11		Log	472.34**	−73.35	−73.46
	Mono	13.53**	−48.06	−48.17		Mono	11.63*	−48.75	−48.86
MCBC1	Exp	470.43**	−63.74	−63.85	QH1176	Exp	811.59**	−89.15	−89.25
	Gomp	3760.74**	−78.26	−78.37		Gomp	3460.02**	−99.28	−99.39
	Log	529.02**	−64.56	−64.67		Log	831.96**	−89.32	−89.43
	Mono	11.99*	−39.35	−39.45		Mono	6.07*	−57.27	−57.38
MCBC10	Exp	75.95**	−79.09	−79.2	QH22	Exp	709.75**	−78.53	−78.64
	Gomp	105.92**	−81.35	−81.46		Gomp	3835.78**	−90.32	−90.43
	Log	76.40**	−79.13	−79.24		Log	749.92**	−78.91	−79.02
	Mono	9.79*	−66.11	−66.22		Mono	13.43**	−51.93	−52.04
MCBC11	Exp	215.45**	−61.99	−62.1	QH326	Exp	1927.52**	−79.65	−79.76
	Gomp	611.87**	−69.24	−69.35		Gomp	1045.50**	−75.38	−75.49
	Log	229.30**	−62.42	−62.53		Log	1929.25**	−79.66	−79.77
	Mono	12.22*	−43.13	−43.23		Mono	5.96*	−41.64	−41.75

(continued)

Table 2 (continued)

Genotype	Model	*F value*	AIC	BIC	Genotype	Model	*F value*	AIC	BIC
MCBC12	Exp	307.86**	−69.53	−69.64	QH37	Exp	37.58**	−53.46	−53.56
	Gomp	729.19**	−75.53	−75.64		Gomp	50.44**	−55.4	−55.51
	Log	319.25**	−69.78	−69.89		Log	38.27**	−53.58	−53.68
	Mono	5.54 ns	−43.96	−44.07		Mono	9.04*	−44.74	−44.85
MCBC13	Exp	79.78**	−99.66	−99.76	QH441	Exp	156.20**	−53.48	−53.59
	Gomp	100.62**	−101.24	−101.34		Gomp	369.80**	−59.45	−59.56
	Log	79.87**	−99.66	−99.77		Log	168.71**	−54.01	−54.12
	Mono	7.99*	−83.96	−84.07		Mono	8.61*	−34.87	−34.97
MCBC14	Exp	431.44**	−77.15	−77.26	QH968	Exp	311.31**	−63.96	−64.07
	Gomp	1664.98**	−86.57	−86.68		Gomp	1148.58**	−73.06	−73.16
	Log	449.47**	−77.43	−77.54		Log	334.94**	−64.47	−64.57
	Mono	14.69**	−54.55	−54.66		Mono	11.07*	−41.97	−42.08
MCBC2	Exp	148.42**	−69.86	−69.97	RP1	Exp	413.88**	−64.47	−64.58
	Gomp	294.93**	−74.61	−74.72		Gomp	1815.58**	−74.79	−74.9
	Log	152.01**	−70.03	−70.13		Log	453.73**	−65.11	−65.22
	Mono	16.42**	−55.33	−55.43		Mono	10.30*	−40.1	−40.2

Note Exp.—Exponential Model; Gomp.—Gompertz model; Log.—Logistic model; Mono.—Monomolecular model; ns—Not Significant at the 95% confidence level; *—Significant at the 95% confidence level; **—Significant at the 99% confidence level

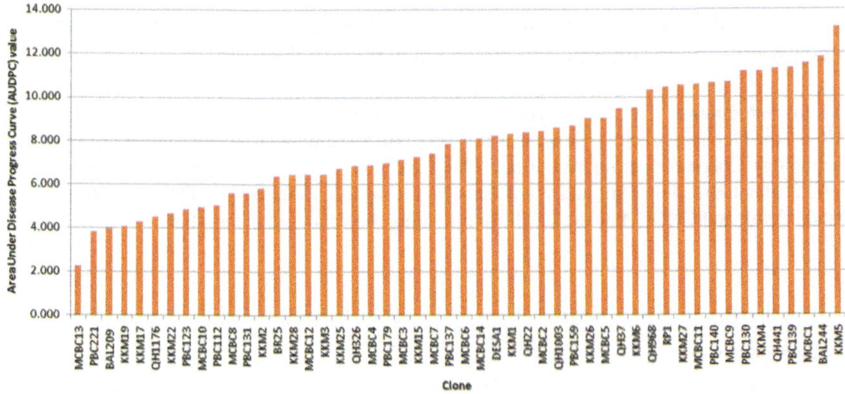

Fig. 1 The AUDPC values arranged in ascending order for 50 cocoa genotypes against cocoa black pod rot disease

3.3 *Estimate the Area Under Disease Progress Curve*

The AUDPC has been widely used in two major areas such as crop loss assessment and field assessment of quantitative resistance as it integrate all aspects of disease progress related to host development and growth [19]. Figure 1 shows the estimated AUDPC from severity disease progress curve fitted with Gompertz models for 50 genotypes. The AUDPC values of each genotype depend on the growth rate of the disease severity over the time, the slow growth rate of the disease severity will see the AUDPC value is lower than the fast growth rate. The lower AUDPC value also indicated that the tolerance level is higher in that genotype compared to the genotype with higher AUDPC value. Result in Fig. 1 showed the AUDPC values ranged from 2.246 to 13.194 with MCBC13 (2.246), PBC221 (3.827), BAL209 (3.992), KKM19 (4.075) and KKM17 (4.297) have the lowest AUDPC values while KKM5 (13.194), BAL244 (11.809), MCBC1 (11.550), PBC139 (11.317) and QH441 (11.290) have the highest AUDPC values.

The selection of AUDPC method as screening method in assessing the tolerance level of cocoa genotypes is better than field assessment which necessitates waiting until a tree bears pods before its level of resistance to black pod infection can be determined [20]. The AUDPC method able to save time and cost in order to provide information about pod resistance to the cocoa black pod disease.

4 Conclusion

In conclusion, the Gompertz models developed for the disease severity on the cocoa black pod rot disease over the 50 genotypes planted in Malaysia by the smallholders was useful in estimating the AUDPC which can be practically used to assess the cocoa genotypes for tolerance to cocoa black pod rot disease.

Acknowledgements The author wishes to thank the Director−General of Malaysian Cocoa Board (MCB), the Deputy Director−General of Research and Development and the Director of Upstream Technology Cocoa for permission of publishing and reviewing this paper. Thanks are also due to staffs at Plant Breeding Unit and Plant Pathology Unit in CRDC Jengka, Pahang, CRDC Bagan Datuk, Perak and CRDC Tawau, Sabah for providing the data used in this study.

References

1. Guest, D.I.: Black pod: diverse pathogens and the greatest global impact on cocoa yield. Phytopathology **97**(12), 1650–1653 (2007)
2. Clias, C., Despréaux, D.: Improvement of cocoa tree resistance to Phytophthora diseases. CIRAD (2004)
3. Macchiavelli, R.E.: Non-linear mixed models for disease incidence and severity: modeling plant diseases in tropical crops. In: Proceedings of the 59th ISI World Statistics Congress, International Statistical Institute, Hong Kong (2013)
4. Madden, L.V., Campbell, C.L.: Nonlinear disease progress curves. In: Kranz, J. (ed.) Epidemics of Plant Diseases: Mathematical Analysis and Modeling, pp. 181–229. Springer, Berlin, Germany (1990)
5. Mohapatra, N.K., Mukherjee, A.K., Suriya Rao, A.V., Nayak, P.: Disease progress curves in the rice blast pathosystem compared with the Logistic and Gompertz models. J. Agric. Biol. Sci. **3**(1), 28–37 (2008)
6. Momol, M.T., Norelli, J.L., Aldwinckle, H.S., Zeller, W.: Use of the area under the disease progress curve for quantification of resistance of apple and pear varieties and rootstocks to *Erwinia amylovora*. Acta Hort. **411**, 373–374 (1996)
7. Haynes, K.G., Weingartner, D.P.: The use of area under the disease progress curve to assess resistance to late blight in potato germplasm. Am. J. Potato Res. **81**, 137–141 (2004)
8. Iwaro, A.D., Sreenivasan, T.N., Umaharan, P.: Foliar resistance to *Phytophthora palmivora* as an indicator of pod resistance in *Theobroma cacao*. Plant Dis. **81**, 619–624 (1997)
9. Jessop, N.H., Awudzi, G., Bateman, R.P.: How best to spray cocoa with motorised mistblowers. Ann. Appl. Biol. **99**, 191–196 (2010)
10. Ten Hoopen, G.M., Deberdt, P., Mbenoun, M., Cilas, C.: Modelling cacao pod growth: implications for disease control. Ann. Appl. Biol. **160**(3), 260–272 (2012)
11. Campbell, A. Ploetz, R., Gutierrez, O.: Using digital image analysis to evaluate disease resistance. In: The Americas Cacao Breeders' Working Group 2nd Annual Meeting, San Salvador (2015)
12. Madden, L.V.: The study of plant disease epidemics. APS Press, St. Paul, MN, USA (2007)
13. SAS for windows, version 9.3. SAS Institute, Cary NC, USA (2011)
14. Burnham, K.P., Anderson, D.R.: Model selection and inference: a practical information−theoretic approach, 2nd edn. Springer, New York, USA (2003)
15. Schwarz, G.: Estimating the dimension of a model. Ann. Stat. **6**, 461–464 (1978)
16. Yeh, S.T.: Using trapezoidal rule for the area under a curve calculation. SAS Conference Proceedings: SAS Users Group International 27, SAS Institute Inc, Cary, NC, USA (2007)
17. Plaut, J.L.: Epidemic progress of three pathosystems as affected by initial disease severity. M.S. thesis, University of Florida, Gainesville (1980)
18. Berger, R.D.: Comparison of the Gompertz and logistic functions to describe plant disease progress. Phytopathology **71**, 716–719 (1981)

19. Jeger, M.J., Viljanen-Rollinson, S.L.H.: The use of the area under the disease−progress curve (AUDPC) to assess quantitative disease resistance in crop cultivars. Theor. Appl. Genet. **102**, 32–40 (2001)
20. Nyadanu, D., Assuah, M.K., Adomako, B., Asiama, Y.P., Opoku, I.Y., Adu-Ampomah, Y.: Efficacy of screening methods used in breeding for black pod disease resistance varieties in cocoa. Afr. Crop Sci. J. **17**(4), 175–186 (2009)

Monotonicity Preserving Interpolation Using Rational Quartic Said-Ball Function

Samsul Ariffin Abdul Karim and Van Thien Nguyen

Abstract A curve interpolation scheme with C^1 continuity is developed by using rational quartic Said-Ball function with quartic numerator and linear denominator. The scheme has a unique representation, with one shape parameter that can be used to change the shape of the curve. Sufficient conditions for the monotonicity of the rational interpolant have been derived. Several numerical results are given to show the capability of the presented scheme.

Keywords Interpolation · Rational quartic · Said-Ball function · Monotonicity

1 Introduction

Data interpolation and approximation are important in sciences and engineering applications. It is important that the interpolant must be able to produce the interpolating curve that preserves its geometric properties such as monotonicity, convexity and positivity. Monotonicity problems arise in our daily lives. Pricing models in economics and finance, the level of uric acid in the body of patients with gout are examples of monotone data [11, 17]. Cubic spline interpolating scheme with continuous is not able to produce the monotone curves. Unwanted oscillations may appear and it will destroy the characteristics of the data. Cubic Hermite function with continuity also cannot preserve the shape of the data. Thus, shape preserving interpolation can be used to produce monotonic curves [2, 3, 8, 9]. There exist many methods for monotonicity preserving by using various types of spline interpolant. For instances,

S. A. Abdul Karim (✉)
Fundamental and Applied Sciences Department and Centre for Systems Engineering (CSE), Institute of Autonomous System, Universiti Teknologi PETRONAS (UTP), Bandar Seri Iskandar, 32610 Seri Iskandar, Perak DR, Malaysia
e-mail: samsul_ariffin@utp.edu.my

V. T. Nguyen
Department of Mathematics, FPT University, Education Zone, Hoa Lac High Tech Park, Thach That Ward, Hanoi, Vietnam
e-mail: thiennv15@fe.edu.vn

S. A. Abdul Karim et al. (eds.), *Proceedings of the 6th International Conference on Fundamental and Applied Sciences*, Springer Proceedings in Complexity,
https://doi.org/10.1007/978-981-16-4513-6_49

Fristch and Butland [4] and Fristch and Carlson [5] discussed interpolation of monotonic data using piecewise cubic spline polynomials. In the later article, the authors have derived the necessary and sufficient conditions for the cubic interpolant to be monotonic on entire given interval. But both methods are required the modification of the first derivative in order to ensure the monotonicity of the data will be preserved completely.

Besides the use of cubic spline for shape preserving interpolation, there exist many schemes involving rational splines such as piecewise rational quadratic functions by Gregory and Delbourgo [8] and Delbourgo and Gregory [3], rational cubic with quadratic denominator by Delbourgo and Gregory [2] and Gregory [9], rational cubic spline with cubic denominator in Sarfraz [18, 19] and Sarfraz et al. [17]. Duan et al. [6, 7] have proposed rational cubic spline with linear denominator in order to interpolate data set from given function. Meanwhile Abbas et al. [1] have proposed new rational cubic spline with quadratic denominator to preserves the monotone data by utilizing two free parameters to refine the final shape of the curves. Shukri et al. [20] have study the use of cubic Bézier curve for data constrained interpolation. They derived the simple sufficient condition when the constrained are subject to circle, an ellipse and straight line. Our work in this paper is slightly different from the work of Shukri et al. [20]. In this paper, the new rational quartic spline (quartic/linear) has been used for monotonicity preserving. For survey articles on shape preserving interpolation and approximation methods, see Goodman [10] and Kvasov [15] and the listed references in their respective articles.

This paper discussed the monotonicity preserving for a given data sets. In Wang and Tan [21], the authors have used Bernstein-Bezier quartic rational interpolant (quartic numerator and linear denominator) with two shape parameters to preserves the monotonic data. The derivation of calculus to obtain first and second-order derivatives is easier when rational spline with linear denominator is used. Thus, it is easy to construct or continuous curve by using rational spline interpolant in the form of quartic/linear. Motivated by Wang and Tan [21], in this paper, a new rational spline interpolant is proposed. It is based from rational Said-Ball curves of degree 4 (quartic/linear) with one shape parameter to generate the monotonic interpolating curves. One of the advantages of the proposed scheme is that, it will produce monotonic interpolating curves for the monotonic data sets. Meanwhile the Wang and Tan [21] scheme produces monotonic curves. Furthermore, we give the method on how to choose the suitable values of first derivative, that satisfy the sufficient condition for monotonicity. The scientific contributions of this paper are: (1) the new rational spline interpolant based on Said-Ball basis function has been introduced for monotonicity preserving; (2) the method on how to calculate the first derivative for monotonicity preserving is discussed in details while in the work of Wang and Tan [21], the authors do not discuss in details on how to calculate the first derivative values, (3) an algorithm to generate the rational quartic spline for monotonicity preserving also has been given and our final contribution is, the numerical comparison between the proposed rational spline scheme with the existing method namely Sarfraz et al. [17] has been done. This paper is organized as follows. Introduction is given in Sect. 1. Rational quartic Said-Ball (quartic/linear) with one shape parameter is discussed in Sect. 2. Meanwhile, Sect. 3 is devoted to the monotonicity preserving by using the proposed rational quartic Said-Ball interpolant together with algorithm for computer implementation. In Sect. 4, Results and Discussion will be elaborated in details together with numerical comparison between the proposed scheme and the rational cubic spline (cubic/cubic) by Sarfraz et al. [17]. Section 5 concludes the paper with some future research recommendation.

2 Rational Quartic Said-Ball Interpolant

Suppose $\{(x_i, f_i),\ i = 1, \ldots, n\}$ is a given set of data points, where $x_1 < x_2 < \cdots < x_n$, $h_i = x_{i+1} - x_i$, $\Delta_i = (f_{i+1} - f_i)/h_i$, local variable θ is defined by $\theta_i = (x - x_i)/h_i$ such that $0 \le \theta \le 1$ and shape parameter $\alpha_i > 0$. For $x \in [x_i, x_{i+1}]$, $i = 1, 2, \ldots, n-1$, the rational quartic Said-Ball interpolant is defined by

$$s(x) = s(x_i + h_i\theta) \equiv S_i(\theta) = \frac{P_i(\theta)}{Q_i(\theta)}, \tag{1}$$

where

$$\begin{aligned} P_i(\theta) &= W_{i0}(1-\theta)^3 + W_{i1}\theta(1-\theta)^2 + W_{i2}\theta^2(1-\theta)^2 + W_{i3}\theta^3(1-\theta) + W_{i4}\theta^4 \\ Q_i(\theta) &= \alpha_i(1-\theta) + \theta. \end{aligned}$$

Now, to ensure the rational quartic Said-Ball interpolant in (1) is on the following interpolatory properties must be satisfied:

$$\begin{aligned} s(x_i) &= f_i, \quad s(x_{i+1}) = f_{i+1} \\ s^{(1)}(x_i) &= d_i, \quad s^{(1)}(x_{i+1}) = d_{i+1} \end{aligned} \tag{2}$$

Hence, the unknowns W_{ij}, $j = 0, 1, 2, 3, 4$ can be shown to have the following values:

$$\begin{aligned} &W_{i0} = \alpha_i f_i,\ W_{i1} = (2\alpha_i + 1)f_i + \alpha_i h_i d_i,\ W_{i2} = 3\alpha_i f_{i+1} + 3f_i \\ &W_{i3} = (\alpha_i + 2)f_{i+1} - h_i d_{i+1},\ W_{i4} = f_{i+1}. \end{aligned} \tag{3}$$

The first derivative values are denoted by d_i. The rational quartic interpolant in (1) is $C^1[x_1, x_n]$. The value of shape parameter α_i can be utilized to changes the final shape of the interpolating curve according to what the user needs [12]. Obviously when $\alpha_i = 1$, then the rational quartic reduce to the non-rational quartic Said-Ball [16] interpolant given as follows:

$$s(x) = f_i(1-\theta)^3 + (3f_i + h_i d_i)\theta(1-\theta)^2 + (3f_i + f_{i+1})\theta^2(1-\theta)^2 \\ + (3f_{i+1} - h_i d_{i+1})\theta^3(1-\theta) + f_{i+1}\theta^3. \quad (4)$$

3 Monotonicity-Preserving Said-Ball Interpolation

Given the monotonic increasing set of data so that $f_1 \leq f_2 \leq \cdots \leq f_n$ or equivalently

$$\Delta_i \geq 0, \text{ for } i = 1, \ldots, n-1 \text{ and } d_i \geq 0 \text{ for } i = 1, \ldots, n. \quad (5)$$

After some simplification, it can be shown that, the first derivative of the rational quartic Said-Ball interpolant $s(x)$ is given by

$$s^{(1)}(x) = \frac{\sum_{j=0}^{4} A_{ij}(1-\theta)^{4-j}\theta^j}{[\alpha_i(1-\theta)+\theta]^2}, \quad (6)$$

where

$$A_{i0} = \alpha_i^2 d_i, \; A_{i1} = 2\alpha_i^2(3\Delta_i - d_i)$$
$$A_{i2} = 3\alpha_i(4\Delta_i - d_i - d_{i+1}), \; A_{i3} = 2(3\Delta_i - d_{i+1}), \; A_{i4} = d_{i+1}.$$

Now, if $s^{(1)}(x)$ then the rational interpolant $s(x)$ will be monotonic increasing for all $x \in [x_1, x_n]$. For $x \in [x_i, x_{i+1}]$, the denominator in (4) is always positive, thus we only consider the numerator in (4) to find the sufficient condition for monotonicity. Since $s^{(1)}(x)$ is non-negative if and only if $A_{ij} \geq 0$. It is obvious that the sufficient conditions for $A_{i1} \geq 0$, $A_{i2} \geq 0$ and $A_{i3} \geq 0$ are

$$d_i \leq 3\Delta_i, \quad d_{i+1} \leq 3\Delta_i, \quad d_i + d_{i+1} \leq 4\Delta_i. \quad (7)$$

Therefore, $s^{(1)}(x) \geq 0$ if (7) holds. The following proposition gives the sufficient condition for monotonicity preserving by using rational quartic Said-Ball interpolant (quartic/linear).

Proposition 1 *Given a monotonic increasing set of data satisfying (5), there exists a class of monotonic rational (of the form quartic/linear) interpolating spline $s(x) \in C^1[x_1, x_n]$ involving free parameter α_i provided that (7) holds.*

Equation (7) can be rewritten as

$$\frac{d_i}{\Delta_i} \leq 3, \quad \frac{d_{i+1}}{\Delta_i} \leq 3, \quad \frac{d_i}{\Delta_i} + \frac{d_{i+1}}{\Delta_i} \leq 4. \quad (8)$$

Now, by following the same idea from Fristch and Carlson [5], the sufficient conditions in (8) can be represented in the form of monotonicity region. Figure 1 shows

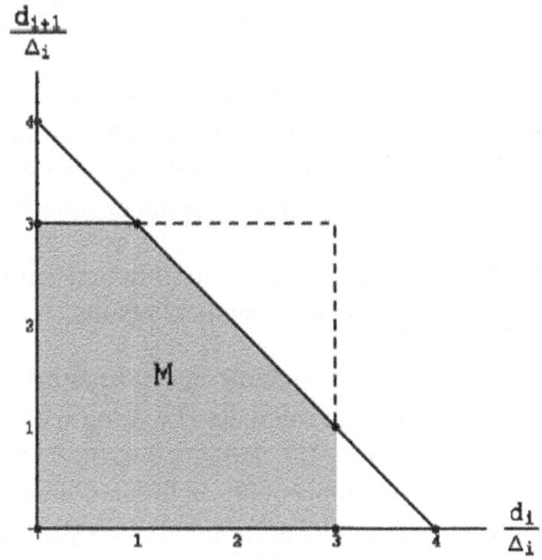

Fig. 1 Monotonicity region for rational quartic Said-Ball interpolant

the monotonicity region for rational quartic Said-Ball (quartic/linear) with one shape parameter (shaded region denoted by M). The monotonicity region is a trapezium with vertices (0, 0), (3, 0), (3, 1), (1, 3) and (0, 3). One question still needs to be answered: How do we choose the first derivative value that will assure the monotonicity preservation is achieved? In order to use the sufficient conditions in (8) for monotonicity preserving, and from Fig. 1, the following choices will guarantee the existence of monotonc rational quartic Said-Ball interpolant for all given strictly monotone data sets

$$0 \leq d_i \leq 2\Delta_i, \quad 0 \leq d_{i+1} \leq 2\Delta_i. \tag{9}$$

Condition (9) satisfies conditions in (7) and (8). This sufficient condition is simple and easy to use. Furthermore, as long as the first derivatives fall inside the monotonicity region in (9), the rational interpolant will produce monotonic interpolating curves for all monotone data sets.

Finally, the first derivative can be chosen as follows:

$$0 \leq d_1 \leq 2\Delta_1, \quad 0 \leq d_n \leq 2\Delta_{n-1} \tag{10}$$

and

$$0 \leq d_i \leq \min\{2\Delta_{i-1}, 2\Delta_i\}, i = 2, 3, \ldots, n-1. \tag{11}$$

Thus, the derivatives value in (10) and (11) will ensure the sufficient condition for monotonicity is met and the user may choose any positive value of shape parameter to generate monotonic interpolating curves.

Remark 1 The sufficient condition in (7) is the same as the sufficient condition for monotonicity in Wang and Tan [21]. The main difference is that in this paper the author has used rational quartic Said-Ball (quartic/linear) with one shape parameter while Wang and Tan [21] have used rational quartic spline (quartic/linear) with two shape parameters. Furthermore, we give complete idea on how to calculate the first derivative values that will produces monotonicity preserving interpolation curves.

Remark 2 In Karim [12] the sufficient condition in (7) and (8) is modified to produce more visual pleasing graphical results.

To generate the default curves (quartic Said-Ball), the first derivative value is the original value that calculated by using arithmetic mean method (AMM) [3] with $\alpha_i = 1$. Figure 2 shows the defaults quartic Said-Ball without shape preserving property. When the sufficient condition for monotonicity is met (by adjusting the first derivative value in which the monotonicity is not preserved), then the monotonic interpolating curves can be generated by choosing any shape parameter value $\alpha_i > 0$. Thus, for same value of shape parameter, i.e. $\alpha_i = 1$, we may have two set of interpolating curves where one interpolating curve having shape preserving property while the other interpolating curve does not have shape preserving property. Figure 3 shows this example clearly. Below is an algorithm that can be used to generate C^1 rational quartic Said-Ball for monotonicity-preserving curves using the main result in Proposition 1.

Algorithm 1

1. Input data points $\{(x_i, f_i)\}_{i=1}^n$ and first derivative at end points $d_1 \in [0, 2\Delta_1]$, $d_n \in [0, 2\Delta_{n-1}]$.
2. For $i = 2, \ldots, n-1$
   Compute the first derivative value: $0 \le d_i \le \min(2\Delta_{i-1}, 2\Delta_i)$.
3. For $i = 1, \ldots, n-1$
   (a) Find h_i and Δ_i
   (b) Assign $\alpha_i > 0$ (we set $\alpha_i = 1$ for the default curve)
   (c) Calculate the inner control ordinates W_{i1}, W_{i2} and W_{i3} in (3)
4. For $i = 1, \ldots, n-1$ By using (1), generates the piecewise rational quartic Said-Ball (quartic/linear) to produces monotonic interpolating curve everywhere.

4 Result and Discussion

We test the proposed method for monotonicity preserving by using monotone data taken from Sarfraz et al. [17]

Figure 2 shows the default interpolating curve when (with no derivative modification), clearly the rational interpolant fails to preserve the monotonicity property especially on interval [0, 2] and [3, 9]. Figure 3 shows the shape preserving interpolation with shape parameter values as follows: (solid), (dashed) and (gray).

Table 1 Monotone data [17]

i	1	2	3	4	5
x_i	0	2	3	9	11
f_i	0.5	1.5	7	9	13

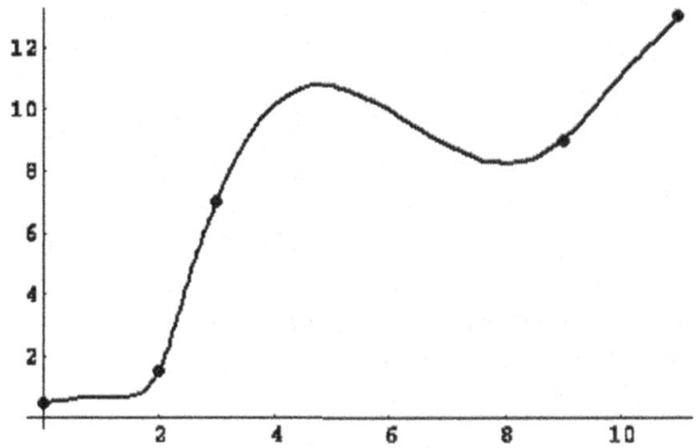

Fig. 2 Quartic Said-Ball curve ($\alpha_i = 1$) without shape preserving (no derivative modification)

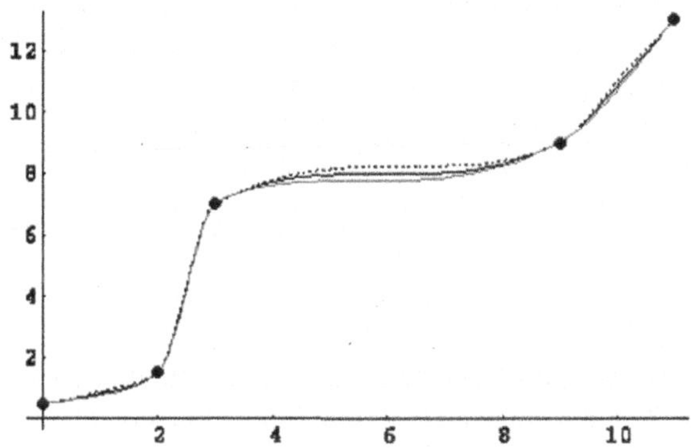

Fig. 3 Monotonic interpolating curves with various values of shape parameter

Meanwhile, Fig. 4 shows the visual pleasing shape preserving monotonic interpolating curve. From the figure, it can be seen clearly that, the proposed rational interpolant Said-Ball function (quartic numerator and linear denominator) with one shape parameter gives an acceptable result. Furthermore, the monotonic interpolating curves have C^1 continuity at their common points. When sufficient conditions for the monotonicity in (6) or (7) have been satisfied, the user can modify the shape of the curve by assigning different shape parameter values. We may have variety types of monotonic interpolating curves for the same data sets. This is one of the unique of the proposed scheme. Furthermore, these properties might be useful for some applications.

Finally, Fig. 5 shows the shape preserving by using the rational cubic spline (cubic/cubic) with continuity proposed by Sarfraz et al. [17]. From Figs. 2, 3, 4 and 5 we could see that both shape preserving interpolation schemes produces very visual pleasing result. But in term of easy to use and less calculation required, the

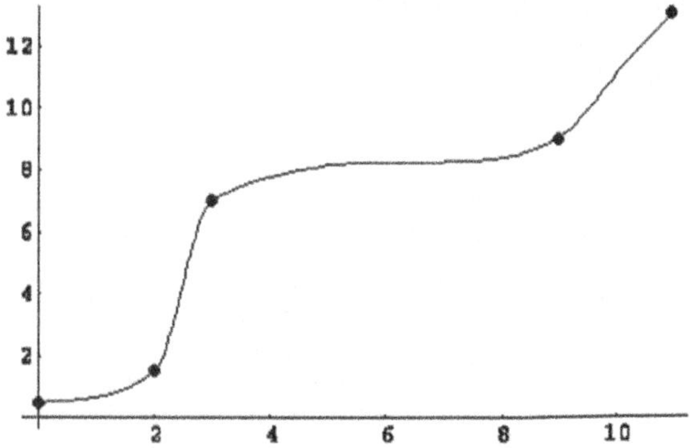

Fig. 4 Shape preserving interpolation (visual pleasing) using rational quartic Said-Ball (quartic/linear) with $\alpha_1 = 0.1, \alpha_i = 2, i = 2, 3, 4$

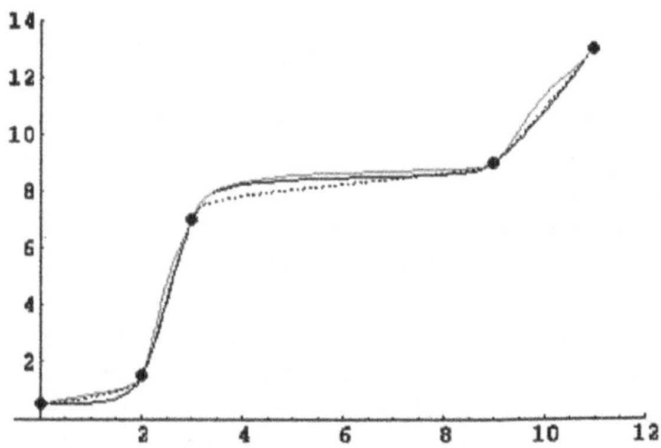

Fig. 5 Shape preserving interpolation using Sarfraz et al. [17] with $r_i = q_i = 1$ (solid), $r_i = 1, q_1 = 2$ (dashed) and $r_i = q_i = 2$ (gray) for data in Table 1

rational quartic Said-Ball function (quartic/linear) with one shape parameter is the best option compare to rational cubic function of Sarfraz et al. [17].

More complete results on shape preserving monotonic data by using rational quartic Said-Ball (quartic/linear) together with shape control analysis were reported in Karim [12].

Remark 3 The first derivative values, $d_i, i = 1, 2, \ldots, n$ can be estimated by using Butland's method [4]. This is since the simpler conditions to guarantee the monotonicity in (6) is to use region inside the box $[0, 2] \times [0, 2]$ as shown in Fig. 1. This is the sufficient condition in original Butland's construction of monotonic interpolation curves by using cubic spline [4]. It should be noted that, one of the main difference between our scheme and Wang and Tan [21] is that, we gives complete methods on how to choose the suitable derivative values that satisfy the sufficient conditions in (7) and (8). Furthermore, Wang and Tan [21] discussed their rational quartic scheme for C^2 monotonicity preserving whiles our scheme utilized rational quartic Said-Ball interpolant with one shape parameter for C^1 monotonicity preserving.

5 Conclusion

This paper has proposed new shape preserving interpolation by using rational quartic Said-Ball (quartic numerator and linear denominator) with one shape parameter. The parameter can be used to change the shape of the monotone curves. The sufficient condition for the monotonicity of the rational interpolant also has been derived. The proposed scheme may be used for data visualization purpose since these schemes require less computation, local and it also gives satisfactory results. Future work will focus on shape preserving for data in 2D and 3D. Finally, the work on shape preserving for monotone, positive and convex data by using GC^1 rational quartic Said-Ball spline is underway. Two early results for shape preserving interpolation by using rational quartic spline with GC^1 continuity can be found in [13, 14]. Such of the findings will be reported in future.

Acknowledgements This study was fully supported by Universiti Teknologi PETRONAS and Ministry of Education, Malaysia through Fundamental and Research Grant Scheme (FRGS): FRGS/1/2018/STG06/UTP/03/1/015MA0-020 and YUTP:0153AA-H24.

References

1. Abbas, M., Majid, A.A., Ali, J.M.: Monotonicity-preserving rational cubic spline for monotone data. Appl. Math. Comput. **219**, 2885–2895 (2012)
2. Delbourgo, R., Gregory, J.A.: Shape preserving piecewise rational interpolation. SIAM J. Sci. Stat. Comput. **6**, 967–976 (1985)
3. Delbourgo, R., Gregory, J.A.: The determination of derivative parameters for a monotonic rational quadratic interpolant. IMA J. Numer. Anal. **5**, 397–406 (1985)
4. Fritsch, F.N., Butland, J.: A method for constructing local monotone piecewise cubic interpolants. SIAM J. Sci. Stat. Comput. **5**, 300–304 (1984)
5. Fritsch, F.N., Carlson, R.E.: Monotone piecewise cubic interpolation. SIAM J. Numer. Anal. **17**, 238–246 (1980)
6. Duan, Q., Djidjeli, K., Price, W.G., Twizell, E.H.: Constrained control and approximation properties of a rational interpolating curve. Inf. Sci. **152**, 181–194 (2003)
7. Duan, Q., Wang, X., Cheng, F.: Constrained interpolation using rational cubic spline curve with linear denominators. Korean J. Comput. Appl. Math. **1**, 203–215 (1999)
8. Gregory, J.A., Delbourgo, R.: Piecewise rational quadratic interpolation to monotonic data. IMA J. Numer. Anal. **2**, 123–130 (1982)
9. Gregory, J.A.: Shape preserving interpolation. Comput. Aided Des. **18**(1), 53–57 (1986)
10. Goodman, T.N.T.: Shape preserving interpolation by curves. In: Levesley, J., Anderson, I.J., Mason, J.C. (eds.) Algorithms for Approximation IV. University of Huddersfield Proceedings, Huddersfield, pp. 24–35 (2002)
11. Hussain, M.Z., Hussain, M.: Visualization of data preserving monotonicity. Appl. Math. Comput. **190**, 1353–1364 (2007)
12. Karim, S.A.A.: Rational Said-Ball Bases for Shape Preserving Monotone Interpolation. Preprint (2013)
13. Karim, S.A.A., Kong, V.P.: Monotonicity preserving using rational quartic spline. AIP Conf. Proc. **1482**, 26–31 (2012). https://doi.org/10.1063/1.4757432
14. Karim, S.A.A., Kong, P.V., Hashim, I.: Positivity preserving using rational quartic spline. AIP Conf. Proc. **1522**, 518–525 (2013). https://doi.org/10.1063/1.4801170

15. Kvasov, B.I.: Methods of Shape-Preserving Spline Approximation. World Scientific, Singapore (2000)
16. Said, H.B.: Generalized Ball curve and its recursive algorithm. ACM Trans. Graphics **8**, 360–371 (1989)
17. Sarfraz, M., Butt, S., Hussain, M.Z.: Visualization of shaped data by a rational cubic spline interpolation. Comput. Graphics **25**, 833–845 (2001)
18. Sarfraz, M.: A rational cubic spline for visualization of monotonic data. Comput. Graphics **24**(4), 509–516 (2000)
19. Sarfraz, M.: A rational cubic spline for the visualization of monotonic data: an alternate approach. Comput. Graphics **27**, 107–121 (2003)
20. Shukri, F.N.A., Abbas, M., Bashir, U., Awang, M.N.Hj., Jamal, E., Ali, J.M.: Cubic Bézier constrained curve interpolation. J. Basic Appl. Sci. Res. **2**(4), 3682–3692 (2012)
21. Wang, Q., Tan, J.: Rational quartic involving shape parameters. J. Inf. Comput. Sci. **1**(1), 127–130 (2004)

Some Inequalities for σ-Convex Functions

Ibrahima Faye and Van Thien Nguyen

Abstract The main purpose of this article is to establish Karamata and Ostrowski type inequalities for σ-convex functions. An alternative definition of σ-convexity is also proposed in this paper. With the alternative definition, simpler proofs of Hermite-Hadamard and Karamata type inequalities are established.

Keywords σ-convex function · Karamata inequality · Ostrowski inequality · Hermite-Hadamard inequality

1 Introduction

The authors in [1] gave a new class of σ-convex set and σ-convex functions. Let σ be a strictly monotonic continuous function. A set $I \subset \mathbb{R}$ is said to be σ-convex set if

$$\sigma^{-1}[(1-t)\sigma(x) + t\sigma(y)] \in I, \forall x, y \in I, t \in [0, 1].$$

And a function $f : I \to \mathbb{R}$ is said to be σ-convex function if

$$f[\sigma^{-1}((1-t)\sigma(x) + t\sigma(y))] \leq (1-t)f(x) + tf(y),$$

$\forall x, y \in I, t \in [0, 1]$.

The notion σ-convex functions generalizes the concepts of geometric convexity, harmonic convexity, p-convexity, log-exponential convex functions that have been

I. Faye
Fundamental and Applied Sciences Department, Universiti Teknologi PETRONAS (UTP), Seri Iskandar, 32610 Seri Iskandar, Perak DR, Malaysia
e-mail: samsul_ariffin@utp.edu.my; ibrahima_faye@petronas.com.my

V. T. Nguyen (✉)
Department of Mathematics, FPT University, Education Zone, Hoa Lac High Tech Park, Thach That Ward, Hanoi, Vietnam
e-mail: thiennv15@fe.edu.vn

S. A. Abdul Karim et al. (eds.), *Proceedings of the 6th International Conference on Fundamental and Applied Sciences*, Springer Proceedings in Complexity,
https://doi.org/10.1007/978-981-16-4513-6_50

investigated before (see [2–4]). For each type of convexity considered, various integral inequailities have been investigated (see [5–7] for some examples. Similarly, the authors in [1] showed the integral inequality of Hermite-Hadamard type for σ-convex functions as it follows.

Theorem 1 *Suppose that* $f : [a, b] \to \mathbb{R}$ *is an integrable* σ*-convex function, then we have the following inequalities:*

$$f\left(\sigma^{-1}\left(\frac{\sigma(a)+\sigma(b)}{2}\right)\right) \leq \frac{1}{\sigma(b)-\sigma(a)} \int_a^b f(x)\sigma'(x)dx$$

$$\leq \frac{f(a)+f(b)}{2}.$$

In this note, we focus on proving the Karamata's inequality and Ostrowski's inequality for σ-convex functions.

2 Karamata's Inequality

Definition 1 A tuples $(x_1, x_2, \ldots, x_n)$ is σ-majorant of the tuples $(y_1, y_2, \ldots, y_n)$ ((x_i) is σ-majorant of (y_i) for short) if the following conditions are satisfied:

(i) $\sigma(x_1) \geq \sigma(x_2) \geq \cdots \geq \sigma(x_n)$, $\sigma(y_1) \geq \sigma(y_2) \geq \cdots \geq \sigma(y_n)$
(ii) $\sum_{k=1}^{i} \sigma(x_k) \geq \sum_{k=1}^{i} \sigma(y_k)$, $\forall i \in \{1, 2, \ldots, n-1\}$
(iii) $\sum_{k=1}^{n} \sigma(x_k) = \sum_{k=1}^{n} \sigma(y_k)$.

We have the Karamata's inequality type for σ-convex function

Theorem 2 *Suppose that* $x_i, y_i \in I, i = 1, \ldots, n$, (x_i) *is* σ*-majorant of* (y_i) *and a function* f *is* σ*-convex on* I*. Then we have*

$$\sum_{i=1}^{n} f(x_i) \geq \sum_{i=1}^{n} f(y_i). \tag{1}$$

To prove this theorem, we need the following lemma.

Lemma 1 *Let* $f : I \to \mathbb{R}$ *is* σ*-convex. Then the different quotient*

$$\frac{f(x)-f(y)}{\sigma(x)-\sigma(y)}$$

is non-decreasing with respect to $\sigma(x)$ *for fixed* y.

Proof We prove for σ is increasing on I (the case σ is decreasing can be proved similarly). Fix $y \in I$. We must show that for $x_1, x_2 \in I$ such that $\sigma(x_1) > \sigma(x_2)$ then

$$\frac{f(x_1) - f(y)}{\sigma(x_1) - \sigma(y)} \geq \frac{f(x_2) - f(y)}{\sigma(x_2) - \sigma(y)}. \tag{2}$$

We consider the following cases.
Case 1. $\sigma(x_1) > \sigma(y) > \sigma(x_2)$.
We have $\sigma(y) = t\sigma(x_1) + (1-t)\sigma(x_2)$ for some $t \in (0, 1)$.
Since f is σ-convex, we obtain that

$$\begin{aligned}(t+1-t)f(y) = f(y) &= \left[f\sigma^{-1}(\sigma(y))\right] \\ = f\left[\sigma^{-}1(t\sigma(x_1) + (1-t)\sigma(x_2))\right] &\leq tf(x_1) + (1-t)f(x_2).\end{aligned}$$

Hence

$$t(f(y) - f(x_1)) \leq (1-t)(f(x_2) - f(y)). \tag{3}$$

Moreover, we have $(t+1-t)\sigma(y) = t\sigma(x_1) + (1-t)\sigma(x_2)$. This shows that

$$t(\sigma(y) - \sigma(x_1)) = (1-t)(\sigma(x_2) - \sigma(y)). \tag{4}$$

Divide (3) by (4), inequality (2) is followed.
Case 2. $\sigma(x_1) > \sigma(x_2) > \sigma(y)$.
We have $\sigma(x_2) = t\sigma(x_1) + (1-t)\sigma(y)$. As above,

$$f(x_2) = f\left[\sigma^{-1}(t\sigma(x_1) + (1-t)\sigma(y))\right]$$

$$\leq tf(x_1) + (1-t)f(y).$$

This implies that

$$f(x_2) - f(y) \leq t(f(x_1) - f(y)). \tag{5}$$

In plus, since $\sigma(x_2) = t\sigma(x_1) + (1-t)\sigma(y)$,

$$\sigma(x_2) - \sigma(y) = t(\sigma(x_1) - \sigma(y)). \tag{6}$$

Divide (5) by (6), we obtain inequality (2).
Case 3. $\sigma(y) > \sigma(x_1) > \sigma(x_2)$.
We have $\sigma(x_1) = t\sigma(y) + (1-t)\sigma(x_2)$.

By σ-convexity of f,

$$\begin{aligned} f(x_1) =& f\left[f(\sigma^{-1}(\sigma(x_1)))\right] \\ =& f\left[\sigma^{-1}(t\sigma(y) + (1-t)\sigma(x_2))\right] \\ \leq& tf(y) + (1-t)f(x_2). \end{aligned}$$

Hence,

$$t(f(x_1) - f(y)) \leq (1-t)(f(x_2) - f(y)). \tag{7}$$

Since $\sigma(x_1) = t\sigma(y) + (1-t)\sigma(x_2)$, we have

$$t(\sigma(x_1) - \sigma(y) = (1-t)(\sigma(x_2) - \sigma(y)). \tag{8}$$

Devide (7) by (8), we get inequality (2).

Proof (Proof of Theorem 2) Set

$$a_i = \sum_{k=1}^{i} \sigma(x_k),\ b_i = \sum_{k=1}^{i} \sigma(y_k) \text{ for } i \in \{1, \ldots, n\}, \tag{9}$$

$$a_0 = b_0 = 0. \tag{10}$$

By the majorization assumption, we have

$$a_i \geq b_i \text{ for } i \in \{1, \ldots, n-1\} \text{ and } a_n = b_n.$$

For each $i \in \{1, \ldots, n\}$, put

$$c_i = \frac{f(x_i - f(y_i))}{\sigma(x_i) - \sigma(y_i)} \text{ for } i \in \{1, \ldots, n\}.$$

By Lemma 1, we have $c_{i+1} \leq c_i$, for each $i \in \{1, \ldots, n-1\}$. Hence

$$\begin{aligned}
&\sum_{k=1}^{n}(f(x_k) - f(y_k)) = \sum_{k=1}^{n} c_k(\sigma(x_k) - \sigma(y_k)) \\
&= \sum_{k=1}^{n} c_k\left[(a_k - a_{k-1}) - (b_k - b_{k-1})\right] \\
&= \sum_{k=1}^{n} c_k(a_k - b_k) - \sum_{k=1}^{n} c_k(a_{k-1} - b_{k-1}) \\
&= c_n(a_n - b_n) + \sum_{k=1}^{n-1}(c_k - c_{k+1})(a_k - b_k) - c_1(a_0 - b_0) \\
&= \sum_{k=1}^{n-1}(c_k - c_{k+1})(a_k - b_k) \geq 0.
\end{aligned}$$

Inequality (1) is proved.

Remark 1 If we take $\sigma(x) = x$ in Theorem 2, then we obtain the classical Karamata's theorem.

3 Ostrowski Inequality

The next inequality we want to mention here is Ostrowski inequality.

Theorem 3 *Let $f : I \to \mathbb{R}$ be an integrable function, and σ is a strictly monotonic function. If f is differentiable on the interior of I and if $|f'(t)| \leq M$ for all $t \in [a, b]$, where $a, b \in I^{\circ}$ then the following inequalities holds*

$$\left| f(x) - \frac{1}{\sigma(b) - \sigma(a)} \int_a^b f(t)\sigma'(t)dt \right| \leq \tag{11}$$

$$\frac{M}{\sigma(b) - \sigma(a)} \left[\int_a^x \sigma(t)dt + \int_b^x \sigma(t)dt + \sigma(a)(a - x) + \sigma(b)(b - x) \right]$$

Proof The proof follows the approach suggested in [8]. Assume that σ is strictly increasing. Using the fact that $-M < f' < M$ we obtain

$$(\sigma(t) - \sigma(a))(f'(t) + M) \geq 0 \tag{12}$$

and

$$(\sigma(b) - \sigma(t))(M - f'(t)) \geq 0 \tag{13}$$

By (12), we get

$$I = \int_a^x (\sigma(t) - \sigma(a))(f'(t) + M)dt \geq 0$$

Integrating by parts for above integral with $u = \sigma(t) - \sigma(a), du = \sigma'(t)$; $dv = (f'(t) + M)dt, v = f(t) + Mt$,

$$\begin{aligned}
I &= [(\sigma(t) - \sigma(a)] \left[f(t) + Mt)\right] \Big|_{t=a}^{a} - \int_a^x \sigma'(t)(f(t) + Mt)dt \\
&= [\sigma(x) - \sigma(a)] \left[f(x) + Mx\right] - \int_a^x \sigma'(t)f(t)dt - M \int_a^x t\sigma'(t)dt \\
&= [\sigma(x) - \sigma(a)]f(x) + Mx\sigma(x) - Mx\sigma(a) - \int_a^x \sigma'(t)f(t)dt \\
&\quad - M(t\sigma(t))\Big|_{t=a}^{x} + M \int_a^x \sigma(t)dt \\
&= [\sigma(x) - \sigma(a)]f(x) + Mx\sigma(x) - Mx\sigma(a) - \int_a^x \sigma'(t)f(t)dt \\
&\quad - Mx\sigma(x) + Ma\sigma(a) + M \int_a^x \sigma(t)dt \geq 0.
\end{aligned}$$

Using (13), we get

$$J = \int_x^b (\sigma(b) - \sigma(t))(M - f'(t))dt \geq 0.$$

Integrating by parts for J with $u = \sigma(b) - \sigma(t), du = -\sigma'(t)dt$; $dv = (M - f'(t))dt, v = Mt - f(t)$, we get

$$
\begin{aligned}
J =& [\sigma(b) - \sigma(t)]\left[Mt - f(t)\right]\Big|_{t=x}^{b} + \int_x^b Mt\sigma'(t)dt \\
& - \int_x^b \sigma'(t)f(t)dt \\
=& [\sigma(x) - \sigma(b)][Mx - f(x)] + M[t\sigma(t)]\Big|_{t=x}^{b} \\
& - M\int_x^b \sigma(t)dt - \int_x^b \sigma'(t)f(t)dt \\
=& Mx\sigma(x) - Mx\sigma(b) + f(x)(\sigma(b) - \sigma(x)) + Mb\sigma(b) \\
& - Mx\sigma(x) - M\int_x^b \sigma(t)dt - \int_x^b \sigma'(t)f(t)dt \geq 0.
\end{aligned}
$$

Hence

$$
\begin{aligned}
I + J =& f(x)(\sigma(b) - \sigma(a)) \\
& + M[\sigma(a)(a - x) + \sigma(b)(b - x)] \\
& + M\left[\int_a^x \sigma(t)dt + \int_b^x \sigma(t)dt\right] - \int_a^b \sigma'(t)f(t)dt \geq 0.
\end{aligned}
$$

This implies that

$$
\begin{aligned}
& f(x) - \frac{1}{\sigma(b) - \sigma(a)}\int_a^b \sigma'(t)f(t)dt \\
& \geq -\frac{M}{\sigma(b) - \sigma(a)}[\sigma(a)(a - x) + \sigma(b)(b - x) + p], \text{ where} \\
& p = \int_a^x \sigma(t)dt + \int_b^x \sigma(t)dt.
\end{aligned}
\tag{14}
$$

Again from $-M \leq f'(t) \leq M$ and σ is increasing we get

$$
K = \int_a^x (\sigma(t) - \sigma(a))(M - f'(t))dt \geq 0 \tag{15}
$$

and

$$L = \int_x^b (\sigma(b) - \sigma(t))(f'(t) + M)dt \geq 0. \tag{16}$$

As above, we are going to estimate K and L. Take integration by parts for K with $u = \sigma(t) - \sigma(a)$, $du = \sigma'(t)dt$; $dv = (M - f'(t))dt$, $v = Mt - f(t)$,

$$\begin{aligned}
K =& [\sigma(t) - \sigma(a)][Mt - f(t)]\Big|_{t=a}^{x} - M \int_a^x t\sigma'(t)dt \\
&+ \int_a^x \sigma'(t)f(t)dt \\
=& [\sigma(x) - \sigma(a)][Mx - f(x)] - M[t\sigma(t)]\Big|_{t=a}^{x} \\
&+ M \int_a^x \sigma(t)dt + \int_a^x \sigma'(t)f(t)dt \\
=& Mx\sigma(x) - Mx\sigma(a) - (\sigma(x) - \sigma(a))f(x) - Mx\sigma(x) \\
&+ Ma\sigma(a) + M \int_a^x \sigma(t)dt + \int_a^x \sigma'(t)f(t)dt \geq 0.
\end{aligned}$$

For L, integration by parts with $u = \sigma(b) - \sigma(t)$, $du = -\sigma'(t)$; $dv = (f'(t) + M)dt$, $v = f(t) + Mt$,

$$\begin{aligned}
L =& [\sigma(b) - \sigma(t)][f(t) + Mt]\Big|_{t=x}^{b} + \int_x^b \sigma'(t)f(t)dt + M \int_x^b t\sigma'(t)dt \\
=& [\sigma(x) - \sigma(b)][f(x) + Mx] - Mx\sigma(b) + \int_x^b \sigma'(t)f(t)dt \\
&+ M[t\sigma(t)]\Big|_{t=x}^{b} - M \int_x^b \sigma(t)dt \\
=& (\sigma(x) - \sigma(b))f(x) + Mx\sigma(x) - Mx\sigma(b) + Mb\sigma(b) - Mx\sigma(x) \\
&+ \int_x^b \sigma'(t)f(t)dt - M \int_x^b \sigma(t)dt \geq 0.
\end{aligned}$$

Hence,

$$K + L =(\sigma(a) - \sigma(b))f(x) + \int_a^b \sigma'(t)f(t)dt + M\,[\sigma(a)(a - x) + \sigma(b)(b - x)] + M\int_a^x \sigma(t)dt + M\int_b^x \sigma(t)dt \geq 0.$$

This lead us to the following inequality

$$f(x) - \frac{1}{\sigma(b) - \sigma(a)}\int_a^b \sigma'(t)f(t)dt \leq \frac{M}{\sigma(b) - \sigma(a)}\left[\sigma(a)(a - x) + \sigma(b)(b - x) + \int_a^x \sigma(t)dt + \int_b^x \sigma(t)dt\right]. \tag{17}$$

Combining (14) and (17) we get the inequality (11).
For σ strictly decreasing, using the facts

$$(\sigma(a) - \sigma(t))(f'(t) + M) \geq 0,$$
$$(\sigma(t) - \sigma(b))(M - f'(t)) \geq 0,$$
$$(\sigma(a) - \sigma(t))(M - f'(t)) \geq 0,$$
$$(\sigma(t) - \sigma(b))(f'(t) + M) \geq 0,$$

Theorem 3 could be proven following the same previous steps.

Remark 2 For $\sigma(x) = x$, by Theorem 3

$$\begin{aligned}
&\left|f(x) - \frac{1}{b-a}\int_a^b f(t)dt\right| \\
\leq&\frac{M}{b-a}\left[\frac{1}{2}t^2\Big|_a^x + \frac{1}{2}t^2\Big|_b^x + a^2 - ax + b^2 - bx\right] \\
\leq&\frac{M}{b-a}\left[\frac{1}{2}(x^2 - a^2 + b^2 - b^2) + \frac{1}{2}(2a^2 - 2ax + 2b^2 - 2bx)\right] \\
=&\frac{M}{2(b-a)}\left[(x-a)^2 + (x-b)^2\right],
\end{aligned}$$

which is the classic Ostrowski inequality.

4 Other Comments

We can define σ-convex function in a easy way as follows

Definition 2 Let $\sigma : I \to \mathbb{R}$ is a strictly monotonic function. A function $f : I \to \mathbb{R}$ is called σ-convex if the composition function

$$f \circ \sigma^{-1} : \sigma(I) \to \mathbb{R}$$

is convex function on $\sigma(I)$.

Based on this definition, we are going to give an easy way to prove Theorems 1 and 2.

Proof (Proof of Theorem 1) By the classical Hermite-Hadamard inequality,

$$(f \circ \sigma^{-1})\left(\frac{\sigma(a)+\sigma(b)}{2}\right) \leq \frac{1}{\sigma(b)-\sigma(a)} \int_{\sigma(a)}^{\sigma(b)} f \circ \sigma^{-1}(t)dt$$

$$\leq \frac{f \circ \sigma^{-1}(\sigma(a)) + f \circ \sigma^{-1}(\sigma(b))}{2}$$

By changing variable $t = \sigma(x)$,

$$\int_{\sigma(a)}^{\sigma(b)} f \circ \sigma^{-1}(t)dt = \int_{a}^{b} f(t)\sigma'(t)dt.$$

Hence, Theorem 1 is established.

Proof (Proof of Theorem 2) By the classical Karamata inequality, we obtain that

$$\sum_{i=1}^{n} f \circ \sigma^{-1}(\sigma(x_i)) \geq \sum_{i=1}^{n} f \circ \sigma^{-1}(\sigma(y_i)),$$

which shows Theorem 2.

Acknowledgements This research was partly done during V. T. Nguyen's visit to UTP in Malaysia, in 2019. He wishes to thank the members of the Department of Fundamental and Applied Sciences for their kind hospitality.

References

1. S. Wu, M. Awan, M. Noor, K. Noor, S. Iftikhar, On a new class of convex functions and integral inequalities. J. Inequalities Appl. **2019**, Article number 131 (2019)
2. Anderson, G.D., Vamanamurthy, M.K., Vuorinen, M.: Generalized convexity and inequalities. J. Math. Anal. Appl. **335**(2), 1294–1308 (2007)
3. Iscan, I.: Hermite-Hadamard type inequalities for harmonically convex functions. Hacet. J. Math. Stat. **43**(6), 935–942 (2014)
4. Zhang, K.S., Wan, J.P.: p-convex functions and their properties. Pure Appl. Math. **23**(1), 130–133 (2007)
5. Set, E., Akdemir, A.O., Ozdemir, M.E.: Simpson type inequalities for convex functions via Riemann-Liouville integrals. Filomat **31**(14), 4415–4420 (2017)
6. Alomari, M., Darus, M., Drogomir, S.S., Cerone, P.: Ostrowski type inequalities for functions whose derivatives are s-convex in the second sense. Appl. Math. Lett. **23**, 1071–1076 (2010)
7. Bai, Y.M., Wu, S.H., Wu, Y.: Hermite-Hadamard type integral inequalities for functions whose second-order mixed derivatives are coordinated $(s, m) - P$-convex. J. Funct. Spaces article ID 1693075 (2018)
8. Farid, G.: Straightforward proofs of Ostrowski inequality and some related results. Int. J. Anal. 3918483 (2016)

On the Space of m-subharmonic Functions

Samsul Ariffin Abdul Karim and Van Thien Nguyen

Abstract In this study, we investigate the space of m-subharmonic functions. We introduce the Mabuchi space of strictly m-subharmonic functions in an m-hyperconvex domain. We also study the metric properties of this space using Mabuchi geodesics. Some theoretical results are given in this study. We prove the connection between geodesics and homogeneous Hessian equations. The main result is important in the geometry of m-subharmonic functions.

Keywords Pluripotential theory · m-subharmonic function · Metric space

1 Introduction

The pluripotential theory is a developed nonlinear complex counterpart of classical potential theory. The basic objects of pluripotential theory are plurisubharmonic functions.

Caffarelli, Nirenberg and Spruck introduced a more general notions than sh and psh functions, to help us study both of them, which called m-subharmonic function. A smooth function u defined in open subset Ω of $\mathbb{C}^n$ is m-subharmonic if

$$\sigma_k(u) = \sum_{1\leq j_1<\cdots<j_k\leq n} \lambda_{j_1}\cdots\lambda_{j_k} \geq 0, \quad \forall k = 1, \ldots, m$$

S. A. Abdul Karim (✉)
Fundamental and Applied Sciences Department and Centre for Systems Engineering (CSE), Institute of Autonomous System, Universiti Teknologi PETRONAS (UTP), Bandar Seri Iskandar, 32610 Seri Iskandar, Perak, Malaysia
e-mail: samsul_ariffin@utp.edu.my

V. T. Nguyen
Department of Mathematics, FPT University, Education Zone, Hoa Lac High Tech Park Thach That Ward, Hanoi, Vietnam
e-mail: thiennv15@fe.edu.vn

S. A. Abdul Karim et al. (eds.), *Proceedings of the 6th International Conference on Fundamental and Applied Sciences*, Springer Proceedings in Complexity,
https://doi.org/10.1007/978-981-16-4513-6_51

where $(\lambda_{j_1}, \ldots, \lambda_{j_n})$ is the eigenvalue vector of the complex Hessian matrix of u. For non-smooth function u, m-subharmonicity of u can be defined if its regularization $u \star \rho_\epsilon$ is m-sh in $\Omega_\epsilon = \{z \in \Omega : distance(x, \partial\Omega) > \epsilon\}$. The complex Hessian operator $H_m(u) = (dd^c u)^m \wedge (dd^c|z|^2)^{n-m}$ of a locally bounded m-subharmonic function u is a Radon measure (see [1]). This operator generalized both the classical Laplacian and the complex Monge-Ampère operators. It has many applications in complex geometry (see e.g. [1–7]). The complex Hessian operator is much harder to study since it is neither linearly nor holomorphically invariant. For recent investigation of m-subharmonic function, please see [8–21].

We shall define a space of m-subharmonic functions along with the Mabuchi metric. We also introduce a geodesics between two points in this space and give an equivalent condition when a curve on the space is geodesics.

2 Space of Strictly m-subharmonic Functions

In this section, we shall define the metric in the space of strictly m-subharmonic functions on the m-hyperconvex domain. We introduction the readers to nice results [22, 23] for plurisubharmonic functions and Kähler manifolds settings.

Let Ω be an m-hyperconvex domain in $\mathbb{R}^n$. We are going to define the Mabuchi space of m-subharmonic functions on Ω.

Definition 1 The Mabuchi space of m-subharmonic functions in Ω is

$$\mathcal{M}_m := \{\varphi \in C^\infty(\Omega, \mathbb{R}) \cap SH_{m,+}(\Omega) : \varphi = 0 \text{ on } \partial\Omega\}, \tag{1}$$

where $SH_{m,+}(\Omega)$ is the set of all strictly m-subharmonic functions defined on Ω. And the tangent space of $\mathcal{M}_m$ at a point φ, denoted by $T_\varphi\mathcal{M}_m$ as followed

$$T_\varphi\mathcal{M}_m = \{\gamma'(0) : \gamma : [-\epsilon, \epsilon] \to \mathcal{M}_m, \gamma(0) = \varphi\}. \tag{2}$$

Proposition 1 *The tangent space of $\mathcal{M}_m$ at φ can be identified with*

$$T_\varphi\mathcal{M}_m \equiv \{u \in C^\infty(\Omega, \mathbb{R}) : u = 0 \ on \ \partial\Omega\}.$$

Proof Let $v \in C^\infty(\Omega, \mathbb{R})$, $v = 0$ on $\partial\Omega$. Put $\gamma(s) := \varphi + sv$. Then we have $\gamma(0) = \varphi$ and $\gamma'(0) = v$. Moreover, since φ is strictly m-subharmonic, we can choose s is close enough to 0 that $\gamma(s) \in \mathcal{M}_m$, which implies that $v \in T_\varphi\mathcal{M}_m$.

Conversely, let $\gamma : [-\epsilon, \epsilon] \to \mathcal{M}_m$. We have that $\gamma(s) \restriction_{\partial\Omega} = 0$ for every $s \in [-\epsilon, \epsilon]$. This implies that $\gamma'(0) \restriction_{\partial\Omega} = 0$. Therefore,

$$v = \gamma'(0) \in \{u \in C^\infty(\Omega, \mathbb{R}) : u = 0 \text{ on } \partial\Omega\}.$$

Definition 2 We define the Mabuchi metric on $T_\varphi \mathcal{M}_m$ for each $\varphi \in \mathcal{M}_m$ as follows

$$\langle v_1, v_2\rangle_\varphi : = \int_\Omega v_1 v_2 H_m(\varphi),$$

where $v_1, v_2 \in T_\varphi \mathcal{M}_m$.

Definition 3 The energy functional on $\mathcal{M}_m$ is defined by

$$E(\varphi_t) = \frac{1}{2}\int_0^1 \int_\Omega (\dot{\varphi}_t)^2 H_m(\varphi_t),$$

where φ_t is a path in $\mathcal{M}_m$ joining φ_0 and φ_1.

Definition 4 Geodesics between two points φ_0, φ_1 in $\mathcal{M}_m$ are defined as the extremals of the energy functional of a path joining φ_0 to φ_1.

Theorem 1 *The geodesics equation is obtained by computing the Euler-Lagrange equation of the functional E as following*

$$\ddot{\varphi}_t H_m(\varphi_t) = m d\dot{\varphi}_t \wedge d^c \dot{\varphi}_t \wedge (dd^c \varphi_t)^{m-1} \wedge \beta^{n-m}. \tag{3}$$

Proof Let $(\phi_{s,t})$ be a variation of φ_t, it means that

$$\phi_{0,t} = \varphi_t, \phi_{s,0} = \varphi_0, \phi_{s,1} = \varphi_1 \quad \text{and} \quad \phi_{s,t} = 0 \quad \text{on } \partial\Omega.$$

We put $\psi_t := \frac{\partial \phi}{\partial s} \restriction_{s=0}$. We can see that $\psi_0 \equiv \psi_1 \equiv 0$ and $\psi_t = 0$ on $\partial\Omega$. Thus,

$$\phi_{s,t} = \varphi_t + s\psi_t + o(s), \quad \frac{\partial \phi_{s,t}}{\partial t} = \dot{\varphi}_t + s\dot{\psi}_t + o(s),$$

and

$$\begin{aligned} H_m(\phi_{s,t}) &= [dd^c(\varphi_t + s\psi_t)]^m \wedge \beta^{n-m} + o(s) \\ &= H_m(\varphi_t) + smdd^c\psi_t \wedge (dd^c\varphi_t)^{m-1} \wedge \beta^{n-m} + o(s). \end{aligned}$$

We can compute

$$E(\phi_{s,t}) = \frac{1}{2}\int_0^1\int_\Omega \left(\frac{\partial \phi_{s,t}}{\partial t}\right)^2 H_m(\phi_{s,t})$$
$$= \frac{1}{2}\int_0^1\int_\Omega (\dot{\varphi}_t + s\dot{\psi}_t)^2\left[H_m(\varphi_t) + smdd^c\psi_t \wedge (dd^c\varphi_t)^{m-1} \wedge \beta^{n-m}\right]dt + o(s) \tag{4}$$
$$= E(\varphi_t) + s\int_0^1\int_\Omega \dot{\varphi}_t\dot{\psi}_t H_m(\varphi_t)dt$$
$$+ \frac{sm}{2}\int_0^1\int_\Omega (\dot{\varphi}_t)^2 dd^c\psi_t \wedge (dd^c\varphi_t)^{m-1} \wedge \beta^{n-m}dt + o(s).$$

We have known that $\psi_0 \equiv \psi_1 \equiv 0$. Using the integration by parts, we obtain that

$$\int_0^1\int_\Omega \dot{\varphi}_t\dot{\psi}_t H_m(\varphi_t)dt$$
$$= -\int_0^1\int_\Omega \psi_t\left[d\dot{\varphi}_t \wedge d^c\dot{\varphi}_t + \dot{\varphi}_t dd^c\dot{\varphi}_t\right] \wedge (dd^c\varphi_t)^{m-1} \wedge \beta^{n-m}dt. \tag{5}$$

Moreover, by Stokes formula,

$$\int_0^1\int_\Omega (\dot{\varphi}_t)^2 dd^c\psi_t \wedge (dd^c\varphi_t)^{m-1} \wedge \beta^{n-m}dt$$
$$= 2\int_0^1\int_\Omega \psi_t(d\dot{\varphi}_t \wedge d^c\dot{\varphi}_t + \dot{\varphi}_t dd^c\dot{\varphi}_t) \wedge (dd^c\varphi_t)^{m-1} \wedge \beta^{n-m}dt. \tag{6}$$

Combining (4), (5) and (6), we get that

$$E(\phi_{s,t}) = E(\varphi_t) + s\int_0^1\int_\Omega \psi_t\left[-\ddot{\varphi}_t(dd^c\varphi_t)^m + md\dot{\varphi}_t \wedge d^c\dot{\varphi}_t\right.$$
$$\left.\wedge (dd^c\varphi_t)^{m-1}\right] \wedge \beta^{n-m}dt + o(s).$$

Hence,

$$\lim_{s\to 0}\frac{E(\phi_{s,t})-E(\varphi_t)}{s}$$
$$=\int_0^1\int_\Omega \psi_t\left[-\ddot{\varphi}_t(dd^c\varphi_t)^m+md\dot{\varphi}_t\wedge d^c\dot{\varphi}_t\wedge(dd^c\varphi_t)^{m-1}\right]\wedge\beta^{n-m}dt.$$

Therefore, (φ_t) is a critical point of E if and only if

$$\ddot{\varphi}_t H_m(\varphi_t)=md\dot{\varphi}_t\wedge d^c\dot{\varphi}_t\wedge(dd^c\varphi_t)^{m-1}\wedge\beta^{n-m}.$$

3 Geodesics and Homogeneous Hessian Equations

In this section, we shall give a bridge connecting geodesics and homogeneous Hessian equations.

For each point $(\varphi_t), t\in[0,1]$ in $\mathcal{M}_m$, we set

$$\Phi(z,\zeta)=\varphi_t(z), z\in\Omega$$

and

$$\zeta=e^{t+is}\in T=\{\xi\in\mathbb{C}\colon 1<|\xi|<e\}.$$

We shall prove that

Theorem 2 *A curve $(\varphi_t)_{0\le t\le 1}$ is a geodesic if and only if $[dd^c_{z,\zeta}\Phi(z,\zeta)]^{m+1}\wedge\beta^{n-m}(z,\zeta)=0$, where $\beta(z,\zeta)=dd^c|(z,\zeta)|^2$.*

Proof We see that

$$d_{z,\zeta}\Phi(z,\zeta)=d_z\Phi+d_\zeta\Phi,\ d^c_{z,\zeta}\Phi=d^c_z\Phi+d^c_\zeta\Phi.$$

Now, computing by hands,

$$\begin{aligned}dd^c_{z,\zeta}\Phi&=(d_z+d_\zeta)(d^c_z\Phi+d^c_\zeta\Phi)\\&=d_zd^c_z\Phi+d_zd^c_\zeta\Phi+d_\zeta d^c_z\Phi+d_\zeta d^c_\zeta\Phi\\&=d_zd^c_z\Phi+R(z,\zeta),\end{aligned}$$

where $R(z,\zeta) = d_z d^c{}_\zeta \Phi + d_\zeta d^c_z \Phi + d_\zeta d^c{}_\zeta \Phi$. Thus,

$$\begin{aligned}
&(dd^c_{z,\zeta}\Phi)^{m+1} \wedge \beta^{n-m}(z,\zeta) \\
&= [d_z d^c_z \Phi + R(z,\zeta)]^{m+1} \wedge \beta^{n-m}(z,\zeta) \\
&= \sum_{k=0}^{m+1} \binom{m+1}{k} (d_z d^c_z \Phi)^k \wedge R^{m+1-k} \wedge \beta^{n-m}(z,\zeta) \\
&= (d_z d^c_z \Phi)^{m+1} \wedge \beta^{n-m}(z,\zeta) \\
&\quad + (m+1)(d_z d^c_z \Phi)^m \wedge R \wedge \beta^{n-m}(z,\zeta) \\
&\quad + \frac{m(m+1)}{2}(d_z d^c_z \Phi)^{m-1} \wedge R^2 \wedge \beta^{n-m}(z,\zeta).
\end{aligned} \tag{7}$$

We have used the fact that $R^3 = R \wedge R \wedge R = 0$.

We shall using the techniques in [22] to show that the right hand side of (7) is zero.

The first term in the right hand side of (7) equal to 0 by the bi-degree computation. Otherwise, we have

$$\begin{aligned}
d_\zeta \Phi &= \partial_\zeta \Phi + \bar{\partial}_\zeta \Phi = \frac{\partial \Phi}{\partial \zeta} d\zeta + \frac{\partial \Phi}{\partial \bar{\zeta}} = \dot{\varphi}_t(z)(d\zeta + d\bar{\zeta}), \\
d^c_\zeta \Phi &= \frac{i}{2}\left[\frac{\partial \Phi}{\partial \bar{\zeta}} d\bar{\zeta} - \frac{\partial \Phi}{\partial \zeta} d\zeta\right] = \frac{i}{2}\dot{\varphi}_t(z)(d\zeta - d\bar{\zeta}), \\
d_\zeta d^c_\zeta \Phi &= i\ddot{\varphi}_t(z) d\zeta \wedge d\bar{\zeta}.
\end{aligned}$$

These imply that

$$\begin{aligned}
R = R(z,\zeta) &= i\ddot{\varphi}_t d\zeta \wedge d\bar{\zeta} + \frac{i}{2} d_z \dot{\varphi}_t \wedge d\bar{\zeta} \\
&\quad - \frac{i}{2} d_z \dot{\varphi}_t \wedge d\zeta + d^c_z \dot{\varphi}_t \wedge d\zeta + d^c_z \dot{\varphi}_t \wedge d\bar{\zeta}, \\
R^2 &= 2i d_z \dot{\varphi}_t \wedge d^c_z \dot{\varphi}_t \wedge d\zeta \wedge d\bar{\zeta}.
\end{aligned}$$

Hence the second term in the right hand side of (7) is computed as

$$\begin{aligned}
&(d_z d^c_z \Phi)^m \wedge R \wedge \beta^{n-m}(z,\zeta) \\
&= \left[d_z d^c_z \varphi_t(z)\right]^m \wedge \left[i\ddot{\varphi_t}(z) d\zeta \wedge d\bar{\zeta} + \frac{i}{2} d_z \dot{\varphi}_t \wedge d\bar{\zeta}\right. \\
&\quad \left. - \frac{i}{2} d_z \dot{\varphi}_t \wedge d\zeta + d^c_z \dot{\varphi}_t \wedge d\zeta + d^c_z \dot{\varphi}_t \wedge d\bar{\zeta}\right] \wedge \beta^{n-m}(z,\zeta) \\
&= i\ddot{\varphi_t}(z)(d_z d^c_z \varphi_t(z))^m \wedge d\zeta \wedge d\bar{\zeta} \wedge \beta^{n-m}(z).
\end{aligned} \tag{8}$$

For the third term, we compute

$$\begin{aligned}(d_z d_z^c \Phi)^{m-1} &\wedge R^2 \wedge \beta^{n-m}(z,\zeta)\\ &= (d_z d_z^c \varphi_t(z))^{m-1} \wedge 2i d_z \dot{(\varphi_t)} \wedge d_z^c \dot{\varphi}_t \wedge d\zeta \wedge d\bar{\zeta}\\ &= -2i(d_z d_z^c \varphi_t(z))^{m-1} \wedge d_z \dot{\varphi}_t \wedge d_z^c \dot{\varphi}_t \wedge d\zeta \wedge d\bar{\zeta} \wedge \beta^{n-m}(z).\end{aligned} \tag{9}$$

From (8) and (9), expression (7) is rewritten by

$$\begin{aligned}(dd^c_{z,\zeta}\Phi)^{m+1} &\wedge \beta^{n-m}(z,\zeta)\\ &= (m+1)i\big[\ddot{\varphi}_t (d_z d_z^c \varphi_t)^m \wedge \beta^{n-m}(z)\\ &\quad - m(d_z d_z^c \varphi_t)^{m-1} \wedge d_z \dot{\varphi}_t \wedge d_z^c \dot{\varphi}_t \wedge \beta^{n-m}(z)\big] \wedge d\zeta \wedge d\bar{\zeta}.\end{aligned}$$

By Theorem 1, we obtain

$$(dd^c_{z,\zeta}\Phi)^{m+1} \wedge \beta^{n-m}(z,\zeta) = 0$$

if and only if the curve $(\varphi_t)_{0\le t\le 1}$ is a geodesic. These concludes the prove of Theorem 2.

4 Conclusion

In this study we have established a connection between geodesics and homogeneous Hessian equations. This connection is important because the homogeneous Hessian equations has been investigated and it is easy to check when an m-subharmonic function is maximal, while geodesics notation in the space of m-subharmonic functions belongs to geometric category. This connection helps us more understand about m-subharmonic functions as well as their geometry.

References

1. Błocki, Z.: Weak solutions to the complex Hessian equation. Ann. Inst. Fourier (Grenoble) **55**(5), 1735–1756 (2005)
2. Åhag, P., Czyż, R., Hed, L.: The geometry of m-hyperconvex domains. J. Geo. Anal. **28**(4), 3196–3222 (2018)
3. Åhag, P., Czyż, R., Hed, L.: Extension and approximation of m-subharmonic functions. Complex. Var. Elliptic. Equ. **63**(6), 783–801 (2018)
4. Dinew, S., Kołodziej, S.: A priori estimates for the complex Hessian equations. Anal. PDE. **1**, 227–244 (2014)
5. Dinew, S., Kołodziej, S.: Liouville and Calabi-Yau type theorems for complex Hessian equations. Am. J. Math. **139**, 403–415 (2017)

6. Li, S.Y.: On the Dirichlet problems for symmetric function equations of the eigenvalues of the complex Hessian. Asian J. Math. **8**, 87–106 (2004)
7. Lu, H.C.: Complex Hessian equations. Doctoral thesis. University of Toulouse III Paul Sabatier (2012)
8. Czyż, R., Nguyen, V.T.: On a constant in the energy estimate. C. R. Math. **355**(10), 1050–1054 (2017)
9. Lu, H.C.: Solutions to degenerate complex Hessian equations. J. Math. Pures Appl. (9) **100**(6), 785–805 (2013)
10. Lu, H.C.: Viscosity solutions to complex Hessian equations. J. Funct. Anal. **264**(6), 1355–1379 (2013)
11. Nguyen, N.C.: Subsolution theorem for the complex Hessian equation. Univ. Iagel. Acta Math. **50**, 69–88 (2013)
12. Nguyen, N.C.: Hölder continuous solutions to complex Hessian equations. Potential Anal. **41**(3), 887–902 (2014)
13. Nguyen, V.T.: On delta m-subharmonic functions. Ann. Polon. Math **118**, 25–49 (2016)
14. Nguyen, V.T.: Maximal m-subharmonic functions and the Cegrell class $\mathcal{N}_m$. Indag. Math. **30**(4), 717–739 (2019)
15. Nguyen, V.T.: A characterization of Cegrell's classes and generalized m-capacities. Ann. Polon. Math **121**, 33–43 (2018)
16. Nguyen, V.T.: Hessian boundary measures. Int. J. Math. 30(3), 1950016 (2019)
17. Nguyen, V.T.: The convexity of radially symmetric m-subharmonic functions. Complex. Var. Elliptic. Equ. **63**(10), 1396–1407 (2018)
18. Sadullaev, A., Abdullaev, B.: Potential theory in the class of m-subharmonic functions. Tr. Mat. Inst. Im. V. A. Steklova **279**, 166–192 (2012)
19. Yong, H., Lu, X.: Regularity of radial solutions to the complex Hessian equations. J. Partial Differ. Equ. **23**(2), 147–157 (2010)
20. Wan, D., Wang, W.: Complex Hessian operator and Lelong number for unbounded m-subharmonic functions. Potential. Anal. **44**(1), 53–69 (2016)
21. Wang, X.-J.: The k-Hessian equation. Lect. Not. Math. **1977** (2009)
22. Abja, S.: Geometry and topology of the space of plurisubharmonic functions. J. Geo. Anal. **29**(1), 510–541 (2019)
23. Darvas, T.: The Mabuchi completion of the space of Kähler potentials. Am. J. Math. **139**(5), 1275–1313 (2017)

Unsteady Nanofluid Flow Over Exponentially Stretching Sheet with Vertical Throughflow

Abdullah Al-Yaari, Hamzah Sakidin, Nooraini Zainuddin, and Ishak Hashim

Abstract Owing to its excellent thermal conductivity, the use of nanofluid as a heat transfer agent has become more popular recently. The study of heat transfer is critical in keeping up with the rapid development of industries such as cooling and advanced the nuclear technology. The purpose of this research is to investigate the unsteady flow, heat, and mass transfer of nanofluid over exponentially stretching sheet. This research's mathematical model manages the nanoparticles passively at the boundary, rather than actively making it physically more realistic. A convective term is also added to the energy equation due to the vertical throughflow. The system of partial differential equations is reduced to nonlinear ordinary differential equations using similarity transformation and then solved using bvp4c solver in MATLAB numerically. The skin friction coefficient, heat transfer rate, velocity, temperature, and nanoparticle volume fraction are computed for various values of relevant parameters involved. Results show that the Brownian motion parameter has no effect on the heat transfer rate. The reduced Nusselt number shows decreasing trend for the thermophoresis parameter and Scmidt number but increases for Prandtl number.

Keywords Unsteady · Nanofluid · Passive control · Vertical throughflow

---

A. Al-Yaari · H. Sakidin (✉) · N. Zainuddin
Fundamental and Applied Sciences Department, Universiti Teknologi PETRONAS, 32610 Seri Iskandar, Perak, Malaysia
e-mail: hamzah.sakidin@utp.edu.my

A. Al-Yaari
e-mail: abdullah_20001447@utp.edu.my

N. Zainuddin
e-mail: aini_zainuddin@utp.edu.my

I. Hashim
School of Mathematical Sciences, Universiti Kebangsaan Malaysia, 43600 Bangi, Selangor, Malaysia
e-mail: ishak_h@ukm.edu.my

S. A. Abdul Karim et al. (eds.), *Proceedings of the 6th International Conference on Fundamental and Applied Sciences*, Springer Proceedings in Complexity,
https://doi.org/10.1007/978-981-16-4513-6_52

1 Introduction

The use of nanofluids to improve heat transfer over the last decade has caught the attention of many scientists. Choi [1] was the first to use the term' nanofluid' referring to the dispersion of particles of nanometer size (less than 100 nm in diameter) in a base fluid such as water, ethylene glycol, propylene glycol, toluene, and oil to increase their thermal conductivity. This makes nanofluids very useful for many engineering applications including cooling and heat exchange. Nanofluid applications in engineering include such areas as microelectronics, coolant, fuel cells, pharmaceutical processes, household refrigerator, and chiller. The ultrafine particles are usually formed from a conductive material via a high-energy-pulsed process. These nanoparticles typically consisting of metals, oxides, carbides, or carbon nanotubes, Oztop et al. [2]. Research has shown that the dispersal of different types of nanoparticles into a base fluid can enhance the fluid's thermal properties, Das et al. [3]. Today, with the rapid progress in nanomanufacturing, various economic combinations of base fluid/nan particles are available. Das et al. [4] stated that the addition to standard heat transfer fluids of a small amount (less than 1 percent by volume) of nanoparticles approximately doubled their thermal conductivity.

Some explications for the improved nanofluid heat transfer characteristics have been proposed. Pak et al. [5] stated that the heat transfer enhancement discovered in nanofluids was due to the dispersion of suspended particles in the base fluid while Xuan et al. [6] suggested that the heat transfer coefficient increase was due to the intensified turbulence produced by the movement of nanoparticles. On the other hand, experimental data on water-glycerin nanofluid has shown that the heat transfer enhancement is the result of rotation of nanoparticles, Ahuja [7]. Nevertheless, after a thorough evaluation, Buongiorno [8] found that the high heat transfer coefficients of nanofluids were not adequately explained by dispersion of nanoparticles, increase in turbulence, or rotation of nanoparticles. Buongiorno [8] proposed a nanofluid convective transport model that considers Brownian diffusion and thermophoresis and explained the high heat transfer coefficient is due to a substantial reduction in the viscosity of the fluid due to a large difference in the boundary layers temperature.

With the recent advancements in nanotechnology, there are two effective models for convective heat transfer in nanofluids. The first modelling was developed by Buongiorno [8] who discovered that absolute velocity of nanoparticles could be regarded as the sum of the base fluid velocity and a slip velocity, with a total of seven slip mechanisms involved: inertia, Brownian diffusion, thermophoresis, diffusiophoresis, Magnus effect, fluid drainage, and gravity settling. Only Brownian motion and thermophoresis are found vital to the development of the convective heat transfer model in nanofluids, of all the slip mechanisms. Tiwari et al. [9] pioneered the second nanofluid model which proposed a model incorporating the effect of the nanoparticle volume fraction on heat transfer enhancement. Many researchers have applied those two well-known nanofluid models. For example, the model proposed by Buongiorno [8] was used by Nield et al. [10] and Kuznetsov et al. [11], while Bachok et al. [12] and Bachok et al. [13] applied the model proposed by Tiwari et al. [9]. Many nanofluid papers, such as the papers by Xuan et al. [6], Buongiorno [8], Tiwari et al. [9], Xuan et al. [14] and Eastman et al. [15], have also been published. It is also worth mentioning that references to nanofluids can be found in Das et al. [3] and Neild et al. [16] articles, and Trisaksri et al. [17]-Mahian et al. [18] review papers, etc. Ellahi et al. [19] recently studied Copper–water nanofluid entropy generation with variable shapes of nanosized particles. More recently, by applying the mesoscopic method, Sheikholeslami [20] found numerical solution for the flow of magnetohydrodynamic (MHD) in copper–water nanofluid past a permeable canal.

The model from Buongiorno [8] has been used by a considerable number of researchers to boost heat transfer in nanofluid with various physical aspects. They nevertheless constantly tried to control the concentration of nanoparticles on the surface. Kuznetsov et al. [21] were the first to eradicate this limitation by changing the fractional boundary state of the nanoparticles to passive rather than active control and named it "revised model." They also claimed that the changed boundary condition is more physically feasible because in real-life application, it is difficult to actively monitor the nanoparticle fraction at the boundary. In this revised model, the surface nanoparticles or mass flux are weighted to be zero and the fraction of the nanoparticles on the boundary is passively controlled rather than actively controlled, independent of the imposed temperature. Later, they reviewed and revised their previous works by implementing the same approach for various physical aspects as in Neild et al. [22] and Kuznetsov et al. [23]. In their studies it was found that due to zero nanoparticle flux on the boundary, the Nusselt number or heat transfer rate became negligible to the Brownian motion parameter. This founding means that Brownian motion's contribution to the equation of thermal energy is now tending to zero. They also reported a decreasing function of the hat transfer rate for the thermophoresis parameter.

Following Kuznetsov et al. [21], different researchers revisited their works and revised their models using the Buongiorno model. A couple of significant works are listed here. Zaimi et al. [24] revised his work on nanofluid flow as a result of a suction stretching / shrinking surface. Rana et al. [25] and Chand et al. [26] revised their work for the elastic Rivlin-Ericksen and the viscoelastic nanofluid models of Walter, respectively. Rana et al. [27] used the revised model to passively control the surface volume fraction of the nanoparticles and reported that the heat transfer rate became almost independent at given Prandtl and Lewis numbers of values. Yadav et al. [28] revisited their rotating nanofluid layer thermal instability model and found that zero mass flux at the boundary has a more destabilizing effect than fixed boundary conditions for alumina-water nanofluid and the opposite effect for copper–water nanofluid is observed. Recently Jahan et al. [29] revised the flow model due to moving surface in a nanofluid by considering passive regulation of concentration of nanoparticles at the boundaries. Uddin et al. [30] recently used the updated model to investigate the effect of buoyancy on MHD stagnation point flow due to a convectively heated stretching layer in a nanofluid.

Work on throughflow is important as it offers the opportunity to monitor possible convective instability through a modification of the throughflow. Wooding [31] was the founder of the principle of throughflow. He contemplated instability of the Rayleigh convection in a permeable, throughflow medium. Sutton [32] subsequently considered the convective flow in a two-dimensional, permeable channel when a net fluid flow through the channel occurs. Nield [33] subsequently studied the effect of throughflow on convection with sloped temperature slopes. Nield et al. [34] subsequently made a reference to their previous work Nield [33]. He then called the correction that there should be a convective portion of mass flux in the presence of throughflow. He thus modified the equation of concentration conservation by adding a convective term to it. Barletta et al. [35] recently investigated the effect of vertical throughflow of a non-Newtonian power law fluid on the initiation of convection in a horizontally saturated permeable layer of fluid. Recently, Pop et al. [36] adopted the work of Nield et al. [34] and updated the work of Khan et al. [37] by adding the convective to the conservation equation of nanoparticles and finding that there are two solutions to this equation.

Buongiorno [8] developed a model combining the impact of parameter Brownian motion and thermophoresis. Since then, several researchers have used the Buongiorno model [8] to study the various flows and heat transfer problems in nanofluid. Nonetheless, the model has one drawback which is the active regulation of nanoparticles flux at the boundary which is very difficult to implement realistically. Nield et al. [34] subsequently made a correction to their previous work Kuznetsov et al. [21] and included the convective term in the conservation of the mass equation, as it is found that a convective term of the mass flux of nanoparticles should exist in the presence of throughflow.

Therefore, the goal of this work is to build a mathematical model for unstable flow, heat and mass transfer of a nanofluid through an exponentially stretching surface, using a more practical boundary condition with vertical throughflow. This work will examine the effect of vertical throughflow and zero mass flux boundary conditions on various flow parameters, including velocity, temperature and concentration profiles as well as the skin friction coefficient and the reduced Nusselt number.

2 Mathematical Formulation

Consider an unsteady three-dimensional boundary layer flow of a viscous and incompressible nanofluid past a permeable stretching/shrinking surface with suction/injection. A set of coordinates (x, y, z) is measured normal to the plate. The x- and y-coordinates are in the plane of the surface, while the z-coordinate is perpendicularly measured to the surface with the flow confined to $z \geq 0$. It is assumed that the plate is stretching continuously in both x- and y-directions with the velocities $u_w(x, t)$ and $v_w(y, t)$ respectively. It is also assumed that T_w is the surface wall temperature and T_∞ is the ambient temperature while C_w and C_∞ are the nanoparticle volume fraction on the surface and away from the surface respectively. Figure 1 shows the physical model and coordinate system for a flow past a stretching sheet.

The governing equations for conservation of mass, momentum, energy, and concentration can be written in vector form as:

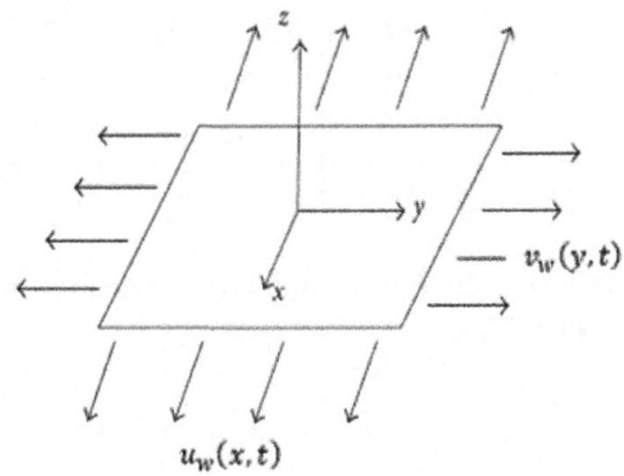

Fig. 1 Physical model and coordinate system

$$\nabla.(v) = 0 \tag{1}$$

$$\frac{\partial v}{\partial t} + (\nabla.v)v = \frac{1}{\rho}\nabla P + \upsilon\nabla^2 v \tag{2}$$

$$\frac{\partial T}{\partial t} + (\nabla.v)T = \alpha\nabla^2 T + \beta\left[D_B\nabla C + (C - C_\infty)v + \frac{D_T}{T_\infty}\nabla\right]\nabla.T \tag{3}$$

$$\frac{\partial C}{\partial t} + (v.\nabla)C = D_B\nabla^2 C + \frac{D_T}{T_\infty}\nabla^2 T \tag{4}$$

where $\boldsymbol{v}$ is the velocity vector, T is the nanofluid temperature, C is the nanoparticle volume fraction, t is time, p is the pressure, ρ is the density, α is the thermal diffusivity of the nanofluid, υ is the kinematic viscosity of the fluid, D_B is the Brownian diffusion coefficient, D_T is the thermophoresis diffusion coefficient, $\beta = (\rho c_p)_p/(\rho c_p)_f$, where $(\rho c_p)_p$ is the effective heat capacity of the nanoparticle material and $(\rho c_p)_f$ is the heat capacity of the fluid. It should be mentioned that, according to [36], the expression inside the square brackets in (3) represents the nanoparticle flux. It is noted the inclusion of a convective term $(C - C_\infty) \cdot v\nabla T$ is due to the vertical throughflow, which was ignored in most of previously published papers on stretching/shrinking sheets using the mathematical nanofluid model proposed by [8].

Under the boundary layer assumptions, (1)–(4) can be written in the Cartesian coordinates x, y and z as:

$$\frac{\partial u}{\partial x} + \frac{\partial v}{\partial y} + \frac{\partial w}{\partial z} = 0 \tag{5}$$

$$\frac{\partial u}{\partial t} + u\frac{\partial u}{\partial x} + v\frac{\partial u}{\partial y} + w\frac{\partial u}{\partial z} = \upsilon\frac{\partial^2 u}{\partial z^2} \tag{6}$$

$$\frac{\partial v}{\partial t} + u\frac{\partial v}{\partial x} + v\frac{\partial v}{\partial y} + w\frac{\partial v}{\partial z} = \upsilon\frac{\partial^2 v}{\partial z^2} \tag{7}$$

$$\frac{\partial T}{\partial t}+u\frac{\partial T}{\partial x}+v\frac{\partial T}{\partial y}+w\frac{\partial T}{\partial z}=\alpha\frac{\partial^2 T}{\partial z^2}$$
$$+\beta\left[D_B\frac{\partial C}{\partial z}\frac{\partial T}{\partial z}+(C-C_\infty)v\frac{\partial T}{\partial z}+\frac{D_T}{T_\infty}\left(\frac{\partial T}{\partial z}\right)^2\right] \tag{8}$$

subject to the following boundary conditions:

For $t < 0$;

$$\begin{aligned}&u=0,\quad v=0,\quad w=0,\\ &T=T_\infty,\quad C=C_\infty,\quad \text{for any } x, y, z\end{aligned} \tag{9}$$

For $t \geq 0$;

$$\begin{aligned}&u=\frac{\lambda U_0}{1-\omega t}e^{\frac{x+y}{L}},\quad v=\frac{\lambda V_0}{1-\omega t}e^{\frac{x+y}{L}}\\ &w=0,\quad T=T_w\\ &D_B\frac{\partial C}{\partial z}+\frac{D_T}{T_\infty}\frac{\partial T}{\partial z}=0\quad \text{at } z=0\\ &u\to 0,\quad v\to 0,\quad w\to 0,\\ &T\to T_\infty,\quad C\to C_\infty\quad \text{as}\quad z\to\infty\end{aligned} \tag{10}$$

where λ is a stretching/shrinking parameter with $\lambda > 0$ corresponds to stretching while $\lambda < 0$ corresponds to shrinking surface. U_0 and V_0 are constants while L is the reference length. ω is the parameter showing the unsteadiness of the problem. The boundary condition $D_B\frac{\partial C}{\partial z}+\frac{D_T}{T_\infty}\frac{\partial T}{\partial z}=0$ at $z=0$ implies that, with Brownian motion and thermophoresis taken into account, the nanoparticles flux at the boundary is zero.

A similarity transformation is introduced as

$$\begin{aligned}&u=\frac{\lambda U_0}{1-\omega t}e^{\frac{x+y}{L}}f'(\eta),\quad v=\frac{\lambda V_0}{1-\omega t}e^{\frac{x+y}{L}}g'(\eta)\\ &w=-\left(\frac{\upsilon U_0}{2L(1-\omega t)}\right)^{\frac{1}{2}}e^{\frac{x+y}{2L}}\left[f+\eta f'+g+\eta g'\right]\\ &\theta=\frac{T-T_\infty}{T_w-T_\infty},\quad \phi=\frac{C-C_\infty}{C_\infty}\quad \eta=\left(\frac{U_0}{2\upsilon L(1-\omega t)}\right)^{\frac{1}{2}}e^{\frac{x+y}{2L}}z\end{aligned} \tag{11}$$

where primes denote differentiation with respect to η. Using the similarity variables (11), (5) is automatically satisfied, while (6)–(9) are reduced to a system of nonlinear ordinary differential equations as follows:

$$f'''-2(f'+g')f'+(f+g)f''-A(\eta f''+2f')=0 \tag{12}$$

$$g''' - 2(f' + g')g' + (f + g)g'' - A(\eta g'' + 2g') = 0 \tag{13}$$

$$\theta'' + Pr(f + g)\theta' + \eta Pr(B - A)\theta' + Nb\theta'\varphi' + Nt\theta'2 = 0 \tag{14}$$

$$\phi'' + Sc(f + g)\phi' - A\eta Sc\phi' + \frac{Nt}{Nb}\theta'' = 0 \tag{15}$$

and the boundary conditions (10) are reduced to:

$$\begin{aligned}
&f'(0) = \lambda, \quad g'(0) = \lambda, \quad f(0) = 0, \quad g(0) = 0, \\
&\theta(0) = 1, \quad Nb\phi'(0) + Nt\theta'(0) = 0 \\
&f'(\eta) \to 0, \quad g'(\eta) \to 0, \quad \theta(\eta) \to 0, \quad \phi(\eta) \to 0 \ \text{ as } \ \eta \to \infty
\end{aligned}$$

where $A = \frac{\omega L}{U_0} e^{-\frac{x+y}{L}}$, $B = \beta \frac{2\upsilon L(C_w - C_\infty)(1-\omega t)}{U_0} \eta e^{-\frac{x+y}{L}}$, $Pr = \frac{\upsilon}{\alpha}$, $Sc = \frac{\upsilon}{D_B}$,

$$Nb = \beta D_B \frac{(C_w - C_\infty)}{\alpha} \quad \text{and} \quad Nt = \beta \frac{D_T}{T_\infty} \frac{T_w - T_\infty}{\alpha}$$

and is solved using the bvp4c solver in MATLAB. The effects of various parameters on the skin friction coefficient, local Nusselt number as well as the velocity, temperature and concentration profiles are discussed.

3 Results and Discussions

The nonlinear partial differential (5)–(9) together with boundary conditions (10) is converted into a set of ordinary differential Eq. (12)–(15). They are then reduced to first order ODEs and is solved using the bvp4c solver in MATLAB. In this method, suitable initial guess for the solution needs to be chosen. These guesses must satisfy the boundary conditions and keep the solution behavior. The convergence criteria for the solution is set to be 10−8. The impact of the governing parameters such as the Prandtl number (Pr), Brownian motion parameter (Nb), thermophoresis parameter (Nt), stretching parameter (λ) and Schmidt number (Sc) on the skin friction coefficient, the reduced Nusselt number, the velocity, temperature, and concentration profiles is discussed. The effects of Nb and Nt on the reduced Nusselt number are observed with active and passive control of nanoparticles by keeping other parameter fixed.

Table 1 shows that an increase in Nb does not give any effect on the heat transfer rate for passive control of nanoparticles. This is because of the implementation of the realistic boundary condition which resulted in zero nanoparticle flux at the boundary. As for active control, the heat transfer rate decreases as Nb increases. This happens because of frequent collisions that occurred with the increase inNb, thus resulting

Table 1 Numerical results of reduced Nusselt number when $Pr = 5$, $Nt = 0.5$, $\lambda = 0.5$, $Sc = 3$ and $A = -1$

Nb	$-\theta'(0)$ Active control of nanoparticle	$-\varphi'(0)$ Passive control of nanoparticle
0.1	1.89751	1.92713
0.3	1.88089	1.92713
0.5	1.86431	1.92713
0.7	1.84777	1.92713
0.9	1.83125	1.92713

more condensed nanoparticles. From Table 2, a decreasing trend is seen for the reduced Nusselt number as value of Nt increases for both active and passive control of nanoparticle. This is due to thermophoresis exhilarates the diffusion of nanoparticles and causing increase in thermal boundary layer thickness. Table 3 shows the numerical results for the heat transfer rate without and with vertical throughflow taking into account. It can be seen that there is difference of values of heat transfer rate between with and without vertical throughflow. This is because when the vertical throughflow is considered, the energy equation included the convective component. This result is more accurate as supported by the work of [34] where they found that in the presence of throughflow there is convective term that is previously ignored by researchers. From Table 4 and Fig. 2, the increase in Nt will decrease the local Nusselt number and an increase in thermal boundary layer thickness. This result coincides with Table 2. Table 5 and Fig. 3 show that Nb has no effect on the heat transfer rate and therefore negligible. This is because of the implementation of the realistic

Table 2 Numerical results of reduced Nusselt number when $Pr = 5$, $Nb = 0.5$, $\lambda = 0.5$, $Sc = 3$ and $A = -1$

Nt	$-\theta'(0)$ Active control of nanoparticle	$-\varphi'(0)$ Passive control of nanoparticle
0.1	2.27185	2.32773
0.3	2.05421	2.11431
0.5	1.86431	1.92713
0.7	1.69821	1.76259
0.9	1.55254	1.61759

Table 3 Numerical results of reduced Nusselt number when $Nb = Nt = 0.5$, $\lambda = 0.5$, $Sc = 3$ and $A = -1$

Pr	$-\theta'(0)$ Without vertical throughflow	$-\varphi'(0)$ With vertical throughflow
5	1.92451	1.92713
10	2.74789	2.75271
15	3.37940	3.38507
20	3.91169	3.91772
25	4.38060	4.38678

Table 4 Values of the reduced Nusselt number and the Sherwood number for some values of Nt when $Pr = 5$, $Nb = 0.5$, $\lambda = 0.5$, $Sc = 3$ and $A = -1$

Nt	$\theta'(0)$	$-\varphi'(0)$
0.1	2.76106	−0.55221
0.2	2.63023	−1.05209
0.3	2.50780	−1.50468
0.4	2.39317	−1.91453
0.5	2.28577	−2.28577

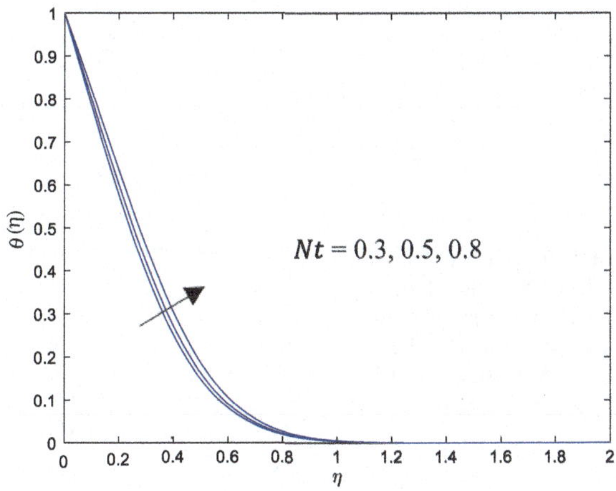

Fig. 2 Temperature distribution for several values of Nt when = 5, $Nb = 0.5$, $\lambda = 0.5$, $Sc = 3$ and $A = -1$

Table 5 Values of the reduced Nusselt number and the Sherwood number for some values of Nb when $Pr = 5$, $Nt = 0.5$, $\lambda = 0.5$, $Sc = 3$ and $A = -1$

Nb	$-\theta'(0)$	$-\varphi'(0)$
0.1	2.28577	−11.42887
0.2	2.28577	−5.71444
0.3	2.28577	−3.80962
0.4	2.28577	−2.85722
0.5	2.28577	−2.28577

boundary condition which resulted in zero nanoparticle flux at the boundary. The result coincides with Table 1. As shown in Table 6 and Fig. 4, when Pr increases, the thermal boundary layer thickness decreases which led to increase in the heat transfer rate. This is due to decreasing fluid thermal conductivity as Pr increases. Table 7 shows that as Sc increases, the heat transfer rate decreases. This is because higher Sc will increase the momentum diffusivity which resulted the widening of thermal boundary layer thickness and increase in temperature. From Fig. 5, the increase in λ will widen the range of the local Nusselt number which then delayed the thermal boundary layer separation. Parabolic behavior is noticed for the case of nanoparticle volume fraction. It is quite interesting to observe the concentration profiles because the similarity variable over-shoots and attains negative values in the neighborhood

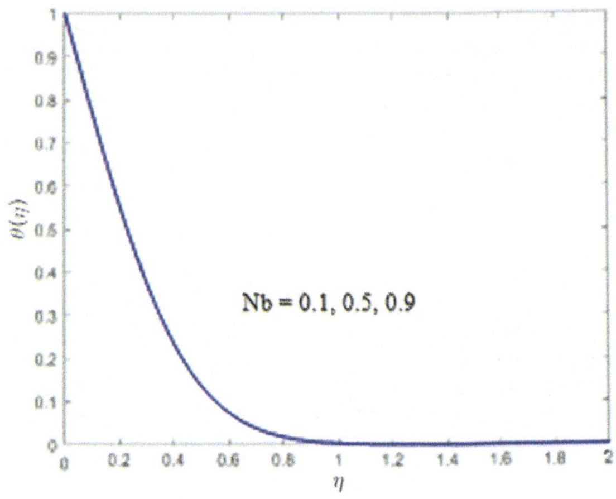

Fig. 3 Temperature distribution for several values of Nb when $Pr = 5$, $Nt = 0.5$, $\lambda = 0.5$, $Sc = 3$ and $A = -1$

Table 6 Values of the reduced Nusselt number and the Sherwood number for some values of Pr when $Nb = Nt = 0.5$, $\lambda = 0.5$, $Sc = 3$ and $A = -1$

Pr	$-\theta'(0)$	$-\varphi'(0)$
5	2.28577	−2.28577
10	3.29495	−3.29495
15	4.06855	−4.06855
20	4.72034	−4.72034
25	5.29439	−5.29439

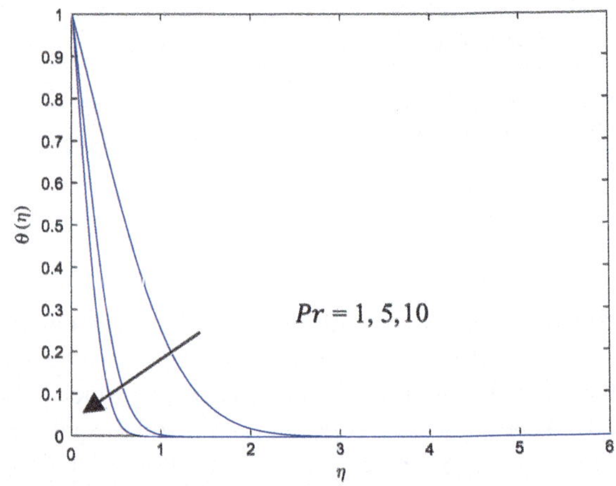

Fig. 4 Temperature distribution for several values of Pr when $Nb = Nt = 0.5$, $\lambda = 0.5$, $Sc = 3$ and $A = -1$

of the surface as shown in Figs. 6, 7, 8, 9 and 10. This interesting behavior is also observed by [21, 22]. This is due to the zero-nanoparticle flux condition (accounting for both Brownian motion and thermophoresis parameters) on the surface, which means that the nanoparticle flux at the surface is suppressed.

Table 7 Values of the reduced Nusselt number and the Sherwood number for some values of Sc when $Pr = 5$, $Nb = Nt = 0.5$, $\lambda = 0.5$, and $A = -1$

Sc	$-\theta'(0)$	$-\varphi'(0)$
1	2.29545	−2.29545
2	2.28892	−2.28892
3	2.28577	−2.28577
4	2.28396	−2.28396
5	2.28282	−2.28282

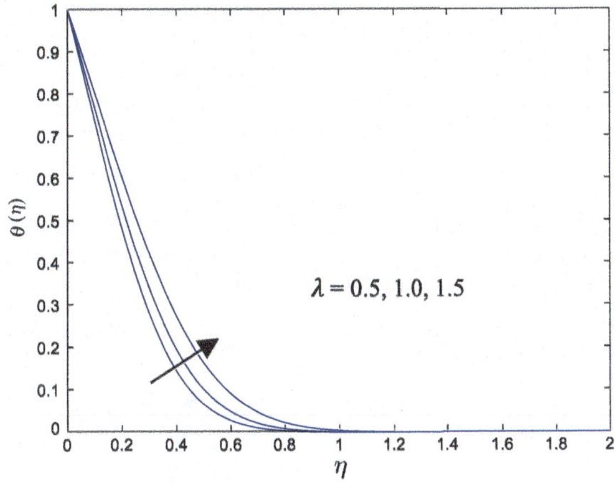

Fig. 5 Temperature distribution for several values of λ when $Pr = 5$, $Nb = Nt = 0.5$, $Sc = 3$ and $A = -1$

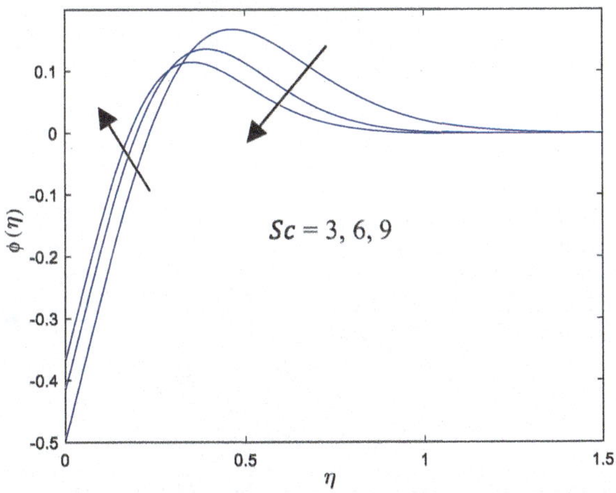

Fig. 6 Concentration distribution for several values of Sc when $Pr = 5$, $Nb = Nt = 0.5$, $\lambda = 0.5$ and $A = -1$

4 Conclusion

The aim of this research is to study the impact of vertical throughflow on unsteady three-dimensional laminar boundary layer flow, heat and mass transfer of nanofluid

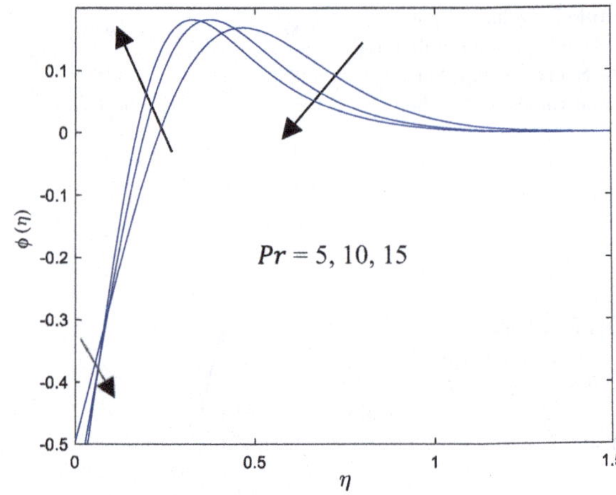

Fig. 7 Concentration distribution for several values of Pr when $Nb = Nt = 0.5$, $\lambda = 0.5$, $Sc = 3$ and $A = -1$

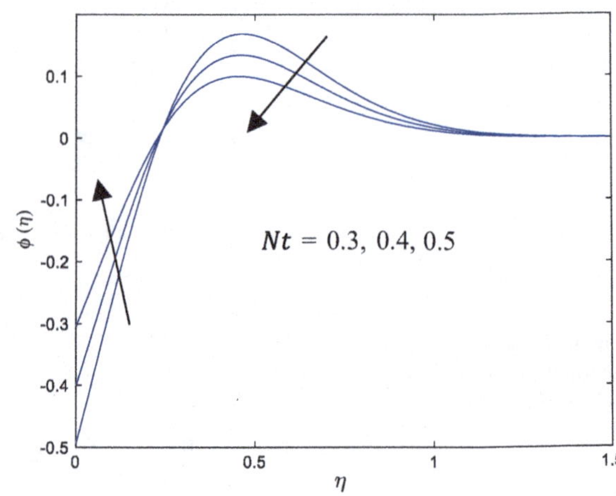

Fig. 8 Concentration distribution for several values of Nt when $Pr = 5$, $Nb = 0.5$, $\lambda = 0.5$, $Sc = 3$ and $A = -1$

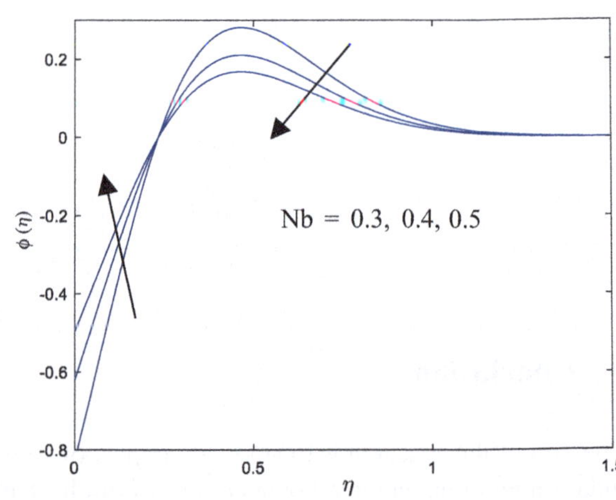

Fig. 9 Concentration distribution for several values of Nb when $Pr = 5$, $Nt = 0.5$, $\lambda = 0.5$, $Sc = 3$ and $A = -1$

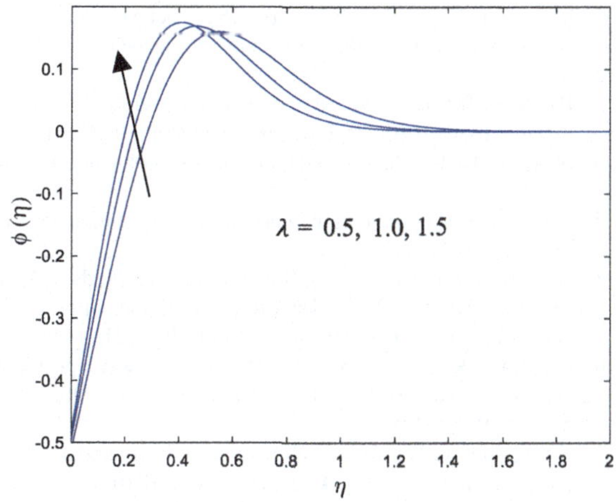

Fig. 10 Concentration distribution for several values of λ when $Pr = 5$, $Nb = Nt = 0.5$, $Sc = 3$ and $A = -1$

towards an exponentially deforming sheet by using the realistic approach. The nanoparticle volume fraction at the boundary is passively controlled to get a more realistic results and a convective term is added to the conservation equation of nanoparticles. The system of governing partial differential equations together with its boundary conditions are reduced to nonlinear ordinary differential equations and then solved numerically using bvp4c solver in MATLAB software. It was found that an increase in *Nb* does not give any effect on the heat transfer rate for passive control of nanoparticles but decreases the heat transfer rate for active control of nanoparticle. It can also be seen that increase in Nt decreases the local Nusselt number for both passive and active control. It is found as Pr increases, the heat transfer rate increases. On the other hand, it was found that increase in λ widen the range of the local Nusselt number. As Sc increases, the heat transfer rate also decreases. Lastly, Parabolic behavior is noticed for the case of nanoparticle volume fraction.

Acknowledgements Authors would like to thank Department of fundamental and applied science, Universiti Teknologi PETRONAS UTP for their support. This work was supported by Yayasan YUTP under grant cost center 015LC0-272.

References

1. Choi, S.U.S.: Enhancing thermal conductivity of fluids with nanoparticles. In: Proceedings of the 1995 ASME International Mechanical Engineering Congress and Exposition, vol. 66, pp. 99–105 (1995).
2. Oztop, H.F., Abu-Nada, E.: Numerical study of natural convection in partially heated rectangular enclosures filled with nanofluids. Int. J. Heat Fluid Flow **29**(5), 1326–1336 (2008)
3. Das, S.K., Choi, S.U.S., Yu, W., Pradeep, Y.: Nanofluids: Science and Technology. Wiley, New Jersey (2008)

4. Choi, S.U.S., Zhang, Z.G., Yu, W., Lockwood, F.E., Grulke, E.A.: Anomalous thermal conductivity enhancement in nanotube suspensions. Appl. Phys. Lett. **79**(14), 2252–2254 (2001)
5. Pak, C., Cho, Y.I.: Hydrodynamic and heat transfer study of dispersed fluids with submicron metallic oxide particles. Exp. Heat Transfer **11**(2), 151–170 (1998)
6. Xuan, Y., Li, Q.: Heat transfer enhancement of nanofluids. Int. J. Heat Fluid Flow **21**(1), 58–64 (2000)
7. Ahuja, A.S.: Augmentation of heat transfer in laminar flow of polystyrene suspensions. J. Appl. Phys. **46**, 3408–3425 (1975)
8. Buongiorno, J.: Convective Transport in Nanofluids. J. Heat Transfer **128**(3), 240 (2006)
9. Tiwari, R.K., Das, M.K.: Heat transfer augmentation in a two-sided lid-driven differentially heated square cavity utilizing nanofluids. Int. J. Heat Mass Transf. **50**(9–10), 2002–2018 (2007)
10. Nield, D., Kuznetsov, A.: The Cheng-Minkowycz problem for natural convective boundary-layer flow in a porous medium saturated by a nanofluid. Int. J. Heat Mass Transf. **52**(25–26), 5792–5795 (2009)
11. Kuznetsov, A., Nield, D.: Natural convective boundary-layer flow of a nanofluid past a vertical plate. Int. J. Therm. Sci. **49**(2), 243–247 (2010)
12. Bachok, N., Ishak, A., Pop, I.: Unsteady three- dimensional boundary layer flow due to a permeable shrinking sheet. Appl. Math. Mech. **31**(11), 1421–1428 (2010)
13. Bachok, N., Ishak, A., Pop, I.: Unsteady boundary-layer flow and heat transfer of a nanofluid over a permeable stretching/shrinking sheet. Int. J. Heat Mass Transf. **55**(7–8), 2102–2109 (2012)
14. Xuan, Y., Roetzel, W.: Conceptions for heat transfer correlation of nanofluids. Int. J. Heat Mass Transf. **43**(19), 3701–3707 (2000)
15. Eastman, J.A., Choi, S.U.S., Li, S., Yu, W., Thompson, L.J.: Anomalously increased effective thermal conductivities of ethylene glycol-based nanofluids containing copper nanoparticles. Appl. Phys. Lett. **78**(6), 718–720 (2001)
16. Nield, D.A., Bejan, A.: Convection in Porous Media, 4th edn. Springer, New York (2013)
17. Trisaksri, V., Wongwises, S.: Critical review of heat transfer characteristics of nanofluids. Renew. Sustain. Energy Rev. **11**(3), 512–523 (2007)
18. Mahian, O., Kianifar, A., Kalogirou, S.A., Pop, I., Wongwises, S.: A review of the applications of nanofluids in solar energy. Int. J. Heat Mass Transf. **57**, 582–594 (2013)
19. Ellahi, R., Hassan, M., Zeeshan, A.: Shape effects of nanosize particles in Cu–H_2O nanofluid on entropy generation. Int. J. Heat Mass Transf. **81**, 449–456 (2015)
20. Sheikholeslami, M.: Numerical investigation for CuO–H_2O nanofluid flow in a porous channel with magnetic field using mesoscopic method. J. Mol. Liq. **249**, 739–746 (2018)
21. Kuznetsov, A.V., Nield, D.: The Cheng-Minkowycz problem for natural convective boundary layer flow in a porous medium saturated by a nanofluid: a revised model. Int. J. Heat Mass Transf. **65**, 682–685 (2013)
22. Nield, D., Kuznetsov, A.V.: Thermal instability in a porous medium layer saturated by a nanofluid: a revised model. Int. J. Heat Mass Transf. **68**, 211–214 (2014)
23. Kuznetsov, A.V., Nield, D.: Natural convective boundary-layer flow of a nanofluid past a vertical plate: a revised model. Int. J. Therm. Sci. **77**, 126–129 (2014)
24. Zaimi, K., Ishak, A., Pop, I.: Boundary layer flow and heat transfer over a nonlinearly permeable stretching/shrinking sheet in a nanofluid. Sci. Rep. **4**, 4404 (2014)
25. Rana, G., Chand, R.: Stability analysis of double- diffusive convection of Rivlin-Ericksen elastico-viscous nanofluid saturating a porous medium: a revised model. Forsch. Ingenieurwes. **79**, 87–95 (2015)
26. Chand, R., Rana, G.: Thermal instability in a horizontal layer of walter's (model B′) vico-elastic nanofluid—a more realistic approach. Appl. Appl. Math. **10**, 1027–1042 (2015)
27. Rana, P., Kumar, L.: Two phase boundary layer nanofluid flow and heat transfer analysis over a stretching sheet: a realistic approach. J. Comput. Theor. Nanosci. **12**, 3090–3095 (2015)
28. Yadav, D., Agrawal, G., Lee, J.: Thermal instability in a rotating nanofluid layer: a revised model. Ain Shams Eng. J. **7**, 431–440 (2015)

29. Jahan, S., Sakidin, H., Nazar, R., Pop, I.: Boundary layer flow of nanofluid over a moving surface in a flowing fluid using revised model with stability analysis. Int. J. Mech. Sci. **131**, 1073–1081 (2017)
30. Uddin, I., Khan, M.A., Ullah, S., Islam, S., Israr, M., Hussain, F.: Characteristics of buoyancy force on stagnation point flow with magneto-nanoparticles and zero mass flux condition. Res. Phys. **8**, 160–168 (2018)
31. Wooding, R.A.: Rayleigh instability of a thermal boundary layer in flow through a porous medium. J. Fluid Mech. **9**(2), 183–192 (1960)
32. Sutton, F.M.: Onset of convection in a porous channel with net through flow. Phys. Fluids **13**(8), 1931–1934 (1970)
33. Nield, D.A.: Modelling the effect of surface tension on the onset of natural convection in a saturated porous medium. Transp. Porous Media **31**(3), 365–368 (1998)
34. Nield, D.A., Kuznetsov, A.V.: The effect of vertical throughflow on thermal instability in a porous medium layer saturated by a nanofluid: a revised model. J. Heat Transfer **137**(5), 52–61 (2015)
35. Barletta, A., Storesletten, L.: Linear instability of the vertical throughflow in a horizontal porous layer saturated by a power-law fluid. Int. J. Heat Mass Transf. **99**, 293–302 (2016)
36. Pop, I., Naganthran, K., Nazar, R., Ishak, A.: The effect of vertical throughflow on the boundary layer flow of a nanofluid past a stretching/shrinking sheet: a revised model. Int. J. Numer. Meth. Heat Fluid Flow **27**(9), 1910–1927 (2017)
37. Khan, W.A., Pop, I.: Boundary-layer flow of a nanofluid past a stretching sheet. Int. J. Heat Mass Transf. **53**(11–12), 2477–2483 (2010)

Pitt's Inequality Associated with Fractional Wavelet Transform

Mawardi Bahri and Samsul Ariffin Abdul Karim

Abstract The fractional wavelet transform is an extension of the conventional wavelet transform in the context of the fractional Fourier transform. In current work, we present the natural link between the fractional Fourier transform and conventional wavelet transform. We apply this relation to provide the different proof of some fundamental properties of the fractional wavelet transform such as the orthogonality relation, inversion formula and reproducing kernel. Based on these properties and relation, we formulate Pitt's inequality associated with the fractional Fourier transform.

Keywords Fractional wavelet transform · Fractional Fourier transform · Pitt's inequality

1 Introduction

As is well known, the conventional wavelet transform (WT) is a useful mathematical tool, which has been widely applied in quantum physics, signal processing and many other fields of science and engineering (see, e.g., [3, 8, 13]). In recent years, a number of efforts have been increasing interest in extending the conventional wavelet transform using the fractional Fourier transform (FRFT). In [5, 6, 12, 14], the authors have studied an extension of the WT in the FRFT domain, the so-called the fractional wavelet transform (FRWT). The generalized transform is obtained by including the Fourier kernel with the FRFT kernel in the definition of the WT. They also have

M. Bahri (✉)
Department of Mathematics, Hasanuddin University, Makassar 90245, Indonesia
e-mail: mawardibahri@gmail.com

S. A. Abdul Karim
Fundamental and Applied Sciences Department and Centre for Systems Engineering (CSE), Institute of Autonomous System, Universiti Teknologi PETRONAS (UTP), Bandar Seri Iskandar, Seri Iskandar, Perak Darul Ridzuan 32610, Malaysia
e-mail: samsul_ariffin@utp.edu.my

S. A. Abdul Karim et al. (eds.), *Proceedings of the 6th International Conference on Fundamental and Applied Sciences*, Springer Proceedings in Complexity,
https://doi.org/10.1007/978-981-16-4513-6_53

derived its essential properties like linearity, orthogonality relation, reconstruction formula and the inequalities.

On the other hand, the author [10, 11] has discussed that the fractional Fourier transform can be changed to the Fourier transformation. Due to this idea, some properties of the fractional Fourier transform can be easily obtained using the basic connection between the fractional Fourier transform and Fourier transform. In this work, we develop this approach within the framework of the fractional wavelet transform. We provide the other proof of the fundamental properties of the fractional Fourier transform such as the orthogonality relation and inversion formula using the basic relationship between the FRWT and the WT. We also build the reproducing kernel related to the fractional Fourier transform. Finally, based on the properties and Pitt's inequality for the fractional Fourier transform we explore Pitt's inequality related to the fractional Fourier transform.

The paper is arranged as follows. In Sect. 2 we shortly recall basic knowledge about the fractional Fourier transform and notations that will be needed in next section. Section 3 is concerned with introducing the definition of the fractional wavelet transform. Some of its essential properties are investigated in some detail in this section. We also establish Pitt's inequality related to the fractional wavelet transform. In Sect. 4 we conclude this work.

2 Preliminaries

This part recalls the definition of the fractional Fourier transform (FRFT) and basic properties (see [1, 7, 9]). We also introduce a natural link between the fractional Fourier transform and conventional Fourier transform. Let us start with the following definitions.

Definition 1 For $1 \leq p < \infty$, $L^p(\mathbb{R})$ is the space of measurable functions on $\mathbb{R}$ with the norm

$$\|f\|_{L^p(\mathbb{R})} = \left(\int_{\mathbb{R}} |f(t)|^p \, dt \right)^{1/p} < \infty. \tag{1}$$

In particular, for $p = 2$, $L^2(\mathbb{R})$ is a Hilbert space with the inner product

$$\langle f, g \rangle_{L^2(\mathbb{R})} = \int_{\mathbb{R}} f(t)\overline{g(t)} \, dt,$$

and $\|f\|_{L^2(\mathbb{R})} = \sqrt{\langle f, f \rangle_{L^2(\mathbb{R})}}$.

Denote by $C^\infty(\mathbb{R})$, the set of all complex-valued smooth functions on $\mathbb{R}$ and by $\mathbb{Z}_+$, the set of non-negative integers.

Definition 2 The Schwartz space is then defined by

$$\mathcal{S}(\mathbb{R}) = \left\{ f \in C^{\infty}(\mathbb{R}) \mid \sup_{t \in \mathbb{R}} |t^{\alpha} D^{\beta} f(t)| < \infty, \ \forall \alpha, \beta \in \mathbb{Z}_{+} \right\}. \tag{2}$$

Elements of the dual space $\mathcal{S}'(\mathbb{R})$ of $\mathcal{S}(\mathbb{R})$ are called tempered distributions.

The Fourier transform of the function $f \in L^1(\mathbb{R})$ is defined by [2]

$$\mathcal{F}\{f\}(\xi) = \hat{f}(\xi) = \int_{\mathbb{R}} f(t) e^{-i\xi t}\, dt. \tag{3}$$

Let $\theta \in \mathbb{R}$ be a real parameter called the transform order. The kernel $K^{\theta}(t, \xi)$ of the fractional Fourier transform is defined by

$$K^{\theta}(t, \xi) = \begin{cases} C^{\theta} e^{i(t^2+\xi^2)\frac{\cot\theta}{2} - it\xi \csc\theta}, & \theta \neq n\pi \\ \frac{1}{\sqrt{2\pi}} e^{-it\xi}, & \theta = \frac{\pi}{2} \\ \delta(t-\xi), & \theta = 2n\pi \\ \delta(t+\xi), & \theta = (2n+1)\pi, n \in \mathbb{Z}, \end{cases} \tag{4}$$

where δ is a Dirac delta function and

$$C^{\theta} = \frac{e^{i(\frac{\pi}{4} - \frac{\theta}{2})}}{\sqrt{2\pi \sin\theta}} = \sqrt{\frac{1 - i\cot\theta}{2\pi}}. \tag{5}$$

Definition 3 The FRFT of $f \in L^1(\mathbb{R})$ is defined in following way

$$\mathcal{F}^{\theta}\{f\}(\xi) = \hat{f}^{\theta}(\xi) = \int_{\mathbb{R}} f(t) K^{\theta}(t, \xi)\, dt. \tag{6}$$

Remark 1 Note that $\mathcal{F}^{\theta}$, $\theta = \frac{\pi}{2}$ becomes precisely the definition of the Fourier transform $\mathcal{F}\{f\}$ and $\mathcal{F}^{\theta}, \theta = -\frac{\pi}{2}$ is the definition of the inverse Fourier transform.

Definition 4 The inverse FRFT of $\hat{f}^{\theta} \in L^1(\mathbb{R})$ is defined by the integral

$$(\mathcal{F}^{\theta})^{-1}\{\hat{f}^{\theta}\}(t) = \int_{\mathbb{R}} \hat{f}^{\theta}(\xi) \overline{K^{\theta}(t, \xi)}\, d\xi, \tag{7}$$

where $\overline{K^{\theta}(t, \xi)}$ is the complex conjugate of $K^{\theta}(t, \xi)$.

In the sequel, we describe a direct connection between the Fourier transform and fractional Fourier transform. It follows from the FRFT definition that

$$\begin{aligned}\mathcal{F}^{\theta}\{f\}(\xi) &= \int_{\mathbb{R}} f(t)\, K^{\theta}(t,\xi)\, dt \\ &= C^{\theta} \int_{\mathbb{R}} f(t) e^{i(t^2+\xi^2)\frac{\cot\theta}{2} - it\xi \csc\theta}\, dt \\ &= C^{\theta} e^{i\xi^2 \frac{\cot\theta}{2}} \int_{\mathbb{R}} f(t) e^{it^2 \frac{\cot\theta}{2} - it\xi \csc\theta}\, dt \\ &= C^{\theta} e^{i\xi^2 \frac{\cot\theta}{2}} \mathcal{F}\{e^{it^2 \frac{\cot\theta}{2}} f(t)\}(\xi \csc\theta). \end{aligned} \tag{8}$$

Now set

$$f_{\theta}(t) = e^{it^2 \frac{\cot\theta}{2}} f(t), \tag{9}$$

we obtain

$$\frac{e^{-i\xi^2 \frac{\cot\theta}{2}}}{\sqrt{1 - i\cot\theta}} \mathcal{F}^{\theta}\{f\}(\xi) = \mathcal{F}\{f_{\theta}\}\,(\xi \csc\theta)\,. \tag{10}$$

Hence,

$$\frac{e^{-i\xi^2 \frac{\cot\theta}{2}}}{\sqrt{1 - i\cot\theta}} \mathcal{F}^{\theta}\{e^{-it^2 \frac{\cot\theta}{2}} f\}(\xi) = \mathcal{F}\{f\}\big(\xi \csc\theta\big). \tag{11}$$

As is known that the FRFT Parseval identity for $f, g \in L^1(\mathbb{R}) \cap L^2(\mathbb{R})$ is given by

$$\langle f, g\rangle_{L^2(\mathbb{R})} = \langle \mathcal{F}^{\theta}\{f\}, \mathcal{F}^{\theta}\{g\}\rangle_{L^2(\mathbb{R})}. \tag{12}$$

and

$$\|f\|^2_{L^2(\mathbb{R})} = \|\mathcal{F}^{\theta}\{f\}\|^2_{L^2(\mathbb{R})}. \tag{13}$$

3 Continuous Fractional Wavelet Transform (CFRWT)

We begin this section by recalling the wavelet transform. Let the shifting parameter be $b \in \mathbb{R}$ and scaling parameter be $a \in \mathbb{R}_+$. We denote the shifting and scaling of the function f by $\tau_b f(t) = f(t-b)$ and $f_a(t) = a^{-1/2} f(t/a)$, respectively. The family of wavelets are functions of the form

$$\psi_{a,b}(t) = \tau_b \psi_a(t) = \frac{1}{\sqrt{a}} \psi\left(\frac{t-b}{a}\right), \tag{14}$$

that are generated from the mother (basic) wavelets $\psi \in L^2(\mathbb{R})$. Further, we obtain the definition of the continuous wavelet transform of $f \in L^2(\mathbb{R})$ with respect to the mother wavelet ψ as

$$\begin{aligned} T_\psi f(a,b) &= \int_{\mathbb{R}} f(t)\overline{\psi_{a,b}(t)}\,dt \\ &= \frac{1}{\sqrt{a}} \int_{\mathbb{R}} f(t)\overline{\psi\left(\frac{t-b}{a}\right)}\,dt. \end{aligned} \tag{15}$$

Let us consider the family of fractional mother wavelets described by

$$\begin{aligned} \psi_{a,b,\theta}(t) &= \frac{1}{\sqrt{a}} \psi\left(\frac{t-b}{a}\right) e^{-\frac{i}{2}(t^2-b^2)\cot\theta} \\ &= \psi_{a,b}(t) e^{-\frac{i}{2}(t^2-b^2)\cot\theta}. \end{aligned} \tag{16}$$

The most important property of family of fractional mother wavelets (16) is the admissibility condition as follows.

Definition 5 We say a mother wavelet $\psi \in L^2(\mathbb{R})$ associated with the FRFT is admissible if and only if the following admissibility condition is satisfied:

$$C_{\psi,\theta} = \int_{\mathbb{R}^+} \left|\mathcal{F}^\theta\{e^{-\frac{i}{2}(\cdot)^2\cot\theta}\psi\}(a\xi)\right|^2 \frac{da}{a} < \infty. \tag{17}$$

In this case, $C_{\psi,\theta}$ is a real positive constant independent of ξ satisfying $|\xi| = 1$.

Lemma 1 *Suppose that*

$$C_\psi = \int_{\mathbb{R}^+} |\mathcal{F}\{\psi\}(a\xi)|^2 \frac{da}{a} < \infty. \tag{18}$$

is the admissibility condition for the conventional wavelet transform. The relationship between (17) *and* (18) *is described by*

$$C_\psi = \sin\theta\, C_{\psi,\theta}. \tag{19}$$

Proof A direct computation using (11) results in

$$\begin{aligned} C_\varphi &= \left(\int_{\mathbb{R}+} \left|\mathcal{F}\{\psi\}(a\xi)\right|^2 \frac{da}{a}\right) \\ &= \left(\int_{\mathbb{R}+} \left|\mathcal{F}\{\psi\}(a\xi \csc\theta)\right|^2 \frac{da}{a}\right) \\ &= \left(\int_{\mathbb{R}+} \left|\frac{e^{-i\xi^2 \frac{\cot\theta}{2}}}{\sqrt{1-i\cot\theta}} \mathcal{F}^\theta\{e^{-it^2 \frac{\cot\theta}{2}}\psi\}(a\xi)\right|^2 \frac{da}{a}\right) \\ &= \left(\int_{\mathbb{R}+} \left|e^{-i\xi^2 \frac{\cot\theta}{2}} \sqrt{i\sin\theta}\, e^{-i\frac{\theta}{2}} \mathcal{F}^\theta\{e^{-it^2 \frac{\cot\theta}{2}}\psi\}(a\xi)\right|^2 \frac{da}{a}\right) \\ &= \sin\theta \left(\int_{\mathbb{R}+} \left|\mathcal{F}^\theta\{e^{-it^2 \frac{\cot\theta}{2}}\psi\}(a\xi)\right|^2 \frac{da}{a}\right), \end{aligned} \tag{20}$$

and the proof is complete.

Note that the Fourier transform of (14) takes the form

$$\hat{\psi}_{a,b}(\xi) = \sqrt{a}\, e^{-ib\xi} \mathcal{F}\{\psi\}(a\xi). \tag{21}$$

The following Lemma 2 gives a relationship between (16) and its FRFT.

Lemma 2 *Let ψ be a mother wavelet. Then, the family of the fractional mother wavelets* (16) *can be expressed in terms of the FRFT as*

$$\begin{aligned} \mathcal{F}^\theta\{\psi_{a,b,\theta}\}(\xi) &= \hat{\psi}^\theta_{a,b,\theta}(\xi) \\ &= \sqrt{a}\, e^{\frac{i}{2}(b^2+\xi^2)\cot\theta - ib\xi\csc\theta - \frac{i}{2}a^2\xi^2\cot\theta} \mathcal{F}^\theta\{e^{-\frac{i}{2}(\cdot)^2\cot\theta}\psi\}(a\xi). \end{aligned} \tag{22}$$

Let us define the continuous fractional wavelet transform (CFRWT).

Definition 6 (CFRWT) The CFRWT of $f \in L^2(\mathbb{R})$ with respect to mother wavelet $\psi \in L^2(\mathbb{R})$ associated with the FRFT is defined by

$$\begin{aligned} T^\theta_\psi f(a,b) &= \int_{\mathbb{R}} f(t)\overline{\psi_{a,b,\theta}(t)}\, dt \\ &= \frac{1}{\sqrt{a}} \int_{\mathbb{R}} f(t)\overline{\psi\left(\frac{t-b}{a}\right)}\, e^{\frac{i}{2}(t^2-b^2)\cot\theta}\, dt. \end{aligned} \tag{23}$$

The following lemma explains an interesting connection between the CFRWT and the CWT (see [12] for more details).

Lemma 3 *Let $\psi, f \in L^2(\mathbb{R})$. Then,*

$$T^\theta_\psi f(a,b) = e^{-\frac{i}{2}b^2\cot\theta} T_\psi \tilde{f}(a,b), \tag{24}$$

where

$$\tilde{f}(t) = f(t)e^{\frac{i}{2}t^2 \cot\theta}. \tag{25}$$

Lemma 4 *Let $\psi \in L^2(\mathbb{R})$. Then, the CFRWT (23) has a fractional Fourier representation form*

$$T_\psi^\theta f(a,b) = \sqrt{a} \int_{\mathbb{R}} e^{-\frac{i}{2}(b^2+\xi^2)\cot\theta + ib\xi\csc\theta + \frac{i}{2}a^2\xi^2\cot\theta} \overline{\mathcal{F}^\theta\{e^{-\frac{i}{2}(\cdot)^2\cot\theta}\psi\}(a\xi)}\hat{f}^\theta(\xi)\, d\xi. \tag{26}$$

In view of the inverse transform of the FRFT defined by (7), we have

$$\mathcal{F}_b^\theta\{T_\psi^\theta f(a,b)\}(\xi) = \frac{\sqrt{a}}{\overline{C^\theta}} e^{\frac{i}{2}a^2\xi^2\cot\theta} \overline{\mathcal{F}^\theta\{e^{-\frac{i}{2}(\cdot)^2\cot\theta}\psi\}(a\xi)}\hat{f}^\theta(\xi). \tag{27}$$

The following results are devoted to the derivation of inversion formula and orthogonality relation related to the CFRWT by applying direct connection between the fractional wavelet transform and the conventional wavelet transform.

Theorem 1 *Let $\psi \in L^2(\mathbb{R})$ satisfy the admissibility condition defined by* (17). *Then, for $f \in L^2(\mathbb{R})$, the following inversion formula holds:*

$$f(t) = \frac{1}{2\pi \sin\theta\, C_{\psi,\theta}} \int_{\mathbb{R}^+}\int_{\mathbb{R}} \psi_{a,b,\theta}(t)\, T_\psi^\theta f(a,b)\, db\, \frac{da}{a^2}. \tag{28}$$

Proof Since $\tilde{f}(t) \in L^2(\mathbb{R})$, the inversion formula for the WFT can be expressed as [3]

$$\tilde{f}(t) = \frac{1}{2\pi\, C_\psi} \int_{\mathbb{R}^+}\int_{\mathbb{R}} \psi_{a,b}(t)\, T_\psi \tilde{f}(a,b)\, db\, \frac{da}{a^2}. \tag{29}$$

The above expression is equal to

$$f(t)e^{\frac{i}{2}t^2\cot\theta} = \frac{1}{2\pi\, C_\psi} \int_{\mathbb{R}^+}\int_{\mathbb{R}} \psi_{a,b}(t)\, T_\psi \tilde{f}(a,b)\, db\, \frac{da}{a^2}. \tag{30}$$

We can write equation (30) in the form

$$f(t) = \frac{1}{2\pi\, C_\psi} \int_{\mathbb{R}^+}\int_{\mathbb{R}} e^{-\frac{i}{2}(t^2-b^2)\cot\theta} \psi_{a,b}(t)\, e^{-\frac{i}{2}b^2\cot\theta} T_\psi \tilde{f}(a,b)\, db\, \frac{da}{a^2}. \tag{31}$$

Using (19) and (24) we see that

$$f(t) = \frac{1}{2\pi \sin\theta\, C_{\psi,\theta}} \int_{\mathbb{R}^+} \int_{\mathbb{R}} e^{-\frac{i}{2}(t^2-b^2)\cot\theta} \psi_{a,b}(t)\, e^{-\frac{i}{2}b^2\cot\theta} T_\psi \tilde{f}(a,b)\, db \frac{da}{a^2}$$
$$= \frac{1}{2\pi \sin\theta\, C_{\psi,\theta}} \int_{\mathbb{R}^+} \int_{\mathbb{R}} e^{-\frac{i}{2}(t^2-b^2)\cot\theta} \psi_{a,b}(t)\, T_\psi^\theta f(a,b)\, db \frac{da}{a^2}. \tag{32}$$

This gives the required result.

Theorem 2 *Assume that the basic wavelet $\varphi \in L^2(\mathbb{R})$ related to the FRFT satisfies the admissibility condition* (17). *For $f, g \in L^2(\mathbb{R}) \cap L^1(\mathbb{R})$ we have*

$$\int_{\mathbb{R}^+} \int_{\mathbb{R}} T_\psi^\theta f(a,b)\, \overline{T_\psi^\theta g(a,b)}\, db \frac{da}{a^2} = 2\pi \sin\theta\, C_{\psi,\theta} (f, g)_{L^2(\mathbb{R})}, \tag{33}$$

and

$$\int_{\mathbb{R}^+} \int_{\mathbb{R}} \left|T_\psi^A f(a,b)\right|^2 db \frac{da}{a^2} = \int_{\mathbb{R}^+} \|T_\psi^\theta f(a,b)\|_{L^2(\mathbb{R})}^2 \frac{da}{a^2} = 2\pi \sin\theta\, C_{\psi,\theta} \|f\|_{L^2(\mathbb{R})}^2. \tag{34}$$

Proof Because $\tilde{f}(t)$ and $\tilde{g}(t)$ defined by (25) are in $L^2(\mathbb{R})$, then the orthogonality relation for the conventional wavelet transform implies that

$$\int_{\mathbb{R}^+} \left(\int_{\mathbb{R}} T_\psi \tilde{f}(a,b)\, \overline{T_\psi \tilde{g}(a,b)}\, db \right) \frac{da}{a^2} = 2\pi\, C_\psi (\tilde{f}, \tilde{g})_{L^2(\mathbb{R})}. \tag{35}$$

where C_ψ is given by (18). The above identity can be rewritten in the form

$$\int_{\mathbb{R}^+} \left(\int_{\mathbb{R}} e^{\frac{i}{2}b^2\cot\theta} e^{-\frac{i}{2}b^2\cot\theta} T_\psi \tilde{f}(a,b)\, \overline{T_\psi \tilde{g}(a,b)}\, db \right) \frac{da}{a^2} = 2\pi\, C_\psi (\tilde{f}, \tilde{g})_{L^2(\mathbb{R})},$$

which gives

$$\int_{\mathbb{R}^+} \left(\int_{\mathbb{R}} e^{\frac{i}{2}b^2\cot\theta} T_\psi \tilde{f}(a,b)\, \overline{e^{\frac{i}{2}b^2\cot\theta} T_\psi \tilde{g}(a,b)}\, db \right) \frac{da}{a^2}$$
$$= 2\pi\, C_\psi \int_{\mathbb{R}} f(t) e^{\frac{i}{2}t^2\cot\theta}\, \overline{g(t) e^{\frac{i}{2}t^2\cot\theta}}\, dt.$$

Using (19) we infer that

$$\int_{\mathbb{R}^+} \left(\int_{\mathbb{R}} T_\psi^\theta f(a,b)\, \overline{T_\psi^\theta g(a,b)}\, db \right) \frac{da}{a^2} - 2\pi \sin\theta\, C_{\psi,\theta} \int_{\mathbb{R}} f(t) \overline{g(t)}\, dt, \tag{36}$$

and the proof is complete.

The next result is reproducing kernel related to the CFRWT, which is an extension of the reproducing kernel of the conventional wavelet transform (see Mallat [3]).

Theorem 3 *[Reproducing kernel] Let ψ be an admissible wavelet. If*

$$K_{\psi}(a,b,\theta;a',b',\theta') = \frac{1}{2\pi \sin\theta\, C_{\psi,\theta}} \left(\psi_{a,b,\theta}, \psi_{a',b',\theta'}\right)_{L^2(\mathbb{R})}, \tag{37}$$

then $K_{\psi}(a,b,\theta;a',b',\theta')$ is a reproducing kernel, i.e.,

$$T_{\psi}^{\theta'} f(a',b') = \int_{\mathbb{R}^+}\int_{\mathbb{R}} T_{\psi}^{\theta} f(a,b) K_{\psi}(a,b,\theta;a',b',\theta')\, db \frac{da}{a^2}. \tag{38}$$

Proof It directly follows from the reproducing kernel for the CWT that

$$\begin{aligned} T_{\psi}\tilde{f}(a',b') &= \int_{\mathbb{R}^+}\int_{\mathbb{R}} T_{\psi}\tilde{f}(a,b) K_{\psi}(a,b;a',b')\, db \frac{da}{a^2} \\ e^{-\frac{i}{2}b'^2 \cot\theta'} T_{\psi}\tilde{f}(a',b') &= \int_{\mathbb{R}^+}\int_{\mathbb{R}} e^{-\frac{i}{2}b'^2 \cot\theta'} T_{\psi}\tilde{f}(a,b) K_{\psi}(a,b;a',b')\, db \frac{da}{a^2}. \end{aligned} \tag{39}$$

With help of (24) and (25) we may write the above identity in the form

$$\begin{aligned} T_{\psi}^{\theta'} f(a',b') &= \int_{\mathbb{R}^+}\int_{\mathbb{R}} T_{\psi}^{\theta} f(a,b) e^{-\frac{i}{2}(t^2-b^2)\cot\theta} e^{\frac{i}{2}(t^2-b'^2)\cot\theta} K_{\psi}(a,b;a',b')\, db \frac{da}{a^2} \\ &= \int_{\mathbb{R}^+}\int_{\mathbb{R}} T_{\psi}^{\theta} f(a,b) K_{\Psi}(a,b,\theta;a',b',\theta')\, db \frac{da}{a^2}. \end{aligned} \tag{40}$$

This proves that K_{ψ} is a reproducing kernel for $T_{\psi}^{\theta} f$ in $L^2(\mathbb{R})$. This achieves the proof.

The next result will be useful in proving the main result of this work.

Theorem 4 *[FRFT Pitt's Inequality] Let $f \in \mathcal{S}(\mathbb{R})$ and $0 \le \alpha < 1$. Then the following inequality holds:*

$$\int_{\mathbb{R}} |\xi|^{-\alpha} \left|\mathcal{F}^{\theta}\{f\}(\xi)\right|^2 d\xi \le |\csc\theta|^{\alpha} C_{\alpha} \int_{\mathbb{R}} |t|^{\alpha} |f(t)|^2\, dt, \tag{41}$$

where

$$C_{\alpha} = \pi^{\alpha}\left[\Gamma\left(\frac{1-\alpha}{4}\right) \Big/ \Gamma\left(\frac{1+\alpha}{4}\right)\right]^2.$$

Here $\Gamma(\cdot)$ is the gamma function.

Proof Based on the Pitt's inequality for the Fourier transformation we have

$$\int_{\mathbb{R}} |\xi|^{-\alpha} |\mathcal{F}\{f\}(\xi)|^2 d\xi \le C_{\alpha} \int_{\mathbb{R}} |t|^{\alpha} |f(t)|^2\, dt. \tag{42}$$

Now inserting f_{θ} into both sides of relation (42) above we see that

$$\int_{\mathbb{R}} |\xi|^{-\alpha} |\mathcal{F}\{f_\theta\}(\xi)|^2\, d\xi \leq C_\alpha \int_{\mathbb{R}} |t|^\alpha |f_\theta(t)|^2\, dt. \tag{43}$$

This implies that

$$\int_{\mathbb{R}} |\xi \csc\theta|^{-\alpha} |\mathcal{F}\{f_\theta\}(\xi \csc\theta)|^2\, d\xi \csc\theta \leq C_\alpha \int_{\mathbb{R}} |t|^\alpha |f_\theta(t)|^2\, dt. \tag{44}$$

Applying (10) we may write the above identity as

$$\int_{\mathbb{R}} |\xi|^{-\alpha} |\csc\theta|^{-\alpha} \left| \frac{e^{-i\xi^2 \frac{\cot\theta}{2}}}{\sqrt{1 - i\cot\theta}} \mathcal{F}^\theta\{f\}(\xi) \right|^2 d\xi \csc\theta \leq C_\alpha \int_{\mathbb{R}} |t|^\alpha |e^{it^2 \frac{\cot\theta}{2}} f(t)|^2\, dt. \tag{45}$$

Hence,

$$\int_{\mathbb{R}} |\xi|^{-\alpha} \left| \mathcal{F}^\theta\{f\}(\xi) \right|^2 d\xi \leq |\csc\theta|^\alpha C_\alpha \int_{\mathbb{R}} |t|^\alpha |f(t)|^2\, dt.$$

This is the desired result.

A generalization of Pitt's inequality in the framework of CFRWT is discussed in the following result.

Theorem 5 *(CFRWT Pitt's Inequality) Let $\psi \in \mathcal{S}(\mathbb{R})$ be a fractional admissible wavelet. Then, for any $f \in \mathcal{S}(\mathbb{R})$ such that $T_\psi^\theta f(a,b) \in \mathcal{S}(\mathbb{R})$, one has*

$$\int_{\mathbb{R}} |\xi|^{-\alpha} \left| \mathcal{F}^\theta\{f\}(\xi) \right|^2 d\xi \leq \frac{|\csc\theta|^\alpha C^\theta C_\alpha}{C_{\psi,\theta}} \int_{\mathbb{R}^+} \int_{\mathbb{R}} |b|^\alpha |T_\psi^\theta f(a,b)|^2\, db \frac{da}{a^2}. \tag{46}$$

Proof Now replacement of the function $f(t)$ by $T_\psi^\theta f(a,b)$ on both sides of Theorem 4 we see that

$$\int_{\mathbb{R}} |\xi|^{-\alpha} \left| \mathcal{F}^\theta\{T_\psi^\theta f(a,b)\} \right|^2 d\xi \leq |\csc\theta|^\alpha C_\alpha \int_{\mathbb{R}} |b|^\alpha |T_\psi^\theta f(a,b)|^2\, db. \tag{47}$$

By integrating both sides of (47) with respect to $\frac{da}{a^2}$, we get

$$\begin{aligned} &\int_{\mathbb{R}^+} \int_{\mathbb{R}} |\xi|^{-\alpha} \left| \mathcal{F}^\theta\{T_\psi^\theta f(a,b)\} \right|^2 d\xi \frac{da}{a^2} \\ &\quad \leq |\csc\theta|^\alpha C_\alpha \int_{\mathbb{R}^+} \int_{\mathbb{R}} |b|^\alpha |T_\psi^\theta f(a,b)|^2\, db \frac{da}{a^2}. \end{aligned} \tag{48}$$

According to relations (17), (27), (34) and the Fubini's theorem, the left-hand side of (48) above may be expressed in the form

$$\int_{\mathbb{R}^+}\int_{\mathbb{R}} |\xi|^{-\alpha} \frac{a}{C^\theta} \left|\overline{\mathcal{F}^\theta\{e^{-\frac{i}{2}(\cdot)^2\cot\theta}\psi\}(a\xi)}\right|^2 \left|\mathcal{F}^\theta\{f\}(\xi)\right|^2 d\xi \, \frac{da}{a^2}$$

$$\leq |\csc\theta|^\alpha C_\alpha \int_{\mathbb{R}^+}\int_{\mathbb{R}} |b|^\alpha |T_\psi^\theta f(a,b)|^2 \, db \, \frac{da}{a^2}$$

$$\int_{\mathbb{R}^+} \left|\mathcal{F}^\theta\{e^{-\frac{i}{2}(\cdot)^2\cot\theta}\psi\}(a\xi)\right|^2 \frac{da}{a} \int_{\mathbb{R}} |\xi|^{-\alpha} \left|\mathcal{F}^\theta\{f\}(\xi)\right|^2 d\xi$$

$$\leq |\csc\theta|^\alpha C^\theta C_\alpha \int_{\mathbb{R}^+}\int_{\mathbb{R}} |b|^\alpha |T_\psi^\theta f(a,b)|^2 \, db \, \frac{da}{a^2}$$

$$C_{\psi,\theta} \int_{\mathbb{R}} |\xi|^{-\alpha} \left|\mathcal{F}^\theta\{f\}(\xi)\right|^2 d\xi \leq |\csc\theta|^\alpha C^\theta C_\alpha \int_{\mathbb{R}^+}\int_{\mathbb{R}} |b|^\alpha |T_\psi^\theta f(a,b)|^2 \, db \, \frac{da}{a^2},$$

which is the desired result.

As a special case of Pitt's inequality related to the CFRWT above we get the following remark.

Remark 2 It should be observed that by choosing $\alpha = 0$ in Theorem 5, we obtain the following inequality

$$\int_{\mathbb{R}} \left|\mathcal{F}^\theta\{f\}(\xi)\right|^2 d\xi \leq \frac{C^\theta}{C_{\psi,\theta}} \int_{\mathbb{R}^+}\int_{\mathbb{R}} |T_\psi^\theta f(a,b)|^2 \, db \, \frac{da}{a^2}. \tag{49}$$

4 Conclusion

In this work, we have investigated some fundamental properties of the fractional Fourier transform like he orthogonality relation and inversion formula using direct interaction between the FRWT and the WT. Based on Pitt's inequality related to the fractional Fourier transform we have derived Pitt's inequality in context of the fractional wavelet transform.

Acknowledgements The work is partially supported by Grant from Ministry of Research, Technology and Higher Education, Indonesia under WCR scheme.

References

1. Almeida, L.B.: The fractional Fourier transform and time-frequency representations. IEEE Trans. Signal Process. **42**(11), 3084–3091 (1994)
2. Bracewell, R.: The Fourier Transform and its Applications. McGraw Hill, Boston, MA (2000)
3. Mallat, S.: A Wavelet Tour of Signal Processing. Academic, San Diego, CA (2001)
4. Namias, V.: The fractional order Fourier transform and its application to quantum mechanics. IMA J. Appl. Math. **25**(3), 241–265 (1980)
5. Shi, J., Zhang, N., Liu, X.: A novel fractional wavelet transform and its applications. Sci. China Inf. Sci **55**(6), 1270–1279 (2012)

6. Prasad, A., Manna, S., Mahato, A., Singh, V.K.: Generalized continuous wavelet transform associated with the fractional Fourier transform. J. Comput. Appl. Math. **259**, 660–671 (2014)
7. Ozaktas, H.M., Zalevsky, Z., Kutay, M.A.: The Fractional Fourier Transform with Application in Optics and Signal Processing. Wiley, Chichester, England (2001)
8. Rieder, A.: The wavelet transform on Sobolev spaces and its approximation properties. Numer. Math. **58**(8), 875–894 (1990)
9. Shi, J., Liu, X., Zhang, N.: On uncertainty principle for signal concentrations with fractional Fourier transform. Signal Process. **92**(12), 2830–2836 (2012)
10. Zayed, Z.I.: On the relationship between the Fourier and fractional Fourier transforms. IEEE Signal Process. Lett. **3**(12), 310–311 (1996)
11. Zayed, Z.I.: A convolution and product theorem for the fractional Fourier transform. IEEE Signal Process. Lett. **5**(4), 101–103 (1998)
12. Bahri, M., Ashino, R.: Logarithmic uncertainty principle, convolution theorem related to continuous fractional wavelet transform and its properties on a generalized Sobolev space. Int. J. Wavelets Multiresolution Inf. Process. **15**(5), 1750050 (2017)
13. Bahri, M., Shah, F.A., Tantary, A.Y.: Uncertainty principles for the continuous shearlet transforms in arbitrary space dimensions. Integral Transforms Special Funct. **31**(7), 538–555 (2020)
14. Guo, Y., Li, B.Z., Yang, L.D.: Novel fractional wavelet transform: principles. MRA and application. Digit. Signal Process. **110**, 102937 (2021)

A Comparison Study on Developed FSWGARCH, SWGARCH and GARCH Models in Time Series Forecasting: An Application to Airline Passenger Volume

Amiratul Liyana Mohamad Hanapi, Mahmod Othman, Rajalingam Sokkalingam, and Hamzah Sakidin

Abstract Several time series data consist of fluctuating information such as risks and uncertainties, arising from instability of the series data. The most popular model for these data is Generalized Auto Regressive Conditional Heteroskedasticity (GARCH) model. However, the GARCH model does not capture the influence of each variance in the observation because the model uses long-run average variance. The computation of the long-run average variance only considers on the entire series, so it loses information on different effects of the variances in each observation. This study therefore develops a new forecasting model using fuzzy window variance to replace the long-run average variance to incorporate more recent returns, which will yield greater weight of forecast. The concept of fuzzy sliding window was embedded in GARCH model to capture the influence of each variance in the observation. This study is aimed at improving the effectiveness of forecasting time series, which in turn increases forecast accuracy. A monthly airline passenger volume dataset is used for evaluation purposes. The accuracy of the proposed model is compared with Sliding Window GARCH (SWGARCH) and GARCH. From the results, the proposed model produces forecasts that are almost accurate as the actual data and outperforms the benchmark models. The proposed model is significantly fitted and reliable for time series forecasting.

Keywords GARCH · Time series · Forecasting · Sliding window · Fuzzy

A. L. M. Hanapi (✉) · M. Othman · R. Sokkalingam · H. Sakidin
Fundamental and Applied Sciences Department, Universiti Teknologi PETRONAS, 32610 Seri Iskandar, Perak, Malaysia
e-mail: amiratul_16005788@utp.edu.my

M. Othman
e-mail: mahmod.othman@utp.edu.my

R. Sokkalingam
e-mail: raja.sokkalingam@utp.edu.my

H. Sakidin
e-mail: hamzah.sakidin@utp.edu.my

S. A. Abdul Karim et al. (eds.), *Proceedings of the 6th International Conference on Fundamental and Applied Sciences*, Springer Proceedings in Complexity,
https://doi.org/10.1007/978-981-16-4513-6_54

1 Introduction

Airline passenger volume forecasting is considered to be of key importance to both airlines and stakeholders, as investment performance is greatly influenced by the precision and adequacy of the estimate carried out [1]. Forecasts also assist airlines in minimizing airport risks and in decision-making on the construction of infrastructure facilities such as availability of seats, thereby improving their passengers' services [2]. Globally, the number of passengers boarding the airline industry is increasing significantly from 2004 to 2017 [3]. This number is expected to increase each year; however, the airline passenger data engage in some seasonal periodicity with an upward trend, making the data non-stationary and nonlinear [4]. Therefore, it is crucial that this airline passenger volume to get forecast accurately by handling the fluctuating information with suitable forecasting models.

Various forecasting techniques have been used by researchers in the airline forecasting industry such as worldwide airline network [5], multitude of decision trees [6], neural network [7], time series models [8] and others. Although these forecasting techniques are mostly used by analysts, the techniques are difficult and unsuitable to handle the nonlinearity of the data that contains changes of variance over time. These changes are related to the dispersion of returns for a given factor and the use of prior values in the observed series. The returns are defined as the change of measurements at the ending time. The selection of a good model is important to handle these changes using standard deviation or variance between the returns of the same factors. These models of nonlinear time series should be able to differentiate assumptions about the effects of volatility by treating the weights of each estimated parameter value.

The most common model for analyzing nonlinear time series data is Generalized Auto Regressive Conditional Heteroskedasticity (GARCH) [9]. Being autoregressive, GARCH processes depend on past square observations and past variances to model current variance. GARCH seeks to reduce forecast errors by taking into account past forecast errors and thereby improving the reliability of ongoing forecasts. GARCH processes are commonly used in finance [10, 11], merely found in medicine [12], physics [13], and agriculture [14]. In forecasting time series data, the GARCH model is more accurate compared to the considered benchmark models, both in long term forecasting and short-term forecasting [15, 16]. Extensive research is being undertaken to improve the GARCH model by combining mathematical and statistical models into the new models. These hybrid models are more efficient than the classic forecasting models [17–20].

However, the GARCH model uses long run average variance which does not reflect the influence of each observation because the long run average variance is calculated from the entire series. Hence, it is essential to improve the long run average variance in order to capture the influence of each observation differently. The hybrid forecasting models involving GARCH are complicated in nature as there are long process of computation, especially the hybrid of Artificial Neural Network (ANN) and GARCH models. The ANN are not easily understandable and only concern on the adjustment of weights in order to get correct output from the given input [21].

Furthermore, the common neural networks are ineffective in detecting nonlinear or dynamic time series behaviors that have average moving conditions and therefore low predictability [22].

This study has interest in fuzzy logic to improve the GARCH model. The applications of fuzzy logic are important to deal with the uncertainties data, as ANN cannot deal with it singlehandedly [23]. The fuzzy models have resulted in higher accuracy compared to other methods [24, 25]. The models can estimate nonlinear continuous functions with arbitrary accuracy uniformly, allowing the model to generate more accurate results than GARCH model [26]. The application of fuzzy theory is robust to outliers and noise data points and offers positive forecast results even in cases where the number of observations is small and entails uncertainties [27].

In order to achieve a higher level of accuracy of the forecasting performance, it is important to decide how much training data is sufficient to adjust weights of the forecast. Sliding window technique is found to be beneficial in determining the adjustment weights and identifying a set cut of points to partition the range into a small number of intervals [28]. The sliding window technique is usually used for the concept of data mining as well as for forecasting [29]. Since the variation in the prior data is crucial in time series forecasting, [21] used the sliding window algorithm with the implementation of fuzzy time series model to forecast weather conditions. They fragmented the data into twelve sliding windows and considered a two-week span of the variation of the previous year to determine the dependence of the variation of the current year. This resulted in high weather prediction accuracy, with the exception of the highly unpredictable conditions in some seasonal months. In addition, the performance of the sliding window technique allows the researchers to understand how cases behave and has an effect on changing the efficiency pattern [30].

The sliding window technique was combined in GARCH model to produce Sliding Window GARCH (SWGARCH) model [31]. This model had overcome one of the weakness of GARCH model. The long run average variance of GARCH was replaced by a window variance to determine weight of forecast. The window variance computation and determination of forecast weight were based on Principal Component Analysis (PCA). However, the determined weights based on PCA did not produce the best accuracy performance. As stated in [32] any PC is prone to error because the original features of the dataset will be converted to a PC that is a linear combination of the original features. As a result, some information is missing as well as not readable and interpreted in comparison to the original list of features. The performance of forecasting using PCA is discussed in next section. Therefore, this study aims to address these drawbacks by proposing Fuzzy Sliding Window GARCH for time series forecasting model (FSWGARCH).

The proposed model is developed by imposing fuzzy window variance, which automatically replaces the long run average variance in the GARCH model. The appropriate forecast weight estimation in the FSWGARCH model allow the prediction model to be efficiently used for time series forecasting. The performance of the FSWGARCH model is then validated using statistical error analysis techniques, which are mean square error, mean absolute percentage error and linear regression model.

2 Methodology

In this section, we present the methodology of GARCH, SWGARCH and FSWGARCH for comparison purpose. The GARCH and SWGARCH models are discussed in Sects. 2.1 and 2.2, respectively, while the proposed FSWGARCH model is explained in Sect. 2.3. More details of GARCH model can be founded in [9–11] and for SWGARCH see [31], for example.

2.1 Generalized Auto Regressive Conditional Heteroskedasticity (GARCH)

Modelling and forecasting of nonlinear time series data is of interest to some practitioners and researchers. GARCH is most common used in the modelling variable and high volatile data. GARCH model is written as GARCH (p, q) where p is the number of auto-regressive terms and q is the number of moving-average terms. GARCH (1, 1) was developed where the parameters of the model were estimated by using maximum likelihood estimation (MLE). The GARCH (1, 1) model is represented as below.

$$\sigma_n^2 = V_L \times \gamma + u_{n-1}^2 \times \alpha + \sigma_{n-1}^2 \times \beta \tag{1}$$

where V_L is calculated from a long-run average variance rate, as well as from return rate u_{n-1}^2 and recent variance σ_{n-1}^2. Parameters γ, α and β are estimated using maximum likelihood method and must sum to one. The return is defined as the continuously compounded return during time t (between the end of time $t-1$ and end of time t). In other words, return is the gain or loss in a particular period.

2.2 Sliding Window Generalized Auto Regressive Conditional Heteroskedasticity (SWGARCH)

The long-run average variance of GARCH does not reflect the influence of each observation because its calculation is from the entire series. Thus, SWGARCH is introduced to overcome the limitation of GARCH model. The SWGARCH replace the component of long-run average variance into component of window variance. The SWGARCH model is represented as below.

$$\sigma_n^2 = V_W \times \gamma + u_{n-1}^2 \times \alpha + \sigma_{n-1}^2 \times \beta \tag{2}$$

where V_W is a window variance that is calculated from a sliding window technique based on PCA, as well as from u_{n-1}^2 and σ_{n-1}^2. The PCA is used to determine the weight of forecast by estimating the size of the window where there is a massive drop in the scree plot of the variance. The weight of each observation in the window consists of two steps. The first step is to calculate the total weight of the window size (*TF*) using Eq. (3) and normalized using Eq. (4).

$$TF = w_n + w_{n-1} + w_{n-2} + \cdots + w_1 \tag{3}$$

$$W_i = \frac{w_i}{TF} \tag{4}$$

where w_1 is the first weight, n is the window size and w_i is the weight for each observation. Multiplying each weight by the return, and computing the sum of the multiplied values as shown in Eq. (5) are performed to obtain the result for the window variance.

$$V_w = u_t^2 \times W_1 + u_{t-1}^2 \times W_2 + \cdots + u_{t-(n-1)}^2 \times W_i \tag{5}$$

where u_t^2 is the return for time t, and W_n ($n = 1, 2, 3, \ldots$) is the window weight subject to $\sum_{i=1}^{n} W_i = 1$.

2.3 The Proposed Fuzzy Sliding Window Generalized Auto Regressive Conditional Heteroskedasticity (FSWGARCH)

The window variance of SWGARCH is calculated based on PCA. Due to some information may miss and the linear combination of PC of the original data features is not readable and interpreted compared to the original list of features, a fuzzy logic is introduced to address these drawbacks to improve the performance of modelling and forecasting the nonlinear data. Applications of fuzzy logic are important for dealing with data uncertainty that is a major issue in the nonlinear data. The proposed FSWGARCH model is improved by replacing the window variance component with a component called fuzzy window variance. Fuzzy sliding window algorithm is developed in the model to compute the fuzzy window variance which based on fuzzy logic and sliding window knowledge. The fuzzy window variance can capture the influence of each variance in the observation thus the information on different effects of the variances in each observation is preserved. The proposed FSWGARCH calculates the variance (σ_n^2) using fuzzy window variance (F_w) as well as from the recent return (u_{n-1}^2) and recent variance (σ_{n-1}^2) as shown as below.

$$\sigma_n^2 = F_w \times \gamma + u_{n-1}^2 \times \alpha + \sigma_{n-1}^2 \times \beta \tag{6}$$

where F_w is a fuzzy window variance calculated from a fuzzy sliding window algorithm, as well as from u_{n-1}^2 and σ_{n-1}^2. The fuzzy sliding window algorithm is used to determine the weight of forecast by estimating the size of window. The algorithm is composed of six simple steps:

(i) load and fuzzify data into triangular fuzzy number (TFN),
(ii) divide the data into a matrix size of 12 × 1 for the current year (Y) and a matrix size of 24 × 1 for the previous year (PY) consisting of 24 months from the preceding two years,
(iii) construct twelve sliding windows (X) with matrix sized of 12 × 1 from PY,
(iv) calculate the difference between current values in Y and previous values in each X using algebraic operation on TFN,
(v) find Euclidean distance (Ed) of each difference in step (iv), and
(vi) compare the Euclidean distance of the 12 sliding windows. The minimum mean of Euclidean distance estimates the size of window thus determines the weight of forecast.

Fuzzification process is done to change a real scalar value into a fuzzy value. This is achieved using triangular fuzzy number (TFN). The definition of TFN is:

Definition 1 Let X be a universe of discourse, $\tilde{A}$ is a fuzzy subset of X, that if for all. If $\tilde{A} = (a_1, a_2, a_3), 0 \le a_1 \le a_2 \le a_3$ then $\tilde{A}$ is defined as TFN.

A matrix size of 12 × 1 is then divided and constructed for the current year (Y) and a matrix size of 24 × 1 is constructed for the previous year (PY) consisting of 24 months from the preceding two years, as adopted from the study by Kapoor and Bedi [21]. The matrix PY is used to partition 12 slices of sliding windows. The sliding windows are labelled as X_i and have the same size as the matrix Y. The matrix Y and the partitions of the sliding windows X_i and are given in Eqs. (7) and (8), respectively.

$$Y = \{Y_{n-11}, X_{n-10}, X_{n-9}, \ldots, X_n\} \tag{7}$$

$$X_i = \left\{X_{n-(i+22)}, X_{n-(i+21)}, X_{n-(i+20)}, \ldots, X_{n-(i-11)}\right\} \tag{8}$$

where $X_{n-(i+11)}$ $(i = 1, 2, 3, \ldots, 12)$ is the previous observation of the first element in matrix of Y. Both matrices X and Y are in the TFN form where $X_i = (x_1, x_2, x_3)$ and $Y = (y_1, y_2, y_3)$. The weakness of SWGARCH is addressed where the elements in current year and 12 sliding window matrices are considered in fuzzy values. Then, Euclidean distance, Ed_i, for every 12 fuzzified sliding windows is obtained first by calculating the difference between current and previous values as shown in Eq. (9).

$$\tilde{Y}_j - \tilde{X}_j = (y_1 - x_3, y_2 - x_2, y_3 - x_1) = (z_1, z_2, z_3) \tag{9}$$

where $j = 1, 2, 3, \ldots, 12$ is the row of the elements in the both matrices of Y and X. Then, the TFN computed in Eq. (9) is defuzzified as in Eq. (10).

$$Ed_{i,j} = \frac{z_1 + (2 \times z_2) + z_3}{4} \tag{10}$$

The Euclidean distance is calculated using Eq. (11).

$$Ed_i = \sqrt{\sum_{j=1}^{12} \left(Ed_{i,j}\right)^2} \tag{11}$$

Next, by comparing the Euclidean distance of each sliding window, the window size can be estimated based on the minimum Euclidean distance Ed_i produced by the sliding window. Thus, the forecast weight can be determined by following the two steps in SWGARCH which are the calculation of the total weight of the window size (TF) using Eq. (3) and normalized using Eq. (4). Further, fuzzy window variance is computed by summing up the multiplication of each weight with the returns of the observation as in Eq. (12).

$$F_w = u_t^2 \times W_1 + \cdots + u_{t-(n-1)}^2 \times W_n \tag{12}$$

where u_t^2 is the return for time t, and W_n ($n = 1, 2, 3, \ldots$) is the window weight subject to $\sum_{i=1}^{n} W_i = 1$.

2.4 Forecasting

In the last step of the GARCH, SWGARCH and FSWGARCH model, the future values are calculated using expected forecast variance, which uses the long-run average variance as in Eq. (13), window variance as in Eq. (14) and fuzzy window variance as in Eq. (15), which act as the forecast weight as shown below.

$$E\left[\sigma_{n+t}^2\right] = V_L + (\alpha + \beta)^t + \left(\sigma_n^2 - V_L\right) \tag{13}$$

$$E\left[\sigma_{n+t}^2\right] = V_w + (\alpha + \beta)^t + \left(\sigma_n^2 - V_w\right) \tag{14}$$

$$E\left[\sigma_{n+t}^2\right] = F_w + (\alpha + \beta)^t + \left(\sigma_n^2 - F_w\right) \tag{15}$$

where n is the measured variance measurement and where t is the additional time representing the forecast value. Eventually, as shown below, the expected value is obtained.

$$X_{n+1} = X_n + \left(X_n \times E\left[\sigma_n^2\right]\right) \tag{16}$$

where X_n is the current observation.

2.5 Model Validation

The performance of the proposed model is validated using error analysis. The error analysis used in this study is in terms of root mean square error (RMSE) and mean absolute percentage error (MAPE). RMSE is a standard deviation of the prediction errors and determines the concentrated data is around the line of best fit. MAPE measures the accuracy of forecast in terms of percentage. The RMSE and MAPE are shown in Eqs. (17) and (18), respectively.

$$RMSE = \sqrt{\frac{1}{n}\sum_{i=1}^{n}\left(\hat{Y}_i - Y_i\right)^2} \tag{17}$$

$$MAPE = \frac{1}{n}\sum_{t-1}^{n}\frac{\left|\hat{Y}_i - Y_i\right|}{Y_i} \times 100 \tag{18}$$

where Y_i is the actual value, $\widehat{Y}_i$ is the forecast value and n is the number of periods. According to previous study, if the proposed model produces smaller value of RMSE and MAPE than the benchmark models considered, it is interpreted as producing higher accuracy of forecast thus becoming a better model [33]. In this study, the performance of the proposed FSWGARCH is compared to the performance of the SWGARCH and GARCH models.

Furthermore, regression analysis according to R-squared can also be used to verify the performance of the proposed model [34]. The analysis determines the relationship between the variance of FSWGARCH and its covariates, which are the fuzzy window variance, recent returns and recent variances. The R-squared analysis tests on a comfortable scale of 0–100% the strength of the relationship between the model and its covariates. A hypothesis testing is also developed as stated below.

$$\begin{aligned} H_0&: \mu_1 = \mu_2 \\ H_1&: \mu_1 \neq \mu_2 \end{aligned} \tag{19}$$

The test statistic used to test the significance of the proposed model is $F_0 = MS_R/MS_E$, where $MS_R = SSR/k$ and $MS_E = SSE/(n-p)$ with the rejection criteria of $F_0 > f_{\frac{\alpha}{2},k,n-p}$ where $f_{\frac{\alpha}{2},k,n-p}$ is the critical value. If H_0 is rejected, then there is enough evidence to say that the proposed model is significant. Other than that, two-sample t-tests allowing unequal variance is also used to compare the differences in the performance of the proposed model and the previous models. The null hypothesis is rejected in favor of the alternative if, $t < t_{\alpha/2}$ or $t > t_{1-\alpha/2}$. If H_0 is not rejected, then there is insufficient evidence to suggest that the proposed model yields completely different results compared to previous ones.

3 Data Analysis and Results

This section comprises the evaluation of the proposed model using a monthly dataset which is airline passenger volume dataset at United State of America (USA) airports, retrieved from the Bureau Transport Statistics repository, https://www.transtats.bts.gov/Data_Elements.aspx?Data=1 for a period of January 2003 to August 2019. A plot of part of the monthly USA airline passenger volumes that are used in the study (see Fig. 1). The x-axis is the time numbered from January 2003 to August 2019 while y-axis is the volume of airline passenger.

Increasing trend and cyclical component are apparent from the plot, indicating the necessity of suitable model for nonlinear time series forecasting, such as GARCH, SWGARCH and FSWGARCH models. From the analysis, the GARCH model forecast the USA airline passenger volume data with optimal estimated parameters $\gamma = 0.0174$, $\alpha = 0.0583$, $\beta = 0.9243$ and the long run variance is $V_L = 0.01063$. A part of forecast values by GARCH are tabulated in Table 1 in the fifth column. The percentage errors produced by GARCH model are low and most of them are within 1 and 3%.

For the SWGARCH, the total weight of window size is $1 + 2 + 3 = 6$ because there is a big drop at component 3 in the scree plot of PCA, which resulted six forecast weight were produced. Hence, the SWGARCH model forecast the USA airline passenger volume data with optimal estimated parameters $\gamma = 0.8138$, $\alpha = 0.0307$ and $\beta = 0.1555$. A part of forecast values by SWGARCH are tabulated in Table 1 in the fourth column. The percentage errors produced by SWGARCH model are low and most of them are within 1 and 4%.

For the proposed FSWGARCH model, SW_{12} produced the minimum mean of Euclidean distance, hence, the total weight of window size is $1 + 2 + 3 + 4 + 5 + 6 + 7 + 8 + 9 + 10 + 11 + 12 = 78$ and resulted twelve forecast weight. The FSWGARCH forecast the USA airline passenger volume data with optimal

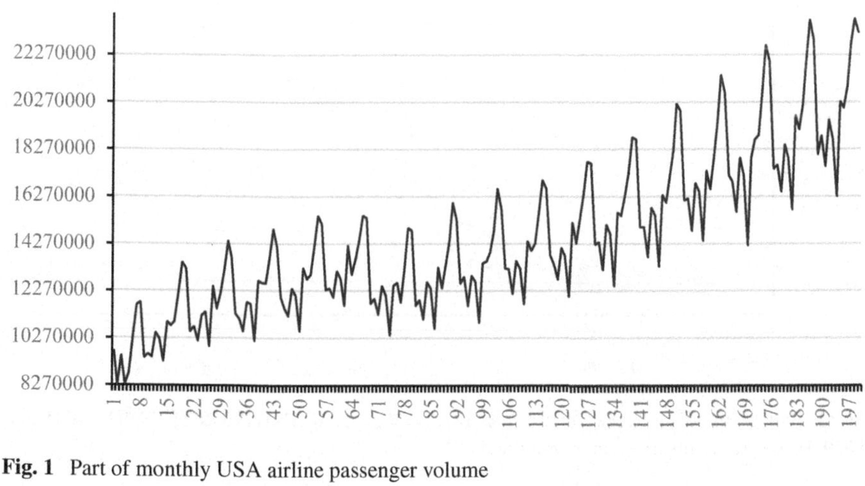

Fig. 1 Part of monthly USA airline passenger volume

Table 1 Comparison results of forecasting USA airlines passenger volume using FSWGARCH, SWGARCH and GARCH

Time	Actual	FSWGARCH	SWGARCH	GARCH
160	18,051,374	18,251,574	18,127,781	18,379,118
161	18,830,854	19,125,724	19,268,080	19,270,583
162	17,552,376	17,805,293	17,840,514	17,934,898
163	19,520,179	19,767,406	19,726,736	19,924,592
164	18,690,817	18,922,832	18,844,133	19,073,183
165	16,276,419	16,450,755	16,373,679	16,587,778
166	20,290,541	20,513,935	20,510,793	20,694,237
167	20,019,304	20,381,540	20,728,048	20,499,970
168	20,911,015	21,260,057	21,404,605	21,375,580
169	22,742,565	23,070,992	23,005,092	23,214,629
170	23,788,306	24,109,024	23,901,549	24,264,150
171	23,208,590	23,483,242	23,301,876	23,643,180

estimated parameters $\gamma = 0.8100$, $\alpha = 0.0497$ and $\beta = 0.1402$. A part of forecast values by FSWGARCH are tabulated in Table 1 in the third column. The percentage errors produced by FSWGARCH model are low and most of them are within 1 and 2%.

The performance of the proposed model is compared to the benchmark models of SWGARCH as shown in Table 1. The comparison is illustrated in Fig. 2, which starts with time 166 for an easy interpretation and the readers can see the differences between the lines.

Figure 2 illustrates the blue line as the actual airline passenger volume, the red line as the forecast value using the FSWGARCH model, the green line as the forecast

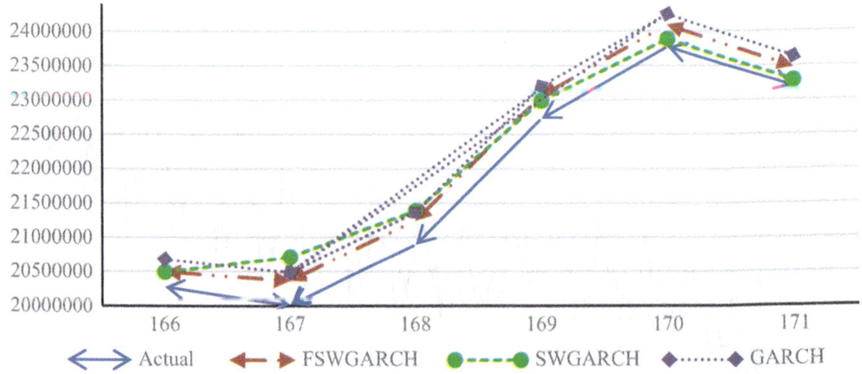

Fig. 2 Plot of the actual data versus forecasted values using FSWGARCH, SWGARCH and GARCH of USA airline passenger volume data

Table 2 Comparison results of the proposed model with the benchmark models for USA airlines passenger volume in terms of RMSE and MAPE

Model	RMSE	MAPE
FSWGARCH	**2.02E+05**	**1.2447**
SWGARCH	2.31E+05	1.2700
GARCH	2.68E+05	1.9868

value using the SWGARCH model and the orange line as the forecast value using the GARCH model. The red line of FSWGARCH is the nearest to the blue line of the actual data. The RMSE and MAPE produced by the proposed and the benchmark models is summarized in Table 2.

From the above table, the FSWGARCH model has the lowest RMSE and MAPE compared to the SWGARCH and GARCH models. This means the proposed model is able to reduce forecasting errors and outperforms the benchmark models in forecasting the airline passenger volume. Since the proposed model is a new model, the model undergoes a goodness-of-fit test using regression analysis. The results of the analysis are tabulated in Tables 3, 4 and 5 to prove the efficacy of the proposed model.

Table 3 Regression statistics of the model

Regression statistics	
Multiple R	0.959384
R square	**0.920417**
Adjusted R square	0.918836
Standard error	0.000751
Observations	155

Table 4 ANOVA of the model

Model	*df*	*F*	*Significance F* *9.72E−83*
Regression	3	582.13	**9.72E-83**
Residual	151		
Total	154		

Table 5 t-test of two models (assuming unequal variances)

	FSWGARCH	*GARCH*	*FSWGARCH*	*SWGARCH*
Mean	1.2447	1.9531	1.2447	1.2699
Variance	0.0624	0.0464	0.0624	0.7545
Hypothesized Mean difference	0		0	
t Stat	−29.2934		−0.3805	
P(T ≤ t) two-tail	**0.0000**		**0.7039**	
t Critical two-tail	1.9665		1.9711	

From Table 2, the R^2 value is equal to 92.04% which is a very strong intensity of the model's relationship.

Referring Table 4, the F value is 582.13 that is bigger than $F_{0.025,3,186} = 3.2269$, with p-value 9.72E-83. The F value is from Table 4, which is higher than the value of $F_{0.025,3,151}$. Thus, H_0 is rejected and there is enough evidence to say that the proposed model is statistically significant.

From Table 5, firstly the t-test of FSWGARCH and GARCH models show that P(T ≤ t) two-tail (0.0000) gives us the probability that a value of the t-Statistic (−29.29) would be observed that is smaller in absolute value than t Critical two-tail (1.97). Since the p-value is smaller than our Alpha (0.05), we reject the null hypothesis that there is statistically significant difference in the means of each FSWGARCH and GARCH models. Secondly, the t-test of FSWGARCH and SWGARCH models show that P(T ≤ t) two-tail (0.7039) gives us the probability that a value of the t-Statistic (−0.38) would be observed that is larger in absolute value than t Critical two-tail (1.97). Since the p-value is larger than our Alpha (0.05), we cannot reject the null hypothesis that there is no statistically significant difference in the means of each FSWGARCH and SWGARCH models. This result is inline with the result of MAPE, that there is small difference between the MAPE of FSWGARCH and MAPE of SWGARCH. Therefore, we can say that the proposed FSWGARCH model is significantly fitted and effective as the SWGARCH model [31].

4 Conclusion

This study presented an enhanced model for forecasting time series data. The enhanced model implemented a component called fuzzy window variance to replace the long run average variance component in the GARCH model. The fuzzy window variance is the combination of fuzzy theory and sliding window technique. Based on the results, by producing lower mean errors, the FSWGARCH model produced higher accuracy and outperformed the benchmark models. The findings have shown that the proposed model is superior, robust and well adapted to forecast time series data. The importance of this work lies in the application of fuzzy sliding window algorithm that overcomes the limitation of GARCH and SWGARCH models.

The FSWGARCH model is capable of capturing the influence of variance in each observation and can identify the true window selection for forecast weight. The true identification of forecast weight yields the best performance in accuracy. In view of the need for a model that is easily understood by most people, particularly forecasters, the proposed model is computationally easy and highly practical. Therefore, the application of the FSWGARCH model will facilitate the airlines in minimizing airport risks and in decision-making on the construction of infrastructure facilities, thereby improving their passengers' services. To extend this research, an improvement in estimating parameters of the model is suggested to improve the accuracy performance.

Acknowledgements This research was funded by Ministry of Higher Education Malaysia (FRGS/1/2018/STG06/UTP/02/2), under Fundamental Research Grant Scheme (FRGS) with grant number 015MA0-021. The authors would like to convey their gratitude to the reviewers for their significant remarks and suggestion for this paper.

References

1. Blinova, T.O.: Analysis of possibility of using neural network to forecast passenger traffic flows in Russia. Aviation **11**(1), 28–34 (2007)
2. Abed, S.Y., Ba-Fail, A.O., Jasimuddin, S.M.: An econometric analysis of international air travel demand in Saudi Arabia. J. Air Transp. Manag. **7**(3), 143–148 (2001)
3. Statista: Airline industry—passenger traffic worldwide 2018. Statistic. https://www.statista.com/statistics/564717/airline-industry-passenger-traffic-globally/, Last accessed 12 May 2018
4. Adhikari, R., Agrawal, R.K.: An Introductory Study on Time Series Modeling and Forecasting. Germany (2013)
5. Mao, L., Wu, X., Huang, Z., Tatem, A.J.: Modeling monthly flows of global air travel passengers: an open-access data resource. J. Transp. Geograph. **48**, 52–60 (2015)
6. Laik Ma, N., Choy, M., Sen, P.: Predicting Airline Passenger Load: A Case Study (2014)
7. Sun, X.S., Brauner, E., Hormby, S.: A large-scale neural network for airline forecasting in revenue management. Oper. Res. Airline Ind. **9**, 46–67 (1998)
8. Kan Tsui, W.H., Balli, H., Gilbey, A., Gow, H.: Forecasting of Hong Kong airport's passenger throughput. Tour. Manag. **42**, 62–76 (2014)
9. Bollerslev, T.: Generalized autoregressive conditional heteroskedasticity. J. Econom. **31**(3), 307–327 (1986)
10. Hanapi, A.L.M., Othman, M., Sokkalingam, R., Sakidin, H.: Developed a hybrid sliding window and GARCH model for forecasting of crude palm oil prices in Malaysia. J. Phys. Conf. Ser. **1123**, 1–8 (2018)
11. Rita, M.R., Wahyudi, S., Muharam, H.: Bad Friday, Monday effect and political issue: application of ARCH-GARCH model to analyze seasonal pattern of stock return. Int. J. Eng. Technol., 38–47 (2018)
12. Brás, S., Nunes, C.S., Amorim, P.: GARCH models for drug effects on patient heart rate during general anaesthesia. IFAC Proc. **39**(18), 441–446 (2006)
13. Zou, Y., Donner, R.V., Marwan, N., Donges, J.F., Kurths, J.: Complex network approaches to nonlinear time series analysis. Phys. Rep., 1–97 (2019)
14. Shiferaw, Y.: Time-varying correlation between agricultural commodity and energy price dynamics with Bayesian multivariate DCC-GARCH models. Phys. A Stat. Mech. Appl. **526**, July 2019
15. Mosalaosi, M., Afullo, T.J.O.: Prediction of asynchronous impulsive noise volatility for indoor powerline communication systems using GARCH models. In: Progress in Electromagnetic Research Symposium (PIERS), Shanghai, China (2016)
16. Guo, H.: Estimating volatilities by the GARCH and the EWMA model of PetroChina and TCL in the stock exchange market of China. In: 6th International Scientific Conference Managing and Modelling of Financial Risks, Ostrava (2012)
17. Lu, X., Que, D., Cao, G.: Volatility forecast based on the hybrid artificial neural network and GARCH-type models. Proc. Comput. Sci. **91**, 1044–1049 (2016)
18. Hajizadeh, E.: A hybrid modeling approach for forecasting the volatility of S&P 500 index return. Exp. Syst. Appl. **39**(1), 431–436 (2012)
19. Narendra, B., Reddy, E.: Selected Indian stock predictions using a hybrid ARIMA-GARCH model. In: 2014 International Conference on Advances in Electronics Computers and Communications, pp. 1–6 (2014)

20. Glosten, L.R., Jagannathan, R., Runkle, D.E.: On the relation between the expected value and the volatility of the nominal excess return on stocks. J. Financ. **48**(5), 1779–1801 (1993)
21. Kapoor, P., Bedi, S.S.: Weather forecasting using sliding window algorithm. ISRN Sig. Process., 1–5 (2013)
22. Tealab, A., Hefny, H., Badr, A.: Forecasting of nonlinear time series using ANN. Future Comput. Inf. J. **2**(1), 39–47 (2017)
23. Singh, P.: Indian summer monsoon rainfall (ISMR) forecasting using time series. Geosci. Front. **9**, 1243–1257 (2018)
24. Dincer, N.G.: Akkuş, Ö.: A new fuzzy time series model based on robust clustering for forecasting of air pollution. Ecol. Inf., 157–164 (2018)
25. Chang, J.-R., Wei, L.-Y., Cheng, C.-H.: A hybrid ANFIS model based on AR and volatility for TAIEX forecasting. Appl. Soft Comput. **11**, 1388–1395 (2011)
26. Maciel, L.: A hybrid fuzzy GJR-GARCH modeling. Rev. Bras. Financas **10**(3), 337–367 (2012)
27. Singh, P., Gaurav, D.: A hybrid fuzzy time series forecasting model based on granular computing and bio-inspired optimization approaches. J. Comput. Sci. **27**, 370–385 (2018)
28. Bingham, E., Gionis, A., Haiminen, N., Hiisila, H., Mannila, H., Terzi, E.: Segmentation and dimensionality reduction. In: SIAM International Conference on Data Mining (2006)
29. Datar, M., Gionis, A., Indyk, P., Motwani, R.: Maintaining stream statistics over sliding windows. Soc. Ind. Appl. Math. **31**(6), 1794–1813 (2002)
30. Ferreira, P., Dionísio, A., Guedes, E.F., Zebendee, G.F.: A sliding windows approach to analyse the evolution of bank shares in the European Union. Phys. A Stat. Mech. Appl., 1355–1367 (2018)
31. Shbier, M.Z.D., Ku-Mahamud, K.-R., Othman, M.: SWGARCH model for time series forecasting. In: IML'17 Proceedings of the 1st International Conference on Internet of Things and Machine Learning. Liverpool, United Kingdom, Oct 2017
32. Raykov, T., Marcoulides, G.A., Li, T.: On the fallibility of principal components in research. Educ. Psychol. Measur. **77**(1), 165–178 (2017)
33. Kapila, M.T.R., Seneviratna, D., Jianguo, W., Arumawadu, H.I.: A hybrid statistical approach for stock market forecasting based on artificial neural network and ARIMA time series models. In: 2015 International Conference on Behavioral, Economic and Socio-Cultural Computing (BESC) (2015)
34. Barucci, E., Reno, R.: On measuring volatility and the GARCH forecasting performance. J. Int. Financ. Mark. Inst. Money **12**, 183–200 (2002)

Optimization of PV/T Solar Collector Performance with Fuzzy If-Then Rules Generation

Noran Nur Wahida Khalili, Mahmod Othman, Hamzah Sakidin, Mohd Nazari Abu Bakar, and Lazim Abdullah

Abstract Solar energy has been a great resource of energy that help to fulfill the world's energy demand. However, the surrounding conditions affect the performance of a photovoltaic/thermal solar collector greatly. Since the operation of air photo-voltaic/thermal solar collector involves energy input to generate the air flow, it is important to ensure balance in the input and output energy. Based on the experimental data collected, membership functions have been developed and a set of rules has been generated using graphical method to ensure the optimization of the solar collector performance under changes of environmental factor such as solar radiation and ambient temperature.

Keywords Solar energy · Membership functions · Fuzzy rules · Optimization

N. N. W. Khalili (✉) · M. Othman · H. Sakidin
Fundamental and Applied Sciences Department, Universiti Teknologi PETRONAS, Seri Iskandar, Perak, Malaysia
e-mail: noran_17005800@utp.edu.my

M. Othman
e-mail: mahmod.othman@utp.edu.my

H. Sakidin
e-mail: hamzah.sakidin@utp.edu.my

M. N. A. Bakar
Faculty of Applied Sciences, Universiti Teknologi MARA Perlis, Arau Campus, Arau, Perlis, Malaysia
e-mail: mohdnazari@uitm.edu.my

L. Abdullah
School of Informatics and Applied Mathematics, Universiti Malaysia Terengganu, Kuala Terengganu, Terengganu, Malaysia
e-mail: lazim_m@umt.edu.my

S. A. Abdul Karim et al. (eds.), *Proceedings of the 6th International Conference on Fundamental and Applied Sciences*, Springer Proceedings in Complexity,
https://doi.org/10.1007/978-981-16-4513-6_55

Fig. 1 Combination of solar thermal and solar PV system

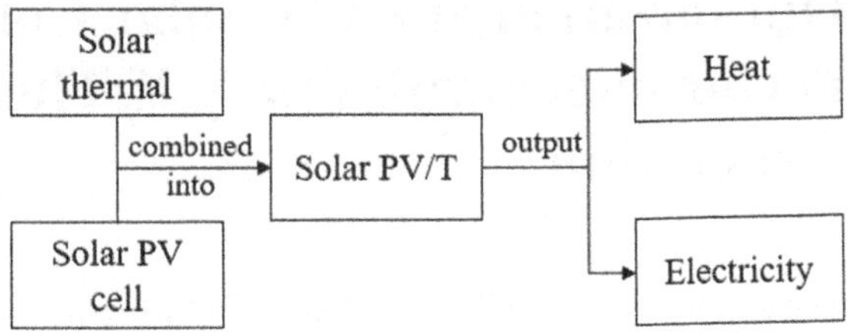

1 Introduction

The increase in demand of the world energy is significantly fulfillable with one of the renewable energies, the solar energy. Solar energy is seen to be high potential because it does not pollute the environment and reliable. Since mid-1970s, researchers have focused their interest on the study of photovoltaic/thermal (PV/T) solar collectors [1], and it continues until today. Research have been done on various collector system aspects, which includes the experimental and theoretical studies of the performance of the solar collectors, and also the designs. PV/T is a combination of a solar photovoltaic (PV) system and a thermal (T) system, which the output of each system is electricity and heat respectively (see Fig. 1).

The operation of air PV/T solar collector is much depending on the parameters affecting its efficiency. Factors such as solar radiation, PV temperature, air mass flowrate affect the collector's performance in significant way. Previously, the velocity of air that is forced into the collector is done manually by the researcher. However, it does not necessarily for the fan to be operated at full speed at every time to remove the heat. There must be a balance between the input and output energy used in the collector system. Therefore, this study is to analyze the effect of parameters on the operation of the PV/T solar collector. From the data collected through the experiment, Mamdani fuzzy if-then rule is used to generate fuzzy rules that represent the effect of the parameters on the collector's efficiency [2].

The objective of this paper is to generate a set of rules from the experimental data collected to predict the optimum speed of fan to be operated to ensure optimum efficiency of the solar collector.

2 Methodology

2.1 Collection of Experimental Data

The factors affecting the operation of the PV/T solar collector are numerous since it depends greatly on the surrounding condition. However, the experiment is carried out under the assumptions that: (a) the system has achieved steady state of energy transfer; (b) heat capacity effects of the solar cells and the back plate have been neglected; (c) the temperature of the collector varies in the air flow direction and (d) the system side losses are neglected [3]. The parameters that have been collected are the PV temperature, ambient temperature, output temperature, solar radiation, voltage and current output and air mass flowrate. The data were collected from the experiment carried out in Universiti Teknologi PETRONAS.

Experimental Setup. The experiment has been carried out using a 50 W monocrystalline silicon solar panel with an aluminium backplate attached with a set of fins, also made of aluminium. The fins act as heat transfer enhancement factor, installed vertically in the direction of air flow. Two DC fans of 12 V have been used to create the air flow in transferring the heat out of the collector system. A pyranometer, air flowmeter, and temperature sensors are used to measure solar radiation, air mass flowrate and temperature of the solar collector. All parameter data collected is recorded by a data logger via a PC (see Fig. 2). However, for this study, only solar radiation and ambient temperature are taken into consideration in generation of the fuzzy rules.

For the experiment set-up, the following conditions have been followed. PV modules can be installed where power source is available, on ground; but in this set-up, the effects of heat capacity of the solar cells and back plate have been neglected. Since shading can adversely affect the photovoltaic arrays the PV system needs clear and uninterrupted access to the solar radiation. The orientation of PV modules is ideally towards true south; based on the latitude angle of the location where it is mounted. The collector system is installed in the compound of Universiti Teknologi PETRONAS, hence the tilt angle is approximately 4.3590°. The detailed description of the experiment setup and results can be found in [4].

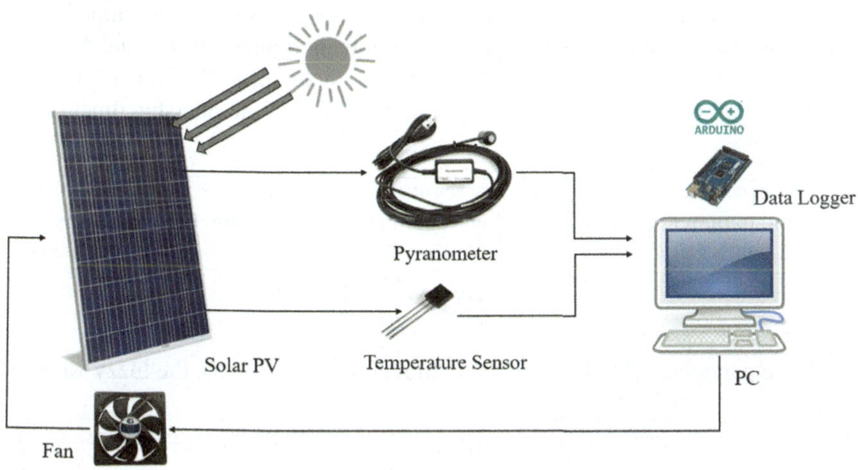

Fig. 2 The experimental setup of the air PV/T solar collector with the installation of fans, temperature sensors, pyranometer and data logger

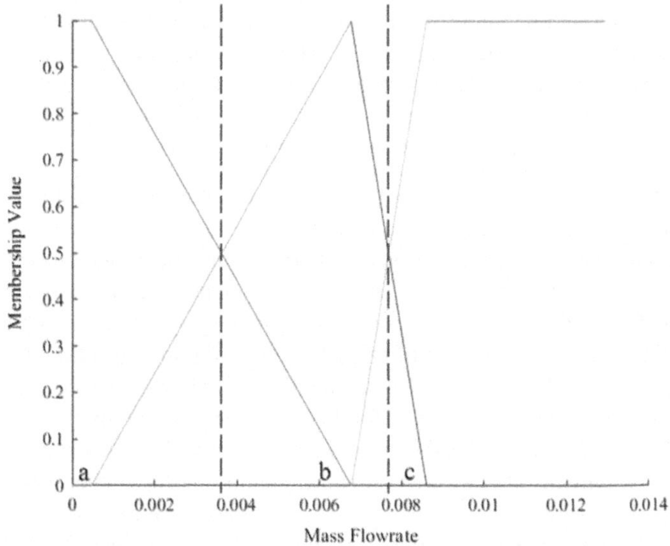

Fig. 3 Membership function for air mass flowrate

2.2 *Membership Functions Development*

From the experiments carried out, the data collected has been used to develop membership functions for the parameters involved, air mass flowrate and total efficiency of the solar collector. The membership functions have been developed using entropy method [5] and are presented in the following figures (Figs. 3 and 4).

Both parameters have been categorized into low, medium, and high in the membership function. The calculations that represent the graphs of membership function are as follows [6]:

1. If the parameter, P falls in the low region ($P \leq a$), the fuzzy value is equal to 1.
2. If the parameter, P falls in the low region ($a < P \leq b$), the fuzzy value is

$$\mu(P) = \frac{b - P}{b - a}$$

3. If the parameter, P falls in the medium region ($a < P \leq b$), the fuzzy value is

$$\mu(P) = \frac{P - a}{b - a}$$

4. If the parameter, P falls in the medium region ($b < P \leq c$), the fuzzy value is

$$\mu(P) = \frac{c - P}{c - b}$$

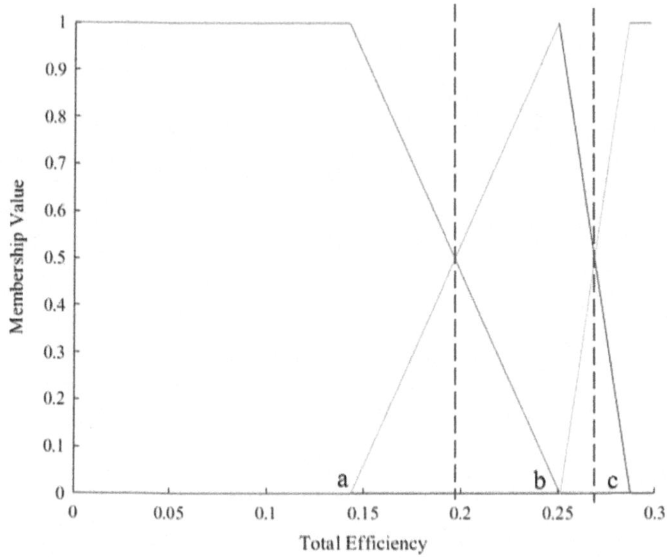

Fig. 4 Membership function for total efficiency of PV/T solar collector

5. If the parameter, P falls in the high region ($b < P \leq c$), the fuzzy value is

$$\mu(P) = \frac{P - b}{c - b}$$

6. If the parameter, P falls in the high region ($P > c$), the fuzzy value is 1.
7. Elsewhere, the fuzzy value is equal to 0.

Then, as shown in Figs. 3 and 4 the values of air mass flowrate and total efficiency of the PV/T solar collector have been categorized into three; depicted by the dotted lines. The category is then plotted into the graphs of the effect of changes in air mass flowrate on the total efficiency (Figs. 5 and 6).

In Fig. 5, when the ambient temperature is 301 K under various solar radiation values, it can be seen that the total efficiency increases with air mass flowrate. However, it is observed that the fan does not necessarily be operated at full speed every time.

For example, when solar radiation is at 100 W/m^2, the total efficiency does not increase significantly even when the air mass flowrate increase. Hence, it can be concluded that the optimum fan speed for the condition of ambient temperature of 301 K and solar radiation of 100 W/m^2 is low speed. Observing another situation, when the solar radiation is 600 W/m^2, the optimum speed is high, since there is more heat to be removed to maintain the collector's efficiency.

The process is repeated for every condition, in which the ambient temperature varies from 301 to 310 K, and solar radiation varies from 100 to 600 W/m^2. The

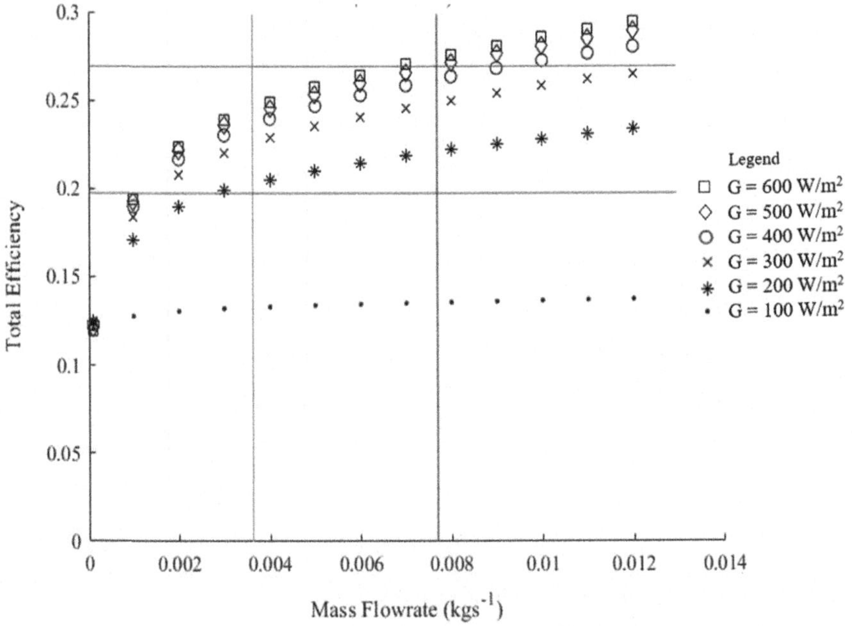

Fig. 5 The effect of changes in air mass flowrate on the total efficiency of the PV/T solar collector performance under ambient temperature 301 K and various solar radiation G

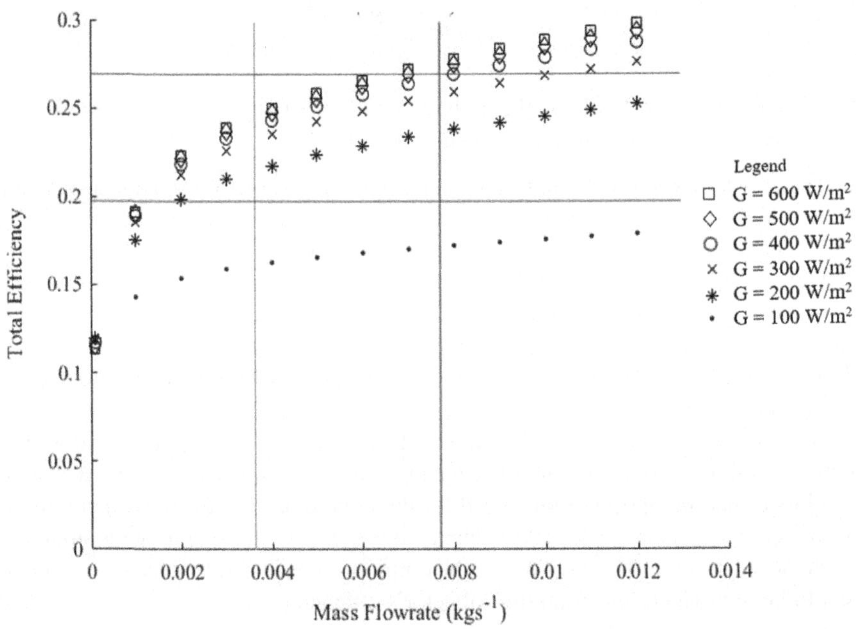

Fig. 6 The effect of changes in air mass flowrate on the total efficiency of the PV/T solar collector performance under ambient temperature 310 K and various solar radiation G

Table 1 Optimum air mass flowrate for various ambient temperature and solar radiation

		Ambient temperature (K)									
		301	302	303	304	305	306	307	308	309	310
Solar radiation	100	L	L	L	L	L	L	L	L	L	L
	200	M	M	M	M	M	M	M	M	M	M
	300	M	M	M	M	M	M	H	H	H	H
	400	M	H	H	H	H	H	H	H	H	H
	500	H	H	H	H	H	H	H	H	H	H
	600	H	H	H	H	H	H	H	H	H	H

final classification of optimum fan speed is tabulated in Table 1. In the table, the categories are written in symbols: Low (L), Medium (M) and High (H).

3 Results and Discussion

From the experiment data collected, it can be seen that the performance of an air PV/T solar collector, calculated from its efficiency is dependent on various factors. By smaller scope, solar radiation and ambient temperature are considered important factors that could significantly changes the collector's performance. From Table 1 shown, the set of rules to optimize the performance of PV/T solar collector can be listed as follows:

Rule 1: IF solar radiation is 100 W/m^2 or below,
THEN the optimum fan speed is Low.

Rule 2: IF solar radiation is in between 100 and 200 W/m^2,
THEN the optimum fan speed is Medium.

Rule 3: IF solar radiation is in between 200 and 300 W/m^2,
AND ambient temperature is 306 K or below,
THEN the optimum fan speed is Medium.

Rule 4: IF solar radiation is in between 200 and 300 W/m^2,
AND ambient temperature is more than 306 K,
THEN the optimum fan speed is High.

Rule 5: IF solar radiation is more than 300 W/m^2,
THEN the optimum fan speed is High.

It is shown here that the fan does not need to be in full speed during the operation of the solar collector. While reducing the speed of fan, the input energy required is also reduced, hence improving the balance with the output energy.

4 Conclusion

This study has presented a fuzzy if-then set of rules to optimize the performance of an air PV/T solar collector. The data has been collected from experiments carried out and membership functions of each parameter has been developed based on the data. Fuzzy rules generated could help with decision making by researchers or industry in optimization of solar collectors. The results showed that the performance of collectors can be predicted, and it can be optimized to ensure balance in input and output energy from the system.

Acknowledgements This research was supported by Yayasan Universiti Teknologi PETRONAS, grant number 0153AA-H27. The authors would like to convey their gratitude to the reviewers for their significant remarks and suggestion for this paper.

References

1. Chow, T.T., Pei, G., Fong, K.F., Lin, Z., Chan, A.L.S., Ji, J.: Energy and exergy analysis of photovoltaic–thermal collector with and without glass cover. Appl. Energy **86**(3), 310–316 (2009)
2. Izquierdo, S., Izquierdo, L.R.: Mamdani Fuzzy Systems for Modelling and Simulation: A Critical Assessment. Available at SSRN 2900827 (2017)
3. Jarimi, H., Abu Bakar, M.N., Othman, M., Din, M.H.: Bi-fluid photovoltaic/thermal (PV/T) solar collector: experimental validation of a 2-D theoretical model. Renew. Energy **85**, 1052–1067 (2016)
4. Khalili, N.N.W., Othman, M., Abu Bakar, M.N., Abdullah, L.: Modelling of a single passage air PV/T solar collector: experimental and simulation design. Processes **8**(7), 763 (2020)
5. Ross, T.J.: Fuzzy Logic with Engineering Applications, vol. 2. Wiley, New York (2004)
6. Ruspini, E.H., Bonissone, P.P., Pedrycz, W.: Handbook of Fuzzy Computation (1998)

Predicting Household Income Due to Fuel Price Fluctuations

Norhana Abd. Rahim and Yumn Suhaylah Yusoff

Abstract The paper mainly aims to predict the household income due to the fuel price fluctuation in Malaysia as resulted from the subsidy removal. Other macroeconomic variables namely household income, household expenditure, household size, inflation rate and gross domestic product (GDP) also added to increase the efficiency of the model. The data is simulated based on the historical data ranges from 2006 to 2016. The Granger causality test is used to examine the causal relationship between variables. The results revealed that there is a causal relationship between the selected variables. Then, the household income is predicted using the Vector Autoregressive (VAR) model. The prediction takes place for the next ten periods. The findings showed that the predicted income is decreasing as the predicted fuel price decreases. This indicate that the future income will be enough as the predicted fuel price decreases. However, if the inverse trend of predicted fuel price occurs, the future income will not be enough to compensate the increase in the fuel price. Thus, this model could be beneficial in helping individual to plan their spending of income.

Keywords Household income · Fuel price · Fluctuation · Prediction

1 Introduction

Starting in November 2014, the government of Malaysia has stopped the fuel subsidy. Since then, the fuel price is depending on the world crude oil price. Initially, the fuel subsidy is provided by the government of Malaysia with the aim to help the poor. However, the fuel subsidy scheme is failed to meet its main objective. The subsidy is most benefited by the rich group instead of helping the poor. In Malaysia, the middle-income group significantly affected by the increased in the fuel price [1].

N. Abd. Rahim (✉) · Y. S. Yusoff
Faculty of Science and Technology, Universiti Sains Islam Malaysia, Nilai, Malaysia
e-mail: norhana@usim.edu.my

Y. S. Yusoff
e-mail: suhaylah@usim.edu.my

S. A. Abdul Karim et al. (eds.), *Proceedings of the 6th International Conference on Fundamental and Applied Sciences*, Springer Proceedings in Complexity,
https://doi.org/10.1007/978-981-16-4513-6_56

The same situation occurs in other countries. Coady et al. [2] found that the gasoline subsidies were mostly enjoyed by the richest 40% of households in Africa and Latin America. In Gabon, about one third of the total subsidy was received by the richest 10% group. However, only 13% of all subsidies were received by the poorest 30% of individuals in Gabon [3]. While in Nigeria, the richest 20% households received double the benefit of kerosene and petrol subsidies as compared to the poorest 20% households [4]. The negative effect of fuel subsidy is affected the poor households in China as they share the largest portion of their income on fuel [5]. This shows that the fuel subsidy is mostly benefited by the rich group. The study on the poverty impact of fuel price adjustment in Indonesia found that the increase in petroleum price by 29% will increase poverty rate by 1.95% which proved that the fuel price increment did had a significant impact on the poor households [6].

Consequently, this situation has also created many problems to Malaysian such as the increase in the price of necessities, food, transportation and others. All these problems have caused the insufficient income among Malaysian households. Several studies have pointed out this significant cause in Malaysia. Saari et al. [7] conducted a study on the impact of fuel price fluctuation across various ethnicities in Malaysia. This study found that Malays faced the biggest impact compared to Chinese and Indians. At the same time, another study by Roslan et al. [1] also investigated the impact of increasing fuel price on three different income groups in Malaysia. This study found that the middle-income group was significantly affected by the increase in fuel price. Moreover, other countries around the world also faced the problem of insufficient income due to the fuel price increment. Oktaviani et al. [8] found that the removal of fuel subsidy has resulted in the reduction of household income in Indonesia. Moreover, a study by Coady et al. [9] stated that when the fuel price increased by 50% on average, the average real income decreased by 4.60%. This is in line with the study by Godek and Murray [10] that showed the overall expenditure was reduced which was a result of the decrease in income due to the increase in the fuel price.

Malaysians nowadays have started taking initiatives to reduce the monthly expenditure in order to compensate the increase in fuel consumption cost. Rohani and Pahazri [11] found that there was a reduction in the frequency of personal vehicle usage. The reduction was about 2–7%, which was based on the trips that were frequently observed. This study also showed that the number of respondents using public transportation has increased when the fuel price increased. Another study by Ahmat [12] showed that individuals had changed their travel patterns because of income insufficiency, which was due to the increase in fuel price.

Therefore, this study undertakes with main aim to predict the household income for the next ten periods. Subsequently, this study analyses the effects of fuel price fluctuation on the Malaysian household income. Hence, this study could be beneficial for individual to plan their income and lives in better life.

2 Methodology

2.1 Data and Variables

The variables used in this study are household income, household expenditure, fuel price, household size, inflation, and gross domestic product (GDP). These variables are selected based on the literature review of the previous studies. These selected variables are proven to have relationship with the household income. This study used yearly data from 2006 to 2016. Based on this data, we run simulation to generate 10,000 simulated values. The simulated values are used to achieve the objective of this study due to limited availability of historical data.

2.2 Stationary Test

Firstly, the data will be tested for stationary property of each variable. The stationary property will be enabling the prediction of time series data. The Augmented Dickey-Fuller Test (ADF) is used. The ADF models are defined as follows:

(i) No constant and no trend (Model 1)

$$\Delta y_t = \beta_1 y_{t-1} + \sum_{s=1}^{m} u_s \Delta y_{t-s} + \varepsilon \tag{1}$$

(ii) Constant, no trend (Model 2)

$$\Delta y_t = \alpha + \beta_1 y_{t-1} + \sum_{s=1}^{m} a_s \Delta y_{t-s} + \varepsilon \tag{2}$$

(iii) Constant and trend (Model 3)

$$\Delta y_t = \alpha + \beta_1 y_{t-1} + \mu_t + \sum_{s=1}^{m} a_s \Delta y_{t-s} + \varepsilon \tag{3}$$

where y is the variable of interest, Δy_{t-1} is the difference between y_{t-1} and y_{t-2}, β_i is a set of parameters and ε is the error term.

The hypothesis testing is performed to test the stationary property of the variable of interest. The hypotheses are defined as follows:

H_0: *The series is not stationary.*
H_1: *The series is stationary.*

If the p-value is less than 5%, the null hypothesis is rejected and hence, the data is stationary.

2.3 Causal Relationship

After the stationary test, the study also examined the causal relationship between variables. The Granger causality test is used to examine the relationship between variables. The Granger causality test is expressed as follows:

$$Y_t = \gamma_0 + \sum_{z=1}^{p} \gamma_z Y_{t-z} + \sum_{i=1}^{q} \lambda_i X_{t-1} + \mu_t \tag{4}$$

$$X_t = \varphi_0 + \sum_{z=1}^{p} \delta_z X_{t-z} + \sum_{i=1}^{q} \varphi_i Y_{t-1} + \varepsilon_t \tag{5}$$

where Y_t and X_t are the tested variables, μ_t and ε_t are error terms, t is time period and z and i are the number of lags.

The hypothesis testing is performed to test the granger causality between variables. The hypotheses are defined as follows:

H_0: *No instantaneous causality between* y *and* x_i.
H_1: *There is instantaneous causality between* y *and* x_i.

If the p-value is less than 5% level of significance, the null hypothesis is rejected. Then, the variable Y is significantly contributed to forecast the value of another variable X, thus Y has a Granger causal relationship with X and vice versa. After investigating the Granger causal relationship between variables, the study will be forecasting all variables for the next ten periods.

2.4 Lag Selection

Firstly, the lag length should be chosen in order to run the test. This is done to ensure that the residuals are not serially correlated. This study used the Akaike's information criterion (AIC) to select the lag length. The Akaike information criterion (AIC) is as follows:

$$AIC = -2\ln(L) + 2k \tag{6}$$

where L is the value of the likelihood and k is the number of estimated parameters.

2.5 Prediction

Following that, the VAR (1) is performed. The VAR (1) is conducted to capture the independences among multiple time series and forecasting the future values of the selected variables. However, the focus of this study is to analyse the household income prediction due to the fuel price fluctuation. The VAR (1) models are defined as follows:

$$y_{1,t} = \beta_{1,1}y_{1,t-1} + \beta_{1,2}y_{2,t-1} + \beta_{1,3}y_{3,t-1} + \beta_{1,4}y_{4,t-1} + \beta_{1,5}y_{5,t-1} + \beta_{1,6}y_{6,t-1} + \varepsilon \quad (7)$$

$$y_{2,t} = \beta_{2,1}y_{1,t-1} + \beta_{2,2}y_{2,t-1} + \beta_{2,3}y_{3,t-1} + \beta_{2,4}y_{4,t-1} + \beta_{2,5}y_{5,t-1} + \beta_{2,6}y_{6,t-1} + \varepsilon \quad (8)$$

$$y_{3,t} = \beta_{3,1}y_{1,t-1} + \beta_{3,2}y_{2,t-1} + \beta_{3,3}y_{3,t-1} + \beta_{3,4}y_{4,t-1} + \beta_{3,5}y_{5,t-1} + \beta_{3,6}y_{6,t-1} + \varepsilon \quad (9)$$

$$y_{4,t} = \beta_{4,1}y_{1,t-1} + \beta_{4,2}y_{2,t-1} + \beta_{4,3}y_{3,t-1} + \beta_{4,4}y_{4,t-1} + \beta_{4,5}y_{5,t-1} + \beta_{4,6}y_{6,t-1} + \varepsilon \quad (10)$$

$$y_{5,t} = \beta_{5,1}y_{1,t-1} + \beta_{5,2}y_{2,t-1} + \beta_{5,3}y_{3,t-1} + \beta_{5,4}y_{4,t-1} + \beta_{5,5}y_{5,t-1} + \beta_{5,6}y_{6,t-1} + \varepsilon \quad (11)$$

$$y_{6,t} = \beta_{6,1}y_{1,t-1} + \beta_{6,2}y_{2,t-1} + \beta_{6,3}y_{3,t-1} + \beta_{6,4}y_{4,t-1} + \beta_{6,5}y_{5,t-1} + \beta_{6,6}y_{6,t-1} + \varepsilon \quad (12)$$

where y_1 is household income, y_2 is household expenditure, y_3 is fuel price, y_4 is household size, y_5 inflation rate and y_6 is gross domestic product (GDP).

3 Results and Discussions

3.1 Stationary Test

The stationary property of all variables was tested using ADF test to avoid the possibility of estimating spurious regressions. Three models were used in this study. Model 1 does not include constant and trend. While Model 2 included only the constant and Model 3 included both the constant and trend. The results of ADF test for all variables from three different models are shown in Tables 1, 2 and 3. The *p*-values of all

Table 1 Augmented Dickey-Fuller (ADF) results for Model 1

Variables	*p*-values	Conclusions
Household income	0.000	Stationary at level
Household expenditure	0.000	Stationary at level
Fuel price	0.000	Stationary at level
Household size	0.000	Stationary at level
Inflation	0.000	Stationary at level
GDP	0.000	Stationary at level

Table 2 Augmented Dickey-Fuller (ADF) results for Model 2

Variables	*p*-values	Conclusions
Household income	0.000	Stationary at level
	0.000	There is a constant term
Household expenditure	0.000	Stationary at level
	0.000	There is a constant term
Fuel price	0.000	Stationary at level
	0.000	There is a constant term
Household size	0.000	Stationary at level
	0.000	There is a constant term
Inflation	0.000	Stationary at level
	0.000	There is a constant term
GDP	0.000	Stationary at level
	0.000	There is a constant term

Table 3 Augmented Dickey-Fuller (ADF) results for Model 3

Variables	*p*-values	Conclusions
Household income	0.000	Stationary at level
	0.354	There is no trend
Household expenditure	0.000	Stationary at level
	0.243	There is no trend
Fuel price	0.000	Stationary at level
	0.057	There is no trend
Household size	0.000	Stationary at level
	0.195	There is no trend
Inflation	0.000	Stationary at level
	0.197	There is no trend
GDP	0.000	Stationary at level
	0.479	There is no trend

variables are 0.000 for Model 1. This indicates that the null hypothesis is rejected. Thus, all variables are stationary at level. Model 2 also showed the same results as Model 1 whereby all variables are stationary at level. The constant term that included in Model 2 shows significant result as *p*-values for all variables are less than 5%. Hence, there is a constant term. However, the constant term is not considered in this study. For the Model 3, the *p*-values are less than 5% for all variables. This indicates that all variables are stationary at level. However, the trend term shows insignificant result and there is no trend term existed. Thus, the trend term will not be included to predict the future values. Since the data are stationary, it is appropriate to proceed with the Granger causality test.

3.2 Granger Causality Relationship

Since all variables are stationary at level, the Granger causality test is performed to investigate the causal relationship between the selected variables. The results are presented in Table 4. The results indicated that household income is Granger cause by other selected variables namely household expenditure, fuel price, household size, inflation and GDP. This means that the changes in household income are mostly driven by the household expenditure, fuel price, household size, inflation and GDP. Household expenditure also found to be Granger cause by household income, fuel price, household size, inflation and GDP. This indicates that the household expenditure influences household income, fuel price, household size, inflation and GDP. Then, the other variables namely fuel price, household size, inflation and GDP also showed the same results. These results are similar to other studies which found there is short run and long run relationship between household expenditure, income and other macroeconomics variables [13–17]. The findings of this study suggest that changes in fuel price and other macroeconomics variables would affect the household income. Since there is a relationship between these selected variables, the VAR can be performed.

Table 4 Granger causality relationship results

Variables	*p*-value	Decision
Household income	0.0000	Reject H_o
Household expenditure	0.0000	Reject H_o
Fuel price	0.0000	Reject H_o
Household size	0.0000	Reject H_o
Inflation	0.0000	Reject H_o
GDP	0.0000	Reject H_o

3.3 Lag Selection and Prediction

Before the VAR was conducted, the lag order selection criteria were performed using Akaike's information criterion. The results (not included in this paper) showed that the most suitable lag length for all variables is lag one. This indicates that the time series for all variables are not serially correlated. Thus, we can proceed with the VAR (1) and predict all variables for the next ten periods. After the lag selection is done, the VAR (1) model without trend and constant is performed. The estimated equations for each variable are as follows:

$$\begin{aligned}\hat{y}_{1,t} &= 0.090y_{1,t-1} + 0.0413y_{2,t-1} + 1105y_{3,t-1} + 25.26y_{4,t-1} \\ &\quad + 22620y_{5,t-1} + 1.624y_{6,t-1} + \varepsilon \end{aligned} \tag{13}$$

$$\begin{aligned}\hat{y}_{2,t} &= 0.053y_{1,t-1} + 0.024y_{2,t-1} + 481y_{3,t-1} + 3.169y_{4,t-1} \\ &\quad + 13040y_{5,t-1} + 0.644y_{6,t-1} + \varepsilon \end{aligned} \tag{14}$$

$$\begin{aligned}\hat{y}_{3,t} &= 0.0000446y_{1,t-1} + 0.0000118y_{2,t-1} + 0.593y_{3,t-1} + 0.0098y_{4,t-1} \\ &\quad + 10.97y_{5,t-1} + 0.000675y_{6,t-1} + \varepsilon \end{aligned} \tag{15}$$

$$\begin{aligned}\hat{y}_{4,t} &= 0.000121y_{1,t-1} + 0.0000137y_{2,t-1} + 0.8481y_{3,t-1} + 0.0212y_{4,t-1} \\ &\quad + 21.77y_{5,t-1} + +0.00129y_{6,t-1} + \varepsilon \end{aligned} \tag{16}$$

$$\begin{aligned}\hat{y}_{5,t} &= 0.000000702y_{1,t-1} + 0.000000139y_{2,t-1} + 0.0103y_{3,t-1} \\ &\quad + 0.0001658y_{4,t-1} + 0.2045y_{5,t-1} + 0.0000128y_{6,t-1} + \varepsilon \end{aligned} \tag{17}$$

$$\begin{aligned}\hat{y}_{6,t} &= 0.00588y_{1,t-1} + 0.00195y_{2,t-1} + 83.55y_{3,t-1} + 1.702y_{4,t-1} \\ &\quad + 1448y_{5,t-1} + 0.07207y_{6,t-1} + \varepsilon \end{aligned} \tag{18}$$

From the estimated equations above, all variables have a positive coefficient and hence indicated the positive relationship. The results for VAR (1) (not included in this paper) showed that all variables have p-value less than 0.000. This indicates that all variables have a significant relationship. These findings similar to other studies which found a significant relationship between household expenditure, income and other macroeconomics variables [15–21].

Next, the forecasted values for all variables are calculated for ten periods ahead. The results showed in Table 5. Based on the results, the fuel price showed a decreasing trend for the next ten periods which cause the decrease in expenditure. Hence, the decreasing trend reflected the household income as well. This finding revealed that lower household income will be enough if the fuel price keeps decreasing. However,

Table 5 Forecasted values for next 10 years

Household income	Household expenditure	Fuel	Household size	Inflation	GDP
4057.554	1935.910	2.059	3.533	0.0363	281.229
4086.512	1917.166	2.048	3.490	0.0359	278.435
4062.885	1906.708	2.036	3.472	0.0357	276.878
4039.919	1895.894	2.025	3.452	0.0355	275.313
4017.071	1885.171	2.013	3.433	0.0353	273.756
3994.761	1874.509	2.002	3.413	0.0351	272.207
3971.761	1863.907	1.990	3.394	0.0349	270.668
3949.298	1853.366	1.979	3.375	0.0347	269.137
3926.962	1842.884	1.968	3.356	0.0345	267.615
3904.752	1832.461	1.957	3.337	0.0343	266.101

when the fuel price increases, the higher household income will be required. Moreover, other macroeconomics factors such as inflation and GDP also predicted to be decreasing over the next ten periods which also contributed to decrease in the household income.

4 Conclusion

This study was carried out with main objective to predict the household income due to the fuel price fluctuation in Malaysia. The data were simulated based on the historical data from 2006 to 2016. The simulation is carried out due to the limited data availability. The variables used in this study are household income, household expenditure, fuel price, household size, inflation rate and gross domestic product (GDP). Firstly, all variables are tested for stationary property and the results showed that all variables are stationary at level. Then, the Granger causality test is performed to investigate the causal relationship between the selected variables. This study found that there is a significant causal relationship between the selected variables. Following that, the prediction is performed using VAR (1) model. All variables showed a decreasing trend over the forecast periods. This study revealed that when the fuel price decreased, the household income will be decreased. The other selected variables also showed the same effect in predicting the household income. In conclusion, this prediction model could be beneficial in helping individual to plan their income efficiently.

Acknowledgements The authors would like to express sincere gratitude to the Universiti Sains Islam Malaysia for this research under grant number PPPI/FST/0217/051000/10718.

References

1. Roslan, A.H., Russayani, I., Nor Azam, A.R.: Fuel subsidy reform in Malaysia: an assessment on the direct welfare impact on consumers. Int. Postgrad. Bus. J. **8**(1), 26–36 (2016)
2. Coady, D., Gillingham, R., Ossowski, R., Piotrowski, J., Tareq, S., Tyson, J.: Petroleum product subsidies: costly, inequitable and rising. IMF Staff Position Note SPN/10/05. International Monetary Fund, Washington, DC (2010)
3. El-Said, M., Leigh, D.: Fuel price subsidies in Gabon: fiscal cost and distributional impact. IMF Working Paper 06 (243). International Monetary Fund, Washington, DC (2006)
4. Soile, I., Mu, X.: Who benefit most from fuel subsidies? Evidence from Nigeria. Energy Policy, 314–324 (2015)
5. Jiang, Z., Ouyang, X., Huang, G.: The distributional impacts of removing energy subsidies in China. China Econ. Rev. **33**, 111–122 (2015)
6. Sugema, I., Hassan, M., Oktaviani, R., Viliani, A., Ritonga, H.: Dampak kenaikan harga BBM dan efektivitas program kompensasi. INDEF Working Paper (2005)
7. Saari, M.Y., Dietzenbacherc, E., Los, B.: The impacts of petroleum price fluctuations on income distribution across ethnic groups in Malaysia. Ecol. Econ. **130**, 25–36 (2016)
8. Oktaviani, R., Hakim, D.B., Siregar, H.: Impact of a lower oil subsidy on Indonesian macroeconomic performance, agricultural sector and poverty incidences: a recursive dynamic computable general equilibrium analysis. MPIA Working Paper, No. 2007-28, 2007. Available at SSRN: https://ssrn.com/abstract=1086380
9. Coady, D., El-Said, M., Gillingham, R., Kpodr, K., Medas, P., Newhouse, D.: The magnitude and distribution of fuel subsidies: evidence from Bolivia, Ghana, Jordan, Mali and Sri Lanka. IMF Working Paper, 06 (247). International Monetary Fund, Washington, DC (2006)
10. Godek, J., Murray, K.: Effects of spikes in the price of gasoline on behavioral intentions: a mental accounting explanation. Behav. Decis. Making **3**(3), 1–15 (2011)
11. Rohani, M.M., Phazri, N.: Survey on how fluctuating petrol prices are affecting Malaysian large city dwellers in changing their trip patterns. IOP Conf. Ser. Earth Environ. Sci. **140** (2018)
12. Ahmat, N.: Perubahan Harga Petrol dan Risiko Keselamatan Jalan Raya **2**(VII), 1435–1444 (2012)
13. Guisan, M.C.: A comparison of causality tests applied to the bilateral relationship between consumption and GDP in the USA and Mexico. Int. J. Appl. Econ. Quant. Stud. **1**(1), 115–130 (2004)
14. Mallik, L., Pradhan, K.C.: Per capita consumption expenditure and personal disposal income in India—an econometric analysis. Int. J. Econ. Res. **3**(2), 96–102 (2012)
15. Ikwuagwu, E.B., Ariwa, F.O., Onyele, K.O.: Determinants of aggregate consumption expenditure in Nigeria. Int. J. Econ. Finan. Manag. **3**(2) (2017)
16. Bonsu, D.O., Muzindutsi, P.F.: Macroeconomics determinants of household expenditure in Ghana: a multivariate cointegration approach. Int. J. Econ. Financ. Issues **7**(4), 737–745 (2017)
17. Amin, S.B.: Causal relationship between consumption expenditure and economic growth in Bangladesh. World J. Soc. Sci. **1**(2), 158–169 (2011)
18. Akekere, J., Yousuo, P.O.J.: Empirical analysis of change in income on private consumption expenditure in Nigeria from 1981 to 2010. Int. J. Acad. Res. Bus. Soc. Sci. **2**(12), 312–331 (2010)
19. Ofwona, A.C.: An estimation of the consumption function for Kenya using Keynes' absolute income hypothesis for the period 1992–2011. J. Emerg. Trends Econ. Manag. Sci. **4**(1), 103–105 (2013)
20. Ezeji, C.E., Ajudua, E.I.: Determinants of aggregate consumption expenditure in Nigeria. J. Econ. Sustain. Dev. **6**(5), 164–168 (2015)
21. Abayomi, O., Michael, O., Olaronke, O.: Income and household consumption expenditure in Nigeria. MPRA Paper No. 83334 (2015). Available at https://mpra.ub.uni-muenchen.de/83334/

Review on Cardiovascular Disease Risk Factors Among Selected Countries in Asia

Yumn Suhaylah Yusoff, Norhana Abd. Rahim, and Muhammad Hariz Hasmi

Abstract Cardiovascular disease is known as the principal cause of death worldwide. Among all the diseases related to cardiovascular disease, myocardial infarction and stroke are the most common contributors to the increased death rate. This paper aims to study the risk factors affecting cardiovascular disease, especially in the region of Asia. This paper reviews past studies regarding cardiovascular disease to investigate the contributing risk factors for this disease. This study found that most of the cardiovascular disease risk factors are related to unhealthy lifestyles. Blood pressure level, smoking status, body mass index, cholesterol level and blood glucose level are among the most significant risk factors for cardiovascular disease. Promoting a healthy lifestyle towards all citizens is necessary to reduce the mortality rate due to cardiovascular disease. A comprehensive prevention plan by the government is essential to reduce the risk and the mortality rate of cardiovascular disease.

Keywords Cardiovascular disease · Myocardial infarction · Stroke · Risk factors

1 Introduction

The heart is an organ in the human body, which ensures continuous blood flow in the blood vessel without any stop. Heart functions to pump the blood to the entire human body. This process is vital to ensure an adequate supply of oxygen to cells throughout the body. The process of blood circulation begins with oxygenated blood cells, which are blood cells that carry oxygen from the lung to the heart. Then, the heart will pump the oxygenated blood cell to the entire body. The oxygenated blood cells will supply the body's tissue with the oxygen needed by the body. Lastly, after the exchange of oxygen and carbon dioxide at the body tissue, the deoxygenated blood cell, which

Y. S. Yusoff (✉) · N. Abd. Rahim · M. H. Hasmi
Faculty of Science and Technology, Universiti Sains Islam Malaysia, 71800 Nilai, Negeri Sembilan, Malaysia
e-mail: suhaylah@usim.edu.my

N. Abd. Rahim
e-mail: norhana@usim.edu.my

S. A. Abdul Karim et al. (eds.), *Proceedings of the 6th International Conference on Fundamental and Applied Sciences*, Springer Proceedings in Complexity,
https://doi.org/10.1007/978-981-16-4513-6_57

carries carbon dioxide, will be transported to the heart before being pump back to the lung. This process will continuously occur until the human dies. The blood in the human body does not only carries oxygen and carbon dioxide, but it also carries nutrients, hormones, waste materials, and interstitial fluid that are needed by the body [1]. Approximately, there are 72 heartbeats per minute. The human heart weight was estimated at around 300–350 g for men and 250–300 g for women [2]. Like other organs in the human body, the heart also requires oxygen to function continuously. The artery that provides the blood to the heart is called the coronary artery [1].

Humans' routine changed from being active and focused on farming for food to being inactive with high consumption of fast food in the last decades due to advancing technology. Besides that, a worldwide increase in cigarette consumption increases the risk of cardiovascular disease (CVD). CVD is a phrase used to describe diseases related to heart and blood vessels. CVD occurs when a blood vessel has narrowed or blocked, disturbing the oxygen supply to the organ's tissue [3]. This condition is known as atherosclerosis, a condition when plaques from fats, cholesterol and other substances developed at the artery's inner wall and disturb the blood circulation system. The plaques' development starts with the formation of tiny cholesterols crystals inside the intima and its underlying smooth muscle. The plaque will then expand with the development of fibrous tissues and its surrounding smooth muscle. The plaque will bend within the arteries and lower the flow of blood. Fibroblast connective tissue development and calcium accumulation in the lesion will cause the arteries to harden. The arteries' irregular surface results in the growth of clots and thrombosis, leading the blood flow being suddenly obstructed [4].

There are two types of CVD, which describes the diseases of the heart and blood circulation. The most common cause for the heart disease is myocardial infarction (MI) or heart attack, and the most common cause for the disease of the circulation is cerebrovascular accident or stroke [2].

Ischaemic Heart Disease (IHD) is a disease due to blockage of the blood flow in the coronary artery. MI is the results of IHD. MI is the condition when there is a blood clot that blocks the coronary artery from delivering blood, which supplies the heart muscle with oxygen. MI also refers to the condition when the myocardial cell dies due to shortage of oxygen [5]. Based on the MI definition, not all MI cases would be due to the presence of a blood clot. The oxygen supply to the heart muscle must be equal to the oxygen demand by the heart muscle. Any imbalance ratio of supply and demand will lead to a rapid heart rate or increased blood pressure. This condition can lead to MI without the existence of blood clot [6].

Stroke is also the most frequent cause of death for patients with CVD. A stroke occurs when there is a sudden decrease in blood flow to the brain and causes brain functions loss due to lack of oxygen and nutrients supply to the brain. A stroke usually presents the symptoms of hemiparesis, vomiting, drowsiness, and loss of consciousness. The most common type of stroke is the ischemic stroke due to blockage of blood vessel towards the brain, while haemorrhagic stroke is due to the spontaneous rupture of blood vessel [7].

CVD was the principal factor of mortality and morbidity worldwide. The total number of deaths due to CVD had increased universally from 14.4 million death in 1990 to 17.5 million death in 2012. In 2012, 31% of worldwide deaths were caused by CVD, and it has been the largest single contributor to global mortality and was projected to maintain its dominance of the world mortality trends in the future [8]. CVD had also contributed to more than 30% of all deaths worldwide, and 82% of CVD deaths happen in low and middle-income countries. It is also predicted around 23.6 million people would die annually due to CVD during 2030, and the most significant rise of CVD deaths would be in South East Asia countries [3]. CVD treatment is the most expensive, with 11% or 5.4 billion dollars of the total expenditure on health was from CVD cases in 2000–2001. Other than that, CVD survivors were mostly disabled and are not able to work. These had become a tremendous financial burden for the patients [3].

Many risk factors can contribute to the trends of CVD cases. This paper aims to review the trend of risk factors on CVD among selected countries in Asia: China, India, Indonesia, Japan, Malaysia, Singapore, and Thailand. Section 2 will review the distribution of CVD among selected countries. Following this, Sect. 3 will discuss the prevalence of the risk factors for CVD among selected countries and the relationship between each country's death trend. Section 4 will discuss the prevention of CVD. Section 5 will discuss the review's findings, and Sect. 6 will conclude all the finding in the research.

2 Distribution of Cardiovascular Disease Among Selected Countries

Type of CVD that are fatal differs among countries. China, Cambodia, North Korea, South Korea, Vietnam, Bangladesh, and Myanmar have a higher mortality rate due to stroke than MI, while in all the other Asian countries the mortality rate due to MI is higher than mortality resulted from stroke [9]. Figure 1 shows the proportion of MI and stroke deaths among the selected countries and shows Malaysia, India, and China to have the highest mortality due to MI in 2016. Other than that, these three countries show an increasing rate of MI mortality in the 16 years. Figure 2 shows that stroke mortality is highest in China and has an increasing trend in China, Indonesia, and India.

Studying the mortality rate of CVD will allow a glance at the mortality rate of MI and stroke since most CVD were due to MI and stroke. Among all CVD cases, MI and stroke comprise 87.8% of all CVD deaths worldwide [9]. Figure 3 shows the CVD mortality rate of the selected countries. Based on the figure, China had the highest CVD mortality rate compared to other countries. China has also shown a 10% increase in CVD mortality rate from 2000 to 2016. Many studies reported a decline in physical activity among individuals in China. During 2002, the Chinese National Nutrition and Health Survey had proved that only 34% of adults in a large city and 45% adults in the small and medium cities have an active lifestyle. Only 18.7% of Chinese adults have regular physical activity. The study believes that the

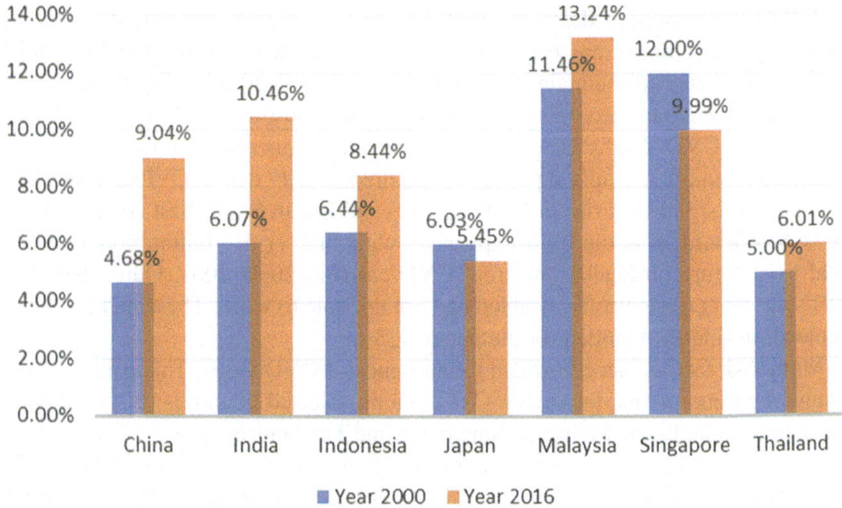

Fig. 1 Percentage of myocardial infarction death in selected countries [9]

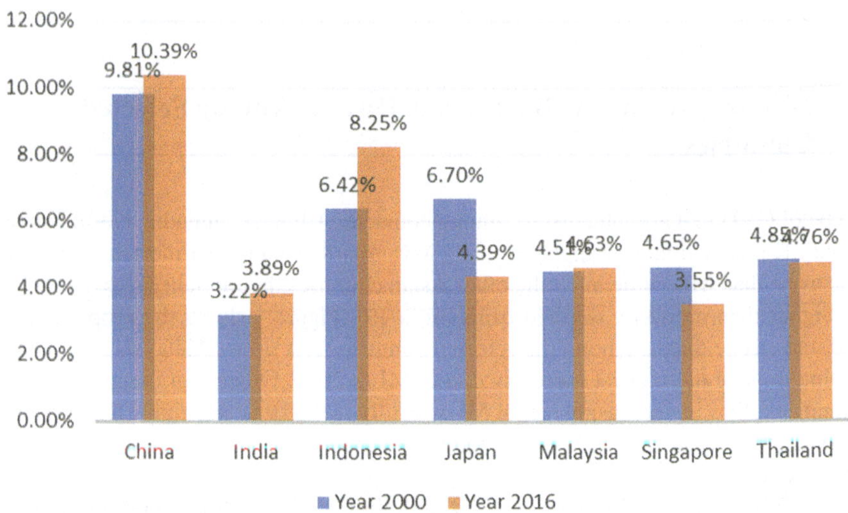

Fig. 2 Percentage of stroke death in selected countries [9]

decline in the number of active Chinese adults will contribute to China's rising cases of CVD deaths [10].

The graph also shows that Japan and Singapore have a declining trend in CVD's mortality from 2000 to 2016. Another study discovered that Japan also reported the lowest MI death worldwide, with 47 deaths per 100,000 populations [11]. Although there was a declining in the statistics of CVD death in Singapore, CVD is still the main

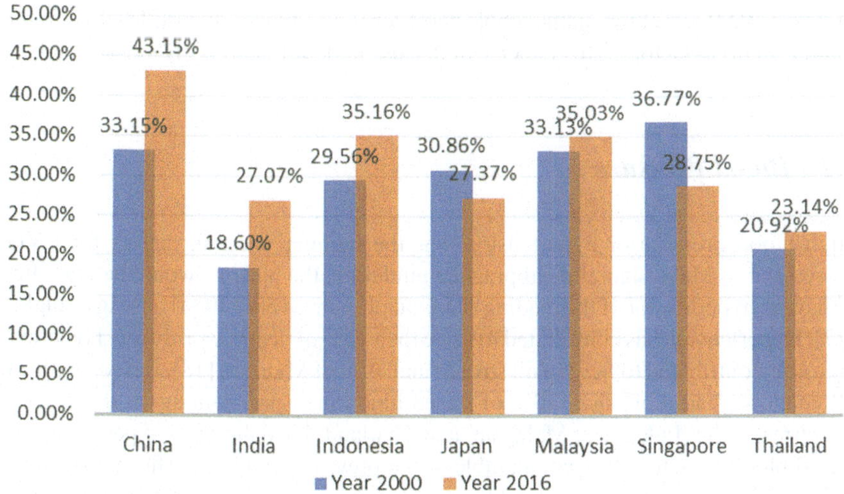

Fig. 3 Percentage of cardiovascular disease death in selected countries [9]

contributor to Singapore's total death because its population was moving towards the ageing population [12]. The graph also shows that China, India, Indonesia, Malaysia, and Thailand have an increased CVD mortality rate. The risk factors which contribute to the different CVD trends would be discussed further in the next section.

3 Prevalence of Risk Factors Among Selected Countries

Many risk factors contribute to CVD. There are two types of risk factors: controllable risk factors and uncontrollable risk factors. The uncontrollable risk factors are the risk factors that could not be modified or avoided. Family history, age, ethnicity, and gender are among the uncontrollable risk factors of CVD. Whereas the controllable risk factors were the risk factors that could be modified or avoided. Dyslipidaemia, hypertension, hyperglycaemia, low physical activities, low nutrition consumption, smoking habits and BMI were the controllable CVD risk factors [13]. A study reported that current smoking behaviour, psychosocial factors, diabetes mellitus, hypertension history, abdominal obesity, alcoholic drinks consumption, reduced regular activities and low daily consumption of fruits and vegetables are the significant CVD risk factors [14]. Another study discovered that diabetes mellitus, hypertension, CVD family history, renal disease, Percutaneous Coronary Intervention (PCI), Killip class and age were the main contributing CVD risk factors [15]. National Health and Morbidity Survey 2019 reported hypertension, diabetes, and high cholesterol as the main risk factors of CVD in Malaysia. 1.7 million people in Malaysia have these three risk factors, while 3.4 million people have two of these three risk factors [16]. Based on a lot of research, it can be concluded that blood pressure, BMI, smoking status, cholesterol level and blood glucose level were the main CVD risk factors. This paper will discuss in detail these CVD's risk factors.

3.1 Blood Pressure

High blood pressure or hypertension was the number one risk factors for CVD. According to Malaysian clinical practice guideline, the patients would be classified as having hypertension if the reading of blood pressure were 140/90 mmHg or higher [17]. Hypertension has contributed to more than 15% of deaths worldwide. Hypertension also contributed to 62% of all strokes and 49% of MI cases [18]. Based on Fig. 4, all countries had more than 10% of raised blood pressure. Among all these seven countries, India, Indonesia, Malaysia, and Thailand, have the highest percentage of raised blood pressure. It is comparable to the previous graph that shows that India, Indonesia, Malaysia, and Thailand have an increasing trend of CVD mortality. This finding concludes that hypertension has a high impact on CVD mortality.

Hypertension is a disease which is related to the increase in age. Increasing age would increase the risk of having hypertension. It was approximated that 60% of people aged 60 years and above have hypertension. Since the global population is moving towards an ageing population, it is predictable that most of the worldwide population aged 60 years and above would develop hypertension by 2030 [19]. One of the factors that contribute to the increasing rate of hypertension cases was excessive salt consumption. Current salt consumption in the world population is alarming. Although sodium has many benefits for the human body, consuming too much sodium

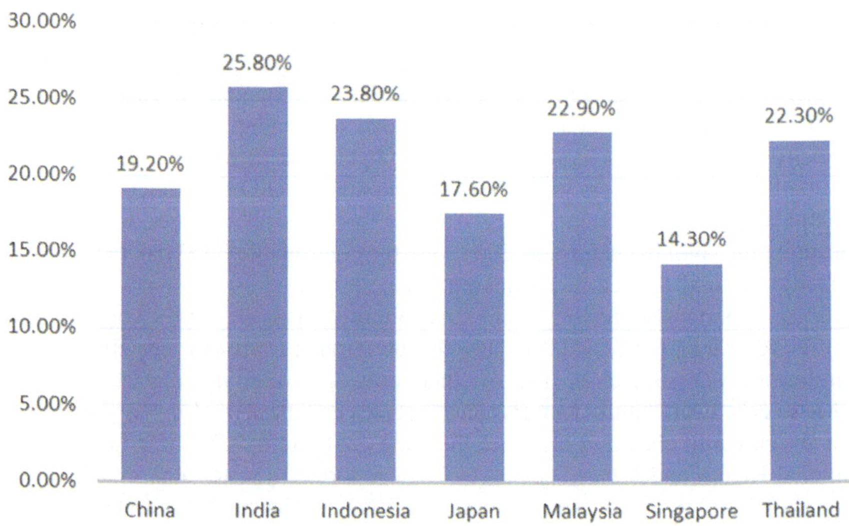

Fig. 4 Percentage of raised blood pressure in selected countries [9]

than what is needed by the body would overload the metabolic system. The overload would increase blood pressure and cause millions of deaths, mostly related to CVD [18].

3.2 Smoking Status

Cigarette smoking was found to be the largest risk factor for premature death among developing countries. Premature death is a condition when a person dies before the average age of the country. It was estimated that approximately 3 million deaths were reported per year worldwide due to smoking-related diseases, and 10 million deaths were forecasted to occur due to smoking-related diseases [20]. Smoking was the second leading risk factor for CVD after high blood pressure. According to the World Health Organization (WHO), more than 1 billion smokers exist worldwide, and the statistics keep increasing [21].

Figure 5 shows the percentage of smokers amongst the selected countries. By analyzing the graph, most of the countries have nearly half of their population that were smokers. India and Singapore recorded the smallest number of smokers' percentage. Indonesia, China, and Malaysia recorded the highest number of smokers in these countries. These countries showed the highest statistics for CVD mortality during 2016. This finding proves that smoking has a high impact on CVD mortality trends.

There are many reasons for the youths to be involved in smoking, such as peer pressure, desire to look mature, kill boredom, and to help relax and release stress.

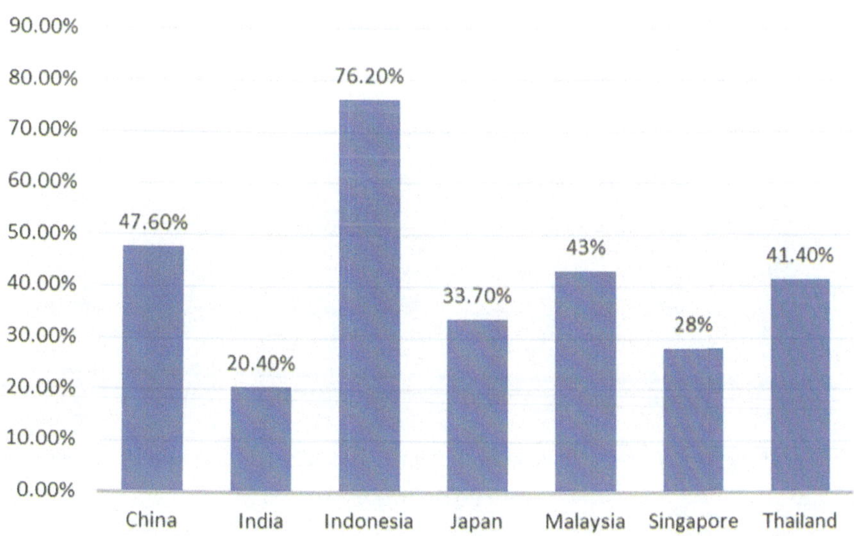

Fig. 5 Percentage of smokers in selected countries [9]

Usually, this would be learned from their friends, influenced by family members who are also smokers or tried by themselves because of curiosity [22]. There are many reasons for people to start smoking either due to the environment or genetic factors. Many smokers would try to stop smoking or reduce their cigarette consumption, but many tend to fail due to strong nicotine addiction. Since majority of smokers are men, it is found that men are two to five times at greater chance of developing CVD [21].

3.3 Body Mass Index

BMI is another significant risk factor for the development of MI and stroke. In Malaysian clinical practice guideline, individuals with BMI reading between 18.5 and 22.9 kg/m^2 were classified as normal. Individuals being above these values are considered obese [23]. Some studies reported that BMI does not directly affect CVD, but it has a second-order CVD effect. High BMI value would increase the risk of developing hypertension, diabetes and hypercholesterolaemia. Smoking had a little impact on BMI. Quitting smoking could raise the BMI level while continuing to smoke cigarette could reduce the BMI level. Studies have proved that reducing BMI would result in the reduction of CVD occurrence risk [24].

Figure 6 demonstrates that most of the countries showed an increase in the BMI average from the year 2000–2016. As for overweight individuals, their BMI values would be in the range of 25–40. Individuals passing the value of 40 would be classed as being obese. Most countries were nearly approaching the average BMI range of

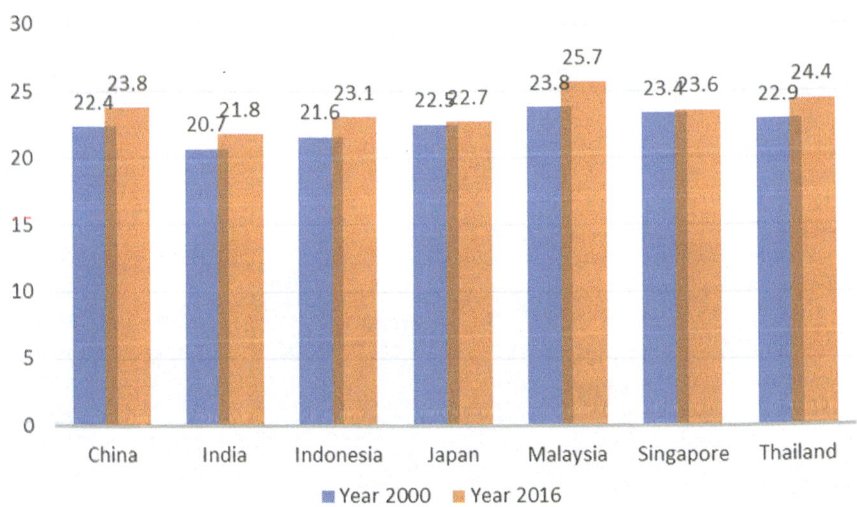

Fig. 6 Average body mass index in selected countries [9]

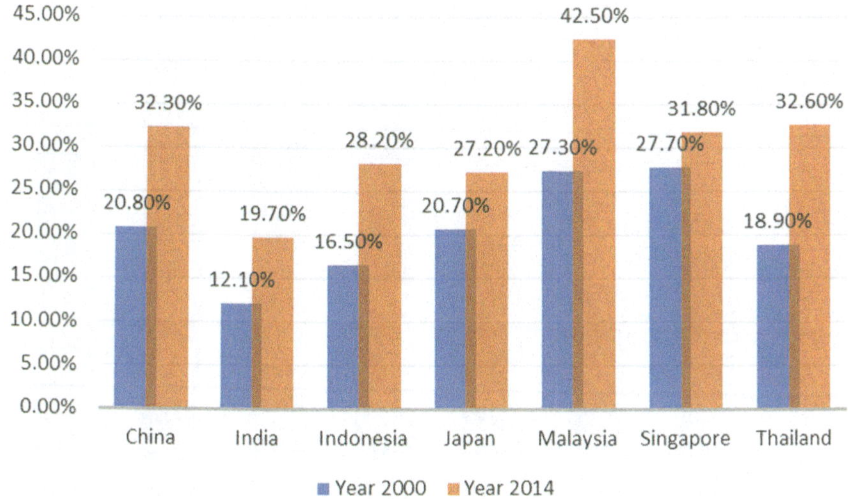

Fig. 7 Percentage of overweight citizens in selected countries [9]

overweight while Malaysia had already passed this range. This finding shows that most of the countries are moving towards the average BMI of being overweight.

Figure 7 also exposed that Malaysia has 42.5% of individuals categorised as being overweight. The graph also shows that the percentage of overweight individuals in all countries has increased from 2000 to 2016. This finding is alarming since the increase in BMI could trigger other CVD risk factors.

Physical inactiveness is the main factor for the increase in BMI level, which can lead to obesity. Increase in BMI would also increase the risk of CVD. Individuals with high physical activity or fitness have the potential to counter the risk of mortality caused by obesity [25].

3.4 Cholesterol

Dyslipidaemia is the condition in which the blood lipid or cholesterol levels are increased. Increase in the blood cholesterol level could cause ischemia or blood vessel blockage in the heart [26]. According to clinical practice guideline in Malaysia, patient with a total cholesterol level higher than 5.2 mmol/L would be diagnosed with dyslipidaemia [27]. Figure 8 shows that Japan, Malaysia, Singapore, and Thailand have more than 50% of the individuals with raised total cholesterol level. Although China and Indonesia recorded the highest CVD deaths, the percentage of individuals with raised total cholesterol is lower. The effect of blood cholesterol level towards CVD is antagonistic in this study, suggesting that the cholesterol level may not have a high impact on CVD risk in these countries.

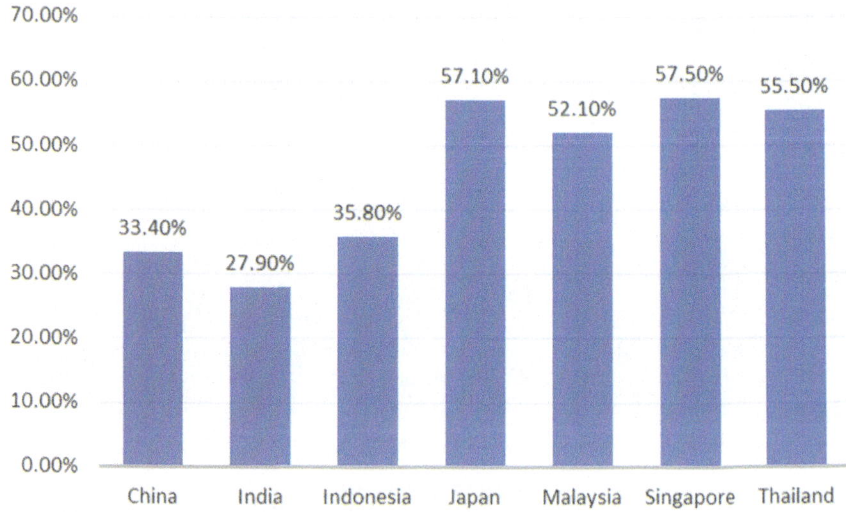

Fig. 8 Percentage of raised total cholesterol in selected countries [9]

Increased cholesterol level would increase the CVD risk, and a decrease in cholesterol level would reduce CVD risk. There are many strategies to reduce the blood cholesterol level, which in turn would reduce the risk of CVD, such as by consuming statins [28].

3.5 *Blood Glucose Level*

Blood glucose level has a significant positive effect on CVD risk. Increase in blood glucose level would increase the risk of CVD [23, 24]. For diabetic patients, the clinical practice guideline in Malaysia had set the reading for venous plasma glucose would be higher than 7.0 mmol/L for fasting patient and 11.1 mmol/L for non-fasting patients [29]. Figure 9 shows that Malaysia has the highest percentage of raised blood glucose, and Japan has the lowest percentage. Findings for other countries varied among each other. The impact of blood glucose level on CVD mortality may not be high, since the percentage of raised blood glucose level is lower than other CVD risk factors that had been discussed.

Diabetes is a well-recognized risk factor for CVD. Diabetic patients who consume insulin should monitor their blood glucose level routinely to assure their diabetes is controlled with the correct diet, exercise, and medication. Taking good care of blood glucose level would reduce the risk of CVD [30].

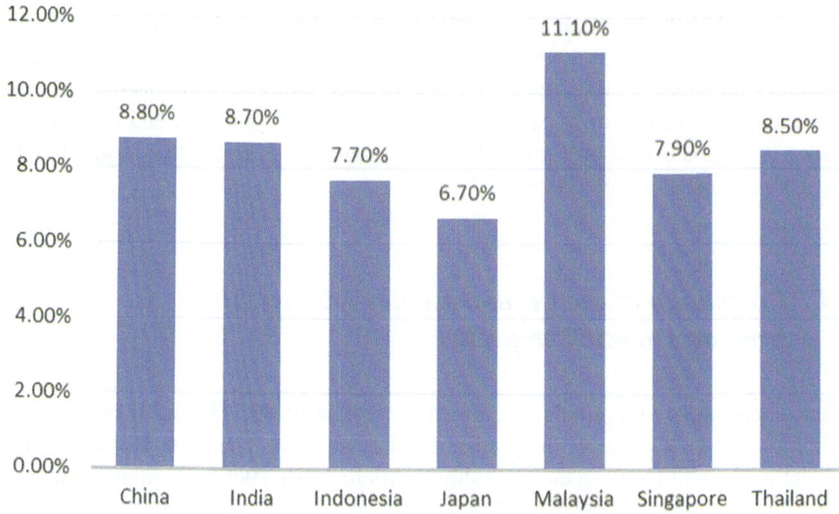

Fig. 9 Percentage of raised fasting blood glucose in selected countries [9]

4 Methods of Preventing the Risk of Suffering from Cardiovascular Disease

Based on the risk factors mentioned above, most of the risk factors mentioned were related to an unhealthy lifestyle. Replacing the bad habit of lifestyles with better habits would reduce the CVD risk, especially for developing MI and stroke. There are many ways to reduce the risk of MI and stroke. Quitting or reducing smoking, having a healthy diet, and being active physically by playing sports and jogging could help reduce this risk. Implementing a healthy lifestyle would lead to cessation of deaths related to MI and stroke [14]. According to the National Health and Morbidity survey 2019, CVD was considered a non-communicable disease. Some ways to help reduce the risk of having the non-communicable disease are by controlling the blood pressure to be below 140/90 mmHg, practicing a healthy diet, maintaining a healthy weight, having regular exercises, stop smoking and reducing the consumption of alcohol [16].

Lessening salt intake would assist to lower the high blood pressure Reducing the number of individuals having high blood pressure may lead to excellent health gain. WHO had recommended reducing the consumption of salt to 5 g per day [18]. Other than that, quitting smoking could also be beneficial for CVD risk reduction. About 40% of smokers try to quit smoking, but only 4–6% would be successful. This failure is due to high nicotine addiction and lack of motivation. Having great motivation would influence someone to stop smoking and having a better life [31]. Having an active lifestyle would reduce the risk of being overweight and obese. People who are obese but have an active lifestyle tend to have a lower risk of CVD than those with an inactive lifestyle. People who are active in sports also less likely to be smokers [25]. Taking statin may result in reducing blood cholesterol or lipid levels. With the exception for individuals with extreme cholesterol panel, controlling diet or dietary therapy may help reduce the risk of developing hyperlipidaemia. Having a balanced diet, proper exercise, consuming fruits and vegetables, and fish oil may reduce the blood cholesterol level [32]. Reducing sugar consumption would reduce the risk of diabetes development [33].

4.1 Prevention of the Cardiovascular Disease Risk from an Islamic Perspective

Islam encourages individuals to take care of their health. Having a good health condition not only prevents individuals from developing dangerous diseases, but it also helps them to focus on their worship to Allah. Allah SWT had said in the Quran:

One of the two women said, "O my father, take him as a hired man (shepherd of sheep), surely the best person you choose is the one who is strong and trustworthy" (Al-Qasas: 26).

From Abi Hurairah RA, Rasulullah SAW said: "A strong believer is better and more favoured by Allah than a weak believer. Even when both are good. Do your utmost to obey Allah and seek His help and do not be discouraged. If you are hit by disaster, do not say 'if I had not done that', but say 'Allah had destined it and what Allah desires would surely happen.' Indeed, the word 'if' could open the door to the devil" (Sahih Muslim).

Both verses from Al-Quran and Al-Hadith explain that it is essential to be strong since Allah love strong believers. 'Strong' means having a healthy life. Healthy people tend to have more energy to work efficiently and perform worship to Allah calmly. Allah SWT also Said in the Quran:

And We say: "Eat of the good things which We have given to you and do not cross the limit because who are over the limit would face My anger and whoever faces My anger would certainly perish" (Taha: 81).

This verse explains that Allah has permitted us to eat all the good foods provided on the earth, but do not consume excessively until the limit is crossed. Anything that is over the limit would usually bring disaster. For example, consuming too much salt would increase the chance of developing high blood pressure, and too much sugar would lead to diabetes. Hence, we need to be balanced and not crossover the limit. Rasulullah SAW had said:

> And indeed, every human body has lumps of blood that if it is good then the body would be good, and if it is damaged the body would be damaged. You know, it is Al-Qalb. (Sahih Bukhari)

From this verse, it can be understood that to have a healthy body, not only having an intense physical activity is required, but the spiritual needs should also be complete. Having a good spiritual condition may help individuals to overcome stress and tension, hence avoiding them from harming themselves by taking dangerous substances to calm themselves, such as by smoking.

Therefore, it can be concluded that Islam encourages individuals to take care of themselves to maintain physical and mental health. Having a balanced physical and spiritual activities would reduce the risk of being unhealthy. Eating from good sources and not consuming over the limit, would provide sufficient nutrients to maintain a healthy body, thus reducing the risk of developing MI, stroke, and other diseases.

5 Discussion

Based on the study above, only China had a lower rate of MI deaths compared to stroke. This previous study also reported that China's stroke mortality was lower than MI, with 157 deaths related to stroke per 100,000 individuals, while MI contributed to 63 deaths per 100,000 individuals [34]. This study also discovered that Japan and Singapore showed a declining mortality rate related to both MI and stroke. A study reported that Singapore's CVD mortality rate had declined from 99 deaths per 100,000 population in 1976 to 59 deaths per 100,000 population in 1994 [34]. They reported that Japan mortality rate due to stroke had decreased from 974 deaths per 100,000 in 1964–1971 to 231 deaths per 100,000 in 1996–2003. The study also found that there is no change in MI mortality rate in Akita, Hiroshima, and Nagasaki from 1960 to 1990. It is estimated that the trend of stroke mortality events in Singapore decreases 3–5% annually. Japan showed a 70% decrease in stroke-related deaths from the year 1960–1990 [35]. There was also a study that discovered a different result from the previous discussion. The study reported that Japan's MI mortality rate had increased by 7.4 deaths per 100,000 individuals in 1979 to 27.0 deaths per 100,000 in 2008 due to the increase in Japan's ageing population. The results were contradictory and could be due to many factors, such as the different analysis method used [36].

Many risk factors were contributing to the trends of CVD mortality rate. Each risk factors have a different level of impact on the death rate of CVD. Blood pressure is the most significant risk factors of CVD, especially for MI and stroke. It was reported that 54% of stroke and MI cases were contributed by high blood pressure. In the Asia Pacific region, hypertension contributes 665 CVD cases. Reduction in blood pressure will give a consistent effect on CVD reduction [37].

Smoking is also an important risk factor for CVD, strongly related to the increase in morbidity and mortality of CVD patients. Smoking cessation would result in declining CVD mortality rates [38]. People who are smokers have 3.436 times more chances to develop MI and 2.158 more chances to develop stroke, compared to those who never smoke. Smoking also increases the chance of developing other risk factors such as diabetes, hypertension and hypercholesterolaemia. Smokers who suffered MI and stroke were 1.766 times more likely to die than people who never smoked. Reducing the number of cigarette smokes could increase the BMI of the individual while resuming smoking could decrease the BMI level [39].

Increase in BMI was directly related to developing MI. Being overweight and obese would affect the health and other independent risk factors of MI [40]. Some studies argued that BMI not an important risk factor of MI and stroke but has a second-order effect since BMI is an important risk factor for hypertension, hypercholesterolaemia and diabetes [24].

Dyslipidaemia is another significant risk factor of CVD. Many patients of dyslipidaemia developed MI. High level of total cholesterol increases the chance of having CVD. Improving blood cholesterol level to manage dyslipidaemia better could reduce the risk of developing MI [40]. Dyslipidaemia in Asian countries is generally than lower than the United States of America and other western countries. Among Asian countries, Singapore recorded with the highest dyslipidaemia cases [41].

Diabetic patients have a higher probability of developing CVD. More than 70% of patients with type 2 diabetes died from CVD [37]. People with type 2 diabetes have a higher probability of injury and death due to CVD. Diabetes raises the risk of CVD by 2 to 4 times. Individuals with diabetes' life expectancy are reduced almost by eight years since the chance of mortality increases. Diabetes also increases the probability of developing MI because diabetes increases the rate of atherosclerotic progression and mitigate the formation of atherosclerotic plaque [40].

6 Conclusion

CVD is the number one cause of mortality in the world. Among all the disease in CVD, MI and stroke are the most common contributor to CVD-related deaths worldwide. There are many risk factors that contribute to the high number of deaths due to CVD. There are controllable and uncontrollable risk factors. Some of the uncontrollable risk factors are family history with CVD, increase in age, ethnicity, and being a male. High blood pressure and smoking are the most important risk factors since it can be reflected in the rate of death in selected countries. The BMI, cholesterol and blood glucose are also significant risk factors for CVD; even when it does not reflect too much on the CVD mortality trends of the selected countries. There are many ways to reduce the risk of CVD such as by quitting or reducing smoking, having healthy food with consistent intake of fruits and vegetables, and having active physicals activities such as sport and jogging. Islam also encourages individuals to have a healthy physical and spiritual and taking sufficient nutrient to have a healthy body.

Acknowledgements We would like to thank the Ministry of Higher Education, Malaysia for supporting this study under the research grant FRGS/1/2019/STG06/USIM/02/1.

References

1. Weinhausr, A.J., Roberts, K.P.: Anatomy of the human heart. In: Iaizzo, P.A. (ed.) Handbook of Cardiac Anatomy, Physiology, and Devices, 2nd edn., pp. 51–79. Springer, US (2009)
2. Gaze, D.C.: Introduction to ischemic heart disease. In: Gaze, D.C. (ed.) Ischemic Heart Disease, pp. 1–13. IntechOpen (2013)
3. Biglu, M., Ghavami, M., Biglu, S.: Cardiovascular diseases in the mirror of science. Tabriz Univ. Med. Sci. **8**(4), 158–163 (2016). https://doi.org/10.15171/jcvtr.2016.32
4. Rafieian-kopaei, M., Setorki, M., Doudi, M., Baradaran, A., Nasri, H.: Atherosclerosis: process, indicators, risk factors and new hopes. Int. J. Prev. Med. **5**(8), 927–946 (2020)
5. Mendis, S., et al.: World health organization definition of myocardial infarction: 2008–09 revision. Int. J. Epidemiol. **40**(1), 139–146 (2011). https://doi.org/10.1093/ije/dyq165
6. Saleh, M., Ambrose, J.A.: Understanding myocardial infarction [version 1; referees: 2 approved]. F1000 Res. **7**(0), 1–8 (2018). https://doi.org/10.12688/f1000research.15096.1
7. Wittenauer, R., Smith, L.: Ischaemic and Haemorrhagic Stroke, Dec 2012
8. Ahmad, M.H., Nishi, N., Mohd Yusoff, M.F., Aris, T.: Cardiovascular disease risk and its association with body mass index in Malaysians based on the world health organization/international society of hypertension risk prediction chart. Heal. Sci. J. **12**(1), 1–7 (2018). https://doi.org/10.21767/1791-809x.1000550
9. OECD/WHO, Health at a Glance: Asia/Pacific 2018: Measuring Progress Towards Universal Health Coverage. OECD Publishing (2018)
10. Wu, Y., Benjamin, E.J., MacMahon, S.: Prevention and control of cardiovascular disease in the rapidly changing economy of China. Circulation **133**(24), 2545–2560 (2016). https://doi.org/10.1161/CIRCULATIONAHA.115.008728
11. Seong, A.C., Chb, M.B., Kok, C., John, M., Cth, F.: A Review of Coronary Artery Disease Research in Malaysia, vol. 71, pp. 42–57 (2016)
12. S. Epidemiology & Disease Control Division, Ministry of Health: The Burden of Disease in Singapore, 1990–2017: An Overview of the Global Burden of Disease Study 2017 Results. Institute for Health Metrics and Evaluation (2017)
13. Mucheru, D., Hanlon, M.C., Campbell, L.E., McEvoy, M., MacDonald-Wicks, L.: Cardiovascular disease lifestyle risk factors in people with psychosis: a cross-sectional study. BMC Public Health **18**(1), 1–14 (2018). https://doi.org/10.1186/s12889-018-5649-5
14. Oliveira, G.B.F., Avezum, A., Roever, L.: Cardiovascular disease burden: evolving knowledge of risk factors in myocardial infarction and stroke through population-based research and perspectives in global prevention. Front. Cardiovasc. Med. **2**, 1–4 (2015). https://doi.org/10.3389/fcvm.2015.00032
15. Juhan, N., Khalid, Z.M., Zubairi, Y.Z., Syadi, A., Zuhdi, M.: Risk Factors of Cardiovascular Disease Infarction Male Patients in Malaysia. J. Teknol. Sci. Eng. **81**(3), 145–150 (2019). https://doi.org/10.11113/jt.v81.12194
16. National Health and Morbidity Survey: Non-communicable Diseases: Risk Factors and Other Health Problems, vol. 1. National Institutes of Health (NIH), Ministry of Health Malaysia (2019)
17. Clinical Practice Guidelines: Management of Hypertension, 5th edn (2018). https://doi.org/10.1080/00325481.1947.11691709
18. Abreu, D., Sousa, P., Matias-Dias, C., Pinto, F.J.: Cardiovascular disease and high blood pressure trend analyses from 2002 to 2016: after the implementation of a salt reduction strategy. BMC Publ. Health **18**(1), 1–9 (2018). https://doi.org/10.1186/s12889-018-5634-z
19. Wu, C.Y., Hu, H.Y., Chou, Y.J., Huang, N., Chou, Y.C., Li, C.P.: High blood pressure and all-cause and cardiovascular disease mortalities in community-dwelling older adults. Med. (United States) **94**(47), e2160 (2015). https://doi.org/10.1097/MD.0000000000002160
20. Lim, K.H., Teh, C.H., Pan, S., Ling, M.Y., Yusoff, M.F.M., Sumarni, M.: Prevalence and factors associated with smoking among adults in Malaysia: findings from the National Health and Morbidity Survey (NHMS) 2015. EUEP Eur. Publ. Behalf Int. Soc. Prev. Tob. Induc. Dis. **16**, 1–12 (2018). https://doi.org/10.18332/tid/82190

21. Keto, J., et al.: Cardiovascular disease risk factors in relation to smoking behaviour and history: a population-based cohort study. Open Hear. **3**, e000358 (2016). https://doi.org/10.1136/openhrt-2015-000358
22. Azad, S.M.N., Hossain, M.M., Parveen, R.: Impacts of smoking habit by young generation in our society impacts of smoking habit by young generation in our society. AIUB J. Bus. Econ. **10**(Jan 2011), 45–66 (2014)
23. Clinical Practice Guidelines: Management of Obesity (2004)
24. Chatterjee, T., Macdonald, A.S., Waters, H.R.: A model for ischaemic heart disease and stroke II: modelling obesity. Ann. Actuar. Sci. **3**(1–2), 83–103 (2008). https://doi.org/10.1017/s1748499500000464
25. Hamer, M., Stamatakis, E.: Low-dose physical activity attenuates cardiovascular disease mortality in men and women with clustered metabolic risk factors. Circ. Cardiovasc. Qual. Outcomes **5**(4), 494–499 (2012). https://doi.org/10.1161/CIRCOUTCOMES.112.965434
26. Ariyanti, R., Besral, B.: Dyslipidemia associated with hypertension increases the risks for coronary heart disease: a case-control study in Harapan Kita Hospital, National Cardiovascular Center, Jakarta. J. Lipids 6 (2019) [Online]. Available: https://doi.org/10.1155/2019/2517013
27. Clinical Practice Guidelines: Management of Dyslipidaemia, vol. 5. Ministry of Health Malaysia, Academy of Medicine Malaysia, National Heart Association of Malaysia (2017)
28. Jeong, S.M., et al.: Effect of change in total cholesterol levels on cardiovascular disease among young adults. J. Am. Heart Assoc. **7**(12) (2018). https://doi.org/10.1161/JAHA.118.008819
29. Clinical Practice Guidelines: Management of Type 2 Diabetes Mellitus, 5th edn (2015)
30. Frankum, S., Ogden, J.: Estimation of blood glucose levels by people with diabetes: a cross-sectional study. Br. J. Gen. Pract. **55**(521), 944–948 (2005)
31. Hughes, J.R.: Motivating and helping smokers to stop smoking. J. Gen. Intern. Med. **18**(12), 1053–1057 (2003). https://doi.org/10.1111/j.1525-1497.2003.20640.x
32. Rosenthal, R.L.: Effectiveness of altering serum cholesterol levels without drugs. Baylor Univ. Med. Cent. Proc. **13**(4), 351–355 (2000). https://doi.org/10.1080/08998280.2000.11927704
33. Park, C., et al.: Fasting glucose level and the risk of incident atherosclerotic cardiovascular diseases. Diab. Care **36**(7), 1988–1993 (2013). https://doi.org/10.2337/dc12-1577
34. Ohira, T., Iso, H.: Cardiovascular disease epidemiology in Asia—an overview. Circ. J. **77**(7), 1646–1652 (2013). https://doi.org/10.1253/circj.CJ-13-0702
35. Khor, G.L.: Cardiovascular epidemiology in the Asia–Pacific region. J. Clin. Nutr. **10**(2), 76–80 (2001). https://doi.org/10.1111/j.1440-6047.2001.00230.x
36. Takii, T., et al.: Trends in acute myocardial infarction incidence and mortality over 30 years in Japan: Report from the MIYAGI-AMI registry study. Circ. J. **74**(1), 93–100 (2010). https://doi.org/10.1253/circj.CJ-09-0619
37. Ministry of Health: Primary and Secondary Prevention of Cardiovascular Disease 2017 (2017)
38. Hasdai, D., Garratt, K.N., Grill, D.E., Leman, A., Holmes, D.R.: Effect of smoking status on the long-term outcome after successful percutaneous coronary revascularization. N. Engl. J. Med. **336**(11), 755–761 (1997)
39. Chatterjee, T., Macdonald, A.S., Waters, H.R.: A model for ischaemic heart disease and stroke III: applications. Ann. Actuar. Sci. **3**(1–2), 105–119 (2008). https://doi.org/10.1017/s1748499500000476
40. Rathore, V., Singh, N., Mahat, R.K.: Risk factors for acute myocardial infarction: a review. Eur. J. Med. Investig. **2**(1), 1–7 (2018). https://doi.org/10.14744/ejmi.2018.76486
41. Ueshima, H., et al.: Cardiovascular disease and risk factors in Asia: a selected review. Circulation **118**(25), 2702–2709 (2008). https://doi.org/10.1161/CIRCULATIONAHA.108.790048. Cardiovascular

Forecasting Petroleum Fuel Price in Malaysia by ARIMA Model

Rajalingam Sokkalingam, Richard M. N. Y. Sarpong-Streetor, Mahmod Othman, Hanita Daud, and Derrick Asamoah Owusu

Abstract Over the years the retail price of petroleum fuel in Malaysia, Ron95, Ron97 and Diesel have been controlled by the governments using the Automatic Price Mechanism (APM) which made the price of fuel in Malaysia relatively stable up until 2004. Beyond the year 2004, the price of petroleum fuel has been volatile even with the APM still being implemented. Even after changing the Policy to Managed float system in 2016 fuel prices have still not been stable. Reasons that have been attributed to the volatilities are the international crude oil price and foreign exchange volatilities and reduction of subsidies to improve government fiscal space. Predicting fuel Prices has become difficult. In this paper we apply the Time Series method Autoregressive Integrated Moving Average (ARIMA) to model and forecast fuel price.

Keywords ARIMA · Equation · Forecasting · Fuel · Price · Malaysia · Managed float system (MFS) · Ron97

R. Sokkalingam · R. M. N. Y. Sarpong-Streetor (✉) · M. Othman · H. Daud
Fundamental and Applied Sciences Department, Universiti Teknologi PETRONAS, 32610 Seri Iskander, Perak, Malaysia
e-mail: richard_17004029@utp.edu.my

R. Sokkalingam
e-mail: raja.sokkalingahm@utp.edu.my

M. Othman
e-mail: mahmod.othman@utp.edu.my

H. Daud
e-mail: hanitadaud@utp.edu.my

D. A. Owusu
Mathematics Department, Kwame Nkrumah University of Science and Technology, Kumasi, Ghana

S. A. Abdul Karim et al. (eds.), *Proceedings of the 6th International Conference on Fundamental and Applied Sciences*, Springer Proceedings in Complexity,
https://doi.org/10.1007/978-981-16-4513-6_58

1 Introduction

Efficient Transport systems depends on cheap fuel prices. One aspect of obtaining cheap fuel is the planning of the fuel price. The fuel price is announced weekly by The government's body Ministry of Domestic Trade and Consumer Affairs in conjunction with the Ministry of Finance based on the Managed Floating System (MFS) Policy [1–5]. Transport owners, pump station managers and other stakeholders of petroleum fuel would want to forecast the fuel price to ensure there is an optimized fuel budget. Based on the MFS policy, the fuel price can be modelled by the formular (1) [2, 6].

$$P = A + B + C + D + E + F \tag{1}$$

Here, (A) is the Refined fuel price, Mean of Platts Singapore (MOPS), published by Platts [7]. (B) is Alpha, the difference between the MOPS and actual purchasing price from the refinery's companies, (C) is Tax/Subsidy [8], (D) is Operational cost at bulk storage for transportation and advertisement, (E) is Bulk distribution company's margin and (F) is Fuel station margins. Using the formula (1) is quiet and difficult and expensive as the MOPS value s is not accessible to all. (A) MOPS is a subscription-based Time series. Medium to small scale and entities are unwilling to subscribe to it, because of the expense liability, and technicalities associated with its Assessment Methodology. To solve this problem, ARIMA modelling is explored. The historical weekly data on fuel price is modelled. This paper focuses on modelling Ron97 a variant of petrol that that has the oxidation number 97. The MFS policy allows Ron97 price to float at the international market price without subsidies. The pricing Policy of 97 has remained consistent. It is a premium product as compared to Ron 95 and Diesel sold on the Malaysian Market. The sometimes subsidized [9–11].

2 Data

Data source for the study is the weekly price of Ron97 announced by the Ministry of trade and Consumerism, Malaysia [3, 4, 12] starting from 7 April 2017 to 6 March 2020, Fig. 1. Validation of the forecasting model is done using Fuel price from 13 march 2020 to 7 August 2020. There are some weeks in the time series where the Policy MFS was suspended for the APM Policy or the period of review of fuel price was changed to one month [1, 13]. Where there is such problem linear interpolation is applied to fill Missing weekly fuel price [14, 15] . The weekly day for announcing Fuel Price, has changed in the timeline of the time series. The dates are aligned to a common day point by taking a weekly average of the historical price of Ron97. Saturday is assigned as the common date point. The compiled data can be found on [16].

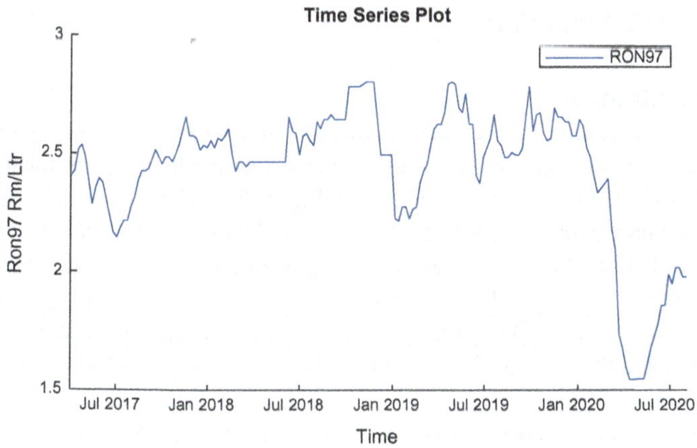

Fig. 1 Time series of Ron97 price in Rm/Ltr

3 Methodology

ARIMA or the Box–Jenkins methodology of order ARIMA (p, d, q) model is a time series forecasting method for non-stationary data series. The future value of a variabley_t, is assumed to be a linear function of several past observations (Ron97 price) and random errors, ε_t as represented by Eq. (2). ARIMA modelling is segmented into three-part process; model identification, parameter estimation and diagnostic checking. Here, p represents the autoregressive order, q, represents the moving average and d is the differencing order of the Time series, fuel price [17, 18].

$$\phi(1 + L)^p (1 + L)^d y_t = c + \theta(1 + L)^q \varepsilon_t \quad (2)$$

MATLAB Econometric Modeler [19] is used to do the model. We implement the models in MATLAB to forecast. Validation and forecast performance are assessed by comparing the Ron97 price with an 18-week horizon forecast Results.

4 Results

Ron97 fuel price has been modelled in a three-step procedure; ARIMA modelling.

4.1 ARIMA Modelling

Model Identification

The first step in identifying the ARIMA model is to check for stationarity of the time series, Ron97. It is not stationary as the fuel price does not fluctuate uniformly around a mean with a constant variance as shown in Fig. 1. The Time series is differenced to achieve stationarity time series. Figure 2a–c are the Sample autocorrelation (SAC) of Ron97 Sample partial autocorrelation (SPAC) of Ron97 of the first difference. They are stationary. The identified order for the SAC and SPAC is 14 and 14 respectively as the relevant leading spiking Lags in the correlograms are found at the 14th Lags respectively, and the rest the rest of the correlations die down. ARIMA (14, 1, 14) is the Identified model.

The Autoregressive order obtained from the SPAC, and Moving Average order obtained from the SAC are too high for the first differencing. Generally lower orders are preferred. The second differencing correlograms, Fig. 2d–f are assessed to determine its order. The SPAC has an order of 0 as the correlation Lags do not die done. The SAC has spikes at Lag 1 and relatively die down after it. Hence ARIMA (0, 2, 1) is identified at the second differencing.

Model Estimation

The Models identified are estimated using the Econometric Modeler or Econometric Tool box in MATLAB. Ron97 has a *t* distribution and from the estimation process it is observed that the models are more parsimonious when the constant, term, c is zero

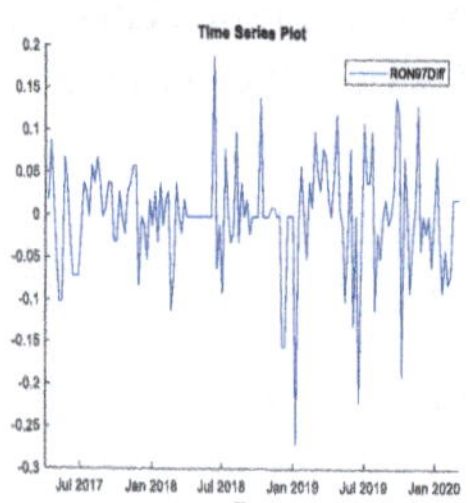

(a) First difference

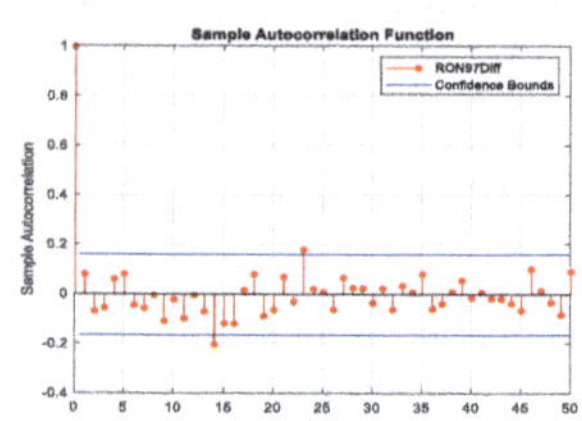

(b) SAC first difference

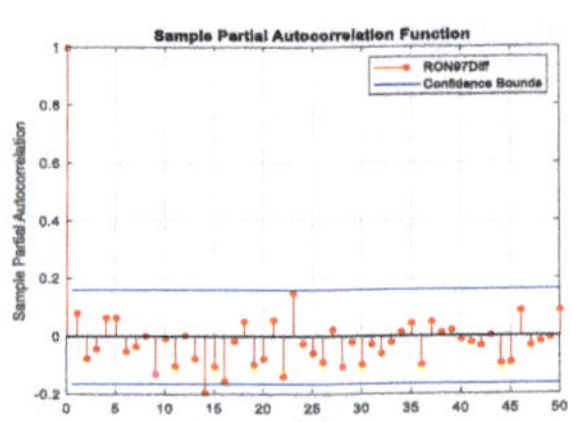

(c) SPAC of first difference

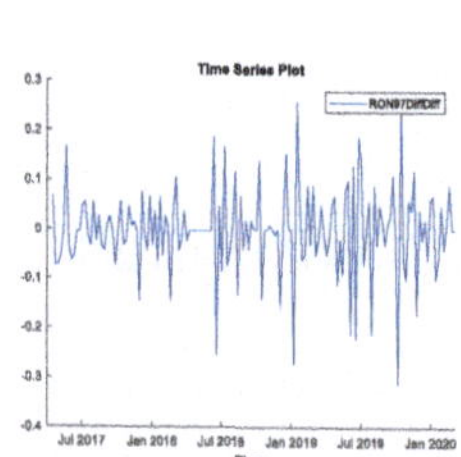

(d) Second difference

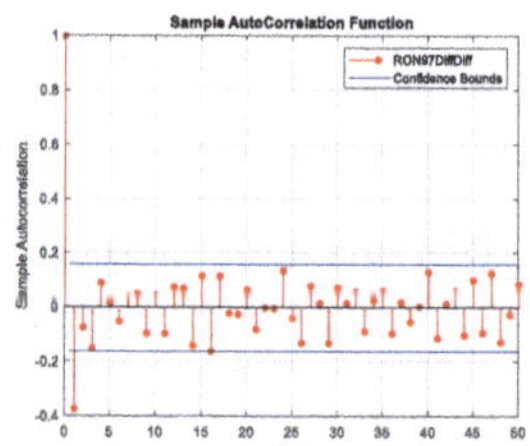

(e) SAC second difference

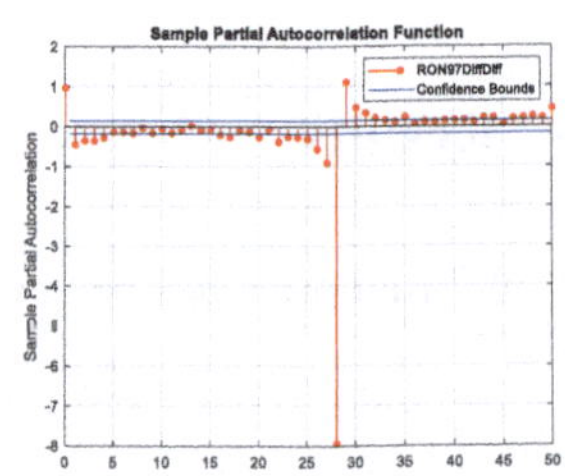

(f) SPAC of second difference

Fig. 2 Correlographs of Ron97

(0). These conditions are specified in the software during the estimating process. The best model is the one with the least Akaike Information Criteria (AIC) or Bayesian Information Criteria (BIC). Tables 1, 2 and 3 are summary of estimates of the model parameters. Comparing the AIC and BIC respectively, it is established that ARIMA (14, 1, 14) is the most parsimonious and the best fit model. Equation (3) is the ARIMA (14, 1, 14). Model.

$$\left(1 - \varnothing_{14}L^{14}\right)(1 - L)y_t = \left(1 + \theta_{14}L^{14}\right)\varepsilon_{14} \tag{3}$$

Equation (3) is expanded to give Eq. (4)

$$y_t = y_{(t-1)} + \phi_{(14)}y_{(t-14)} - \phi_{(14)}y_{(t-15)} + \theta_{14}\varepsilon_{(t-14)} + \varepsilon_t \tag{4}$$

If the parameters are substituted in Eq. (4), with the parameter estimates in Table 1. Estimation Results for ARIMA (14, 1, 14) , the tentative forecasting model is achieved as Eq. (5)

$$y_t = y_{t-1} - 0.7084y_{t-14} + 0.7084y_{t-15} + 0.6308\varepsilon_{t-14} + \varepsilon_t \tag{5}$$

Table 1 Estimation results for ARIMA (14, 1, 14)

Parameter	Value	Standard error	t statistic	*P*-value
Constant	0	0		
AR{14}	−0.7084	0.0620	−11.4199	3.3269e−30
MA{14}	0.6308	0.0817	7.7248	1.1201e−14
DoF	2.2366	0.7204	3.1049	0.0019
Variance	0.0114	0.0291	0.3901	0.6965

Table 2 Estimation results for ARIMA (0, 1, 2)

Parameter	Value	Standard error	t statistic	*P*-value
Constant	0	0		
MA{1}	−0.9834	0.0108	−91.0724	0
DoF	2.2273	0.7612	2.9261	0.0034
Variance	0.0134	0.0382	0.3518	0.7250

Table 3 Goodness of fit, of models

Model	AIC	BIC
ARIMA(14, 1, 14)	−439.1807	−427.4717
ARIMA(0, 2, 1)	−420.5204	−411.4686

Table 4 Ljung-box Q-test

	Null rejected	*P*-value	Test statistic	Critical value	Lags	DOF	Significance level
1	True	0.0191	13.5066	11.0705	15	5	0.05
2	True	0.0012	20.1843	11.0705	20	5	0.05
3	True	3.2044e−05	28.2830	11.0705	25	5	0.05
4	True	8.5631e−06	31.1974	11.0705	30	5	0.05
5	True	2.2489e−06	34.1245	11.0705	35	5	0.05
6	True	5.8798e−07	37.0400	11.0705	40	5	0.05
7	True	4.1995e−07	37.7686	11.0705	45	5	0.05
8	True	7.1972e−09	46.4961	11.0705	50	5	0.05
9	True	3.173e−09	48.2400	11.0705	55	5	0.05
10	True	1.9315e−10	54.1725	11.0705	60	5	0.05

Model Diagnostic Checking

The model is assessed for its adequacy in forecasting the fuel price, Ron97. The model residual is assessed, using the LJung-Box Q-test and residual correlation plot. The Ljung-Box Q-test is a quantitative way to test for autocorrelation at multiple lags jointly. In perfuming the Ljung-Box test, the we specify the degree of freedom to 5 as there are only five independent variables in the model. The number of possible lags is increased by 5 lags from the 15th lag onward till the 150th lag, Table 4. The *p*-value of the Ljung test remains less than 5% for the Null Hypothesis: "The first m autocorrelations of the residuals of ARIMA_RON97 are jointly 0". This implies the are no autocorrelation in the ARIMA residual. The Residual correlation plot Fig. 3 shows only one feeble spike at the 23rd lag which is not that significant. Hence the Model is adequate for the forecasting of Ron97 price in the neighbourhood of the period considered.

4.2 *Model Forecast Performance and Validations*

The forecast for ARIMA(14, 1, 14) performed better than ARIMA(0, 1, 2) as shown in the circled area in the plot. It is the second best in the graph compared to observed data. NARNET, Nonlinear autoregressive neural network model was even better (Fig. 4).

5 Conclusion

The main objective of this paper is to assess the ability to forecast Ron97 fuel price using ARIMA models. It is possible to forecast accurately fuel price using

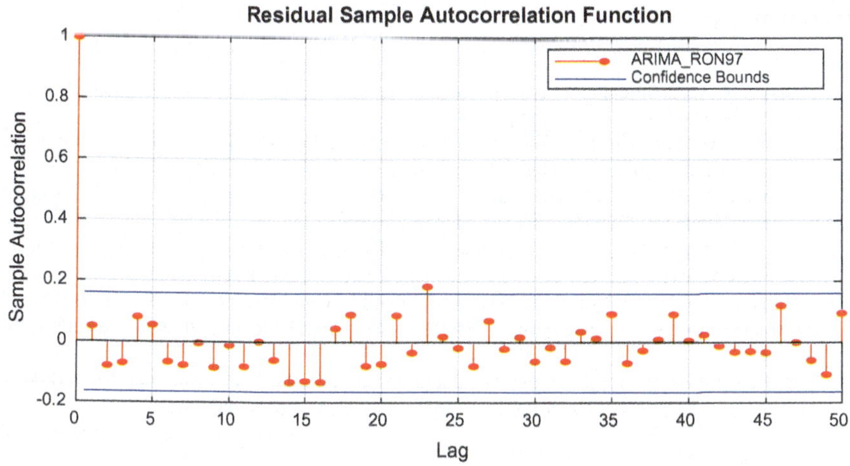

Fig. 3 Residual SAC of ARIMA_ Ron97

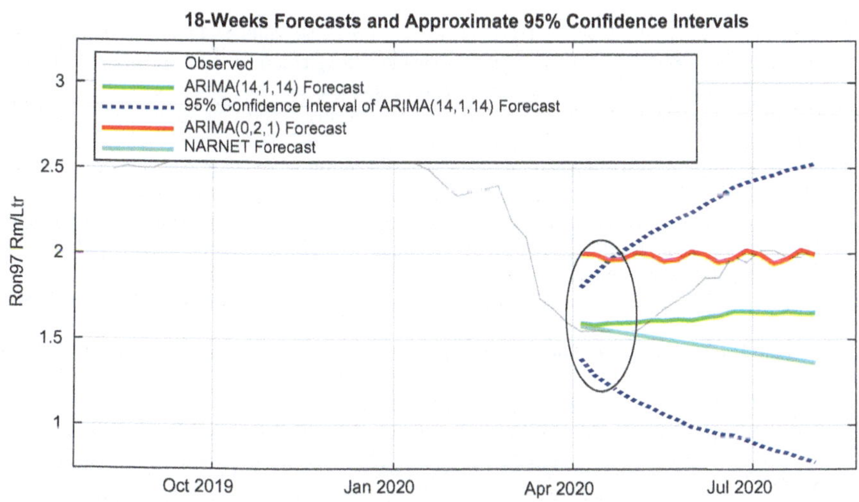

Fig. 4 Forecast performance of the ARIMA models

the ARIMA models but limited to a shorter period, month. Improvement can thought be made to the model to increase the length of forecast periods.

Acknowledgements We would like to thank anonymous referees for many the helpful suggestions which improved the paper considerably. We appreciate the FUNDAMENTAL RESEARCH GRANT SCHEME (FRGS) Reference code FRGS/1/2018/STG06/UTP/02/4, Malaysia for sponsoring the research. Lastly, we appreciate Universiti Teknologi PETRONAS where the research was conducted. The views expressed are those of the authors and do not reflect any other authority.

References

1. Babulal, V.: Government goes back to weekly managed float system to determine fuel prices. In: New Straits Times, Kuala Lumpur, 04 Jan 2019
2. Corporate-Malaysia: System, Why govt go for manage-float. In: The Malaysian Reserve, p. Business, Kuala Lumpor, 17 Mar 2017
3. The Ministry of Domestic Trade and Consumer Affairs: Fuel prices. In: The Ministry of Domestic Trade (2020). https://www.kpdnhep.gov.my/ms/
4. M. of F. Malaysia: Retail price of petroleum products. In: Retail Price of Petroleum Products (2021). https://www.mof.gov.my/en/component/tags/tag/price-of-petroleum%0A
5. Bofinger, P., Wollmershäuser, T.: Managed Floating: Understanding the New International Monetary Order W.E.P. Würzburg, 30, 2001. [Online]. Available: http://hdl.handle.net/10419/48479
6. Tan, P.: APM: How fuel prices are calculated in Malaysia. In: Malaysian Fuel Prices (2009). https://paultan.org/2009/02/15/how-fuel-prices-are-calculated-in-malaysia/
7. P. S & P Global: Methodology and specifications guide Asia Pacific & Middle East refined oil products. In: Platts, p. 47 (2017)
8. Chambers, A.G.: Sales Tax Regulations, p. 86. Malaysia Sales and Service Tax'Royal, Malaysia (2018)
9. Razak, Y.D.S.M.N.T.H.A.: The 2014 Budget Speech, pp. 1–52. Dewan Rakyat, Putrajaya (2013)
10. Ramasamy, M., Koon, C.P.: Malaysia raises fuel prices as Najib seeks to trim budget gap. In: Bloomberg Business (2013). https://www.bloomberg.com/news/articles/2013-09-02/malaysia-raises-fuel-prices-as-najib-seeks-to-trim-budget-gap
11. Hakim, R.A., Ismail, R., Razak, N.A.A.: Fuel subsidy reform in Malaysia: an assessment on the direct welfare impact on consumers. IPBJ **8**(1), 26–36 (2016) [Online]. Available: http://oyagsb.uum.edu.my/images/2018/List_of_Issues/Fuel_Subsidy_Reform_in_Malaysia_An_Assessment_on_the_Direct_Welfare_Impact_on_Consumers.pdf
12. CompareHero.my: Latest Petrol Price for Ron95, Ron97 & Diesel in Malaysia. In: Money Tips, Transportation (2020). https://www.comparehero.my/transportation/articles/latest-petrol-price-ron95-ron97-diesel
13. NST: PH fulfils pledge to stabilise fuel prices. In: New Straits Times, Kuala Lumpur, News, 17 Aug 2018
14. Burden, R.L., Faires, J.D.: Interpolation and polynomial approximation. In: Ostedt, G. (ed.) Numerical Analysis, 6th edn, pp. 104–165. Brooks/Cole Publishing Company, USA (1997)
15. Moler, C.: Numerical Computing with MATLAB (2015)
16. Sarpong-Streetor, R.M.N.Y.: ARIMAX Modelling of Ron97 Price with Crude Oil Price as an Exogenous Variable in Malaysian, Mendeley Data, V1 (2021). https://doi.org/10.17632/zxjnrpmwd8.1
17. O'Connell, R.T., Koehler, A.B., Bowerman, B.L., O'Connell, R.T., Koehler, A.B.: Forecasting, Time Series, and Regression: An Applied Approach, vol. 4. Thomson Brooks/Cole (2005)
18. Box, G.E.P., Jenkins, G.M., Reinsel, G.C., Ljung, G.M.: Time Series Analysis: Forecasting and Control. Wiley (2015)
19. Manjon, J.: Econometric Modeler. MathWorks Inc., Massachusetts, United States, 20AD

ARIMAX Modelling of Ron97 Price with Crude Oil Price as an Exogenous Variable in Malaysian

Richard M. N. Y. Sarpong-Streetor, Rajalingam Sokkalingam, Mahmod Othman, Hanita Daud, and Derrick Asamoah Owusu

Abstract Malaysia's petroleum fuel pricing policy has shuffled between Automatic Adjusted Formula (APM) and Managed Float System (MFS). One major problem in forecasting fuel prices is the use of the structural model. The model uses an input or independent variable, the refined fuel price also known as Mean of Platts Singapore (MOPS). MOPS is published regularly but is sold, which makes it inaccessible to a section of forecasters in Malaysia. In this paper, we suggest an alternative method of forecasting Ron97 price by applying the time series method, Autoregressive Integrated Moving Average with Explanatory Variable (ARIMAX) to model and forecast Ron97 price in Malaysia using publicly available data.

Keywords Automatic price mechanism (APM) · ARIMAX · ARIMA · Forecast · Fuel · Price · OPEC · Ron97

R. M. N. Y. Sarpong-Streetor (✉) · R. Sokkalingam · M. Othman · H. Daud
Fundamental and Applied Sciences Department, Universiti Teknologi PETRONAS, 32610 Seri Iskandar, Perak, Malaysia
e-mail: richard_17004029@utp.edu.my

R. Sokkalingam
e-mail: raja.sokkalingahm@utp.edu.my

M. Othman
e-mail: mahmod.othman@utp.edu.my

H. Daud
e-mail: hanitadaud@utp.edu.my

D. A. Owusu
Mathematics Department, Kwame Nkrumah University of Science and Technology, Kumasi, Ghana

S. A. Abdul Karim et al. (eds.), *Proceedings of the 6th International Conference on Fundamental and Applied Sciences*, Springer Proceedings in Complexity,
https://doi.org/10.1007/978-981-16-4513-6_59

1 Introduction

Over the years the retail price of petroleum fuel in Malaysia, Ron95, Ron97 and Diesel have been controlled by the governments using the Automatic Price Mechanism (APM) which made the price of fuel in Malaysia relatively stable up until 2004. Beyond the year 2004, the price of petroleum fuel has been volatile even with APM, still being implemented. After changing the Policy to Managed float system in 2016 fuel price have still not been stable. Reasons that have been attributed to the volatilities are the international crude oil price and foreign exchange volatilities and reduction of subsidies to improve government fiscal space [1].

The unstable nature of the fuel price demands the need for forecasting of the fuel price. Modelling and forecasting fuel price for a section of forecasters have become difficult. This because the APM model, Eq. (1), the popular model used for fuel price forecasting uses an input variable MOPS, (A), which is published regularly but sold to forecasters, making it difficult to access. Equation (1), have the inputs (A) as the Refined fuel price, Mean of Platts Singapore (MOPS), published by Platts [2], (B) Alpha, the difference between the MOPS and actual purchasing price from the refinery's companies, (C) Tax/Subsidy [3, 4], (D) Operational cost at bulk storage for transportation and advertisement, (E) Bulk distribution company's margin and (F) Fuel station [5, 6].

$$P = A + B + C + D + E + F \tag{1}$$

In this paper, we apply the time series method Autoregressive Integrated Moving Average with Explanatory Variable (ARIMAX) to model and forecast the petroleum fuel pricee for Ron97 using available public data.

2 Data

The primary data used in the paper are the published weekly price of Ron97 [7–9], daily crude oil price in barrels (WTI, BRENT and OPEC) [10, 11] and daily foreign exchange rate (Selling rate) of the Ringgit per US dollars [12]. The data is pre-processed to clean the data and standardize the data for the modelling process. The crude oil price and foreign exchange rates are converted to weekly averages. Daily missing data are replaced with weekly averages. The foreign exchange rates, (G) in Ringgits per US dollar (RM/USD) [12], is multiplied with the international crude oil price (X) to convert ringgits per barrel (RM/blur), (I), using Eq. (2). The crude oil price are converted to ringgit per litre using the Barrel to Litre Metric Conversion (M), [13], as shown in Eq. (3). The compiled data can be found on the doi link: https://doi.org/10.17632/zxjnrpmwd8.1 [14].

$$I = X \times G \tag{2}$$

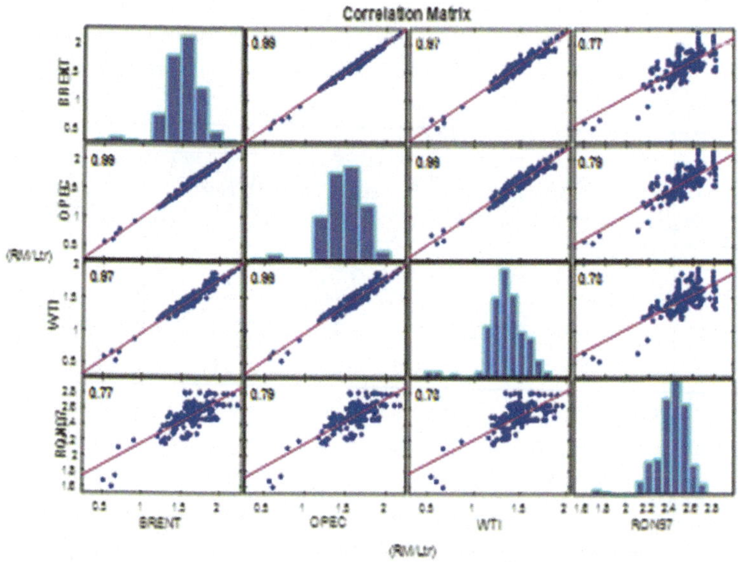

Fig. 1 Correlation matrix of Ron97 and crude oil price in Malaysia

$$Q = I \times M \tag{3}$$

Data used for the model, range from 7 April 2017 to 6 March 2020. Data used for the validation process, range from 13 March 2020 to 7 August 2020. We apply the correlation matrix to select the best crude oil price time series to model Ron97. Figure 1 is the Correlation Matrix of the three crude oil prices being assessed to model the Ron97 price. OPEC crude oil price have a better correlation of 0.79 with Ron97 compared with that of WTI and Brent crude oils. OPEC crude oil price is therefore selected as the exogenous variable to apply in modelling Ron97 price. Figure 2 are the selected Data for Ron97 price modelling.

3 Methodology

An Autoregressive Integrated Moving Average with Explanatory Variable model (ARIMAX), Eq. (4), can be viewed as a multiple regression model with one or more autoregressive (AR) terms and/or one or more moving average (MA) terms.

$$\phi(1 + L)^p(1 + L)^d y_t = c + \sum_{i=1}^{n} x_i \beta_i + \theta(1 + L)^q \varepsilon_t \tag{4}$$

Fig. 2 Selected data for Ron97 modelling

This method is suitable for forecasting when data is stationary/nonstationary, and multivariate with any type of data pattern. ARIMAX is related to the ARIMA technique but, while ARIMA is suitable for datasets that are univariate, ARIMAX is suitable for analysis where there are additional explanatory variables (multivariate) in categorical and/or numeric format [15–18]. The ARIMAX modelling is in three stages, the Model Identification, Model estimation and Diagnostic checking [19]. The Modelling and 18-week forecast are implemented in MATLAB. The model is validated by plotting the forecast of the ARIMAX model with ARIMA, NARNET and actual data recorded over the time [20–22].

4 Results

The result has been sectioned into two parts the modelling and Evaluation of the forecast.

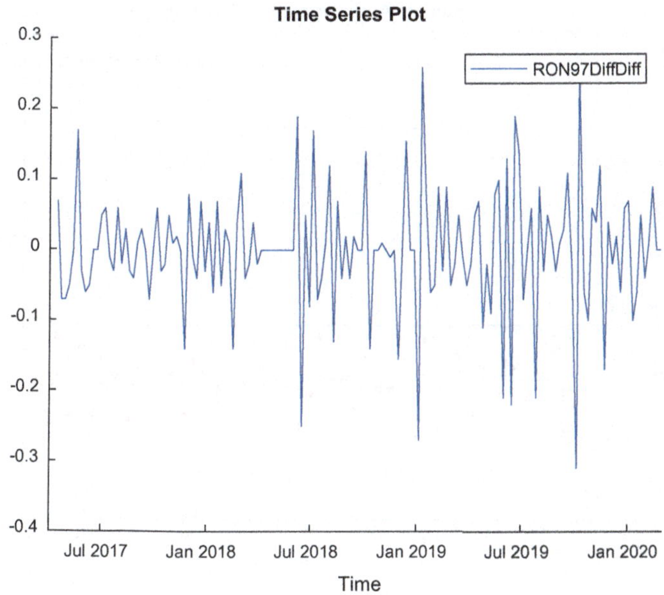

Fig. 3 Second difference of Ron97

4.1 *ARIMAX Modelling*

Ron97 fuel price has been modelled in a three-step procedure;

Model Identification

The precondition for the ARIMAX model to be identified is for the time series to be stationary or there exist no unit roots in the time series [23]. Thus for multiple time series they should be cointegrated [24]. We consider two Time Series, Ron97 and OPEC time series for the ARIMAX modelling. A suitable ARIMA model is identified for the dependent variable, Ron97, then the best regression is done on the ARIMA model and the independent variable, OPEC (Fig. 3). The Sample autocorrelation (SAC) of Ron97 dies down sharply at lag 1 on the second differencing as shown in Fig. 4; thus, stationarity is at difference order of 2 and moving average order is 1. The sample partial autocorrelation (SPAC) at the second differencing, Fig. 5 does not die down, which implies the autoregression order is 0 at the second differencing. The analysis deduced from the above information implies the ARIMA model have a differencing order of 2 and the ARIMA model is set at ARIMA (0, 2, 1) for Ron97. The ARIMA model is modified to ARIMAX by introducing the exogenous variable, OPEC (Fig. 6). OPEC must not be unit root time series. Based on Phillips-Perron tests [25], OPEC at lags 0, 1 and 2, (OPEC, OPECDiff, OPECDiffDiff) with significance level of (p-value) of 0.05 do not contain unit root as shown in Table 1. We will need the ARIMA model of OPEC to forecast the exogenous value in

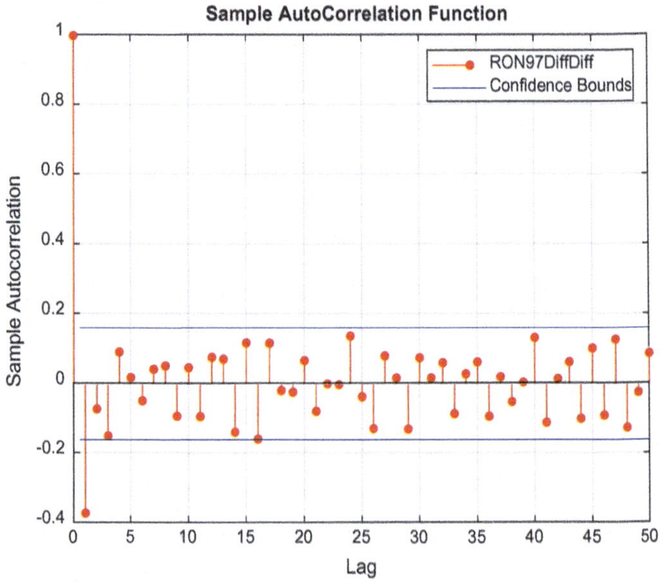

Fig. 4 Sample autocorrelation (SAC) of Ron97

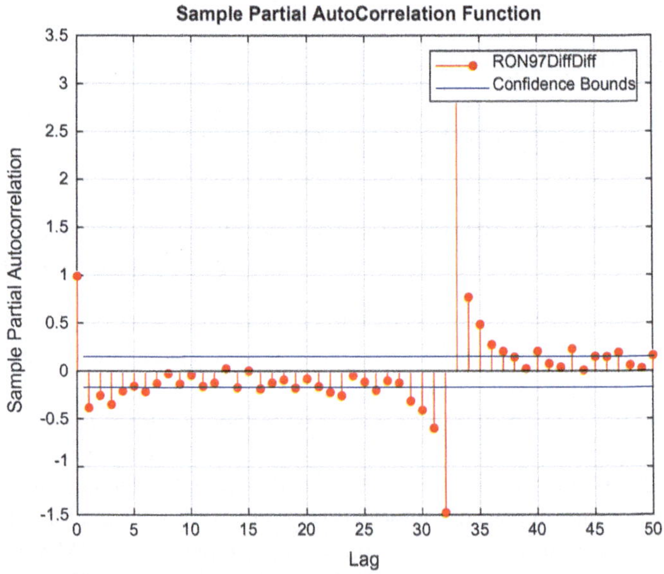

Fig. 5 Sample partial autocorrelation (SPAC) of Ron97

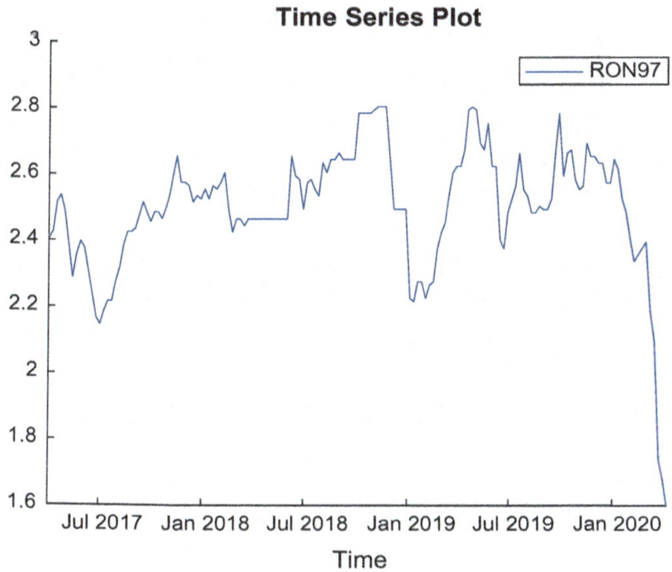

Fig. 6 Time series of OPEC

Table 1 Phillips-Perron test on OPEC time series

Null hypothesis: OPEC contains a unit root								
Test parameters					Test results			
S/N	Lags	Model	Test statistic	Sign. level	Null rejected	*P*-value	*T*-statistic	Critical value
1	0	ARD	t1	0.05	False	0.994	0.829	−2.881
2	1	ARD	t1	0.05	False	0.974	0.233	−2.881
3	2	ARD	t1	0.05	False	0.953	−0.037	−2.881

the ARIMAX model. Assessing the SAC and SPAC of OPEC as shown in Figs. 7 and 8 respectively, the SAC does not die down, but the SPAC is considered to die down at lag 2. Thus, the ARIMA (2, 0, 0) is chosen as the model for OPEC. The distributions of Ron97 and OPEC are also considered. Observing the histograms from the correlation matrix table in Fig. 1, Ron97 and OPEC have skewed tail on the left side which is characteristics of the *t*-distribution, hence the *t*-distribution is specified over the Gaussian for all ARIMA and ARIMAX models that will be estimated.

Model Estimation

MATLAB Econometric Medullar Application [26] is applied to estimate the model parameters. The Model Identification section identified the tentative model,

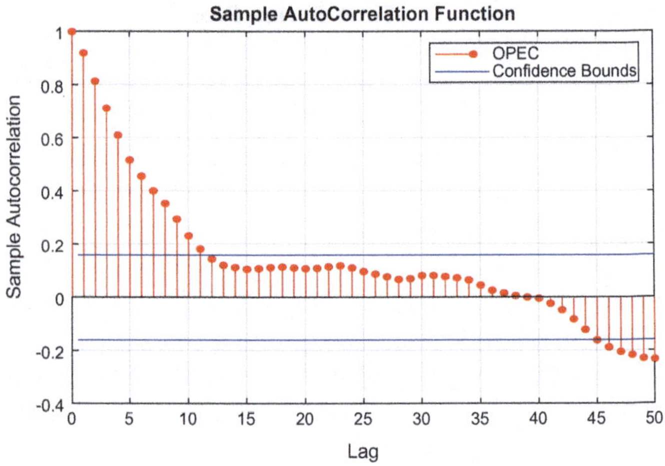

Fig. 7 Sample autocorrelation (SAC) of OPEC

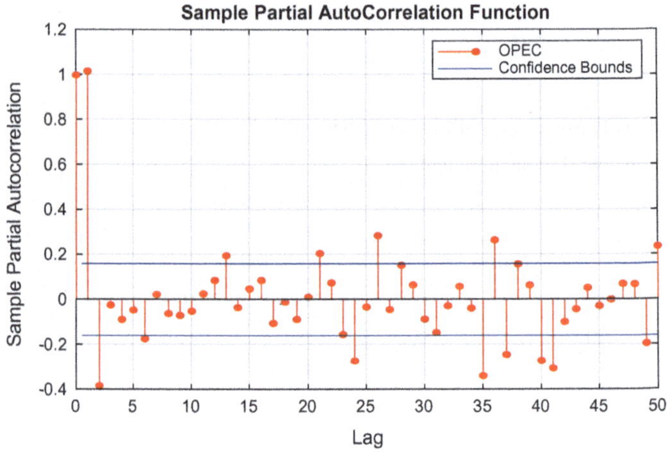

Fig. 8 Sample partial autocorrelation (SPAC) of OPEC

ARIMAX (0, 2, 1) with *t*-distribution and beta innovations of the exogenous predictors' time series (OPEC, OPECDiff, OPECDiffDiff). The ARIMA model associated with the ARIMAX is estimated first, then the exogenous predictors are introduced. Table 2 is the estimated parameters of ARIMA (0, 2, 1) model, Eq. (5). Equation (5) is expanded to give Eq. (6) and the estimated parameters is substituted into Eq. (6) to give Eq. (7)

$$(1 - L)^2 y_t = (1 + \theta_1 L)\varepsilon_t \tag{5}$$

Table 2 Estimation results ARIMA (0, 2, 1) (t-distribution) for Ron97

Parameter	Value	Standard error	T-statistic	P-value
Constant	0	0		
MA {1}	−0.9801	0.0126	−77.8371	0
DoF	2.031	0.5462	3.7183	0.0002
Variance	0.0910	1.5674	0.0581	0.9537

Table 3 Estimation results ARIMAX (0, 21) (t-distribution) for Ron97 with OPEC, OPECDiff and OPECDiffDiff as exogeneous variables

Parameter	Value	Standard error	T-statistic	P-value
Constant	0	0		
MA{1}	−1	0.017	−57.67	0
Beta (OPEC)	−0.00019	7.48e−05	−2.473	0.013
Beta (OPECDiff)	0.03361	0.020	1.661	0.097
Beta (OPECDiffDiff)	−0.04975	0.074	−0.668	0.504
DoF	2.42790	0.826	2.940	0.003
Variance	0.00945	0.013	0.702	0.483

$$y_t = 2y_{t-1} - y_{t-2} + \varepsilon_t + \theta_1\varepsilon_{t-1} \tag{6}$$

$$y_t = 2y_{t-1} - y_{t-2} + \varepsilon_t - 0.9801\varepsilon_{t-1} \tag{7}$$

$$(1-L)^2 y_t = X_1\beta_1 + X_2\beta_2 + X_3\beta_3 + (1+\theta_1 L)\varepsilon_t \tag{8}$$

Table 3 presents the tentative model estimates of the parameters of the ARIMAX which have the exogeneous time series' variables OPEC, OPECDiff and OPECDiffDiff with its associated model, is Eq. (8). We want the most parsimonious model for the ARIMAX Ron97 model. From Table 3, the p-value of the regression coefficients OPECDiff and OPECDiffDiff are statistically not significant at significance level of 5%. They are 9.7% and 50.4% respectively. On the other-hand, the p-value of OPEC is significant with a value of 1.3%. Estimating the ARIMAX model again with OPEC as the only exogenous variable Table 4, gives the p-value, 2.51%, Table 4.

Re-estimating the model, Eq. (8) as Eq. (9) makes the model more parsimonious. The Akaike information criterion (AIC) or the Bayesian information criterion (BIC) of the Eq. (9) is smaller than that of Eq. (8). The errors of Eq. (9) have been minimized more than that of Eq. (8), as shown. Figure 5. Hence the final model, ARIMAX (0, 2, 1) model with OPEC as the exogeneous variable is chosen as shown in Eq. (9).

$$(1-L)^2 y_t = X_1\beta_1 + (1+\theta_1 L)\varepsilon_t \tag{9}$$

Table 4 Parameter estimation results, ARIMAX (0, 2, 1) (t-distribution) for Ron97 with OPEC as exogeneous variable

Parameter	Value	Standard error	T-statistic	P-value
Constant	0	0		
MA{1}	−1	0.0138	−72.565	0
Beta (OPEC)	−0.00014	6.0668e−05	−2.24	0.0251
DoF	2.0983	0.6364	3.2972	0.0010
Variance	0.0313	0.1883	0.1663	0.8680

$$y_t = 2y_{t-1} - y_{t-2} + X_1\beta_1 + \varepsilon_t + \theta_1\varepsilon_{t-1} \tag{10}$$

$$y_t = 2y_{t-1} - y_{t-2} - 0.00014\,X_1 + \varepsilon_t - \varepsilon_{t-1} \tag{11}$$

Note that the constant term was omitted when specifying the model. This helped achieved a more parsimonious model. Expanding Eq. (9) into Eq. (10), and substituting the parameters from Table 4 into Eq. (10), we obtain Eq. (11) (Table 5)

Lastly, we estimate the OPEC ARIMA model which is needed to update the exogenous variable X_t in the forecast horizon interval in the ARIMAX model. Table 6 is the model parameters and Eq. (12) is it associated equation. Expanding Eq. (12) and substituting the parameters from Table 6 into Eq. (13), we obtain Eq. (14), the exogenous variable simulator for the ARIMAX model.

$$\left(1 - \phi_1 L - \phi_2 L^2\right) y_t = \varepsilon_t \tag{12}$$

Table 5 Goodness of fit for ARIMAX

Model	Exogenous variable	AIC	BIC
ARIMAX (0, 2, 1)	OPEC, OPECDif, OPECDifDif	−399.98	−381.75
ARIMAX (0, 21)	OPEC	−406.45	−394.25

Table 6 Estimation results of ARIMA(2, 0, 0) (t-distribution) with OPEC

Parameter	Value	Standard error	T-statistic	P-value
Constant	0	0		
AR{1}	1.324	0.068	19.423	4.964e−84
AR{2}	−0.324	0.068	−4.785	1.706e−06
DoF	6.751	1.674	4.033	5.503e−05
Variance	0.003	0.0005	6.696	2.150e−11

$$y_t = \phi_1 y_{t-1} + \phi_2 y_{t-2} + \varepsilon_t \quad (13)$$

$$y_t = 1.324 y_{t-1} - 0.324 y_{t-2} + \varepsilon_t \quad (14)$$

In the current scenario a one step ahead forecast can be done using Eqs. (15) and updating the exogenous variable with Eq. (16). These models are deduced from Eqs. (13) and (14)

$$y_{t+1} = 2y_t - y_{t-1} - 0.00014\, X_{1+1} + \varepsilon_{t+1} - \varepsilon_t \quad (15)$$

$$X_t = 1.324 X_{t-1} - 0.324 X_{t-2} + \varepsilon_t \quad (16)$$

Diagnostic Checking
After fitting the ARIMAX model we are left with residual Time series of Ron97, We assess the residual to confirm the adequacy of the ARIMAX model. We apply the residual autocorrelation plot tests and the Ljung Box Q test in the diagnostic check [27]. The sample autocorrelation function is very weak as the correlation plot does not spike. This is corroborated by the Ljung Box Q Test. The null Hypothesis is accepted, 'The first *m* autocorrelations of the residuals of ARIMAX_RON97 are jointly zero '0''. This implies the ARIMAX model is adequate for forecasting the fuel price in the neighbourhood of the period considered.

4.2 *Model Forecast Performance and Validations*

The ARIMAX was better at forecasting Ron97 for the first month than the bench marks ARIMA and NARNET as can be seen in Fig. 9.

5 Conclusion

This paper has examined the ability of ARIMAX to model the fuel price of Ron97 using times series data of Ron97 and another exogenous time series, the crude oil price, in this case OPEC in Malaysia. It is possible to do the modelling and forecasting accurately using the ARIMAX.

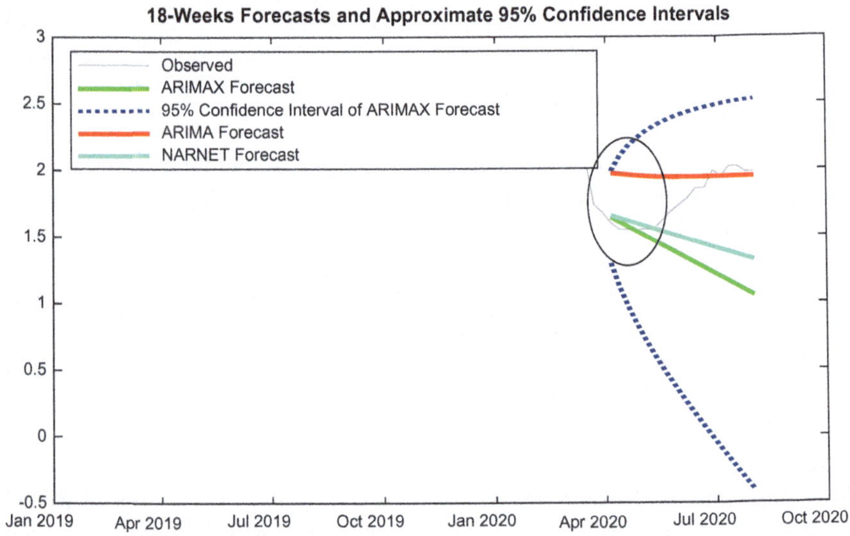

Fig. 9 Forecast performance of the ARIMAX models

Acknowledgements We would like to thank anonymous referees for many the helpful suggestions which improved the paper considerably. We appreciate the FUNDAMENTAL RESEARCH GRANT SCHEME (FRGS) Reference code FRGS/1/2018/STG06/UTP/02/4, Malaysia for sponsoring the research. Lastly, we appreciate Universiti Teknologi PETRONAS where the research was conducted. The views expressed are those of the authors and do not reflect any other authority.

References

1. Bridel, A., Lontoh, L.: Lessons Learned: Malaysia's 2013 Fuel Subsidy Reform (2014) [Online]. Available: http://www.iisd.org/gsi
2. S&P Global Platts: Methodology and Specifications Guide Asia Pacific and Middle East Refined Oil Products, p. 47. Platts (2017)
3. Royal Malaysian Customs Department: Guide on Goods and Service Tax on Downstream Petroleum, vol. 1, 2016 edn., p. 24 (2019)
4. Chambers, A.G.: Sales Tax Regulations, p. 86. Malaysia Sales and Service Tax' Royal, Malaysia (2018)
5. Wagner, A.: Fuel Price Policies Reading List Update 2001–2015. Dtsch. Ges. Int. Zusammenarbeit, 40 (2015) [Online]. Available: http://www.giz.de/fuelprices
6. Tan, P.: APM: How fuel prices are calculated in Malaysia. In: Malaysian Fuel Prices (2009). https://paultan.org/2009/02/15/how-fuel-prices-are-calculated-in-malaysia/
7. M. of F. Malaysia.: Retail Price of Petroleum Products (2021). https://www.mof.gov.my/en/component/tags/tag/price-of-petroleum%0A
8. CompareHero.my.: "Latest Petrol Price for Ron95, Ron97 & Diesel in Malaysia," Money Tips, Transportation (2020). https://www.comparehero.my/transportation/articles/latest-petrol-price-ron95-ron97-diesel
9. The Ministry of Domestic Trade and Consumer Affairs.: "Fuel Prices," The Ministry of Domestic Trade (2020). https://www.kpdnhep.gov.my/ms/

10. OPEC: OPEC basket reference price data. In: Organization of the Petroleum Exporting Countries (2020). http://www.opec.org/opec_web/en/data_graphs/40.htm
11. EIA: Spot prices for crude oil and petroleum products. In: Energy Information Administration (2020). https://www.eia.gov/dnav/pet/hist/LeafHandler.ashx?n=pet&s=rbrte&f=w
12. Bank Negara Malaysia: Exchange rate, ringgit to dollar, Selling rate @ 12 Noon (2020). https://www.bnm.gov.my/index.php?ch=statistic&pg=stats_spot_rates%0A
13. W. H. Ltd.: US Barrels (Oil) to Litres. In: Metric Conversions (2020). https://www.metric-conversions.org/volume/us-oil-barrels-to-liters.html
14. Sarpong-Streetor, R.M.N.Y.: ARIMAX Modelling of Ron97 Price with Crude Oil Price as an Exogenous Variable in Malaysian, Mendeley Data, V1 (2021). https://doi.org/10.17632/zxjnrpmwd8.1
15. Yang, M., Xie, J., Mao, P., Wang, C., Ye, Z.: Application of the ARIMAX model on forecasting freeway traffic flow. In: CICTP 2017 Transportation Reform and Change—Equity, Inclusiveness, Sharing, Innovation—Proceedings of the 17th COTA International Conference of Transportation Professionals, vol. 2018, pp. 593–602, July 2017. https://doi.org/10.1061/9780784480915.061
16. Anggraeni, W., Vinarti, R.A., Kurniawati, Y.D.: Performance comparisons between Arima and Arimax method in Moslem kids clothes demand forecasting: case study. Proc. Comput. Sci. **72**, 630–637 (2015). https://doi.org/10.1016/j.procs.2015.12.172
17. Andrews, B.H., Dean, M.D., Swain, R., Cole, C.: Building ARIMA and ARIMAX models for predicting long-term disability benefit application rates in the public/private sectors sponsored by society of actuaries health section prepared by University of Southern Maine. Soc. Actuar. Aug 2013
18. Kongcharoen, C., Kruangpradit, T.: Autoregressive integrated moving average with explanatory variable (ARIMAX) model for Thailand export. In: 33rd International Symposium on Forecasting. South Korea, pp. 1–8, June 2013 [Online]. Available: http://www.researchgate.net/publication/255731345_Autoregressive_Integrated_Moving_Average_with_Explanatory_Variable_(ARIMAX)_Model_for_Thailand_Export/file/9c9605209ac48013f6.pdf
19. Mathworks: ARIMA model including exogenous covariates. In: Mathworks Help Centre (2021). https://ch.mathworks.com/help/econ/arima-model-including-exogenous-regressors.html
20. Yang, Z., Mehmed, E.E.: Artificial neural networks in freight rate forecasting. Marit. Econ. Logist. **21**(3), 390–414 (2019). https://doi.org/10.1057/s41278-019-00121-x
21. Mathworks: narxnet. In: Deep Learning Toolbox Documentation (2010) https://ch.mathworks.com/help/deeplearning/ref/narxnet.html;jsessionid=4351ba4802441e46dae0d9a3a7f1. Accessed 30 June 2020
22. Durrani, A., Khurram, M., Khan, H.R.: Smart weather alert system for dwellers of different areas. In: Proceedings of 2019 16th International Bhurban Conference on Applied Sciences and Technology (IBCAST 2019), pp. 333–339 (2019). https://doi.org/10.1109/IBCAST.2019.8667190
23. Diebold, F.X., Kilian, L.: Unit-root tests are useful for selecting forecasting models. J. Bus. Econ. Stat. **18**(3), 265–273 (2014)
24. Escribano, Á., Wang, D.: Mixed random forest, cointegration, and forecasting gasoline prices. Int. J. Forecast. (2021). ISSN 0169-2070. https://doi.org/10.1016/j.ijforecast.2020.12.008. (https://www.sciencedirect.com/science/article/pii/S0169207020301941)
25. Kwiatkowski, D., Phillips, P.C.B., Schmidt, P., Shin, Y.: Testing the null hypothesis of stationarity against the alternative of a unit root. How sure are we that economic time series have a unit root? J. Econom. **54**(1–3), 159–178 (1992). https://doi.org/10.1016/0304-4076(92)90104-Y
26. Manjon, J.: Econometric Modeler. MathWorks Inc., Massachusetts, USA
27. Ljung, G.M., Box, G.E.P.P.: On a measure of lack of fit in time series models. Biometrika **65**(2), 297–303 (1978). https://doi.org/10.1093/biomet/65.2.297

Application of Newton-GS Iterative Method with Second-Order Quadrature Scheme in Solving Nonlinear Fredholm Integral Equations

Labiyana Hanif Ali, Jumat Sulaiman, Azali Saudi, and Xu Ming Ming

Abstract In this study, we discuss the application of Newton-GS iterative method with quadrature schemes in solving nonlinear Fredholm integral equations. This study proposes the application of Newton-Gauss–Seidel (NGS) iteration with the second-order quadrature scheme in getting the approximate solution of nonlinear Fredholm integral equations of the second kind (NFIE-2) in comparison with the first-order quadrature scheme. The main idea of this study is to apply the second-order quadrature scheme to discretize the NFIE-2 to form a system of nonlinear integral equations. Then we convert the nonlinear system into the corresponding linear system by imposing the Newton approach. By having this large-scale and sparse linear system, the numerical implementation of Newton-Jacobi (NJ) and NGS iteration approaches alone with first- and second-quadrature schemes have recorded their number of iterations, computational time, and maximum absolute error. As a result of these measured parameters, the comparative study can be performed to gauge the effectiveness of NGS with second-order quadrature scheme when compared with the numerical results of first-order quadrature scheme and NJ iteration. Based on numerical experiments, it can be important to highlight that the implementation of NGS iteration with second-order quadrature scheme has significantly improved the accuracy of its approximate results.

Keywords Nonlinear Fredholm integral equations · Quadrature scheme · Simpson's 1/3 rule · Gauss–Seidel iteration · Newton linearization

---

L. H. Ali (✉) · J. Sulaiman · X. M. Ming
Faculty of Science and Natural Resources, Universiti Malaysia Sabah, Sabah, Malaysia

J. Sulaiman
e-mail: jumat@ums.edu.my

X. M. Ming
e-mail: xmmzg@sina.com

A. Saudi
Faculty of Computing and Informatics, Universiti Malaysia Sabah, Sabah, Malaysia
e-mail: azali@ums.edu.my

S. A. Abdul Karim et al. (eds.), *Proceedings of the 6th International Conference on Fundamental and Applied Sciences*, Springer Proceedings in Complexity,
https://doi.org/10.1007/978-981-16-4513-6_60

1 Introduction

Academically, the problem of NFIE-2 has been solved using several analytical and numerical methods such as generalized extrapolation method [1], collocation method [2–4], multi projection method [5], Homotopy perturbation method [6], Adomian decomposition method [7], parameter continuation method [8], successive approximation method [9], etc. However, due to its important application in science and engineering, the studies of solving this problem is continued until now. Recently, the researchers have proposed several methods including Nyström method [10], Nyström-quasilinearization method [11], and trigonometric basis functions [12] to solve the problem of NFIE-2.

Inspired by the accuracy of the data through implementation of higher order quadrature scheme, this study is focusing on the implementation of the second-order quadrature scheme to solve the NFIE-2 in the following form

$$u(t) = f(t) + \int_a^b k(t, x, u(x))dx, \quad x \in [a, b], \tag{1}$$

where k is continuous on interval $[a, b]$, $f(t)$ is known function and $u(t)$ is the unknown function [13]. Family of quadrature schemes is one of numerical integration schemes which used widely in numerical studies due to its useful properties. A system of linear or nonlinear equations can be generated from a single approximation equation through discretization process using a particular quadrature scheme. Many studies have been conducted with implementing these quadrature schemes to discretize the problem of differential and integral equations such as in [14–18]. The implementation of different type or order of quadrature schemes have influenced the accuracy of the approximate solutions.

Since one of the advantages of the second-order quadrature rule, in comparison with first-order quadrature rule is that it provides more accurate approximation equations by considering more grid points on interval $[a, b]$, subsequently, the implementation of the composite Simpson's 1/3 (CS1) scheme is expected to be resulting more accurate solutions in terms of maximum absolute error. Prior to progressively utilizing these quadrature schemes, we design the methodology of solving nonlinear Fredholm integral equations using the NGS and NJ iterative methods with the composite Simpson's 1/3 (CS1) scheme namely NGS-CS1 and NJ-CS1. As for comparative effect, we also establish the formulation and implementation of NGS and NJ iterative methods with the composite Trapezoidal (CT) scheme namely NGS-CT and NJ-CT. Moreover, we just use NJ-CT as a control method to test the efficiency of other three proposed iterative methods particularly NGS-CS1, NGS-CT and NJ-CS1 in solving NFIE-2.

This paper is organized as follows: In Sect. 2, we will discuss the methodology of this study by implementing the second-order quadrature scheme mainly on the composite Simpson's 1/3 (CS1) scheme towards NFIE-2 to form a system of nonlinear integral equations. Then, we discuss the Newton-GS iteration on solving the corresponding nonlinear system. In Sect. 3, we present some numerical examples from the previous studies to illustrate the effectiveness of the proposed approach. Then, we discuss the numerical findings of this studies. Finally, we make some conclusions and suggestion for further studies on solving NFIE-2 in Sect. 4.

2 Methodology

The methodology of this study is design by two main parts: in part one, we discretize the nonlinear Fredholm integral equations using first- and second-order quadrature schemes to form nonlinear integral system. In part two, using NJ and NGS iterative methods, firstly, we need to do the linearization process over the generated nonlinear system using Newton's method to get the corresponding linear system. Later, the linear system can be solved iteratively by using NGS-CS1, NGS-CT, NJ-CS1 and NJ-CT iterative methods respectively to get their numerical solutions.

2.1 *Discretization of NFIE-2 Using First- and Second-Order Quadrature Schemes*

The formulation of integration function in solving NFIE-2 in (1) using quadrature scheme can be defined as follows

$$\int_a^b f(x)dx = \sum_{j=0}^{n} A_j f(x_j) + \epsilon_n(f) \tag{2}$$

where $t_j, (j = 0, 1, \ldots, n)$ are abscissas of the partition points of the integration interval on interval $[a, b]$, $A_j, j = 0, 1, 2, \ldots, n$ are the numerical coefficients and $\epsilon_n(y)$ is the truncation error. Constant $A_j, j = 0, 1, 2, \ldots, n$ for Trapezium rule is defined as [16, 17]

$$A_j = \begin{cases} \frac{1}{2}h, & j = 0, n \\ h, & otherwise \end{cases} \tag{3}$$

whereas the value of $A_j, j = 0, 1, 2, \ldots, n$ for Simpson's 1/3 is denoted in the following expression

$$A_j = \begin{cases} \frac{1}{3}h, & j = 0, n \\ \frac{4}{3}h, & j = 1, 3, 5, \ldots, n-1 \\ \frac{2}{3}h, & otherwise \end{cases} \tag{4}$$

where

$$h = \frac{b - a}{n}. \tag{5}$$

The differences between Trapezoidal and Simpson's 1/3 rule can be illustrated as follows.

Figure 1 shows the comparison of finite grid networks between Trapezoidal rule and Simpson's 1/3 rule on interval $[a, b]$. For illustration, let the total number of subintervals to be n. The implementation of Trapezoidal rule will consider as much $(n + 1)$ node points while the Simpson's 1/3 will also consider $(n + 1)$ total node point. This means Simpson's 1/3 rule consider more node points on interval $[a, b]$ resulting the neighboring point between the node points to be increased. Thus, it will

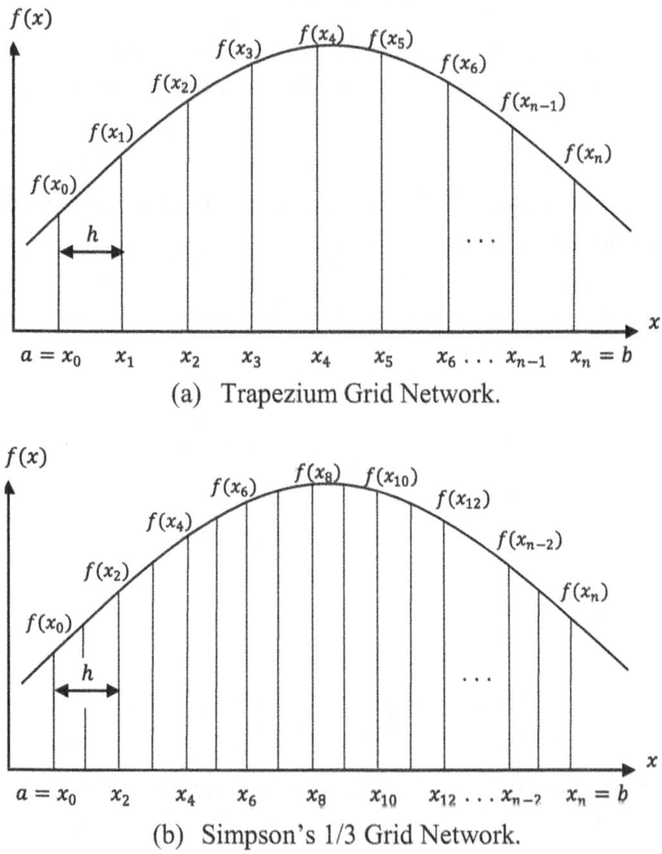

Fig. 1 The comparison of finite grid networks for **a** Trapezium and **b** Simpson's 1/3 on interval $[a, b]$

help to reduce the neighboring distance between the node points and provide more accurate approximate solution of the problem.

In general, we can form the following nonlinear approximation equations based on the implementation of quadrature scheme in integration part of (1) for $i, j = 0, 1, 2, \ldots, n$, as follows

$$\begin{aligned} u_i - A_j hk(t, x_0, u_0) - A_j k(t, x_1, u_1) - A_j k(t, x_2, u_2) - \cdots \\ - A_j hk(t, x_n, u_n) = f_i \end{aligned} \tag{6}$$

The nonlinear approximation (6) can be easily specified by replacing the A_j based on the given values in (3) and (4) for first- and second-order quadrature schemes. The nonlinear function of (6) can be defined as

$$\begin{aligned} G_i(u_0, u_1, u_2, \ldots, u_n) = u_i - A_j hk(t, x_0, u_0) - A_j hk(t, x_1, u_1) \\ - A_j hk(t, x_2, u_2) - \cdots - A_j hk(t, x_n, u_n) - f_i \end{aligned} \tag{7}$$

Thus, we can form the following nonlinear system of NFIE-2 in the following form

$$G_i(u_0, u_1, u_2, \ldots, u_n) = 0, \quad i = 0, 1, 2, 3, \ldots, n \tag{8}$$

2.2 Formulation of NGS Iteration with Quadrature Schemes

In the second part, we will discuss the formulation of NGS iteration with quadrature schemes to solve the generated nonlinear system in Part A. Using Newton's method, we can represent the corresponding nonlinear system (8) into a linear system as follows [19]

$$J\left(\underline{u}^{(k)}\right)\Delta u^{(k)} = -G\left(\underline{u}^{(k)}\right) \tag{9}$$

where

$$J\left(\underline{u}^{(k)}\right) = \begin{bmatrix} \frac{dG_0}{du_0} & \frac{dG_0}{du_1} & \frac{dG_0}{du_2} & \cdots & \frac{dG_0}{du_n} \\ \frac{dG_1}{du_0} & \frac{dG_1}{du_1} & \frac{dG_1}{du_2} & \cdots & \frac{dG_1}{du_n} \\ \frac{dG_2}{du_0} & \frac{dG_2}{du_1} & \frac{dG_2}{du_2} & \cdots & \frac{dG_2}{du_n} \\ \vdots & \vdots & \vdots & \ddots & \vdots \\ \frac{dG_n}{du_0} & \frac{dG_n}{du_1} & \frac{dG_n}{du_2} & \cdots & \frac{dG_n}{du_n} \end{bmatrix}_{(n+1)\times(n+1)},$$

$$\Delta u^{(k)} = \left[\Delta u_0 \ \Delta u_1 \ \Delta u_2 \ \cdots \ \Delta u_n\right]^T,$$

and $\underline{u}^{(k)}$ is determined by $u_i^{(k+1)} = u_i^{(k)} + \Delta u_i$.

To solve the linear system of (9), Jacobian matrix, $J\left(\underline{u}^{(k)}\right)$ needs to be decompose into $J\left(\underline{u}^{(k)}\right) = D - L - U$, where D is diagonal matrix, L is strictly lower matrix, and U is strictly upper matrix so we can formulate the formulation of Gauss–Seidel iteration to solve NFIE-2 in (9) as follows [20]

$$\Delta\underline{u}^{(k+1)} = (D - L)^{-1} U \Delta\underline{u}^{(k)} + (D - L)^{-1} \underline{f} \tag{10}$$

Algorithm 1 Implementation of NGS-CS1 Iteration.

i. Let $\nabla \underline{u}^{(k)} = 0$, $k = 0$ and $\varepsilon = 10^{-10}$.
ii. Set $q = 0$ and compute matrix $J\left(\underline{u}^{(k)}\right)$ and $G\left(\underline{u}^{(k)}\right)$.
iii. Compute the current value, $\Delta u_i^{(k+1)}$
   a. For $i = 0, 1, 2, \ldots, n$ and $j = 0, 1, 2, \ldots, n$, calculate

$$\Delta u_i^{(k+1)} \leftarrow \frac{1}{A_{i,i}} \left(\begin{array}{c} f_i - \frac{1}{A_{i,i}} \left(\sum\limits_{j=0}^{i-1} A_{i,j} \Delta u_j^{(k+1)} \right) \\ -\frac{1}{A_{i,i}} \left(\sum\limits_{j=i+1}^{n} A_{i,j} \Delta u_j^{(k)} \right) \end{array} \right).$$

   b. Conduct the convergence test, $\left| \Delta u_i^{(k+1)} - \Delta u_i^{(k)} \right| \leq \varepsilon$. If satisfied, continue to step iv, otherwise repeat step iii(a).
iv. Conduct the convergence test, $\left| G\left(\underline{u}^{(k+1)}\right) \right| \leq \varepsilon$. If satisfied, display the approximate solution, and otherwise repeat step iii.
v. Display the output.
vi. Stop.

3 Numerical Experiments and Discussion

This study considers five large mesh size which are 512, 1024, 2048, 4096, and 8192. Using three main parameters, number of iteration (Iter), computational time (Time) and maximum absolute error (Err), we compared the data obtained for four suggested iterative methods, NGS-CS1, NGS-CT, NJ-CS1 and NJ-CT on three numerical examples as stated here.

Example 1 Consider the following NFIE-2 problem [21]

$$u(t) = 1 - \frac{5}{12} t + \int_0^1 t x [u(x)]^2 dx \tag{11}$$

where the exact solution for this problem is $u(t) = 1 + \frac{1}{3}t$.

Example 2 Consider the following NFIE-2 problem [22]

$$u(t) = -\frac{t}{9} - \frac{t^2}{8} + t^3 + \int_0^1 (t^2x + tx^2)u^2(x)dx \tag{12}$$

where the exact solution for this problem is $u(t) = t^3$.

Example 3 Consider the following NFIE-2 problem [23]

$$u(t) = t + \frac{\cos\left(e^{(1)} + t\right) - \cos(1+t)}{20} + \int_0^1 \frac{\sin\left(e^{(x)} + t\right)}{20} e^{(u(x))}dx \tag{13}$$

where the exact solution for this problem is $u(t) = t$.

Tables 1, 2 and 3 show the implementation of NJ and NGS using Trapezium and Simpson's 1/3 does not bring much different in terms of number of iteration and iteration time. But in terms of maximum absolute error, the approximate solutions for all proposed problems recorded more accurate approximate error using Simpson's 1/3 rule compared to Trapezium rule. Figures 2, 3 and 4 demonstrates th3e comparison of Newton-iterative methods in terms of maximum absolute error graphically for Examples 1, 2, and , respectively. The figures show NGS-CS1 iterative method have provided more accurate solutions compared to methods with Trapezium rule. This is

Table 1 Numerical results of NJ and NGS iterative methods using Trapezium and Simpson's 1/3 for Example 1

		Mesh size				
		512	1024	2048	4096	8192
Iter	NJ-CT	329	330	330	330	330
	NGS-CT	183	183	183	184	184
	NJ-CS1	329	330	330	330	330
	NGS-CS1	183	183	183	184	184
Time	NJ-CT	1.66	6.65	26.56	106.12	423.31
	NGS-CT	0.94	3.76	14.97	60.23	240.11
	NJ-CS1	1.68	6.62	26.45	105.7	422.29
	NGS-CS1	0.96	3.75	14.96	60.15	241.12
Err	NJ-CT	3.18E−06	7.94E−07	1.98E−07	4.93E−08	1.20E−08
	NGS-CT	3.18E−06	7.95E−07	1.99E−07	4.95E−08	1.23E−08
	NJ-CS1	3.89E−10	3.92E−10	3.94E−10	3.95E−10	3.95E−10
	NGS-CS1	1.23E−10	1.24E−10	1.25E−10	1.25E−10	1.25E−10

Table 2 Numerical results of NJ and NGS iterative methods using Trapezium and Simpson's 1/3 for Example 2

		Mesh size				
		512	1024	2048	4096	8192
Iter	NJ-CT	105	105	105	105	105
	NGS-CT	64	64	64	64	64
	NJ-CS1	105	105	105	105	105
	NGS-CS1	64	64	64	64	64
Time	NJ-CT	0.55	2.19	8.78	35.05	138.82
	NGS-CT	0.35	1.41	5.54	22.00	87.79
	NJ-CS1	0.58	2.19	8.70	34.75	138.96
	NGS-CS1	0.36	1.39	5.48	21.91	87.93
Err	NJ-CT	1.13E−05	2.82E−06	7.04E−07	1.76E−07	4.40E−08
	NGS-CT	1.13E−05	2.82E−06	7.04E−07	1.76E−07	4.40E−08
	NJ-CS1	4.47E−11	5.57E−11	6.28E−11	6.36E−11	6.39E−11
	NGS-CS1	9.81E−11	6.99E−12	1.14E−11	1.18E−11	1.19E−11

Table 3 Numerical results of NJ and NGS iterative methods using Trapezium and Simpson's 1/3 for Example 3

		Mesh size				
		512	1024	2048	4096	8192
Iter	NJ-CT	22	22	22	22	22
	NGS-CT	18	18	18	18	18
	NJ-CS1	22	22	22	22	22
	NGS-CS1	18	18	18	18	18
Time	NJ-CT	1.69	6.73	26.94	108.01	428.01
	NGS-CT	1.42	5.66	22.72	90.89	360.65
	NJ-CS1	1.69	6.76	26.99	107.97	441.91
	NGS-CS1	1.43	5.69	23.01	91.07	363.81
Err	NJ-CT	1.45E−07	3.63E−08	9.08E−09	2.27E−09	5.68E−10
	NGS-CT	1.45E−07	3.63E−08	9.08E−09	2.27E−09	5.68E−10
	NJ-CS1	2.84E−13	3.83E−14	2.82E−14	2.76E−14	2.78E−14
	NGS-CS1	2.95E−13	1.84E−14	5.27E−15	4.80E−15	5.12E−15

due to the implementation of high order quadrature scheme which helps to increase the neighborhood distance between each node points in interval $[a, b]$ which resulting more accurate results. Moreover, when comparing both NJ-CS1 and NGS-CS1, NGS-CS1 iteration recorded more accurate results compared to NJ-CT.

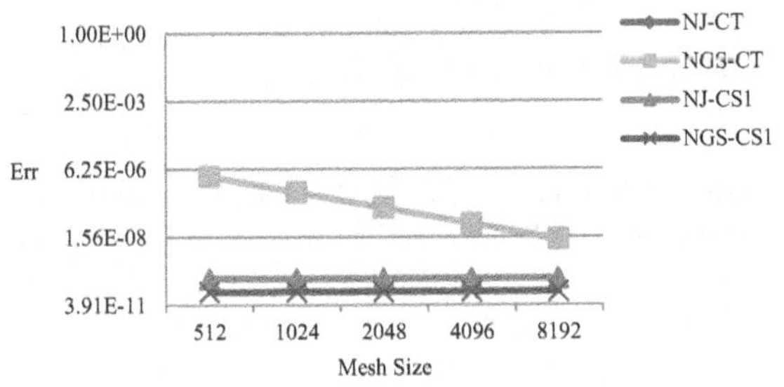

Fig. 2 Plots of maximum absolute error for of NJ and NGS iterative methods using Trapezium and Simpson's 1/3 (Example 1)

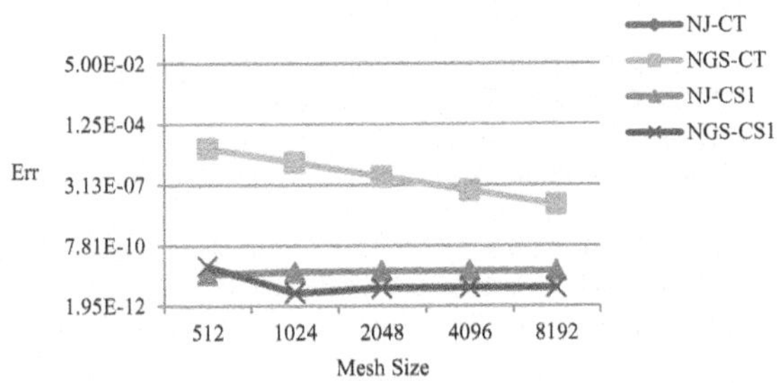

Fig. 3 Plots of maximum absolute error for of NJ and NGS iterative methods using Trapezium and Simpson's 1/3 (Example 2)

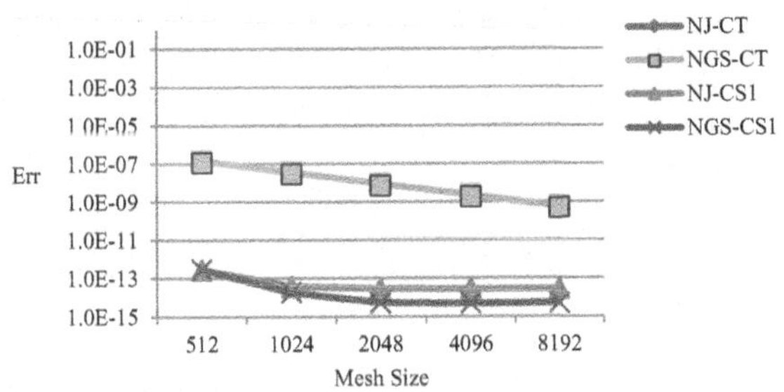

Fig. 4 Plots of maximum absolute error for of NJ and NGS iterative methods using Trapezium and Simpson's 1/3 (Example 3)

4 Conclusions

In this study, we discuss the application of Newton-iterative method with first- and second-order quadrature schemes in getting the approximate solution of NFIE-2. Based on the data obtained, we can conclude that the implementation of second-order quadrature scheme can improve the accuracy of the output compared to the first-order quadrature scheme. This means the approximate solution obtained for NFIE-2 is very close to the exact solutions. Thus, we conclude that Newton-GS iteration with second-order quadrature scheme, NGS-CS1 to be the most efficient method in solving NFIE-2 compared to the rest tested methods in this study. In the future study, this finding had a wide potential to be extended using the combination of the half-sweep iteration concept together with the weighted parameter iteration family, specifically to weighted parameter [16, 17], modified weighted parameter

[24, 25] and Accelerated parameter [26, 27] iteration families to reduce the iteration number and computational time and to improve the accuracy of the approximate solution.

Acknowledgements All authors would like to thank Universiti Malaysia Sabah for funding this research under UMS Research Grant Scheme: GUG0485-1/2020.

References

1. Farzi, J.: Generalized extrapolation methods for solving nonlinear Fredholm integral equations. Math. Commun. **19**, 375–390 (2014)
2. Ebrahimi, N., Rahidinia, J.: Collocation method for linear and nonlinear Fredholm and Volterra integral equations. Appl. Math. Comput. **270**, 156–164 (2015)
3. Bazm, S.: Bernoulli polynomials for the numerical solution of some classes of linear and nonlinear integral equations. J. Comput. Appl. Math. **275**, 44–60 (2015)
4. Maleknejad, K., Rostami, Y., Kalalagh, H.S.: Numerical solution for first kind Fredholm integral equations by using Sinc collocation method. Int. J. Appl. Phys. Math. **6**(3), 120–128 (2016)
5. Das, P., Nelakanti, G.: Error analysis of discrete Legendre multi-projection methods for nonlinear Fredholm integral equations. Numer. Funct. Anal. Optim., 1–26 (2017)
6. Hasan, M.M., Matin, M.A.: Approximate solution of nonlinear integral equations of the second kind by using Homotopy perturbation method. Dhaka Univ. J. Sci. **65**(2), 151–155 (2017)
7. Mohedul, H.M., Abdul, M.M.: Solving nonlinear integral equations by using Adomian decomposition method. J. Appl. Comput. Math. **6**(2), 1–4 (2017)
8. Binh, N.T., Ninh, K.V.: Parameter continuation method for solving nonlinear Fredholm integral equations of the second kind. Bull. Malays. Math. Sci. Soc. **42**, 3379–3407 (2019)
9. Maturi, D.A.: The successive approximation method for solving nonlinear Fredholm integral equation of the second kind using Maple. Adv. Pure Math. **9**, 823–843 (2019)
10. Awawdeh, F., Smail, L.: Convergence analysis of a highly accurate Nyström scheme for Fredholm integral equations. Appl. Numer. Math. **152**, 231–242 (2020)
11. Najafi, E.: Nyström-quasilinearization method and smoothing transformation for the numerical solution of nonlinear weakly singular Fredholm integral equations. J. Comput. Appl. Math. **368**, 1–13 (2020)
12. Amiri, S., Hajipour, M., Baleanu, D.: On accurate solution of the Fredholm integral equations of the second kind. Appl. Numer. Math. **150**, 478–490 (2020)
13. Nadjafi, F.S., Heidari, M.: Solving nonlinear integral equations in the Urysohn form by Newton Kantorovich quadrature method. Appl. Numer. Math. **150**, 478–490 (2020)
14. Janodi, M.R., Majid, Z.A., Ismail, F., Senu, N.: Numerical solution of Volterra integro-differential equations by hybrid block with quadrature rules method. Malays. J. Math. Sci. **14**(2), 191–208 (2020)
15. Muthuvalu, M.M., Sulaiman, J.: Comparison of quadrature schemes with arithmetic mean iterative method for second kind linear Fredholm integral equations. J. Math. Comput. Sci. **3**, 174–186 (2010)
16. Ali, L.H., Sulaiman, J., Hashim, S.R.M.: SOR iterative method with Simpson's 1/3 rule for the numerical solution of fuzzy second kind Fredholm integral equations. J. Phys. Conf. Ser. **1123**, 012030 (2018)
17. Ali, L.H., Sulaiman, J., Hashim, S.R.M.: Numerical solution of SOR iterative method for fuzzy Fredholm integral equations of second kind. In: Proceeding of the International Conference on Mathematics, Engineering and Industrial Applications, AIP Conference Proceedings **2013**, 020016 (2018)

18. Emamzadeh, M.J., Kajani, M.T.: Nonlinear Fredholm integral equation of the second kind with quadrature methods. J. Math. Ext. **4**(2), 51–58 (2010)
19. Sulaiman, J., Hasan, M.K., Othman, M., Karim, S.A.A.: Newton-EGMSOR methods for solution of second-order two-point nonlinear boundary value problems. J. Math. Syst. Sci. **2**, 185–190 (2012)
20. Youssef, I.K.: On the successive overrelaxation method. J. Math. Stat. **8**(2), 176–184 (2012)
21. Maleknejad, K., Nediasl, K.: Application of Sinc-collocation method for solving a class of nonlinear Fredholm integral equations. Comput. Math. Appl. **62**, 3292–3303 (2011)
22. Sahu, P.K., Ray, S.S.: Numerical approximate solutions of nonlinear Fredholm integral equations of second kind using B-Spline wavelets and variational iteration method. Comput. Model. Eng. Sci. **93**(2), 91–112 (2013)
23. Allahviranloo, T., Ghanbari, M.: Discrete homotopy analysis method for the nonlinear Fredholm integral equations. Ain Shams Eng. J. **2**, 133–140 (2011)
24. Akhir, M.K.M., Othman, M., Sulaiman, J., Majid, Z.A., Suleiman, M.: The four point-EDGMSOR iterative method for solution of 2D Helmholtz equations. Commun. Comput. Inf. Sci. CCIS **253**(PART 3), 218–227 (2011)
25. Akhir, M.K.M., Othman, M., Sulaiman, J., Majid, Z.A., Suleiman, M.: Numerical solution of Helmholtz equation using a new four point EGMSOR iterative method. Appl. Math. Sci. **5**(80), 3991–4004 (2011)
26. Sunarto, A., Sulaiman, J., Saudi, A.: Implicit finite difference solution for time-fractional diffusion equations using AOR method. J. Phys. Conf. Ser. **495**, 012032 (2014)
27. Dahalan, A.A., Saudi, A., Sulaiman, J.: Autonomous navigation on modified AOR iterative method in static indoor environment. J. Phys. Conf. Ser. **1366**(1), 012020 (2019)

Gauss-Newton and L-BFGS Methods in Full Waveform Inversion (FWI)

Samsul Ariffin Abdul Karim, Mudassar Iqbal, Afza Shafie, and Muhammad Izzatullah

Abstract Full waveform inversion (FWI) is a recent powerful method in the area of seismic imaging where it used for reconstructing high-resolution images of the subsurface structure from local measurements of the seismic wavefield. This method consists in minimizing the distance between the predicted and the recorded data. The predicted data are computed as the solution of a wave-propagation problem. In this study, we investigate two algorithms Gauss-Newton and L-BFGS for solving FWI problems. We compare these algorithms in terms of its robustness and speed of convergence. Also, we implement the Tikhonov regularization for assisting convergence. Numerical results show that Gauss-Newton method performs better than L-BFGS method in terms of convergence of l_2-norm of misfit function gradient since it provides better convergence as well as the quality of high resolution constructed images. Yet, L-BFGS outperforms Gauss-Newton in terms of computationally efficiency and feasibility for FWI.

S. A. Abdul Karim (✉)
Fundamental and Applied Sciences Department and Centre for Systems Engineering (CSE), Institute of Autonomous System, Universiti Teknologi PETRONAS, Seri Iskandar, Perak Darul Ridzuan, Malaysia
e-mail: samsul_ariffin@utp.edu.my

M. Iqbal · A. Shafie
Department of Fundamental and Applied Sciences, Universiti Teknologi PETRONAS, Seri Iskandar, Perak Darul Ridzuan 32610, Malaysia
e-mail: mudassar.iqbal@buitms.edu.pk

A. Shafie
e-mail: afza@utp.edu.my

M. Iqbal
Department of Mathematical Sciences, Balochistan University of Information Technology, Engineering and Management Sciences (BUITEMS),Quetta, Pakistan

M. Izzatullah
Seismic Modeling and Inversion Group (SMI), King Abdullah University of Science and Technology, Thuwal, Saudi Arabia
e-mail: muhammad.izzatullah@kaust.edu.sa

S. A. Abdul Karim et al. (eds.), *Proceedings of the 6th International Conference on Fundamental and Applied Sciences*, Springer Proceedings in Complexity,
https://doi.org/10.1007/978-981-16-4513-6_61

Keywords Full waveform inversion · Gauss-Newton · L-BFGS · Seismic wavefield · Tikhonov regularization

1 Introduction

Full-waveform inversion (FWI) is a recent powerful method based on based on nonlinear optimization technique in the area of seismic imaging. FWI was proposed by [1–3] back in the early of 1980s for reconstructing high-resolution images of the subsurface structure from local measurements of the seismic wavefield by minimizing the distance between the predicted and the recorded data [4–6]. Since then there are many numerical studies and new implementation of algorithms have been done [7, 8].

In this study, we investigate two algorithms Gauss-Newton and L-BFGS for solving frequency domain FWI as proposed in [7]. We compare these algorithms in terms of its robustness and speed of convergence via realistic synthetic model with marine exploration seismic setting. Also, we implement the Tikhonov regularization for assisting convergence.

2 Problem Formulation

We will formulate the FWI problem in the frequency domain as proposed by **PRatt**. Consider the slowness-squared as model parameters $\mathbf{m} \in \mathbb{R}^{n_{grid}}$ and the measurement vector $\mathbf{d} \in \mathbb{C}^{n_{data}}$ are related through a known but nonlinear relationship denoted as

$$\mathbf{d} = F(\mathbf{m}) + \epsilon, \tag{1}$$

where $\epsilon \sim \mathcal{N}(0, \mathbf{C}_D)$ is additive, normally distributed noise with zero mean and covariance $\mathbf{C}_D \in \mathbb{C}^{n_{data} \times n_{data}}$.

The nonlinear forward modeling map $F(\mathbf{m})$ can be desribed as

$$F(\mathbf{m}) = \mathbf{PA}(\mathbf{m})^{-1}\mathbf{q}, \tag{2}$$

where $\mathbf{q} \in \mathbb{C}^{n_{grid}}$ is the discretized source term which considered known. The operator $\mathbf{A}(\mathbf{m}) \in \mathbb{C}^{n_{grid} \times n_{grid}}$ represents the discretized Helmholtz operator $(\nabla^2 + \omega^2\mathbf{m})$ where $\omega = 2\pi f$ is the angular frequency. The operator $\mathbf{P} \in \mathbb{R}^{n_{data} \times n_{grid}}$ denotes the sampling operator which samples the data $\mathbf{d}$ from the field vector variables $\mathbf{u}$, which is the solution of the Helmholtz equation $\mathbf{u} = \mathbf{A}(\mathbf{m})^{-1}\mathbf{q}$.

By choosing the matrix that $\mathbf{L}$ as the first order finite difference operator which commonly referred to as roughening matrix, we can define the least-square misfit function with Tikhonov regularization as

$$V(\mathbf{m}) = \frac{1}{2}\left|\left|F(\mathbf{m}) - \mathbf{d}\right|\right|_2^2 + \frac{\alpha}{2}\left|\left|\mathbf{Lm}\right|\right|_2^2, \tag{3}$$

where α is the regularization coefficient. The optimal model $\mathbf{m}$ can be sought by minimizing the misfit function $V(\mathbf{m})$ in 3. The resulting optimization problem is typically solved using a gradient-based method which generates iterates of the form

$$\mathbf{m}_{k+1} = \mathbf{m}_k - B_k \nabla V(\mathbf{m}_k), \tag{4}$$

where B_k includes appropriate scaling/smoothing of the gradient. In this study the matrix B_k could be represented either as the inverse of the Gauss-Newton approximation or the L-BFGS approximation of Hessian which will be explained in details in the following sections. For the gradient of the misfit function, it can be computed through adjoint-state method [9] and the explicit formula can be described as

$$\nabla V(\mathbf{m}) = \mathbf{J}^T(F(\mathbf{m}) - \mathbf{d}) + \alpha(\mathbf{L}^T\mathbf{L})\mathbf{m}, \tag{5}$$

with $\mathbf{J}$ the Jacobian of $\mathbf{F}(\mathbf{m})$.

3 Gauss-Newton Method

Gauss-Newton method is a method derived from Newton method for solving the nonlinear optimization problem. The issue with Newton method in solving the nonlinear optimization problem especially FWI is the computation of full Hessian. In Eq. 4, the matrix B_k has two terms based on Newton method which can be presented as

$$\mathbf{H} = \mathbf{J}^T\mathbf{J} + \frac{\partial \mathbf{J}}{\partial \mathbf{m}}(F(\mathbf{m}) - \mathbf{d}). \tag{6}$$

Commonly, the computation of the second term is avoided due to its tedious calculation and which in any case should be small by assuming the problem is approximately linear, which, in practice, implies that the starting model is sufficiently close to the true model. This is where the Gauss-Newton method is being derived from. The difference between Newton and Gauss-Newton method is the negligence of the second term in the Hessian computation. Based on [7, 10], we can safely dropped off the second term in the Eq. 6 because of its value is too small and it is only important if changes in the parameters cause a change in the partial derivative of the Helmholtz equation's solution.

The Gauss-Newton method and its approximation of Hessian can be presented as

$$\mathbf{m}_{k+1} = \mathbf{m}_k - \mathbf{H}_{GN} \nabla V(\mathbf{m}_k), \tag{7}$$

$$\mathbf{H}_{GN} = \mathbf{J}^T \mathbf{J}, \tag{8}$$

where the matrix $\mathbf{H}_{GN}$ is assumed to have full column rank, and is thus invertible. See [11] for more details regarding to this algorithm.

4 L-BFGS Method

The limited- memory BFGS method (L-BFGS) is a quite successful modification of the quasi-Newton methods [11, 12]. In this method, no Hessian approximation is ever actually formed, but rather a collection of the last several (s_k, y_k) pairs is stored and used to compute the step. Let m, the memory size, be the number of (s, y) pairs stored. Then, given an initial matrix H_0, the matrix H_k can be defined as follows:

$H \leftarrow H_0$
for $m, \ldots, 1$ **do**
  $H \longleftarrow V_{k-i}^T H V_{k-i} + \rho_{k-i} s_{k-i} s_{k-i}^T$
end for
$H_k \leftarrow H.$

The notation is simplified by eliminating the iteration counter k and choosing to store the most recent value of s, that is, $s_k - 1$, in $s_m - 1$ and the oldest value, $s_k - m$, in s_0. The vectors $y_i, i = 0, \ldots, m-1$, are stored similarly. With these values, it can be shown that the search direction in Eq. 4 can be represented as

$$B_k \nabla V(\mathbf{m}_k) = H_k \nabla V(\mathbf{m}_k), \tag{9}$$

where the matrix H_k is the L-BFGS approximation to the inverse Hessian and can be computed through the algorithm presented above.

5 Numerical Examples

In these numerical examples, we illustrate the performance of Gauss-Newton and L-BFGS algorithms through solving the frequency domain FWI problem. We solve two FWI problems with two different velocity models with an objective to compare these two algorithms in reconstructing the velocity models from the recorded data.

For first numerical example, we use a homogeneous velocity model with an inclusion in the centre which acts as an reflector, depicted in the Fig. 1a. A standard

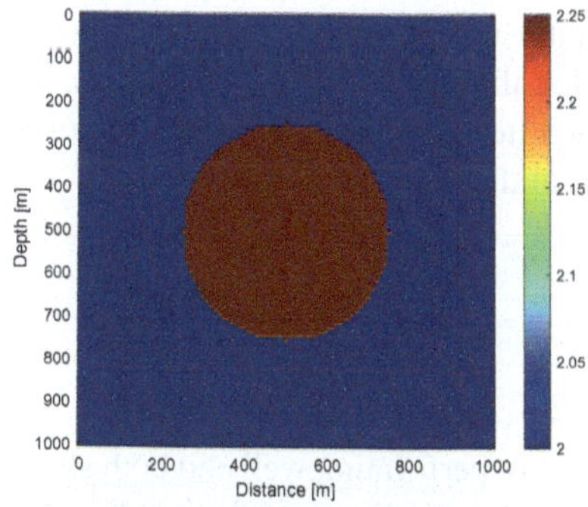

(i) True Model

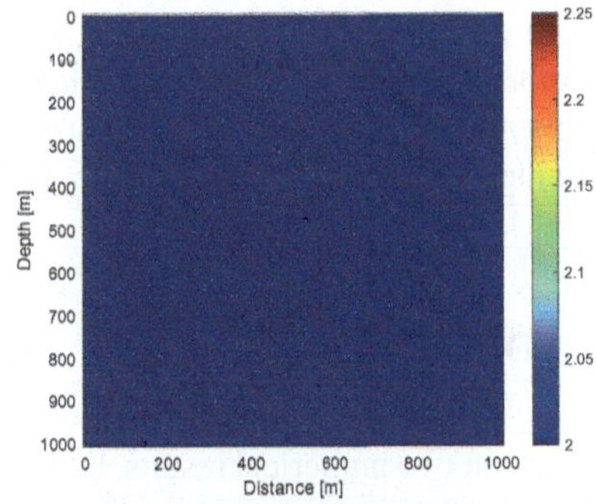

(ii) Initial Model

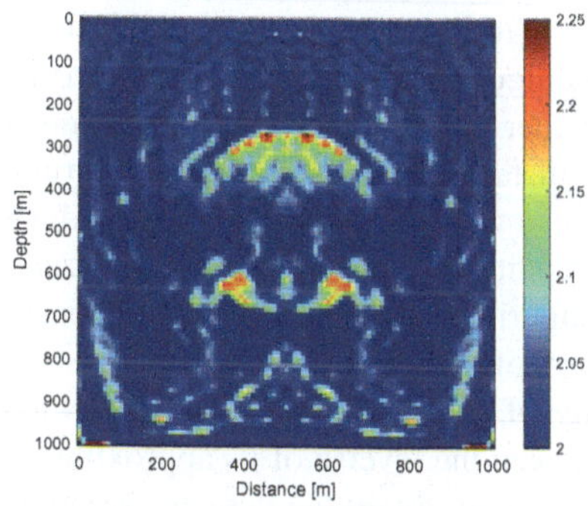

(iii) L BFGS

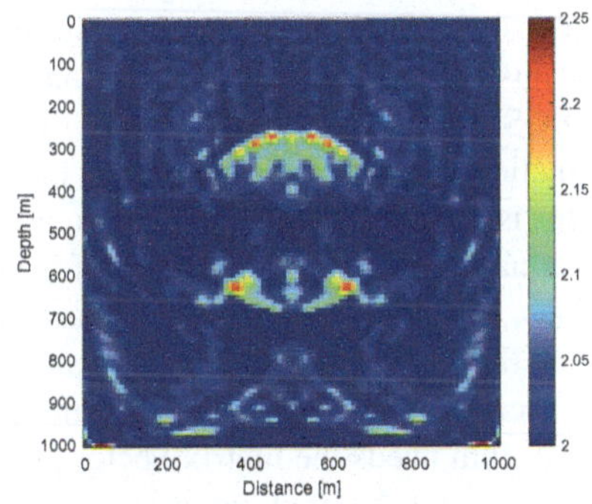

(iv) Gauss-Newton

Fig. 1 Numerical example 1: reconstruction of homogeneous velocity model with an inclusion in the centre

finite-difference method is used to solve the Helmholtz equation. The grid size is 100×100, and grid spacing is 10×10 m. In this numerical example we consider collocated sources-receivers setting with sources-receivers are located at every 20m. We use frequency content 5 to 25 Hz with frequency sampling of 3.33 Hz.

In the second numerical example, we use the Marmousi model as depicted in the Fig. 3a to perform the numerical studies. A standard finite-difference method is used to solve the Helmholtz equation. The grid size is 61×220, and grid spacing is 50×50 m. 50 shots at every 100 and 100 receivers at every 50m are used in this numerical example. This sources-receivers setting is resembling the marine exploration seismic setting. We use frequency content from 0.5 Hz to 3.95 Hz with frequency sampling of 0.5 Hz.

For both numerical examples, we performed 100 Gauss-Newton and L-BFGS iterations each, starting from the initial model depicted in the Figs. 1b and 3b respectively to obtain the optimal model $\mathbf{m}$ as shown in the bottom row of Figs. 1 and 3. As regularization, we use the Tikhonov regularization method with regularization operator $\mathbf{L}$ as first order derivative operator and regularization parameter α equals to 0.01.

In practice, the Hessian is not store explicitly in memory and only its matrix-vectors product are being computed. Thus, for the Gauss-Newton iterations, we are solving a system of linear equations at each iteration using the preconditioned conjugate gradient (PCG) to estimate the descent direction.

6 Discussions

Based on two numerical results, both algorithms are performing well and both showing a good convergence of misfit values and the values of l_2-norm of misfit function gradient as illustrated in Figs. 2 and 4, respectively. As we can observe, the misfit values of L-BFGS is better than Gauss-Newton algorithm, yet the values of l_2-norm of misfit function gradient for Gauss-Newton algorithm is lower compared to L-BFGS algorithm. In practice, we should consider the values of l_2-norm of misfit function gradient as it represents the optimal distance of the solution to the truth. This is because the true solution could be obtained when the misfit function gradient is equal to zero or in the vicinity of l_2-norm of misfit function gradient closes to zero. Thus, based on this practice, Gauss-Newton algorithm is perform better compared to L-BFGS because of its lower value in l_2-norm of misfit function gradient.

Here we also should discuss the feasibility of each algorithm. Gauss-Newton algorithm needs the matrix-vector product between the inverse of its approximated Hessian and the gradient at each iteration in order to obtain the descent direction. This computation is computationally intensive thus it takes longer time per iteration to solve the optimization problem. Meanwhile, in L-BFGS algorithm no Hessian approximation is ever actually formed, but rather a collection of the last several (s_k, y_k) pairs is stored and used to compute the step. This makes L-BFGS algorithm is computationally efficient compared to the Gauss-Newton algorithm.

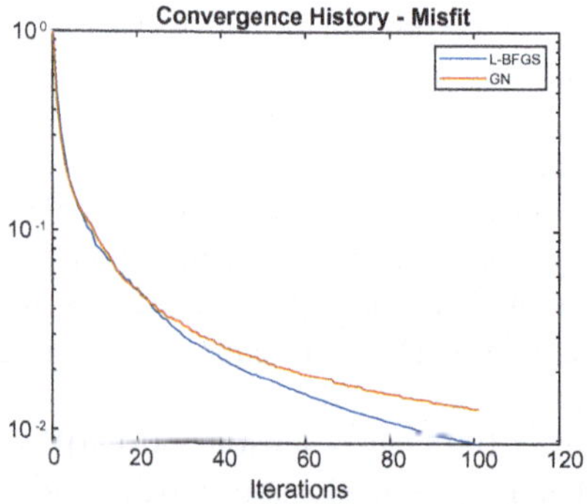

(i) Misfit

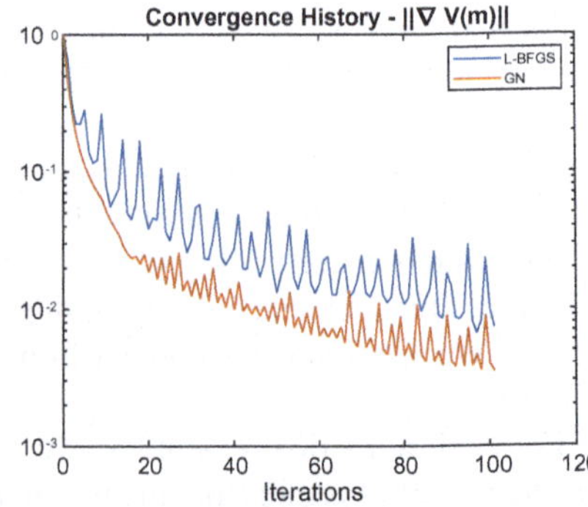

(ii) Norm of gradient

Fig. 2 Numerical example 1: misfit and norm of gradient values at each iteration

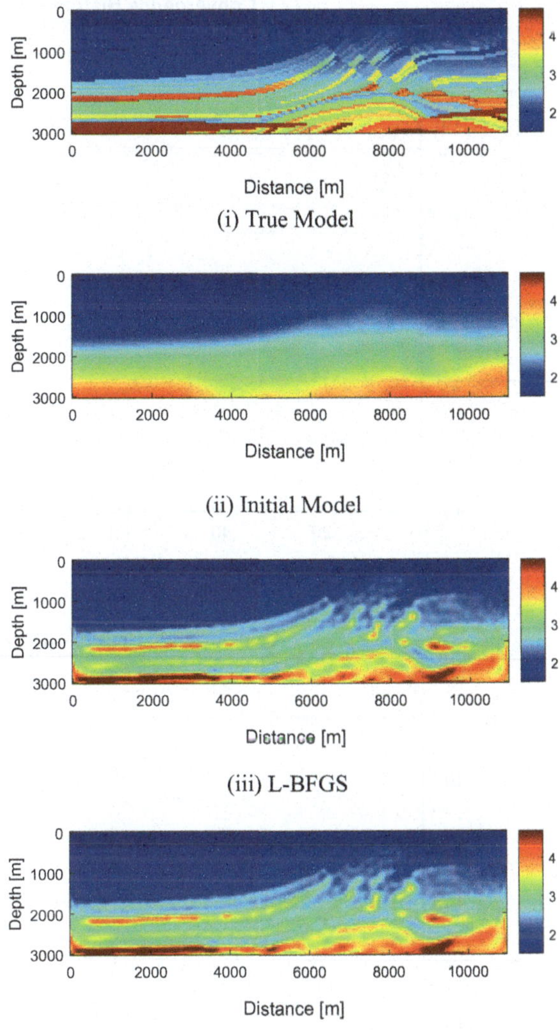

Fig. 3 Numerical example 2: reconstruction of marmousi model

(i) True Model

(ii) Initial Model

(iii) L-BFGS

(iv) Gauss-Newton

7 Conclusion

In conclusion, both algorithms, L-BFGS and Gauss-Newton are comparable to each other in terms of performance. Gauss-Newton algorithm gives a better result in the convergence of l_2-norm of misfit function gradient sense, yet it is computationally intensive. Meanwhile, L-BFGS performance is comparable to the Gauss-Newton and in terms of computationally efficiency and feasibility, L-BFGS is outperformed the Gauss-Newton for the large scale optimization problems especially in FWI.

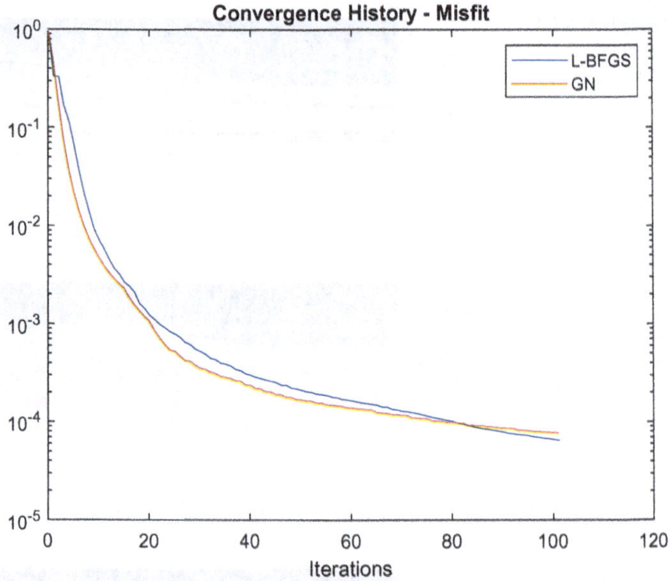

(i) Misfit

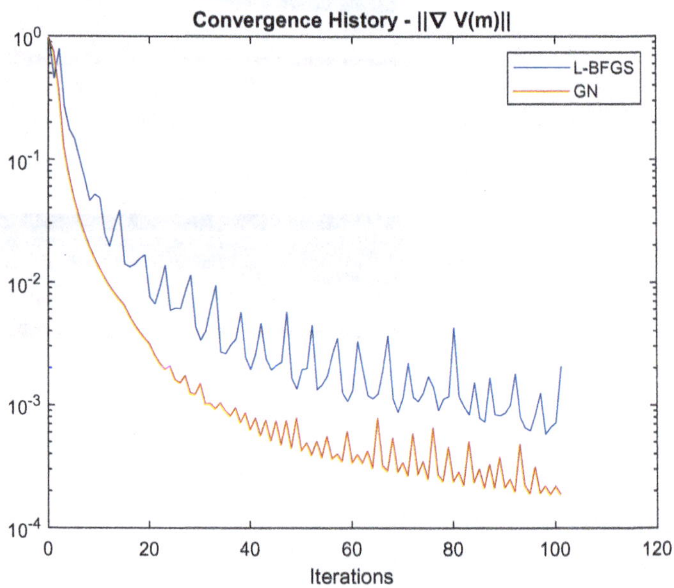

(ii) Norm of gradient

Fig. 4 Numerical example 2: misfit and norm of gradient values at each iteration

Acknowledgements This research was fully supported by Universiti Teknologi PETRONAS (UTP) through a research grant YUTP: 015LC0-315 (Uncertainty estimation based on Quasi-Newton methods for Full Waveform Inversion (FWI)).

References

1. Tarantola, A.: Inversion of seismic reflection data in the acoustic approximation. Geophysics **49**(8), 1259–1266 (1984)
2. Tarantola, A.: Linearized inversion of seismic reflection data. Geophys. Prospect. **32**(6), 998–1015 (1984)
3. Lailly, P.: The seismic inverse problem as a sequence of before stack migrations. In: Conference on Inverse Scattering, Theory and Applications, Society for Industrial and Applied Mathematics (1983)
4. Valette, B., Tarantola, A.: Generalized nonlinear inverse problems solved using the least squares criterion. Rev. Geophys. **20**(2), 219 (1982)
5. Virieux, J., Operto, S.: An overview of full-waveform inversion in exploration geophysics. Geophysics **74**(6), WCC1–WCC26 (2009)
6. Virieux, J., Asnaashari, A., Brossier, R., Métivier, L., Ribodetti, A., Zhou, W.: An introduction to full waveform inversion (2017)
7. Gerhard Pratt, R., Shin, C., Hick, G.J.: Gauss–Newton and full Newton methods in frequency–space seismic waveform inversion. Geophys. J. Int. **133**(2), 341–362 (1998)
8. Métivier, L., Brossier, R., Operto, S., Virieux, J.: Full waveform inversion and the truncated Newton method. SIAM Rev. **59**(1), 153–195 (2017)
9. Plessix, R.E.: A review of the adjoint-state method for computing the gradient of a functional with geophysical applications. Geophys. J. Int. **167**(2), 495–503 (2006)
10. Tarantola, A.: Inverse Problem Theory and Methods for Model Parameter Estimation. Society for Industrial and Applied Mathematics (2005)
11. Nocedal, J., Wright, S.J.: Numerical Optimization, 2nd edn. New York, Springer (2006)
12. Liu , D.C., Nocedal, J.: On the limited memory BFGS method for large scale optimization. Math. Program. **45**(1), 503–528 (1989)

Removing the Blurring from X-Ray Image Using BM3D Technique

Ariful Islam, Nooraini Zainuddin, and Samsul Ariffin Bin Abdul Karim

Abstract The x-ray image quality of normal patient is needed to enhance to diagnose accurately. For this reason, block-matching 3D (BM3D) technique is chosen for denoising the x-ray images. The currently the best BM3D denoising system utilizes a white Gaussian noise (WGN) design. The similar 2D x-ray image is converted to 3D data arrays by grouping to improve the sparsity and it is called grouping. Collaborative filtering is a unique method for dealing with these three-dimensional groups. The collaborative filtering reduces noise, demonstrating even the details of image shared by grouped blocks while preserving the crucial unique characteristics from every individual block. After that, the shifted blocks are replaced with new positions. As these blocks coincide, we get a variety of special predictions with each pixel, which we have to combine. The Wiener filtering process is implemented in the transform coefficients to a post-thresholding signal in the present BM3D algorithm for improved noise removal. Wiener filtering of transform domain co-efficient is used based on the properties of x-ray images in terms of PSNR and SNR value. The hard thresholding system is used in previous step to denoise the x-ray image in the utter lack of a ground-truth signal. The performance of BM3D technique is compared with wavelet transform to evaluate image quality.

Keywords Collaborative filtering · Image denoising · Block matching · BM3D

A. Islam (✉) · N. Zainuddin
Department of Fundamental and Applied Sciences, Universiti Teknologi PETRONAS, 32610 Seri Iskandar, Perak Darul Ridzuan, Malaysia

N. Zainuddin
e-mail: aini_zainuddin@utp.edu.my

S. A. B. A. Karim
Department of Fundamental and Applied Sciences and Centre for Systems Engineering (CSE), Institute of Autonomous System, Universiti Teknologi PETRONAS, 32610 Seri Iskandar, Perak, Malaysia
e-mail: samsul_ariffin@utp.edu.my

S. A. Abdul Karim et al. (eds.), *Proceedings of the 6th International Conference on Fundamental and Applied Sciences*, Springer Proceedings in Complexity,
https://doi.org/10.1007/978-981-16-4513-6_62

1 Introduction

CT (computed tomography) has indeed been widely used in biological imaging in therapeutic applications as a non-invasive imaging method [1, 2]. CT scans are known to be the main standard for thoracic disease assesment. The low dose of CT screening is investigated for emphysema. Potential emphysema is defined in locations with extremely low attenuation coefficients in this diagnosis, and noise can heavily weaken this choice [3, 4]. Quantitative emphysema scoring's reliability on CT images was intensively reviewed in [5, 6], demonstrating that adaptive image processing can improve image quality. To improve the arrangement between full-dose and low-dose sufferers, reconstruction techniques is used.

The block-matching 3D (BM3D) denoising algorithm is investigated in this study to overcome reduced CT denoising complications [7, 8]. When natural/synthetic images are perverted by white Gaussian noise, the BM3D strategy has revealed state-of-the-art efficiency. It consists of two synchronous denoising stages, the second of which uses transform-domain collaborative Wiener filtering to enhance quality over the estimate with the first hard-thresholding denoising. When signal/noise total independence rules are met, the Wiener filter's efficacy in the current BM3D mechanism is called into question. Due to its inefficacy in attempting to deal with control noise that deviated from the Gaussian system, Wiener filtering in BM3D was abandoned out during [9]. Pre-whitening and a Wiener filtering tailored to noise variances were used to modify the BM3D process in order to minimise associated noise in [10]. Similarly, the WGN assumptions limit the application of the simple BM3D to low-dose CT denoising. In the existing BM3D approach, the hard-thresholding denoising module is being used as a plug-in spectrum evaluator for the ground-truth amplitude [11].

In this study, the proposed BM3D technique is used to remove noise from the x-ray image of patients. The noise variance is computed exactly in collaborative transform to compute fast. The image quality is evaluated using SNR and PSNR value after applying this technique. The SNR and PSNR value of BM3D technique is compared with wavelet technique. This process can be helpful to diagnose the patients accurately.

2 Methodology

In first step, the x-ray images are taken and the SNR, PSNR value is calculated from x-ray images. These values are gathered in multiplexer operator. Then, BM3D technique is applied for x-ray images to reduce noise (Fig. 1).

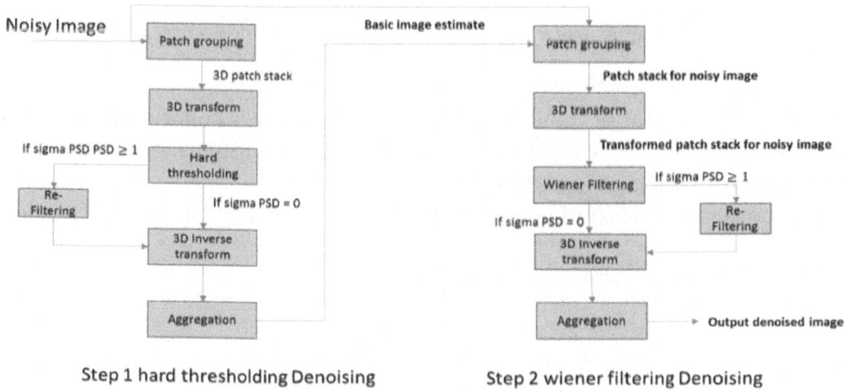

Fig. 1 The Block Diagram of Proposed Framework

2.1 BM3D Technique

Grouping: Grouping is the first sub-step. Patches M similar to the reference patch L are found in a P-centered neighbourhood of the noisy image u. The set of patches that are similar is simply defined by

$$\underset{\wedge}{L}(L) = \{M : d(L, M) \leq \tau^{hard}\} \tag{1}$$

The distance threshold for d below that two patches are supposed to be closely related is called τ^{hard}.$d(L, M) = \frac{\gamma'(L) - \gamma'(M)}{(k^{hard})^2}$ is the quadratic distance among patches. The hard thresholding operator is γ'. The variance of Gaussian noise is σ^2.

For collaborative filtering, the inverse linear transform can be referred as

$$\underset{\wedge}{L}(L)^{hard} = \tau_{3D}^{hard^{-1}}\left(\gamma\left(\tau_{3D}^{hard}(L)\right)\right) \tag{2}$$

where, a hard thresholding operator is set as γ and threshold value is set as $\lambda_{3D}^{hard}\sigma$:

$$\gamma(x) = \begin{cases} 0\ if x \leq \lambda_{3D}^{hard}\sigma \\ x\ otherwise \end{cases}. \tag{3}$$

For practical purposes, the 3D transform τ_{3D}^{hard} consists of two transforms. The 2D and 1D transform represented as τ_{2D}^{hard} and τ_{1D}^{hard} that is used in every patch $\underset{\wedge}{L}(L)$.

Aggregation: Whenever the collaborative filtering is complete, each used patch will receive an estimate, as well as a different set of predictions by each pixel.

$$\forall M \in L, \forall x \in M, \begin{cases} \vartheta(x) = \vartheta(x) + \omega_L^{hard} u_{M,L}^{hard}(x) \\ \delta(x) = \delta(x) + \omega_L^{hard} \end{cases} \tag{4}$$

The numerator part of significant investment of the image collected at the end of such grouping stage is denoted by ϑ (resp. δ) $u_{M,L}^{hard}(x)$ is estimated from patch M. The retained coefficients are referred by N_L^{hard}.

In second step, Wiener filter used the basic estimate as u^{basic} from original image u.

The set of similar patches can be written as

$$\overset{basic}{\underset{\wedge}{L}}(L) = \left\{M : d(L, M) \leq \tau^{wien}\right\} \tag{5}$$

Collaborative Filtering: Collaborative filtering can begin once the two 3D groups have been obtained. In order to achieve this, empirical Wiener coefficients are defined as follows:

$$\omega_L(\varepsilon) = \frac{\left|\tau_{3D}^{wien}\left(\overset{basic}{\underset{\wedge}{L}}(L)\right)(\varepsilon)\right|^2}{\left|\tau_{3D}^{wien}\left(\overset{basic}{\underset{\wedge}{L}}(L)\right)(\varepsilon)\right|^2 + \sigma^2} \tag{6}$$

The sub-step is estimated from 3D group that is obtained by

$$\overset{basic}{\underset{\wedge}{L}}(L) = \tau_{3D}^{wien-1}\left(\omega_L . \tau_{3D}^{wien}\left(\underset{\wedge}{L}(L)\right)\right) \tag{7}$$

Aggregation: The assumptions at each pixel are saved in a buffer because once collaborative filtering is completed:

$$\forall M \in L, \forall x \in M, \begin{cases} \vartheta(x) = \vartheta(x) + \omega_L^{wien} u_{M,L}^{wien}(x) \\ \delta(x) = \delta(x) + \omega_L^{wien} \end{cases} \tag{8}$$

$u_{M,L}^{wien}(x)$ is estimated from patch M during the collaborative filtering. To decrease boundary impact, $K^{wien} \times K^{wien}$ Kaiser window is implemented in first step. The second step can be given by

$$u^{final}(x) = \frac{\sum_L \omega_L^{wien} \sum_{M \in L} XM(x) u_{M,L}^{wien}(x)}{\sum_L \omega_L^{wien} \sum_{M \in \underset{\wedge}{L}(L)} XM(x)} \tag{9}$$

which is simply obtained by dividing both buffers (numerator and denominator) element-by element. Here $XM(x) = 1$ if and only if $X \in M$, 0 otherwise.

PSNR Equation:

$$PSNR = 10 \log_{10} \frac{R^2}{MSE} \tag{10}$$

The highest variability in the given input data type is denoted by R. If the input image is double highly precise inflatable type of data, for example, R equals 1. R is 255 if the data type is an 8-bit unreleased imaginary number.

MSE Equation:

$$\text{MSE} = \frac{1}{\text{MN}} \sum_{i=1}^{M} \sum_{j=1}^{N} (x(i, j) - y(i, j))^2 \tag{11}$$

SNR Equation: Signal S can be defined in a variety of ways. S can be a single patch pixel level or the pixel difference corresponding to a specific scene density range. SNR can be expressed as

$$\text{SNR (db)} = 20 \log_{10}\left(\frac{S}{N}\right) \tag{12}$$

3 Experimental Results

Here, The SNR and PSNR value of x-ray images is represented in Table 1 before implementing the BM3D method. And Table 2 represents the PSNR and SNR value after applying the BM3D technique.

It is also clear that the image quality increases after using the BM3D framework and it is mentioned in Fig. 2. It is shown that the PSNR and SNR value is increased after applying the BM3DBlock-Matching 3D (BM3D) technique and it is mentioned in Figs. 3 and 4.

Table 1 Parameters for deblurring the X-ray image

2D_HT_name 'dst'		Wiener filtering parameters	
pro.lambda_thr3D	2.9	N1_wiener	8
pro.N1_wiener	8	Nstep_wiener	3
pro.Nstep_wiener	2	N2_wiener	32
pro.N2_wiener	16	Ns_wiener	39
pro.Ns_wiener	39	tau_match_wiener	400
pro.tau_match_wiener	800	beta_wiener	2
pro.beta_wiener	0	decLevel	0

Table 2 The PSNR and SNR value before applying the BM3D technique

X-ray images (before applying BM3D)		X-ray images (after applying BM3D)	
PSNR	SNR	PSNR	SNR
22.29	18.6899	38.45	31.61
22.03	18.8076	33.75	30.46
22.25	18.0765	31.46	29.85
22.38	17.0019	30.04	29.44
22.08	18.0291	29.04	29.14
21.52	15.9502	28.2	28.88
22.37	17.4062	27.49	28.66
22.35	17.3868	25.68	28.06
21.64	14.5841	24.69	27.71
22.03	18.716	24.18	27.53
21.91	18.4518	23.89	27.42
21.75	18.7977	23.6	27.31
21.76	18.2033	23.09	27.12
21.35	17.462	22.63	26.94
22.27	17.1885	22.2	26.77
22.28	17.1931	21.84	26.63

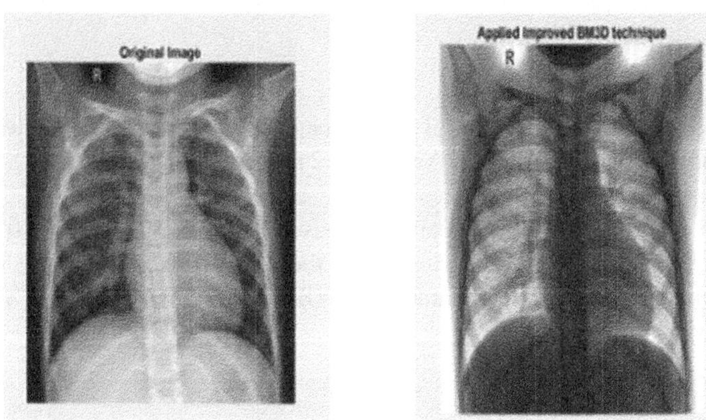

Fig. 2 The image quality enhancement is shown after applying BM3D technique

4 Conclusion

By considering the relative sizes of the adjusted blocks, we described a technique that requires with both exact computation and effective estimations of the output signal in nonlocal collaborative transforms. The exact deviations, in particular, perform more precise shrinkage and avoid matching blocks that are closely linked in noise but not

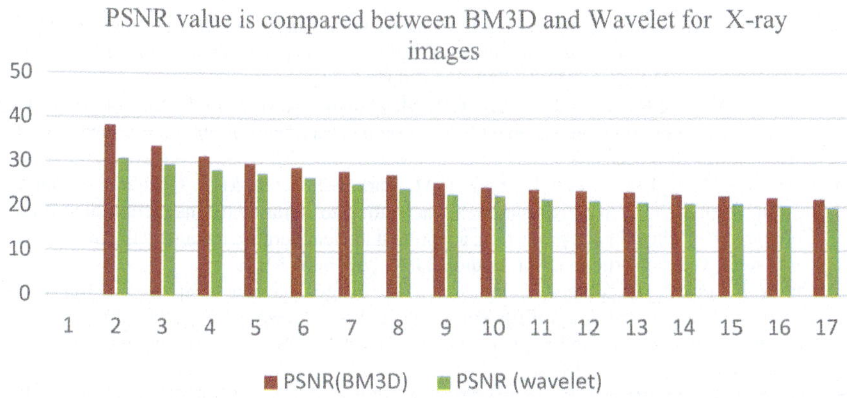

Fig. 3 The comparison of PSNR value between BM3D and Wavelet technique for x-ray images

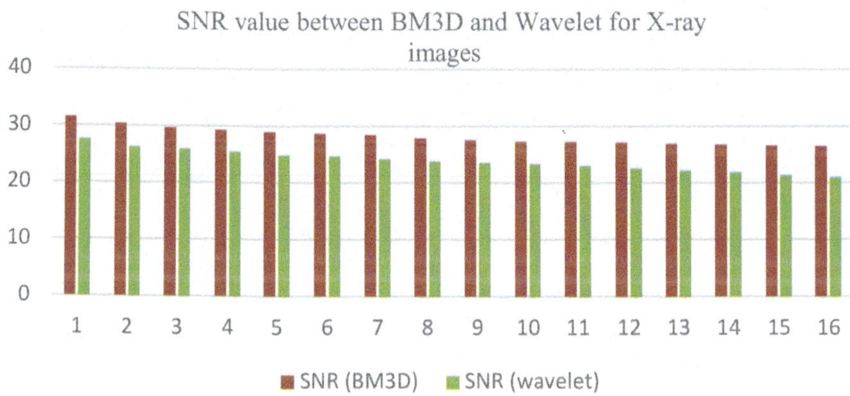

Fig. 4 The comparison of SNR value between BM3D and Wavelet technique for x-ray images

related in the noise-free image. Re-filtering the image approximation with an added thresholded residual can significantly reduce the block's small size. Then, the PSNR and SNR value is calculated of X-ray images using BM3D technique. Image quality depends on the SNR and PSNR value. Experimental result show that SNR and PSNR value is increased after applying the BM3D technique.

References

1. Brenner, D.J.: Radiation risks potentially associated with low-dose CT screening of adult smokers for lung cancer. Radiology **231**(2), 440–445 (2004)
2. Brenner, D.J., Hall, E.J.: Computed tomography-an increasing source of radiation exposure. N. Engl. J. Med. **357**(22), 2277–2284 (2007)

3. Mayo, J.R., Kim, K.I., MacDonald, S.L., Johkoh, T., Kavanagh, P., Coxson, H.O., Vedal, S.: Reduced Radiation Dose Helical Chest CT: Effect on Reader Evaluation of Structures and Lung Findings. Radiology **232**(3), 749–756 (2004)
4. Yuan, R., Mayo, J.R., Hogg, J.C., Pare, P.D., McWilliams, A.M., Lam, S., Coxson, H.O.: The effects of radiation dose and CT manufacturer on measurements of lung densitometry. Chest J. **132**(2), 617–623 (2007)
5. Choo, J.Y., Goo, J.M., Lee, C.H., Park, C.M., Park, S.J., Shim, M.S.: Quantitative analysis of emphysema and airway measurements according to iterative reconstruction algorithms: Comparison of filtered back projection, adaptive statistical iterative reconstruction and model-based iterative reconstruction. Eur. Radiol. **24**(4), 799–806 (2014)
6. Dabov, K., Foi, A., Katkovnik, V., Egiazarian, K.: Image denoising by sparse 3-D transform-domain collaborative filtering. IEEE Trans. Image Process. **16**(8), 2080–2095 (2007)
7. Lebrun, M.: An analysis and implementation of the BM3D image denoising method. Image Process. On Line **2**, 175–213 (2012)
8. Gao, J., Chen, Q., Blasch, E.: Image denoising in the presence of non-Gaussian, power-law noise. In: 2012 IEEE National Aerospace and Electronics Conference (NAECON), pp.103–108 (2012)
9. Matrecano, M., Poggi, G., Verdoliva, L.: Improved BM3D for correlated noise removal. VISAPP **1**, 129–134 (2012)
10. Hoffman, J., Young, S., Noo, F., McNitt-Gray, M.: Technical note: FreeCT wFBP: A robust, efficient, open-source implementation of weighted filtered backprojection for helical, fan-beam CT. Med. Phys. **43**(3), 1411–1420 (2016)
11. Young, S., Kim, H.G., Ko, M.M., Ko, W.W., Flores, C., McNitt-Gray, M.: Variability in CT lung-nodule volumetry: Effects of dose reduction and reconstruction methods. Med. Phys. **42**(5), 2679–2689 (2015)

Empirical Formulation of (n, p) Nuclear Cross-Section at 14–15 meV

Mohamad Amin Bin Hamid, Beh Hoe Guan, Xiao Yan Chew, and Saba Ayub

Abstract The empirical equation of (n, p) nuclear cross-section is formulated at 14–15 meV incident neutron energy. The data used for the construction of the empirical formula is obtained from EXFOR with a mass range of $9 < A < 111$. From the data, we fit empirical formula with exponential dependency of $\left(\frac{N-Z+A}{A}\right)^n$ where n is determined from the non-linear fitting. The empirical formulation is then compared using the chi-square coefficient as a quality criterion. The derived empirical formulation is compared with five previously published formulation. We show that the new formulation derived in this work gives good agreement with the experimental nuclear cross-section value.

Keywords Nuclear cross-section data · Empirical formulation · (n · p) nuclear reaction

M. A. B. Hamid (✉) · B. H. Guan
Centre of Innovative Nanostructure and Nanodevices, Universiti Teknologi PETRONAS, 32610 Seri Iskandar, Perak, Malaysia

B. H. Guan
Department of Fundamental & Applied Sciences, Universiti Teknologi PETRONAS, 32610 Seri Iskandar, Perak, Malaysia
e-mail: beh.hoeguan@utp.edu.my

X. Y. Chew
Department of Physics Education, Pusan National University, Busan 46241, Republic of Korea

S. Ayub
Research Center for Dielectric and Advanced Matter Physics, Pusan National University, Busan 46241, Republic of Korea
e-mail: saba_20000009@utp.edu.my

S. A. Abdul Karim et al. (eds.), *Proceedings of the 6th International Conference on Fundamental and Applied Sciences*, Springer Proceedings in Complexity,
https://doi.org/10.1007/978-981-16-4513-6_63

1 Introduction

Neutron-induced nuclear cross-section data is important in the study of nuclear structure, radiation damage, nuclear heating and structural transmutation of nuclear reactors [1]. Nuclear cross-section data is also valuable in determining the possible nuclear reaction channels which are important in the production of radionuclide [2]. Experimental nuclear cross-sections can be obtained from the EXFOR database, obtained from the measurement of various authors [3–5]. However, there are problems of the discrepancy between measurements done by different authors. This discrepancy is especially large at neutron incident energy around 14 meV. Some nuclear reaction is very hard to experiments on, the experimental cross-section data is scarce.

The empirical and semi-empirical formulation is crucial in theoretical calculations of nuclear cross-sections. For (n, p) nuclear reaction, the various empirical and semi-empirical formulation has been developed at neutron incident energy around 14 meV. In all of the empirical formulation for (n, p) nuclear cross-section, the nuclear cross-section is assumed to have exponential dependence on the number of nucleons in the target nucleus [6]. In this work, we proposed a new empirical formulation of (n, p) nuclear reaction by the non-linear fitting of EXFOR nuclear cross-section data for 61 nuclear isotopes with atomic mass range $9 < A < 111$ at neutron incident energy around 14–15 meV.

2 Methodology and Data

Generally, the statistical model defined the (n, p) nuclear reaction cross-section as

$$\sigma_{(n,p)} = \sigma_C(n) * G(p) \tag{1}$$

where $\sigma_C(n)$ is the cross-sections for the compound nucleus form during the collisions of the nucleus with incident neutron [7]. $\sigma_C(n)$ is generally known as the non-elastic cross sections. $G(p)$ is the probability of the compound nucleus to decays, where a proton is ejected from the nucleus. $G(p)$ can be defined as

$$G(p) = \frac{\Gamma_{\mathrm{p}}}{\Gamma_{\mathrm{n,p}}} \tag{2}$$

where Γ_{p} is the partial decay width of emitted proton and $\Gamma_{\mathrm{n,p}}$ is the total decay width. Due to $\Gamma_{\mathrm{n,p}}$ to be the total decay width, it is given as the sum of partial decay width of emitted proton (Γ_p), neutron (Γ_n), deuteron (Γ_d), triton (Γ_t), hellion (Γ_{He}) and alpha particles (Γ_α). However, all of the modes of decays is assumed as negligible excepts for Γ_n as shown below.

$$\Gamma_{n,p} = \Gamma_p + \Gamma_n + \Gamma_d + \Gamma_t + \Gamma_{He} + \Gamma_\alpha$$

$$\Gamma_{n,p} \approx \Gamma_n$$

The Γ_p can be defined as follows

$$\Gamma_\mathrm{p} = h \int \frac{d^2W}{d\varepsilon_p * dt} * d\varepsilon_p$$

where according to Weisskopf-Ewing theory, the emission rate of the proton, $\frac{d^2W}{d\varepsilon_p * dt}$ can be defined as

$$\frac{d^2W}{d\varepsilon_p * dt} \approx \frac{2I_p + 1}{2\pi^2 h} k_p^2 \sigma_C \exp - \frac{\varepsilon_p + S_p}{T}$$

Here, k_p and I_p are the wavenumber and the spin of the emitted proton while T and S_p is the nuclear temperature and separation energy of proton []. At around 14 meV (fast neutrons energy range), the nuclear cross-section for charged can be approximated as

$$\sigma(n, p) = C * \sigma_C * \exp a\left(\frac{N - Z}{A}\right) \quad (3)$$

This is the empirical nuclear cross-section proposed by Levkovski [6], where the nuclear cross-section has an exponential dependence on asymmetry parameter, $\frac{N-Z}{A}$. From this empirical formulation, many other authors have subsequently modified and developed other version of empirical formulation that has exponential dependent on $\frac{N-Z}{A}$. Other forms of exponential dependency have also been developed.

In Table 1, we tabulated previously proposed empirical nuclear cross sections by other authors. In this table, we tabulated the expression proposed, the mass region, the incident neutron energy and the chi-square value (χ^2) of the fitting. Here we defined the chi-square value (χ^2) as

$$\chi^2 = \frac{1}{N - n_0} \sum_{i=1}^{N} \frac{|\sigma_{exp} - \sigma_{emp}|}{\sigma_{exp}} \quad (4)$$

where N is the total number of cross-sections data points, n_0 is the number of free parameters, σ_{exp} is the experimental nuclear cross-sections and σ_{emp} is the empirical nuclear cross-sections. In this work, we proposed an empirical formulation in the form of 3 free parameters.

$$\sigma(n, p) = C_1 * \left(A^{\frac{1}{3}} + 1\right) * \exp C_2\left(\frac{N - Z + A}{A}\right)^n \quad (5)$$

Table 1 Empirical formulation of (n, p) nuclear cross-section at around 14 meV incident energy

Authors	Energy (meV)	Mass region	Formula (mb)	χ^2
Ait Tahar [8]	14	$40 \leq A \leq 239$	$\sigma = 107.98\left(A^{\frac{1}{3}}+1\right)^2 \exp(-36.749\frac{N-Z+1}{A})$	7.08
Kasugai et al. [9]	14	$19 \leq A \leq 188$	$\sigma = 1830(N-Z+1)\exp(-50.7\frac{N-Z+1}{A})$	8.23
Doczi et al. [10]	14.7	$31 \leq A \leq 181$	$\sigma = 23.659\left(A^{\frac{1}{3}}+1\right)^2 \exp\left(-23.041\left(\frac{N-Z}{A}+\left(\frac{N-Z}{A}\right)^2\right)\right)$	4.24
Luo et al. [11]	14.5	$46 \leq A \leq 196$	$\sigma = 62.98\left(A^{\frac{1}{3}}+1\right)^2 \exp(-34.45\frac{N-Z}{A})$	2.64
Yigit [7]	14–15	$9 \leq A \leq 239$	$\sigma = 48.367\left(A^{\frac{1}{3}}+1\right)^2 \exp(-31.859\frac{N-Z}{A})$	0.29

where C_1, C_2 and n is obtained from fittings. The fitting is done by using a dataset of nuclear isotopes obtained from the EXFOR database. The dataset contains the list of 61 nuclear isotopes with atomic mass range $9 < A < 111$ at neutron incident energy around 14–15 meV. Then, we compare the empirical formulation we obtained with previous formulations in terms of chi-square value (χ^2).

3 Results and Discussions

From Fig. 1, we plotted the experimental nuclear cross-sections (orange dot) and the fitted empirical formulation (blue dot) that we obtained as a function of $\left(\frac{N-Z+A}{A}\right)$. From the patterns of the plot in Fig. 1, there is an exponential dependence of $\sigma(n, p)$ with $\left(\frac{N-Z+A}{A}\right)$. From non-linear fitting, we obtained that the value of C_1, C_2 and n that minimized the chi-square value (χ^2) are 7, 4.3 and -19.6. We tabulated the fitting parameters in Table 2. The complete empirical formulation proposed in this work for (n, p) nuclear reaction with mass range $9 \leq A \leq 111$ is as follows:

$$\sigma(n, p) = 7 * \left(A^{\frac{1}{3}} + 1\right) * \exp 4.3\left(\frac{N - Z + A}{A}\right)^{-19.6} \tag{6}$$

The minimum chi-square value (χ^2) obtained for our empirical formulation is 1.23, which is better than the results from Ait Tahar [8], Kasugai et al. [9], Doczi et al. [10] and Luo et al. [11]. Our results is higher compared to Yigit [7]. However, we find that our result is acceptable and the empirical formulation fit strong enough to the experimental nuclear cross-section data which can be observed in Fig. 2.

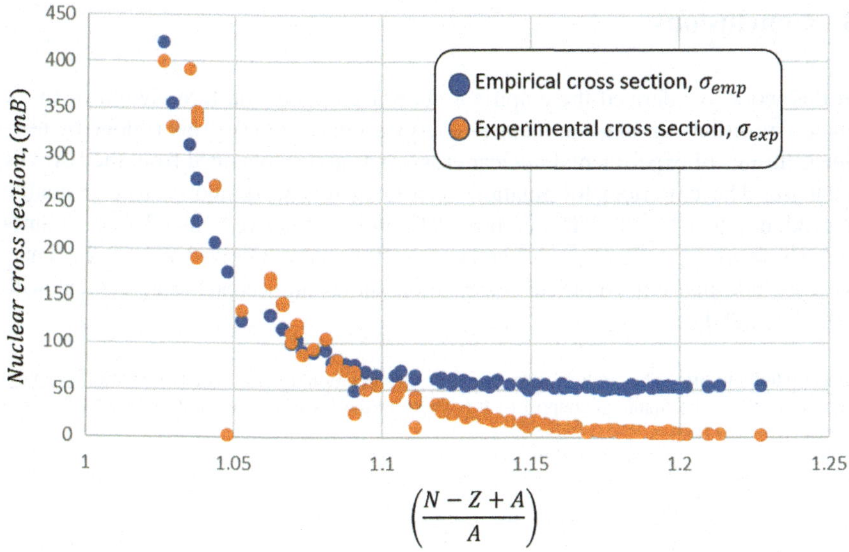

Fig. 1 Non-linear fitting of experimental cross-sections. The blue dots are the empirical nuclear cross-section while the orange dots are the experimental nuclear cross-section at incident neutron energy around 14 meV

Table 2 Fitting parameter obtained for the empirical formulation of (n, p) nuclear cross-section at around 14 meV incident energy

C_1	C_2	n	χ^2
7	4.3	− 19.6	1.230680986

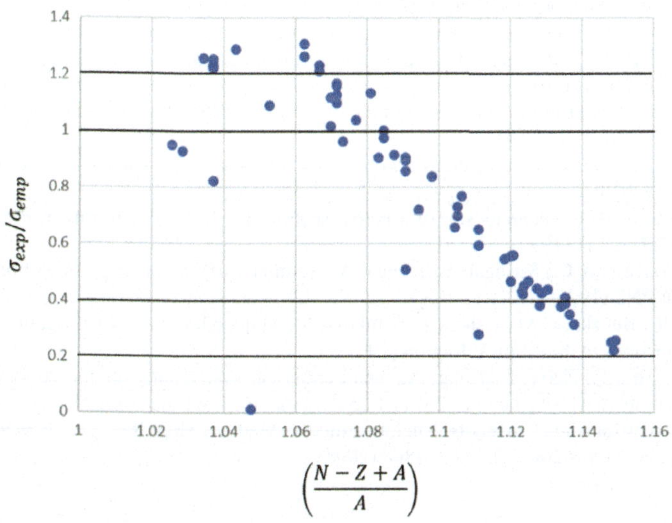

Fig. 2 Ratio of $\sigma_{exp}/\sigma_{emp}$ as a function of $\left(\frac{N-Z+A}{A}\right)$

4 Conclusions

In this work, we derived the empirical nuclear cross-section formulation of (n, p) nuclear reactions for neutron incident energy around 14 meV. This is done by non-linear fittings of experimental nuclear cross-section data obtained from the EXFOR database. The empirical formulation derived in this work shows an exponential dependency to $\left(\frac{N-Z+A}{A}\right)$. The empirical formula we derived have 3 free parameters which are $C_1 = 7$, $C_2 = 4.3$ and $n = -19.6$ that minimized $\chi^2 = 1.23$. Thus, we conclude that our empirical formulation shows an acceptable agreement with experimental data.

Acknowledgements The authors express their appreciation to Yayasan Universiti Teknologi PETRONAS for the financial support of this study in the form of a research grant (YUTP Cost Centre: 015LC0-063).

References

1. Kara, A., Yiğit, M., Korkut, T., Tel, E.: Cross section calculations of neutron induced reactions on 124,126,128,134,136Xe. J. Fusion Energy **34**(4), 882–886 (2015)
2. Qaim, S.M.: Nuclear data relevant to the production and application of diagnostic radionuclides. Radiochim. Acta **89**(4–5/2001), 223 (2001)
3. Zerkin, V.V., Pritychenko, B.: The experimental nuclear reaction data (EXFOR): Extended computer database and web retrieval system (2018)
4. Semkova, V., Otuka, N. Simakov, S.P., Zerkin, V.: Experimental nuclear reaction data collection EXFOR. In: 2011 2nd International Conference on Advancements in Nuclear Instrumentation, Measurement Methods and their Applications, pp. 2–4 (2011)
5. Otuka, N. et al.: Towards a more complete and accurate experimental nuclear reaction data library (EXFOR): International collaboration between nuclear reaction data centres (NRDC). Nucl. Data Sheets (2014)
6. Levkovskii, V.N.: Empirical behavior of the (n, p) cross-section for 14–15 meV neutrons. Sov. Phys. JETP **18**, 213 (1964)
7. Yiğit, M.: Analysis of (n, p) cross sections near 14 meV. Appl. Radiat. Isot. **135**(January), 115–122 (2018)
8. Ait-Tahar, S.: The systematics of (n, p) cross sections for 14 meV neutrons. J. Phys. G Nucl. Phys. **13**(7), L121 (1987)
9. Ikeda, Y., Konno, C., Kasugai, Y., Kumar, A.: Summary of activation cross section measurements at FNS (1996)
10. Doczi, R., Buczko, C.M., Csikai, J., Semkova, V., Majdeddin, A.D.: Investigations on (n, p) Cross Sections in the 14 meV Region (1997)
11. Luo, J., Tuo, F., Zhou, F., Kong, X.: Semi-empirical systematics for the cross-sections of the reactions (n,α), (n,p) and (n,2n) at 14.5 meV neutrons on the basis of experimental data measured by Lanzhou University. Nucl. Instruments Methods Phys. Res. Sect. B Beam Interact. with Mater. Atoms **266**(22), 4862–4868 (2008)

Outlier Detection in Subsurface Modeling of 2D Electrical Resistivity Imaging by Using Boxplot

Siti Nur Athirah Mazlan, Hanita Daud, Khairul Ariffin Mohd Noh, and Muhammad Naeim Mohd Aris

Abstract The 2D electrical resistivity method has huge applications in environmental, engineering, and shallow subsurface investigations. This electrical resistivity imaging (ERI) survey obtains the subsurface distribution by injecting current into the ground using two current electrodes (C1 and C2) while another two-potential electrode (P1 and P2) is injected to measure the resulting voltage difference. The resistivity value is calculated from the current and voltage differences obtained from this survey. For 2D electrical resistivity imaging, a large set of data is required to effectively map the complex resistivity distribution of the subsurface structure. However, due to the nature of the measurements, noise is detected present in this 2D ERI survey. This noise may affect the quality of the data obtained and will contribute to the quality of the model. A good quality model must have good quality acquired data and must have minimum impact from the presence of the noise. This work aims to conduct noise detection mechanism by using statistical tool, called boxplot. Noise detected by the boxplot was removed. 2D electrical resistivity imaging (ERI) survey was replicated by using Geotomo software to generate synthetic data that is used in developing the forward and inverse models. The developed models were analyzed by comparing their respective Root Mean Square (RMS) values before and after the removal of the noise. The subsurface model after noise removal has shown higher RMS value if compared to the model without noise detection as the outlier is

S. N. A. Mazlan (✉) · H. Daud · M. N. M. Aris
Department of Fundamental & Applied Sciences, Universiti Teknologi PETRONAS, Seri Iskandar, 32610 Perak, Malaysia
e-mail: siti_19001034@utp.edu.my

H. Daud
e-mail: hanita_daud@utp.edu.my

M. N. M. Aris
e-mail: naeim_16000991@utp.edu.my

K. A. M. Noh
Department of Geosciences, Universiti Teknologi PETRONAS, Seri Iskandar, 32610 Perak, Malaysia
e-mail: khairula.nmoh@utp.edu.my

S. A. Abdul Karim et al. (eds.), *Proceedings of the 6th International Conference on Fundamental and Applied Sciences*, Springer Proceedings in Complexity,
https://doi.org/10.1007/978-981-16-4513-6_64

replaced. This indicates that the proposed noise detection mechanism has managed to improve the current practice of manually removing the outliers.

Keywords 2D resistivity imaging · Outliers · Boxplot

1 Introduction

Electrical resistivity imaging (ERI) is widely used in geophysical studies such as pre-construction investigation, detection of voids or buried archeological structures and landslide investigation [1–3]. The ground resistivity is related to various geological parameters such as the mineral and fluid content, porosity and degree of water saturation in the rock. Electrical surveys aims to obtain the subsurface distribution by making measurements on the ground surface. The measurement is conducted by injecting current into the ground through two current electrodes (C1 and C2) and the resulting voltage difference is measured at two potential electrodes (P1 and P2). The resistivity value calculated from the current and voltage difference is not the true resistivity of the subsurface but it is the resistivity of a homogeneous ground which will give the same resistance value for the same electrode arrangement. The calculated resistivity value is referred to as apparent resistivity value [3].

From the current (I) and voltage (V) values, an apparent resistivity (p_a) value is calculated by

$$p_a = k\left(\frac{V}{I}\right) \quad (1)$$

where k is the geometric factor which depends on the arrangement of the four electrodes. Resistivity meters normally give a resistance value,

$$R = \frac{V}{I} \quad (2)$$

The apparent resistivity and the true resistivity has a complex relationship. Hence, an inversion of the measured apparent resistivity value must be carried out to determine the true subsurface resistivity [3].

In ERI, Root Mean Squared (RMS) value is used to find a good fit model for the apparent resistivity value in the inversion method of 2D electrical imaging. The inversion subroutine tries to reduce the square of difference between the measured and calculated apparent resistivity value. The RMS error is defined as

$$RMS = \sqrt{\frac{\sum_{i=1}^{N}\left(\frac{d_i^{Pred} - d_i^{Meas}}{d_i^{Meas}}\right)}{N}} \times 100 \quad (3)$$

where N is the total number of measurements, d^{Pred} is the predicted data and d^{Meas} is the measured data [4]. However, the presence of the noise outliers will drive the inversion and the RMS value [5]. In order to attain a preferable inversion model, the data must be in good quality and not affected by any noise. With that, the outliers are removed from the analysis [1]. Conflictingly, removing the bad data of 2D subsurface resistivity imaging based on user's perspective is a poor mechanism in controlling the noise. Outliers are not necessarily "errors": they can also correspond to unusual circumstances or be members of a different population [6].

A good statistical analysis usually begins at first with the exploration of outliers [7]. Outliers are also known as anomalies in machine learning [6]. Outliers are often present in real data but may go unnoticed because nowadays much data are produced by computer without careful inspection or screening [8]. Different behavior of some cases from the majority of the data often happens in real world dataset which might spoil the analysis but might also contain valuable information [6]. Outliers may be mistakes in data entry or otherwise, accurate but unexpected observations which could shed new light on the phenomenon under study [8].

An outlier is a value that significantly differs from rest of the data. Detection of outliers is both science and art. Science because there are set principles those have to be followed in order to decide about outliers and art because without the sound understanding of the background knowledge of data collection; it is difficult to confidently declare a value as outlier. Identification of outliers plays an important role for further analysis and estimation of the parameters. The presence of outlier(s) is an indication towards re-examination of the collected data. In order to proceed further for statistical analysis of data and modeling; it is recommended to thoughtfully decided about outliers [7].

If quality of data is not improved before the data analysis, then the result will not come as expected [9]. Outlier can cause us to misinterpret patterns in plots; it can affect visual resolution of remaining data in plot (forces observations into clusters) and may indicate that model fails to capture important characteristics of the data [8]. In other words, data cleaning is introduced as a process of removing typographical error or correcting the wrong values against the known values [9]. Data cleaning is said to be essential in geotechnical engineering datasets [10] and auctions field uses the double smoothing method [8]. Importance of proper data cleaning in geophysical study is gaining researcher's attention as Principal Component Analysis (PCA) and Bayesian Learning is introduced [10, 11].

In data preparation, outlier detection and processing is one of the important preprocessing steps [12]. It is crucial to identify and remove outliers since their presence in a data set produces deceptive results in the modeling [13]. However, outliers can provide both useless and interesting information. An outlying data must be carefully studied in order to decide whether to keep or delete it before making any decision. Those outlying values carrying important information should be kept while those that impedes data analytics are better be removed [14]. In 2D subsurface electrical imaging, every data points does carry important information of their respective geological parameter of the homogenous ground. Hence, the current practice of removing those outlying values manually by the users is a poor mechanism. Therefore, this study aims to address this shortcoming by introducing the statistical tools known as boxplot in order to detect the outliers. Next, the outliers will be replaced by the mean of the data set until no outliers are present. Finally, the model's RMS values are compared.

2 Materials and Methods

2.1 Data Acquisition

The data used in this study is the landfill survey in Germany by Buro Fur Geophysik, a synthetic data obtained from the Geotomo software with $3m$ electrode spacing by Wenner array. The data set contains 392 data points of 16 depths. The data set with no added outliers (0%), 20% and 40% added outliers is used. Geotomo software is used to generate the model while MATLAB is used to construct the box plot.

2.2 Boxplot in Outlier Identification

The most common way in outlier detection is box plot also called box and whisker plot method [13]. Box plot is the form of summary of a given dataset which includes, the median, the inter quartile range and the computation and the meaning each of the values are described below.

The median
Median is the middle data observation in a ranked of any dataset and as a measure for central tendency of the data and the same as the 50th percentile of a data.

The inter quartile range
The box represents the middle 50% of the ranked data and is drawn from the lower quartile value to the upper quartile value which is the 25th until 75th percentile.

The outer range
The whiskers represent an outer range and are drawn as a vertical lines extending outward from the end of the box. This whisker represents the maximum and minimum of said data.

The box plot consist of lower boundary and upper boundary defined as:

$$[L\ U] = [Q_1 - 1.5 \times IQR\ \ Q_3 + 1.5 \times IQR] \quad (5)$$

where L is the lower critical value and U denotes the upper critical value, Q_i is the i-th quartile and IQR is the difference of third and first quartile. Data points which lies outside the L and U are considered as outliers which will be replaced by the mean of the data set.

3 Result and Discussions

After box plot construction, it is observed that outliers are present in both data sets. Figure 1 shows the red marks are the outliers present in Depth 6 and Depth 11. Figures 2 and 3 illustrates the outliers in the 20 and 40% noise added data set. Only one outlier present in Fig. 2 while there are two outliers in Fig. 3. These outliers are removed and replaced by the mean of the respective data set.

Next, the subsurface model is generated. Figure 4a shows the model sensitivity section for 0% added outliers where the RMS value is 0.22%. After box plot is applied to the data set the RMS values is given by 0.56% as shown by Fig. 4b. For 0% added outliers it can be seen that the number of iterations has been reduced after outliers are replaced. Figure 5a shows the sensitivity section of the 20% added noise with RMS value 5.6% while the RMS value remains the same after outlier is identified (refer Fig. 5b). Figure 6v shows RMS value of 35.1% before outliers are replaced by the mean and Fig. 6b show RMS value of 35.6% after outliers are identified.

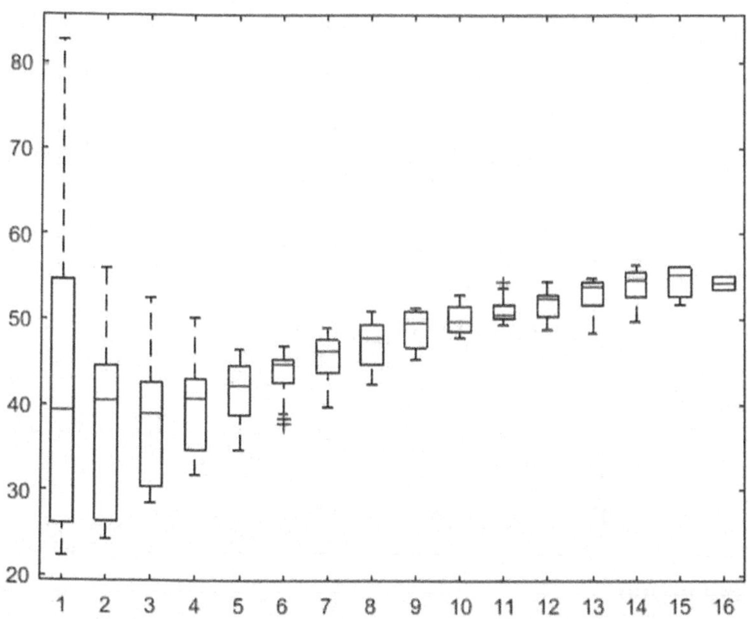

Fig. 1 Box plot displaying the outlier present in the 0% noise added data set

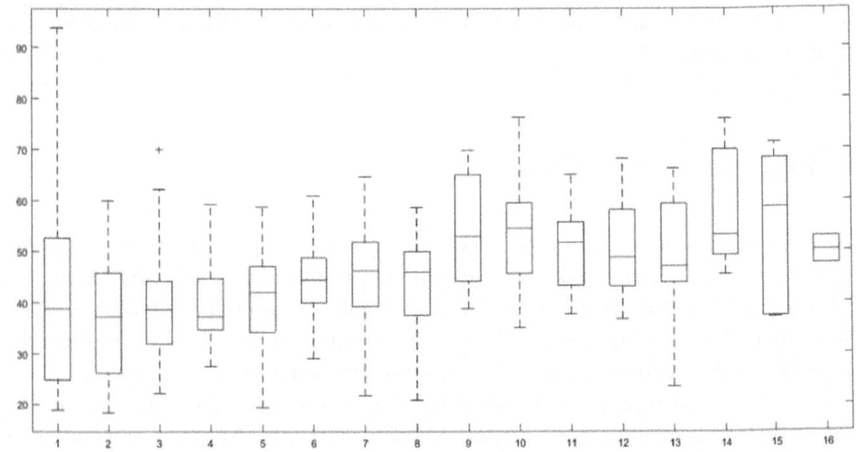

Fig. 2 Box plot displaying the outlier present in the 20% noise added data set

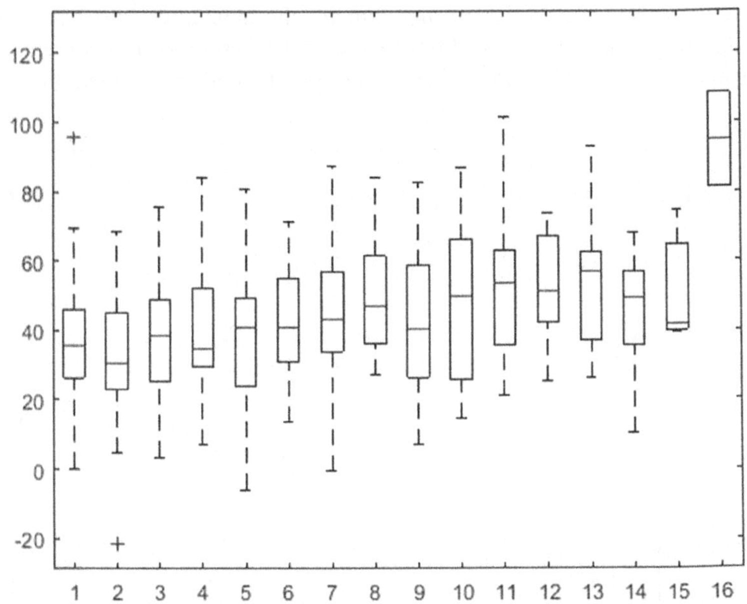

Fig. 3 Box plot displaying the outlier present in the 40% noise added data set

4 Conclusion

It is observed that the RMS value after outliers are replaced is higher than those before box plot is applied. This happens as the outlier is the calculated apparent resistivity itself replaced by the mean of the data set which obviously has different

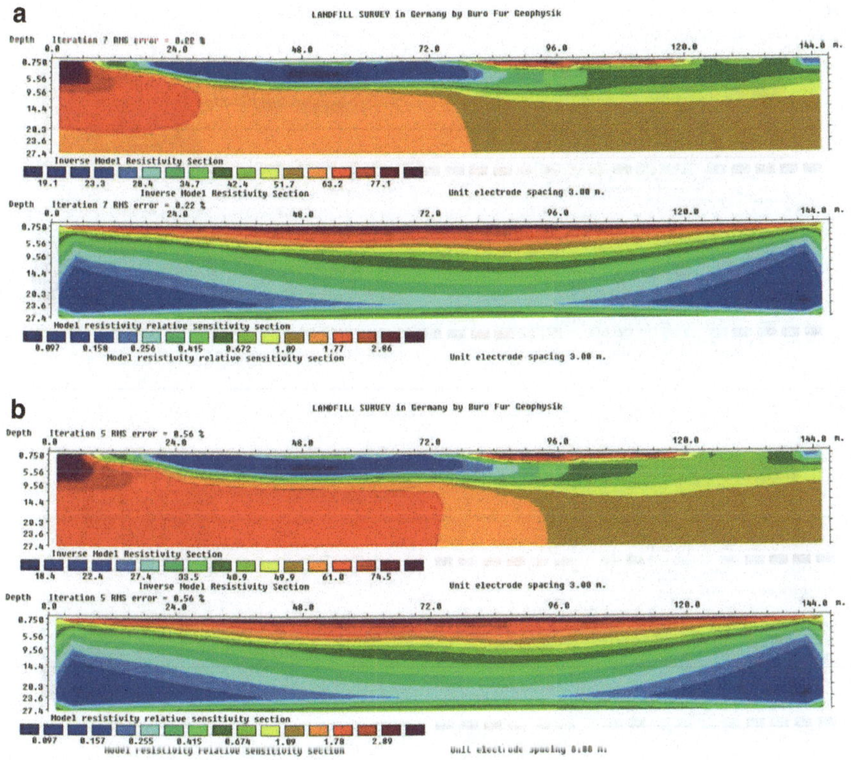

Fig. 4 **a** The model resistivity relative sensitivity section for 0% added noise. **b** The model resistivity relative sensitivity section for 0% added noise after outlier identification

value as compared to the outlier. By referring to Eq. (3), replacing the calculated apparent resistivity with another value far from its original value (the outlier) will definitely drive the RMS value. However, this paper manages to prove that using box plot in detecting outlier in 2D subsurface electrical resistivity modeling is applicable and is a proper mechanism.

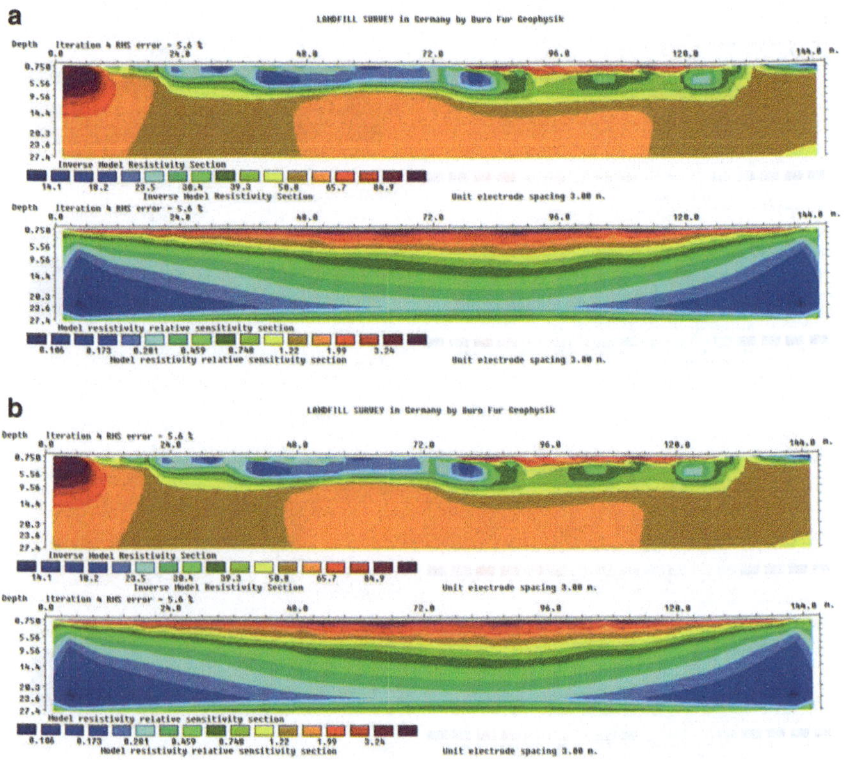

Fig. 5 **a** The model resistivity relative sensitivity section for 20% added noise. **b** The model resistivity relative sensitivity section for 20% added noise after outlier identification

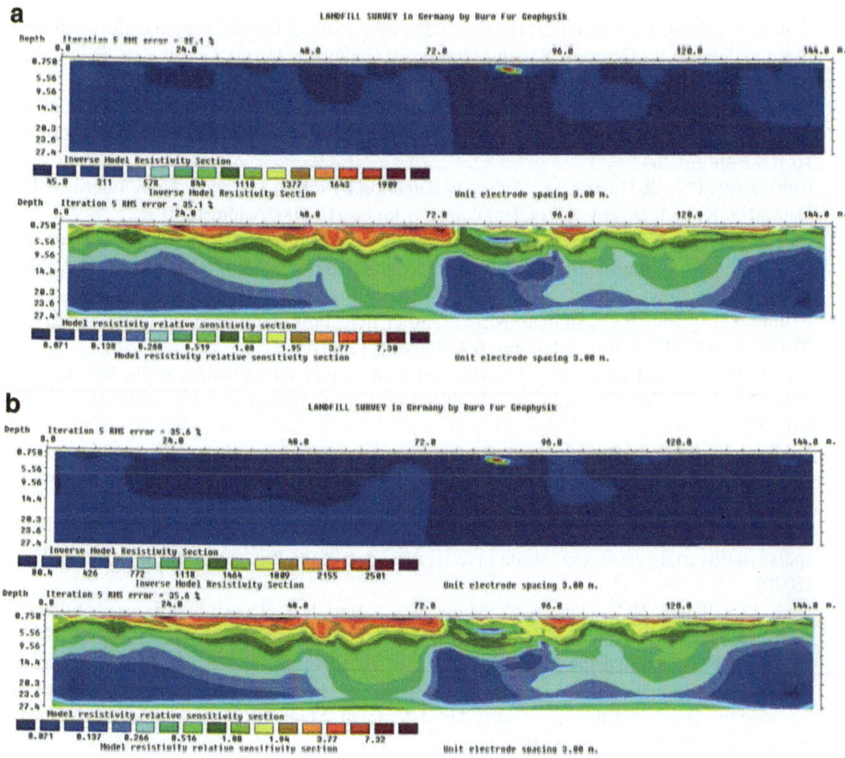

Fig. 6 **a** The model resistivity relative sensitivity section for 40% added noise. **b** The model resistivity relative sensitivity section for 40% added noise after outlier identification

Acknowledgements The authors would like to express their gratitude to those who have contributed to this paper. This study is funded by Yayasan Universiti Teknologi PETRONAS Fundamental Research Grant (YUTP-FRG) which cost center is 015LC0-114.

References

1. Osinowo, O.O., Falufosi, M.O.: 3D electrical resistivity imaging (ERI) for subsurface evaluation in pre-engineering construction site investigation. NRIAG J. Astron. Geophys. **7**(2), 309–317 (2018). https://doi.org/10.1016/j.nrjag.2018.07.001
2. Ungureanu, C., Priceputu, A., Bugea, A.L., Chirică, A.: Use of electric resistivity tomography (ERT) for detecting underground voids on highly anthropized urban construction sites. Procedia Engi. **209**, 202–209 (2017). https://doi.org/10.1016/j.proeng.2017.11.148
3. Perrone, A., Lapenna, V., Piscitelli, S.: Electrical resistivity tomography technique for landslide investigation: A review. Earth Sci. Rev. **135**, 65–82 (2014). https://doi.org/10.1016/j.earscirev.2014.04.002

4. Wahab, S., Saibi, H., Mizunaga, H.: Groundwater aquifer detection using the electrical resistivity method at Ito Campus, Kyushu University (Fukuoka, Japan). Geosc. Lett. **8**(1) (2021). https://doi.org/10.1186/s40562-021-00188-6
5. Schnaidt, S., Heinson, G.: Bootstrap resampling as a tool for uncertainty analysis in 2-D magnetotelluric inversion modelling. Geophys. J. Int. **203**(1), 92–106 (2015). https://doi.org/10.1093/gji/ggv264
6. Rousseeuw, P. J., & Hubert, M.: Anomaly detection by robust statistics. Wiley Interdisc. Rev. Data Min. Knowl. Discov. **8**(2) (2017). https://doi.org/10.1002/widm.1236
7. Iqbal, M., Habib, S., Khan, M., Kashif, M.: Comparison of different techniques for detection of outliers in case of multivariate data. Pak. J. Agri. Sci. **57**(3) (2020). https://doi.org/10.21162/PAKJAS/20.9369
8. Arimie, C.O., Biu, E.O., Ijomah, M.A.: Outlier detection and effects on modeling. OALib **07**(09), 1–30 (2020). https://doi.org/10.4236/oalib.1106619
9. Kumar Jha, A., Lal Saini, M.: A review on data quality enhancement using data cleaning algorithm. UGC Care Group I Listed J. **10**(6) (n.d.). https://doi.org/10.46528/JK.2020.V10I06N08.07
10. Siddiqui, F., Sargent, P., Montague, G.: The use of PCA and signal processing techniques for processing time-based construction settlement data of road embankments. Adv. Eng. Inform. **46**, 101181 (2020). https://doi.org/10.1016/j.aei.2020.101181
11. Zheng, S., Zhu, Y., Li, D., Cao, Z., Deng, Q., Phoon, K.: Probabilistic outlier detection for sparse multivariate geotechnical site investigation data using Bayesian learning. Geosci. Front. (2020)
12. Nam, D.V., Oanh, N.T., Hoai, N.X., Manh, N.V., Hien, N.T.: Detect and process outliers for temperature data at 3h monitoring stations in Vietnam. J. Min. Earth Sci. **61**(1), 132–146 (2020). https://doi.org/10.46326/jmes.2020.61(1).15
13. Hasan, M.F., Sobhan, M.A.: Describing Fuzzy membership function and detecting the outlier by using five number summary of data. Am. J. Comput. Math. **10**(03), 410–424 (2020). https://doi.org/10.4236/ajcm.2020.103022
14. Smiti, A.: A critical overview of outlier detection methods. Comput. Sci. Rev. **38**, 100306 (2020). https://doi.org/10.1016/j.cosrev.2020.100306

Numerical Investigations of Nano-fluid Flow in Square Porous Cavity: Buongiorno's Mathematical Model

Mudasar Zafar, Hamzah Sakidin, Iskandar Dzulkarnain, and Farkhanda Afzal

Abstract Buongiorno's mathematical Nanofluid model is used to investigate the steady-state natural convection process and heat generation in a two-dimensional (2D) porous square cavity dripped in Nanofluid. The problem focuses on the dimensionless momentum equations, energy equations, and nanoparticle concentration. The Rayleigh number, Lewis number, Darcy number, buoyancy-ratio parameter, Brownian motion parameter, thermophoresis parameter, Prandtl number, and aspect ratio are all investigated in this paper. The governing system of partial differential equations was discretized using the finite volume method (FVM), the continuity and momentum equations were discretized using the SIMPLE algorithm, and the generated algebraic equations are solved iteratively. The generated algorithm has been evaluated in terms of Nusselt number and Sherwood number by comparing it to earlier published work, and the findings are in good agreement.

Keywords Nano fluid · Porous medium · Square cavity · Finite volume method · Buongiorno's mathematical model

H. Sakidin · I. Dzulkarnain
Universiti Teknologi PETRONAS, 32610 Seri Iskandar, Perak, Malaysia
e-mail: hamzah.sakidin@utp.edu.my

I. Dzulkarnain
e-mail: iskandar_dzulkarnain@utp.edu.my

M. Zafar (✉)
Fundamental and Applied Science Department, Universiti Teknologi PETRONAS, Seri Iskandar, Perak, Malaysia
e-mail: mudasar_20000296@utp.edu.my

F. Afzal
MCS, National University of Science and Technology (NUST), Islamabad, Pakistan
e-mail: farkhanda@mcs.edu.pk

S. A. Abdul Karim et al. (eds.), *Proceedings of the 6th International Conference on Fundamental and Applied Sciences*, Springer Proceedings in Complexity,
https://doi.org/10.1007/978-981-16-4513-6_65

1 Introduction

Increased heating/cooling in an industrial process can save energy, shorten process time, increase thermal rating, and extend equipment life. Enhanced heat transfer has a limited impact on a small number of processes. The use of high-execution thermal systems for heat transfer enrichment is becoming more common. The transfer of heat energy is used in a wide variety of industrial processes. These days as science and technology progress very rapidly, there is a need to save energy and lower down the production time and save the production cost. This is the main target of every industry for example oil and gas, nuclear reactors, biomedical industries, automobiles, etc. When we are working with Nanofluids, they provide us a source of energy recovery. Nanofluid has special physical properties which attract researchers to use them for enhancement of heat transfer, because of their high thermal conductivity we can achieve the heat transfer enhancement very easily and it also saves time, energy, lower down the production cost and meet the requirement of the daily life. The researcher is employing nanofluids to increase heat transmission by using nanoscale particles to improve fluid thermal characteristics [1–4].

There are numerous applications for convection heat transfer processes with porous medium filled with nanofluids, including thermal and porous catalyst, oil recovery, food processing geothermal energy, nuclear reactor cooling, fuel cells, and so on [5–7]. The concept of nanoparticles is first developed in 1995 by Choi [8], which are extremely useful in conventional heat transfer. Nanoparticles are extra stable when dispersed in fluids and enhance the thermal properties of the fluids. Some other characteristics of nanoparticles that make them appropriate in heat transfer comprise the Brownian motion of the particles, particles/fluid nanolayers, and their decrease pump power when associated with pure liquids to attain strengthened heat transfer. It is very small in size and strong thermal conductivity attracts researchers to use and utilize nanofluids in new advanced technologies. Nanotechnology is applied in nearly every discipline, including heat and mass transfer, medical sciences, renewable energy sources, food processing, fertilizer industries, heat exchangers, oil and gas industry, nuclear reactors, and so on [9–11].

Since the beginning of the year 2000, the use of nanofluids has been a new and very interesting topic in the domain of heat transfer enhancement. The classic books by Nield and Bejan Ingham [12] and Pop [13] cover a broad variety of physical and mathematical aspects of convection in porous media. Over the past few years, several mathematical models to explain heat transfer in nanofluids are being used by researchers such as Tiwari and Dass, Brinkmann, and Buongiorno models [14, 15]. Numerous studies that combine the use of the above-mentioned models with different cavities in nanofluids to improve heat transfer have been published [16–18]. Marina et al. [19] studied the Brinkman-extended Darcy model for porous medium filled with iron oxide nanoparticles in the trapezoidal cavity. Mehmood et al. [20] consider the Koo-Kleinstreuer Lee model with alumina-water nanofluids in a square cavity.

In this paper, we investigated natural convection in the square cavity dripping with the porous medium in the occupancy of heat generation, as mentioned in [21] by considering the influence of pressure in fluid flow. To discretize the above-mentioned model the finite volume method (FVM) is used and the governing equations are solved using the numerical method. The following is a summary of the content of this article. The problem's geometry and mathematical structure are presented in Sect. 2. A numerical method is discussed in Sect. 3, and the result and discussion are described in Sect. 4. The closing remarks were found in Sect. 5.

2 Geometry and Mathematical Model

2.1 *Physics of the Problem*

We investigate Nanofluid flow in a two-dimensional (2D) steady-state, square cavity drip in a porous medium. In this problem, natural convection heat transfer is investigated, and the problem is depicted in Fig. 1. The rectangular coordinates are x and y for the cavity.

The following assumptions are made for this problem,

a. The temperature at solid walls are T_o so that the flux of nanoparticles is zero.
b. The temperature at horizontal walls is adiabatic.

We denote the fluid temperature, Brownian motion, thermophoresis parameter, and volume fraction of nanoparticles in this problem by T, D_B, D_T and ϕ respectively.

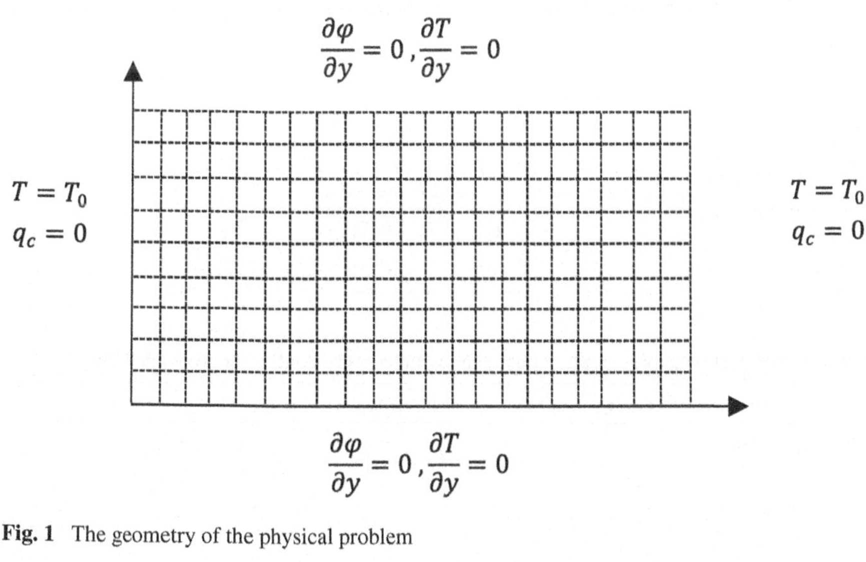

Fig. 1 The geometry of the physical problem

2.2 Governing Equations

We combined two different models to develop a mathematical model for our problem, namely Buongiorno's model and the Brinkmann-Boussinesq approximation.

$$\nabla.\mathbf{v} = 0 \tag{1}$$

$$0 = -\nabla p - \frac{\mu_{mnf}}{K} v + (\rho\beta)_{nf}(T - T_O)g \tag{2}$$

$$(\rho C)_{mnf}\frac{\partial T}{\partial t} + (\rho c)_{nf}(v.\nabla)T = k_{mnf}\nabla^2 T + \varepsilon(\rho c)_p\left(D_B\nabla\varphi + \frac{D_T}{T_o}\right).\nabla T + q_o''' \tag{3}$$

$$\frac{\partial\varphi}{\partial t} + \frac{1}{\varepsilon}(v.\nabla)\varphi = \nabla\left(D_B\nabla\varphi + \frac{D_T}{T_o}\nabla T\right) \tag{4}$$

$$T = T_O, D_B\frac{\partial\varphi}{\partial y} + \frac{D_T}{T_o}\frac{\partial T}{\partial y}, x = 0, L \tag{5}$$

$$\frac{\partial\varphi}{\partial y} = 0, \frac{\partial T}{\partial y} = 0, y = 0, L \tag{6}$$

In this case, Laplacian vector, Darcian velocity, permeability, and porosity is denoted by, ∇^2, v, T, ε, and K respectively. The physical properties of Nanofluids and nanofluids dripped in a porous medium are shown in Tables 1 and 2 respectively.

The subscripts nf, f, n, mnf, s, $m$denotes nanofluids, fluids, nanoparticles, nanoparticles dripped in a porous medium, a solid matrix with a porous medium, and

Table 1 Physical properties of nanofluids [22]

Viscosity	$\mu_{nf} = \frac{\mu_f}{(1-\varphi_0)^{2.5}}$
Heat capacitance	$(\rho C)_{nf} = (1-\varphi_0)(\rho C)_f + \varphi_0(\rho C)_p$
Thermal conductivity	$\frac{k_{nf}}{k_f} = \frac{(k_p+2k_f)-2\varphi_o(k_f-k_p)}{(k_p+2k_f)+\varphi_o(k_f-k_p}$
Buoyancy coefficient	$(\rho\beta)_{nf} = \varphi_0(\rho\beta)_p + (1-\varphi_0)(\rho\beta)_f$

Table 2 Physical properties of the copper nanoparticles, pure fluid, and porous medium [22]

Heat capacitance	$(\rho C)_{mnf} = \varepsilon(\rho C)_{nf} + 1 - \varepsilon)(\rho C)_s = (\rho C)_m[1 - \varepsilon\varphi_o\frac{(\rho C)_f-(\rho C)_p}{(\rho C)_m}$
Thermal conductivity	$k_{mnf} = \varepsilon k_{nf} + (1-\varepsilon)k_s = k_m\left\{1 - \frac{3\varepsilon\varphi_o k_f(k_f-k_p)}{k_m[(k_p+2k_f)+\varphi_o(k_f-k_p)]}\right\}, \alpha_{mnf} = \frac{k_{mnf}}{(\rho C)_{nf}}$

a solid matrix with pure fluid (water) dripped in a porous medium, respectively and φ_0 denotes the initial uniform concentration of nanoparticles in the square cavity.

We also include the dimensionless variables listed below.

$\dot{X} = \frac{x}{L}$, $\dot{Y} = \frac{y}{L}$, $\dot{U} = \frac{uL}{\alpha_{mnf}}$, $\dot{V} = \frac{vL}{\alpha_{mnf}}$, $\theta = \frac{T-T_o}{\frac{q_o''' L^2}{k_{mnf}}}$, $P = \frac{pL^2}{\rho_f \alpha_{mnf}^2}$, $\Phi = \frac{\varphi}{\varphi_o}$, $\tau = \frac{\alpha_{mnf} t}{\sigma L^2}$, $\sigma = \frac{(\rho C)_m}{(\rho C)_f}$, and $\alpha_m = \frac{k_m}{(\rho C)_f}$.

$$\left(\frac{\partial \dot{U}}{\partial \dot{X}} + \frac{\partial \dot{V}}{\partial \dot{Y}}\right) = 0 \tag{7}$$

$$\frac{\partial p}{\partial \dot{X}} = \frac{\Pr}{Da}\left[-M(\varphi_o)\dot{U} + RaH(\varphi_o)\theta)\right] \tag{8}$$

$$\frac{\partial p}{\partial \dot{Y}} = \frac{\Pr}{Da}\left[-M(\varphi_o)\dot{V} + RaH(\varphi_o)\theta)\right] \tag{9}$$

$$\left[Le\frac{\partial \theta}{\partial \tau} + \dot{U}\frac{\partial \theta}{\partial \dot{X}} + \dot{V}\frac{\partial \theta}{\partial \dot{Y}}\right] = \left(\frac{\partial^2 \theta}{\partial \dot{X}^2} + \frac{\partial^2 \theta}{\partial \dot{Y}^2}\right) + \frac{\mathrm{I}(\varphi_o)}{Le}\left(\frac{\partial \Phi}{\partial \dot{X}}\frac{\partial \theta}{\partial \dot{X}} + \frac{\partial \Phi}{\partial \dot{Y}}\frac{\partial \theta}{\partial \dot{Y}}\right) + \frac{N_{BT\mathrm{I}(\varphi_o).\mathrm{J}(\varphi_O)}}{Le}\left[\left(\frac{\partial \theta}{\partial \dot{X}}\right)^2 + \left(\frac{\partial \theta}{\partial \dot{Y}}\right)^2\right] + 1 \tag{10}$$

$$\left[\frac{1}{\sigma}\frac{\partial \Phi}{\partial \tau} + \frac{1}{\varepsilon}\left(\dot{U}\frac{\partial \Phi}{\partial \dot{X}} + \dot{V}\frac{\partial \Phi}{\partial \dot{Y}}\right)\right] = \frac{\mathrm{F}(\varphi_O)}{Le}\left(\frac{\partial^2 \Phi}{\partial \dot{X}^2} + \frac{\partial^2 \Phi}{\partial \dot{Y}^2}\right) + \frac{N_{BT}\,\mathrm{F}(\varphi_O)}{\mathrm{J}(\varphi_O)}\left(\frac{\partial^2 \theta}{\partial \dot{X}^2} + \frac{\partial^2 \theta}{\partial \dot{Y}^2}\right) \tag{11}$$

$$\begin{aligned} &\dot{X} = 0, 1, \dot{U} = 0, \dot{V} = 0, \frac{\partial \varphi}{\partial \dot{X}} + N_{BT}J(\varphi_0)\frac{\partial \theta}{\partial \dot{X}} = 0 \\ &\dot{Y} = 0, 1, \dot{U} = 0, \dot{V} = 0, \frac{\partial \theta}{\partial \dot{Y}} = 0, \frac{\partial \varphi}{\partial \dot{Y}} = 0 \end{aligned} \tag{12}$$

where $\dot{X}$, $\dot{Y}$,$\dot{U}$ and $\dot{V}$ defines the dimensionless rectangular coordinates and velocity components respectively.

Rayleigh numbers, Lewis numbers, the dimensionless ratio of thermophoresis to Brownian coefficients, the Darcy number, and the Prandtl number are also represented by Ra, Le, N_{BT}, Da and Pr, $H(\varphi_0)$, $M(\varphi_0)$, $L(\varphi_o)$, $I\ (\varphi_o)$ and $J(\varphi_O)$ are defined below (Table 3),

$$Ra = \frac{L^3(\rho\beta)_f g q_o'''.K}{\mu_f \alpha_m.k_m}, Le = \frac{\alpha_m}{D_B}, N_{BT} = \frac{L^2 q_o''' D_T}{T_0 k_m D_B}, \Pr = \frac{\mu_{nf}}{\rho_f \alpha_{mnf}}, Da = \frac{K}{L^2} \tag{13}$$

$$H(\varphi_0) = \frac{\left[(1-\varphi_0) + \frac{\varphi_0(\rho\beta)_p}{(\rho\beta)_f}\right].\left[(1-\varphi_0) + \frac{\varphi_0(\rho C)_p}{(\rho C)_f}\right](1-\varphi_0)^{2.5}}{\left\{1 - \frac{3\varepsilon\varphi_o k_f(k_f-k_p)}{k_m\left[(k_p+2k_f)+\varphi_o(k_f-k_p)\right]}^2\right\}}, \mathrm{M}(\varphi_0) = \frac{1}{(1-\varphi_0)^{2.5}} \tag{14}$$

Table 3 Thermophysical properties of the copper nanoparticles, pure fluid, and porous medium [23]

Physical properties	Head pure fluid (water)	Copper nanoparticles	Aluminum foam
Specific heat capacitance	4179	385	897
Density	997.1	8933	2700
Thermal conductivity	0.613	1163.1	205
$\beta \times 10^{-5}$	21	1.67	2.22

$$L(\varphi_o) = \frac{\left[1 - \frac{\varepsilon\varphi_0(\rho C)_p - (\rho C)_p}{(\rho c)_m}\right]}{\left[(1-\varphi_0) + \frac{\varphi_0(\rho C)_p}{(\rho C)_f}\right]}, \mathrm{I}(\varphi_0) = \frac{\varepsilon(\rho c)_p}{\left\{1 - \frac{3\varepsilon\varphi_o k_f(k_f - k_p)}{k_m[(k_p + 2k_f) + \varphi_o(k_f - k_p)]}\right\}(\rho C)_f} \tag{15}$$

$$J(\varphi_O) = \frac{1}{\varphi_O\left\{1 - \frac{3\varepsilon\varphi_o k_f(k_f - k_p)}{k_m[(k_p + 2k_f) + \varphi_o(k_f - k_p)]}\right\}} \tag{16}$$

$$Nu_L = -\frac{k_{mnf}}{k_m}\left(\frac{\partial\theta}{\partial x}\right)_{X=0}, \quad Sh_L = -\frac{k_{mnf}}{k_m}\left(\frac{\partial\varphi}{\partial x}\right)_{X=0}$$

$$Nu_{avg} = \int_0^1 Nudx, \quad Sh_{avg} = \int_0^1 Shdx \tag{17}$$

3 Numerical Method

The finite volume method is used to solve the system of governing equations, which includes partial differential Eqs. (8)–(11) and the Neumann boundary condition (12). The velocity–pressure coupling equation is often discretized using the concept of a staggered grid. A block diagram of staggered mesh, as well as a flow chart to illustrate the definition, can be found in Figs. 2 and 3.

We use the SIMPLE algorithm to discretize the continuity and momentum equations, and its working principle is briefly described below,

Step #1 Determine (Guess) the amount of pressure filed.
Step #02 Using momentum equations, calculate the values u^*, v^*, and w^*.
Step #3 Determine the pressure correction equation (p').
Step #4 Using the equation $p = p^* + p'$, calculate the pressure values.
Step #5 Using the equations below, calculate the values of u, v, and w.

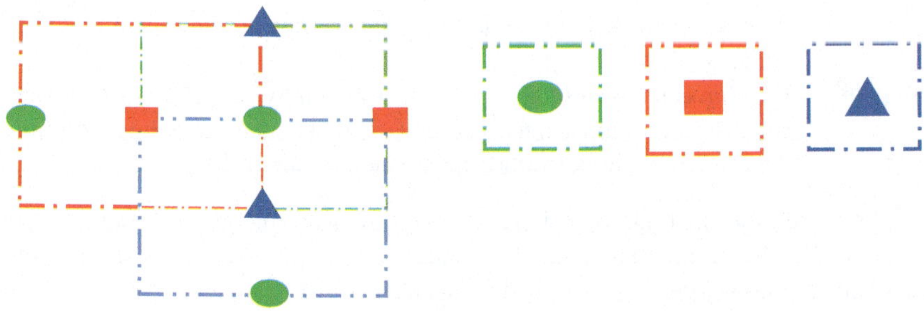

Fig. 2 The block diagram of staggered mesh with staggered mesh for pressure and scalars as green color, the staggered grid for x momentum as red color, and staggered mesh for y momentum as blue color

Fig. 3 Flow chart for staggered mesh

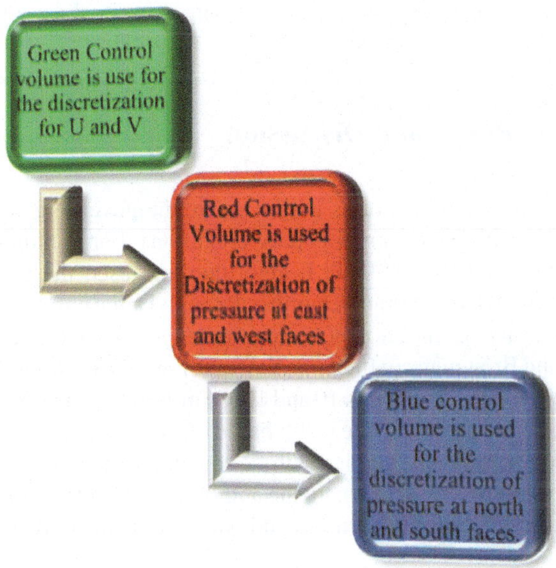

$$u_e = u_e^* + d_e\left(p_p' - p_E'\right)$$

$$v_e = v_e^* + d_e\left(p_p' - p_N'\right)$$

$$w_e = w_e^* + d_e\left(p_p' - p_S'\right)$$

Step #6 Solve the discretization equation for the remaining (such as temperature, concentrations, and turbulence quantities). If they influence the flow field via fluid properties, source terms, etc. (If a particular does not affect the flow field, it is better to calculate it after the flow field has been converged).

Step #07 If a converged solution is obtained, stop; otherwise, return to step 2 and solve the algorithm again because we did not obtain a converged solution. Also, as a new initial value, update the pressure value.

The system of algebraic equations is solved by using the gauss seidel iterative method. We also utilize the Neumann boundary condition for the solution to our problem. The convergence is obtained by using Eq. (18),

$$\left\| \frac{new\ approximation - old\ approximation}{new\ approximation} \right\| \leq \delta \tag{18}$$

Here we described δ as the error and dependents on the values of the examined parameters. Where the value of δ lies in the range of $10^{-5} - 10^{-10}$.

4 Result and Discussion

FVM discretization is used to solve the governing equation derived from Eqs. (7–11) with Neumann boundary conditions. For continuity and momentum equations we utilized the SIMPLE algorithm and used the staggered mesh. The governing parameters are Prandtl number $Pr = 1$, Darcy number $Da = 10^{-2}$, porosity $\varepsilon = 0.9\varepsilon$, initial concentration of Nanoparticles $\varphi_o = 0.01\varphi_o$, the ratio between thermophoresis and Brownian coefficients $N_{BT} = 0.1$ and dimensionless time $\tau = 0.16$. The value of Rayleigh's number is 10 and 100, similarly, Lewis's number is 1 and 10. Table 4 gives the comparison of the result based on Nusselt number Nu and Sherwood number Sh. We perform the studies with finite volume method and use cooper nanoparticles and compare with [21] where the author used the finite difference method and use carbon nanotube as nanoparticles and our result is in good agreement with the literature.

Table 4 Values of Nusselt number and Sherwood number compared to the previous result

Physical properties		Present		Previous [21]	
Ra	Le	Nu	Sh	Nu	Sh
10	1	0.548934	0.890285	0.550278	0.892496
	10	0.548940	0.890288	0.550287	0.892510
100	1	0.549030	0.890376	0.550320	0.892563
	10	0.549032	0.890379	0.550322	0.892567

5 Conclusion

In this paper, we investigated the natural convection phenomena for Nanofluid flow in a square cavity dripped with a porous medium numerically. We consider the Aluminum foam as a porous medium and water-based Copper as nanoparticles. We solve our problem iteratively and discretize the set of governing equations using the Finite volume method. For velocity pressure coupling equations SIMPLE algorithm is used. The result is in good agreement with the previous work for the enhancement of heat transfer.

Acknowledgements The authors gratefully acknowledge financial support from Yayasan Universiti Teknologi PETRONAS (YUTP) Malaysia, which was given as a Graduate Research Assistantship (GRA) Scheme under the Grant cost center 015LC0-272.

References

1. Das, S.K., Choi, S.U., Patel, H.E.: Heat transfer in nanofluids—A review. Heat Transfer Eng. **27**(10), 3–19 (2006)
2. Wang, X.Q., Mujumdar, A.S.: Heat transfer characteristics of nanofluids: a review. Int. J. Therm. Sci. **46**(1), 1–19 (2007)
3. Kakaç, S., Pramuanjaroenkij, A.: Review of convective heat transfer enhancement with nanofluids. Int. J. Heat Mass Transf. **52**(13–14), 3187–3196 (2009)
4. Awais, M., Ullah, N., Ahmad, J., Sikandar, F., Ehsan, M.M., Salehin, S., Bhuiyan, A.A.: Heat transfer and pressure drop performance of Nanofluid: A state-of-the-art review. Int. J. Thermofluids 100065 (2021)
5. Khanafer, K., Vafai, K.: Applications of nanofluids in porous medium. J. Therm. Anal. Calorim. **135**(2), 1479–1492 (2019)
6. Ebrahimnia-Bajestan, E., Moghadam, M.C., Niazmand, H., Daungthongsuk, W., Wongwises, S.: Experimental and numerical investigation of nanofluids heat transfer characteristics for application in solar heat exchangers. Int. J. Heat Mass Transf. **92**, 1041–1052 (2016)
7. Alomair, O. A., Matar, K. M., Alsaeed, Y. H.: Nanofluids application for heavy oil recovery. In: SPE Asia Pacific oil & gas conference and exhibition. Society of Petroleum Engineers (Oct, 2014).
8. Choi, S.U., Eastman, J.A.: Enhancing Thermal Conductivity of Fluids with Nanoparticles (No. ANL/MSD/CP-84938; CONF-951135-29). Argonne National Lab., IL (United States) (1995)
9. Reddy, N., Murugesan, K.: Numerical investigations on the advantage of nanofluids under DDMC in a lid-driven cavity. Heat Transf.—Asian Res. **46**(7), 1065–1086 (2017)
10. Snoussi, L., Ouerfelli, N., Chesneau, X., Chamkha, A.J., Belgacem, F.B.M., Guizani, A.: Natural convection heat transfer in a nanofluid filled U-shaped enclosures: Numerical investigations. Heat Transf. Eng. **39**(16), 1450–1460 (2018)
11. Liu, S., Zhu, Q.Y.: Experimental and numerical investigations on combined Buoyancy–Marangoni convection heat and mass transfer of power-law nanofluids in a porous composite with complex surface. Int. J. Heat Mass Transf. **138**, 825–843 (2019)
12. Nield, D.A., Bejan, A.: Convection in Porous Media, vol. 3. Springer, New York (2006)
13. Pop, I., & Ingham, D. B. Transport phenomena in porous media II. Elsevier (2002).
14. Tiwari, R.K., Das, M.K.: Heat transfer augmentation in a two-sided lid-driven differentially heated square cavity utilizing nanofluids. Int. J. Heat Mass Transf. **50**(9–10), 2002–2018 (2007)
15. Buongiorno, J.: Convective transport in nanofluids. J. Heat Transf. **128**(3), 240–250 (2006)

16. Selimefendigil, F., Chamkha, A.J.: MHD mixed convection of Ag–MgO/water nanofluid in a triangular shape partitioned lid-driven square cavity involving a porous compound. J. Therm. Anal. Calorim. **143**(2), 1467–1484 (2021)
17. Raizah, Z.A.S., Aly, A.M., Ahmed, S.E.: Natural convection flow of a power-law non-Newtonian nanofluid in inclined open shallow cavities filled with porous media. Int. J. Mech. Sci. **140**, 376–393 (2018)
18. Rashad, A.M., Rashidi, M.M., Lorenzini, G., Ahmed, S.E., Aly, A.M.: Magnetic field and internal heat generation effects on the free convection in a rectangular cavity filled with a porous medium saturated with Cu–water nanofluid. Int. J. Heat Mass Transf. **104**, 878–889 (2017)
19. Astanina, M.S., Sheremet, M.A., Oztop, H.F., Abu-Hamdeh, N.: MHD natural convection and entropy generation of ferrofluid in an open trapezoidal cavity partially filled with a porous medium. Int. J. Mech. Sci. **136**, 493–502 (2018)
20. Mehmood, K., Hussain, S., Sagheer, M.: Numerical simulation of MHD mixed convection in alumina–water nanofluid filled square porous cavity using KKL model: Effects of non-linear thermal radiation and inclined magnetic field. J. Mol. Liq. **238**, 485–498 (2017)
21. Groşan, T., Revnic, C., Pop, I., Ingham, D.B.: Free convection heat transfer in a square cavity filled with a porous medium saturated by a nanofluid. Int. J. Heat Mass Transf. **87**, 36–41 (2015)
22. Sheremet, M.A., Grosan, T., Pop, I.: Free convection in a square cavity filled with a porous medium saturated by nanofluid using Tiwari and Das' nanofluid model. Transp. Porous Media **106**(3), 595–610 (2015)
23. Zhuang, Y.J., Zhu, Q.Y.: Numerical study on combined buoyancy–Marangoni convection heat and mass transfer of power-law nanofluids in a cubic cavity filled with a heterogeneous porous medium. Int. J Heat Fluid Flow **71**, 39–54 (2018)

Forecasting Electricity Consumption in Malaysia by Hybrid ARIMA-ANN

Nur Ezzati Mohd Izudin, Rajalingam Sokkalingam, Hanita Daud, Hermiza Mardesci, and Abdullah Husin

Abstract Forecasting electricity consumption is of national interest to any country. Electricity forecast is not only required for short-term and long-term power planning activities but also in the structure of the national economy. Electricity consumption time series data consists of linear and non-linear patterns. Thus, the patterns make the forecasting difficult to be done. Neither autoregressive integrated moving average (ARIMA) nor artificial neural networks (ANN) can be adequate in modeling and forecasting electricity consumption. The ARIMA cannot deal with non-linear relationships while a neural network alone is unable to handle both linear and non-linear pattern equally well. This research is an attempt to develop ARIMA-ANN hybrid model by considering the strength of ARIMA and ANN in linear and non-linear modeling. The Malaysian electricity consumption data is taken to validate the performance of the proposed hybrid model. The results will show that the proposed hybrid model will improve electricity consumption forecasting accuracy by compare with other models.

Keywords Forecasting · Time series · ARIMA · ANN · Hybrid method

N. E. M. Izudin (✉) · R. Sokkalingam · H. Daud
Fundamental and Applied Sciences Department, Universiti Teknologi PETRONAS, 32610 Seri Iskandar, Perak, Malaysia
e-mail: nur_17008265@utp.edu.my

R. Sokkalingam
e-mail: raja.sokkalingam@utp.edu.my

H. Daud
e-mail: hanitadaud@utp.edu.my

H. Mardesci
Fakultas Pertanian, Universitas Islam Indragiri, Riau, Indonesia
e-mail: hermizamardesci@unisi.ac.id

A. Husin
Program Studi Sistem Informasi, Universitas Islam Indragiri, Riau, Indonesia

S. A. Abdul Karim et al. (eds.), *Proceedings of the 6th International Conference on Fundamental and Applied Sciences*, Springer Proceedings in Complexity,
https://doi.org/10.1007/978-981-16-4513-6_66

1 Introduction

In Malaysia, the demand for electricity increases by 4.7 percent per year over the outlook period, to reach 274 TWh in 2030 [1]. The growth in electricity demand is heavily influenced by strong demand from the industrial and residential sectors, which is projected to increase annually at 5.4 and 4.9 percent respectively over the outlook period. The past 15 years (2000–2015) record show that the electricity consumption trend in Malaysia is continuously increasing. Therefore, electricity consumption forecast has fundamental importance in the energy planning of Malaysia. A good forecasting technique is critically important to estimate the level of electricity's demand accurately, thus proper planning could be made by the utility company and the government to meet the country's future. Several techniques are being used for electricity consumption planning to accurately forecast future electricity consumption in Malaysia.

This paper reviews previous research works related to ARIMA, ANN and Hybrid method only which were applied in various applications [2]. Using statistics rule-based approach to forecast peak load electricity demand and they modified the basic regression model using Box-Jenkins autoregressive error. It was produced an adequate model with 2.41% forecasting error. While Haiges et al. forecast using ARIMA method that provides a projection that relies on past historical data to reach a state of statistical equilibrium [3]. Besides, Mohamed Othman et al. using ANN-based forecast the electricity consumption in Malaysia [4]. The authors developed ANN model which involved the creation of numerous feed-forward backpropagation network in MATLAB and select the best ANN model via cross-validation method.

Yan and Zou using Hybrid ARIMA-ANN techniques to forecast water quality [5]. The ARIMA models were first used to do the water quality forecasting and then with the obtained errors ANNs were built taking into the nonlinear patterns. For modeling linear and nonlinear components of a time series. The hybrid ARIMA-ANN combines linear and non-linear models to improve the forecasting performance of the price of Robusta Coffee in India [6]. Mohan and Reddy apply hybrid ARIMA-ANN model to predict the resource usage in server virtualization [7]. The authors using the measurement base approach with time series for prediction and reviewed the effectiveness of the ARIMA model. They compared hybrid ARIMA-ANN to ARIMA and ANN and the result shows that the accuracy has improved 5 times. Mucaj and Sinaj presented three models, ARIMA, NAR, and ARIMA-ANN. The hybrid model proposed to improve the forecasting accuracy and the result shows that the hybrid model has the best results among the three models [8].

Many empirical studies [1, 2, 4, 9–11] has been conducted to forecast electricity consumption using multivariate techniques and time-series analysis such as autoregressive (AR), moving average (MA) algorithm, general exponential smoothing algorithm, ARMA algorithm, and AR integrated MA (ARIMA). The above methods have their own advantages and drawbacks in the variable selection, model selection, segmenting and lack of comparison with a combination of different models to improve the forecasting accuracy.

This paper presents the forecast electricity consumption by using hybrid ARIMA-ANN based on historical data. Therefore, the authors would like to propose a hybrid ARIMA-ANN model for forecasting electricity consumption in Malaysia.

2 Methodology

Figure 1 shows the flowchart model development of hybrid ARIMA-ANN which consists of five stages.

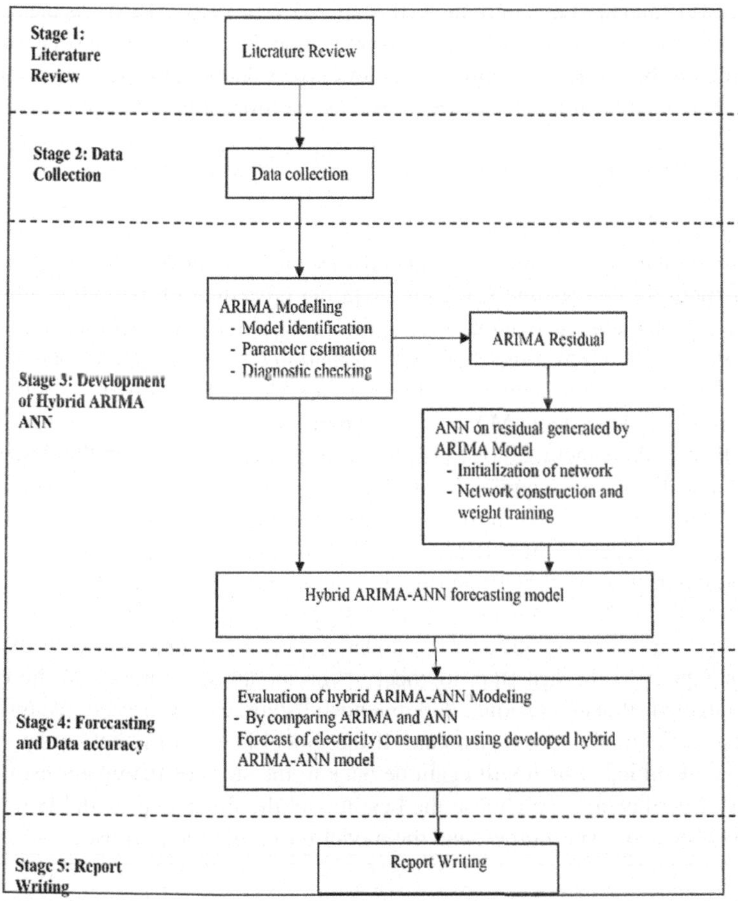

Fig. 1 Model development hybrid ARIMA-ANN

2.1 Data Collection

The dataset was obtained from the Malaysian Energy Commission on a yearly basis for 40 years, between the years 1978 and 2017. Interpolation process was used to obtain more data.

2.2 The ARIMA Model

Box-Jenkins method or ARIMA model was one of the most popular approaches of forecasting and has been introduced this approach in 1976. ARIMA is the combination of the autoregressive and moving average models, the future value of a variable is assumed to be a linear function of random errors and several past observations. The mathematical formula can be expressed as the form of Eq. (1):

$$y_t = \theta_0 + \varphi 1 y_{t-1} + \varphi 1 y_{t-2} + \ldots + \varphi_p y_{t-p} + \varepsilon_t - \theta_1 \varepsilon_{t-1} - \theta_{12} \varepsilon_{t-2} - \ldots - \theta_q \varepsilon_{t-q} \tag{1}$$

where p; number of lags of the considered variable, q; number of lags of the error term, y_t; actual value at time t, y_{t-i}; series in the preceding ith period, φ; ith autoregressive coefficient, ε_t; random error at time t, ε_{t-1}; preceding error term at the ith period, θ_i; ith moving average coefficient. Equation (1) is an important special case of the ARIMA family of models. (1) becomes an AR model of order p when $q = 0$, and the model reduces to an MA model of order q if $p = 0$.

The Box-Jenkins methodology [12] has a fundamental impact on the time series analysis and forecasting application to building ARIMA model. It includes three iterative steps model identification, parameter estimation, and diagnostic checking. To identify the order of the ARIMA model, Box and Jenkins proposed to use the autocorrelation function (ACF) and the partial autocorrelation function (PAFC) of the sample data. The parameter estimated such that an overall measure of errors is minimized by using a nonlinear optimization procedure. Diagnostic checking of model adequacy is the last step, to check if the model assumptions of the errors, ε_t, are satisfied. Plots of residual and diagnostic static can be used to examine the goodness of the model to the past data. If the model is not sufficient, a new model should be identified, which will again be back to the steps of parameter estimation and model verification. To choose the best model the three-step model building is typically repeated several times, and the model for prediction purpose.

2.3 The ANN Model

The model is characterized by a network of three layers of simple processing units connected by acyclic links. The relationship between the output (y_t) and the inputs $(y_{t-1}, \ldots, y_{t-p})$ has the following mathematical representation:

$$y_t = \omega_0 + \sum_{j=1}^{q} \omega_j g\left(\omega_{0j} + \sum_{j=1}^{p} \omega_{i,j} y_{t-i}\right) + e_t \tag{2}$$

where ω_j $(j = 1, 2, .., q)$ and $\omega_{i,j}$ $(i = 0, 1, 2, .., p; j = 1, 2, .., q)$ are the model parameters often called connection weights; p is the number of input nodes and q is the number of hidden nodes. The sigmoid function is often used as the hidden layer transfer function, that is,

$$sig(x) = \frac{1}{(1 + \exp(-x))} \tag{3}$$

Hence, the ANN model, in fact performs a nonlinear functional mapping from the past observations $(y_{t-1}, \ldots, y_{t-p})$ to the future value y_t,

$$y_t = f\left(y_{t-1}, \ldots, y_{t-p,}\omega\right) + e_t \tag{4}$$

where ω is a vector of all parameters and f is a function determined by a network structure and connection weights. Thus the neural network is equivalent to a nonlinear autoregressive model.

2.4 The Hybrid ARIMA-ANN Model

Both ARIMA and ANN models have achieved successes in their own linear o nonlinear domains. However, none of them is a universal model that is suitable for all circumstances. The approximation of ARIMA models to complex nonlinear problems may not be adequate. On the other hand, using ANNs to model linear problems have yield mixed results. Hence, it is not wise to apply ANNs blindly to any type of data. Since it is difficult to completely know the characteristic of the detain a real problem, hybrid model that has both linear and nonlinear modeling capability can be a good choice for forecasting electricity consumption. Different aspect of the underlying patterns may be captured to combining different models. Linear autocorrelation structure and a nonlinear component:

$$Y_t = L_t + N_t \tag{5}$$

where L_t denotes the linear component and N_t denotes nonlinear component. Both of these two parameters have to be estimated from the time series data.

$$e_t = Y_t - YF_t \tag{6}$$

where YF_t is the predicted value of the ARIMA model at time t. With n inputs nodes, the ANN model for the residuals will be:

$$e_t = f(e_{t-1}, e_{t-2}, \ldots, e_{t-n}) + u_t \tag{7}$$

where f denotes nonlinear function and u_t denotes random error. NF_t as the forecast from above equation, the combine prediction will be:

$$YF_t = LF_t + NF_t \tag{8}$$

Hybrid ANN and ARIMA system consist of two-step, first ARIMA model is used to analyze the linear part of the problem and the second is neural network model is developed to model the residuals from the ARIMA model.

3 Result and Discussion

3.1 Data Set

The data used in this study were load data of electricity consumption in Malaysia obtained from Malaysia Energy Information hub [13].

3.2 ARIMA Modelling

The historical data of the electricity consumption of Malaysia from 1978 to 2017 was increasing over time as shown in Fig. 2.

Through Fig. 3 of the original series of ACF and PACF, we note that it is non-stationary in the data of the original series as there are some values outside the confidence interval. And to make the series stationary we make differences. The best fit AR parameters and MA parameters should be estimated according to ACF and PACF respectively. Figure 5 show the ACF and PACF. It was obvious that the ACF died off smoothly at a geometric rate after one lag and PACF declined geometrically after one lagged. Therefore, the parameters of AR and MA can be chosen as 1 for the ARIMA model (Fig. 4).

However, in the practical fitting process, any other AR/MA parameters could be selected. For instance, the AR parameters defined as 1 and the MA parameters can

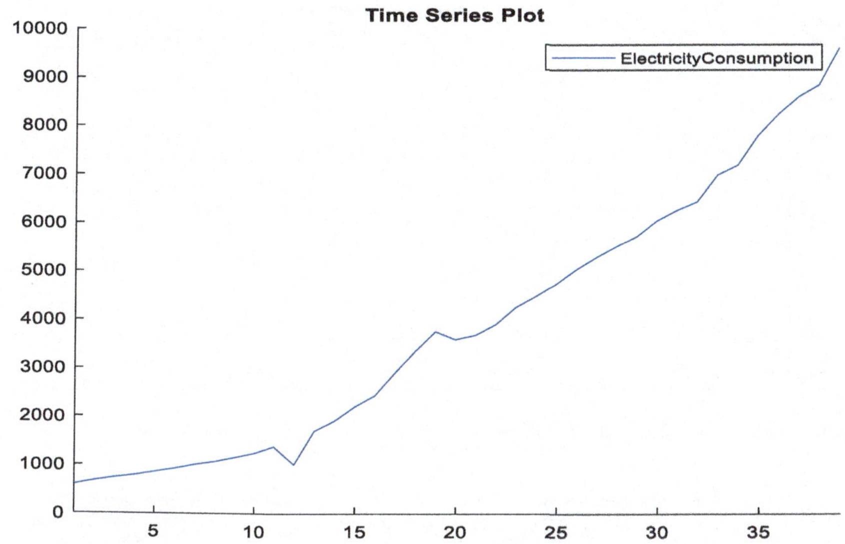

Fig. 2 Electricity consumption of Malaysia from 1978 to 2017 (Kilotonne of oil equivalent)

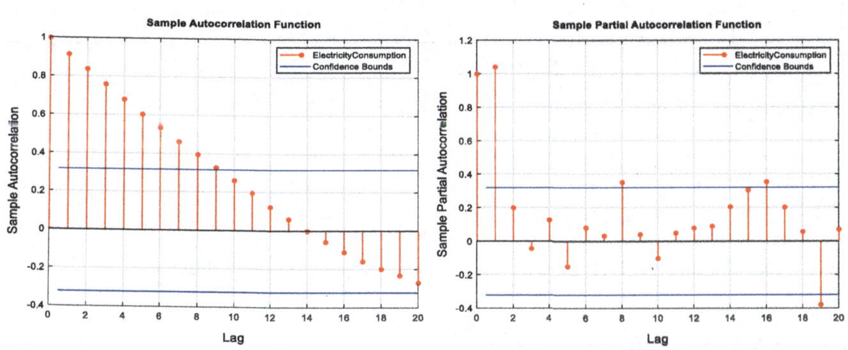

Fig. 3 The ACF and PACF of the series of electricity consumption

be defined as 1 or 3. After fitted, ARIMA(1,2,1) has been found to be the best model among all ARIMA models. Once the ultimately fittest model was identified, the equation form of the model could be obtained:

$$y_t = 2y_{t-1} - 1 - 2y_{t-2} + 10.9595 + e_t + \theta e_t \tag{9}$$

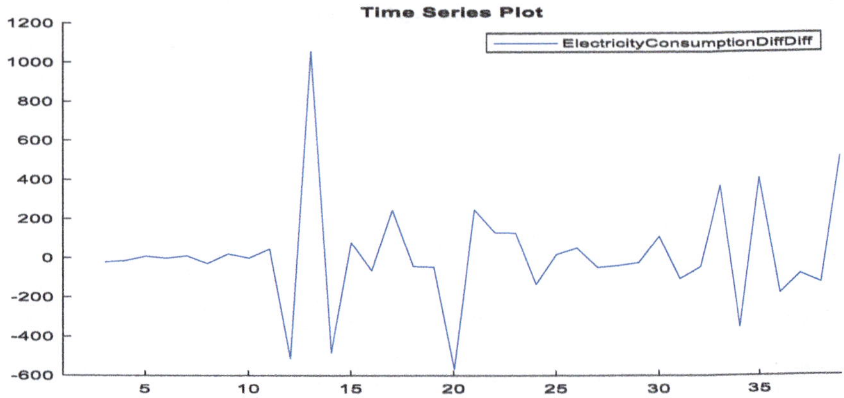

Fig. 4 Graphical representation of the series difference in electricity consumption

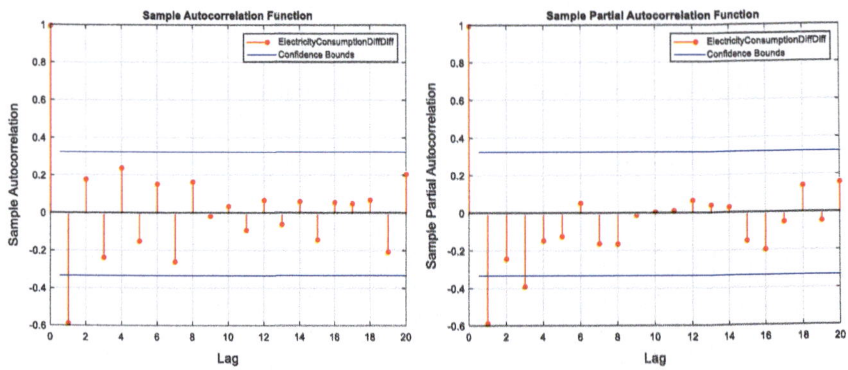

Fig. 5 The ACF and PACF of the series of electricity consumption after difference

3.3 *ANN Modelling*

A three-layer feedforward neural network model was developed for the forecasting of electricity consumption using an optimized Levenberg–Marquardt training algorithm. The data for the period between 1978 and 2017 were available for the modeling purposes. Electricity consumption time series data were divided into two independent data sets. The first data set of 1978–2013 was used for model training and the other datasets from 2014 to 2017 were used for model verification purposes. In the ANN modeling process, the input and output electricity consumption data sets for each parameter were normalized to the range of [0,1].

The number of neurons in the input and output layers have been set as 5 and 1 respectively. A series of different topologies were used to determine the optimum number of hidden nodes. Compared with the training results, it was found that the training set had the lowest error value when the number of hidden units was 10. 10

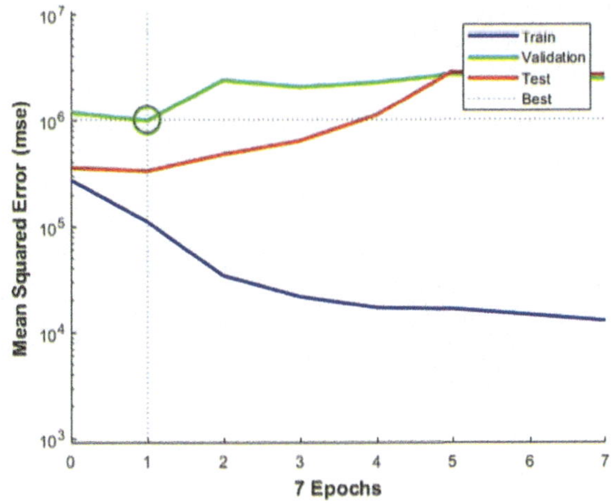

Fig. 6 Training performance of ANN model

is chosen as the number of hidden nodes. Thus the number of each layer's neurons in the network was 5-10-1 respectively. The parameters of the network were chosen as follows: the transformation function of hidden neuron was "tansig" and "logsig" was the output layer function. The stop criterion of error function was set to 0.001 and the maximum of number of iteration was 1000. Computer program has been performed under MATLAB 2019a. Figure 6 demonstrate the ANN model training performance for electricity consumption parameter.

3.4 *Hybrid ARIMA-ANN Modeling*

The proposed algorithm of the hybrid system consisted of two steps. In the first step, to analyze the linear part of the problem, an ARIMA model was employed. In the second step, the residuals from the ARIMA model were modeled by using a neural network model. Since the ARIMA model cannot detect the nonlinear structure of the electricity consumption time series data, the residuals of linear model will contain information about nonlinearity. The output from the neural networks can be used as forecasts of the error terms of the ARIMA model.

The hybrid model utilizes the unique feature and strength of ARIMA model as well as ANN model in determining different patterns. Therefore, it may be favorable to model linear and nonlinear patterns separately by using different models and then combine the forecasts to improve the overall modeling and forecasting performance. In the hybrid modeling algorithm, the input and output electricity consumption data sets for each parameter were normalized to the range of [0,1]. In the modeling process, the hybrid model was trained to adjust the model so that the model forecasted electricity consumption parameters match well with observed data. The verification results of 2014–2017 listed in Fig. 7 indicates that the hybrid model forecast results

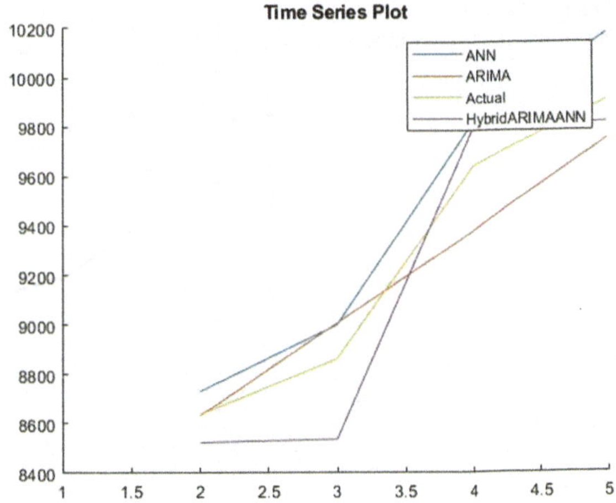

Fig. 7 Comparing the predicted results with the actual value

reasonable match the observed electricity consumption data. By this model, the forecasting electricity consumption of each year was calculated.

3.5 Comparison of Model Performance

Measure of forecast accuracy should always be evaluated as part of a model validation effort. When more than one forecasting accuracy technique seems reasonable. To evaluate the performance of the forecasting accuracy, the three evaluation statistics which is mean absolute error (MAE), root mean square error (RMSE) and mean absolute percentage forecast error (MAPE) are used to evaluate each model.

$$e_t = y_t - y_{t-1} \tag{10}$$

$$MAE = \frac{1}{n}\sum_{t=1}^{n}|e_t| \tag{11}$$

$$RMSE = \frac{1}{n}\sqrt{\sum_{t=1}^{n}(e_t)^2} \tag{12}$$

$$MAPE = \frac{100\%}{n}\sum_{t=1}^{n}|\frac{e_t}{y_{t-1}}| \tag{13}$$

Table 1 reports the MAE, RMSE and MAPE for the year 2014–2017 from the ARIMA, ANN and hybrid AIMA-ANN models. It can be seen that the error levels

Table 1 Forecasting performance of different model

	ARIMA	ANN	Hybrid ARIMA ANN
MAE	2734.9756	239.5	171.3417
RMSE	14,226.0899	275.0873	194.6846
MAPE	0.3930	0.0251	0.0188

in the case of hybrid model are lower than ARIMA and ANN, which leads to the conclusion that the hybrid model show the better consequently and reasonable results.

4 Conclusion

The main objective of this paper is to provide accurate electricity usage prediction models to increase power system reliability. Comparison between the results showed that hybrid ARIMA-ANN model produced better results. The performance of each model is assessed by three statistical measures: RMSE, MAE, and MAPE. The results of the statistical measures will select the best model and will be an effective tool to improve forecasting accuracy. Consequently, forecast of the electricity consumption can be successfully done.

Acknowledgements We would like to thank anonymous referees for many the helpful suggestions which improved the paper considerably. We appreciate the **International Collaborative Research Fund (Universiti Teknologi PETRONAS—Universitas Islam Indragiri) 2021, with cost centre 015-ME0-223** and Fundamental Research Grant Scheme (FRGS) Reference code FRGS/1/2018/STG06/UTP/02/4, Malaysia for sponsoring the research. Lastly, we appreciate Universiti Teknologi PETRONAS where the research was conducted. The views expressed are those of the authors and do not reflect any other authority.

References

1. Aman, S., Ping, H.: Modelling and forecasting electricity consumption of Malaysian large steel mills. Sci. Res. [Internet] **6**(8), 1817–30 (2011). Available from: http://academicjournals.org/sre/PDF/pdf2011/18Apr/Amanetal.pdf
2. Ismail, Z., Yahya, A., Mahpol, K.A.: Forecasting peak load electricity demand using statistics and rule based approach. Am. J. Appl. Sci. **6**(8), 1618–1625 (2009)
3. Haiges, R., Wang, Y.D., Ghoshray, A., Roskilly, A.P.: Forecasting Electricity Generation Capacity in Malaysia: An Auto Regressive Integrated Moving Average Approach. Energy Procedia [Internet] (2017) **105**, 3471–3478 (Oct 2016). Available from: https://doi.org/10.1016/j.egypro.2017.03.795
4. Mohamed Othman, M.S., Johari, D., Musirin, I., Abdul Rahman, T.K., Nik Ismail, N.F.: Artificial neural network-based forecast for electricity consumption in Malaysia. In: 2010 IEEE International Conference on Power and Energy, pp. 24–28 (2010)
5. Yan, H., Zou, Z.: Application of a hybrid ARIMA and neural network model to water quality time series forecasting. J. Converg. Inf. Technol. **8**(4), 59–70 (2013)

6. Naveena, K., Singh, S., Rathod, S., Singh, A.: Hybrid ARIMA-ANN modelling for forecasting the price of robusta coffee in India. Int. J. Curr. Microbiol. Appl. Sci. **6**(7), 1721–1726 (2017)
7. Mohan, B.R., Reddy, G.R.M.: A hybrid ARIMA-ANN model for resource usage prediction. Int. J. Pure. Appl. Math. **119**(12), 12633–12642 (2018)
8. Mucaj, R., Sinaj, V.: Exchange rate forecasting using ARIMA. NAR and ARIMA-ANN Hybrid Model **4**(10), 8581–8586 (2017)
9. Kaytez, F., Taplamacioglu, M.C., Cam, E., Hardalac, F.: Forecasting electricity consumption: A comparison of regression analysis, neural networks and least squares support vector machines. Int. J. Electr. Power Energy Syst. [Internet] **67**(May), 431–438 (2015). Available from https://doi.org/10.1016/j.ijepes.2014.12.036
10. Haiges, R., Wang, Y.D., Ghoshray, A., Roskilly, A.P.: Forecasting Electricity Generation Capacity in Malaysia: An Auto Regressive Integrated Moving Average Approach. Energy Proc. [Internet] **105**, 3471–3478 (2017). Available from: https://doi.org/10.1016/j.egypro.2017.03.795
11. Jifri, M.H., Hassan, E.E., Binti Wan Abd Razak, I.A., Miswan, N.H.: Time series performance for electricity load demand in Johor. IEEE Reg. 10 Ann. Int. Conf. Proceedings/TENCON (2017), 998–1003 (2017 Dec)
12. Box, G.E.P., Jenkins, G.M., Reinsel, G.C., Ljung, G.M.: Time Series Analysis : Forecasting and Control [Internet] [cited 2019 Aug 26]. Available from: https://books.google.com.my/books?hl=en&lr=&id=rNt5CgAAQBAJ&oi=fnd&pg=PR7&dq=G.E.P&ots=DJ75AUq1Wz&sig=1K-wN4sDuGNEohthwSvuTfr4d6o&redir_esc=y#v=onepage&q=G.E.P&f=false
13. Statistics—Malaysia Energy Information Hub [Internet] [cited 2018 Apr 8]. Available from: https://meih.st.gov.my/statistics;jsessionid=4BC29F2B42BE4A1C11E4F34DBF2A4A1C?p_auth=Fl82AN7u&p_p_id=Eng_Statistic_WAR_STOASPublicPortlet&p_p_lifecycle=1&p_p_state=maximized&p_p_mode=view&p_p_col_id=column-1&p_p_col_pos=1&p_p_col_count=2&_Eng_Statistic_WA

Fractional Model for the Flow of Casson Nanofluid using the Generalized Fourier's Law for Heat Transfer

Nadeem Ahmad Sheikh, Dennis Ling Chuan Ching, Hamzah Sakidin, and Ilyas Khan

Abstract The present work used a fractional model of Casson fluid by utilizing a generalized Fourier's Law to construct Caputo Fractional model. A flow of nanofluid flowing in a channel is considered with free convection and electrical conduction. A novel transformation is applied for the energy equation and then solved using integral transforms, combined, the Fourier and Laplace transformations. The results are shown in the form of the Mittag–Leffler function. The influence of physical parameters has been presented in graphs and tables and discussed in this work.

Keywords Fractional model · Nanofluid · Exact solutions

1 Introduction

Nanofluids are used in different engineering and industrial sectors to overcome heat transfer problems in conventional fluids. Casson fluid is one of the essential industrial fluids as it has remarkable properties and applications. Heat transfer is a vital segment in engineering, petroleum, and industrial sectors that necessitates the detailed examination of surface area for heat exchange mechanisms. However, the values are unrealistic when there is a surge in heat transfer. This mechanism can be resolved by teaching the true soul and precise thermo-physical parameters for various fluids

N. A. Sheikh (✉) · D. L. C. Ching · H. Sakidin
Fundamental and Applied Sciences Department, Universiti Teknologi PETRONAS, Bandar Seri Iskandar, 32610 Seri Iskandar, Perak, Malaysia
e-mail: nadeem_18000052@utp.edu.my

D. L. C. Ching
e-mail: dennis.ling@utp.edu.my

H. Sakidin
e-mail: hamzah.sakidin@utp.edu.my

I. Khan
Department of Mathematics, Majmaah University, Majmaah 11952, Saudi Arabia

S. A. Abdul Karim et al. (eds.), *Proceedings of the 6th International Conference on Fundamental and Applied Sciences*, Springer Proceedings in Complexity,
https://doi.org/10.1007/978-981-16-4513-6_67

like water, alcohol, oil, or ethylene glycol. This challenge can be solved by incorporating the change in thermal conductivity of fluid with the heat transfer process [1]. Nanofluids exhibiting nanometer-sized metal oxides (silica, carbon nanotubes or polymers) suspended in the conventional fluids pose a significant development in resolving this issue [2]. Choi [3] proposed this application of nanofluids in 1995 that explains the mechanism of inhibiting the clogging phenomenon that occurred along with the walls of heat transfer equipment to enhance the efficiency, performance, and cost-effectiveness [4]. Ethylene glycol-based nanofluid was discussed by Saqib et al. [5] through exact solutions using the integral transformation. They discussed the Casson nanofluids by considering Tiwari and Das model Molybdenum disulfide (MoS_2) nanoparticles suspended in the fluid.

Temperature, velocities and skin friction coefficient for an incompressible Casson fluid were examined in the presence porous media with MHD boundary layer conditions. The temperature increases with an increase in the heat generation parameter, and the higher Casson fluid parameter is associated with the skin friction coefficient [6]. The Lagrangian equation computationally analyzed the dynamics of submarine debris flow in viscoplastic fluids to compare various rheological models. The downslope movement of high-density fluid was discussed, keeping the fluid volume as constant to describe the transition of fluid between viscous and plastic nature of flow [7]. The numerical analysis of viscosity and yield stress parameters for submarine debris flows' rheology was studied using the plastic Bingham model. The yield surface is widely determined by the shear rate and viscosity of fluid [8]. The flow of viscoelastic fluid, through a uniform magnetic field over an infinite accelerated plate through a porous medium was examined in this study. Laplace transformation technique (LTT) was applied to study the velocity parameter and skin friction. The velocity of fluid has positively influenced by elasticity and permeability, while skin friction also increases with an increase in medium permeability [9]. A Caputo fractional model of MHD Casson fluid flowing in a channel was numerically analyzed by applying both Fourier and Laplace transformations. The Casson fluid behaves like a Newtonian fluid by increasing the values of the Casson fluid parameter [10].

2 Mathematical Modelling

We have considered the motion of Casson nanofluid is a vertical channel. The flow is assumed to be in the direction of x-axis while the y-axis is taken perpendicular to the plates. With ambient temperature T_1, both the fluid and plates are at rest when $t \leq 0$. At $t = 0^+$, the plate at $y = d$ begin to move in its plane with velocity $Uh(t)$. At $y = d$, the plate temperature level raised to $\Theta_1 + (\Theta_2 - \Theta_1)f(t)$ with time t.

We suppose that the rheological equation for an incompressible Casson fluid is [11, 12]:

Table 1 Thermophysical properties of nanoparticles and base fluid

Properties	Engine oil	Cadmium telluride (CdTe)
$\rho\ (\mathrm{kg\,m^{-3}})$	863	5855
$c_p\ (\mathrm{Jkg^{-1}K^{-1}})$	2048	209
$k\ (\mathrm{Wm^{-1}K^{-1}})$	0.1404	7.5
$\beta(\mathrm{K^{-1}})$	0.00007	0.00005
$\sigma\ (\mathrm{Sm^{-1}})$	0.0000055	0.0000007

$$\tau_{ij} = \begin{cases} 2\left(\mu_\gamma + \frac{p_y}{\sqrt{2\pi}}\right)e_{ij}, \pi > \pi_c, \\ 2\left(\mu_\gamma + \frac{p_y}{\sqrt{2\pi}}\right)e_{ij}, \pi_c > \pi \end{cases} \tag{1}$$

The free convection flow of Casson nonfluid together with heat and mass transfer is governed by the corresponding partial differential equations using the possibly the best Boussinesq's approximation [13, 14]:

$$\rho_{nf}\frac{\partial u(y,t)}{\partial t} = \mu_{nf}\left(1+\frac{1}{\gamma_C}\right)\frac{\partial^2 u(y,t)}{\partial y^2} - \sigma_{nf}B_0^2 u(y,t) + (\rho\beta_\Theta)_{nf}g(\Theta-\Theta_1), \tag{2}$$

$$\left(\rho c_p\right)_{nf}\frac{\partial \Theta(y,t)}{\partial t} = -\frac{\partial q(y,t)}{\partial y}, \tag{3}$$

$$q(y,t) = -k_{nf}\frac{\partial \Theta(y,t)}{\partial y}, \tag{4}$$

For the properties of the nanofluids with a subscript (nf), refer to [15]. The thermophysical properties of nanoparticles and base fluid are given in Table 1.

In the dimensionless form the initial and boundary conditions are:

$$\begin{aligned} &u(y,0)=0,\ u(0,t)=0,\ u(d,t)=Uh(t), \\ &\Theta(y,0)=\Theta_1,\ \Theta(0,t)=\Theta_1,\ \Theta(d,t)=\Theta_1+(\Theta_2-\Theta_1)f(t). \end{aligned} \tag{5}$$

Introducing the following dimensionless variables:

$$\begin{aligned} &\upsilon = \frac{u}{U}, \xi = \frac{y}{d}, \tau = \frac{\nu_f}{d^2}t, \theta = \frac{\Theta-\Theta_1}{\Theta_2-\Theta_1}, \delta = \frac{qd}{k_f(\Theta_2-\Theta_1)}, \\ &f(\tau) = f\left(\frac{d^2}{\nu_f}t\right),\ h(\tau) = h\left(\frac{d^2}{\nu_f}t\right), \end{aligned}$$

into Eqs. (3), (4) and (5) we get:

$$\frac{\partial v(\xi,\tau)}{\partial \tau} = \beta_1 \frac{\partial^2 v(\xi,\tau)}{\partial \xi^2} - \beta_2 v(\xi,\tau) + \beta_3 \theta(\xi,\tau), \tag{6}$$

$$\frac{\partial \theta(\xi,\tau)}{\partial \tau} = -\frac{1}{\vartheta_5 \Pr} \frac{\partial \delta(\xi,\tau)}{\partial \xi}, \tag{7}$$

$$\delta(\xi,\tau) = -\vartheta_6 \frac{\partial \theta(\xi,\tau)}{\partial \xi} \tag{8}$$

$$\begin{aligned} &v(\xi,\,0) = 0,\; v(0,\,\tau) = 0,\; v(1,\,\tau) = h(\tau), \\ &\theta(\xi,\,0) = 0,\; \theta(0,\,\tau) = 0, \theta(1,\,\tau) = f(\tau), \end{aligned} \tag{9}$$

where $M = \frac{\sigma_f B_0^2 d^2}{\mu_f}$ is the Hartman number, $Gr = \frac{g d^2 \beta_\Theta}{\nu_f U}(\Theta_2 - \Theta_1)$ is the thermal Grashof number, and $\Pr = \frac{(\rho c_p)_f \upsilon_f}{k_f}$ is the Prandtl number.

3 Fractional Model

To develop a fractional model for the mentioned flow problem, the generalized Fourier's law is used as under:

$$^{C}\wp_\tau^\alpha \theta(\xi,\tau) = \beta_4 \frac{\partial^2 \theta(\xi,\tau)}{\partial^2 \xi}, \tag{10}$$

4 Methodology and Solution of the Problem

4.1 Energy Field

Using the following transformation

$$\chi(\xi,\tau) = \theta(\xi,\tau) - \xi f(\tau), \tag{11}$$

Equation (10) takes the form

$$^{C}\wp_\tau^\alpha \chi(\xi,\tau) + \xi\, {}^{C}\wp_\tau^\alpha f(\tau) = \beta_4 \frac{\partial^2 \chi(\xi,\tau)}{\partial^2 \xi}, \tag{12}$$

with the corresponding initial and boundary conditions as:

$$\chi(\xi,0) = 0,\;\; \chi(0,\tau) = 0,\;\; \chi(1,\tau) = 0. \tag{13}$$

Applying the Laplace and Fourier sine transform, we get

$$\overline{\chi}_F(n,s) = s\overline{f}(s)\frac{(-1)^n}{n\pi}\frac{s^{\alpha-1}}{s^\alpha+\beta_6}, \tag{14}$$

inverting the integral transformations of Eq. (14), we have

$$\chi(\xi,\tau) = 2\sum_{n=1}^{\infty}\frac{(-1)^n\sin(\xi n\pi)}{n\pi}\int_0^\tau \dot{f}(\tau-t)E_{\alpha,\alpha-1}(-\beta_6 t^\alpha)dt, \tag{15}$$

therefore, the solution of the energy equation is

$$\theta(\xi,\tau) = \chi(\xi,\tau) + \xi f(\tau). \tag{16}$$

4.2 *Velocity Profile*

Applying the Laplace and Fourier transforms to Eq. (6) using Eq. (9) we arrived at

$$\begin{aligned}\overline{v}_F(n,s) &= \frac{(-1)^{n+1}\overline{h}(s)}{n\pi} + \left(\frac{\beta_7}{s} + \frac{\beta_8}{s+\beta_5}\right)\frac{(-1)^n s\overline{h}(s)}{n\pi}\\ &+\frac{\beta_3}{s+\Re_1}\left(s\overline{f}(s)\frac{(-1)^n}{n\pi}\frac{s^{\alpha-1}}{s^\alpha+\beta_6} + \overline{f}(s)\frac{(-1)^{n+1}}{n\pi}\right),\end{aligned} \tag{17}$$

where

$$\beta_0 = 1+\frac{1}{\gamma_C},\ \beta_1 = \frac{\vartheta_2}{\beta_0\vartheta_1},\ \beta_2 = \frac{\vartheta_3 M}{\vartheta_1},\ \beta_3 = \frac{\vartheta_4 Gr}{\vartheta_1},\ \beta_4 = \frac{\vartheta_6}{\vartheta_5 \Pr},\ \beta_5 = \beta_2 + \beta_1(n\pi)^2,$$

$$\beta_6 = \beta_4(n\pi)^2,\ \beta_7 = \frac{\beta_2}{\beta_5},\ \beta_8 = \frac{\beta_5-\beta_2}{\beta_5},\ \vartheta_1 = (1-\phi)+\phi\left(\frac{\rho_p}{\rho_f}\right),\ \vartheta_2 = \frac{1}{(1-\phi)^{2.5}},$$

$$\vartheta_3 = 1+3\frac{(\sigma-1)\phi}{(\sigma+2)-(\sigma-1)\phi},\ \vartheta_4 = (1-\phi)+\phi\left(\frac{\rho_p\beta_{\Theta p}}{\rho_f\beta_{\Phi f}}\right),$$

$$\vartheta_5 = (1-\phi)+\phi\frac{\rho_p c_{pp}}{\rho_f c_{pf}},\ \vartheta_6 = \frac{k_{nf}}{k_f},$$

inverting the Laplace and Fourier sine transformations of Eq. (17) we have:

$$v(\xi,\tau) = h(\tau)\xi + 2\sum_{n=1}^{\infty}\frac{(-1)^n}{n\pi}\dot{h}(\tau) * (\beta_7 H(\tau) + \beta_8 \exp(-\beta_5\tau))\sin(\xi n\pi)$$
$$+2\beta_3\sum_{n=1}^{\infty}\frac{(-1)^n}{n\pi\exp(\beta_5\tau)} * \begin{pmatrix} f(\tau) + \int_0^\tau \dot{f}(\tau - q) \\ \times E_{\alpha,\alpha-1}(-\beta_6 q^\alpha)dq \end{pmatrix} \sin(\xi n\pi) \tag{18}$$

Here $H(\tau)$ is the unit step function and $E_{a,b}(.)$ is the Mittag Leffler function [16].

5 Results and Discussion

The exact solutions for the MHD flow of Casson nanofluid in a channel embedded in a porous media with heat transfer are obtained in this study. The associated energy equation is fractionalized using generalized Fourier's law. The obtained exact solutions are plotted through graphs, and the effects of different physical parameters on the flow and heat transfer are presented.

The variations in the velocity of the nanofluid for different values of the fractional parameter is displayed in Fig. 1. From this figure, it is noticed that four different velocity profiles are obtained for four different values of fractional parameter keeping all the other physical parameters constant. This is showing that the fractional parameter has a significant influence on the obtained solutions; even this is not a physical parameter and is a purely mathematical parameter. These variations are due to the memory effect, which cannot be studied through integers order derivatives.

An increasing trend is noticed in the velocity of the Casson nanofluid for increasing values of the Casson fluid parameter in Fig. 2. Physically, the viscosity of the fluid is increased for smaller values of the Casson fluid parameter. Another impressive result

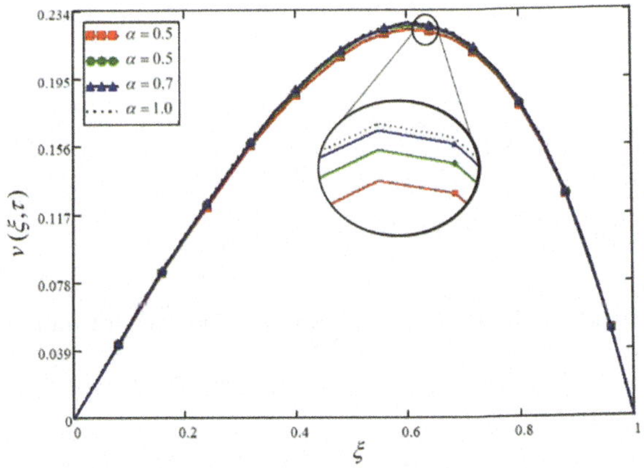

Fig. 1 Influence of the fractional parameter on the nanofluid velocity

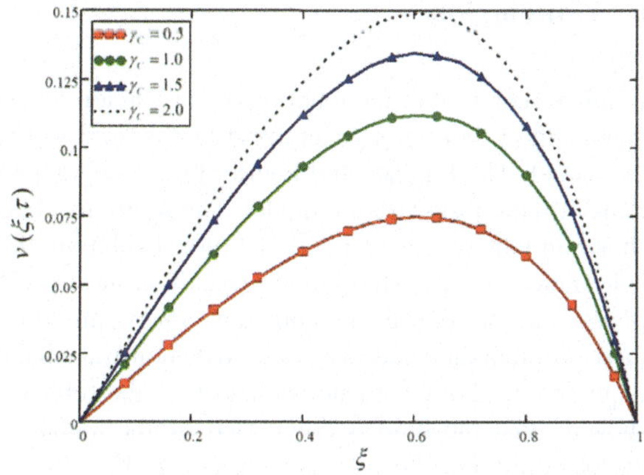

Fig. 2 Influence of the Casson fluid parameter on the nanofluid velocity

can be drawn through this graph that Casson fluid is more viscous than Newtonian fluid and when $\gamma_C \to \infty$, the fluid behaves like a Newtonian fluid.

In this study, we have considered engine oil as a base fluid and Cadmium Telluride (CdTe) as nanoparticles.

Figure 3 is drawn to show the effect of the volume fraction of nanoparticles on the fluid velocity. The fluid velocity is decreasing with the higher values of the volume fraction of nanofluid. This means the fluid will become more viscous with the addition of nanoparticles, and as a result, the lubrication of the engine oil will be improved. For the interest of the readers.

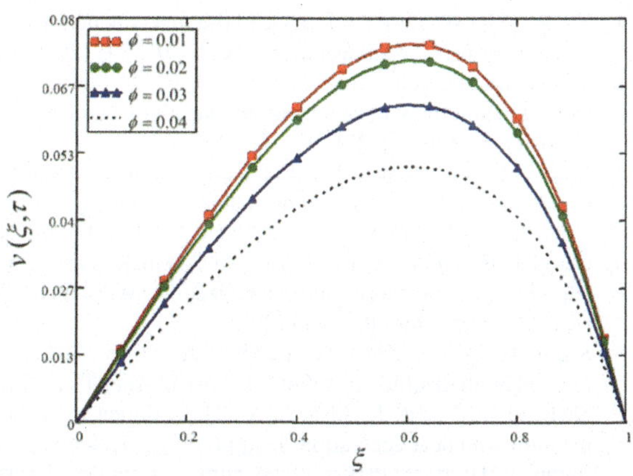

Fig. 3 Influence of nanoparticles volume fraction on the nanofluid velocity

6 Conclusion

In this research, the fractional model of Casson nanofluid is being developed by a new method. Fourier's generalized law is used to obtain the generalized form of the model. The Laplace and Fourier processing approaches are used for a modern transformation to solve the model. The solutions obtained are drawn in graphs and shown in tables. For the fractional model solution, the new transformation is more believable. This transformation makes it easier to solve the fractional model. This transformation reduces the computer time required to identify precise solutions to these problems and makes it easier to demonstrate that the solutions meet the limiting requirements. For the greater values of γ_C, the velocity of the Casson fluid is greater, showing that the fluid is comported by the higher values of γ_C as a Newtonian viscous fluid. For the various values of α, the differences in all profiles are shown. It is important to point out here that for one value of time we have distinct lines. This effect demonstrates the fluid's memory effect, which the integer model cannot describe.

Acknowledgements The authors acknowledge Universiti Teknologi PETRONAS for supporting this research work. This research work was sponsored by Yayasan UTP (YUTP) grants with cost center: CRG 015MC0-011 awarded to Dr. Dennis Ling Chuan Ching.

References

1. Yu, W., Xie, H., Bao, D.: Enhanced thermal conductivities of nanofluids containing graphene oxide nanosheets. Nanotechnology **21**(5), 055705 (2009)
2. Reddy, J.R., Sugunamma, V., Sandeep, N.: Impact of nonlinear radiation on 3D magneto-hydrodynamic flow of methanol and kerosene based ferrofluids with temperature dependent viscosity. J. Mol. Liq. **236**, 93–100 (2017)
3. Choi, S.U., Eastman, J.A.: "Enhancing thermal conductivity of fluids with nanoparticles," presented at the International mechanical engineering congress and exhibition, San Francisco (1 Oct 1995) ANL/MSD/CP-84938; CONF-951135-29 ON: DE96004174; TRN: 96:001707
4. Öztop, H.F., Estellé, P., Yan, W.-M., Al-Salem, K., Orfi, J., Mahian, O.: A brief review of natural convection in enclosures under localized heating with and without nanofluids. Int. Commun. Heat Mass Transfer **60**, 37–44 (2015)
5. Saqib, M., Ali, F., Khan, I., Sheikh, N.A., Shafie, S.B.: Convection in ethylene glycol-based molybdenum disulfide nanofluid. J. Therm. Anal. Calorim. **135**(1), 523–532 (2019)
6. Shehzad, S., Hayat, T., Alsaedi, A.: Three-dimensional MHD flow of Casson fluid in porous medium with heat generation. J. Appl. Fluid Mech. **9**(1), 215–223 (2016)
7. Imran, J., Harff, P., Parker, G.: A numerical model of submarine debris flow with graphical user interface. Comput. Geosci. **27**(6), 717–729 (2001)
8. Jeong, S.W.: Determining the viscosity and yield surface of marine sediments using modified Bingham models. Geosci. J. **17**(3), 241–247 (1 Sept 2013). https://doi.org/10.1007/s12303-013-0038-7
9. Eldabe, N.T.M., Moatimid, G.M., Ali, H.S.: Magnetohydrodynamic flow of non-Newtonian visco-elastic fluid through a porous medium near an accelerated plate. Can. J. Phys. **81**(11), 1249–1269 (1 Nov 2003). https://doi.org/10.1139/p03-092

10. Sheikh, N.A., Ching, D.L.C., Khan, I., Kumar, D., Nisar, K.S.: A new model of fractional Casson fluid based on generalized Fick's and Fourier's laws together with heat and mass transfer. Alexandria Eng. J. (in press) (31 Dec 2019). https://doi.org/10.1016/j.aej.2019.12.023
11. Aman, S., Khan, I., Ismail, Z., Salleh, Alshomrani, and M. S. Alghamdi, "Magnetic field effect on Poiseuille flow and heat transfer of carbon nanotubes along a vertical channel filled with Casson fluid. AIP Adv. **7**(1) (2017). https://doi.org/10.1063/1.4975219
12. Ali, F., Sheikh, N.A., Khan, I., Saqib, M.: Magnetic field effect on blood flow of Casson fluid in axisymmetric cylindrical tube: A fractional model. J. Magn. Magn. Mater. **423**, 327–336 (2017)
13. Khan, A., Khan, D., Khan, I., Ali, F., Karim, F.U., Imran, M.: MHD flow of sodium alginate-based casson type nanofluid passing through a porous medium with newtonian heating. Sci. Rep. **8**(1), 8645 (5 June 2018). https://doi.org/10.1038/s41598-018-26994-1
14. Khalid, A., Khan, I., Khan, A., Shafie, S.: Unsteady MHD free convection flow of Casson fluid past over an oscillating vertical plate embedded in a porous medium. Eng. Sci. Technol. Int. J. **18**(3), 309–317 (2015)
15. Sheikh, N.A., Ching, D.L.C., Khan, I.: A comprehensive review on theoretical aspects of nanofluids: Exact solutions and analysis. Symmetry **12**(5), 725 (2020)
16. Ali, F., Saqib, M., Khan, I., Ahmad Sheikh, N.: Heat transfer analysis in ethylene glycol based molybdenum disulfide generalized nanofluid via Atangana–Baleanu fractional derivative approach. In: Fractional Derivatives with Mittag-Leffler Kernel (Studies in Systems, Decision and Control, 2019, Chap. 13), pp. 217–233

Author Index

S. A. Abdul Karim et al. (eds.), *Proceedings of the 6th International Conference on Fundamental and Applied Sciences*, Springer Proceedings in Complexity,
https://doi.org/10.1007/978-981-16-4513-6

Subject Index

S. A. Abdul Karim et al. (eds.), *Proceedings of the 6th International Conference on Fundamental and Applied Sciences*, Springer Proceedings in Complexity,
https://doi.org/10.1007/978-981-16-4513-6

N

O

The manufacturer's authorised representative in the EU is Springer Nature Customer Service Centre GmbH, Europaplatz 3, 69115 Heidelberg, Germany. If you have any concerns regarding our products, please contact ProductSafety@springernature.com

Printed and bound by CPI Group (UK) Ltd, Croydon, CR0 4YY

30/09/2026

02225254-0018